# 中国高等植物

## Higher Plants of China in Colour

《中国高等植物彩色图鉴》编委会　主编

Edited by
Editorial Committee of
Higher Plants of China in Colour

科 学 出 版 社
北 京

# 中国高等植物彩色图鉴

## Higher Plants of China in Colour

《中国高等植物彩色图鉴》编委会　主编
Edited by Editorial Committee of Higher Plants of China in Colour

### 第 5 卷　被子植物　大戟科－山茱萸科
### Volume Ⅴ　Angiosperms　Euphorbiaceae—Cornaceae

卷编辑　刘　博　林秦文
Edited by Bo LIU
Qinwen LIN

## 内容简介

本套图鉴精选中国境内野生高等植物和重要栽培植物1万余种，配以图片近2万张，每一物种以中英文形式简要介绍植物的中文名、拉丁学名、形态特征、花果期、生境和分布。图鉴共分为9卷，收载苔藓植物100科、蕨类植物40科、裸子植物11科、被子植物232科，共计383科，且除苔藓植物之外，已收全所有科。本套图鉴是继《中国高等植物图鉴》、《中国植物志》、Flora of China之后，又一部大型植物分类学巨著。本卷为第5卷。

本书适合植物学领域的科研人员、管理人员及爱好植物学的普通大众阅读和收藏。

This set of pictorial books contains nearly 20 thousand photographs, presenting the cream of wild higher plants and important cultivated plants in China, the species of which number more than 10 thousand. Each of the species is concisely introduced in both Chinese and English from such aspects as Chinese name, Latin name, morphological features, flowering and fruiting season, habitat and distribution. Divided into nine volumes, this work includes 100 bryophyte families, 40 pteridophyte families, 11 gymnosperm families and 232 angiosperm families, 383 families altogether; the inclusion of all the said families is complete except for the bryophytes. The set of pictorial books is another monumental work on plant taxonomy, after *Iconographia Cormophytorum Sinicorum*, *Flora Reipublicae Popularis Sinicae*, and *Flora of China*. This is volume V of the series.

This work is intended for scientific researchers and administrators in the field of botany and also for botany enthusiasts. As well as for reading, the work can be a classic collection.

**图书在版编目（CIP）数据**

中国高等植物彩色图鉴＝Higher Plants of China in Colour. 第5卷，被子植物. 大戟科—山茱萸科：汉英 /《中国高等植物彩色图鉴》编委会主编；刘博，林秦文分册主编. —北京：科学出版社，2016.1

ISBN 978-7-03-047065-2

Ⅰ. ①中… Ⅱ. ①中… ②刘… ③林… Ⅲ. ①高等植物-中国-图集 ②大戟科-中国-图集 ③山茱萸科-中国-图集 Ⅳ. ①Q949.4-64

中国版本图书馆CIP数据核字（2016）第013518号

责任编辑：王 静 付 聪 马 俊 / 责任校对：李 影
责任印制：肖 兴 / 书籍设计：北京美光设计制版有限公司

科学出版社 出版

北京东黄城根北街16号
邮政编码：100717
http://www.sciencep.com

北京汇瑞嘉合文化发展有限公司 印刷

科学出版社发行 各地新华书店经销

*

2016年1月第 一 版 开本：787×1092 1/16
2016年1月第一次印刷 印张：35 1/4
字数：1 613 000

**定价：520.00元**

（如有印装质量问题，我社负责调换）

# 《中国高等植物彩色图鉴》编委会




## 作者

费　勇　刘怡涛　刘　冰　陶国达　成　晓　陈　彬　陈又生　傅立国　何国生
金效华　李振宇　林秦文　刘　博　王文采　吴声华　徐松芝　涂晔春　于胜祥
喻勋林　张　力　张代贵　张树仁　张宪春　朱鑫鑫　左　勤

# 丛书图片主要拍摄者

(按姓氏汉语拼音排序)

阿不都拉·阿巴斯　白鹭　白重炎　毕延超　邴艳红　曹同　车晋滇　陈彬　陈高
陈丽　陈庆　陈鑫　陈炳华　陈世品　陈贤兴　陈又生　陈志雄　成晓　程文达
迟敏杰　戴攀峰　邓涛　邓云飞　丁炳扬　丁学欣　董仕勇　杜诚　杜巍　杜玉芬
段长虹　段士民　方振东　方振兴　费勇　冯君茹　傅连中　甘啟良　高贤明　高信芬
高云东　葛斌杰　耿玉英　古训铭　顾余兴　管开云　郭世伟　韩国营　郝加琛　郝云庆
何海　何理　何春梅　何国生　和兆荣　侯元同　胡光万　胡国雄　华国军　黄健
黄江华　黄圣卓　黄向旭　黄俞淞　惠肇祥　季定乾　贾渝　姜林　蒋宏　蒋蕾
蒋日红　金伟涛　金孝锋　金效华　康世昌　赖阳均　郎楷永　黎斌　黎兴江　李东
李恒　李凯　李敏　李攀　李不言　李策宏　李东辉　李家湘　李建民　李剑武
李良千　李文军　李先源　李晓东　李小杰　李新华　李新伟　李学东　李泽贤　李振宇
李志奇　李智选　李中阳　梁同军　廖明林　廖云标　林敏　林祁　林维　林广旋
林建勇　林俊杰　林茂祥　林秦文　林哲丽　刘冰　刘博　刘静　刘军　刘坤
刘夙　刘翔　刘鑫　刘演　刘莹　刘大伟　刘光裕　刘海桑　刘红梅　刘伦辉
刘全儒　刘晟源　刘怡涛　刘正宇　刘宗才　柳永红　卢刚　卢元　罗柳青　骆适
吕碧凤　吕志学　马林　马炜梁　马文章　马欣堂　莫水松　牟善杰　沐先运　慕泽泾
南程慧　倪静波　倪素碧　农东新　潘勃　潘建斌　彭博　彭镜毅　彭日成　乔明明
秦卫华　秦祥堃　覃海宁　邱志敬　仁琛　任飞　任丽华　任明波　任昭杰　尚策
邵剑文　沈阳肇　施忠辉　石硕　寿海洋　税玉民　宋纬文　宋柱秋　买买提明·苏来曼
苏丽飞　苏享修　孙航　孙苗　孙观灵　孙明洲　孙卫邦　孙小美　谭运洪　陶国达
田乾福　田新民　童毅　童毅华　汪远　王辰　王东　王弘　王晖　王健
王进　王强　王颖　王耘　王喆　王长荣　王钧杰　王慷林　王清隆　王秋美
王文卿　王雅琼　王亚玲　王英伟　王玉兵　王正元　王祝年　韦宏金　韦毅刚　韦玉梅
卫然　魏来　温韩东　温九良　翁茂伦　吴丰　吴磊　吴双　吴棣飞　吴凤琴
吴光弟　吴国晞　吴林芳　吴声华　吴望辉　吴问舫　吴永红　吴增源　伍凯　武全安
武素功　武玉东　夏念和　肖翠　肖亮　肖艳　肖红菊　谢磊　辛夷　辛晓伟
辛益群　辛宇明　熊源新　徐徭　徐锦泉　徐克学　徐连升　徐申健　徐文斌　徐晔春
徐永福　许为斌　寻路路　严新富　严岳鸿　阳文静　杨浩　杨永　杨成梓　杨建昆
杨金财　杨科明　杨青山　杨世雄　杨奕绯　杨增宏　姚永飚　叶德平　叶建飞　叶喜阳
叶幸儿　易思荣　殷建涛　尹志坚　于胜祥　郁文彬　喻勋林　袁彩霞　曾孝濂　曾云保
张力　张良　张强　张伟　张莹　张勇　张彩飞　张重岭　张代贵　张凤秋
张海华　张宏伟　张金龙　张金政　张守君　张淑梅　张树仁　张维柱　张宪春　张霄林
张志翔　赵宏　赵伟　赵大昌　郑宝江　郑希龙　郑小明　钟智明　周繇　周重建
周海成　周家宝　周兰平　周喜乐　周小林　周浙昆　朱弘　朱大海　朱仁斌　朱淑霞
朱维明　朱鑫鑫　David E. Boufford　Dmitry Sokoloff　Jan Thomas Johansson
Jozef Lemmens　Kirill Tkachenko　Pavel Novák　Ralf Knapp　Richard Ree
Susan Kelley

香港植物标本室(免费提供)

## Major Photographers of the Series

(in the order of Chinese pinyin)

| | | | | | |
|---|---|---|---|---|---|
| Abdulla ABASI | Lu BAI | Chongyan BAI | Yanchao BI | Yanhong BING | Tong CAO |
| Jindian CHE | Bin CHEN | Gao CHEN | Li CHEN | Qing CHEN | Xin CHEN |
| Binghua CHEN | Shipin CHEN | Xianxing CHEN | Yousheng CHEN | Chihhsiung CHEN | Xiao CHENG |
| Wenda CHENG | Minjie CHI | Panfeng DAI | Tao DENG | Yunfei DENG | Bingyang DING |
| Xuexin DING | Shiyong DONG | Cheng DU | Wei DU | Yufen DU | Changhong DUAN |
| Shimin DUAN | Zhendong FANG | Zhenxing FANG | Yong FEI | Junru FENG | Lianzhong FU |
| Qiliang GAN | Xianming GAO | Xinfen GAO | Yundong GAO | Binjie GE | Yuying GENG |
| Xunming GU | Yuxing GU | Kaiyun GUAN | Shiwei GUO | Guoying HAN | Jiachen HAO |
| Yunqing HAO | Hai HE | Li HE | Chunmei HE | Guosheng HE | Zhaorong HE |
| Yuantong HOU | Guangwan HU | Guoxiong HU | Guojun HUA | Jian HUANG | Jianghua HUANG |
| Shengzhuo HUANG | Xiangxu HUANG | Yusong HUANG | Zhaoxiang HUI | Dingqian JI | Yu JIA |
| Lin JIANG | Hong JIANG | Lei JIANG | Rihong JIANG | Weitao JIN | Xiaofeng JIN |
| Xiaohua JIN | Shihchang KANG | Yangjun LAI | Kaiyong LANG | Bin LI | Xingjiang LI |
| Dong LI | Heng LI | Kai LI | Min LI | Pan LI | Buyan LI |
| Cehong LI | Donghui LI | Jiaxiang LI | Jianmin LI | Jianwu LI | Liangqian LI |
| Wenjun LI | Xianyuan LI | Xiaodong LI | Xiaojie LI | Xinhua LI | Xinwei LI |
| Xuedong LI | Zexian LI | Zhenyu LI | Zhiqi LI | Zhixuan LI | Zhongyang LI |
| Tongjun LIANG | Minglin LIAO | Yunbiao LIAO | Min LIN | Qi LIN | Wei LIN |
| Guangxuan LIN | Jianyong LIN | Junjie LIN | Maoxiang LIN | Qinwen LIN | Zheli LIN |
| Bing LIU | Bo LIU | Jing LIU | Jun LIU | Kun LIU | Su LIU |
| Xiang LIU | Xin LIU | Yan LIU | Ying LIU | Dawei LIU | Guangyu LIU |
| Haisang LIU | Hongmei LIU | Lunhui LIU | Quanru LIU | Shengyuan LIU | Yitao LIU |
| Zhengyu LIU | Zongcai LIU | Yonghong LIU | Gang LU | Yuan LU | Liuqing LUO |
| Shi LUO | Pifong LU | Zhixue LÜ | Lin MA | Weiliang MA | Wenzhang MA |
| Xintang MA | Shuisong MO | Shannjye MOORE | Xianyun MU | Zejing MU | Chenghui NAN |
| Jingbo NI | Subi NI | Dongxin NONG | Bo PAN | Jianbin PAN | Bo PENG |
| Ching-I PENG | Richeng PENG | Mingming QIAO | Weihua QIN | Xiangkun QIN | Haining QIN |
| Zhijing QIU | Chen REN | Fei REN | Lihua REN | Mingbo REN | Zhaojie REN |
| Ce SHANG | Jianwen SHAO | Yangzhao SHEN | Zhonghui SHI | Shuo SHI | Haiyang SHOU |

| | | | | | |
|---|---|---|---|---|---|
| Yumin SHUI | Weiwen SONG | Zhuqiu SONG | Mamtimin SULAYMAN | Lifei SU | Xiangxiu SU |
| Hang SUN | Miao SUN | Guanling SUN | Mingzhou SUN | Weibang SUN | Xiaomei SUN |
| Yunhong TAN | Guoda TAO | Qianfu TIAN | Xinmin TIAN | Yi TONG | Yihua TONG |
| Yuan WANG | Chen WANG | Dong WANG | Hong WANG | Hui WANG | Jian WANG |
| Jin WANG | Qiang WANG | Ying WANG | Yun WANG | Zhe WANG | Changrong WANG |
| Junjie WANG | Kanglin WANG | Qinglong WANG | Chiumei WANG | Wenqing WANG | Yaqiong WANG |
| Yaling WANG | Yingwei WANG | Yubing WANG | Zhengyuan WANG | Zhunian WANG | Hongjin WEI |
| Yigang WEI | Yumei WEI | Ran WEI | Lai WEI | Handong WEN | Jiuliang WEN |
| Maolun WENG | Feng WU | Lei WU | Shuang WU | Difei WU | Fengqin WU |
| Guangdi WU | Guoxi WU | Linfang WU | Shenghua WU | Wanghui WU | Wenfang WU |
| Yonghong WU | Zengyuan WU | Kai WU | Quan'an WU | Sugong WU | Yudong WU |
| Nianhe XIA | Cui XIAO | Liang XIAO | Yan XIAO | Hongju XIAO | Lei XIE |
| Yi XIN | Xiaowei XIN | Yiqun XIN | Yuming XIN | Yuanxin XIONG | Yao XU |
| Jinquan XU | Kexue XU | Liansheng XU | Shenjian XU | Wenbin XU | Yechun XU |
| Yongfu XU | Weibin XU | Lulu XUN | Hsinfu YEN | Yuehong YAN | Wenjing YANG |
| Hao YANG | Yong YANG | Chengzi YANG | Jiankun YANG | Jincai YANG | Keming YANG |
| Qingshan YANG | Shixiong YANG | Yifei YANG | Zenghong YANG | Yongbiao YAO | Deping YE |
| Jianfei YE | Xiyang YE | Xing'er YE | Sirong YI | Jiantao YIN | Zhijian YIN |
| Shengxiang YU | Wenbin YU | Xunlin YU | Caixia YUAN | Xiaolian ZENG | Yunbao ZENG |
| Li ZHANG | Liang ZHANG | Qiang ZHANG | Wei ZHANG | Ying ZHANG | Yong ZHANG |
| Caifei ZHANG | Chongling ZHANG | Daigui ZHANG | Fengqiu ZHANG | Haihua ZHANG | Hongwei ZHANG |
| Jinlong ZHANG | Jinzheng ZHANG | Shoujun ZHANG | Shumei ZHANG | Shuren ZHANG | Weizhu ZHANG |
| Xianchun ZHANG | Xiaolin ZHANG | Zhixiang ZHANG | Hong ZHAO | Wei ZHAO | Dachang ZHAO |
| Baojiang ZHENG | Xilong ZHENG | Xiaoming ZHENG | Zhiming ZHONG | You ZHOU | Chongjian ZHOU |
| Haicheng ZHOU | Jiabao ZHOU | Lanping ZHOU | Xile ZHOU | Xiaolin ZHOU | Zhekun ZHOU |
| Hong ZHU | Dahai ZHU | Renbin ZHU | Shuxia ZHU | Weiming ZHU | Xinxin ZHU |
| David E. Boufford | Dmitry Sokoloff | Jan Thomas Johansson | | Jozef Lemmens | Kirill Tkachenko |
| Pavel Novák | Ralf Knapp | Richard Ree | Susan Kelley | | |

Hong Kong Herbarium (free of charge)

## 丛书文字主要编写者

(按姓氏汉语拼音排序)

| | | | | | | | | | |
|---|---|---|---|---|---|---|---|---|---|
| 陈世龙 | 陈又生 | 陈之端 | 成　晓 | 崔逸群 | 邓云飞 | 杜　宁 | 段士民 | 樊　杰 | 方瑞征 |
| 费　勇 | 傅立国 | 高　凡 | 高　乞 | 谷粹芝 | 郭　慧 | 韩　宇 | 郝加琛 | 何　理 | 侯学良 |
| 侯元同 | 胡光万 | 黄向旭 | 黄俞淞 | 蒋　宏 | 金孝锋 | 金效华 | 赖阳均 | 雷立公 | 黎　斌 |
| 李　恒 | 李　嵘 | 李秉滔 | 李宏哲 | 李剑武 | 李梦华 | 李文军 | 李锡文 | 李晓贤 | 李新华 |
| 李新伟 | 李章海 | 李振宇 | 李中阳 | 廖文波 | 林秦文 | 刘　冰 | 刘　博 | 刘　演 | 刘大伟 |
| 刘海桑 | 刘全儒 | 刘衍男 | 卢金梅 | 马欣堂 | 潘　勃 | 覃海宁 | 邱志敬 | 萨　仁 | 尚　策 |
| 税玉民 | 孙　苗 | 孙久琼 | 万　涛 | 王　东 | 王　晖 | 王　健 | 王　强 | 王德艺 | 王文采 |
| 王英伟 | 卫　然 | 魏　来 | 吴　磊 | 吴鹏程 | 吴声华 | 向巧萍 | 向小果 | 谢　磊 | 徐松芝 |
| 徐晓婷 | 薛大伟 | 闫瑞亚 | 严岳鸿 | 杨世雄 | 叶建飞 | 游旨价 | 于胜祥 | 袁　慷 | 张　力 |
| 张　梅 | 张　强 | 张重岭 | 张钢民 | 张红瑞 | 张树仁 | 张宪春 | 张志翔 | 赵　宏 | 赵存峰 |
| 周兰平 | 左　勤 | | | | | | | | |

## Major Textwriters of the Series

(in the order of Chinese pinyin)

| | | | | |
|---|---|---|---|---|
| Shilong CHEN | Yousheng CHEN | Zhiduan CHEN | Xiao CHENG | Yiqun CUI |
| Yunfei DENG | Ning DU | Shimin DUAN | Jie FAN | Ruizheng FANG |
| Yong FEI | Liguo FU | Fan GAO | Qi GAO | Cuizhi GU |
| Hui GUO | Yu HAN | Jiachen HAO | Li HE | Xueliang HOU |
| Yuantong HOU | Guangwan HU | Xiangxu HUANG | Yusong HUANG | Hong JIANG |
| Xiaofeng JIN | Xiaohua JIN | Yangjun LAI | Ligong LEI | Bin LI |
| Heng LI | Rong LI | Bingtao LI | Hongzhe LI | Jianwu LI |
| Menghua LI | Wenjun LI | Xiwen LI | Xiaoxian LI | Xinhua LI |
| Xinwei LI | Zhanghai LI | Zhenyu LI | Zhongyang LI | Wenbo LIAO |
| Qinwen LIN | Bing LIU | Bo LIU | Yan LIU | Dawei LIU |
| Haisang LIU | Quanru LIU | Yannan LIU | Jinmei LU | Xintang MA |
| Bo PAN | Haining QIN | Zhijing QIU | Ren SA | Ce SHANG |
| Yumin SHUI | Miao SUN | Jiuqiong SUN | Tao WAN | Dong WANG |
| Hui WANG | Jian WANG | Qiang WANG | Deyi WANG | Wentsai WANG |
| Yingwei WANG | Ran WEI | Lai WEI | Lei WU | Pengcheng WU |
| Shenghua WU | Qiaoping XIANG | Xiaoguo XIANG | Lei XIE | Songzhi XU |
| Xiaoting XU | Dawei Xue | Ruiya YAN | Yuehong YAN | Shixiong YANG |
| Jianfei YE | Zhijia YOU | Shengxiang YU | Qian YUAN | Li ZHANG |
| Mei ZHANG | Qiang ZHANG | Chongling ZHANG | Gangmin ZHANG | Hongrui ZHANG |
| Shuren ZHANG | Xianchun ZHANG | Zhixiang ZHANG | Hong ZHAO | Cunfeng ZHAO |
| Lanping ZHOU | Qin ZUO | | | |

# 丛书前言

中国是世界上植物最丰富的国家之一，已知有三万五千多种野生和重要栽培的高等植物，其中特有种达一万五千多种，形成复杂而独具特色的植物区系。中国的先人们创造了古老而辉煌的农业文明，选育出水稻、大豆、茶、枣、桃、柿等重要作物，其中水稻的栽培历史可追溯到约七千年前新石器时期的河姆渡文化，如今稻米已成为世界上近一半人口的粮食。丰富的植物资源和灿烂的历史文化，使中国成为“花园之母”和世界农作物七大起源中心之一。

中国植物学家为了系统地展示中国植物的多样性，历经艰辛，相继编研了《中国高等植物图鉴》和《中国植物志》，并与外国专家合作出版Flora of China等大型志书，这些著作在国内外应用广泛、影响巨大，客观地展现了不同时期的植物分类学研究和植物资源调查的成果，成为植物分类学领域最重要的大型经典著作。但是，它们都有一个共同的缺憾，即仅有黑白线条图，难以充分表达植物各器官的质地和颜色等自然状态下的外貌特征，其效果难以满足部分读者鉴赏植物的需要。

大多数发达国家都有自己的植物彩色图鉴，这些图鉴不仅展示了本国的生物多样性，还兼备工具书功能和富有感染力的艺术效果，具有很高的应用和收藏价值。迄今为止，国内出版的植物彩色图书多为地区性的，或局限于某一类植物的，如观赏植物、栽培作物和药用植物。作为世界生物多样性大国，中国应当拥有一套全面体现本国野生植物多样性的的大型鉴赏类彩色图册。

将灿烂的瞬间变为永恒是广大植物爱好者和摄影爱好者的追求。为了填补上述空白，台湾吴声华研究员策划并启动了这项工作。在海峡两岸学者的共同努力下，本书的规模在不断扩大，从最初的云南植物写真集扩展到全国性大型彩色植物图鉴。中国科学院植物研究所王文采院士出任丛书编委会主任，吴声华研究员和中国科学院植物研究所李振宇研究员任副主任。编委会遴选国内从事植物分类学研究的专家担任各卷卷编辑，邀请中国大陆、台湾和香港近200位植物学家承担各科的编写和审稿工作，卷编辑在专家审稿的基础上，再次对本卷内容进行核查。近400位摄影作者提供了大量精美的植物彩色照片。丛书还采用了著名的动植物科学画大师曾孝濂先生绘制的20余幅优雅而灵动的彩色图片。

本丛书划分为九卷，共收录中国高等植物1万余种，种类以野生植物为主，同时收载重要的栽培植物，精选图片近2万张。本丛书中科的系统排列如下：苔藓植物主要参考《中国苔藓志》中的系统；蕨类植物按张宪春2015年在《石松类和蕨类名词及名称》提出的系统；裸子植物和被子植物的系统排列按第尔斯(L. Diels, 1936)于A. Engler's Syllabus der Pflanzenfamilien中采用的系统。仅第三卷将毛茛科分为星叶草科、毛茛科和芍药科，将木兰科分为木兰科、八角科、五味子科和水青树科。全书收载中国高等植物383科，其中苔藓植物100科，占全国苔藓科总数的大多数；其余是蕨类植物40科，裸子植物11科，被子植物232科，分别代表了国产三大门类所有的科。本丛书收载的植物中有一些是Flora of China出版后发表的国产新种，如香港鹅耳枥(*Carpinus insularis*)、球柱楼梯草(*Elatostema globosostigmatum*)和西藏小囊兰(*Micropera tibetica*)，以及中国分布新记录，如轮叶三棱栎(*Trigonobalanus verticillata*)和格力兜兰(*Paphiopedilum gratrixianum*)。

为了方便更多的读者阅读，本丛书的文字采用中英文，简要介绍各种植物的中文名、拉丁学名、形态特征、花果期、生境和分布。

本书在编写过程中，承中国科学院植物研究所中国植物图像库和中国自然标本馆提供了许多方便和帮助，在此向他们表示衷心的感谢。

感谢国家出版基金和科学出版社对本丛书出版的大力支持。

由于编著者的业务水平有限、错漏之处，欢迎批评指正。

《中国高等植物彩色图鉴》编委会
2015年10月31日

## Preface to the Series

As one of the countries with the richest diversity of plant species in the world, China has more than 35 000 known species of wild and important cultivated higher plants, among which there are over 15 000 endemic species, forming a complex and unique flora. The ancestors of the Chinese people created an ancient and splendid agricultural civilization. They selected and cultivated significant crops like rice, soya bean, tea, jujube, peach, and persimmon. Among these crops, the cultivation history of rice can be traced back to the Hemudu culture of the Neolithic Period around 7000 years ago. Nowadays, rice has become the staple food for nearly half of the world's population. With abundant plant resources and a long history and great culture, China is renowned as 'the mother of gardens' and is one of the seven important centers of origin for crops in the world.

In order to present the diversity of China's plants systematically, botanists from China have made pain-taking efforts to compile a series of large-volume floras including *Iconographia Cormophytorum Sinicorum*, *Flora Reipublicae Popularis Sinicae* (Chinese version) by themselves, and *Flora of China* (English version) with the collaboration of international specialists. These books are well known both in China and abroad and have been used extensively for studying Chinese plants and plants from adjacent areas, these works present the results of plant taxonomic study and study of plant resources in China at different periods, and constitute some of the most important large classic volumes in the field of plant taxonomy. However, in all these works the plants are only partly illustrated by black and white line-drawings, unable to present the texture and colour of flowers and leaves fully in their natural state, and they hardly reveal the spectacular beauty and fascination of the wealth of plant species.

Most developed countries have colour pictorial books of their plants, which form greatly desirable works, because they are not only a presentation of the plant diversity of the countries, but are also an attractive record of the beauty of the nature. So far, most of the Chinese colour pictorial books of plants are regional treatments, or concentrate on particular groups, such as ornamental plants, cultivated crops and medicinal plants or certain taxonomic groups. As a country with a high level of biodiversity, China merits a large-scale colour pictorial book with high appreciation value featuring the wild plants that occur within its territory.

It is the goal of every lover of plants and plant photography to capture the essence of plant beauty and make it permanent. In order to fill the above-mentioned gap, Professor Shenghua WU from Taiwan, planned and launched the present project. With the joint effort of specialists from all over China, the scale of the book has expanded from the initial pictorial book of plants of Yunnan to a many-volume colour pictorial book of plants of the whole country. Academician Wentsai WANG of the Institute of Botany, Chinese Academy of Sciences, took up the post of the chairman of the editorial committee, and the positions of vice chairmen of the editorial committee were assumed by Prof. Shenghua WU, Taiwan, and Prof. Zhenyu LI of the Institute of Botany, Chinese Academy of Sciences. The editorial committee then selected experienced plant taxonomists as volume editors for each volume, and invited nearly 200 botanists from mainland China, Taiwan and Hong Kong to undertake the compilation and reviewing work for each plant family by volumes. Nearly 400 photographers provided numerous beautiful full colour plant photos. The well known zoological and botanical artist, Xiaolian ZENG, kindly allowed the use of more than twenty of his elegant and vivid plant portraits in this series.

This whole work is divided into nine volumes, depicting more than 10 thousand species of higher plants from China, dealing mainly with wild plants, but also including some important cultivated plants, and has involved the careful selection of nearly 20 thousand photographs. The system arrangement for plant families are as follows: bryophytes are mainly arranged according to the system used in *Flora Bryophytorum Sinicorum*; pteridophytes are arranged according to the system proposed by Professor Xianchun ZHANG in *A Glossary of Terms and Names of Lycopods and Ferns* (2015); gymnosperms and angiosperms are arranged according to the system used in *A. Engler's Syllabus der Pflanzenfamilien* (L. Diels, 1936), with the difference that in Volume III, Ranunculaceae is divided into Circaesteraceae, Ranunculaceae, and Paeoniaceae, and Magnoliaceae is divided into Magnoliaceae, Illiciaceae, Schisandraceae, and Tetracentraceae. The higher plants of China included in this work comprise 383 families; with 100 families of bryophytes, which represent the majority of the bryophyte families in China; the others are 40 pteridophyte families, 11 gymnosperm families and 232 angiosperm families, which represent all the families distributed in China respectively. The work includes some new additions of species published since *Flora of China*, such as *Carpinus insularis*, *Elatostema globosostigmatum*, *Micropera tibetica*, and new distribution records for China, such as *Trigonobalanus verticillata* and *Paphiopedilum gratrixianum*.

To facilitate and attract readers from both China and abroad, the text of this book series is bilingual in Chinese and English, providing the Chinese name, Latin name, morphological features, flowering and fruiting season, habitat and distribution.

In the process of compiling this work, Plant Photo Bank of China (PPBC) and Chinese Field Herbarium, both of which are under the Institute of Botany of Chinese Academy of Sciences, provided great help with the selection of photographs, for which we express our gratitude.

We also thank National Publication Foundation and Science Press, Beijing, for their great support for the publication of this book series.

It will be appreciated if mistakes and omissions are brought to our attention.

Editorial Committee of *Higher Plants of China in Colour*
31 October, 2015

# 关于本图鉴

1995年夏天，我参加由中国科学院昆明植物研究所臧穆教授带领的云南野外工作，同行的还有国际真菌学会理事长德籍的Franz Oberwinkler教授与法国的学者。臧教授爽朗好客，外国人都喜欢他的热情。那年去丽江，再去南部的西双版纳。西双版纳热带植物园的陶国达先生带领我们的野外工作，他是当地植物鉴定首席专家，知道好的树林在何处。一天，在傣族传统农家的木架房子吃中饭。臧教授建议陶先生既然喜欢摄影，何不出一本版纳植物图鉴，问我能不能帮忙在台湾找出版。我答应回去问问。

先问自然科学博物馆的李家维馆长，他对植物研究及保育充满热诚，对这项工作有兴趣。但未久他感觉这项工作所需时间过久，博物馆经费也不足以出版。我又问其他出版公司，没有得到响应。我想应该先有成果再问出版吧，就请陶先生持续植物拍摄。臧教授和夫人黎兴江教授推荐了费勇帮忙这项工作。1997年夏天，我在昆明机场与臧教授和费勇会合，一同飞去版纳。费勇年纪与我相当，长得瘦黑，话不太多。陶先生带领我们野外工作。回程时费勇说他想找几个同事一起负责滇西北的植物拍摄工作，与陶先生滇南的工作结合成为云南植物图鉴。回台后看陶先生给我的幻灯片，感觉质量不是太好，问他才知道所用的相机是正牌，镜头却是小厂牌。我汇钱请他购买一套相机，以利拍摄质量。

1998年，我到昆明植物所，和陶国达、费勇及孙航，讨论植物图鉴工作。费勇对此工作充满兴趣，人缘也好，决定由他征集昆明植物所人员拍摄的植物照片，并且中、英文字也由他撰写。翌年臧穆夫妇介绍昆明植物所的著名画家曾孝濂先生。曾先生长期考察云南山野的植物与动物，画作结合了科学性与艺术美感，是中国写实花鸟画得最好的。

2000年，费勇在日本富山县中央植物园半年，其间拍摄植物园栽培的中国植物。那年秋天费勇带我去大理点苍山和楚雄紫溪山。一天，我们在大理古城一间白族旧庭院吃风味晚餐，他兴致好，畅所欲言。费勇起初给我的印象是有些木讷，几次往来后就把我当熟人。几次的讨论，感觉他满心想做好这件事，并不在意条件。大理巷弄中有摊贩卖当地特产乳扇，他说闺女爱吃，买了两大张带走。

2001年年初一个早上，臧教授发来邮件，通知我费勇前一日在丽江不幸去世。一个年轻健康生命的突然离去，令人难以承受。出席完上午的会议后即打电话到昆明。黎教授说费勇到丽江出差，半夜室友听到声响，见他口吐白沫急送医院。地方医院初以为是癫痫，到清晨就不治了。

几个月后我有事联络曾孝濂先生，他告诉我费勇太太想与我联系。费勇太太姓向，我们称小向。她电话中希望植物图鉴工作能够继续，而且费勇的几个同事愿意帮忙。当年夏天在昆明的一个晚上，小向同昆明植物所的成晓、孙航、周浙昆一起和我见面，商讨后续的工作。成晓说他与费勇是同学，同时毕业，同时上班，他一定会帮忙。他确实尽力后续工作的联系与推动。2002年在昆明，几个朋友见面，小向带初中的女儿同来。女儿乖巧懂事，我说长得像费勇，她眼眶微微红了。成晓研究蕨类，他的岳父武素功先生及岳母方瑞征女士也是昆明植物所学者，两位在图片提供及文稿修订均提供协助。昆明植物所李锡文教授对植物分类的造诣比较全面，负责图片和文稿审查。昆明植物所还有多位专家对本书工作做出贡献，不在此逐一罗列。

2001年，曾孝濂先生介绍昆明植物所的画家刘怡涛先生。刘先生在版纳热带植物园待过，建立独特的版纳风光绘画风格，也喜好摄影，带过我几次野外工作。他建议我把植物图鉴工作扩大到全中国。艺术家天生具有美感，曾、刘两位画家拍摄的植物图，构图与取景皆有独到之处。2003年我到河北与吉林进行野外工作，2004年到新疆与吉林时决定把植物图鉴范围扩大到全中国。我和小向说明书的分量和质量要到位，才能彰显费勇的努力精神。费勇原本即有中国植物图鉴的梦想，干脆一次到位。

2004年，中国科学院植物研究所覃海宁博士来台，我们是1994年在英国邱园认识的。中国植物图像库在海宁领导下建立得有声有色。海宁总是满脸笑容，热诚谦虚，听我说植物图鉴的事，立即寻思找人帮忙。他人面广，介绍不少人，拍摄较好的有福建的何国生、四川的吴光弟、广西的刘演、广东的李泽贤。刘演的图片色彩饱满令人赞叹。我去爱丁堡皇家植物园时知道David Chamberlain博士是杜鹃花科专家，他同意审查杜鹃花科及小檗科图片。彭镜毅介绍哈佛大学David E. Boufford博士，他的图片是从中国西南的横断山脉植物调查工作所拍摄。David又推荐Susan Kelley及Richard Ree提供植物图片。

中科院植物所吴鹏程教授是苔藓专家，1990年我在芬兰赫尔辛基大学即将取得博士学位时他在赫大待了几个月。吴教授介绍几位中科院植物所的专家帮忙图片审查及文字撰写。台湾真菌学前辈吕理燊博士介绍昆明市农业局副局长惠肇祥先生提供杂草图片，惠先生又介绍北京的车晋滇先生提供华北的杂草图片。台湾赖明洲教授介绍上海自然博物馆的秦祥堃先生提供华东植物图片，又介绍中国科学院沈阳应用生态研究所的赵大昌先生提供长白山植物图片。我2004年到乌鲁木齐开会，组织会议的新疆大学阿不都拉教授拍了不少新疆植物图片，也提供给我。

大学同学康世昌是植物及计算机高手，拍摄的植物图片也提供给我。他早预想到网络世界的影响力，不推荐大部头实体书的出版构想。多年前他写个网址要我去看，那是我不知道的“Google”，可以查询信息。网络上图片的数量越来越多，趋势是如此。我在芬兰的指导教授Tuomo Niemelä出版过大型真菌的小书，亲自编排，图片与文字搭配得美感十足，我每翻阅总是心情愉悦。我向Tuomo请教对这套植物实体书的意见。他说网络的数据有时会消失，且许多没经过审查。我想这套书终要完成，无法顾及趋势与新世代人类的想法。

2007年年初，我在网络发现中科院植物所的中国植物图像库有影像部分。负责的是李敏，我问他图片提供者，他推荐几位拍摄较好的。多数是中科院植物所的年轻人，有刘冰、林秦文、于胜祥、李敏、高贤明、郦艳红，还有陕西的王萙。我当时已收集中国植物5000种的图片，工作超过10年理应收尾。然不加入这批有许多北方植物的图片实在不舍。刘冰是植物分类奇葩，这么年轻就拍到数量惊人的植物图片。刘冰和刘怡涛是给这项工作提供图片最多的两位。刘博帮忙不少文字撰写及图片审查，工作积极。当年年底，图片收集到6500种以上，接着准备文字、图片审查等出书的各项工作。

2010年在台湾“中央研究院”召开一项研讨会，覃海宁和李敏也来了，他们的报告显示中国植物图像库已收到数十万张图片。我如果再搜寻一次图片，能收到更好及更多的图片，但面对许多人殷盼这套书问世，时间的延长，压力更大。终究，我相信费勇会支持这最后一批图片的征求。湖南喻勋林及张代贵两位教授寄来许多华中植物图片，浙江张宏伟先生及安徽施忠辉先生也送来图片。吉林通化的周繇教授寄来他辛苦拍摄的长白山植物图片。近三年送来较多图片的还有朱鑫鑫、陈又生、徐晔春、陈彬、陈世品、何海、周喜乐，以及蕨类的张宪春和兰科的金效华。好友张力负责苔藓部分。

“中央研究院”彭镜毅博士提供了许多秋海棠科图片，也修订这科的文稿。牟善杰是台湾的蕨类学家，提供一些蕨类图片给本书，也审查过蕨类图片及文字。我在台湾大学念博士时，善杰是大学生，见他圆圆的笑脸，成天在标本馆研究。2010年11月，44岁的他突然中风走了，令人感慨！吕碧凤小姐是台湾优秀的业余蕨类专家，提供一些好的蕨类图片。还要感谢提供及审查图片的几位同事：王秋美、陈志雄、胡维新、黄俊霖、严新富和邱少婷。

早期收到的是幻灯片及少数印好的照片，2005年以后送来的是数码影像。数码图片干净，缺点是饱和度、清晰度和锐利度表现稍差，绿色部分有时偏黄。图效调整可改善这些问题。幻灯片的影像则会受到底片、冲洗、保存、扫描等质量的影响，好质量的并不多。图片须裁切出重点部位，再调整影像效果，这些工作大多是我处理。商请到一批人分别撰写文字。虽然有范本给撰写人参考，但各人的写法与仔细程度难免不一，有疏漏或小错误的情形普遍存在。起初我自己参考文献逐一查核，修订了约两千种的文字，但工作量太大，无法继续亲为。文字工作贡献较多的有费勇、刘博、萨仁、谷粹芝、成晓、杜宁、李锡文、徐晓婷和方瑞征等。

吴鹏程教授与科学出版社生物分社社长王静女士提及这项工作，王静有兴趣了解出版的可能，我们2010年在北京见面。自己过于深入这项工作，甚至如排版形式、字形等都亲自研究。像是自己养大的小孩，不放心交给他人处理，而且书稿已经在台湾找设计公司开始排版了。王静有毅力，持续两年逐渐消减我的疑虑。二十年来两岸的社会经济形势改变，使得这套书在大陆出版成为自然。德高望重的王文采院士及植物分类权威李振宇教授鼎力相助、组织动员，国家出版基金给予资助，促使整体工作能顺利完成。

吴声华

2015年10月20日

## About the Pictorial Series

In the summer of 1995, I took part in the Yunnan fieldwork led by Prof. Mu ZANG from Kunming Institute of Botany, Chinese Academy of Sciences. Joining us were German professor Franz Oberwinkler, director general of International Mycological Association, and some French scholars. Prof. ZANG was candid, cordialand hospitable, which impressed everyone, especially the foreign guests. We first went to Lijiang and then Sipsongpanna in the south. During this fieldwork, we were guided by Mr. Guoda TAO from Xishuangbanna Tropical Botanical Garden, Chinese Academy of Sciences. He was the chief expert of plant identification in the area, knowing which areas of the woods were worth this field inspection of ours. One day, when we were having lunch together in a traditional wooden house of an ethnic Dai family, Prof. ZANG proposed to Mr. TAO: "Since you are so fond of photography, why not compile a pictorial book of Banna's plants?" Prof. ZANG then turned to me, asking whether I could give help in getting the book published in Taiwan, and I promised to give it a try after returning to Taiwan.

I first contacted Dr. Chiawei LI, director of Museum of Natural Science, who was passionate about plant research and conservation and interested in the project. But before long, his passion faded due to his sense that the project was likely to take too long a time, and the Museum did not have sufficient fund to support the publishing. I then inquired of other publishing companies, but none of them gave a positive response. These setbacks sent me thinking that perhaps we should make some tangible achievements first before our work could be accepted for publication. So I asked Mr. TAO to proceed with shooting plants. Prof. ZANG and his wife Prof. Xingjiang LI recommended Yong FEI to provide assistance to the work. In the summer of 1997, I met Prof. ZANG and Yong FEI at Kunming Airport, and we flew to Sipsongpanna together for the fieldwork led by Mr. TAO. Of the same age as mine, Yong FEI was a thin and swarthy man, not very talkative. On our way back, Yong FEI said he was considering asking several of his colleagues to join him in shooting plants of the northwest of Yunnan so that the pictures taken in the two areas (the SouthYunnan and the Northwest Yunnan) could be combined to make a single pictorial book that could be called "Plants of Yunnan". After returning to Taiwan, I browsed the slides given by Mr. TAO, feeling that their quality was not ideal. Having asked Mr. TAO about this, I learned that it had been caused by his camera whose main body was of good brand and quality but whose lens was made by a mediocre producer. I remitted money to him for purchasing a new camera set, hoping that the quality of photos could be ensured by a high-quality camera.

In 1998, I visited Kunming Institute of Botany to discuss with Guoda TAO, Yong FEI and Hang SUN about the work of pictorial book for plants. Yong FEI was full of enthusiasm on the work, and had good relations with people, so we decided to commission him to collect plant photos taken by staff from the Kunming Institute, and to compose text both in English and Chinese. The next year, Prof. Mu ZANG and Prof. LI introduced to me Mr. Xiaolian ZENG, who had been engaging in the investigation of plants and animals in the wilds of Yunnan Province for a long time and was also a famous painter from the Kunming Institute of Botany. His paintings are the best realistic bird-and-flower works in China, blending scientificity with artistic beauty.

In 2000, Yong FEI spent half a year in Botanic Gardens of Toyama (Japan), taking photos of Chinese plants grown in the Gardens. In autumn of the same year, with Yong FEI as my guide, we went to Diancang Mountain in Dali and Zixi Mountain in Chuxiong. One day, when we were having local delicacies for super in an old courtyard of Bai nationality, Yong FEI got into high spirit and chatted with me without restraint. My first impression of Yong FEI was that he was a bit unapproachable, but after several rounds of conversations, he regarded me as his close friend. After several discussions with him, I felt that he very much concentrated on doing the work well, paying no attention to remuneration. In a lane of Dali, we found a vendor selling milk fan cake, a kind of local specialty, and he bought two big pieces, saying that they were for his daughter who liked such food.

One early morning in the early 2001, Prof. ZANG sent me an email, saying sadly that Yong FEI passed away in Lijiang the day before. It was really unbearable to hear of the sudden passing of such a young life. As soon as the meeting in that morning ended, I called to Kunming. The call was answered by Prof. LI who said that Yong FEI had been on a working trip at the time. At midnight, his roommates were awakened by some noises and found him foaming at the mouth. He was rushed to a local hospital and initially diagnosed as only having a fit of epilepsy, but no amount of treatment took effect on him; he passed away just as dawn came.

Several months later, when contacting Mr. Xiaolian ZENG, I was told that Yong FEI's wife was looking for me. The family name of Yong FEI's wife was XIANG, so we called her Little XIANG, a traditional way of Chinese people addressing their acquaintances who were younger than themselves. Little XIANG expressed her wish over phone that the project of the pictorial book should go on as usual and she also said that several of Yong FEI's colleagues were willing to help. One summer evening of the same year in Kunming, Little XIANG, together with Xiao CHENG, Hang SUN and Zhekun ZHOU all from the Kunming Institute of Botany, had a meeting with me to discuss about the remaining work of the project. Xiao CHENG said he and Yong FEI were classmates, graduating and first getting employed at the same time, so he would definitely offer his help. And in fact he did try his best to facilitate the progress of the work through networking. In 2002, we had a gathering in Kunming. Little

XIANG brought her daughter there, who was then a junior-secondary-school student. The girl was both clever and well-behaved, and when I said to her that she looked like her father, her eyes moistened slightly. Xiao CHENG was a fern researcher. His father-in-law Mr. Sugong WU and mother-in-law Mrs. Ruizheng FANG were also scholars of the Kunming Institute of Botany, both of whom offered their assistance in providing photos and editing texts. Prof. Xiwen LI also from the Kunming Institute of Botany, who had comprehensive attainments in plant taxonomy, was responsible for examining photos and texts. There were many other experts from the Kunming Institute of Botany who made contributions to this book, but due to space constraint, their names are not listed here one by one.

In 2001, Mr. Xiaolian ZENG introduced to me Mr. Yitao LIU, another painter from Kunming Institute of Botany. Mr. LIU used to stay in Xishuangbanna Tropical Botanical Garden, where he developed his distinctive painting style with which to depict typical Sipsongpanna's landscape. He was also a lover of photography, and used to be my fieldwork guide for several times. Mr. LIU suggested that I expand the pictorial plant book project to cover the whole territory of China. Due to the innate aesthetic sense of artist, the plant photos taken by the two painters - Mr. ZENG and Mr. LIU - had unique characteristics both in picture composing and view finding. My fieldwork in Hebei and Jilin in 2003 and then my travelling in Xinjiang and Jilin in 2004 prompted my final decision to expand the pictorial plant book project to the whole country. I explained to Little XIANG that only when the book was comprehensive enough and of high quality, could Yong FEI's hardworking spirit and aspiration in this regard be fully manifested. And only in this way could Yong FEI's cherished dream of compiling a pictorial book on plants of China be realized without unnecessary pre-steps.

2004 saw Dr. Haining QIN's visit to Taiwan. Dr. QIN was from Institute of Botany, Chinese Academy of Sciences, and we got to know each other at British Kew Gardens in 1994. Under the leadership of Haining, the construction of Plant Photo Bank of China was making marvelous progress. Haining was a cordial and modest man, with his face always shining with smile. Upon knowing that I was conducting the project of pictorial plant book, he offered to give help. Taking advantage of his wide network, he brought in many talents, among whom Guosheng HE from Fujian, Guangdi WU from Sichuan, Yan LIU from Guangxi, and Zexian LI from Guangdong were all good at photography. In terms of color, Yan LIU's photos were particularly good, which was admirable. In addition, Dr. David Chamberlain, an expert in Ericacea, whom I got to know when I visited Royal Botanic Garden Edinburgh, agreed to review the photos of Ericaceae and Berberidaceae. Besides, Dr. Ching-I PENG introduced Dr. David E. Boufford from Harvard University who provided photos taken when he was investigating the plants of the Hengduan Mountains in the southwest of China. And David also recommended Susan Kelley and Richard Ree who both offered their plant photos.

Prof. Pengcheng WU from Institute of Botany, Chinese Academy of Sciences. was an expert in bryophytes. In 1990, he stayed in University of Helsinki, Finland for a few months when I was about to obtain my doctorate awarded by the University. Prof. WU introduced several experts from Institute of Botany to help review photos and write text. Dr. Liisin LEU, a Taiwan veteran in mycology, recommended Mr. Zhaoxiang HUI, deputy director of Kunming Municipal Bureau of Agriculture, to provide photos of weeds. And Mr. HUI invited Mr. Jindian CHE from Beijing to provide photos of weeds in Northern China. Prof. Mingjou LAI from Taiwan involved Mr. Xiangkun QIN from Shanghai Natural History Museum in contributing photos of plants in Eastern China, and then recommended Mr. Dachang ZHAO from Shenyang Institute of Applied Ecology, Chinese Academy of Sciences to offer photos of plants in Changbai Mountains. In 2004, I went to Urumqi to attend a conference whose organizer, Prof. Abdulla from Xinjiang University, gave me many photos of Xinjiang plants taken by himself.

My college classmate Shihchang KANG, an expert in plants and computer, also sent me plant photos taken by himself. Having long foreseen the power of internet, he did not quite agree with the idea of publishing a bulky physical book. Years ago, he wrote down a website address and asked me to visit it. The website, which I had never heard of before, was 'Google', a 'search engine' enabling us to search for information easily. And it turned out that this became a strong upward trend, with more and more photos being uploaded onto internet for people to view or download. However, Prof. Tuomo Niemelä, my Finnish adviser, had a different view on this phenomenon. He had published a handbook about large fungi, whose formatting was done by himself. The photos and text were arranged so well that a full sense of beauty permeated the entire book, and this always made me in a good mood each time I read it. When being consulted about the idea of publishing a physical plant book like this one, he encouraged me to continue doing so, saying that sometimes online data and materials would vanish for no reason and many online materials could not be said to be authentic because they had not undergone necessary review and approval. With this encouragement, I decided to carry out this project through to the end, paying no attention to the trends and fashionable ideas of new generations.

At the beginning of 2007, I found on internet that the Plant Photo Bank of China owned by Institute of Botany contained image data being managed by Min LI. So I asked him for sources of these photos.

Min LI recommended several persons whose photos in the Database were regarded as excellent. Most of these photo-takers were young people from the Institute of Botany. They were Bing LIU, Qinwen LIN, Shengxiang YU, Min LI, Xianming GAO, Yanhong BING. Besides, Yun WANG from Shaanxi was also added to the list of recommendation. By that time, I had already collected photos of 5000 species of plants in China through over 10 years of my hardwork which could have very well wound up. However, it would have been regrettable if I had not added so many fine photos of plants in Northern China to this important book. Bing LIU was a wonder in plant taxonomy - so young as he was, he had taken astonishingly large number of plant photos. It was Bing LIU and Yitao LIU who provided the largest number of photos for this work. Bo LIU, who was very active in work, helped a lot in writing text and reviewing photos. By the end of the same year, we had collected photos of more than 6500 species, paving the way for doing other publication-related preparatory work such as text writing and photo reviewing.

In 2010, "Academia Sinica" held a seminar in Taiwan, at which Haining QIN and Min LI delivered their reports which revealed that the Plant Photo Bank of China had collected hundreds of thousands of photos. In this circumstance, one more round of photo searching and collecting would certainly make more and better photos available for this upcoming book. Only, it would take more time. With so many people looking forward to the publication of the book, the longer time we took in publishing, the heavier pressure we would face. But in final analysis, I believed that Yong FEI, if he were still alive, would support this last round of photo searching and collecting. Prof. Xunlin YU and Prof. Daigui ZHANG from Hunan sent me many photos of plants of Central China. Mr. Hongwei ZHANG from Zhejiang and Mr. Zhonghui SHI from Anhui also sent photos to me. Prof. You ZHOU from Tonghua of Jilin contributed the photos of plants of Changbai Mountain that he took with great efforts. In the recent three years, a lot of photos were also provided by Xinxin ZHU, Yousheng CHEN, Yechun XU, Bin CHEN, Shipin CHEN, Hai HE, and Xile ZHOU. Xianchun ZHANG offered many photos of ferns. Xiaohua JIN submitted many photos of Orchidaceae plants. My good friend Li ZHANG was responsible for bryophytes.

Dr. Ching-I PENG from "Academia Sinica" provided a lot of pictures of Begoniaceae and edited the draft for this family. Shannjye MOORE, an expert of ferns from Taiwan, contributed some pictures of ferns to this book, and reviewed the pictures and text for the fern part. When I studied for doctorate in Taiwan University, Shannjye was still an undergraduate of the University. With a lovely round face often with smile, he was always seen studying in herbarium. In November 2010, however, he suddenly died of a stroke at the age of 44, making us very sad and regretful. Miss Pifong LU, an excellent amateur expert of ferns from Taiwan, contributed some good pictures of ferns. I also would like to express my thanks to the following colleagues who provided and reviewed pictures for me Chiumei WANG, Chihhsiung CHEN, Weihsin HU, Chunlin HUANG, Hsinfu YEN and Shauting CHIU.

What we received in earlier stages were slides and a small number of prints,and after 2005, all contributions were in the form of digital image. Digital photos are clean, but their saturation, definition and sharpness are not very ideal, with green parts tending to turn slightly yellowish. Fortunately, these problems could be solved through photo-effect modification. As to the slides, high-quality ones were not many, as the quality of such images hinged on such factors as: quality of the film, developing process, storage condition and scanning, etc. The photos first needed some trimming so as to highlight their essential parts and then required modification to the image effect; most of the work was done by myself. In the meantime, we engaged a group of people to do text writing. Although templates were provided to text writers, inconsistency still appeared in some places due to different writing styles of different writers. There were also not a few oversights or slips caused by some writers who were not conscientious enough. At first, I myself did the correction and revision one by one against reference literature, finishing the work on about two thousand species, but as the amount of this kind of work was so big that I could not continue to do it all by myself. Here, I would like to list those who made greater contributions to the text. They are: Yong FEI, Bo LIU, Ren SA, Cuizhi GU, Xiao CHENG, Ning DU, Xiwen LI, Xiaoting XU and Ruizheng FANG, etc.

Prof. Pengcheng WU mentioned this work to Ms. Jing WANG, director of Biological Division of Science Press, who was interested in exploring the possibility of publishing the work, so we met each other in Beijing in 2010. Before this, I had devoted myself to the work so deeply that even small details like typesetting and font were studied and arranged by myself. Therefore, the work was like a child brought up by myself, so I would feel uneasy if I put it in the care of someone else, and moreover, we had already commissioned a design company in Taiwan to start typesetting the draft. However, my concern and worry were gradually dispelled by Jing WANG's sincerity and her perseverance in persuasion and explanation over two successive years. And the changes in social and economic situations across the Straits also made it natural for the book to be published in the Mainland. Also worthy of mentioning are: the generous support, organization and mobilization given or conducted by both Wentsai WANG, a renowned academician and Prof. Zhenyu LI, an expert on plant taxonomy, as well as the funding by National Publication Foundation. All this facilitated the smooth completion of the entire work.

Shenghua WU

20 October, 2015

# 第5卷编审者分工

| 科名 | 编审者 |
|---|---|
| 大戟科 | 李秉滔　李章海 |
| 虎皮楠科 | 刘　博　林秦文 |
| 水马齿科 | 王　东 |
| 黄杨科 | 林秦文　彭　华 |
| 岩高兰科 | 张重岭 |
| 马桑科 | 刘　博 |
| 漆树科 | 刘　博　林秦文 |
| 五列木科 | 刘大伟 |
| 冬青科 | 刘　博　林秦文　陈书坤 |
| 卫矛科 | 刘　博　刘全儒 |
| 翅子藤科 | 刘　博 |
| 刺茉莉科 | 刘　冰 |
| 省沽油科－茶茱萸科 | 刘　博　林秦文　彭　华 |
| 槭树科 | 陈又生　刘　博　徐松芝 |
| 七叶树科 | 刘　博　叶建飞 |
| 无患子科－清风藤科 | 刘　博　林秦文 |
| 凤仙花科 | 于胜祥　刘克明 |
| 鼠李科 | 刘　博　林秦文　王秋美 |
| 葡萄科 | 刘　博　陈之端 |
| 杜英科－椴树科 | 刘　博　林秦文 |
| 锦葵科 | 刘　博　林秦文　于胜祥 |
| 木棉科－五桠果科 | 刘　博　林秦文 |
| 猕猴桃科 | 李新伟　刘　博　张红瑞 |
| 金莲木科 | 刘　博　林秦文 |
| 山茶科 | 杨世雄　邓云飞 |
| 藤黄科 | 刘　博　林秦文　张代贵 |
| 龙脑香科 | 李剑武 |
| 沟繁缕科 | 刘　博　林秦文 |
| 瓣鳞花科 | 金效华 |
| 柽柳科 | 段士民　彭　华 |
| 半日花科 | 林秦文 |
| 红木科 | 王　晖 |
| 堇菜科 | 陈又生 |
| 大风子科－旌节花科 | 刘　博　林秦文　彭　华 |
| 时钟花科 | 刘　冰 |
| 西番莲科－四数木科 | 刘　博　林秦文 |
| 秋海棠科 | 彭镜毅　税玉民 |
| 钩枝藤科 | 刘　博　林秦文 |
| 仙人掌科 | 李振宇 |
| 瑞香科 | 刘　博　林秦文　方瑞征 |
| 胡颓子科 | 孙　苗　彭　华 |
| 千屈菜科 | 刘　博　林秦文　邓云飞 |
| 海桑科 | 王　晖 |
| 隐翼科－玉蕊科 | 刘　博　林秦文 |
| 红树科 | 王　晖 |
| 紫树科 | 刘　博　林秦文 |
| 八角枫科 | 刘　博　赖阳均 |
| 使君子科 | 税玉民 |
| 桃金娘科 | 蒋　宏　张代贵 |
| 野牡丹科 | 税玉民　邓云飞 |
| 菱科 | 于胜祥 |
| 柳叶菜科 | 陈家瑞　刘　博　林秦文 |
| 小二仙草科 | 王　东 |
| 杉叶藻科 | 刘　博 |
| 假繁缕科 | 徐松芝 |
| 锁阳科 | 刘　博 |
| 五加科 | 李　嵘　刘　博 |
| 伞形花科 | 刘　博　刘启新　林秦文 |
| 山茱萸科 | 刘　博　于胜祥 |

## Authors and Reviewers of Volume Ⅴ

| | | | |
|---|---|---|---|
| Euphorbiaceae | Bingtao LI | Zhanghai LI | |
| Daphniphyllaceae | Bo LIU | Qinwen LIN | |
| Callitrichaceae | Dong WANG | | |
| Buxaceae | Qinwen LIN | Hua PENG | |
| Empetraceae | Chongling ZHANG | | |
| Coriariaceae | Bo LIU | | |
| Anacardiaceae | Bo LIU | Qinwen LIN | |
| Pentaphylacaceae | Dawei LIU | | |
| Aquifoliaceae | Bo LIU | Qinwen LIN | Shukun CHEN |
| Celastraceae | Bo LIU | Quanru LIU | |
| Hippocrateaceae | Bo LIU | | |
| Salvadoraceae | Bing LIU | | |
| Staphyleaceae—Icacinaceae | Bo LIU | Qinwen LIN | Hua PENG |
| Aceraceae | Yousheng CHEN | Bo LIU | Songzhi XU |
| Hippocastanaceae | Bo LIU | Jianfei YE | |
| Sapindaceae—Sabiaceae | Bo LIU | Qinwen LIN | |
| Balsaminaceae | Shengxiang YU | Keming LIU | |
| Rhamnaceae | Bo LIU | Qinwen LIN | Chiumei WANG |
| Vitaceae | Bo LIU | Zhiduan CHEN | |
| Elaeocarpaceae—Tiliaceae | Bo LIU | Qinwen LIN | |
| Malvaceae | Bo LIU | Qinwen LIN | Shengxiang YU |
| Bombacaceae—Dilleniaceae | Bo LIU | Qinwen LIN | |
| Actinidiaceae | Xinwei LI | Bo LIU | Hongrui ZHANG |
| Ochnaceae | Bo LIU | Qinwen LIN | |
| Theaceae | Shixiong YANG | Yunfei DENG | |
| Guttiferae | Bo LIU | Qinwen LIN | Daigui ZHANG |
| Dipterocarpaceae | Jianwu LI | | |
| Elatinaceae | Bo LIU | Qinwen LIN | |
| Frankeniaceae | Xiaohua JIN | | |
| Tamaricaceae | Shimin DUAN | Hua PENG | |

| | | | |
|---|---|---|---|
| Cistaceae | Qinwen LIN | | |
| Bixaceae | Hui WANG | | |
| Violaceae | Yousheng CHEN | | |
| Flacourtiaceae—Stachyuraceae | Bo LIU | Qinwen LIN | Hua PENG |
| Turneraceae | Bing LIU | | |
| Passifloraceae—Tetramelaceae | Bo LIU | Qinwen LIN | |
| Begoniaceae | Ching-I PENG | Yumin SHUI | |
| Ancistrocladaceae | Bo LIU | Qinwen LIN | |
| Cactaceae | Zhenyu LI | | |
| Thymelaeaceae | Bo LIU | Qinwen LIN | Ruizheng FANG |
| Elaeagnaceae | Miao SUN | Hua PENG | |
| Lythraceae | Bo LIU | Qinwen LIN | Yunfei DENG |
| Sonneratiaceae | Hui WANG | | |
| Crypteroniaceae—Lecythidaceae | Bo LIU | Qinwen LIN | |
| Rhizophoraceae | Hui WANG | | |
| Nyssaceae | Bo LIU | Qinwen LIN | |
| Alangiaceae | Bo LIU | Yangjun LAI | |
| Combretaceae | Yumin SHUI | | |
| Myrtaceae | Hong JIANG | Daigui ZHANG | |
| Melastomaceae | Yumin SHUI | Yunfei DENG | |
| Trapaceae | Shengxiang YU | | |
| Onagraceae | Jiarui CHEN | Bo LIU | Qinwen LIN |
| Haloragaceae | Dong WANG | | |
| Hippuridaceae | Bo LIU | | |
| Theligonaceae | Songzhi XU | | |
| Cynomoriaceae | Bo LIU | | |
| Araliaceae | Rong LI | Bo LIU | |
| Umbelliferae | Bo LIU | Qixin LIU | Qinwen LIN |
| Cornaceae | Bo LIU | Shengxiang YU | |

# 目录 | Contents

第5卷
Volume Ⅴ

# 被子植物
# 大戟科－山茱萸科
# Angiosperms
# Euphorbiaceae－Cornaceae

# 大戟科 Euphorbiaceae

闭花木 *Cleistanthus sumatranus*

假肥牛树 *Cleistanthus petelotii*

## 雀儿舌头
**Leptopus chinensis** (Bunge) Pojark.

直立灌木，雌雄同株。叶膜质至纸质，两面无毛至密具粗毛。花序由单性或两性花组成；雄花每簇1-4朵；雌花萼片基部常具1或2个指状腺体；子房3室。蒴果平滑或具浅网纹。生海拔3000米以下的石山坡、灌丛中或林下。产中国西南、华南、华中、华北、华东和华西。巴基斯坦、缅甸和西南亚亦有。

Erect shrubs, monoecious. Leaves membranous to papery, both surfaces glabrous to densely hirsute. Inflorescences unisexual or bisexual; staminate flowers 1-4 per fascicle; pistillate flowers sepals often with 1 or 2 digitate glands at base; ovary 3-locular. Capsules smooth to faintly reticulate. Stony slopes, thickets or under forests below 3000 m. Distributed in SW, S, C, N, E and W China. Also in Pakistan, Myanmar and SW Asia.

雀儿舌头 *Leptopus chinensis*

## 闭花木
**Cleistanthus sumatranus** (Miq.) Müll. Arg.

常绿乔木，高达18米。叶卵圆形、椭圆形或椭圆状长圆形，纸质或硬纸质；少花簇生于叶腋(多达7朵)。雄花花瓣5，长约0.8毫米；雌花花瓣长约1毫米；子房卵球形至球形。蒴果红色，卵球状三角形。花期3-8月，果期4-10月。生海拔500(-700)米以下的密林中。产云南、广西、广东和海南。泰国、越南、柬埔寨、文莱、马来西亚、新加坡、印度尼西亚和菲律宾亦有。

Evergreen trees, up to 18 m tall. Leaves ovate, elliptic or ovate-oblong, papery or stiffly papery; flowers in axillary few-flowered fascicles (up to 7). Staminate flowers petals 5, ca. 0.8 mm long; pistillate flowers petals ca. 1 mm long; ovary ovoid to globose. Capsules red, ovoid-trigonous. Fl. Mar-Aug. Fr. Apr-Oct. Dense forests below 500(-700) m. Distributed in Yunnan, Guangxi, Guangdong and Hainan. Also in Thailand, Vietnam, Cambodia, Brunei, Malaysia, Singapore, Indonesia and the Philippines.

## 假肥牛树
**Cleistanthus petelotii** Merr. ex Croizat

乔木，雌雄同株，全株无毛。叶革质；侧脉每边6-7条，未达叶缘而联结。雄花数朵组成腋生团伞花序，花丝中部以下合生为圆筒状包围子房基部，花盘腺体倒卵形；雌花花盘坛状或筒状，包围子房；子房平滑。蒴果近圆球状。花期4-6月，果期5-11月。生海拔200-400米的石灰岩山地林中。产广西西部。越南北部亦有。

Trees, monoecious, glabrous throughout. Leaves leathery, lateral veins 6-7 pairs, anastomosing near margin. Male flowers several grouped into axillary glomerules, filaments connate below middle into a cylinder surrounding rudimentary ovary; disk scales obovate, female flowers disk urceolate or cylindric, surrounding ovary; ovary smooth. Capsules subglobose. Fl. Apr-Jun. Fr. May-Nov. Limestone forests at 200-400 m. Distributed in W Guangxi. Also in N Vietnam.

膜叶土蜜树 *Bridelia glauca*

## 膜叶土蜜树
**Bridelia glauca** Blume

乔木，雌雄同株，高达15米。叶椭圆状卵形、长圆形或倒卵形，膜质或薄纸质，上面和叶柄被毛。多花簇生多达50朵花，腋生；雄花白色。核果椭圆体形，基部具宿存萼片。花期5-9月，果期9-12月。生海拔500-1600米的山坡疏林中。产云南、广西、广东和台湾。南亚和东南亚亦有。

Trees, monoecious, up to 15 m tall. Leaves elliptic-ovate, oblong, or obovate, membranous or thickly papery, adaxial surface and petioles hairy. Many-flowered fascicles with up to 50 flowers, axillary; staminate flowers white. Drupes ellipsoidal, base with persistent sepals. Fl. May-Sep. Fr. Sep-Dec. Open forests on slopes at 500-1600 m. Distributed in Yunnan, Guangxi, Guangdong and Taiwan. Also in S and SE Asia.

## 禾串树
**Bridelia balansae** Tutcher

乔木，雌雄同株。叶近革质，椭圆形或窄椭圆形，边缘反卷，上面和叶柄均无毛，下面被毛。团伞花序腋生，多达12花，除花萼和花瓣具淡黄色短柔毛外无毛，子房球形至卵球形。果梗极粗壮；核果长圆状卵球形，成熟后紫黑色。花期5-8月，果期9-11月。生海拔200-1000米的林中。产中国西南和华南。老挝、越南和日本亦有。

Trees, monoecious. Leaves subcoriaceous, elliptic or narrowly elliptic, margin revolute, adaxial surface and petioles glabrous, abaxially hairy. Glomerules axillary, up to 12-flowered, glabrous except for yellowish pubescent sepals and petals; ovary globose to ovoid. Fruiting pedicels very stout; drupes oblong-ovoid, purple-black when mature. Fl. May-Aug. Fr. Sep-Nov. Forests at 200-1000 m. Distributed in SW and S China. Also in Laos, Vietnam and Japan.

禾串树 *Bridelia balansae*

## 网脉核果木
**Drypetes perreticulata** Gagnep.

乔木。小枝具棱。托叶线形，宿存；叶革质，网脉密而明显。雄花萼片4，花盘扁平，雄蕊约25，无退化雌蕊；雌花子房1室。核果长1.8-2.5厘米，宽1.4-1.8厘米，平滑；外果皮革质，中果皮肉质，内果皮木质；1室，具种子1粒。花期1-3月，果期5-10月。生海拔800米以下的常绿林中。产广东、广西、贵州、海南和云南。泰国和越南亦有。

Trees. Branchlets angulate. Stipules linear, persistent; leaves leathery, reticulate veins dense and prominent. Male flowers sepals 4; disk flattened; stamens ca. 25; rudimentary ovary absent. female flowers ovary 1-celled. Drupes 1.8-2.5 × 1.4-1.8 cm, smooth; exocarp leathery; mesocarp fleshy; endocarp woody, 1-celled, 1-seeded. Fl. Jan-Mar. Fr. May-Oct. Evergreen forests below 800 m. Distributed in Guangdong, Guangxi, Guizhou, Hainan and Yunnan. Also in Thailand and Vietnam.

网脉核果木
*Drypetes perreticulata*

柳叶核果木 *Drypetes salicifolia*

## 柳叶核果木
**Drypetes salicifolia** Gagnep.

乔木，雌雄异株，高达10米，除萼片、子房、果实和果梗具柔毛外无毛。叶纸质，条状披针形，下面有黑色小斑点。雄花1-3朵腋生；雄蕊约12；雌花单生。核果球形，黄褐色，常1室，具1粒种子。花期5-10月，果期7-10月。生海拔200-1000米的常绿阔叶林中。产中国西南、华中和华西。老挝和越南亦有。

Trees, dioecious, up to 10 m tall, glabrous except pubescent sepals, ovary, fruits, and fruiting pedicels. Leaves papery, linear-lanceolate, abaxially minutely black-maculate. Staminate flowers 1-3 axillary; stamens ca. 12; pistillate flowers solitary. Drupes globose, fulvous, usually 1-celled, 1-seeded. Fl. May-Oct. Fr. Jul-Oct. Evergreen broad-leaved forests at 200-1000 m. Distributed in SW, C and W China. Also in Laos and Vietnam.

## 黄毛五月茶
**Antidesma fordii** Hemsl.

小乔木，幼枝、叶柄和花序轴密具黄色绒毛。叶长圆形，有时椭圆形，纸质。花序顶生或腋生；雌花花盘环状；子房椭圆体形；花柱2深裂。核果纺锤形，长约7毫米，直径约4毫米。花期3-7月，果期7月至翌年1月。生海拔200-2300米的密林中。产云南、广西、广东、海南和福建。老挝和越南亦有。

Treelets, young twigs, petioles, and Inflorescences axes densely yellow tomentose. Leaves oblong, sometimes elliptic, papery. Inflorescences terminal or axillary; pistillate flowers with annular disk; ovary elliptic; styles 2-parted. Drupes fusiform, ca. 7 mm long, ca. 4 mm diam. Fl. Mar-Jul. Fr. Jul to next Jan. Dense forests at 200-2300 m. Distributed in Yunnan, Guangxi, Guangdong, Hainan and Fujian. Also in Laos and Vietnam.

## 方叶五月茶
**Antidesma ghaesembilla** Gaertn.

乔木。幼枝被柔毛。托叶针形，早落；叶片椭圆形、稀卵形和倒卵形。雄花萼片5(或4-7)，离生；花丝着生于4-7个分离的花盘裂片之间；雌花子房被短柔毛。核果椭圆形。花期3-9月，果期6-12月。生海拔200-1100米的疏林中。产广东、广西、海南和云南。南亚、东南亚、巴布亚新几内亚和澳大利亚南部亦有。

Trees. Young twigs pubescent. Stipules needlelike, caducous; leaves oblong, more rarely ovate or obovate. Male flowers sepals 5(or 4-7), free; disk consisting of 4-7 free alternistaminal obconical lobes; female flowers ovary pubescent. Drupes ellipsoid. Fl. Mar-Sep. Fr. Jun-Dec. Sparse forests at 200-1100 m. Distributed in Guangdong, Guangxi, Hainan and Yunnan. Also in S and SE Asia, Papua New Guinea and S Australia.

黄毛五月茶 *Antidesma fordii*

方叶五月茶 *Antidesma ghaesembilla*

西南五月茶 *Antidesma acidum*

五月茶 *Antidesma bunius*

## 西南五月茶

**Antidesma acidum** Retz.

灌木或小乔木。叶倒卵形至椭圆状长圆形，膜质至纸质，下面被短柔毛。总状花序；雄花花萼杯状至球形，(3或)4裂，开裂至约1/3；子房无毛；雄蕊1-3。核果椭圆体形，近圆柱形至侧扁。花期5-7月，果期6-11月。生海拔100-1500米的山坡疏林中。产云南、四川和贵州。南亚和东南亚亦有。

Shrubs or treelets. Leaves obovate to elliptic-oblong, membranaceous to papery, abaxially pubescent. Inflorescences racemose; staminate flowers calyx cupular to globose, (3 or)4-lobed, divided for ca. 1/3; ovary glabrous; stamens 1-3. Drupes ellipsoid, nearly terete to laterally compressed. Fl. May-Jul. Fr. Jun-Nov. Open forests on slopes at 100-1500 m. Distributed in Yunnan, Sichuan and Guizhou. Also in S and SE Asia.

## 小叶五月茶

**Antidesma montanum** Blume var. **microphyllum** (Hemsl.) Petra ex Hoffmam.

乔木。叶柄长2-3(-5)毫米；叶披针形至条形，纸质或革质，长为宽的4-10倍或叶宽少于2厘米，平均3-6 × 0.4-1.5厘米，先端急尖或钝。花序和果序长1-4厘米，雄性者不分枝至6分枝，雌性者不分枝至2分枝。花期4-6月，果期6-11月。生海拔100-1200米的河边。产中国西南和华南。老挝、泰国和越南亦有。

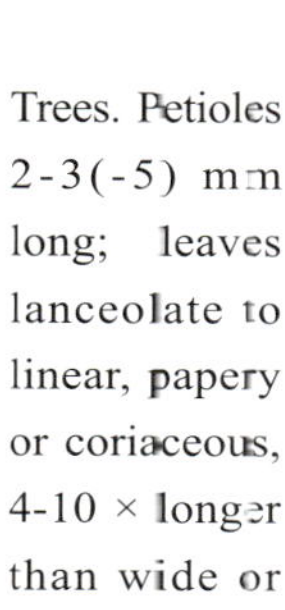

Trees. Petioles 2-3(-5) mm long; leaves lanceolate to linear, papery or coriaceous, 4-10 × longer than wide or less than 2 cm wide, average size 3-6 × 0.4-1.5 cm, apex acute or obtuse. Inflorescences and infructescences 1-4 cm long, males unbranched to 6-branched, females unbranched to 2-branched. Fl. Apr-Jun. Fr. Jun-Nov. Along rivers at 100-1200 m. Distributed in SW and S China. Also in Laos, Thailand and Vietnam.

## 五月茶

**Antidesma bunius** (L.) Spreng.

乔木。叶长圆形、椭圆形或倒卵形，(5-)10-23(-32) × (2-)3-10厘米，革质或厚纸质；无孔穴。花序腋生或顶生；轴粗壮；雄花花萼长1-1.5毫米，杯状。核果椭圆体形，侧扁，5-11 × 4-7毫米，成熟时红色至黑色。花期3-5月，果期6-11月。生海拔200-1800米的林中。产中国西南和华南。南亚、东南亚和大洋洲亦有。

Trees. Leaves oblong, elliptic, or obovate, (5-)10-23(-32) × (2-)3-10 cm, coriaceous or thickly papery; domatia absent. Inflorescences axillary or terminal; axes robust; staminate flowers calyx 1-1.5 mm long, cup-shaped. Drupes ellipsoid, laterally compressed, 5-11 × 4-7 mm, red to black when ripe. Fl. Mar-May. Fr. Jun-Nov. Forests at 200-1800 m. Distributed in SW and S China. Also in S and SE Asia, and Oceania.

小叶五月茶 *Antidesma montanum* var. *microphyllum*

一叶萩 *Flueggea suffruticosa*

## 一叶萩

**Flueggea suffruticosa** (Pall.) Baill.

灌木。雌雄异株，叶纸质，侧脉5-8对。花序腋生，聚伞花序；雄花3-18朵簇生；雄蕊5；雌花萼片5，近全缘；子房卵球形，(2或)3室。蒴果三角状扁球形，成熟后红褐色。花期3-8月，果期6-11月。生海拔500-2500米的山坡灌丛、林缘、溪岸或路边。产中国各地。俄罗斯、蒙古、朝鲜半岛和日本亦有。

Shrubs,dioecious. Leaves papery, lateral veins 5-8 per side. Inflorescences axillary, cymose; staminate flowers 3-18-clustered; stamens 5; pistillate flowers sepals 5, subentire; ovary ovoid, (2 or)3-locular. Capsules triquetrou-soblate, reddish brown when ripe. Fl. Mar-Aug. Fr. Jun-Nov. Thickets on slopes, forest edges, stream banks or roadsides at 500-2500 m. Widely distributed in China. Also in Russia, Mongolia, Korean Peninsula and Japan.

## 白饭树

**Flueggea virosa** (Roxb. ex Willd.) Voigt

灌木，雌雄异株，无毛。叶纸质，边缘全缘。花序腋生，簇生；苞片鳞片状；花无花瓣；雄花萼片5，浅黄色；雄蕊5；雌花序具(1-)3-10花；子房卵球形，3室。蒴果浆果状，近球形，淡白色，不开裂。花期3-8月，果期6-11月。生海拔100-2000米的山坡灌丛中。产中国西南、华南和华东。东南亚、东亚、大洋洲和非洲亦有。

Shrubs, dioecious, glabrous. Leaves papery, margin entire. Inflorescences axillary, fascicled; bracts scaly; flowers without petals; staminate flowers sepals 5, yellowish; stamens 5; pistillate flowers Inflorescences (1-)3-10-flowered; ovary ovoid, 3-locular. Capsules baccate, subglobose, whitish, indehiscent. Fl. Mar-Aug. Fr. Jun-Nov. Thickets on slopes at 100-2000 m. Distributed in SW, S and E China. Also in SE and E Asia, Oceania and Africa.

白饭树 *Flueggea virosa*

## 毛白饭树

**Flueggea acicularis** (Croiz.) Webster

灌木，雌雄异株。树皮淡褐色，初时被柔毛。叶纸质，倒卵形，3-7 × 2-5毫米，侧脉4-5对，具微柔毛。花单生或几朵簇生于短枝顶端；雄蕊6；雌花萼片5；子房3室。浆果球形。花期3-5月，果期6-10月。生海拔300-400米的山坡灌丛中。产云南、四川和湖北。

Shrubs, dioecious. Bark brownish, at first pubescent. Leaves papery, obovate, 3-7 × 2-5 mm, lateral veins 4-5 per side, puberulent. Flowers solitary or several flowers clustered at apex of short branches; stamens 6; pistillate flowers with 5 sepals; ovary 3-celled. Berries globose. Fl. Mar-May. Fr. Jun-Oct. Thickets on slopes at 300-400 m. Distributed in Yunnan, Sichuan and Hubei.

毛白饭树 *Flueggea acicularis*

小果叶下珠 *Phyllanthus reticulatus*

## 小果叶下珠
**Phyllanthus reticulatus** Poir.

灌木，雌雄同株。叶形多变，膜质至纸质。花序腋生，由2-10雄花和1或2雌花簇生组成；雄花萼片5或6，2轮，雄蕊5；子房4-12室，平滑。浆果球形至扁球形，成熟后黑色及深紫色。花期3-6月，果期6-10月。生海拔200-800米的林下或灌木丛中。产中国西南、华南和东南。南亚和东南亚、非洲西部和澳大利亚东北部亦有。

Shrubs, monoecious. Leaves varying in shape, membranous to papery. Inflorescences an axillary fascicle, with 2-10 staminate and 1 or 2 pistillate flowers; male flowers sepals 5 or 6, in 2 series; stamens 5; ovary 4-12-celled, smooth. Berries globose to oblate, black and dark purplish at maturity. Fl. Mar-Jun. Fr. Jun-Oct. Forests or thickets at 200-800 m. Distributed in SW, S and SE China. Also in S and SE Asia, W Africa and NE Australia.

## 落萼叶下珠
**Phyllanthus flexuosus** (Siebold et Zucc.) Müll. Arg.

灌木，雌雄同株，全株无毛。叶椭圆形至卵形，沿枝渐变大，二列。叶腋簇生花序具多达5朵雄花和1朵雌花；雄花萼片5，深紫色；雄蕊5；雌花萼片6，早落。果实呈浆果状，扁球形。花期4-5月，果期6-9月。生海拔700-1500米的疏林或灌丛中。产中国西南、东南、华中和华东。日本亦有。

Shrubs, monoecious, glabrous throughout. Leaves elliptic to ovate, progressively larger along shoot, distichous. Inflorescences an axillary fascicle with up to 5 staminate and 1 pistillate flowers; staminate flowers sepals 5, dark purple; stamens 5; pistillate flowers sepals 6, caducous. Fruits berrylike, oblate. Fl. Apr-May. Fr. Jun-Sep. Open forests or thickets at 700-1500 m. Distributed in SW, SE, C and E China. Also in Japan.

## 余甘子
**Phyllanthus emblica** L.

落叶乔木，雌雄同株。叶条状长圆形，纸质至革质，基部两侧不对称；托叶三角形，褐红色。花序簇生，具多个雄花和有时具1或2个稍大的雌花；雄花萼片6，膜质，黄色；雄蕊3。蒴果呈核果状，圆球形，直径1-1.3厘米。花期4-6月，果期7-9月。生海拔200-2300米的疏林、灌丛、荒地或向阳山坡上。产中国西南、东南和华南。南亚和东南亚亦有。南美洲有栽培。

Trees, monoecious, deciduous. Leaves linear-oblong, papery to coriaceous, base asymmetrical; stipules deltoid, brown-red. Fascicles with many staminate flowers and sometimes 1 or 2 larger pistillate flowers; staminate flowers sepals 6, membranous, yellow; stamens 3. Capsules drupaceous, globose, 1-1.3 cm diam. Fl. Apr-Jun. Fr. Jul-Sep. Open forests, thickets, wastelands or sunny slopes at 200-2300 m. Distributed in SW, SE and S China. Also in S and SE Asia. Cultivated in South America.

落萼叶下珠 *Phyllanthus flexuosus*

余甘子 *Phyllanthus emblica*

## 刺果叶下珠

**Phyllanthus forrestii** W. W. Sm.

灌木，约20厘米高。叶二列，近圆形至长圆形，薄纸质。花序丛生于叶腋，雄花数个，雌花单生；雄花萼片4，紫色；雄蕊2；雌花萼片6，2轮。蒴果球形，具密软刺。花期6-8月，果期8-10月。生海拔300-3300米的山坡灌丛中。产云南、四川、贵州和湖北。

Shrubs, ca. 20 cm tall. Leaves distichous, suborbicular to oblong, thinly papery. Inflorescences an axillary fascicle, staminate flowers several, pistillate flowers solitary; staminate flowers sepals 4, purple; stamens 2; pistillate flowers sepals 6, 2-seriate. Capsules globose, with dense, soft prickles. Fl. Jun-Aug. Fr. Aug-Oct. Thickets on slopes at 300-3300 m. Distributed in Yunnan, Sichuan, Guizhou and Hubei.

## 黄珠子草

**Phyllanthus virgatus** Forst. f.

一年生草本，全株无毛。叶近革质，条状披针形或长圆形。花序两性簇生于叶腋，常由2-4朵雄花和1朵雌花组成；雄花萼片6；雄蕊3；雌花萼片6，反折，紫色，具淡白色膜质边缘，果期宿存。蒴果扁球形，紫色，具突起的鳞片或平滑。花期4-5月，果期6-11月。生海拔200-1400米的草坡、林中或灌丛。产中国西南、华南、东南、华中和华北。热带和亚热带亚洲、澳大利亚(昆士兰)亦有。

黄珠子草 *Phyllanthus virgatus*

Annual herbs, glabrous throughout. Leaves subcoriaceous, linear-lanceolate or oblong. Inflorescences bisexual axillary fascicles usually with 2-4 staminate and 1 pistillate flower; staminate flowers sepals 6; stamens 3; pistillate flowers sepals 6, reflexed, purple with whitish membranous margins, persistent in fruit. Capsules oblate, purple, with raised scales or smooth. Fl. Apr-May. Fr. Jun-Nov. Grassy slopes, forests or thickets at 200-1400 m. Distributed in SW, S, SE, C and N China. Also in tropical and subtropical Asia, and Australia (Queensland).

## 叶下珠

**Phyllanthus urinaria** L.

一年生草本，雌雄同株。叶长圆形或卵形，纸质，边缘具1-3列毛。雄花3-4，簇生于叶腋；雄蕊3；雌花萼片6；子房卵球形，具鳞状突起，花柱分离，顶端2裂。花期4-6月，果期7-11月。生海拔100-600米的盆地、旷野、路边、弃荒地、疏林或山坡上。产中国除东北和西北以外大部分地区。南亚、东南亚、日本和南美洲亦有。

Annual herbs, monoecious. Leaves oblong or ovate, papery, margin with 1-3 rows of hairs. Staminate flowers 3-4-flowered, clustered at leaf axils; stamens 3; pistillate flowers with 6 sepals; ovary ovoid-globose, scaly-verrucose, styles free, apex 2-lobed. Fl. Apr-Jun. Fr. Jul-Nov. Basins, fields, roadsides, wastelands, open forests or slopes at 100-600 m. Distributed in most parts of China, except NE and NW China. Also in S and SE Asia, Japan and South America.

刺果叶下珠 *Phyllanthus forrestii*

叶下珠 *Phyllanthus urinaria*

## 单花水油甘

**Phyllanthus nanellus** P. T. Li

灌木，雌雄同株。生叶小枝扁，两侧具翅。叶二列，薄革质，卵形或长卵形，基部偏斜。花通常单朵腋生；花小，黄绿色。蒴果圆球状，3瓣裂。果期4-5月。生海拔300-400米的山谷林下或溪旁灌木丛中。产海南。

单花水油甘 *Phyllanthus nanellus*

银柴 *Aporosa dioica*

毛银柴 *Aporosa villosa*

Shrubs, monoecious. Leafy branchlets compressed, laterally winged. Leaves distichous, thinly coriaceous, ovate or narrowly ovate, base oblique. Flowers usually solitary, axillary; flowers small, greenish-yellow. Capsules globose, 3-valved. Fr. Apr-May. Forests in valleys or scrubs by streams at 300-400 m. Distributed in Hainan.

## 枝翅珠子木

**Phyllanthodendron dunnianum** Lévl.

灌木或小乔木，雌雄同株。枝条两侧具翅，全株无毛。托叶卵状披针形，长约3毫米；叶革质或厚纸质，侧脉每边6-8条，叶缘前联结。雄花花盘腺体5，雄蕊3，花丝合生成圆柱状；雌花花盘腺体6，子房3室，花柱3。蒴果圆球状。花期5-7月，果期7-10月。生海拔500-1000米的山地阔叶林或石灰岩山地灌木丛。产广西、贵州和云南。

Shrubs or treelets, monoecious. Branches ±2-winged, glabrous throughout. Stipules ovate-lanceolate, ca. 3 mm long; leaves leathery or thickly papery, lateral veins in 6-8 pairs, anastomosing before margins. Male flowers: disk segments 5, stamens 3; filaments connate into a terete column; female flowers: disk segments 6; ovary 3-locular; styles 3. Capsules globose. Fl. May-Jul. Fr. Jul-Oct. Montane broad-leaved forests or limestone scrubs at 500-1000 m. Distributed in Guangxi, Guizhou and Yunnan.

## 银柴

**Aporosa dioica** (Roxb.) Müll. Arg.

乔木。叶柄两侧具2腺体；叶椭圆形、狭卵形、长圆状椭圆形、倒卵形或倒披针形，革质，下面沿脉疏具短柔毛。萼片常4；子房卵球形，密具短柔毛，2室；每室2胚珠。蒴果椭圆体形。花果期几乎全年。生海拔200-1000米的山地疏林、林缘或灌丛中。产广东、广西、海南和云南。南亚和东南亚亦有。

Trees. Petioles apex bilateral with 2 glands; leaves elliptic, narrowly ovate, oblong-elliptic, obovate, or oblanceolate, coriaceous, sparsely pubescent along nerves abaxially. Sepals usually 4; ovary ovoid, densely pubescent, bilocular; ovules 2 per locule. Capsules ellipsoid. Fl. and fr. almost all year. Open forests, forest edges or thickets at 200-1000 m. Distributed in Guangdong, Guangxi, Hainan and Yunnan. Also in S and SE Asia.

## 毛银柴

**Aporosa villosa** (Lindl.) Baill.

灌木或小乔木，植株被铁锈色短柔毛。叶柄先端具2腺体；叶多宽卵形、宽椭圆形，有时圆形至长圆状卵形，革质。雄穗状花序腋生；苞片半圆形；花萼3-6；雄蕊2或3。蒴果椭圆体形。花果期全年。生海拔100-1500米的山坡丛林或灌丛中。产广东、海南、广西和云南。南亚和东南亚亦有。

Shrubs or treelets, plants ferruginous-pubescent. Petioles 2-glandular at apex; leaves mostly broadly ovate or broadly elliptic, sometimes rotund to oblong-ovate, coriaceous. Staminate flowers in axillary spikes; bracts semiorbicular; sepals 3-6; stamens 2 or 3. Capsules ellipsoid. Fl. and fr. all year. Forests on slopes or thickets at 100-1500 m. Distributed in Guangdong, Hainan, Guangxi and Yunnan. Also in S and SE Asia.

枝翅珠子木 *Phyllanthodendron dunnianum*

## 云南银柴

**Aporosa yunnanensis** (Pax et Hoffm.) Metc.

小乔木。叶长圆形、狭椭圆形、狭卵形至披针形，上面被黑色小斑点，下面无毛，或仅幼脉具微柔毛。雄穗状花序腋生；花萼3-5，外面具短柔毛；雄蕊2；子房椭圆体形，无毛，2室。蒴果近球形，成熟后橘红色，无毛。花果期1-10月。生海拔200-1500米的林中或灌丛中。产云南、贵州、广西、广东、海南和江西。印度、缅甸、泰国和越南亦有。

Treelets. Leaves oblong, narrowly elliptic, or narrowly ovate to lanceolate, adaxially black small punctate, abaxially glabrous, or only young nerves puberulent. Staminate flowers in axillary spikes; sepals 3-5, pubescent outside; stamens 2; ovary ellipsoid, glabrous, bilocular. Capsules subglobose, red-yellow when mature, glabrous. Fl. and fr. Jan-Oct. Forests or thickets at 200-1500 m. Distributed in Yunnan, Guizhou, Guangxi, Guangdong, Hainan and Jiangxi. Also in India, Myanmar, Thailand and Vietnam.

木奶果 *Baccaurea ramiflora*

云南银柴 *Aporosa yunnanensis*

## 木奶果

**Baccaurea ramiflora** Lour.

常绿乔木，雌雄异株，高达20米。叶两面无毛。花小，无花瓣，多花，组合呈总状圆锥花序；子房被铁锈色糙伏毛。果卵球形或近球形，2-2.5 × 1.5-2厘米，黄色或成熟后变紫色。花期3-4月，果期6-10月。生海拔100-1300米的林中。产云南、广西、广东和海南。南亚和东南亚亦有。

Evergreen trees, dioecious, up to 20 m tall. Leaves glabrous on both surfaces. Flowers small, apetalous, many-flowered, compound into racemelike panicles; ovary ferruginous-strigose. Capsules ovoid or subglobose, 2-2.5 × 1.5-2 cm, yellow to purple when mature. Fl. Mar-Apr. Fr. Jun-Oct. Forests at 100-1300 m. Distributed in Yunnan, Guangxi, Guangdong and Hainan. Also in S and SE Asia.

## 厚叶算盘子

**Glochidion hirsutum** (Roxb.) Voigt

灌木或小乔木，植株密被长柔毛。叶卵圆形或长圆形，革质，沿叶脉密被短柔毛，老渐无毛，下面密柔毛，托叶披针形。雄花萼片长3-4毫米；雄蕊5-8。蒴果扁球形，被短柔毛，

厚叶算盘子 *Glochidion hirsutum*

具5或6沟。花果期几乎全年。生海拔100-1800米的林中、灌丛或河边。产中国西南和华南。印度亦有。

Shrubs or treelets, plants densely villose. Leaves ovate or oblong, coriaceous, densely pubescent along nerves, glabrous at manturity, abaxially densely pubescent, stipules lanceolate. Staminate flowers sepals 3-4 mm long; stamens 5-8. Capsules depressed globose, pubescent, 5- or 6-grooved. Fl. and fr. almost all year. Forests, thickets or riversides at 100-1800 m. Distributed in SW and S China. Also in India.

## 艾胶算盘子

**Glochidion lanceolarium**

(Roxb.) Voigt

常绿灌木或乔木，雌雄同株，除子房和蒴果外全株无毛。叶革质，椭圆形，长圆形或长圆状披针形，基部急尖或阔楔形。花簇生于叶腋；雄花萼片6，倒卵形或倒卵状长圆形；子房密被短柔毛。蒴果近球形，具6-8沟。花期4-9月，果期7月至翌年2月。生海拔500-1200米的山地疏林中或灌丛中。产云南、广西、广东、海南和福建。印度、老挝、泰国、越南和柬埔寨亦有。

Evergreen shrubs or trees, monoecious, glabrous throughout except for hairy ovary and capsules. Leaves leathery, ellptic, oblong or oblong-lanceolate, base acute or rounded. Flowers fascicled, axillary; staminate flowers sepals 6, obovate or obovate-oblong, base acute or broadly cuneate; ovary densely pubescent. Capsules subglobose, 6-8-grooved. Fl. Apr-Sep. Fr. Jul to next Feb. Montane open forests or thickets at 500-1200 m. Distributed in Yunnan, Guangxi, Guangdong, Hainan and Fujian. Also in India, Laos, Thailand, Vietnam and Cambodia.

## 毛果算盘子

**Glochidion eriocarpum**

Champ. ex Benth.

灌木或小乔木，雌雄同株。小枝密被淡黄色长柔毛。叶片纸质，密被长柔毛；基部钝、截形或圆形。雄花萼片6，雄蕊3；雌花萼片6，子房密被柔毛，4-5室，花柱合生呈圆柱状。蒴果密被长柔毛，具宿存花柱。花果期几乎全年。生海拔100-1700米的山坡、山谷灌丛、草地或林缘。产华东和华南。泰国和越南亦有。

Shrubs or treelets, monoecious. Branchlets densely yellowish or gray-yellow villous. Leaves papery, densely villous, base obtuse, truncate, or rounded. Male flowers sepals 6; stamens 3; female flowers sepals 6; ovary densely pubescent, 4-5-locular; style column cylindric. Capsules densely villous, with persistent style column. Fl. and fr. almost all year. Slopes, valley scrubs, grassy areas, sometimes at forest edges at 100-1700 m. Distributed in E and S China. Also in Thailand and Vietnam.

艾胶算盘子 *Glochidion lanceolarium*

毛果算盘子 *Glochidion eriocarpum*

## 算盘子

**Glochidion puberum** (L.) Hutch.

直立灌木，雌雄同株。小枝密被短柔毛。叶纸质或近革质，下面密被短柔毛。花腋生成簇，具2-5花；萼片6，狭长圆形或长圆状倒卵形；子房球形，密被短柔毛，5-10室。蒴果扁球形。花期4-8月，果期7-11月。生海拔300-2200米的山坡、溪边、灌丛中或林缘。产中国西南、华南、东南、华中、华西和华东。日本亦有。

Erect shrubs, monoecious. Branchlets densely pubescent. Leaves papery or subcoriaceous, densely pubescent abaxially. Flowers in axillary clusters, 2-5-flowered; sepals 6, narrowly oblong or oblong-obovate; ovary globose, densely pubescent, 5-10-locular. Capsules depressed-globose. Fl. Apr-Aug. Fr. Jul-Nov. Slopes, by streams, shrubs or forest edges at 300-2200 m. Distributed in SW, S, SE, C, W and E China. Also in Japan.

## 四裂算盘子

**Glochidion ellipticum** Wight

乔木，雌雄同株。枝条无毛。叶宽椭圆形、卵圆形至披针形，纸质或近革质，两面无毛，托叶三角形。花两性簇生于叶腋，其中多数雄花及少数雌花；萼片6；雄蕊3。蒴果扁球状，常4室。花期5-8月，果期7-11月。生海拔100-1700米的林中、灌丛中或溪边。产中国西南和华南。南亚亦有。

四裂算盘子 *Glochidion ellipticum*

Trees, monoecious. Branches glabrous. Leaves broadly elliptic, ovate to lanceolate, papery or subcoriaceous, both surfaces glabrous, stipules deltoid. Flowers in bisexual axillary clusters, with many staminate flowers and few pistillate flowers; sepals 6; stamens 3. Capsules depressed globose, usually 4-locular. Fl. May-Aug. Fr. Jul-Nov. Forests, thickets or stream banks at 100-1700 m. Distributed in SW and S China. Also in S Asia.

算盘子 *Glochidion puberum*

## 革叶算盘子

**Glochidion daltonii** (Müll. Arg.) Kurz

灌木或乔木，雌雄同株，除叶柄和子房披短柔毛至无毛外均无毛。枝条具棱。叶纸质或近革质，下面灰白色，先端渐尖或短渐尖。花簇生于叶腋，基部具2苞片。蒴果扁球状，具4-6沟。花期3-5月，果期4-10月。生海拔200-1700米的疏林或山坡灌丛。产中国西南、华南、华中和华东。印度、缅甸、泰国、越南和马来西亚亦有。

Shrubs or trees, monoecious, glabrous except for petioles and ovary pubescent to glabrous. Branches angular. Leaves papery or subcoriaceous, abaxially gray-white, apex acuminate or shortly so. Flowers clustered in axillary, with 2 bracts at base. Capsules oblate, 4-6-grooved. Fl. Mar-May. Fr. Apr-Oct. Open forests or thickets on slopes at 200-1700 m. Distributed in SW, S, C and E China. Also in India, Myanmar, Thailand, Vietnam and Malaysia.

革叶算盘子 *Glochidion daltonii*

## 龙脷叶

**Sauropus spatulifolius** Beille

常绿小灌木。茎粗糙。枝蝎尾状弯曲。叶常在枝上部聚生，常弯曲或下垂；叶匙形，倒卵状长圆形或卵形，有时长圆形，鲜时稍肉质。花序茎生，2-5花簇，有时组成短聚伞花序；雄花具3雄蕊，花丝合生呈柱状；子房近球形，3室，花柱3，顶端2裂。花期2-10月。广东、广西和福建有栽培。原产越南北部。泰国、马来西亚和菲律宾有栽培。

Evergreen shrublets. Stems scabrous. Branches scorpioid-curved. Leaves usually clustered apically, often recurved or pendulous; leaves spatulate, obovate-oblong, or ovate, sometimes oblong, ± fleshy when fresh. Inflorescences cauliflorous, 2-5-flowered clusters, sometimes in short cymes; staminate flowers with 3 stamens, filaments connate in column; ovary globose, 3-celled; styles 3, 2-lobed at apex. Fl. Feb-Oct. Cultivated in Guangdong, Guangxi and Fujian. Native to N Vietnam. Cultivated in Thailand, Malaysia and the Philippines.

茎花守宫木 *Sauropus bonii*

## 茎花守宫木

**Sauropus bonii** Beille

灌木，全株无毛。枝条具棱。叶纸质，顶端渐尖或短渐尖；羽状脉。雌雄同序；总状聚伞花序着生于茎的下部或基部，长6-15厘米；具覆瓦状排列的苞片和小苞片；雄花花萼膜质，淡黄色有红色斑纹；雌花花萼钟状，2轮，子房3室。成熟蒴果6爿裂。花期4-8月，果期6-10月。生海拔200-500米的石灰岩山地林下或山坡灌丛。产广西。越南亦有。

Shrubs, glabrous throughout. Branches angular. Leaves papery, apex acuminate or shortly so; venation pinnate. Bisexual; inflorescences cauliflorous, from base to lower part of stem, narrow racemelike thyrses, 6-15 cm long; with imbricate bracts and bracteoles; male flowers: calyx membranous, yellowish, with red streak; female flowers: calyx campanulate, sepals biseriate; ovary 3-locular. Capsules prominently 6-valved when mature. Fl. Apr-Aug. Fr. Jun-Oct. Forests or scrubby slopes on limestone at 200-500 m. Distributed in Guangxi. Also in Vietnam.

## 网脉守宫木

**Sauropus reticulatus** X. L. Mo ex P. T. Li

灌木，全株无毛。托叶三角形，早落；叶柄长约5毫米；叶片革质，顶端渐尖；网脉两面均突起。蒴果单生于叶腋；果梗长约3厘米；宿存萼片6，宽倒卵形，较厚；宿存花柱3，分离，顶端2裂。果期8-11月。生海拔500-800米的石灰岩山地疏林下或灌丛中。产广西西部和云南西部。

Shrubs, glabrous throughout. Stipules triangular, caducous; petioles ca. 5 mm long; leaves leathery, apex acuminate; reticulate veins raised on both surfaces. Infructescences axillary, 1-fruited; fruiting pedicels ca. 3 cm long; persistent sepals 6, broadly obovate, thick; persistent styles 3, free, bifid. Fr. Aug-Nov. Open forests or scrubby slopes on limestone at 500-800 m long. Distributed in W Guangxi and W Yunnan.

龙脷叶 *Sauropus spatulifolius*

网脉守宫木 *Sauropus reticulatus*

## 艾堇

**Sauropus bacciformis** (L.) Airy Shaw

草本或半灌木，全株无毛。叶干后膜质，侧脉不明显。花序簇状腋生；花萼内面具腺槽；雄花萼片6，花盘腺体6；雌花无花盘，子房3室，花柱3，分离，顶端2裂；宿存萼片反卷。蒴果成熟时紫红色。花期4-7月，果期7-11月。生海拔100米以下的海边沙滩特别是含盐黏土中。产广东、广西、海南和台湾等地。南亚和印度洋岛屿亦有。

Herbs or subshrubs, glabrous throughout. Leaves membranous when dried, lateral veins obscure. Inflorescences axillary, flowers in clusters; sepals with adaxial gland-pits; male flowers sepals 6; disk lobes 6; female flowers disk absent, ovary 3-locular; styles 3, free, bifid at apex; fruiting sepals reflexed. Capsules purple when mature. Fl. Apr-Jul. Fr. Jul-Nov. Seashore sandy tracts, especially on brackish clayey soil near sea level to below 100 m. Distributed in Guangdong, Guangxi, Hainan and Taiwan. Also in S Asia and Indian Ocean Islands.

## 黑面神

**Breynia fruticosa** (L.) Hook. f.

直立灌木。小枝压扁状，紫红色。叶卵形、宽卵形或菱状卵形，革质，干后变黑，背面被黑色小斑点。花小，单生或2-4花叶腋簇生；花萼钟形，先端浅6裂，果期膨大至直径约8毫米。蒴果球形，浅黄色至橘黄色。花期全年，果期5-12月。生海拔100-1000米的山坡、灌丛、平坝或林中。产中国西南、华南和东南。老挝、泰国和越南亦有。

Erect shrubs. Branchlets compressed, purple-red. Leaves ovate, broadly ovate, or rhombic-ovate, coriaceous, nigrescent when dry, abaxially black small punctate. Flowers small, solitary or 2-4-flowered in axillary clusters; calyx campanulate, shallowly 6-fid at apex, much enlarged in fruits to ca. 8 mm diam. Capsules globose, yellowish to orange. Fl. year-round. Fr. May-Dec. Slopes, thickets, flat lands or forests at 100-1000 m. Distributed in SW, S and SE China. Also in Laos, Thailand and Vietnam.

黑面神 *Breynia fruticosa*

钝叶黑面神 *Breynia retusa*

## 钝叶黑面神

**Breynia retusa** (Dennst.) Alston

直立灌木。小枝四棱形。叶椭圆形至稍倒卵形，干后灰色，上面近边缘处密被小鳞片，先端圆或近急尖。花常单生；花萼钟形至陀螺状，6裂。蒴果球形，外果皮肉质，缓慢开裂，红色，熟后褐色。花期3-10月，果期翌年2-3月和7-8月。生海拔300-2000米的山地疏林中或山谷灌丛中。产云南、西藏、贵州和广西。南亚亦有。

Erect shrubs. Branchlets tetragonous. Leaves elliptic to slightly obovate, grayish when dry, adaxially densely small scaly near margin, apex rounded to subacute. Flowers usually solitary; calyx campanulate to turbinate, 6-lobed. Capsules globose, exocarps fleshy, tardily dehiscent, red and ripening brown. Fl. Mar-Oct. Fr. next Feb-Mar and Jul-Aug. Open forests on slopes or thickets in valleys at 300-2000 m. Distributed in Yunnan, Xizang, Guizhou and Guangxi. Also in S Asia.

艾堇 *Sauropus bacciformis*

秋枫（重阳木，茄苳）*Bischofia javanica*

## 秋枫（重阳木，茄苳）

**Bischofia javanica** Blume

常绿乔木，雌雄异株，高达40米。茎具红色液体。托叶膜质，披针形；叶为掌状3(-5)小叶。花序腋生，圆锥花序；子房光滑，无毛，3或4室。果实球形或近球形，淡褐色。花期4-5月，果期8-10月。生海拔800米以下的丛林、平坝或河岸；广栽培。产中国西南、华南、东南、华中和华东。南亚、东南亚、日本、澳大利亚和波利尼西亚亦有。

Evergreen trees, dioecious, up to 40 m tall. Stems with red juice. Stipules membranous, lanceolate; leaves palmately 3(-5)-foliolate. Inflorescences axillary, paniculate; ovary smooth, glabrous, 3- or 4-locular. Fruits globose or subglobose, brownish. Fl. Apr-May. Fr. Aug-Oct. Forests, flat lands or riversides below 800 m; widely cultivated. Distributed in SW, S, SE, C and E China. Also in S and SE Asia, Japan, Australia and Polynesia.

## 白叶桐

**Sumbaviopsis albicans**

(Blume) J. J. Sm.

乔木。叶卵状长圆形至长圆形，10-30 × 5-15厘米，厚纸质，狭盾状着生，下面白色或淡黄褐色微绒毛。总状花序长6-30厘米，具微绒毛；雄花2或3朵束生；雄蕊50-70。蒴果直径2.5-3厘米。花果期3-12月。生海拔400-900米的林中。产云南南部。南亚和东南亚亦有。

Trees. Leaves ovate-oblong to oblong, 10-30 × 5-15 cm, thickly papery, narrowly peltate, abaxially white or ochraceous tomentulose. Racemes 6-30 cm long, tomentulose; staminate flowers 2- or 3-fascicled; stamens 50-70. Capsules 2.5-3 cm diam. Fl. and fr. Mar-Dec. Forested limestone valleys and hills at 400-900 m. Distributed in S Yunnan. Also in S and SE Asia.

## 广东地构叶

**Speranskia cantonensis**

(Hance) Pax et Hoffm.

多年生草本。茎少分枝。叶柄长1-3.5厘米；叶纸质，边缘具圆齿或钝锯齿；顶端急尖。雄花花瓣长不及1毫米，花盘腺体5；雄蕊10-12；雌花无花瓣，子房具疣状突起和疏柔毛，花柱3，各2深裂，裂片羽状撕裂。蒴果3爿裂；具瘤状突起。花期2-7月，果期5-12月。生海拔200-1000(-2600)米的草地或山地灌丛。产中国西南、华南和华中。

Perennial herbs. Few branched. Petioles 1-3.5 cm long; leaves papery, margin coarsely crenate or dentate, apex acute. Male flowers petals less than 1 mm long; disk glands 5; stamens 10-12; female flowers petals absent; ovary densely tuberculate and pilose; styles 3, deeply 2-lobed, plumose-lacerate. Capsules 3-lobed, usually tuberculate. Fl. Feb-Jul. Fr. May-Dec. Grassy slopes or mountain thickets at 200-1000(-2600) m. Distributed in SW, S and C China.

白叶桐 *Sumbaviopsis albicans*

广东地构叶 *Speranskia cantonensis*

滑桃树 *Trewia nudiflora*

## 滑桃树
**Trewia nudiflora** L.

乔木，高达25米。叶对生，等大，卵圆形或长圆形，基部具2或4腺体，先端渐尖。雄花序密被长柔毛；子房被微绒毛；花柱常3，基部合生，长2-2.5厘米。核果球形，2-4室。花期12月至翌年3月，果期翌年6-12月。生海拔100-800米的山谷或溪边疏林。产广西、海南和云南。东南亚亦有。

Trees, up to 25 m tall. Leaves opposite, equal in size, ovate or oblong, base with 2 or 4 glands, apex acuminate. Staminate inflorescences densely villous; ovary tomentulose; styles often 3, basally connate, 2-2.5 cm long. Drupes globose, 2-4-locular. Fl. Dec to next Mar. Fr. next Jun-Dec. Open forests in valleys or along streams at 100-800 m. Distributed in Guangxi, Hainan and Yunnan. Also in SE Asia.

## 云南野桐
**Mallotus yunnanensis**
Pax et Hoffm.

灌木，雌雄异株。小枝密被褐色星柔毛。叶对生，叶脉背面具柔毛，脉腋处有芒，疏生颗粒状腺体；侧脉4-6对。雄花：雄蕊35-40；雌花：花柱3，基部合生。蒴果3室，被淡黄色柔毛，散生短软刺和鳞状腺体。花果期4-12月。生海拔100-1400米的山坡或石灰岩灌木丛。产广西、贵州、海南和云南。越南亦有。

云南野桐 *Mallotus yunnanensis*

Shrubs, dioecious. Branchlets densely brownish stellate-pubescent. Leaves opposite, abaxially gray pubescent along veins, and barbate in vein axils, scattered glandular-scaly; lateral veins 4-6 pairs. Male flowers stamens 35-40; female flowers: styles 3, base connate. Capsules 3-locular, yellowish pubescent, sparsely shortly softly spiny and glandular-scaly. Fl. and fr. Apr-Dec. Mountain slopes or limestone, thickets at 100-1400 m. Distributed in Guangxi, Guizhou, Hainan and Yunnan. Also in Vietnam.

## 贵州野桐
**Mallotus millietii** Lévl.

攀援灌木，小枝、叶柄和花序被黄色星状绒毛及长柔毛。叶纸质或革质，下面被暗黄色绒毛，散生黄色腺状鳞片。雄花序不分枝；雄花2-5朵束生；子房密被橙黄色绒毛。蒴果3室，密被橙黄色绒毛。种子直径约6毫米。花期5-6月，果期8-10月。生海拔500-1400米的石灰岩区域、丘陵山坡、林中或灌丛中。产云南、贵州、广西、湖北和湖南。

Climbing shrubs, branchlets, petioles, and inflorescences yellow stellate-tomentose and villous. Leaves chartaceous or coriaceous, abaxially dull yellow tomentose, scattered yellow glandular-scaly. Staminate inflorescences unbranched; staminate flowers 2-5-fascicled; ovary densely orange-yellow tomentose. Capsules 3-locular, densely orange-yellow tomentose. Seeds ca. 6 mm diam. Fl. May-Jun. Fr. Aug-Oct. Limestone regions, hill slopes, forests or thickets at 500-1400 m. Distributed in Yunnan, Guizhou, Guangxi, Hubei and Hunan.

贵州野桐 *Mallotus millietii*

粗糠柴 *Mallotus philippensis*

## 石岩枫

**Mallotus repandus** (Willd.) Müll. Arg.

攀援灌木，小枝、叶柄和花序具暗黄褐色星状微绒毛。叶互生，纸质，三角状卵形或卵形，下面被星状短柔毛，散生浅黄色颗粒状腺体。花萼裂片3或4；雄蕊40-75。蒴果2(-3)个果瓣，被黄褐色微绒毛，散生腺状鳞片。花期3-5月，果期6-9月。生海拔100-1000米的疏林、灌丛或山谷中。产中国西南、华南、东南、华中、华西和华东。南亚、东南亚、澳大利亚北部和太平洋岛屿亦有。

Climbing shrubs, branchlets, petioles, and inflorescences dull yellowish-brownish stellate-tomentulose. Leaves alternate, papery, triangular-ovate to ovate, abaxially stellate-pubescent, scatteredly yellowish granular-glandular. Calyx lobes 3 or 4; stamens 40-75. Capsules consist of 2(-3) cocci, yellowish-brownish tomentulose, scattered glandular-scaly. Fl. Mar-May. Fr. Jun-Sep. Sparse forests, thickets or mountain valleys at 100-1000 m. Distributed in SW, S, SE, C, W and E China. Also in S and SE Asia, N Australia and Pacific Islands.

## 粗糠柴

**Mallotus philippensis** (Lam.) Müll. Arg.

灌木或小乔木，小枝、叶柄和花序被黄褐色星状绒毛。叶革质，下面具灰黄色绒毛和疏生红色腺状鳞片，边缘近全缘。蒴果近球形，直径6-8毫米，密被红色颗粒状腺体。花期3-5月，果期6-8月。生海拔300-1600米的疏林、石灰山或河谷。产中国西南、华南、东南、华中和华东。南亚、东南亚和太平洋热带地区亦有。

Shrubs or small trees, branchlets, petiole, and inflorescences yellow-brownish stellate-tomentose. Leaves coriaceous, abaxially gray-yellow tomentulose and sparsely red glandular-scaly, margin subentire. Capsules subglobose, 6-8 mm diam, with dense red granular-glands. Fl. Mar-May. Fr. Jun-Aug. Open forests, limestone hills or river valleys at 300-1600 m. Distributed in SW, S, SE, C and E China. Also in S and SE Asia, and tropical Pacific regions.

石岩枫 *Mallotus repandus*

## 毛桐
**Mallotus barbatus** Müll. Arg.

灌木或小乔木，小枝、叶柄和花序密被淡褐色丛卷绒毛或浅褐色或浅黄色绒毛。叶纸质，下面被星状绒毛，散生黄色腺状鳞片；基出脉5-7。雌花序长15-25厘米。蒴果排列稀疏，密被淡黄色星状毛和紫红色软刺。花期4-5月，果期9-10月。生海拔200-1300米的疏林、灌丛、石灰岩、路边或常于开阔地。产中国西南、华南和华中。南亚和东南亚亦有。

Shrubs or small trees, branchlets, petioles, and inflorescences densely brownish floccose-tomentose or brownish or yellowish tomentose. Leaves papery, abaxially stellate-tomentose, scattered yellow glandular-scaly; basal veins 5-7. Pistillate inflorescences 15-25 cm long. Capsules sparsely arranged, densely yellowish stellate-hairy and purple-red soft spiny. Fl. Apr-May. Fr. Sep-Oct. Open forests, thickets, limestone hills, roadsides or often in clearings at 200-1300 m. Distributed in SW, S and C China. Also in S and SE Asia.

## 桂野桐
**Mallotus conspurcatus** Croiz.

灌木。小枝密被棕色星状柔毛。叶纸质，下面密被棕色柔毛，散生淡红色颗粒状腺体；基部钝，盾状着生，具斑状腺体8-9个；边全缘或齿状。蒴果3室，直径约1.5厘米，密生星状毛和软刺，软刺钻形。种子卵形，长5毫米，褐色，具瘤状突起。果期8-9月。生海拔400-500米的石灰岩丘陵密林中。产广西西部。

Shrubs. Branchlets densely brown stellate pulveraceous tomentose when young. Leaves papery, abaxially brown tomentose, sparsely reddish granular-glandular; base obtuse, peltate, with 8-9 small basal glands, margin subentire or denticulate. Capsules 3-locular, ca. 1.5 cm diam, densely stellate-pubescent and softly spiny, spines subulate. Seeds ovoid, 5 mm long, brown, verruculose. Fr. Aug-Sep. Limestone hills or forests at 400-500 m. Distributed in W Guangxi.

桂野桐 *Mallotus conspurcatus*

## 白背叶
**Mallotus apelta** (Lour.) Müll. Arg.

灌木或小乔木。小枝幼时被淡白色和浅褐色星状微绒毛。叶互生，下面被淡白色星状微绒毛和散生橘黄色腺状鳞片，下面几无毛或散生星状小微柔毛。雄花序不分枝，萼片4，卵形，雄蕊50-75；雌花萼片3-5，子房3(或4)室，被星状微绒毛。蒴果近球形，密被软刺。花期5-9月，果期8-11月。生海拔100-1000米山坡、林中或山谷灌丛中。产中国西南、华南和东南。越南亦有。

Shrubs or small trees. Branchlets whitish and brownish stellate-tomentulose when young. Leaves alternate, abaxially whitish tomentulose and scattered orange glandular-scaly, adaxially glabrescent or sparsely stellate-pilosulose. Staminate Inflorescences unbranched, calyx lobes 4, ovate, stamens 50-75; pistillate flowers calyx lobes 3-5, ovary 3(or 4)-locular, with stellate-tomentulose hairs. Capsules subglobose, densely softly spiny. Fl. May-Sep. Fr. Aug-Nov. Slopes, forests or shrubs in valleys at 100-1000 m. Distributed in SW, S and SE China. Also in Vietnam.

毛桐 *Mallotus barbatus*

白背叶 *Mallotus apelta*

## 广西白背叶
**Mallotus apelta** (Lour.) Müll. Arg. var. **kwangsiensis** Metc.

小乔木。小枝被白色星状微绒毛。叶互生，下面被白色星状绒毛，上面有散生橘黄色腺体。雄花序疏具分枝；萼片4，卵形；雄蕊50-75；雌花萼片3-5；子房3(或4)室，具星状绒毛。蒴果近球形，密具软刺。花果期7-10月。生海拔200-1000米的石灰岩地区林中或灌丛中。产云南、广西和广东。

Small trees. Branchlets white

广西白背叶 *Mallotus apelta* var. *kwangsiensis*

stellate-tomentulose. Leaves alternate, adaxially white-stellate-tomentose, abaxially sparsely orange-yellow glandular. Staminate Inflorescences laxly branched; calyx lobes 4, ovate; stamens 50-75; pistillate flowers calyx lobes 3-5; ovary 3(or 4)-locular, with stellate-tomentulose hairs. Capsules subglobose, densely softly spiny. Fl. and fr. Jul-Oct. Forests on limestone or thickets at 200-1000 m. Distributed in Yunnan, Guangxi and Guangdong.

## 小果野桐

**Mallotus microcarpus** Pax et Hoffm.

灌木。嫩枝密被白色微柔毛。叶纸质，散生黄色颗粒状腺体；基出脉3-5。雌花序不分枝，12-14厘米；子房密被长柔毛，粗糙，花柱3，具羽状突。蒴果3室，直径6毫米；疏生短软刺，被长柔毛和鳞状腺体。花期8-10月。生海拔200-1000米的山坡或路边灌丛中。产广东、广西、贵州、湖南和江西。越南亦有。

Shrubs. Branchlets densely whitish pubescent. Leaves papery, sparsely yellowish glandular-scaly; basal veins 3-5. Female inflorescences unbranched, 12-14 cm long, ovary puberulent and scabrous; styles 3, plumose. Capsules 3-locular, 6 mm diam, sparsely shortly softly spiny, puberulent and glandular-scaly. Fl. Aug-Oct. Mountain slopes, roadsides or thickets at 200-1000 m. Distributed in Guangdong, Guangxi, Guizhou, Hunan and Jiangxi. Also in Vietnam.

## 野梧桐

**Mallotus japonicus** (L. f.) Müll. Arg.

灌木。小枝幼时具暗褐色星状微绒毛。叶近圆卵形或菱状卵形，有时边缘波状具三尖头，纸质，下面被星状小疏柔毛，疏被淡黄色腺状鳞片，基出脉3-5。雄花序被灰色或淡褐色微绒毛；雄蕊70-100。蒴果直径约8毫米。花期4-6月，果期7-8月。生海拔100-600米的沟谷、林地或林缘。产台湾、浙江南部和江苏。朝鲜半岛和日本亦有。

Shrubs. Branchlets dull brownish stellate-tomentulose when young. Leaves suborbicular-ovate or rhombic-ovate, sometimes margin repand-tricuspidate, papery, abaxially sparsely stellate-pilosulose, sparsely yellowish glandular-scaly, basal veins 3-5. staminate inflorescences gray or brownish tomentulose; stamens 70-100. Capsules ca. 8 mm diam. Fl. Apr-Jun. Fr. Jul-Aug. Valleys, forests or forest edges at 100-600 m. Distributed in Taiwan, S Zhejiang and Jiangsu. Also in Korean Peninsula and Japan.

小果野桐 *Mallotus microcarpus*

野梧桐 *Mallotus japonicus*

红叶野桐 *Mallotus tenuifolius* var. *paxii*

## 红叶野桐

**Mallotus tenuifolius** var. **paxii** (Pamp.) H. S. Kiu

灌木或小乔木。小枝和花序被灰色星柔毛。叶纸质；叶背具灰色星柔毛，散生淡红色颗粒状腺体；上面干后暗褐色或红褐色。雄花3-9朵簇生，花萼裂片4，约3毫米；雌花花柱3或4，长约4毫米。蒴果被星柔毛；脊状突起5-8毫米。花期5-7月，果期7-9月。生海拔300-1200米的山坡山谷灌木丛或次生林及路边。产华西、华中、华东、华南和西南。

Shrubs or small trees. Branchlets and inflorescences gray stellate-tomentulose. Leaves papery, abaxially gray stellate-tomentulose, scattered reddish glandular-scaly, adaxially dull brown or reddishbrownish when dry. Male flowers 3-9-fascicled; calyx lobes 4, ca. 3 mm long; female flowers: styles 3 or 4, ca. 4 mm long. Capsules stellatepilose, spines 5-8 mm long. Fl. May-Jul. Fr. Jul-Sep. Mountain valleys or slopes, thickets, secondary forests or roadsides at 300-1200 m. Distributed in W, C, E, S and SW China.

## 尼泊尔野桐

**Mallotus nepalensis** Müll. Arg.

灌木或小乔木。小枝和花序被黄褐色星状毛。叶纸质，叶基具2斑状腺体；边全缘；基出脉3。雌花序不分枝，10-20厘米，花梗2-5厘米；子房密被软刺和星状毛。蒴果直径约1.5厘米，密被软刺和星状毛。种子具疣突。花果期6-7月。生海拔1700-2500米的山谷或山坡灌木丛。产西藏和云南。不丹、印度、缅甸和尼泊尔亦有。

Shrubs or small trees. Branchlets and inflorescences brownish-yellowish stellate-tomentose. Leaves papery, base with 2 maculate glands, margin entire; basal veins 3. Female inflorescences unbranched, 10-20 cm long; peduncles 2-5 cm; ovary densely softly spiny and stellate-pilose. Capsules ca. 1.5 cm diam, densely softly spiny and stellate-pilose. Seeds verruculose. Fl. and fr. Jun-Jul. Mountain valleys or slopes, thickets at 1700-2500 m. Distributed in Xizang and Yunnan. Also in Bhutan, India, Myanmar and Nepal.

## 血桐

**Macaranga tanarius** (L.) Müll. Arg.

乔木。小枝幼时被淡黄褐色短柔毛。叶卵状圆形或近圆形，纸质，下面具腺状鳞片，沿脉被短柔毛；掌状脉7-11。雄花每苞片内约11；萼片3；雄蕊(4或)5或6(-10)；雌花单生；花萼2或3浅裂；子房2或3室，疏被软刺。花期4-6月，果期6-7月。生沿海低山灌木林或次生林中。产广东和台湾。印度南部、缅甸、泰国、越南、印度尼西亚、菲律宾、日本南部、澳大利亚北部和太平洋岛屿亦有。

Trees. Branchlets yellowish brown pubescent when young. Leaves ovate-orbicular or suborbicular, papery, abaxially glandular-scaly, pubescent along veins; palmate veins 7-11. Staminate flowers ca. 11 per bract; sepals 3; stamens (4 or)5 or 6(-10); pistil-

尼泊尔野桐 *Mallotus nepalensis*

血桐 *Macaranga tanarius*

late flowers solitary; calyx 2- or 3-lobed; ovary 2- or 3-locular, sparsely softly spiny. Fl. Apr-Jun. Fr. Jun-Jul. Thickets below mountains along sea or secondary forests. Distributed in Guangdong and Taiwan. Also in S India, Myanmar, Thailand, Vietnam, Indonesia, the Philippines, S Japan, N Australia and Pacific Islands.

毛丹麻杆 *Discocleidion rufescens*

## 印度血桐

**Macaranga indica** Wight

乔木。小枝幼时淡黄褐色短柔毛。叶薄革质，卵圆形，下面被短柔毛和腺状鳞片，掌状脉9。雄花序有分枝；小枝“之”字形；雄花每苞片内多朵；萼片3，卵形；雌花单生；萼片4；子房1室；花柱1，近钻形。蒴果球形。花期8-10月，果期10-11月。生海拔300-1900米的山谷、溪边，常绿阔叶林中。产云南、西藏、贵州、广西和广东。南亚和东南亚亦有。

Trees. Branchlets yellowish brown pubescent when young. Leaves thinly coriaceous, ovate, abaxially pubescent and glandular-scaly, palmate veins 9. Staminate inflorescences branched; branchlets zigzag; staminate flowers many per bract; sepals 3, ovate; pistillate flowers solitary; sepals 4; ovary 1-locular; styles 1, subulate. Capsules globose. Fl. Aug-Oct. Fr. Oct-Nov. Valleys, by streams or evergreen broad-leaved forests at 300-1900 m. Distributed in Yunnan, Xizang, Guizhou, Guangxi and Guangdong. Also in S and SE Asia.

## 草鞋木

**Macaranga henryi** (Pax et Hoffm.) Rehd.

乔木或灌木。小枝被锈色微绒毛。叶长圆状卵形或长圆状披针形，纸质至厚纸质，下面疏被腺状鳞片，基部具2或4个腺体，侧脉7-10对。雄蕊6-12；花萼坛形，4齿裂或近截形；子房2室；花柱2。花期3-5月，果期7-9月。生海拔300-1400米的山坡、石灰岩丘或林中。产云南、贵州和广西。越南北部亦有。

Shrubs or trees. Branchlets ferruginous tomentulose. Leaves oblong-ovate or oblong-lanceolate, papery to thickly papery, abaxially sparsely glandular-scaly, base with 2 or 4 glands, veins 7-10 pairs. Stamens 6-12; calyx urceolate, 4-denticulate or subtruncate; ovary 2-locular; styles 2. Fl. Mar-May. Fr. Jul-Sep. Mountain slopes, limestone hills or forests at 300-1400 m. Distributed in Yunnan, Guizhou and Guangxi. Also in N Vietnam.

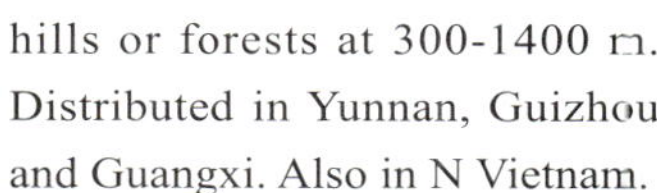

## 毛丹麻杆

**Discocleidion rufescens** (Franch.) Pax et Hoffm.

灌木或小乔木，小枝、叶、花序和花密具浅黄色或淡白色短柔毛。叶卵形或三角状卵形，纸质，下面具绒毛。每苞片具3-15雄花；雄蕊35-60；每苞片具1或2雌花。蒴果直径6-8毫米。花期5-10月，果期7-11月。生海拔200-1000米的山坡、石灰岩林地或灌丛。产中国西南、华南、华中、华北和华西。

Shrubs or small trees. Branchlets, leaves, inflorescences, and flowers densely yellowish or whitish pubescent. Leaves ovate or triangular-ovate, papery, abaxially tomentose. Staminate flowers 3-15 per bract; stamens 35-60; pistillate flowers 1 or 2 per bract. Capsules 6-8 mm diam. Fl. May-Oct. Fr. Jul-Nov. Mountain slopes, usually in limestone forests or thickets at 200-1000 m. Distributed in SW, S, C, N and W China.

印度血桐 *Macaranga indica*

草鞋木 *Macaranga henryi*

## 羽脉山麻杆

**Alchornea rugosa** (Lour.) Müll. Arg.

灌木或小乔木，雌雄异株。叶纸质，侧脉8-12对，基部钝或稍心形，具2腺体，无托叶。雄花序圆锥状，顶生；苞片三角形；萼片5；花柱3，丝状。蒴果近球形。花果期几全年。生海拔600米以下的林中。产云南、广西、广东和海南。东南亚和澳大利亚北部亦有。

Shrubs or small trees, dioecious. Leaves papery, lateral veins 8-12 pairs, base obtuse or slightly cordate, with 2 glands, stipels absent. Staminate inflorescences panicled, terminal; bracts triangular; sepals 5; styles 3, filiform. Capsules subglobose. Fl. and fr. almost all year. Forests below 600 m. Distributed in Yunnan, Guangxi, Guangdong and Hainan. Also in SE Asia and N Australia.

## 山麻杆

**Alchornea davidii** Franch.

落叶灌木，雌雄异株。托叶披针形；叶薄纸质，基部具斑状腺体2或4个，小托叶线状；基出脉3。雄花葇荑花序状，1.5-3.5厘米；雄花：雄蕊6-8枚；雌花：子房被绒毛，花柱3，长10-12毫米，部分合生。蒴果密生柔毛。种子具小瘤体。花期4-5月，果期6-7月。生海拔300-2000米的沟谷、河边坡地或落叶林中。产华南。

Deciduous shrubs, dioecious. Stipules lanceolate; Leaves papery, base with 2 or 4 glands, stipels filiform; basal veins 3. Male inflorescences catkinlike, 1.5-3.5 cm long; male flowers stamens 6-8; female flowers: ovary tomentose; styles 3, 10-12 mm long, partly connate. Capsules densely pubescent. Seeds tuberculate. Fl. Apr-May. Fr. Jun-Jul. Valleys, slopes of streams or rivers, or deciduous forests at 300-2000 m. Distributed in S China.

## 红背山麻杆

**Alchornea trewioides** (Benth.) Müll. Arg.

灌木，雌雄异株。叶薄纸质，背面浅红色；基部腺体4个。雄花序腋生，7-15厘米；雄花萼片4，雄蕊7或8；雌花序顶生，5-6厘米，基部腺体2个；雌花萼片5-8；子房被短绒毛。蒴果被微柔毛。种子具瘤体。花期4-5月，果期6-8月。生海拔1200米以下的平原、山坡矮灌丛或疏林、石灰岩丘陵。产华南。泰国、越南和日本亦有。

Shrubs, dioecious. Leaves papery, abaxially light red, base with 4 glands. Male inflorescences axillary, 7-15 cm long; male flowers sepals 4; stamens 7 or 8; female inflorescences terminal, 5-6 cm long; base with 2 glands; female flowers sepals 5-8; ovary tomentulose. Capsules puberulent. Seeds tuberculate. Fl. Apr-May. Fr. Jun-Aug. Plains, mountains, slopes, thickets, open scrub, limestone hills below 1200 m. Distributed in S China. Also in Thailand, Vietnam and Japan.

羽脉山麻杆 *Alchornea rugosa*

山麻杆 *Alchornea davidii*

红背山麻杆 *Alchornea trewioides*

## 椴叶山麻杆

**Alchornea tiliifolia** (Benth.) Müll. Arg.

灌木或小乔木。叶卵状披针形至阔卵形，纸质，下面具短柔毛，基部楔形或近截形，具4腺体，托叶披针形。雄花疏生短柔毛；雌花萼片5(-6)，近卵形，不等大，长3-4毫米。蒴果椭圆体形。花期3-6月，果期6-9月。生海拔200-1300米的石灰山森林或灌丛中。产云南、贵州和广西。印度、不丹、孟加拉国、缅甸、泰国、越南和马来西亚亦有。

Shrubs or small trees. Leaves ovate-rhombic to broadly ovate, papery, abaxially pubescent, base cuneate or subtruncate, with 4 glands, stipels lanceolate. Staminate flowers sparsely pubescent; sepals of pistillate flowers 5(-6), subovate, unequal, 3-4 mm long. Capsules ellipsoid. Fl. Mar-Jun. Fr. Jun-Sep. Forests or thickets on limestone hills at 200-1300 m. Distributed in Yunnan, Guizhou and Guangxi. Also in India, Bhutan, Bangladesh, Myanmar, Thailand, Vietnam and Malaysia.

灰岩棒柄花 *Cleidion bracteosum*

## 棒柄花

**Cleidion brevipetiolatum** Pax et Hoffm.

小乔木。叶薄革质，下面脉腋具髯毛，基部具2-4个斑状腺体。雄花每苞片有3-7朵；雄蕊(40-)55-65；雌花萼片3，不等大，披针形；花梗粗厚，长2-3.5(-7)厘米。蒴果3浅裂。花果期3-10月。生海拔200-1000米的石灰岩丘陵的林中。产云南、贵州、广西、广东和海南。老挝、泰国和越南亦有。

Small trees. Leaves thinly coriaceous, abaxially bearded in vein axils, base with 2-4 maculate glands. Staminate flowers 3-7 per bract; stamens (40-)55-65; sepals of pistillate flowers 3, unequal, lanceolate; pedicels thick, 2-3.5 (-7) cm long. Capsules 3-lobed. Fl. and fr. Mar-Oct. Forests on limestone hills at 200-1000 m. Distributed in Yunnan, Guizhou, Guangxi, Guangdong and Hainan. Also in Laos, Thailand and Vietnam.

## 灰岩棒柄花

**Cleidion bracteosum** Gagnep.

小乔木，雌雄异株。小枝无毛。叶薄革质，斑状腺体2-4个。雄花单朵疏生于花序轴上，雄蕊100-200；雌花萼片5，三角形，花后几不增大，子房3室，密生黄色毛，花柱2深裂。蒴果直径约1.5厘米。花期12月至翌年2月，果期翌年4-5月。生海拔350-1000米的石灰岩常绿林中。产广西、贵州和云南。越南亦有。

Small trees, dioecious. Branchlets glabrous. Leaves thinly leathery, with 2-4 maculate glands. Male flower solitary, remote along axis; stamens 100-200; female flower sepals 5, triangular, scarcely enlarged in fruit; ovary 3-locular, densely yellow villous; styles deeply 2-cleft. Capsules ca. 1.5 cm diam. Fl. Dec to next Feb. Fr. next Apr-May. Evergreen forests, commonly on limestone hills at 350-1000 m. Distributed in Guangxi, Guizhou and Yunnan. Also in Vietnam.

椴叶山麻杆 *Alchornea tiliifolia*

棒柄花 *Cleidion brevipetiolatum*

## 白桐树

**Claoxylon indicum** (Reinw. ex Blume) Hassk.

灌木或小乔木。叶卵形至阔卵形，纸质，两面疏被柔毛。雄花每苞片中有3-7朵；雄蕊15-25；雌花每苞片通常1朵；蒴果3浅裂，直径7-8毫米。种子近球形，外种皮红色。花果期3-12月。生海拔100-1500米的平坝、沟谷或林中。产云南南部、广西、广东和海南。南亚和东南亚亦有。

Shrubs or small trees. Leaves ovate to broadly ovate, papery, both surfaces sparsely hairy. Staminate flowers 3-7 per bract; stamens 15-25; pistillate flowers often 1 per bract. Capsules 3-lobed, 7-8 mm diam. Seeds subglobose; testas red. Fl. and fr. Mar-Dec. Plains, valleys or forests at 100-1500 m. Distributed in S Yunnan, Guangxi, Guangdong and Hainan. Also in S and SE Asia.

## 轮叶戟

**Lasiococca comberi** Haines var. **pseudoverticillata** (Merr.) H. S. Kiu

乔木或灌木。小枝幼时具灰黄色短柔毛。叶互生、对生或近轮生；叶革质，无毛，基部狭心形；萼片3，卵形，无毛。雄蕊多数；子房密具刚毛状刺。蒴果近球形，直径约1.2厘米，具刺。花期4-6月，果期6-7月。生海拔300-1000米的山坡或山谷林中。产云南和海南。越南北部亦有。

Trees or shrubs. Branchlets gray-yellowish pubescent when younger. Leaves alternate, opposite or subwhorled; leaves coriaceous, glabrous, base narrowly cordate; sepals 3, ovate, glabrous. Stamens many; ovary densely setose-muricate. Capsules subglobose, ca. 1.2 cm diam, muricate. Fl. Apr-Jun. Fr. Jun-Jul. Forested slopes or valleys at 300-1000 m. Distributed in Yunnan and Hainan. Also in N Vietnam.

轮叶戟 *Lasiococca comberi* var. *pseudoverticillata*

## 水柳

**Homonoia riparia** Lour.

灌木，雌雄异株。叶条状长圆形或窄披针形，下面密生鳞片和短柔毛。花序长5-10厘米，具柔毛；苞片近卵形；雄花花萼3浅裂；雌花萼片5，长圆形。蒴果近球形，被短柔毛。花期3-5月，果期4-7月。生海拔1000米以下的河岸边或沙州。产中国西南和华南。南亚和东南亚亦有。

Shrubs, dioecious. Leaves linear-oblong or narrowly lanceolate, abaxially densely scaly and pubescent. Inflorescences 5-10 cm long, puberulent; bracts subovate; staminate flowers calyx 3-lobed; pistillate flowers sepals 5, oblong. Capsules subglobose, pubescent. Fl. Mar-May. Fr. Apr-Jul. River banks or sandbars below 1000 m. Distributed in SW and S China. Also in S and SE Asia.

## 蓖麻

**Ricinus communis** L.

一年生草本，幼嫩部分苍白色，全株常淡红色或淡紫色。托叶合生，长2-3厘米；叶掌状7-11裂，边圆具细锯齿。花序长达30厘米；果梗长达45毫米。蒴果椭圆体形或卵球形，有刺毛。花期6-9月或1-12月。栽培或归化于整个中国。世界各地均有栽培。

白桐树 *Claoxylon indicum*

水柳 *Homonoia riparia*

蓖麻 *Ricinus communis*

Annual herbs, younger parts glaucous, whole plant often reddish or purplish. Stipules connate, 2-3 cm long; leaves palmately 7-11-lobed, margin serrate. Inflorescences to 30 cm long; fruiting pedicels to 45 mm long. Capsules ellipsoid or ovoid, echinate. Fl. Jun-Sep, or Jan-Dec. Cultivated or naturalized throughout China. Also cultivated worldwide.

## 白大凤

**Cladogynos orientalis** Zipp. ex Span.

灌木。小枝密被白色星状毛。托叶基部具1腺体；叶纸质，下面被灰白色绒毛；掌状脉5-7条，侧脉4-5对。花序长约2.5厘米；雄花花萼被星状毛；雌花苞片2，1枚叶状，1枚线形，萼片边缘疏生宿存腺体，子房被绒毛。蒴果被白色短绒毛。花果期3-11月。生海拔200-500米的石灰岩山坡灌丛或林下。产广西。东南亚亦有。

Shrubs. Branchlets white stellate-tomentose. Stipules base with 1 gland; Leaves thickly papery, abaxially white tomentose; palmate veins 5-7, lateral veins 4-5 pairs. Inflorescences ca. 2.5 cm long; male flowers: calyx stellate pubescent; female flowers: bracts 2, 1 leaflike, 1 linear; sepals margin sparsely glandular, persistent; ovary tomentose. Capsules white tomentulose. Fl. and fr. Mar-Nov. Thickets or forests on limestone at 200-500 m. Distributed in Guangxi. Also in SE Asia.

## 风轮桐

**Epiprinus siletianus** (Baill.) Croiz.

灌木或乔木。叶互生，常在枝顶端密集呈假轮状，提琴状椭圆形或匙状披针形，厚纸质，无毛，基部狭心形或耳状心形。花柱顶端常二回叉状裂。蒴果裂片近球形，具微绒毛，果瓣厚革质。花期1-6月，果期6-10月。生海拔100-1000米的河边或山地常绿林中。产云南南部和海南。印度东部、缅甸、老挝和越南亦有。

Trees or shrubs. Leaves alternate, often conferted at branch apices, pseudo-verticillate, panduriform-elliptic or spatulate-lanceolate, thickly papery, glabrous, base narrowly cordate or auriculate-cordate. Styles often dichotomous-divided at apex. Capsules lobes subglobose, tomentulose, valves thickly coriaceous. Fl. Jan-Jun. Fr. Jun-Oct. Evergreen forests on slopes or riversides at 100-1000 m. Distributed in S Yunnan and Hainan. Also in E India, Myanmar, Laos and Vietnam.

白大凤 *Cladogynos orientalis*

风轮桐 *Epiprinus siletianus*

蝴蝶果 *Cleidiocarpon cavaleriei*

肥牛树 *Cephalomappa sinensis*

## 肥牛树

**Cephalomappa sinensis** (Chun et F. C. How) Kosterm.

常绿乔木。嫩枝被短柔毛。叶革质，基部具2细小腺体；叶缘淡紫色。花序长1.5-2.5厘米；雄花花丝基部合生，不育雌蕊柱状，顶部2裂；雌花花萼5深裂，子房粗糙，花柱下半部合生，顶部2浅裂。蒴果3室，密生棱柱状刺。种子具浅褐色斑纹。花期3-4月，果期5-7月。生海拔100-500米的石灰岩山地林中。产广西和云南。越南亦有。

Trees, evergreen. Branchlets pubescent when young. Leaves leathery, base with 2 small glands, margin purplish. Inflorescences 1.5-2.5 cm long; male flowers filaments base connate; pistillode columnar, 2-lobed; female flowers: calyx deeply 5-lobed; ovary muricate; styles basally connate, upper part 2-lobed. Capsules 3-locular, densely prismatic muricate-echinate. Seeds brownish marbled. Fl. Mar-Apr. Fr. May-Jul. Forests on limestone at 100-500 m. Distributed in Guangxi and Yunnan. Also in Vietnam.

## 蝴蝶果

**Cleidiocarpon cavaleriei** (Lévl.) Airy Shaw

乔木。幼嫩枝密被星状毛。叶纸质；基部楔形；钻形小托叶2。圆锥花序密生灰黄色星状毛；雄花不育雌蕊柱状；雌花副萼5-8枚，早落，子房2室，被短绒毛，花柱上部3-5裂，裂片叉裂。核果呈偏斜的卵球形或双球形。花果期5-11月。生海拔100-1000米的石灰岩山、山坡或山谷。产广西、贵州和云南。越南亦有。

Trees. Branchlets sparsely minutely stellate-pubescent when young. Leaves thickly papery, base cuneate, with 2 subulate stipels. Panicles yellowish gray tomentulose; male flowers: pistillode columnar; female flowers epicalyx lobes 5-8, caducous; ovary 2-locular, tomentulose; style upper part spreading, 3-5-lobed, lobes lobed. Drupes obliquely ovoid or 2-lobed. Fl. and fr. May-Nov. Forests on limestone or mountain slopes or in valleys at 100-1000 m. Distributed in Guangxi, Guizhou and Yunnan. Also in Vietnam.

## 铁苋菜

**Acalypha australis** L.

一年生草本，雌雄同株。叶膜质，下面沿脉被小微柔毛。花序腋生，稀顶生，不分枝；雌花苞片近轴生，1或2(-4)朵，卵形、心形，边缘具圆锯齿；花柱3，撕裂5-7条。蒴果3室。花果期4-12月。生海拔100-1900米的草地、山坡或田中。产中国大部分地区，内蒙古和新疆除外。老挝、越南、菲律宾、俄罗斯、朝鲜半岛和日本亦有。

Annual herbs, monoecious. Leaves membranaceous, abaxially pilosulose along veins. Inflorescences axillary, rarely terminal, unbranched; pistillate bracts proximal, 1 or 2(-4) per bract, ovate, cordate, margin crenate; styles 3, 5-7-laciniate. Capsules 3-loculed. Fl. and fr. Apr-Dec. Grasslands, slopes or fields at 100-1900 m. Distributed in most parts of China, except Neimenggu and Xinjiang. Also in Laos, Vietnam, the Philippines, Russia, Korean Peninsula and Japan.

铁苋菜 *Acalypha australis*

裂苞铁苋菜 *Acalypha supera*

## 裂苞铁苋菜
**Acalypha supera** Forssk.

一年生草本，雌雄同株。叶卵形或菱状卵形，膜质，基部心形。花序腋生，1-3个聚生，长不及1厘米；雌花苞片5深裂，具1朵雌花；花柱3，撕裂3-5条。蒴果3室。花期5-12月。生海拔100-1900米的湿润草地、溪边、梯田或山坡路边。产中国西南、华南、华中、华北、华西和华东。南亚、东南亚和热带非洲亦有。

Annual herbs, monoecious. Leaves ovate or rhombic-ovate, membranous, base cordate. Inflorescences axillary, 1-3 together, less than 1 cm long; bracts of pistillate flowers 5-parted, 1-flowered; styles 3, 3-5-laciniate. Capsules 3-locular. Fl. May-Dec. Moist grassy places, streamsides, terraced fields or roadsides on slopes at 100-1900 m. Distributed in SW, S, C, N, W and E China. Also in S and SE Asia, and tropical Africa.

## 卵叶铁苋菜
**Acalypha kerrii** Craib

灌木，雌雄同株。嫩枝密生浅黄色柔毛；小枝红褐色，无毛。托叶披针形，具毛；叶膜质，卵形或长卵形，叶缘具粗锯齿。花序腋生；雄花花梗具疏毛，花萼裂片4，雄蕊8；雌花子房被毛和短刺。蒴果3室，果皮具疏毛和具毛短软刺。花期3-8月。生海拔200-500米的石灰岩地区林下或灌丛。产广西和云南。缅甸、泰国和越南亦有。

Shrubs, monoecious. Branchlets yellowish pubescent when young; branches brownish, glabrescent. Stipules lanceolate, pilose; Leaves membranous, ovate or long ovate, margin coarsely serrate; flowers, axillary; male flowers pedicel pilose; sepals 4; stamens 8. Female flowers ovary hairy, sparsely shortly echinate. Capsules 3-locular, pilose and shortly softly few echinate. Fl. Mar-Aug. Limestone forests, thickets at 200-500 m. Distributed in Guangxi and Yunnan. Also in Myanmar, Thailand and Vietnam.

## 灰岩粗毛藤
**Cnesmone tonkinensis** (Gagnep.) Croiz.

缠绕或攀援状灌木，被黄柔毛和刺毛。叶椭圆形或长圆状卵形，上面被粗毛。总状花序被粗毛；雄花花萼裂片3，被粗毛；雌花花萼裂片6，不等大，大的5-7 × 2-3毫米，小的长2-5毫米，宽0.5-1毫米；子房密被粗毛。蒴果被粗毛。花期4-9月，果期5-10月。生海拔100-600米的山谷、石灰岩山坡、灌丛或林中。产广西和海南。泰国和越南亦有。

Subshrubs, stem twining or climbing, densely yellow villous and with stinging hairs. Leaves elliptic, oblong-ovate, adaxially villous, abaxially tomentose. Racemes villous; male flowers: calyx lobes 3, pilose; female flowers: sepals 6, unequal, larger 5-7 × 2-3 mm, smaller 2-5 × 0.5-1 mm; ovary densely hispid. Capsules villous. Fl. Apr-Sep. Fr. May-Oct. Mountain valleys, limestone rocks, slopes, thickets or forests at 100-600 m. Distributed in Guangxi and Hainan. Also in Thailand and Vietnam.

卵叶铁苋菜 *Acalypha kerrii*

灰岩粗毛藤 *Cnesmone tonkinensis*

石山巴豆 *Croton euryphyllus*

## 石山巴豆

**Croton euryphyllus** W. W. Sm.

灌木或小乔木，被很快脱落的星状柔毛。叶柄顶端有2枚具柄腺体；叶纸质，近圆形至阔卵形；基出脉(3-)5(-7)。花序长达15厘米；雄花萼片长约2.5毫米，花瓣比萼片小，边缘被绵毛；雌花子房密被星状毛，花柱2裂。蒴果密被短星状毛。花期4-5月，果期6-9月。生海拔200-2400米的疏林中。产广西、贵州、四川和云南。

Shrubs or treelets, indumentum stellate-pubescent, mostly very quickly deciduous. Petioles apex with 2 stalked glands; leaves papery, rotund to broadly ovate; basal veins (3-)5(-7). Inflorescences to 15 cm long; male flowers sepals ca. 2.5 mm long; petals smaller than sepals, margins woolly; female flowers ovary densely stellate-pubescent; styles bifid. Capsules densely stellate-pubescent. Fl. Apr-May. Fr. Jun-Sep. Open forests at 200-2400 m. Distributed in Guangxi, Guizhou, Sichuan and Yunnan.

鸡骨香 *Croton crassifolius*

## 鸡骨香

**Croton crassifolius** Geisel.

灌木。老枝近无毛。叶柄顶端或中脉基部具2个带柄的杯状腺体；叶卵形、卵状椭圆体形或长圆形。总状花序顶生；雄花萼片外面被星状绒毛；雄蕊14-20；子房密被黄色星状绒毛；花柱4深裂。果近球形。花期11月至翌年6月，果期翌年2-9月。生海拔100-800米的丘陵地或山坡灌丛中。产广西、广东、海南和福建。缅甸、老挝、泰国和越南亦有。

Shrubs. Older branches subglabrous. Petioles apex or base of midrib with 2 stalked and cupular glands; leaves ovate, ovate-elliptic, or oblong. Racemes terminal; sepals of staminate flowers abaxially stellate-tomentose; stamens 14-20; ovary densely stellate-tomentose; styles 4-parted. Fruits globose. Fl. Nov to next Jun. Fr. next Feb-Sep. Hill regions or thickets on slopes at 100-800 m. Distributed in Guangxi, Guangdong, Hainan and Fujian. Also in Myanmar, Laos, Thailand and Vietnam.

## 毛果巴豆

**Croton lachnocarpus** Benth.

灌木，密被星柔毛，老枝近无毛。叶基或叶柄顶端有2枚具柄杯状腺体；叶纸质，边缘细锯齿间有1枚有柄杯状腺体；叶背密被星柔毛。总状花序顶生；雌雄花萼片均被星状毛；子房被黄色绒毛，花柱2裂。蒴果被毛。花期4-5月，果期6-9月。生海拔100-900米的山地疏林或灌丛中。产广东、广西、贵州、湖南和江西。东南亚亦有。

Shrubs, indumentum densely stellate-pubescent, older branches subglabrous. Petioles apex or base of leaves with 2 stalked and cupular glands; leaves papery,

毛果巴豆 *Croton lachnocarpus*

margin obscurely serrulate, usually with stalked and cupular glands; densely stellate-pubescent abaxially. Inflorescences terminal; male flowers and female flowers sepals stellate-pubescent; ovary yellow tomentose; styles bifid. Capsules hairy. Fl. Apr-May. Fr. Jun-Sep. Sparsely forested slopes, thickets at 100-900 m. Distributed in Guangdong, Guangxi, Guizhou, Hunan and Jiangxi. Also in SE Asia.

巴豆 *Croton tiglium*

## 卵叶巴豆
**Croton caudatus** Geisel.

攀援灌木。枝条近无毛。叶卵圆形，纸质，下面密被星状短柔毛，上面常具一对有柄腺体。花序顶生，长8-16厘米；苞片条形；花瓣边缘被白色绵毛；雄蕊约20。果实球形，密具黄褐色星状糙硬毛。花期5-8月，果期7-10月。生海拔500-600米的疏林中。产云南西南部。南亚、东南亚和澳大利亚北部亦有。

Scandent shrubs. Branches subglabrous. Leaves ovate, papery, abaxially densely stellate-pubescent, adaxially often with pair of stalked glands. Inflorescences terminal, 8-16 cm long; bracts linear; petals margins white woolly; stamens ca. 20. Fruits globose, densely yellow-brown stellate-hispid. Fl. May-Aug. Fr. Jul-Oct. Open forests at 500-600 m. Distributed in SW Yunnan. Also in S and SE Asia, and N Australia.

## 巴豆
**Croton tiglium** L.

小乔木。幼枝绿色，疏被星状毛，成熟后无毛。叶纸质，无毛或几无毛。总状花序顶生；萼片长圆状披针形；子房密被星状短柔毛。蒴果椭圆体形、长圆状卵球形或近球形。花期1-7月，果期5-9月。生海拔300-700米林中、石灰岩灌丛或村边，有时栽培。产中国西南、华南、东南和华东。南亚、东南亚和日本南部亦有。

Treelets. Young branches green, sparsely stellate-hairy, glabrous at maturity. Leaves papery, glabrous or glabrescent. Racemes terminal; sepals oblong-lanceolate; ovary densely stellate-pubescent. Capsules ellipsoidal, oblong-ovoid, or subglobose. Fl. Jan-Jul. Fr. May-Sep. Forests, limestone shrublands or near villages at 300-700 m, sometimes cultivated. Distributed in SW, S, SE and E China. Also in S and SE Asia, and S Japan.

## 光叶巴豆
**Croton laevigatus** Vahl

灌木或乔木，嫩枝、叶柄和花序密被蜡质星状鳞毛。叶聚生茎末端，纸质，下面被疏星状毛，叶干后苍灰色。花序顶部簇生；花萼密具平伏星状毛；子房密被紧贴星状毛。蒴果倒卵球形或三角形。花期10-12月，果期12月至翌年3月。生海拔100-600米的林中。产海南。

Shrubs or trees, young branches, petioles and inflorescences densely waxy stellate-scaly. Leaves clustered at stem apex, papery, abaxially with sparse stellate hairs, gray when dry. Inflorescences terminally clustered; sepals densely and appressed stellate-hairy; ovary with densely appressed stellate hairs. Capsules obovoid or trigonous. Fl. Oct-Dec. Fr. Dec to next Mar. Forests at 100-600 m. Distributed in Hainan.

卵叶巴豆 *Croton caudatus*

光叶巴豆 *Croton laevigatus*

石栗 *Aleurites moluccana*

## 石栗

**Aleurites moluccana** (L.) Willd.

常绿乔木。叶卵形至椭圆状披针形，纸质，全缘或至3-5浅裂。圆锥花序顶生；花瓣狭长圆状匙形，淡黄白色；花盘腺体浅3裂；雄蕊15-20，排成4轮。核果近球形或斜球形。花期4-10月，果期10-12月。生海拔1000米以下的常绿阔叶林或栽培。产云南、广西、广东、海南、台湾和福建。亚洲热带和亚热带、太平洋岛屿亦有；热带地区普遍栽培。

Everygreen trees. Leaves ovate to elliptic-lanceolate, papery, entire or up to shallowly 3-5-lobed. Panicles terminal; petals narrowly oblong-spatulate, yellowish white; disk glands shallowly 3-lobed; stamens 15-20, 4-whorled. Drupes subglobose or obliguely globose. Fl. Apr-Oct. Fr. Oct-Dec. Evergreen forests or cultivated below 1000 m. Distributed in Yunnan, Guangxi, Guangdong, Hainan, Taiwan and Fujian. Also in tropical and subtropical Asia, and Pacific Islands; widely cultivated in the tropics.

## 油桐

**Vernicia fordii** (Hemsl.) Airy Shaw

落叶乔木。叶卵圆形，全缘，稀1-3浅裂，掌状脉5(-7)。花序为平顶聚伞圆锥花序，在新叶发出前长出，常两性；花萼2(或3)分裂；花瓣为倒卵形，基部黄色，粉红色至淡紫色，粉红色脉；雄蕊8-12；子房3-5(-8)室，被短柔毛。核果近球形，平滑。花期3-4月，果期8-11月。栽培于海拔200-2000米的山坡或疏林。产中国西南、华南、东南、华中和华东。越南亦有。

Deciduous trees. Leaves ovate, entire, rarely shallowly 1-3-lobed, palmate veins 5(-7). Inflorescences flat-topped panicles of cymes, appearing generally before new leaves, usually bisexual; calyx 2(or 3)-fid; petals obovate, yellow at base, pink to purplish, pink-veined; stamens 8-12; ovary 3-5(-8)-locular, pubescent. Drupes subglobose, smooth. Fl. Mar-Apr. Fr. Aug-Nov. Cultivated on slopes or open forests at 200-2000 m. Distributed in SW, S, SE, C and E China. Also in Vietnam.

## 木油桐 (千年桐)

**Vernicia montana** Lour.

常绿乔木，雌雄异株。叶宽卵形，掌状5条脉。花序与新叶同时长出，常单性；雄花2或3分裂；花瓣倒卵形，基部白色或紫红色，具紫红色脉纹；雄花雄蕊8-10，花丝被毛；子房密被锈色短柔毛；花柱3，2深裂。核果卵球形，直径3-5厘米。花期4-6月，果期7-10月。生海拔1600米以下

油桐 *Vernicia fordii*

木油桐（千年桐） *Vernicia montana*

麻疯树 *Jatropha curcas*

的疏林中。产中国西南、华南、东南、华中和华东。缅甸、泰国和越南亦有。日本有栽培。

Evergreen trees, dioecious. Leaves broadly ovate, palmately 5-nerved. Inflorescences produced with new leaves, usually unisexual; staminate flowers 2- or 3-fid; petals obovate, white or purple-red at base and with purple-red nerve-stripes; staminate flowers with 8-10 stamens, filaments hairy; ovary densely rusty-pubescent; styles 3, 2-parted. Drupes ovoid, 3-5 cm diam. Fl. Apr-Jun. Fr. Jul-Oct. Sparse forests below 1600 m. Distributed in SW, S, SE, C and E China. Also in Myanmar, Thailand and Vietnam. Cultivated in Japan.

## 东京桐

**Deutzianthus tonkinensis** Gagnep.

乔木，雌雄异株。嫩枝密被星状毛，叶痕明显。叶柄顶端有2腺体；叶全缘。雌花序苞片宿存；雄花花萼钟状，裂片长约1毫米，花盘5深裂，雄蕊7；雌花花萼2-5毫米，花盘杯状，5裂，子房被绢毛，花柱顶端2次分叉。果直径约4厘米，外果皮厚壳质。花期4-6月，果期7-9月。生海拔900米以下的山坡密林。产广西和云南。越南亦有。

Trees, dioecious. Young branches densely stellate-hairy, with prominent leaf scars. Petioles 2-glandular at apex; leaves margin entire. Female inflorescences bracts persistent; male flowers: calyx campanulate, lobes ca. 1 mm long; disk 5-parted; stamens 7; female flowers: calyx 2-5 mm long; disk cupular, 5-fid; ovary sericeous; styles dichotomous, diverging at apex. Fruits ca. 4 cm diam; exocarp thickly crustaceous. Fl. Apr-Jun. Fr. Jul-Sep. Dense forested slopes below 900 m. Distributed in Guangxi and Yunnan. Also in Vietnam.

## 麻疯树

**Jatropha curcas** L.

灌木或小乔木，具水状乳汁。叶全缘或3-5浅裂，纸质，基部心形，先端急尖；掌状脉5-7。花序腋生，长6-10厘米；萼片5；花瓣长圆形，黄绿色，合生至中部；雄蕊10。蒴果椭圆体形或圆球形，直径2.5-3厘米，黄色。花期9-10月，果期10-12月。海拔200-1600米栽培或逸生。产中国西南和华南。其他热带地区亦有栽培。原产热带美洲。

Shrubs or treelets, with watery latex. Leaves entire or 3-5-lobed, papery, base cordate, apex acute; palmate veins 5-7. Inflorescences axillary, 6-10 cm long; sepals 5; petals oblong, green-yellow, connate to middle; stamens 10. Capsules ellipsoid or globose, 2.5-3 cm diam, yellow. Fl. Sep-Oct. Fr. Oct-Dec. Cultivated or naturalized at 200-1600 m. Distributed in SW and S China. Cultivated in other tropical regions of the world. Native to tropical America.

东京桐 *Deutzianthus tonkinensis*

宿萼木
*Strophioblachia fimbricalyx*

叶轮木 *Ostodes paniculata*

## 宿萼木

**Strophioblachia fimbricalyx** Boerl.

灌木。嫩枝灰白色，被疏柔毛；成熟后无毛，具皮孔。叶卵状披针形、卵形至倒卵状披针形。雄花花瓣倒卵形，腺体宽且扁，雄蕊15-30；雌花萼片边缘生腺毛，无花瓣，花盘环状，子房3室，花柱3，2深裂。蒴果具3纵沟。花期(3-)5-10月，果期(6-)7-12月。生海拔200-400米的密林或灌木丛。产广西、海南和云南。印度尼西亚、菲律宾和越南亦有。

Shrubs. Young branches gray-white and sparsely pubescent, glabrous and lenticellate at maturity. Leaves ovate-lanceolate, ovate, or obovate-lanceolate. Male flowers: petals obovate; disk glands broad and flat; stamens 15-30; female flowers: sepals margins with glandular hairs; petals absent; disk annular; ovary 3-locular; styles 3, bipartite. Capsules with 3 longitudinal grooves. Fl. (Mar-)May-Oct. Fr. (Jun-)Jul-Dec. Dense forests or thickets at 200-400 m. Distributed in Guangxi, Hainan and Yunnan. Also in Indonesia, the Philippines and Vietnam.

## 叶轮木

**Ostodes paniculata** Blume

乔木，枝和叶均无毛。叶常在顶端簇生；叶薄革质。花序无毛或被稀疏贴生微柔毛；花瓣5，白色；雄蕊20-35；子房被长硬毛。蒴果微柔毛和密生疣状突起；中果皮硬，木质。花期3-5月，果期8-9月。生海拔400-1400米的常绿林中。产云南和海南。南亚和东南亚亦有。

Trees, branches and leaves glabrous. Leaves usually apically clustered; leaves thinly coriaceous. Inflorescences glabrous or sparsely adpressed-puberulous; petals 5, white; stamens 20-35; ovary hirsute. Capsules tomentulose and densely elevated-verrucose; mesocarps hard, woody. Fl. Mar-May. Fr. Aug-Sep. Evergreen forests at 400-1400 m. Distributed in Yunnan and Hainan. Also in S and SE Asia.

## 云南叶轮木

**Ostodes katharinae** Pax

乔木。叶及叶柄均无毛，或仅在叶背面沿中脉被稀疏毛；叶薄革质。花序总状，花序轴密被绒毛；雄蕊多至40。蒴果稍扁球形，密被褐色微绒毛，疏具不明显疣状凸起。花期4-6月，果期7-8月。生海拔900-2000米的密林中。产云南和西藏。泰国北部亦有。

Trees. Leaves and petioles glabrous, or abaxially sparsely hairy along costa; leaves thinly coriaceous. Inflorescences racemelike, axils densely tomentose; stamens to 40. Capsules slightly depressed globose, densely brown tomentulose, obscurely sparsely elevated-verrucose. Fl. Apr-Jun. Fr. Jul-Aug. Dense forests at 900-2000 m. Distributed in Yunnan and Xizang. Also in N Thailand.

云南叶轮木 *Ostodes katharinae*

长梗三宝木 *Trigonostemon thyrsoideum*

## 长梗三宝木

**Trigonostemon thyrsoideum** Stapf

灌木或小乔木。枝被微柔毛至无毛。叶距一致，倒卵状椭圆形或长圆状椭圆形至披针形，纸质，两面无毛。花序顶生，金字塔状聚伞圆锥花序，长达20厘米，具褐色短柔毛。果柄长约3厘米；蒴果直径约1.5厘米，具纵向3深槽。花期4-7月，果期6-9月。生海拔600-1000米的密林。产云南南部、贵州和广西。缅甸、老挝、泰国北部和越南亦有。

Shrubs or treelets. Branches puberulent to glabrous. Leaves uniformly spaced, obovate-elliptic or oblong-elliptic to lanceolate, papery, both surfaces glabrous. Inflorescences terminal, pyramidal thyrses, to 20 cm long, brown pubescent. Fruiting pedicels ca. 3 cm long; capsules ca. 1.5 cm diam, deeply longitudinally 3-grooved. Fl. Apr-Jul. Fr. Jun-Sep. Dense forests at 600-1000 m. Distributed in S Yunnan, Guizhou and Guangxi. Also in Myanmar, Laos, N Thailand and Vietnam.

## 黄花三宝木

**Trigonostemon fragilis** (Gagnep.) Airy Shaw

灌木，嫩枝、叶柄和花序密被黄褐色柔毛。叶柄长1-3厘米；叶纸质，椭圆形至长圆形；侧脉每边8-10条。圆锥花序顶生，长达25厘米；花瓣黄色，有爪；雄花萼片5，长约2毫米；雌花花盘环状，子房无毛，柱头头状。蒴果近球形，长约1厘米。花期4-5月，果期6-8月。生海拔500-600米的石灰岩山地灌木林。产广西和海南。越南亦有。

Shrubs, young branches, petiole, and inflorescences uniformly densely spreading tawny pubescent. Petioles 1-3 cm long; leaves papery, elliptic to oblong; lateral veins 8-10 side. Panicles terminal, to 25 cm long; petals yellow, hooked; male flowers: sepals 5, ca. 2 mm long; female flowers: disk annular; ovary glabrous; stigmas capitate. Capsules subglobose, ca. 1 cm long. Fl. Apr-May. Fr. Jun-Aug. Thickets on limestone at 500-600 m. Distributed in Guangxi and Hainan. Also in Vietnam.

黄花三宝木 *Trigonostemon fragilis*

勐仑三宝木 *Trigonostemon bonianus*

木薯 *Manihot esculenta*

## 勐仑三宝木

**Trigonostemon bonianus** Gagnep.

灌木或小乔木，高2-4米。幼枝被柔毛。叶纸质，两面均无毛；基部楔形，顶端锐尖至尾尖或钝；侧脉3-5条。总状花序顶生，开展，长达15厘米；苞片线形；雄花直径约6毫米，花瓣黄色；雌花花瓣椭圆形，黄色，子房光滑，花柱3，较短，柱头头状。花期5月。生海拔500-700米的疏林中。产云南南部。越南亦有。

Shrubs or treelets, 2-4 m tall. Young branches puberulent. Leaves papery, glabrous on both surfaces, base cuneate, apex acuminate to caudate-acuminate, acumen obtuse; lateral veins 3-5. Racemes terminal, spreading, to 15 cm long; bracts linear; male flowers ca. 6 mm diam; petals yellow; female flowers: petals elliptic, yellow; ovary glabrous; styles 3, short; stigmas capitate. Fl. May. Open forests at 500-700 m. Distributed in S Yunnan. Also in Vietnam.

## 小盘木

**Microdesmis caseariifolia** Planchon ex Hook.

灌木或小乔木。嫩枝密被柔毛。叶披针形、长圆状披针形至长圆形。花小，黄色，簇生叶腋；雄花花瓣长约2毫米，雄蕊10，2轮，外轮5枚较长；雌花花瓣比雄花的略长，子房2室。核果内具种子2粒。花期3-9月，果期7-11月。生海拔200-800米的山坡密林下、山谷或灌木丛中。产广东、广西、海南和云南。南亚和东南亚亦有。

Shrubs or treelets. Young branches pubescent. Leaves lanceolate, oblong-lanceolate, or oblong. Flowers yellow, small, in axillary fascicles; male flowers: petals ca. 2 mm long; stamens 10, two-wheeled, outer 5 longer; female flowers: petals slightly larger than those of male flowers; ovary 2-celled. Drupes 2-seeded. Fl. Mar-Sep. Fr. Jul-Nov. Dense forests, valleys, slopes, brushwood at 200-800 m. Distributed in Guangdong, Guangxi, Hainan and Yunnan. Also in S and SE Asia.

## 木薯

**Manihot esculenta** Crantz

直立灌木。块根圆柱状。叶柄稍盾状着生；叶掌状深裂，倒披针形至狭椭圆形；全缘。总状花序，苞片条状披针形；雄花花萼紫红色，内面被毛；雌花子房卵形，具6条纵棱，摺扇状柱头外弯。蒴果具6条纵翅；粗糙。花期9-11月。广泛栽培于福建、广东、广西、贵州、海南、台湾和云南。原产巴西；现热带地区广泛栽培。

Erect shrubs. Root tubers terete. Petioles slightly peltate; leaves palmately, lobes oblanceolate to narrowly elliptic, entire. Racemes bracts oblong-lanceolate; male flowers: calyx purple-red; hairy inside; female flowers: ovary ovoid, longitudinally 6-angled; stigmas recurved, plaited. Capsules longitudinally 6-winged, scabrous. Fl. Sep-Nov. Widely cultivated in Fujian, Guangdong, Guangxi, Guizhou, Hainan, Taiwan and Yunnan. Native to

小盘木 *Microdesmis caseariifolia*

Brazil; cultivated throughout the tropics.

## 白树

**Suregada multiflora** (A. Juss.) Baill.

灌木或乔木。枝条灰黄色至灰褐色，无毛。叶长圆形或狭椭圆形，先端急尖或短渐尖，偶有1-2粗齿；叶柄长3-8毫米；小聚伞花序具花序梗；雄蕊30-60。蒴果球形，长11-15毫米。稍肉质，近无毛。花期3-9月，果期6-11月。生海拔100-600米的灌丛中。产云南、广西、广东和海南。东南亚和澳大利亚亦有。

Shrubs or trees. Branches gray-yellow to gray-brown, glabrous. Leaves oblong or narrowly elliptic, apex acute or shortly acuminate, occasionally with 1-2 teeth; petioles 3-8 mm long. Inflorescences pedunculate cymules; stamens 30-60. Capsules globose, 11-15 mm long, slightly fleshy, subglabrous. Fl. Mar-Sep. Fr. Jun-Nov. Thickets at 100-600 m. Distributed in Yunnan, Guangxi, Guangdong and Hainan. Also in SE Asia and Australia.

## 斑籽木

**Baliospermum solanifolium** (Burman) Suresh

灌木，雌雄异株。幼枝绿色，被紧贴短柔毛，成熟后无毛。叶椭圆形、长圆形或倒披针形至阔卵形，有时浅3-5裂，纸质。圆锥花序腋生；雌花腋生或着生于雄花序基部。蒴果下垂，近球形。花期3-5月，果期6-9月。生海拔700米以下的杂木林中。产云南西南部。南亚和东南亚亦有。

Shrubs, monoecious. Young branches green, appressed pubescent, glabrous at maturity. Leaves elliptic, oblong, or oblanceolate to broadly ovate, sometimes shallowly 3-5-lobed, papery. Panicles axillary; pistillate flowers axillary or inserted at base of staminate inflorescences. Capsules pendulous, subglobose. Fl. Mar-May. Fr. Jun-Sep. Mixed woodlands below 700 m. Distributed in SW Yunnan. Also in S and SE Asia.

云南斑籽木 *Baliospermum calycinum*

## 云南斑籽木

**Baliospermum calycinum** Müll. Arg.

灌木，雌雄异株。枝无毛。叶椭圆形，狭长椭圆形至长圆形，长10-15厘米，宽3.5-8厘米。花序腋生；花序梗伸长；花序圆锥状，多花；子房无毛。蒴果直立，稍扁球形。花期6-10月，果期9-12月。生海拔500-2500米的林中。产云南南部。南亚亦有。

Shrubs, dioecious. Branches glabrous. Leaves elliptic, narrowly elliptic to oblong, 10-15 cm long, 3.5-8 cm wide. Inflorescences axillary; peduncles elongated; inflorescences paniculate, many flowered; ovary glabrous. Capsules erect, ± depressed globose. Fl. Jun-Oct. Fr. Sep-Dec. Forests at 500-2500 m. Distributed in S Yunnan. Also in S Asia.

白树 *Suregada multiflora*

斑籽木 *Baliospermum solanifolium*

黄桐 *Endospermum chinense*

地杨桃 *Microstachys chamaelea*

## 黄桐

**Endospermum chinense** Benth.

乔木，高6-35米。叶薄革质，两面无毛或下面被星状毛。花序腋生；花萼杯状；雄蕊5-12，2或3轮；花盘环状，2-4齿裂；子房2或3室。果实近球形，密被灰黄色星状短柔毛。花期5-8月，果期8-11月。生海拔800米以下的常绿林中。产云南、广西、广东、海南和福建。印度、缅甸、泰国和越南亦有。

Trees, 6-35 m tall. Leaves thinly coriaceous, glabrous or abaxially stellate-hairy. Inflorescences axillary; calyx cupular; stamens 5-12, in 2 or 3 series; disk annular, 2-4-serrate-lobed; ovary 2- or 3-locular. Fruits subglobose, densely gray yellow stellate-pubescent. Fl. May-Aug. Fr. Aug-Nov. Evergreen forests below 800 m. Distributed in Yunnan, Guangxi, Guangdong, Hainan and Fujian. Also in India, Myanmar, Thailand and Vietnam.

## 地杨桃

**Microstachys chamaelea** (L.) Müll. Arg.

多年生草本或多分枝灌木。茎基部稍木质化，分枝具棱。叶厚纸质；基部具小腺体。雄花苞片具2匙形腺体，雄蕊3；雌花萼片3，基部向轴面具2腺体，子房具皮刺6列。蒴果分果爿具皮刺。花果期3-11月。生海拔300米以下的沙滩、旱地、路边或溪边。产广东、广西和海南。南亚、东南亚、非洲、澳大利亚和太平洋岛屿亦有。

Perennial herbs to many-stemmed subshrubs. Stem ± woody at base, branching slightly angular. Leaves thickly papery, usually with paired glands. Male flowers: bracts mostly covered by a pair of large spatulate glands; stamens 3; female flowers: sepals 3, sometimes with 2 glands at base inside; ovary with 6 rows of spines. Capsules lobes softly spiny. Fl. and fr. Mar-Nov. Beaches, dry fields, roadsides, along rivers below 300 m. Distributed in Guangdong, Guangxi and Hainan. Also in S and SE Asia, Africa, Australia and Pacific Islands.

## 红背桂

**Excoecaria cochinchinensis** Lour.

常绿灌木，雌雄异株。叶狭椭圆形或长圆形，纸质，无毛，背面紫红色或深红色。总状花序腋生或顶生；雄花苞片阔卵形，具2腺点，每苞具1花；小苞片2，线形；花萼3。蒴果球形，直径约8毫米，基部截形。花期几乎全年。云南、广西、广东、海南、台湾和福建栽培。原产越南；热带和亚热带地区普遍栽培。

Evergreen shrubs, dioecious. Leaves narrowly elliptic or oblong, papery, glabrous, abaxially purple-red or scarlet. Flowers in axillary or terminal racemes; staminate flowers bracts broadly ovate, 2-glandular, each bract 1-flowered; bractlets 2, linear; sepals 3. Capsules globose, ca. 8 mm diam, truncate at base. Fl. almost all year. Cultivated in Yunnan, Guangxi, Guangdong, Hainan, Taiwan and Fujian. Native to Vietnam; widely cultivated in tropical and subtropical areas.

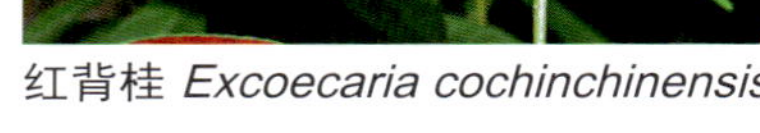

红背桂 *Excoecaria cochinchinensis*

## 鸡尾木

**Excoecaria venenata** S. K. Lee et F. N. Wei

灌木，雌雄同株。叶近革质，上面幼时淡红色或上面仅脉上

鸡尾木 *Excoecaria venenata*

海漆 *Excoecaria agallocha*

紫红色，无毛，先端渐尖，具镰刀状短尖头。花单生；雄花：苞片阔三角形；小苞片条形，基部有2腺体；萼片3，条状披针形；花药球形。蒴果球形。花期8-10月。生石灰岩林中或灌丛中。产广西西南部。

Shrubs, monoecious. Leaves subcoriaceous, reddish or only veins adaxially purple when young, glabrous, apex acuminate, with falcate mucro. Flowers solitary; staminate flowers: bracts broadly triangular; bracteoles linear, base with 2 glands; sepals 3, linear-lanceolate; anther globose. Capsules globose. Fl. Aug-Oct. Limestone forests or thickets. Distributed in SW Guangxi.

## 海漆

**Excoecaria agallocha** L.

落叶乔木，雌雄异株。皮孔多数。叶互生，近革质；叶柄顶端具2腺体。雄花：每一苞片内1朵花，苞片基部具2腺体，雄蕊伸出萼片外；雌花花柱3，分离。蒴果具3沟槽。种子直径约4毫米。花果期1-9月。生海拔100米以下的红树林和潮间林、盐碱地和稻田。产广东、广西、海南和台湾。日本、南亚、东南亚、澳大利亚和太平洋岛屿亦有。

Trees deciduous, dioecious. Branches lenticellate. Leaves alternate throughout; subleathery; with 2 distinct glands at base near junction with petiole. Male flowers 1 per bract; bracts base inside 2-glandular; stamens usually exceeding sepals; female flowers: stigmas 3, free. Capsules trisulcate. Seeds ca. 4 mm diam. Fl. and fr. Jan-Sep. Mangrove and tidal forests, brackish areas and rice fields below 100 m. Distributed in Guangdong, Guangxi, Hainan and Taiwan. Also in Japan, S and SE Asia, Australia and Pacific Islands.

## 云南土沉香

**Excoecaria acerifolia** Didr.

灌木，雌雄同株。叶片卵形、卵状披针形，稀椭圆形。花单性，总状花序顶生及腋生，雌花在下部，雄花在上部；雄花基部两侧各具1圆形腺体，每个苞片具2或3花。蒴果近球状或三角状。花期6-8月。生海拔1200-3000米的林中或灌丛中。产云南、四川、湖北、湖南和甘肃。印度和尼泊尔亦有。

Shrubs, monoecious. Leaves ovate or ovate-lanceolate, rarely elliptic. Flowers unisexual, terminal and axillary racemes, female in lower part, staminate in upper part; stamina flowers base bilateral each with 1 round gland, each bract with 2 or 3 flowers. Capsules subglobose or triangular. Fl. Jun-Aug. Forests or thickets at 1200-3000 m. Distributed in Yunnan, Sichuan, Hubei, Hunan and Gansu. Also in India and Nepal.

云南土沉香 *Excoecaria acerifolia*

乌桕 *Triadica sebifera*

## 乌桕

**Triadica sebifera** (L.) Small

乔木。树皮暗绿色，具纵裂纹。叶互生；叶片菱形、菱状卵形或阔卵形，纸质。花淡黄绿色，组成顶生长3-35厘米的总状花序；雄花具2或稀3雄蕊；子房顶端具3花柱。蒴果近球形至梨状球形，成熟后黑色。花期4-8月，果期8-12月。生海拔100米以下的荒野或疏林中，或栽培。产中国西南、华南、东南、华中和华东。越南和日本亦有。栽培于欧洲、北美洲和非洲。

Trees. Bark dark green, with longitudinal stripes. Leaves alternate; leaves rhomboid, rhomboid-ovate or broadly ovate, papery. Flowers yellowish green in terminal 3-35 cm long racemes; staminate flowers with 2, rarely 3 stamens; ovary with 3 styles at apex. Capsules subglobose to pyriform-globose, black when mature. Fl. Apr-Aug. Fr. Aug-Dec. Wilderness or sparse forests below 100 m, or cultivated. Distributed in SW, S, SE, C and E China. Also in Vietnam and Japan. Cultivated in Europe, North America and Africa.

## 圆叶乌桕

**Triadica rotundifolia** (Hemsl.) Esser

乔木，雌雄同株，全株无毛。叶互生，近革质；叶片近圆形，全缘；顶端圆，稀凸尖。花单性，密集成顶生总状花序；雄花苞片基部具腺体，花萼杯状；雌花花萼3裂，花柱3，基部合生，柱头外卷。蒴果直径8-10毫米。种子具蜡质假种皮。花期4-6月，果期7-10月。生海拔100-500米的石灰岩山地林中。产广东、广西、贵州、湖南和云南。越南亦有。

Trees, monoecious, glabrous. Leaves alternate; subleathery; leaves subrotund, margin entire, apex rounded, rarely acute. Flowers unisexual, densely fascicled in terminal racemes; male flowers: bracts base with gland; calyx cupular; female flowers: calyx 3-lobed; styles 3, connate at base; stigma revolute. Capsules 8-10 mm diam. Seeds with waxy arils. Fl. Apr-Jun. Fr. Jul-Oct. Limestone montane forests at 100-500 m. Distributed in Guangdong, Guangxi, Guizhou, Hunan and Yunnan. Also in Vietnam.

## 山乌桕

**Triadica cochinchinensis** Lour.

乔木，雌雄同株，无毛。叶互生；叶柄顶端具2腺体；叶椭圆形或长圆状卵形，长4-10 × 2.5-5厘米，纸质，幼时淡红色。组成顶生总状花序；花序长4-9厘米，雌花位于下部，雄花位于上部或均有。蒴果黑色，球形，直径7-9毫米。花期4-6月，果期7-10月。生海拔100-1100米的常绿湿润阔叶林、近热带林中、山坡林或灌丛。产中国西南、华南、东南、华中和华东。南亚和东南亚亦有。

Trees, monoecious, glabrous. Leaves alternate; petioles 2-glandular at apex; leaves elliptic or oblong-ovate, 4-10 × 2.5-5 cm, papery, reddish when young. Flowers in terminal racemes;

圆叶乌桕 *Triadica rotundifolia*

山乌桕 *Triadica cochinchinensis*

白木乌桕 *Neoshirakia japonica*

通奶草 *Euphorbia hypericifolia*

inflorescences 4-9 cm long, pistillate flowers in lower part, staminate flowers in upper part or throughout. Capsules black, globose, 7-9 mm diam. Fl. Apr-Jun. Fr. Jul-Oct. Moist broad-leaved evergreen forests, subtropical forests, montane forests or brushwood at 100-1100 m. Distributed in SW, S, SE, C and E China. Also in S and SE Asia.

## 白木乌桕

**Neoshirakia japonica** (Siebold et Zucc.) Esser

小乔木，雌雄同株。叶卵形、卵状长方形或椭圆形，纸质，侧脉8-10对。花序顶生；雌花苞片3，深裂几达基部；萼片3；子房3室；柱头3。蒴果三棱状球形。花期5-6月，果期7-9月。生海拔100-400米的林中湿润处或溪边。产中国西南、东南、华中和华东。朝鲜半岛和日本亦有。

Treelets, monoecious. Leaves ovate, ovate-rectangular, or elliptic, papery, lateral veins 8-10-paired. Inflorescences terminal; bracts of pistillate flowers 3, parted nearly to base; sepals 3; ovary 3-locular; stigmas 3. Capsules trigonous-globose. Fl. May-Jun. Fr. Jul-Sep. Wet forests or by rivers at 100-400 m. Distributed in SW, SE, C and E China. Also in Korean Peninsula and Japan.

## 通奶草

**Euphorbia hypericifolia** L.

一年生草本。叶对生，下面亮绿色，有时紫红色，基部圆形；托叶三角形；合生或分离。总苞陀螺状，5裂；腺体4，附属物白色或亮粉红色。蒴果3棱状，光滑，无毛。花果期8-12月。生旷野荒地、路边、灌丛或田间。产中国西南和华南，北京逸生。亦广布于世界热带和亚热带地区。

Annual herbs. Leaves opposite, light green abaxially, sometimes purple-red, base rounded; stipules triangular, connate or free. Involucres turbinate, 5-lobed; glands 4, appendages white or light pink. Capsules 3-angular, smooth, glabrous. Fl. and fr. Aug-Dec. Waste places, roadsides, thickets or fields. Distributed in SW and S China, naturalized in Beijing. Also widely in tropical and subtropical regions of the world.

## 飞扬草

**Euphorbia hirta** L.

一年生草本。茎被淡褐色粗硬毛。叶卵圆形，上面绿色至红色，有时沿中脉具紫色斑块，基部稍倾斜，先端急尖或钝。杯状聚伞花序密集，常头状；总苞钟形；腺体4，红色，附属物白色至淡红色；花药红色。蒴果被短柔毛。花果期6-12月。生海拔900-2100米的路边、草丛、开阔林地、山坡或灌丛中。产中国西南、华南和东南。热带和亚热带地区亦有。

Annual herbs. Stems brown-hispid. Leaves ovate, adaxially green to red, sometimes with purple blotch along midrib, base slightly oblique, acute or obtuse at apex. Cyathia in dense, often headlike; involucres campanulate; glands 4, red, appendages white to reddish; anthers red. Capsules pubescent. Fl. and fr. Jun-Dec. Roadsides, grasslands, open forests, slopes or thickets at 900-2100 m. Distributed in SW, S and SE China. Also in tropical and subtropical regions of the world.

飞扬草 *Euphorbia hirta*

## 土库曼大戟

**Euphorbia granula** Forsk.

一年生矮小草本。茎基部木质化且极多分枝，斜倚向上或匍匐；节与节间明显。叶对生，近椭圆形，长3-6毫米；托叶钻状。花序单生于叶腋；总苞陀螺状，腺体4，具白色附属物，且不等大，近轴面窄，略宽于腺体，远轴面为腺体宽的2-4倍。蒴果被柔毛或少许疏柔毛。花果期6-9月。生海拔约500米的沙质或石质地。产新疆。中亚、西亚和北非亦有。

Annual small herbs. Stems usually woody at base and much branched, ascending or prostrate; nodes and internodes conspicuous. Leaves opposite, subelliptic, 3-6 mm long; stipules subulate. Inflorescences solitary in leaf axils; involucre turbinate, glands 4, appendages white and unequal, narrow adaxially, but wider than glands, abaxially 2-4 times as wide as gland. Capsules pilose or sparsely pilose. Fl. and fr. Jun-Sep. Sandy or stony fields at ca. 500 m. Distributed in Xinjiang. Also in C and W Asia, and N Africa.

## 地锦草

**Euphorbia humifusa** Willd.

一年生草本。茎多自基部分枝，匍匐或斜升，常红色或淡紫红色。叶对生；托叶膜质，常深裂为线状裂片，早落；叶长圆形或椭圆形。杯状聚伞花序单生于叶腋；总苞螺旋状，腺体4，附属物白色或紫红色。蒴果3棱状卵球形。花果期5-10月。生海拔3000(-3800)米及以下的田中、路边、沙丘、海岸、山坡或生开阔地。产中国(除海南)。欧亚大陆广布。

Herbs, annual. Stems many from base, prostrate or ascending, often red or pinkish red. Leaves opposite; stipules membranous, deeply divided into often thread-like lobes, caducous; leaves oblong or elliptic. Cyathia single, axillary; involucres turbinate, glands 4, appendages white or pink-red. Capsules 3-angular-ovoid-globose. Fl. and fr. May-Oct. Fields, roadsides, sandy hills, seashores, slopes or open situations below 3000(-3800) m. Distributed throughout China, except Hainan. Distributed widely in Eurasia.

地锦草 *Euphorbia humifusa*

## 匍匐大戟

**Euphorbia prostrata** Aiton

一年生草本。根纤细。茎匍匐。叶对生；托叶长三角形，易脱落；叶上面绿色，有时下面亮红色或红色。杯状聚伞花序常单生或少簇生于叶腋；总苞陀螺状；腺体4，附属物白色，极狭。蒴果三棱状。花果期4-10月。生路边、荒地、村庄或灌丛。中国西南、华南和东南逸生。其他热带、亚热带地区亦有归化；原产美洲。

Annual herbs. Roots slender. Stems creeping. Leaves opposite; stipules long triangular, easily fallen; leaves adaxially green, sometimes with light red or red abaxially. Cyathia single, axillary or few clustered; involucres turbinate; glands 4, appendages white, extremely narrow. Capsules 3-angular. Fl. and fr. Apr-Oct. Roadsides, wastelands, villages or thickets. Naturalized in SW, S and SE China. Also in other tropical and subtropical areas; native to America.

土库曼大戟 *Euphorbia granula*

匍匐大戟 *Euphorbia prostrata*

千根草 *Euphorbia thymifolia*

斑地锦 *Euphorbia maculata*

## 千根草
**Euphorbia thymifolia** L.

一年生草本。根纤维状。叶对生，边缘常具细锯齿，两面具短柔毛。杯状聚伞花序单生或簇生于叶腋；总苞狭钟状至陀螺状，外被稀疏短柔毛；腺体4，附属物白色。蒴果被短柔毛。花果期6-11月。生路边、灌丛、草地或疏林中。产中国西南、华南、东南和华东。广布于世界热带和亚热带地区。

Annual herbs. Root fibrous. Leaves opposite, margin usually finely serrulate, both surfaces pubescent. Cyathia solitary, or conferted at leaf axils; involucres narrowly campanulate to turbinate, abaxially sparsely pubescent; glands 4, appendage white. Capsules pubescent. Fl. and fr. Jun-Nov. Roadsides, scrub, grassy places or open forests. Distributed in SW, S, SE and E China. Also in tropical and subtropical regions of the world.

## 斑地锦
**Euphorbia maculata** L.

一年生草本。茎匍匐。叶对生，上面绿色，中部常具一个长圆形的紫色斑点，两面无毛。杯状聚伞花序单生于节上叶腋；总苞狭杯状；腺体4，黄绿色；雌蕊1。蒴果三角状卵形，3瓣裂。花果期4-9月。生草地或路边。产台湾、浙江、湖北、河南、河北、江苏和江西。原产北美洲；归化于欧亚大陆。

Annual herbs. Stems stoloniferous. Leaves opposite, adaxially green, often with an oblong purple spot in middle, both surfaces glabrous. Cyathia simple from nodes at leaf axil; involucres narrowly cup-shaped; glands 4, yellow-green; pistils 1. Capsules triangular-ovate, 3-valved. Fl. and fr. Apr-Sep. Grasslands or by roads. Distributed in Taiwan, Zhejiang, Hubei, Henan, Hebei, Jiangsu and Jiangxi. Native to North America; naturalized in Eurasia.

## 金刚纂
**Euphorbia neriifolia** L.

乔木或灌木。茎圆柱形，具5列小瘤隆起呈螺旋状排列的脊。叶互生，枝顶簇生，宿存，肉质；托叶2，刺状。杯状聚伞花序近顶生呈聚伞状；苞叶膜质，早落；总苞钟形；腺体5，肉质，厚且全缘；雄花多数，苞片线形；雌花稀能育。花期6-9月。云南、广西、广东和海南栽培或逸生。广泛栽培于热带亚洲。原产印度。

Trees or shrubs. Stems terete, with 5 spiral ranks of tubercles. Leaves alternate, apically clustered, persistent, succulent; stipules 2, spiniform. Cyathia in subterminal cymes; cyathophylls membranous, caducous; involucres campanulate; glands 5, succulent, thick and entire; staminate flowers many, bracts linear; piatillate flowers rarely developed. Fl. Jun-Sep. Cultivated or naturalized in Yunnan, Guangxi, Guangdong and Hainan. Widely cultivated in tropical Asia. Native to India.

金刚纂 *Euphorbia neriifolia*

霸王鞭 *Euphorbia royleana*

一品红(圣诞红) *Euphorbia pulcherrima*

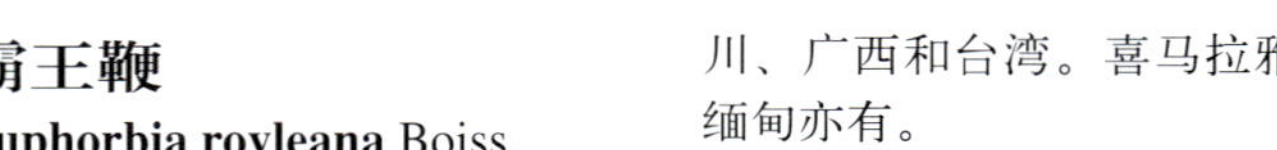

## 霸王鞭
**Euphorbia royleana** Boiss.

小乔木或灌木。叶肉质，互生，集于枝顶；托叶刺状，长3-5毫米，宿存；二歧聚伞花序。苞叶与总苞等长，膜质；总苞杯状，黄色；腺体5，暗绿色。蒴果3瓣裂。花果期5-7月。生亚热带多雨沟谷，在岩坡上形成其群落。产云南、四川、广西和台湾。喜马拉雅和缅甸亦有。

Small trees or shrubs. Leaves fleshy, alternate, aggregated at top of branches; stipules spinecent, 3-5 mm long, persistent; dichasial cymes. Cyathophylls as long as involucre, membranous; involucres cup-shaped, yellow; glands 5, dark green. Capsules 3-lobed. Fl. and fr. May-Jul. Subtropical rainshadow valleys, forming its own communities on rocky slopes. Distributed in Yunnan, Sichuan, Guangxi and Taiwan. Also in Himalaya and Myanmar.

## 一品红 (圣诞红)
**Euphorbia pulcherrima** Willd. ex Klotz.

灌木至小乔木。叶卵状椭圆形至披针形，上部的较狭，朱红色。杯状聚伞花序形成极密的单侧复合花序；总苞坛状，淡绿色，一边有一大的黄色腺体，压扁，二唇形。蒴果为3瓣裂的球形。花果期10月至翌年4月。中国大部分地区有栽培。亦广泛栽培于热带和亚热带地区。原产中美洲。

Shrubs to small trees. Leaves ovate-elliptic to lanceolate, upper ones narrower, scarlet. Cyathia in a very congested, 1-sided synflorescence; involucres urceolate, greenish, bearing a large yellow gland at one side, compressed, 2-lipped. Capsules 3-lobed-globose. Fl. and fr. Oct to next Apr. Cultivated in most parts of China. Also widely cultivated in tropical and subtropical regions of the world. Native to Central America.

猩猩草 *Euphorbia cyathophora*

## 猩猩草
**Euphorbia cyathophora** Murr.

一年生或多年生草本。叶卵形至椭圆形，两面无毛。杯状聚伞花序顶生；基部总苞叶淡红色或基部红色；总苞桶形，无毛；腺体压扁，近二唇形。果无毛。花果期5-11月。中国各地广泛栽培或归化。栽培并归化于泛热带地区。原产美洲。

Annual or perennial herbs. Leaves ovate to elliptic, both surfaces glabrous. Cyathiums terminal; basal bracts reddish or scarlet at base; involucres barrel-shaped, glabrous; glands compressed, somewhat bilabiate. Fruits glabrous. Fl. and fr. May-Nov. Cultivated or naturalized in most parts of China. Cultivated and naturalized in pantropical areas. Native to America.

白苞猩猩草 *Euphorbia heterophylla*

## 白苞猩猩草
**Euphorbia heterophylla** L.

一年生草本。叶形变异大，从卵形至条形，叶缘具红色。杯状聚伞花序顶生；基部苞片与叶相似，浅绿色或基部白色；总苞桶形，无毛；腺体漏斗形，阔圆形，常具红缘。果被柔毛。种子明显具疣。花果期2-11月。中国西南、华南、东南、华中、华北和华东栽培或归化。栽培并归化于泛热带地区。原产美洲。

Annual herbs. Leaves very variable in shape from ovate to linear, margin red. Cyathiums terminal; basal bracts similar to leaves, pale green or white at base; involucres barrel-shaped, glabrous; glands funnel-shaped, opening circular, often red-rimmed. Fruits pubescent. Seeds obviously verrucate. Fl. and fr. Feb-Nov. Cultivated or naturalized in SW, S, SE, C, N and E China. Cultivated and naturalized in pantropical areas. Native to America.

## 续随子
**Euphorbia lathyris** L.

一年生草本。叶交互对生，无托叶，无叶柄，条状披针形，无毛。花序为一顶生的假伞形花序，常复合；初生苞叶稍淡黄绿色；聚伞花序整齐多分叉；总苞近钟形；腺体4，深褐色，横向长圆状肾形，每端具棒状角。蒴果三角状球形，不开裂。花期4-7月，果期6-9月。野生或栽培。产中国大部分地区。全球大部分地区亦有。

Annual herbs. Leaves decussate, stipules absent, petioles absent, linear-lanceolate, glabrous. Inflorescences a terminal pseudumbel, often compound; primary involucral leaves slightly yellowish green; cymes regularly many forked; involucres subcampanulate; glands 4, dark brown, transversely oblong-reniform with a club-shaped horn at each tip. Capsules trigonous-globose, indehiscent. Fl. Apr-Jul. Fr. Jun-Sep. Wild or cultivated. Distributed in most parts of China. Also in most parts of the world.

续随子 *Euphorbia lathyris*

## 泽漆
**Euphorbia helioscopia** L.

草本。根纤维状。叶互生，边缘具锯齿。花序为复假伞形花序，常极紧密；初生总苞叶5，淡黄绿色；苞叶2，倒卵形，边缘具锯齿；杯状聚伞花序近无柄；总苞钟形。蒴果三角状圆柱形。花果期4-10月。生山沟、灌丛、路旁、荒野或山坡。产中国大部分地区。亚洲、北非和欧洲亦有。

Herbs. Root fibrous. Leaves alternate, margin dentate. Inflorescences a compound pseudumbel, usually rather compact; primary involucral leaves 5, yellowish green; cyathophylls 2, obovate, margin dentate; cyathiums subsessile; involucres campanulate. Capsules trigonous-terete. Fl. and fr. Apr-Oct. Valleys, thickets, roadsides, wastelands or slopes. Distributed in most parts of China. Also in Asia, N Africa and Europe.

泽漆 *Euphorbia helioscopia*

黄苞大戟 *Euphorbia sikkimensis*

## 黄苞大戟
**Euphorbia sikkimensis** Boiss.

草本。根圆柱形。叶互生，除最下面外大小一致，无毛，全缘。花序为顶生的假伞形花序；总苞叶和苞叶黄色；苞叶2，卵形；总苞钟形；腺体4，褐色，圆形。蒴果球形。花期4-7月，果期6-9月。生海拔600-4500米的山坡、疏林或灌丛中。产中国西南。喜马拉雅亦有。

Herbs. Roots terete. Leaves alternate, all but lowermost uniform in size, glabrous, margin entire. Inflorescences a terminal pseudumbel; involucral leaves and cyathophylls yellow; cyathophylls 2, ovate; involucres campanulate; glands 4, brown, rounded. Capsules globose. Fl. Apr-Jul. Fr. Jun-Sep. Slopes, open forests or thickets at 600-4500 m. Distributed in SW China. Also in Himalaya.

## 圆苞大戟
**Euphorbia griffithii** Hook. f.

多年生草本。茎多数单一。叶互生，卵状长圆形至椭圆形；无托叶；苞叶2，常红色或紫红色，多变。杯状聚伞花序单生，总苞杯状，裂片圆形，边缘和内面被白色疏柔毛；花柱分离。蒴果球形。花果期6-9月。生海拔2500-4900米的疏林、灌丛或草甸。产云南、四川和西藏。喜马拉雅亦有。

Perennial herbs. Stems mostly single. Leaves alternate, ovate-oblong to elliptic; stipules absent; cyathophylls 2, usually red or red-purple, much varied. Cyathiums solitary; involucres cuplike, lobes rounded, white pilose at margin and inside; styles free. Capsules globose. Fl. and fr. Jun-Sep. Sparse forests, thickets or meadows at 2500-4900 m. Distributed in Yunnan, Sichuan and Xizang. Also in Himalaya.

## 高山大戟
**Euphorbia stracheyi** Boiss.

匍匐草本。叶互生，全缘。花序为顶生的假伞形花序；初生苞叶下面常带黑紫色，有时淡黄褐色；苞叶常3，倒卵形；总苞杯状；腺体4，灰褐色，肾状圆形。蒴果卵状球形，光滑，无毛。花果期5-8月。生海拔1000-4900米的高山草甸、灌丛或杂木疏林下。产云南、四川、西藏、甘肃和青海。印度、尼泊尔和不丹亦有。

Herbs, prostrate. Leaves alternate, entire. Inflorescences a terminal pseudumbel; primary involucral leaves often flushed dark purplish abaxially, sometimes brownish yellow; cyathophylls often 3, obovate; involucres cuplike; glands 4, pale brown, reniform-rounded. Capsules ovoid-globose, smooth, glabrous. Fl. and fr. May-Aug. Alpine meadows, thickets or mixed sparse forests at 1000-4900 m. Distributed in Yunnan, Sichuan, Xizang, Gansu and Qinghai. Also in India, Nepal and Bhutan.

## 大果大戟
**Euphorbia wallichii** Hook. f.

多年生草本。叶互生，无毛。花序为顶生紧密假伞形花序；初生苞叶黄色或黄绿色；总苞宽钟形，外面被褐色短柔毛，4裂；腺体4，肾状圆形。蒴果无毛，球形，9-11 × 9-11毫米。花果期5-9月。生海拔1800-4700米的高山草甸、山坡或林缘。产云南、四川、西藏和青海。印度北部、尼泊尔、不丹、克什米尔地区和阿富汗亦有。

Perennial herbs. Leaves alternate, glabrous. Inflorescences a compact terminal pseudumbel; primary involucral leaves yellow or yellow-green; involucres broadly campanulate, abaxially brown pubescent, 4-lobed; glands 4, reniform-orbicular. Capsules glabrous, globose, 9-11 × 9-11 mm. Fl. and fr. May-Sep. Alpine meadows, slopes or forest edges at 1800-4700 m. Distributed in Yunnan, Sichuan, Xizang and Qinghai. Also in N India, Nepal, Bhutan, Kashmir and Afghanistan.

圆苞大戟 *Euphorbia griffithii*

高山大戟 *Euphorbia stracheyi*

大果大戟 *Euphorbia wallichii*

pale brown, reniform-rounded, entire. Capsules globose, densely long tuberculate. Seeds light brown. Fl. and fr. Mar-Jul. Grasslands, slopes, thickets or sparse forests at 200-3000 m. Distributed in Yunnan, Sichuan and Taiwan. Also in Korean Peninsula and Japan.

湖北大戟 *Euphorbia hylonoma*

## 湖北大戟

**Euphorbia hylonoma** Hand.-Mazz.

直立草本。叶互生，下面有时淡紫色，侧脉6-10对。杯状聚伞花序伞形状，着生于二歧分枝顶端，伞幅3-5；总苞叶2-5，钟形，腺体4，深褐色，肾形；雌蕊1；花柱3，柱头2裂。蒴果球形，3瓣裂。花期4-7月，果期6-9月。生海拔200-3000米的路边、山谷、山坡、灌丛、草地或疏林中。产中国除西北以外各地。俄罗斯和蒙古亦有。

Erect herbs. Leaves alternate, sometimes light purple abaxially, lateral veins 6-10 pairs. Cyathium, umbellate in dichotomy at apex, branches 3-5; involucres 2-5, campanulate; glands 4, dark brown, reniform; pistils 1; styles 3; stigmas 2. Capsules globose, 3-valved. Fl. Apr-Jul. Fr. Jun-Sep. Roadsides, valleys, slopes, thickets, grasslands or spare forests at 200-3000 m. Distributed in most parts of China, except NW China. Also in Russia and Mongolia.

## 大狼毒

**Euphorbia jolkinii** Boiss.

多年生草本。叶互生，全缘；初生苞叶黄色；苞片2或3，黄色。总苞杯状，裂片卵状三角形；腺体4，灰褐色，肾状圆形，全缘。蒴果球形，密具长瘤状突起。种子亮褐色。花果期3-7月。生海拔200-3000米的草地、山坡、灌丛或疏林内。产云南、四川和台湾。朝鲜半岛和日本亦有。

Perennial herbs. Leaves alternate, margin entire; primary involucral leaves yellow; cyathophylls 2 or 3, yellow. Involucres cuplike, lobes ovate-triangular; glands 4,

大狼毒 *Euphorbia jolkinii*

土瓜狼毒 *Euphorbia prolifera*

## 土瓜狼毒
**Euphorbia prolifera** Buch.-Ham. ex D. Don

草本。叶互生，条状长圆形；初生总苞叶4-6，卵状长圆形至宽卵形。总苞阔钟形；腺体4(-8)，褐色，近新月形，先端具2角；花单生：雄蕊多数；雌蕊1。蒴果卵球形。花果期4-8月。生海拔500-2300米的山谷、草坡或疏林下。产云南、四川和贵州。喜马拉雅、缅甸和泰国亦有。

Herbs. Leaves alternate, linear-oblong; primary involucre leaves 4-6, ovate-oblong to broadly ovate. Involucres broadly campanulate; glands 4(-8), brown, subcrescent, apex 2-horned; flowers solitary; stamens numerous; pistils 1. Capsules ovate-globose. Fl. and fr. Apr-Aug. Valleys, sloping grasslands or sparse forests at 500-2300 m. Distributed in Yunnan, Sichuan and Guizhou. Also in Himalaya, Myanmar and Thailand.

## 钩腺大戟
**Euphorbia sieboldiana** C. Morren et Decne.

多年生草本。根状茎粗壮。叶互生，无毛。聚伞花序多二歧状，有时为单歧状；苞叶2，极多变；杯状聚伞花序具短柄；总苞杯状；腺体4，多为黄褐色，鸡冠状，先端伸长为2个纤细的角；雄花多数，雌花1；蒴果三棱状球形。花果期4-9月。生田中、林缘、疏林中或草地。产中国大部分地区。俄罗斯(远东地区)、朝鲜半岛和日本亦有。

Perennial herbs. Rhizomes robust. Leaves alternate, glabrous. Cymes mostly dichasial, sometimes becoming monochasial; cyathophylls 2, very variable; cyathiums shortly stalked; involucres cuplike; glands 4, mainly yellow-brown, crescent-shaped, tips extended into 2 slender horns; staminate flowers many, pistillate flowers 1. Capsules triqueter-globose. Fl. and fr. Apr-Sep. Fields, thickets, forest edges, sparse forests or grasslands. Distributed in most parts of China. Also in Russia (Far East), Korean Peninsula and Japan.

## 乳浆大戟
**Euphorbia esula** L.

多年生草本。叶无柄，常于茎顶较大；叶条形至卵形，极多变。顶生假伞形花序，常于轴下部具侧生聚伞花序；雄花多数，雌花1。蒴果三棱状球形，具3条纵沟。花果期4-10月。生路边、田中、草地、山坡、疏林中或沙地。产中国大部分地区。西南亚、东北亚和欧洲亦有；归化于北美洲。

Perennial herbs. Leaves sessile, often larger toward stem apex; leaves linear to ovate, very variable. Inflorescences a terminal pseudumbel, often with lateral cymes from axils below; cymes mostly dichotomous; staminate flowers numerous, pistillate flowers 1. Capsules triqueter-globose, longitudinally 3-grooved. Fl. and fr. Apr-Oct. Roadsides, fields, grasslands, slopes, sparse forests or sandy areas. Distributed in most parts of China. Also in SW and NE Asia, and Europe; naturalized in North America.

钩腺大戟 *Euphorbia sieboldiana*

乳浆大戟 *Euphorbia esula*

# 虎皮楠科 Daphniphyllaceae

西藏虎皮楠 *Daphniphyllum himalense*

## 交让木
**Daphniphyllum macropodum** Miq.

乔木或灌木。小枝粗壮，深褐色。叶柄紫红色；叶革质；侧脉细而密，12-18对。雄蕊8-10；雌花花序长4.5-8厘米；子房周围退化雄蕊10；花柱极短，2裂，顶端膨大。果椭圆体形，有疣状皱纹。花期3-5月，果期8-10月。生海拔600-1900米的阔叶林中。产中国西南、华南、东南、华中和华东。朝鲜半岛和日本亦有。

Trees or shrubs. Branchlets stout, dark brown. Petioles purplish red; leaves leathery; lateral veins dense and slender, 12-18 pairs. Stamens 8-10; female flowers Inflorescences 4.5-8 cm long; staminodes 10 around ovary; styles short, 2-lobed, inflated at apex. Fruits ellipsoid, verrucose-corrugate. Fl. Mar-May. Fr. Aug-Oct. Broad-leaved forests at 600-1900 m. Distributed in SW, S, SE, C and E China. Also in Korean Peninsula and Japan.

## 西藏虎皮楠
**Daphniphyllum himalense** (Benth.) Müll. Arg.

乔木或小乔木。小枝粗壮，具白色皮孔。叶长圆状披针形、长圆形或长圆状椭圆形，下面具白粉和细小乳突，叶纸质至薄革质，全缘。雄蕊8-12；子房周围退化雄蕊5；果序长7.5-9.5厘米。核果椭圆体形。果期9-11月。生海拔1200-2500米的阔叶林中。产云南西北部和西藏东南部。印度东北部、尼泊尔东部、不丹和缅甸北部亦有。

Trees or small trees. Branchlets stout, white lenticellate. Leaves oblong-lanceolate, oblong, or oblong-elliptic, glaucous and finely papillate abaxially, papery or thinly leathery, entire; stamens 8-12; staminodes 5 around ovary; infructescences 7.5-9.5 cm long. Drupes ellipsoid. Fr. Sep-Nov. Broad-leaved forests at 1200-2500 m. Distributed in NW Yunnan and SE Xizang. Also in NE India, E Nepal, Bhutan and N Myanmar.

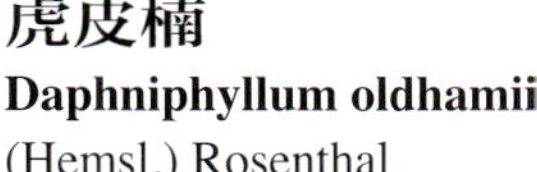

## 虎皮楠
**Daphniphyllum oldhamii** (Hemsl.) Rosenthal

乔木或灌木。小枝细弱，深褐色。叶纸质，下面被白粉和乳突。花萼小，不规则4-6裂；雄蕊7-10；雌花花序长4-6厘米；柱头2，叉裂。果暗褐色至黑色，有疣状突起，椭圆体形或倒卵球形。花期3-5月，果期8-11月。生海拔100-1400米的阔叶林下。产中国西南、华南、东南和华中。朝鲜半岛和日本亦有。

Trees or shrubs. Branchlets slender, dark brown. Leaves papery, abaxially pruinose and papillate. Calyx small, irregularly 4-6-parted; stamens 7-10; female flowers: Inflorescences 4-6 cm long; stigmas 2, furcated; fruits dark-brown to black, verrucose-papillate, ellipsoid or obovoid. Fl. Mar-May. Fr. Aug-Nov. Broad-leaved forests at 100-1400 m. Distributed in SW, S, SE and C China. Also in Korean Peninsula and Japan.

交让木 *Daphniphyllum macropodum*

虎皮楠 *Daphniphyllum oldhamii*

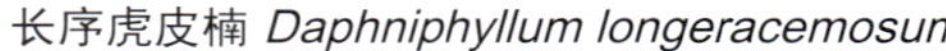
长序虎皮楠 *Daphniphyllum longeracemosum*

牛耳枫 *Daphniphyllum calycinum*

## 长序虎皮楠

**Daphniphyllum longeracemosum** Rosenthal

乔木。小枝粗壮。叶长圆状椭圆形，长16-26厘米，宽6-9厘米，纸质，下面无白粉，无乳突；侧脉12-14对。雄蕊10-16；果序长10-16厘米，直立。核果椭圆体形，具瘤，具白粉。花期4-5月，果期8-11月。生海拔100-1800米的密林中。产云南东南部和广西南部。越南北部亦有。

Trees. Branchlets stout. Leaves oblong-elliptic, 16-26 cm long, 6-9 cm wide, papery, neither glaucous nor papillate abaxially; lateral veins 12-14 pairs. Stamens 10-16; infructescences 10-16 cm long, erect. Drupes ellipsoid, tuberculate, glaucous. Fl. Apr-May. Fr. Aug-Nov. Dense forests at 100-1800 m. Distributed in SE Yunnan and S Guangxi. Also in N Vietnam.

## 大叶虎皮楠

**Daphniphyllum majus** Müll. Arg.

乔木或灌木。小枝粗壮，灰褐色。叶宽大，长圆状椭圆形或倒卵状长圆形，长22-37厘米，宽7-14厘米，纸质，下面具白粉，侧脉15-18对。花萼4裂；雄蕊9-12；果序长约4厘米。核果倒卵球状椭圆体形，具瘤。花期3-5月，果期10-12月。生海拔1100-1500米的林中。产云南南部。印度、缅甸、泰国和越南亦有。

Trees or shrubs. Branchlets stout, grayish brown. Leaves large, oblong-elliptic or obovate-oblong, 22-37 cm long, 7-14 cm wide, papery, glaucous abaxially, lateral veins 15-18 pairs; calyx 4-lobed; stamens 9-12; infructescences ca. 4 cm long. Drupes obovoid-ellipsoid, tuberculate. Fl. Mar-May. Fr. Oct-Dec. Forests at 1100-1500 m. Distributed in S Yunnan. Also in India, Myanmar, Thailand and Vietnam.

## 牛耳枫

**Daphniphyllum calycinum** Benth.

灌木。叶倒卵形或倒卵状椭圆形，纸质，下面具白粉和不明显乳突，侧脉8-11对。雄花花萼盘状；雄蕊9-10；雌花苞片卵形；花柱短，柱头2，外弯；果序长4-5厘米。核果卵球状椭圆体形。花期4-6月，果期8-11月。生海拔(100米以下)200-700米疏林或灌丛中。产广西、广东、福建、湖南和江西。越南北部和日本亦有。

Shrubs. Leaves obovate or obovate-elliptic, papery, glaucous and inconspicuously papillate abaxially, lateral veins 8-11-paired; male flowers with disk sepals; stamens 9-10; female flowers with ovate bracts; styles short, stigmas 2, recurved; infructescences 4-5 cm long. Drupes ovoid-ellipsoid. Fl. Apr-Jun. Fr. Aug-Nov. Sparse forests or thickets at (below 100 m)200-700 m. Distributed in Guangxi, Guangdong, Fujian, Hunan and Jiangxi. Also in N Vietnam and Japan.

大叶虎皮楠 *Daphniphyllum majus*

# 水马齿科 Callitrichaceae

### 水马齿
**Callitriche palustris** L.

一年生水生或湿生小草本。叶对生，在茎顶密集呈莲座状，水中叶舌状线形至稍膨大，顶端圆或具凹缺，常具1中肋；浮水叶椭圆形至近圆形，常具3脉。果倒卵状椭圆体形，上部边缘具翅。花期8-9月，果期10月。生海拔700-3800米的沼泽地或潮湿地。产中国西南、华南、东南、华东、华西和东北。亚洲温带地区、欧洲和北美洲亦有。

Annual submerged herbs, with floating rosettes or on wet mud. Leaves opposite, often in rosette at stem apices, lingulate leaves linear to narrowly expanded, emarginate, expanded leaves elliptic to ± orbicular, usually triplinerved. Fruits obovoid-ellipsoid, winged on upper margin. Fl. Aug-Sep. Fr. Oct. Bogs or moist places at 700-3800 m. Distributed in SW, S, SE, E, W and NE China. Also in temperate Asia, Europe and North America.

水马齿 *Callitriche palustris*

# 黄杨科 Buxaceae

### 杨梅黄杨
**Buxus myrica** Lévl.

灌木，高1-3米。幼枝细弱，具棱，内侧具柔毛。叶长圆状披针形或狭披针形，3-7 × (0.8-)1-2厘米。花序腋生；雄花约10；花柱反折或卷曲。蒴果近球形。生海拔200-2000米的林中、山坡林中、山坡溪边或河岸。产云南东部、四川、贵州南部和西南部、广西西北部、海南和湖南。越南亦有。

Shrubs, 1-3 m tall. Young branches slender, tetragonous, inner sides puberulent. Leaves oblong-lanceolate or narrowly lanceolate, 3-7 × (0.8-)1-2 cm. Inflorescences axillary; male flowers ca. 10; styles recurved or circinate. Capsules subglobose. Forests, forests on slopes, along mountain streams or riverbanks at 200-2000 m. Distributed in E Yunnan, Sichuan, S and SW Guizhou, NW Guangxi, Hainan and Hunan. Also in Vietnam.

杨梅黄杨 *Buxus myrica*

### 匙叶黄杨
**Buxus harlandii** Hance

小灌木。小枝四棱形。叶薄革质，匙形，稀狭长圆形，2-3.5(-4) × 0.5-0.8(-0.9)厘米，中脉两面凸出。可育雄蕊长为花萼的约1/2。蒴果近球形，宿存花柱长3毫米，末端稍外曲。花期5月，果期10月。生溪边或疏林中。产广东和海南。

Small shrubs. Branchlets subtetragonous. Leaves thin leathery, spatulate, rarely narrowly oblong, 2-3.5(-4) × 0.5-0.8(-0.9) cm, mid nerve convex at both sides. Sterile pistils ca. 1/2 of sepal in length. Capsules subglobose. Persistent styles ca. 3 mm long, apex slightly curved. Fl. May. Fr. Oct. By streams or in sparse forests. Distributed in Guangdong and Hainan.

匙叶黄杨 *Buxus harlandii*

雀舌黄杨 *Buxus bodinieri*

黄杨 *Buxus sinica*

## 雀舌黄杨
**Buxus bodinieri** Lévl.

灌木。幼枝具棱，具柔毛或渐无毛。叶常匙形，亦狭卵形或倒卵形，顶端最宽，20-40 × 8-18毫米，中脉基部一半具柔毛。花序腋生，头状，花密集；柱头长约1.5毫米，稍压扁。蒴果卵球形。花期2月，果期5-8月。生海拔400-2700米的山区林中或山坡。产中国西南、华南、华中、华西和华东。

Shrubs. Young branches tetragonous, pubescent or glabrescent. Leaves usually spatulate, also narrowly ovate or obovate, widest in apical part, 20-40 × 8-18 mm, puberulent on basal half of midrib. Inflorescences axillary, capitate, flowers dense; style ca. 1.5 mm long, slightly compressed. Capsules ovoid. Fl. Feb. Fr. May-Aug. Forests in hilly areas or, mountain slopes at 400-2700 m. Distributed in SW, S, C, W and E China.

## 黄杨
**Buxus sinica** (Rehd. et Wils.) M. Cheng

灌木或小乔木。叶革质，阔椭圆形或长圆状椭圆形，长1.5-3.5 × 0.8-2厘米，侧脉可见或不可见，上面具皱。雄花无柄；不育雌蕊为萼片长度的2/3。蒴果近球形。花期3月，果期5-6月。生海拔1200-2600米的山谷、溪边或林下。产中国西南、华南、华西和华东。

Shrubs or small trees. Leaves leathery, broadly elliptic or oblong-elliptic, 1.5-3.5 × 0.8-2 cm, lateral veins visible or not and rugulose adaxially. Staminate flowers sessile; leagth of sterile pistils is as long as 2/3 length of tepals. Capsules almost globose. Fl. Mar. Fr. May-Jun. Valleys, by streams or under forests at 1200-2600 m. Distributed in SW, S, W and E China.

## 长叶柄野扇花
**Sarcococca longipetiolata** M. Cheng

灌木，高1-3米。叶柄长1-1.5厘米；叶片披针形、长圆状披针形或狭披针形。花序腋生和顶生，总状花序或近头状花序组成复总状花序；雄花4-8，生花序轴上部；雌花2-4，生花序轴下部。核果球形；宿存花柱2个。花期9月，果期12月。生海拔300-800米的林中溪边。产广东(乳源县、阳山县)和湖南(宜章县)。

Shrubs, 1-3 m tall. Petioles 1-1.5 cm long; leaves lanceolate, oblong-lanceolate or narrowly lanceolate. Inflorescences axillary and terminal, racemes or subcapitate to compound racemes; staminate flowers 4-8, inserted apically in rachises; pistillate flowers 2-4, inserted basally in rachises. Drupes globose; persistent styles 2. Fl. Sep. Fr. Dec. Streamsides in forests at 300-800 m. Distributed in Guangdong (Ruyuan County, Yangshan County) and Hunan (Yizhang County).

长叶柄野扇花 *Sarcococca longipetiolata*

## 云南野扇花
**Sarcococca wallichii** Stapf

灌木。叶椭圆形、长圆状披针

云南野扇花 *Sarcococca wallichii*

形或披针形，6-10 × 2-3.5(-5)厘米，最下一对侧脉从离叶基2-5毫米处发出上升，成明显的离基三出脉。花序近头状或短穗状；花白色，芳香；雄蕊长3-4毫米。核果近球形或椭圆体形；宿存花柱2或3。花果期9-12月。生海拔1300-2700米的山坡林中或沟谷。产云南南部和西部、西藏。印度东北部、尼泊尔、不丹和缅甸亦有。

Shrubs. Leaves elliptic, oblong-lanceolate, or lanceolate, 6-10 × 2-3.5(-5) cm, basal lateral veins a pair, distance 2-5 mm long from base, to distinctly tri-plinerved. Inflorescences subcapitate or shortly spicate; flowers white, fragrant; male flowers 3-4 mm long. Drupes subglobose or ellipsoid; persistent styles 2 or 3. Fl. and fr. Sep-Dec. Forests on mountain slopes or in valleys at 1300-2700 m. Distributed in S and W Yunnan, Xizang. Also in NE India, Nepal, Bhutan and Myanmar.

## 海南野扇花
**Sarcococca vagans** Stapf

灌木，高1-3米。小枝无毛或被微柔毛。叶纸质，长椭圆形至长圆状披针形，8-16(-20) × 4-6厘米，离基三出脉，距基部1.5-2毫米，先端尖。短总状或近头状花序；雄花7-10；雌花1或2(-5)。核果球形，宿存花柱2。花果期9月至翌年3月。生海拔500-800米的山谷或石灰岩山丘林下。产云南和海南。缅甸和越南亦有。

Shrubs, 1-3 m tall. Young branches glabrous or slightly puberulent. Leaves papery, narrowly elliptic to oblong-lanceolate, 8-16(-20) × 4-6 cm, triplinerved and distance from base 1.5-2 mm long, apex acute. Inflorescences short racemes or subcapitate; staminate flowers 7-10; pistillate flowers 1 or 2(-5). Drupes globose, persistent styles 2. Fl. and fr. Sep to next Mar. Valleys or forests on limestone hills at 500-800 m. Distributed in Yunnan and Hainan. Also in Myanmar and Vietnam.

## 野扇花
**Sarcococca ruscifolia** Stapf

灌木，植株具纤维状根。叶近革质，卵圆形至卵状披针形，近基生三出脉。短总状花序，长1-2厘米；雄花2-7；雌花2-5；花白色，芳香。蒴果球形，猩红色或黑褐色；宿存花柱2或3。花果期10月至翌年2月。生海拔200-2600米的杂木林下、石灰岩地区。产中国西南、华中和华西。

Shrubs, plants with fibrous roots. Leaves subleathery, ovate to ovate-lanceolate, subbasal-trinervious. Inflorescences short racemes, 1-2 cm long; staminate flowers 2-7; pistillate flowers 2-5; flowers white, fragrant. Capsules globose, scarlet or black-brown; persistent styles 2 or 3. Fl. and fr. Oct to next Feb. Mixed forests, prefer to grow on calcareous soils at 200-2600 m. Distributed in SW, C and W China.

海南野扇花 *Sarcococca vagans*

野扇花 *Sarcococca ruscifolia*

## 羽脉野扇花

**Sarcococca hookeriana** Baill.

灌木或小乔木。叶披针形、倒披针形、椭圆状披针形、长圆状披针形或狭披针形，侧脉羽状。花序总状；雄花5-8；雌花1或2。果实球形，蒴果黑色或蓝黑色。生海拔1000-3500米的林中。产云南、四川、西藏东部、重庆、湖北西部和陕西南部。印度东北部、尼泊尔、不丹和阿富汗亦有。

Shrubs or small trees. Leaves lanceolate, oblanceolate, elliptic-lanceolate, oblong-lanceolate, or narrowly lanceolate, lateral veins pinnate. Inflorescences racemes; male flowers 5-8; female flowers 1 or 2. Fruits globose, capsules black or blue-black. Forests at 1000-3500 m. Distributed in Yunnan, Sichuan, E Xizang, Chongqing, W Hubei and S Shaanxi. Also in NE India, Nepal, Bhutan and Afghanistan.

## 双蕊野扇花

**Sarcococca hookeriana** Baill. var. **digyna** Franch.

灌木。叶在小枝顶端对生或近对生，形状和大小多变，窄卵形或长圆形披针形，羽状脉。总状花序；雄花5-8，雌花1或2。蒴果球形，黑色或蓝黑色；宿存花柱2或3。花期1月，果期9月。生海拔1000-3500米的山坡或杂木林下。产云南、四川、湖北和陕西。

Shrubs. Leaves opposite or subopposite at apex of branchlets, more variable in shape and size, narrowly ovate or oblong-lanceolate, pinnate-nerved. Inflorescences racemes; staminate flowers 5-8, pisllate flowers 1 or 2. Capsules globose, black or blue-black; persistent styles 2 or 3. Fl. Jan. Fr. Sep. Slopes or mixed forests at 1000-3500 m. Distributed in Yunnan, Sichuan, Hubei and Shaanxi.

双蕊野扇花 *Sarcococca hookeriana* var. *digyna*

## 板凳果

**Pachysandra axillaris** Franch.

亚灌木。叶卵状长圆形，基部圆形、截形或近心形，背面被柔毛或长柔毛，边缘具较深的粗锯齿。花序腋生，长1-2厘米；花白色、玫瑰红或红色；雄花(5-)10-20；雌花(1-)3-6；花柱3。果球形。花期2-5月，果期9-10月。生海拔600-2500米的山坡、沟边或林下。产中国西南和华南。

Subshrubs. Leaves ovate-oblong, base rounded, truncate, or subcordate, pubescent or villous abaxially, margin thick-serrate.

羽脉野扇花 *Sarcococca hookeriana*

板凳果 *Pachysandra axillaris*

Inflorescences axillary, 1-2 cm long; flowers white, rose, or red; staminate flowers (5-)10-20; pistillate flowers (1-)3-6; styles 3. Fruits globose. Fl. Feb-May. Fr. Sep-Oct. Slopes, by streams or forests at 600-2500 m. Distributed in SW and S China.

### 顶花板凳果
**Pachysandra terminalis** Siebold et Zucc.

半灌木。叶菱状倒卵形，革质，上面沿中脉具柔毛，基部渐狭至叶柄，边缘近顶端具锯齿。花序顶生，长2-4厘米；花白色；雄花多于15；雌花1或2。果卵球形。花果期4-5月。生海拔1000-2600米的林中阴湿地。产四川、浙江、湖北、陕西和甘肃。日本亦有。

Subshrubs. Leaves rhombic-obovate, leathery, puberulent along midrib adaxially, base attenuate into petiole, margin dentate toward apex. Inflorescences terminal, 2-4 cm long; flowers white; staminate flowers more than 15; pistillate flowers 1 or 2. Fruits ovoid. Fl. and fr. Apr-May. Shady and damp lands in forests at 1000-2600 m. Distributed in Sichuan, Zhejiang, Hubei, Shaanxi and Gansu. Also in Japan.

## 岩高兰科 Empetraceae

### 东北岩高兰
**Empetrum nigrum** L. var. **japonicum** K. Koch

常绿灌木。茎匍匐。叶互生，无柄，条形，革质，长3-5毫米，中脉下陷，边缘全缘，稍反卷。雄花具短梗，花瓣3，紫红色，倒卵形，长约2毫米，雄蕊3；雌花无梗，柱头6-9裂；核果球形，直径5-8毫米，紫黑色。花期6-7月，果期8月。生海拔530-1500米的石山。产黑龙江北部和内蒙古东北部。日本、朝鲜半岛、蒙古和俄罗斯东部亦有。

Evergreen shrubs. Stems creeping . Leaves alternate, sessile, linear, leathery, 3-5 mm long, midvein impressed, margin entire, slightly reflexed. Staminate flowers shortly petiolate, peteals 3, purple-red, obovate, ca. 2 mm long, stamens 3; pistillate flowers sessile, stigma 6-9-lobed. Drupes globose, 5-8 mm diam, purple-black. Fl. Jun-Jul. Fr. Aug. Stony hills at 530-1500 m. Distributed in N Heilongjiang and NE Neimenggu. Also in Japan, Korean Peninsula, Mongolia and E Russia.

顶花板凳果 *Pachysandra terminalis*

东北岩高兰 *Empetrum nigrum* var. *japonicum*

# 马桑科 Coriariaceae

草马桑 *Coriaria terminalis*

## 马桑
**Coriaria nepalensis** Wall.

灌木。叶椭圆形或阔椭圆形，先端急尖，3出脉至顶端。花序腋生；雄花序密具多花，花先叶开放，雄花中具退化雌蕊；雌花与叶同期开放；花瓣小，肉质。果成熟后红色至深紫色或紫黑色，近球形。花期2-5月，果期5-8月。生海拔200-3200米的灌丛或山坡。产中国西南、华中、华东和华西。南亚亦有。

Shrubs. Leaves elliptic or broadly elliptic, apex acute, 3-veined to apex. Inflorescences axillary; male Inflorescences densely multi-florous, flowers opening before leaves, staminate flowers with rudimentary pistils; pistillate flowers opening at same time that leaves appear; petals small, fleshy. Fruits red to dark purple or purplish black when mature, subglobose. Fl. Feb-May. Fr. May-Aug. Thickets or mountain slopes at 200-3200 m. Distributed in SW, C, E and W China. Also in S Asia.

## 草马桑
**Coriaria terminalis** Hemsl.

半灌木状草本。下部叶阔卵形或近圆形，纸质，两面具腺状柔毛，基部具(3-)5-9脉。总状花序顶生，花序轴紫红色，具腺柔毛；花瓣肉质，花后伸长。果实成熟后紫红色至黑色。花期5-6月，果期7-9月。生海拔1800-3700米的山坡灌丛。产云南西北部、四川西部、西藏南部和东南部。印度、尼泊尔和不丹亦有。

Herbs subshrubby. Leaves: lower ones broadly ovate or almost orbicular, papery, both surfaces glandular pubescent, basally (3-)5-9-veined. Racemes terminal, rachis purplish red, white glandular pilose; petals fleshy, enlarged after anthesis. Fruits purplish red to black when mature. Fl. May-Jun. Fr. Jul-Sep. Thickets on mountain slopes at 1800-3700 m. Distributed in NW Yunnan, W Sichuan, and S and SE Xizang. Also in India, Nepal and Bhutan.

马桑 *Coriaria nepalensis*

# 漆树科
# Anacardiaceae

豆腐果 *Buchanania latifolia*

### 豆腐果
**Buchanania latifolia** Roxb.

常绿乔木。叶阔长圆形，革质，起初两面具长的锈色毛，后上面近无毛。圆锥花序顶生，花序轴和总梗粗壮，密被锈色长绒毛；花瓣5，白色，外无毛；萼裂片披针形。核果双凸镜状，成熟时棕黑色。花期3-4月，果期5-7月。生海拔100-900米的低地森林中。产海南和云南南部。南亚和东南亚亦有。

Evergreen trees. Leaves broadly oblong, leathery, at first with long ferruginous hairs on both sides, later subglabrous adaxially. Inflorescences paniculate, terminal, peduncles and rachises robust, densely covered with long ferruginous hairs; petals 5, white, outside glabrous; calyx lobes lanceolate. Drupes lens-shaped, brownish black at maturity. Fl. Mar-Apr. Fr. May-Jul. Lowland forests at 100-900 m. Distributed in Hainan and S Yunnan. Also in S and SE Asia.

### 腰果
**Anacardium occidentale** L.

乔木或灌木。叶革质，倒卵形，侧脉12对，革质，全缘。花序圆锥状，长10-20厘米；花黄色；雄蕊7-10。果期花托膨大，肉质，形成棒状或梨状的假果，内果皮肾形。花期3-4月，果期7-8月。栽培于云南、广西、广东、台湾和福建。原产热带美洲。

Trees or shrubs. Leaves leathery, obovate, lateral veins 12 pairs, leathery, margin entire. Inflorescences paniculate, 10-20 cm long; flowers yellow; stamens 7-10. Receptacles inflated, fleshy, forming a pear-shaped or clavate pseudo-carp when mature, endocarps reniform. Fl. Mar-Apr. Fr. Jul-Aug. Cultivated in Yunnan, Guangxi, Guangdong, Taiwan and Fujian. Native to tropical America.

### 杧果
**Mangifera indica** L.

乔木，高10-20米。叶长圆形至长圆披针形，12-30 × 3.5-6.5厘米，革质。花序圆锥形，顶生，长20-35厘米；花瓣浅黄色，花期反折；可育雄蕊1；不育雄蕊4。核果卵球形、长圆形或近肾形，黄色。花期3-4月，果期5-7月。福建南部、广东、广西、台湾和云南南部栽培。原产东南亚大陆；世界热带地区广泛栽培。

Trees, 10-20 m tall. Leaves oblong to oblong-lanceolate, 12-30 × 3.5-6.5 cm, leathery. Inflorescences paniculate, terminal, 20-35 cm long; petals light yellow, recurved at anthesis; fertile stamen 1; staminodes 4. Drupes ovoid, oblong or nearly reniform, yellow. Fl. Mar-Apr. Fr. May-Jul. Cultivated in S Fujian, Guangdong, Guangxi, Taiwan and S Yunnan. Native to continental SE Asia; cultivated in tropical regions of the world.

杧果 *Mangifera indica*

腰果 *Anacardium occidentale*

林生杧果 *Mangifera sylvatica*

扁桃(天桃木)
*Mangifera persiciforma*

## 扁桃(天桃木)

**Mangifera persiciforma** C. Y. Wu et T. L. Ming

乔木。叶狭卵形至披针形，11-20 × 2-2.8厘米，革质，无毛。花序圆锥状，顶生；花瓣黄绿色；可育雄蕊1，无花药。核果圆形至稍压扁，直径4-5厘米。花期4月，果期5-6月。生海拔200-600米的低地林中。产云南东南部、贵州南部和广西南部。

Trees. Leaves narrowly ovate to lanceolate, 11-20 × 2-2.8 cm, leathery, glabrous. Inflorescences paniculate, terminal; petals yellowish green; fertile stamen 1, anthers absent. Drupes rounded to slightly compressed, 4-5 cm diam. Fl. Apr. Fr. May-Jun. Lowland forests at 200-600 m. Distributed in SE Yunnan, S Guizhou and S Guangxi.

## 林生杧果

**Mangifera sylvatica** Roxb.

乔木，高6-20米。叶披针形至长圆披针形，长12-24厘米，宽3-5.5厘米，纸质或薄革质。花序圆锥状，顶生；花瓣白色，花期顶端反折和扭曲；可育雄蕊1；不育雄蕊1或2。核果斜长卵球形，先端延长成尖锐而向下弯曲的喙。花期4-5月，果期6-8月。生海拔600-1900米的低地或山麓林中。产云南南部。印度、尼泊尔、不丹、孟加拉国、缅甸、泰国和柬埔寨亦有。

Trees, 6-20 m tall. Leaves lanceolate to oblong-lanceolate, 12-24 cm long, 3-5.5 cm wide, papery or thinly leathery. Inflorescences paniculate, terminal; petals white, recurved and twisted apically at anthesis; fertile stamen 1; staminodes 1 or 2. Drupes long ovoid, oblique narrowly ovoid decurrent into a curved downward acute beak at apex. Fl. Apr-May. Fr. Jun-Aug. Lowland or hill forests at 600-1900 m. Distributed in S Yunnan. Also in India, Nepal, Bhutan, Bangladesh, Myanmar, Thailand and Cambodia.

## 槟榔青

**Spondias pinnata** (L. f.) Kurz

落叶乔木。奇数羽状复叶具5-11对生小叶。小叶明显具边缘脉，两面无毛。圆锥花序长25-35厘米，先花后叶；萼片无毛。核果卵球形或椭圆体形，横切面四边形。花期4-6月，果期8-9月。生海拔300-1200米的低山丘或山谷疏林。产云南、广西和海南。南亚和东南亚广泛栽培归化。可能原产印度尼西亚和菲律宾。

Deciduous trees. Imparipinnately compound with 5-11 opposite leaflets; leaflets with distinct submarginal nerves, glabrous on both surfaces. Panicles 25-35 cm long, flowers appearing before leaves; sepals glabrous. Drupes ovoid or ellipsoid, quadrangle in transversal section. Fl. Apr-Jun. Fr. Aug-Sep. Low hills or sparse forests in valleys at 300-1200 m. Distributed or cultivated in Yunnan, Guangxi and Hainan. Widely cultivated and naturalized in S and SE Asia. Probably native to Indonesia and the Philippines.

槟榔青 *Spondias pinnata*

岭南酸枣 *Spondias lakonensis*

大果人面子 *Dracontomelon macrocarpum*

## 岭南酸枣
**Spondias lakonensis** Pierre

落叶乔木。奇数羽状复叶长25-35厘米，具11-23小叶。小叶腹面脉上稍被柔毛，侧脉叶腋处稍被簇毛。花序圆锥状，腋生，稍被灰褐色柔毛；花瓣白色。核果成熟时红色，4室，每室1粒种子。花期5-6月，果期9-12月。生林中。产云南、广西、广东、海南和福建。老挝、泰国和越南亦有。

Deciduous trees. Leaves imparipinnate, 25-35 cm long, with 11-23 leaflets; leaflets minutely pubescent abaxially along veins, with tufts of hairs in axils of lateral veins. Inflorescences paniculate, axillary, minutely grayish brown pubescent; petals white. Drupes red at maturity, 4-locular with 1 seed per locule. Fl. May-Jun. Fr. Sep-Dec. Forests. Distributed in Yunnan, Guangxi, Guangdong, Hainan and Fujian. Also in Laos, Thailand and Vietnam.

## 人面子
**Dracontomelon duperreanum** Pierre

乔木，高于20米。小叶两侧多少对称，基部略偏斜，两面沿中脉微具柔毛，下面侧脉脉腋具簇毛。花白色，萼片5，边缘被细睫毛，花瓣5，覆瓦状排列，具3-5暗褐色脉纹。核果扁球形，直径约2.5厘米。花期4-6月，果期7-11月。生海拔100-400米的林中。产云南东南部、广西和广东。越南亦有。

Trees, more than 20 m tall. Leaflets ± symmetric on both sides, less oblique at base, minutely pubescent along midrib on both surfaces, abaxially with white tufts of hair in axils of lateral veins. Flowers white, sepals 5, margin ciliolate, petals 5, imbricate, with 3-5 dull brown veins. Drupes depressed-globose, ca. 2.5 cm diam. Fl. Apr-Jun. Fr. Jul-Nov. Forests at 100-400 m. Distributed in SE Yunnan, Guangxi and Guangdong. Also in Vietnam.

## 大果人面子
**Dracontomelon macrocarpum** H. L. Li

乔木，高约18米。叶长达50厘米，奇数羽状复叶具互生小叶；小叶两侧不对称，基部极偏斜，两面无毛。核果扁球形，直径3.5-4厘米，黑色，5室；中果皮肉质。种子棕色，富含可食油分。果期6月。生海拔约1200米的混交林中。产云南南部。

Trees, ca. 18 m tall. Leaves to 50 cm long, imparipinnately compound with alternate leaflets, leaflets asymmetric, with very oblique base, glabrous on both surfaces. Drupes depressed-globose, 3.5-4 cm diam, black, 5-locular; mesocarp fleshy. Seeds brown, richly containing edible oil. Fr. Jun. Mixed forests at ca. 1200 m. Distributed in S Yunnan.

人面子 *Dracontomelon duperreanum*

藤漆 *Pegia nitida*

南酸枣 *Choerospondias axillaris*

## 藤漆

**Pegia nitida** Colebr.

攀援木质灌木，密被黄色绒毛。奇数羽状复叶，具4-7对生小叶；小叶卵形。花序圆锥状，长20-35厘米，疏具分枝，密具黄色绒毛。核果椭圆体形，倾斜，稍压扁，成熟时黑色。花期1-4月，果期5-7月。生海拔(200-)500-1800米的山谷林中。产云南、贵州和广西。印度、尼泊尔、不丹、缅甸和泰国亦有。

Woody climbers, densely yellow-tomentose. Leaves compound imparipinnately with 4-7 opposite leaflets; leaflets ovate. Inflorescences paniculate, 20-35 cm long, loosely branched, densely yellow tomentose. Drupes ellipsoid, oblique, slightly compressed, black at maturity. Fl. Jan-Apr. Fr. May-Jul. Valley forests at (200-)500-1800 m. Distributed in Yunnan, Guizhou and Guangxi. Also in India, Nepal, Bhutan, Myanmar and Thailand.

## 利黄藤

**Pegia sarmentosa** (Lecte.) Hand.-Mazz.

木本攀援藤本。小枝、叶轴和叶柄近无毛。奇数羽状复叶具11-15对生叶；小叶长圆形。圆锥花序，分枝疏散，稍具柔毛。核果椭圆体形或卵球形，压扁。种子近肾形。花期2-4月，果期4-5月。生海拔200-900米的石土森林或灌丛中。产广东、广西、贵州和云南。老挝、泰国、越南、柬埔寨、马来西亚东部和印度尼西亚(加里曼丹岛)亦有。

Woody climbers. Branchlets, leaf rachis, and leaf Petioles ± glabrous. Leaves imparipinnately compound with 11-15 opposite leaflets; leaflets oblong. Inflorescences paniculate, loosely branched, minutely pubescent. Drupes ellipsoid or ovoid, compressed. Seeds subreniform. Fl. Feb-Apr. Fr. Apr-May. Forests or thickets on rocky soils at 200-900 m. Distributed in Guangdong, Guangxi, Guizhou and Yunnan. Also in Laos, Thailand, Vietnam, Cambodia, E Malaysia and Indonesia (Kalimantan Island).

## 南酸枣

**Choerospondias axillaris** (Roxb.) Burtt et Hill

落叶乔木。叶长25-40厘米，奇数羽状复叶具3-6对小叶。雄花序为聚伞圆锥花序，雌花单生于上部叶腋；花柱分离。核果椭圆体形或倒卵球状椭圆体形，成熟时黄色；内果皮不压扁。花期3-4月，果期7-10月。生海拔300-2000米的山坡、丘陵或山谷林中。产中国西南、华南、华中和东南。南亚和日本亦有。

Deciduous trees. Leaves 25-40 cm long, imparipinnately compound, with 3-6 pairs of leaflets. Male inflorescences racemose cymes, pistillate flowers solitary or axillary at upper part; styles free. Drupes ellipsoid or obovoid-ellipsoid, yellow at maturity; endocarps not compressed. Fl. Mar-Apr. Fr. Jul-Oct. Mountain slopes, hills or valley forests at 300-2000 m. Distributed in SW, S, C and SE China. Also in S Asia and Japan.

利黄藤 *Pegia sarmentosa*

厚皮树 *Lannea coromandelica*

清香木 *Pistacia weinmanniifolia*

## 厚皮树

**Lannea coromandelica** (Houtt.) Merr.

落叶乔木，高5-10米。奇数羽状复叶，常集生小枝顶端；花小，黄色或带紫色，组成顶生圆锥花序或总状花序；子房无毛，4室，常仅1胚珠可育。核果卵球形至稍肾形，成熟后紫红色。花期3月，果期4-6月。生海拔100-1800米的山坡、溪边或旷野林中。产云南、广西和广东。南亚和东南亚亦有。

Deciduous trees, 5-10 m tall. Leaves imparipinnate, often clustered at branch apex; flowers small, yellow or purplish in terminal panicles or racemes; ovary glabrous, 4-locular, usually only 1 ovule fertile. Drupes ovoid to slightly reniform, purplish red at maturity. Fl. Mar. Fr. Apr-Jun. Forests on slopes, along streams or fields at 100-1800 m. Distributed in Yunnan, Guangxi and Guangdong. Also in S and SE Asia.

## 黄连木

**Pistacia chinensis** Bunge

落叶乔木。树皮呈鳞片状剥落。奇数羽状复叶具1-13对生小叶；小叶纸质，披针形或卵状披针形，先端渐尖。花先叶开放；雄花序排列紧密，雄花无退化雄蕊，雌花序排列疏松。核果倒卵球形。花期3-5月，果期8-11月。生海拔100-3600米的石山林中。产中国西南、华南、东南、华中、华北和华东。

Deciduous trees. Bark peeled off by scales. Leaves imparipinnately compound with 1-13 opposite leaflets; leaflets papery, lanceolate or ovate-lanceolate, with acuminate apex. Flowers produced before leafing; male inflorescences dense, male flowers without pistillode, female inflorescences sparse. Drupes obovoid. Fl. Mar-May. Fr. Aug-Nov. Forest in stony mountains at 100-3600 m. Distributed in SW, S, SE, C, N and E China.

## 清香木

**Pistacia weinmanniifolia** J. Poiss. ex Franch.

常绿灌木至小乔木。偶数羽状复叶具8-18对生小叶；小叶革质，长圆形或倒卵状长圆形，顶端截形或钝。圆锥花序被黄棕色和红色腺毛；花小，紫红色；雄花具退化雄蕊。核果近球形。花期3-5月，果期6-8月。生海拔500-2700米的石灰山林下或灌丛。产云南、四川、西藏、贵州和广西。缅甸北部亦有。

Evergreen shrubs to Small trees. Leaves paripinnately compound with 8-18 opposite leaflets; leaflets leathery, oblong or obovate-oblong, with truncate or retuse apex. Panicles with mixed yellowish brown and red glandular pubescent; flowers small, purple-red; male flowers with pistillode. Drupes subglobose. Fl. Mar-May. Fr. Jun-Aug. Forests on limestone hills or thickets at 500-2700 m. Distributed in Yunnan, Sichuan, Xizang, Guizhou and Guangxi. Also in N Myanmar.

黄连木 *Pistacia chinensis*

毛黄栌 *Cotinus coggygria* var. *pubescens*

## 毛黄栌

**Cotinus coggygria** Scop. var. **pubescens** Engl.

灌木，高3-5米。叶宽椭圆形至倒卵形，宽2.5-6厘米，两面被灰色短柔毛。圆锥花序，被短柔毛；花盘5裂，紫褐色；子房近球形。核果肾形，约4.5 × 2.5厘米，无毛。花期2-8月，果期5-11月。生海拔700-2400米的山麓林中、山麓或山地灌丛。产中国西南、东南、华中、华北、华西和华东。西南亚和欧洲亦有。

Shrubs, 3-5 m tall. Leaves broadly elliptic to obovate, 2.5-6 cm wide, gray pubescent on both sides. Inflorescences paniculate, pubescent; disk 5-lobed, purplish brown; ovary subglobose. Drupes reniform, ca. 4.5 × 2.5 mm, glabrous. Fl. Feb-Aug. Fr. May-Nov. Hill forests, hills or mountain thickets at 700-2400 m. Distributed in SW, SE, C, N, W and E China. Also in SW Asia and Europe.

## 盐肤木

**Rhus chinensis** Mill.

灌木至乔木。奇数羽状复叶具(5-)7-13小叶，叶轴具宽翅至无翅，具锈色柔毛。圆锥花序顶生，花白色。核果球形，混生柔毛和腺状柔毛，成熟时红色；中果皮胶质。花期8-9月，果期10月。生海拔100-2800米的山坡、杂木林或灌丛。产中国除东北以外大部分地区。南亚、东南亚和东北亚亦有。

Shrubs to trees. Imparipinnately compound with (5-)7-13 leaflets, rachises broadly winged to wingless, ferruginous pubescent. Panicles terminal, with white flowers. Drupes globose, mixed pilose and glandular-pubescent, red at maturity; mesocarps glutinous. Fl. Aug-Sep. Fr. Oct. Slopes, mixed forests or shrubs at 100-2800 m. Distributed in most parts of China, except NE China. Also in S, SE and NE Asia.

盐肤木 *Rhus chinensis*

## 火炬树
**Rhus typhina** L.

灌木至乔木。小枝红褐色，密具柔毛。奇数羽状复叶；小叶19-25；小叶长圆状卵形至披针形，边缘具锯齿。花序多分枝；花白色。核果球形，压扁，密生粗毛。花期8-9月，果期10月。栽培于中国大部分省区。原产北美洲。

Shrubs to trees. Branchlets red brown, densely pubescent. Imparipinnately compound; leaflets 19-25; leaflets oblong-ovate to lanceolate, margin dentate. Inflorescences many branched; flowers white. Drupes globose, compressed, mixed dense bristles. Fl. Aug-Sep. Fr. Oct. Cultivated in most provinces of China. Native to North America.

火炬树 *Rhus typhina*

## 白背麸杨
**Rhus hypoleuca** Champ. ex Benth.

灌木或小乔木。小枝紫褐色。叶柄和轴具灰色绒毛，无翅；奇数羽状复叶具9-17小叶，叶背密具白绒毛，基部倾斜。圆锥花序长达20厘米。核果被白色柔毛和红色腺毛。花期8月，果期9-10月。生海拔800-1500米的林中。产广东、台湾、福建和湖南。

Shrubs or small trees. Branchlets purplish brown. Petioles and rachis grayish tomentose, wingless; imparipinnately compound with 9-17 leaflets, leaflets abaxially densely white tomentose, base oblique. Panicles up to 20 cm long. Drupes clothed with white pilose and red glandular hairs. Fl. Aug. Fr. Sep-Oct. Forests at 800-1500 m. Distributed in Guangdong, Taiwan, Fujian and Hunan.

## 红麸杨
**Rhus punjabensis** Stewart var. **sinica** (Diels) Rehd. et Wils.

乔木或小乔木。奇数羽状复叶；叶轴上端具狭翅，被微柔毛；小叶7-13，无柄，全缘。花序长15-20厘米；花白色，花瓣于花期反卷。核果近球形，成熟时紫红色，混生柔毛和腺毛。花期6-7月，果期7-9月。生海拔400-3000米的丘陵和山地林中。产中国西南、华中和华西。

Trees or small trees. Leaves imparipinnately compound; rachis narrowly winged distally, minutely pubescent; leaflets 7-13, sessile, margin entire. Inflorescences 15-20 cm long; flowers white, petals revolute at anthesis. Drupes subglobose, purplish red at maturity, mixed pilose and glandular-pubescent. Fl. Jun-Jul. Fr. Jul-Sep Hills and montane forests at 400-3000 m. Distributed in SW, C and W China.

白背麸杨 *Rhus hypoleuca*

红麸杨 *Rhus punjabensis* var. *sinica*

## 青麸杨
**Rhus potaninii** Maxim.

落叶乔木。小枝无毛。奇数羽状复叶具7-11小叶，叶轴无翅；小叶具柄，长圆形至披针形，基部阔楔形至圆形，先端渐尖。圆锥花序长10-20厘米，被微柔毛。核果近球形，略压扁，成熟时红色。花期5-7月，果期7-10月。生海拔900-2500米的山坡疏林或灌丛。产云南西北部、四川、河南、陕西南部、山西南部和甘肃南部。

Deciduous trees. Branchlets glabrous. Leaves imparipinnately compound with 7-11 leaflets, rachises wingless; leaflets petiolulate, oblong to lanceolate, base broadly cuneate to rounded, apex acuminate. Panicles 10-20 cm long, minutely pubescent. Drupes subglobose, slightly compressed, red at maturity. Fl. May-Jul. Fr. Jul-Oct. Sparse montane forests or bushes at 900-2500 m. Distributed in NW Yunnan, Sichuan, Henan, S Shaanxi, S Shanxi and S Gansu.

青麸杨 *Rhus potaninii*

## 黄毛漆
**Toxicodendron fulvum** (Craib) C. Y. Wu et T. L. Ming

乔木。小枝具黄色绒毛；奇数羽状复叶具9-13小叶。小叶片革质，长圆形或长圆状披针形，腹面密被锈色绒毛。圆锥花序腋生，长20-33厘米。果近球形，黄色，光亮，被微柔毛。花期5-6月，果期8-10月。生海拔约1000米的石灰山山丘顶疏林。产云南(勐腊县、勐连县)。泰国北部亦有。

Trees. Branchlets yellow tomentose; imparipinnately compound with 9-13 leaflets. Leaflets leathery, oblong or oblong-lanceolate, abaxially densely rusty-tomentose. Panicles axillary, 20-33 cm long. Fruits subglobose, yellow, lucid, minutely pubescent. Fl. May-Jun. Fr. Aug-Oct. Sparse forests on limestone hill-top at ca. 1000 m. Distributed in Yunnan (Mengla County, Menglian County). Also in N Thailand.

## 绒毛漆
**Toxicodendron wallichii** (Hook. f.) O. Kuntze

乔木。小枝密被锈色绒毛。奇数羽状复叶，小叶7-11；小叶卵形或椭圆形，叶背密被锈色绒毛。圆锥花序腋生，长12-15厘米。果序直立；果密集排列，球形，成熟时不规则开裂；中果皮白色蜡质。花期7-8月，果期9-10月。生海拔1850-2400米的常绿阔叶林中。产西藏(吉隆县)。印度和尼泊尔亦有。

Trees. Branchlets densely rusty-tomentose. Imparipinnately compound leaves with 7-11 leaflets; leaflets ovate or elliptic, abaxially densely rusty-tomentose. Panicles axillary, 12-15 cm long. Infructescences erect; fruits densely clustered, globose, irregularly dehiscent; mesocarp offwhite, waxy. Fl. Jul-Aug. Fr. Sep-Oct. Evergreen forests at 1850-2400 m. Distributed in Xizang (Gyirong County). Also in India and Nepal.

黄毛漆 *Toxicodendron fulvum*

绒毛漆 *Toxicodendron wallichii*

毛漆树 *Toxicodendron trichocarpum*

## 毛漆树

**Toxicodendron trichocarpum** (Miq.) Kuntze

落叶乔木或灌木。叶柄被棕黄色绒毛，基部膨大；奇数羽状复叶，小叶9-15，纸质。圆锥花序密被棕黄色柔毛，花黄绿色。核果扁球形，黄色，稍具刚毛；外果皮薄，分离；中果皮厚，蜡质，具棕色纵向树脂管。花期6月，果期7-9月。生海拔900-2000米的林中或灌丛。产贵州、福建、浙江、湖北、湖南、安徽和江西。朝鲜半岛和日本亦有。

Deciduous trees or shrubs. Petioles yellowish brown pubescent, inflated at base. Leaves imparipinnately compound, leaflets 9-15, papery. Inflorescences paniculate, densely yellowish brown pubescent; flowers yellowish green. Drupes oblate, yellow, minutely setaceous; epicarps thin, separating; mesocarps thick, waxy, with brown longitudinal resin ducts. Fl. Jun. Fr. Jul-Sep. Forests or bushes at 900-2000 m. Distributed in Guizhou, Fujian, Zhejiang, Hubei, Hunan, Anhui and Jiangxi. Also in Korean Peninsula and Japan.

## 漆

**Toxicodendron vernicifluum** (Stokes) F. A. Barkl.

落叶乔木。叶柄基部膨大，叶柄与叶轴稍被柔毛；奇数羽状复叶具9-13小叶。圆锥花序，下垂。核果不对称，无毛；外果皮薄，不裂；中果皮厚，蜡质，具纵向树脂管。花期5-6月，果期7-10月。生海拔800-2800米的林中。产中国除西北以外大部分地区。印度、朝鲜半岛和日本亦有。

Deciduous trees. Petioles inflated at base, petioles and rachis minutely pubescent; imparipinnately compound with 9-13 leaflets. Inflorescences paniculate, grayish yellow minutely pubescent; petals yellowish green; infructescences pendulous. Drupes symmetrical, glabrous; epicarps thin, indehiscent; mesocarps thick, waxy, with brown longitudinal resin ducts. Fl. May-Jun. Fr. Jul-Oct. Forests at 800-2800 m. Distributed in most parts of China, except NW China. Also in India, Korean Peninsula and Japan.

漆 *Toxicodendron vernicifluum*

木蜡树 *Toxicodendron sylvestre*

## 木蜡树

**Toxicodendron sylvestre**

(Siebold et Zucc.) Kuntze

乔木或小乔木。叶柄和叶轴密被黄色绒毛；奇数羽状复叶。小叶7-15，对生。圆锥花序；花黄色；雄蕊反折。核果偏斜，压扁，先端偏于一侧；外果皮薄，分离；中果皮厚，蜡质，具棕色纵向树脂道。花期4-5月，果期6-10月。生海拔100-800(-2300)米的低地或山地林中。产中国西南、华南、东南、华中和华东。朝鲜半岛和日本亦有。

Trees or small trees. Petioles and rachises densely yellow tomentose. Leaves imparipinnately compound; leaflets 7-15, opposite. Inflorescences paniculate; flowers yellow; stamens exserted. Drupes oblique, compressed, apex eccentric; epicarps thin, separating; mesocarps thick, waxy, with brown longitudinal resin canal. Fl. Apr-May. Fr. Jun-Oct. Lowlands or montane forests at 100-800 (-2300) m. Distributed in SW, S, SE, C and E China. Also in Korean Peninsula and Japan.

## 尖叶漆

**Toxicodendron acuminatum**

(DC.) C. Y. Wu et T. L. Ming

小乔木。奇数羽状复叶；小叶5-9，对生；小叶椭圆形或长圆形，纸质，两面无毛，先端长尾尖。花序圆锥状，微具柔毛；花黄绿色；花瓣长圆形。核果扁圆形，对称。花期5-7月，果期8-10月。生海拔1600-2600米的丘陵及山地林中。产云南西南部和西藏南部。印度北部、尼泊尔和不丹亦有。

Small trees. Leaves imparipinnately compound; leaflets 5-9, opposite; leaflets elliptic or oblong, papery, glabrous on both surfaces, apex long caudate. Inflorescences paniculate, minutely pubescent; flowers yellowish green; petals oblong. Drupes oblate, symmetrical. Fl. May-Jul. Fr. Aug-Oct. Hill and montane forests at 1600-2600 m. Distributed in SW Yunnan and S Xizang. Also in N India, Nepal and Bhutan.

尖叶漆 *Toxicodendron acuminatum*

野漆 *Toxicodendron succedaneum*

## 野漆

**Toxicodendron succedaneum**
(L.) Kuntze

落叶灌木或小乔木。叶轴圆柱形或上部具狭翅；奇数羽状复叶具5-15小叶，小叶无毛至两面疏具柔毛。圆锥花序腋生；花小，黄绿色，杂性；花盘5裂。核果较大，不对称，黄色，无毛。花期5月，果期7-10月。生海拔(100-)300-1500(-2500)米的山坡或沟旁。产中国除东北以外大部分地区。南亚、东南亚和东亚亦有。

Shrubs or small trees, deciduous. Rachises terete or narrowly winged distally; imparipinnately compound with 5-15 leaflets, leaflets glabrous to sparsely pubescent on both surfaces. Panicles axillary; flowers small, yellowish green, polygamous; disks 5-lobed. Drupes large, asymmetrical, yellow, glabrous. Fl. May. Fr. Jul-Oct. Slopes or ditch sides at (100-)300-1500(-2500) m. Distributed in most parts of China, except NE China. Also in S, SE and E Asia.

## 小漆树

**Toxicodendron delavayi**
(Franch.) F. A. Barkl.

小灌木。小枝紫红色。奇数羽状复叶具5-11小叶；小叶对生，卵状披针形至条状披针形，纸质。疏散总状花序，无毛；花亮黄色；花盘10裂。核果倾斜至不对称，稍压扁，光亮。花期5-6月，果期9-10月。生海拔1100-2500米的山地林中或向阳坡灌丛。产云南和四川。

Small shrubs. Branchlets purplish red. Imparipinnately compound with 5-11 leaflets; leaflets opposite, ovate-lanceolate to linear-lanceolate, papery. Laxly racemes, glabrous, flowers light yellow; disks 10-lobed. Drupes oblique to symmetrical, slightly compressed, lucid. Fl. May-Jun. Fr. Sep-Oct. Montane forests or thickets on sunny slopes at 1100-2500 m. Distributed in Yunnan and Sichuan.

## 三叶漆

**Terminthia paniculata**
(Wall. ex G. Don) C. Y. Wu et T. L. Ming

灌木或小乔木。掌状3(-5)小叶；小叶倒卵状椭圆形、倒卵状长圆形或倒卵形。花序圆锥状，顶生或腋生；花浅黄色。核果近球形，稍压扁；外果皮成熟时橘红色，最后分离；中果皮胶质。花期9-11月，果期11月至翌年5月。生海拔400-1500米的山坡、稀树草地、灌丛或疏林。产云南。印度东北部、不丹和缅甸北部亦有。

Shrubs or small trees. Leaves palmately 3(-5)-foliolate; leaflets obovate-elliptic, obovate-oblong or obovate. Inflorescences paniculate, terminal or axillary; flowers light yellow. Drupes subglobose, slightly compressed; exocarps orange-red at maturity, finally separating; mesocarps glutinous. Fl. Sep-Nov. Fr. Nov to next May. Slopes, savanas, scrubs or sparse forests at 400-1500 m. Distributed in Yunnan. Also in NE India, Bhutan and N Myanmar.

小漆树 *Toxicodendron delavayi*

三叶漆 *Terminthia paniculata*

大叶肉托果 *Semecarpus gigantifolia*

## 大叶肉托果

**Semecarpus gigantifolia** Vidal.

常绿乔木。小枝粗壮。叶集生小枝顶端，革质，披针形，长可达50厘米。圆锥花序顶生；花小，密集，白色；花萼钟状；花瓣卵状披针形。核果扁球形；果时花托肉质膨大，包于果的中下部。花果期11-12月。产台湾。菲律宾亦有。

Evergreen trees. Branchlets robust. Leaves clustered at top of branchlets, coriaceous, lanceolate, to 50 cm long. Panicles terminal, large; flowers small, dense, white; calyx campanulate; petals ovate-lanceolate. Drupes oblate-globose; receptacles swollen during fruiting, fleshy, surrounding the lower part of fruit. Fl. and fr. Nov-Dec. Distributed in Taiwan. Also in the Philippines.

## 辛果漆

**Drimycarpus racemosus** (Roxb.) Hook. f.

大乔木，高8-18米。叶互生，革质，长圆形或长圆状披针形，全缘，侧脉15-20对，网脉两面突起。圆锥花序腋生；花小，杂性；花瓣卵形，直立。核果椭圆状，直径约2厘米，果肉具树脂。生海拔130-900米的山谷或沟边密林中。产云南东南部。不丹、印度、缅甸和越南亦有。

Large trees, 8-18 m tall. Leaves alternate, coriaceous, oblong or oblong-lanceolate, margin entire; lateral veins 15-20 pairs, reticulate veinlets elevated on both surfaces. Panicles axillary; flowers small, polygamous; petals ovate, erect. Drupes ellipsoid, ca. 2 cm diam, sarcocarp contains rosin. Thickets in valleys or by streams at 130-900 m. Distributed in SE Yunnan. Also in Bhutan, India, Myanmar and Vietnam.

辛果漆 *Drimycarpus racemosus*

羊角天麻 *Dobinea delavayi*

## 羊角天麻

**Dobinea delavayi** (Baill.) Baill.

多年生亚灌木状草本。叶互生，卵状心形，边缘具不整齐锯齿。雄花序顶生，聚伞圆锥状，花梗纤细；花萼4(-5)裂；花瓣4(-5)，匙形，具爪；雄蕊8(-10)。雌花序总状，苞片大，膜质，灰白色，先端圆形。果大，径3-4毫米。生海拔1100-2300米的向阳草坡或灌丛中。产云南和四川。

Perennial suffrutescent herbs. Leaves alternate, ovate-cordate, margin irregularly serrate. Male inflorescences terminal, thyrsiform, pedicels tenuous; calyx 4(-5)-lobed; petals 4(-5), spatulate, clawed; stamens 8(-10). Female inflorescences racemose, bracts large, membranous, grayish white, apex rotundate. Fruit large, 3-4 mm diam. Sunny grass slopes or scrubs at 1100-2300 m. Distributed in Yunnan and Sichuan.

## 九子母

**Dobinea vulgaris** Buch.-Ham. et D. Don

灌木，高1-3米。小枝具柔毛。叶长圆状披针形，纸质，边缘具细齿。圆锥花序顶生，长约18厘米，被微柔毛；花萼钟形，4裂；花瓣4，具爪；雄蕊8。果柄贴生于增大的与花对生苞片，苞片膜质，灰白色，近圆形，直径1-1.3厘米。生海拔1300-1400米的江边林中或山谷。产云南西北部和西藏东南部。印度东北部、不丹和尼泊尔亦有。

Shrubs, 1-3 m tall. Branchlets pubescent. Leaves oblong-lanceolate, papery, margin serrulate. Inflorescences terminal, ca. 18 cm long, minutely pubescent; calyx campanulate, 4-lobed; petals 4, clawed; stamens 8. Fruit pedicels adnate to accrescent floral subtending bract, bract membranous, grayish white, suborbicular, 1-1.3 cm diam. Forests by rivers or in valleys at 1300-1400 m. Distributed in NW Yunnan and SE Xizang. Also in NE India, Bhutan and Nepal.

九子母 *Dobinea vulgaris*

# 五列木科 Pentaphylacaceae

### 五列木

**Pentaphylax euryoides** Gardn. et Champ.

常绿小乔木或灌木。叶互生，革质，卵形至长圆状披针形，先端尾尖，边缘全缘略反卷，无毛。总状花序直立；萼片5，圆形，外面密被灰白色鳞片，里面疏被白色贴伏微柔毛，宿存；花瓣5，白色；雄蕊5，与花瓣互生。蒴果椭圆体形，5瓣裂，中轴宿存。种子具翅。花期6月，果期11月。生海拔600-2000米的密林中。产华南、湖南和江西。东南亚亦有。

Evergreen small trees or shrubs. Leaves alternate, leathery, ovate to oblong-lanceolate, apex caudate, margin entire and slightly revolute, glabrous. Racemes erect; sepals 5, orbicular, outside with dense grayish white scales, inside sparsely white appressed puberulent, persistent; petals 5, white; stamens 5, alternate with petals. Capsules ellipsoid, 5-valved, central axis persistent. Seeds winged. Fl. Jun. Fr. Nov. Dense forests at 600-2000 m. Distributed in S China, Hunan and Jiangxi. Also in SE Asia.

五列木 *Pentaphylax euryoides*

# 冬青科 Aquifoliaceae

### 黄毛冬青

**Ilex dasyphylla** Merr.

常绿灌木或乔木。小枝、叶柄、叶片、花梗及花萼均密被锈黄色瘤基短硬毛。叶革质，全缘或中部以上具稀疏小齿。聚伞花序单生；花4或5基数；宿存柱头厚盘状，凸起。分核4或5。花期5月，果期8-11月。生海拔250-700米的山地疏林、灌丛中或路旁。产广西、广东、福建、湖南和江西。

Evergreen shrubs or trees. Branchlets, petioles, leaves, pedicels, and calyx ferruginous hispidulous, hairs with tuberculate bases. Leaves leathery, margin entire or upper half sparsely serrulate. Cymes solitary; flowers 4-merous or 5-merous; persistent stigmas thickly discoid, convex. Pyrenes 4 or 5. Fl. May. Fr. Aug-Nov. Sparse montane forests, shrubby areas or roadsides at 250-700 m. Distributed in Guangxi, Guangdong, Fujian, Hunan and Jiangxi.

### 显脉冬青

**Ilex editicostata** Hu et Tang

常绿灌木或小乔木。叶绿色，披针形或长圆形，革质，无毛，主脉于叶腹面凸起。聚伞花序或二歧聚伞花序单生。果序具1-3果。果实直径(6-)9-10(-12)毫米；分核4-6。花期5-7月，果期8-12月。生海拔500-1700米的山坡常绿阔叶林中或林缘。产中国西南、华南、东南、华中和华东。

Evergreen shrubs or small trees. Leaves green, lanceolate or oblong, leathery, glabrous, main veins convex in adaxial side of leaves. Cymes or dichotomous cymes solitary. Infructescences 1-3-fruited. Fruits (6-)9-10(-12) mm diam; pyrenes 4-6. Fl. May-Jul. Fr. Aug-Dec. Mountainuos evergreen forests or forest edges at 500-1700 m. Distributed in SW, S, SE, C and E China.

### 木姜冬青

**Ilex litseifolia** Hu et T. Tang

常绿灌木至小乔木。当年生幼枝干后紫黑色，具细棱。叶痕半圆形，隆起；叶全缘，稍反卷，中脉上面具柔毛，网脉模

黄毛冬青 *Ilex dasyphylla*

显脉冬青 *Ilex editicostata*

糊；叶柄稍具柔毛。雄聚伞花序具(3-)5-7花；宿存花萼平展。分核干后光滑，下面具浅槽。花期5-7月，果期8-11月。生海拔700-1100(-2100)米的山地常绿阔叶林中或林缘。产中国西南、华南、东南和华中。

Evergreen shrubs to small trees. Current year's young branchlets purple-black when dry, thinly angular. Leaf scars subcircular, convex; leaves margin entire, slightly recurved, midvein adaxially pubescent, reticulate veins obscure; Petioles ± pubescent. Male inflorescences cymes, (3-)5-7-flowered; persistent calyx explanate. Pyrenes smooth when dry, abaxially shallowly sulcate. Fl. May-Jul. Fr. Aug-Nov. Evergreen broad-leaved forests or forest edges on mountain slopes at 700-1100(-2100) m. Distributed in SW, S, SE and C China.

香冬青 *Ilex suaveolens*

## 香冬青

**Ilex suaveolens** (Lévl.) Loes.

常绿乔木。幼枝褐色，具棱，全株光滑。叶柄长1.5-2厘米，具翅；叶卵形或椭圆形，长5-6.5(-10) × 2-2.5(-4)厘米，革质，两面无毛，侧脉8-10对，两面稍凸起。聚伞状果序单个腋生，具3-7果，有时伞形。果实红色，窄球形，直径约6毫米。花期6月。生海拔600-1600米的常绿阔叶林中。产中国西南、华南、东南、华中和华东。

Trees evergreen. Young branchlets brown, angular, glabrous throughout. Petioles 1.5-2 cm long, winged; leaves ovate or elliptic, 5-6.5(-10) × 2-2.5(-4) cm, leathery, both surfaces glabrous, lateral veins 8-10 pairs, slightly raised on both surfaces. Infructescences: cymes solitary, axillary, 3-7-fruited, sometimes umbelliform. Fruits red, narrowly globose, ca. 6 mm diam. Fl. Jun. Evergreen broad-leaved forests at 600-1600 m. Distributed in SW, S, SE, C and E China.

木姜冬青 *Ilex litseifolia*

冬青 *Ilex chinensis*

广东冬青 *Ilex kwangtungensis*

## 冬青

**Ilex chinensis** Sims

常绿乔木。叶椭圆形或披针形，边缘具圆齿，或有时幼叶具锯齿，无毛，稀上面幼时沿脉具柔毛，干后深褐色。雄花花序三至四回分枝，雌花花序一至二回分枝。果红色，窄球形；分核4或5，内果皮厚革质。花期4-7月，果期7-12月。生海拔2000米以下的常绿阔叶林中或山坡林缘。产中国西南、华南、东南、华中和华东。日本亦有。

Evergreen trees. Leaves elliptic or lanceolate, margin crenate, or sometimes young leaf serrate, glabrous, rarely adaxially pilose on midvein when young, deep brown when dry. Male inflorescences 3-4-branched, female inflorescences 1-2-branched. Fruits red, narrowly globose; pyrenes 4 or 5, endocarps thickly leathery. Fl. Apr-Jul. Fr. Jul-Dec. Evergreen broad-leaved forests or forest edges on mountain slopes below 2000 m. Distributed in SW, S, SE, C and E China. Also in Japan.

## 硬叶冬青

**Ilex ficifolia** C. J. Tseng ex S. K. Chen et Y. X. Feng

常绿灌木或乔木。叶干后紫褐色或黄褐色，4-11.7 × 1.5-4.3厘米，革质，两面无毛。单生聚伞花序腋生。果实干后黑色，球形；分核5粒，背面具一纵沟纹，内果皮革质。花期5-6月，果期9-11月。生海拔400-1200米的丘陵疏林中。产广西、广东、福建、浙江、湖南和江西。

Evergreen shrubs or trees. Leaves purple-brown or yellowish brown when dry, 4-11.7×1.5-4.3 cm, leathery, both surfaces glabrous. Inflorescences cymes, solitary, axillary. Fruits black when dry, globose; pyrenes 5, abaxially longitudinally 1-sulcate, endocarps leathery. Fl. May-Jun. Fr. Sep-Nov. Sparse forests on hills at 400-1200 m. Distributed in Guangxi, Guangdong, Fujian, Zhejiang, Hunan and Jiangxi.

## 广东冬青

**Ilex kwangtungensis** Merr.

常绿灌木或小乔木。叶干后褐色或深橄榄色，卵状椭圆形、长圆形或披针形，中脉上面凹陷。雄花序由具1-3花的聚伞花序簇生，带梗长0.9-1.2厘米；雌花2-3花簇生于次年生枝叶腋内。果椭圆体形，分核4，内果皮革质。花期5-8月，果期9-12月。生海拔300-1200米的山坡常绿阔叶林或灌木丛中。产中国西南、华南、东南和华中。

Evergreen shrubs or small trees. Leaves brown or deep olivaceous when dry, ovate-elliptic, oblong, or lanceolate, midvein impressed adaxially. Male inflorescences

硬叶冬青 *Ilex ficifolia*

铁冬青 *Ilex rotunda*

with 1-3 clustered cymes, with peduncles 0.9-1.2 cm long; pistillate flowers 2-3 clustered axillary on second year's branches. Fruits ellipsoidal, pyrenes 4, endocarps leathery. Fl. May-Aug. Fr. Sep-Dec. Evergreen broad-leaved forests or shrubs on mountain slopes at 300-1200 m. Distributed in SW, S, SE and C China.

## 铁冬青

**Ilex rotunda** Thunb.

常绿灌木或乔木。老枝圆柱形，挺直，粗糙，具纵裂缝。叶卵形、倒卵形或椭圆形，4-9 × 1.8-4厘米，薄革质或纸质，两面无毛。花梗无毛；花白色，5(-7)基数，花萼浅杯状。果大，直径6-8毫米，内果皮木质或近木质。花期4-6月，果期8-12月。生海拔400-1100(-1700)米的常绿阔叶林或山坡林缘。产中国西南、华南、华中和华东。越南、朝鲜半岛和日本亦有。

Shrubs or trees, evergreen. Older branchlets terete, straight, rough, longitudinally fissured. Leaves ovate, obovate, or elliptic, 4-9 × 1.8-4 cm, thinly leathery or papery, both surfaces glabrous. Pedicels glabrous; flowers white, 5(-7)-merous; calyx shallowly cup-shaped. Drupes large, 6-8 mm diam, endocarps woody or subwoody. Fl. Apr-Jun. Fr. Aug-Dec. Evergreen broad-leaved forests or forest edges on mountain slopes at 400-1100(-1700) m. Distributed in SW, S, C and E China. Also in Vietnam, Korean Peninsula and Japan.

## 三花冬青

**Ilex triflora** Blume

灌木或小乔木。小枝常之字形。叶椭圆形、长圆形至卵状椭圆形，下面有点，先端急尖或锐尖。雄聚伞花序具1-3花，稀具多花；花梗与总花梗等长或稍长；退化子房金字塔形。果通常3。花期5-6月，果期9-10月。生海拔(130-)250-1800(-2200)米的林中或灌丛。产中国西南、华南、东南和华中。南亚和东南亚亦有。

Shrubs or small trees. Branchlets usually zigzag. Leaves elliptic, oblong or ovate-elliptic, abaxially punctate, apex acute to acuminate. Male cymes 1-3-flowered, rarely more flowered; pedicels equaling or slightly longer than peduncles; rudimentary ovary pyramidal. Drupes 3. Fl. May-Jun. Fr. Sep-Oct. Forests or thickets at (130-)250-1800(-2200) m. Distributed in SW, S, SE and C China. Also in S and SE Asia.

## 钝头冬青

**Ilex triflora** Blume var. **kanehirai** (Yamam.) S. Y. Hu

本变种与三花冬青的区别在于本变种的叶片倒卵形或长圆状椭圆形，先端圆形或钝，绝不渐尖。花期5-7月，果期8-11月。生海拔200-1100米的山地林中、林缘或山谷灌丛中。产广东、台湾、福建、浙江、湖南和江西。

This variety differs from the typical variety in its leaves obovate or oblong-elliptic, apex rounded, obtuse, or rarely acute, never acuminate. Fl. May-Jul. Fr. Aug-Nov. Forests, forest edges or shrubs in valleys at 200-1100 m. Distributed in Guangdong, Taiwan, Fujian, Zhejiang, Hunan and Jiangxi.

三花冬青 *Ilex triflora*

钝头冬青 *Ilex triflora* var. *kanehirai*

四川冬青 *Ilex szechwanensis*

## 四川冬青

**Ilex szechwanensis** Loes.

灌木或小乔木。小枝被微柔毛。叶革质，卵状椭圆形，卵状长圆形或椭圆形，稀近披针形，腹面有黄褐色腺点。雄聚伞花序具1-7花，单生于当年生小枝上；花白色，4或5数。果熟时黑色。花期5-6月，果期8-10月。生海拔(250-)450-2500米的山地常绿阔叶林、杂木林或溪边。产中国西南、华南、华中和华东。

Shrubs or small trees. Branchlets puberulent. Leaves leathery, ovate-elliptic, ovate-oblong, or elliptic, rarely sublanceolate, abaxially fulvous glandular punctate. Male cymes 1-7-flowered, solitary on current year's branchlets; flowers white, 4- or 5-merous. Fruits black at maturity. Fl. May-Jun. Fr. Aug-Oct. Montane evergreen broad-leaved forests, mixed forests or streamsides at (250-)450-2500 m. Distributed in SW, S, C and E China.

## 齿叶冬青

**Ilex crenata** Thunb.

常绿灌木。幼枝密被短柔毛。托叶钻形，微小；叶缘具圆锯齿。雄花1-7排成聚伞花序，单生于当年枝的鳞片腋内或下部的叶腋内，很少假簇生于二年生枝的叶腋内。果球形，成熟后黑色。花期5-6月，果期8-10月。生海拔700-2100米的丘陵或山地林中或灌丛中。产中国西南、华南、东南、华中和华东。朝鲜半岛和日本亦有。

Evergreen shrubs. Young branches densely pubescent. Stipules subulate, minute; leaves margin crenate-serrate. Male inflorescences cymes, 1-7-flowered, solitary, axillary on scales or lower leaves of current year's branchlets, rarely pseudofasciculate on second year's branchlets. Fruits globose, black at maturity. Fl. May-Jun. Fr. Aug-Oct. Forests or thickets on hills or mountains at 700-2100 m. Distributed in SW, S, SE, C and E China. Also in Korean Peninsula and Japan.

绿叶冬青 *Ilex viridis*

## 绿叶冬青

**Ilex viridis** Champ. ex Benth.

常绿灌木或小乔木。叶倒卵形、倒卵状椭圆形或阔椭圆形，革质，侧脉5-8对，两面明显。雄花序为1-3分枝的聚伞状，具1-5花，单生；花白色，4数；雌花序为单花聚伞状单生。果实黑色，球形或稍扁球形，直径9-11毫米；分核4。花期4-5月，果期10月。生海拔300-1700(-2100)米的常绿阔叶林下或山坡疏林及灌丛中。产中国西南、华南、东南和华东。

Evergreen shrubs or small trees. Leaves obovate, obovate-elliptic, or broadly elliptic, leathery, lateral veins 5-8 pairs, distinct on both surfaces. Male inflorescences: cymes of order 1-3, 1-5-flowered, solitary; flowers white, 4-merous; female inflorescences: 1-flowered cymes, solitary. Fruits black, globose

齿叶冬青 *Ilex crenata*

or slightly depressed globose, 9-11 mm diam; pyrenes 4. Fl. Apr-May. Fr. Oct. Understories of evergreen broad-leaved forests or sparse forests and thickets on mountains at 300-1700(-2100) m. Distributed in SW, S, SE and E China.

## 具柄冬青
**Ilex pedunculosa** Miq.

常绿灌木或乔木。幼枝淡褐色或栗色，近圆柱形，具纵棱角。叶长4-12.5厘米；叶柄长1.5-2.5厘米。聚伞花序单个腋生；雌花梗长4-4.5厘米。果红色或橙色，球形；核背面光滑，具1纵纹。花期5-6月，果期7-11月。生海拔(900-)1200-1900(-3000)米的林中、林缘或灌丛中。产中国西南、东南、华中和华东。日本亦有。

Evergreen shrubs or trees. Young branchlets brownish or castaneous, subterete, longitudinally angular. Leaves 4-12.5 cm long; petioles 1.5-2.5 cm long. Cymes solitary, axillary; pedicels of pistillate flower 4-4.5 cm long. Fruits red or orange, globose; pyrenes abaxially smooth, longitudinally 1-striate. Fl. May-Jun. Fr. Jul-Nov. Forests, forest edges or thickets at (900-)1200-1900 (-3000) m. Distributed in SW, SE, C and E China. Also in Japan.

枸骨 *Ilex cornuta*

## 枸骨
**Ilex cornuta** Lindl. et Paxton

常绿灌木或小乔木。幼枝具纵脊及沟。叶片厚革质，四角状长圆形或卵形，具5或6对侧脉，先端具1枚尖硬刺齿，常反曲。果熟时鲜红色，球形；分核4，石质，具皱纹和纹孔。花期4-5月，果期8-12月。生海拔150-1900米的灌丛、疏林、丘陵、溪旁或路边。产华南、华中、东南、华西和华东。朝鲜半岛亦有。

Evergreen shrubs or small trees. Young branchlets longitudinally ridged and sulcate. Leaves thickly leathery, quadrangular-oblong or rarely ovate, lateral veins 5 or 6 pairs; apex with 1 strong spine, often reflexed. Fruits bright red when matured, globose; drupes 4, stony, rugose and pitted. Fl. Apr-May. Fr. Aug-Dec. Shrubby areas, sparse forests, hillsides, streamsides or roadsides at 150-1900 m. Distributed in S, C, SE, W and E China. Also in Korean Peninsula.

## 浙江冬青
**Ilex zhejiangensis** C. J. Tseng ex S. K. Chen et Y. X. Feng

常绿灌木或小乔木。小枝具纵棱，被柔毛。叶卵状椭圆形或椭圆形，边缘疏具4-7对刺齿，先端急尖，稀钝，边缘具疏离的(2-)4-7枚小刺状黑色锯齿。花序簇生于叶腋；花4基数。果球形；分核背面不具1纵纹。花期4月，果期8-11月。生海拔500-1200米的山地灌丛中或林缘。产浙江。

Evergreen shrubs or small trees. Branches longitudinally angular, pubescent. Leaves ovate-elliptic or elliptic, margin sparsely spinulose-serrate in 4-7 pairs, apices acute, rarely obtuse, margin sparsely (2-)4-7-spinulose-serrate, teeth black. Inflorescences clustered, axillary; flowers 4-merous. Fruit globose; pyrenes abaxially not longitudinally 1-ridged. Fl. Apr. Fr. Aug-Nov. Montane thickets or forest edges at 500-1200 m. Distributed in Zhejiang.

具柄冬青 *Ilex pedunculosa*

浙江冬青 *Ilex zhejiangensis*

猫儿刺 *Ilex pernyi*

## 猫儿刺

**Ilex pernyi** Franch.

常绿灌木或乔木。叶片革质，先端三角形渐尖，渐尖头长达1.2-1.4厘米，终于1粗刺，边缘具1-3对深波状刺齿。花序簇生于二年生枝的叶腋内，多为2-3花聚生成簇；宿存花萼四角形。分核背面具掌状条纹及沟槽。花期4-5月，果期10-11月。生海拔1000-2500米的山谷林中或灌丛中。产中国西南、东南和华中。

Evergreen shrubs or trees. Leaves leathery, apex triangularly acuminate, acumen 1.2-1.4 cm long, with a strong spine, margin sinuate-dentate with 1-3 pairs of spines. Inflorescences 2-3-flowered cymes, fasciculate, axillary on second year's branchlets; persistent calyx quadrangular; pyrenes abaxially palmately striate and sulcate. Fl. Apr-May. Fr. Oct-Nov. Forests or shrubby areas in valleys at 1000-2500 m. Distributed in SW, SE and C China.

## 扣树

**Ilex kaushue** S. Y. Hu

常绿乔木。叶生于第一年至第二年小枝上，长圆形至长圆状椭圆形，10-18 × 4.5-7.5厘米，革质，侧脉14或15对，网脉在两面明显。花序为聚伞状假圆锥或假总状；花瓣4。果实红色，球形，直径9-12毫米。花期4-6月，果期7-10月。生海拔1000-1200米的密林中。产中国西南、华南和华中。

Evergreen trees. Leaves on first to second year's branchlets, oblong to oblong-elliptic, 10-18 × 4.5-7.5 cm, leathery, lateral veins 14 or 15 pairs, reticulate veins evident on both surfaces. Inflorescences: cymes pseudopaniculate or pseudoracemose; petals 4. Fruits red, globose, 9-12 mm diam. Fl. Apr-Jun. Fr. Jul-Oct. Dense forests at 1000-1200 m. Distributed in SW, S and C China.

扣树 *Ilex kaushue*

大叶冬青 *Ilex latifolia*

## 大叶冬青

**Ilex latifolia** Thunb.

常绿乔木，全株无毛。叶片8-19(-28) × 4.5-7.5(-9)厘米，深绿色，有光泽，侧脉在叶腹面明显。花淡黄绿色；柱头盘状，4裂。果球形，直径约7毫米；分核长圆状椭圆体形，背面具纵纹。花期4-5月，果期9-10月。生海拔200-1500米的山坡常绿阔叶林中、灌丛中或竹林中。产中国西南、华南、东南、华中和华东。日本亦有。

Evergreen trees, glabrous throughout. Leaves 8-19(-28) × 4.5-7.5 (-9) cm, adaxially deep green, shiny; lateral veins obvious adaxially. Flowers yellowish green; stigmas discoid, 4-lobed. Fruits globose, ca. 7 mm diam; pyrenes oblong-ellipsoidal, abaxially longitudinally ridged. Fl. Apr-May. Fr. Sep-Oct. Evergreen broad-leaved forests, shrub forests or bamboo forests at 200-1500 m. Distributed in SW, S, SE, C and E China. Also in Japan.

## 康定冬青
**Ilex franchetiana** Loes.

常绿灌木或小乔木。叶倒披针形或长圆状披针形，(3-)6-12.5 × (1.2-)2-4.2厘米，近革质，边缘具锯齿。花序聚伞状、束状，腋生于第二年小枝上；花淡绿色，4数。雄花序具3花；花瓣4，长圆形。果实红色，球形，直径(3-)6-7毫米。花期5-6月，果期9-11月。生海拔800-2300(-2900)米的林中或山坡。产云南东北部、四川、西藏东南部、贵州和湖北。缅甸北部亦有。

Evergreen shrubs or small trees. Leaves oblanceolate or oblong-lanceolate, (3-)6-12.5 × (1.2-)2-4.2 cm, subleathery, margin serrulate. Inflorescences: cymes, fasciculate, axillary on second year's branchlets; flowers greenish, 4-merous. Male inflorescences 3-flowered; petals 4, oblong. Fruits red, globose, (3-)6-7 mm diam. Fl. May-Jun. Fr. Sep-Nov. Forests or hills at 800-2300 (-2900) m. Distributed in NE Yunnan, Sichuan, SE Xizang, Guizhou and Hubei. Also in N Myanmar.

康定冬青 *Ilex franchetiana*

## 狭叶冬青
**Ilex fargesii** Franch.

常绿乔木或灌木。叶近革质，条状倒披针形或条状披针形。聚伞花序簇生于次年生枝条叶腋。果实红色，球形，具纵条纹；分核4，长圆形，背面凹陷，具掌状条纹和沟，内果皮木质。花期5月，果期9-10月。生海拔1500-3000米的林中、山坡树林或灌丛中。产四川、湖北、湖南、陕西和甘肃。

Evergreen trees or shrubs. Leaves subleathery, linear-oblanceolate or linear-lanceolate. Inflorescences cymes, fasciculate, axillary on second year's branchlets. Fruits red, globose, longitudinally striate; pyrenes 4, oblong, abaxially convex, palmately striate and sulcate, endocarps woody. Fl. May. Fr. Sep-Oct. Forests, or forests or shrubs on mountain slopes at 1500-3000 m. Distributed in Sichuan, Hubei, Hunan, Shaanxi and Gansu.

## 弯尾冬青
**Ilex cyrtura** Merr.

常绿乔木。叶近革质，边缘有浅锯齿，先端较长，常镰状尾尖。花黄色；柱头凸起，4浅裂。果球形，直径约6毫米，果柄长5-6毫米；宿存花萼平展，四棱形。花期4月，果期6-9月。生海拔750-1800米山地阔叶林中。产云南西北部、贵州、广西和广东。不丹和缅甸北部亦有。

Evergreen trees. Leaves subleathery, margin shallowly serrate, apex long and often falcate-caudate. Flowers yellow; stigmas very convex, 4-lobed. Fruits globose, ca. 6 mm diam, fruiting pedicels 5-6 mm long; persistent calyx explanate, quadrangular. Fl. Apr. Fr. Jun-Sep. Broad-leaved montane forests at 750-1800 m. Distributed in NW Yunnan, Guizhou, Guangxi and Guangdong. Also in Bhutan and N Myanmar.

狭叶冬青 *Ilex fargesii*

弯尾冬青 *Ilex cyrtura*

假香冬青 *Ilex wattii*

毛冬青 *Ilex pubescens*

## 假香冬青
**Ilex wattii** Loes.

常绿乔木，全株无毛。叶片纸质至薄革质，椭圆形、长圆形或长圆状椭圆形，中脉在叶面凹陷，背面隆起。雄花序为聚伞花序，具1-3花。果近球形，宿存柱头盘形或脐状；分核倒卵球状长圆形。花期(2-)6-7月，果期8-10月。生海拔2100-3000米的山坡林中。产云南西南部。印度东北部亦有。

Evergreen trees, whole plant glabrous. Leaves papery to thinly leathery, elliptic, oblong, or oblong-elliptic, midveins concave at adaxial side, convex at abaxial side. Male cymes with flowers 1-3. Fruits subglobose, persistent stigma discoid or navel-shaped; pyrenes obovoid-oblong. Fl. (Feb-)Jun-Jul. Fr. Aug-Oct. Montane forests at 2100-3000 m. Distributed in SW Yunnan. Also in NE India.

## 榕叶冬青
**Ilex ficoidea** Hemsl.

常绿乔木。小枝无毛。叶片长圆状椭圆形至卵状椭圆形，具不规则细圆齿状锯齿，先端长尾状。聚伞花序或单花簇生于当年生枝的叶腋内；花4基数；雄蕊长于花瓣。宿存花柱薄盘状或脐状。花期3-5月，果期8-11月。生海拔(100-)300-1500米的常绿阔叶林、灌丛、林缘或疏林。产中国西南、华南、东南和华中。日本亦有。

Evergreen trees. Branchlets glabrous. Leaves oblong-elliptic or obovate-elliptic, margin irregularly crenate-serrate, apex abruptly long caudate. Cymes or solitary flowers in axils of current year's branches; flowers 4-merous; stamens longer than petals. Persistent stigmas thinly discoid or navel-like. Fl. Mar-May. Fr. Aug-Nov. Evergreen broad-leaved forests, thickets, forest edges or sparse forests at (100-) 300-1500 m. Distributed in SW, S, SE and C China. Also in Japan.

榕叶冬青 *Ilex ficoidea*

## 毛冬青
**Ilex pubescens** Hook. et Arn.

常绿灌木或小乔木。小枝近4棱形，密被粗硬毛。叶纸质或膜质。雄花序为单个或两分枝，具1-3花的聚伞花序。果实球形；分核(5或)6(或7)，两头尖，具3条纹，内果皮革质或近木质。花期4-5月，果期8-11月。生海拔1000米以下的林中、林缘、灌丛中、溪边或路边。产华南、东南和华中。

Evergreen shrubs or small trees. Branchlet subtetragonous, densely hirsute. Leaves papery or membranous. Male inflorescences: cymes of order 1 or 2, 1-3-flowered. Fruits globose; pyrenes (5 or)6(or 7), pointed at both ends, 3-striate, endocarps leathery or subwoody. Fl. Apr-May. Fr. Aug-Nov. Forests, forest edges, shrubs, streamsides or roadsides below 1000 m. Distributed in S, SE and C China.

## 广西毛冬青
**Ilex pubescens** Hook. et Arn. var. **kwangsiensis** Hand. -Mazz.

本变种与毛冬青的区别在于本变种的叶4-10 × 2-7厘米，先端急渐尖。花白色。花期6月。生海拔500-1000米的常绿阔叶林中。产云南东南部、贵州南部和广西西部。

This variety differs from the typical variety in its leaves 4-10 × 2-7 cm, apex abruptly acuminate. Flowers white. Fl. Jun. Evergreen broad-leaved fo-

广西毛冬青 *Ilex pubescens* var. *kwangsiensis*

rests at 500-1000 m. Distributed in SE Yunnan, S Guizhou and W Guangxi.

## 河滩冬青

**Ilex metabaptista** Loes.

常绿灌木或乔木。叶近革质，披针形或倒披针形，近全缘。聚伞花序簇生于次年生小枝腋。果红色，卵球状椭圆体形；分核5-8，椭圆体形，末端尖，背面与侧面具纵条纹及沟，内果皮革质。花期4-6月，果期7-10月。生海拔300-1200米的林中、河边、山地路边、海滨或山坡。产中国西南和华中。

Evergreen shrubs or trees. Leaves subleathery, lanceolate or oblanceolate, margin subentire. Inflorescences cymes, fasciculate, axillary on second year's branchlets. Fruits red, ovoid-ellipsoidal; pyrenes 5-8, ellipsoidal, ends pointed, abaxially and laterally longitudinally striate and sulcate, endocarps leathery. Fl. Apr-Jun. Fr. Jul-Oct. Forests, stream banks, roadsides on mountains, shores or mountain slopes at 300-1200 m. Distributed in SW and C China.

厚叶冬青 *Ilex elmerrilliana*

## 厚叶冬青

**Ilex elmerrilliana** S. Y. Hu

常绿灌木或小乔木。叶片厚革质，椭圆形或长圆状椭圆形，具光泽。花序簇生或单生于鳞片腋处；宿存花柱明显，柱头头状；分核6或7，长圆体形。花期4-5月，果期7-11月。生海拔(200-)500-1500米的山地常绿阔叶林中、灌丛中或林缘。产中国西南、华南、东南、华中和华东。

Evergreen shrubs or small trees. Leaves thick leathery, elliptic or oblong-elliptic, lustrous. Inflorescences clustered axillary or solitary in scales; persistent styles evident, stigmas capitate; pyrenes 6 or 7, oblong. Fl. Apr-May. Fr. Jul-Nov. Montane evergreen forests, tickets or forest edges at (200-)500-1500 m. Distributed in SW, S, SE, C and E China.

河滩冬青 *Ilex metabaptista*

## 尾叶冬青

**Ilex wilsonii** Loes.

常绿灌木或乔木。小枝灰色，四棱形。叶革质或厚革质，中脉在背面稍隆起，侧脉7或8对，先端急尾尖状渐尖。花序：聚伞状、束状，腋生于第二年生小枝腋处。果小，球形，内果皮革质。花期5-6月，果期8-10月。生海拔400-1900米的山谷林中、灌丛中或山丘上。产中国西南、华南、东南和华中。

Evergreen shrubs or trees. Branchlets gray, quadrangular. Leaves leathery or thick leathery, midvein slightly raised abaxially, lateral veins 7 or 8 pairs, apex abruptly caudate-acuminate. Inflorescences: cymes, fasciculate, axillary on second year's branchlets. Fruits small, globose, endocarps leathery. Fl. May-Jun. Fr. Aug-Oct. Valley forests, among thickets or hills at 400-1900 m. Distributed in SW, S, SE and C China.

矮冬青 *Ilex lohfauensis*

## 福建冬青

**Ilex fukienensis** S. Y. Hu

常绿灌木，全株无毛。小枝圆柱状，纤细。叶革质，基部圆形，上面晦暗，中脉在上面深陷，侧脉9或10对。聚伞花序簇生于次年生枝叶腋内，基部具革质，具三尖头的苞片。果球形；分核2。花期4-6月，果期8-10月。生海拔600-900米的林中或山坡灌丛中。产福建。

Evergreen shrubs, glabrous throughout. Branchlets cylindric, slender. Leaves leathery, base rounded, adaxially opaque, midvein deeply impressed adaxially, lateral veins 9 or 10 pairs. Inflorescences cymes, fasciculate, axillary on second year's branchlets, basal bracts leathery, tricuspidate. Fruits globose; pyrenes 2. Fl. Apr-Jun. Fr. Aug-Oct. Forests or thickets on slopes at 600-900 m. Distributed in Fujian.

## 矮冬青

**Ilex lohfauensis** Merr.

常绿灌木或小乔木。叶纸质或薄革质，长圆形或椭圆形。雄花序由具1-3花的聚伞花序簇生；雌花2-3花簇生于次年生枝叶腋内；宿存柱头厚盘状或头状，4(-5)裂；分核4，广椭圆体形。花期4-8月，果期8-12月。生海拔(100-)200-1000(-1300)米的山坡林中或灌丛中。产中国西南、华南、东南、华中和华东。

Evergreen shrubs or small trees. Leaves papery or thinly leathery, oblong or elliptic. Male inflorescences with 1-3 clustered cymes; pistillate flowers 2-3 clustered axillary on second year's branches; persistent stigmas thickly discoid or capitate, 4(-5)-lobed; pyrenes 4, broadly ellipsoidal. Fl. Apr-Aug. Fr. Aug-Dec. Forests on mountain slopes or thickets at (100-)200-1000(-1300) m. Distributed in SW, S, SE, C and E China.

尾叶冬青 *Ilex wilsonii*

福建冬青 *Ilex fukienensis*

小果冬青 *Ilex micrococca*

## 小果冬青

**Ilex micrococca** Maxim.

落叶乔木。叶膜质或纸质，侧脉5-8对。三或四回复聚伞花序单生；雄花5或6；雌花6-8。果实红色或黄色，球形，直径约3毫米；分核6-8，椭圆体形，具一纵沟，内果皮革质。花期5-6月，果期7-10月。生海拔500-1900米的常绿阔叶林或山地。产中国西南、华南、东南和华中。越南和日本亦有。

Deciduous trees. Leaves membranous or papery, lateral veins 5-8 pairs. Inflorescences: compound cymes, cymules of order 3 or 4, solitary; staminate flowers 5 or 6; pistillate flowers 6-8. Fruits red or yellow, globose, ca. 3 mm diam; pyrenes 6-8, ellipsoidal, longitudinally 1-sulcate, endocarps leathery. Fl. May-Jun. Fr. Jul-Oct. Evergreen broad-leaved forests or mountains at 500-1900 m. Distributed in SW, S, SE and C China. Also in Vietnam and Japan.

## 多脉冬青

**Ilex polyneura** (Hand.-Mazz.) S. Y. Hu

落叶乔木。叶仅见于当年生枝上，长圆状椭圆形或卵状椭圆形，侧脉10-20对。假伞形花序单生于当年生枝条的叶腋内，二级轴通常不发育，若发育，较花梗短。果球形，宿存柱头盘状、凸起。花期5-6月，果期10-12月。生海拔1000-2600米的山地林中或灌丛中。产贵州、四川、西藏和云南。

Deciduous trees. Leaves only visible in annual shoots, oblong-elliptic or ovate-elliptic, lateral veins 10-20 pairs. Inflorescences umbelliform cymes, solitary, axillary on current year's branchlets, secondary axis usually absent or shorter than pedicels. Fruits globose, persistent stigmas disciform, convex. Fl. May-Jun. Fr. Oct-Dec. Forests or among thickets of mountains at 1000-2600 m. Distributed in Guizhou, Sichuan, Xizang and Yunnan.

## 落霜红

**Ilex serrata** Thunb.

落叶灌木。叶椭圆形，膜质，边缘密具锐锯齿，侧脉6-8对，背面凸起。雄花序为二或三歧聚伞花序，雌花序为具1-3花的聚伞花序。果球形；无花柱，柱头碟形；分核4-8；内果皮革质。花期5月，果期10月。生海拔500-1600米的山坡林缘或灌丛中。产四川、福建、浙江、湖南和江西。日本亦有。

Deciduous shrubs. Leaves elliptic, membranous, margin densely sharply serrate, lateral veins 6-8 pairs, raised abaxially. Male cymes dichotomous or trichotomous; female cymes with 1-3 flowers. Fruits globose; styles absent, stigma discoid; pyrenes 4-8, endocarps leathery. Fl. May. Fr. Oct. Forest edges of mountain slopes or shrubs at 500-1600 m. Distributed in Sichuan, Fujian, Zhejiang, Hunan and Jiangxi. Also in Japan.

多脉冬青 *Ilex polyneura*

落霜红 *Ilex serrata*

大果冬青 *Ilex macrocarpa*

沙坝冬青 *Ilex chapaensis*

## 大果冬青
**Ilex macrocarpa** Oliv.

落叶乔木。小枝、叶和花序无毛。叶片卵形或卵状椭圆形，稀长椭圆形，纸质。果梗和叶柄一样长或稍长。核果直径10-14毫米；分核7-9；退化子房垫状，先端稍凹；果梗较短，长12-15毫米。花期4-5月，果期10-11月。生海拔400-4500米的山谷或山坡林中。产中国西南、华南、华中和华东。

Deciduous trees. Branchlets, leaves and inflorescences glabrous. Leaves ovate or ovate-elliptic, rarely oblong-elliptic, papery. Fruiting pedicels as long as or slightly longer than petioles. Drupes 10-14 mm diam; pyrenes 7-9; rudimentary ovary pulvinate, apex slightly retuse; fruiting pedicels short, 12-15 mm long. Fl. Apr-May. Fr. Oct-Nov. Valleys or forests on slopes at 400-4500 m. Distributed in SW, S, C and E China.

## 沙坝冬青
**Ilex chapaensis** Merr.

落叶乔木。叶在长枝上互生，短枝上簇生枝顶端。花白色，雄花序具1-3分枝，每枝具1-5花，假簇生。果球形，成熟后绿色，干后黑色，直径12-20毫米；分核6或7；退化子房圆锥形，具喙，先端浅裂。花期4月，果期10-11月。生海拔500-3000米的山地疏林或杂木林中。产云南、贵州、广西、广东、海南和福建。越南亦有。

Deciduous trees. Leaves alternate at long branches and cluster at top at short branches. Flowers white, male inflorescences cymes, of order 1-3, 1-5-flowered, pseudo-fasciculate. Fruits globose, green when mature, black when dry, 12-20 mm diam; pyrenes 6 or 7; rudimentary ovary conical, rostellate, apex shallowly lobed. Fl. Apr. Fr. Oct-Nov. Sparse forests or mixed forests on mountains at 500-3000 m. Distributed in Yunnan, Guizhou, Guangxi, Guangdong, Hainan and Fujian. Also in Vietnam.

## 秤星树
**Ilex asprella** (Hook. et Arn.) Champ. ex Benth.

落叶灌木。叶卵形或卵状椭圆形，上面具柔毛，先端尾尖状渐尖。雄花序中2-3花呈束状或单生于叶腋或鳞片腋内；雌花序单生于叶腋或鳞片腋内；宿存柱头头状；分核4-6。花期3月，果期4-10月。生海拔400-1000米的山地疏林或路旁灌丛中。产中国西南、华南、东南和华中。菲律宾亦有。

秤星树 *Ilex asprella*

Deciduous shrubs. Leaves ovate or ovate-elliptic, adaxially puberulent, apex caudate-acuminate. 2-3 flowers of male inflorescences bunched or solitary in leaf axils or scale axils; female inflorescences solitary in leaf axils or scale axils; persistent stigma capitates; pyrenes 4-6. Fl. Mar. Fr. Apr-Oct. Mountain sparse forests or among thickets roadsides at 400-1000 m. Distributed in SW, S, SE and C China. Also in the Philippines.

紫果冬青 *Ilex tsoi*

## 满树星

**Ilex aculeolata** Nakai

落叶灌木。叶膜质或薄纸质，边缘具细齿。聚伞花序单生于叶腋或鳞片腋内。果实黑色，球形，干时具纵棱及沟；分核4，轮廓椭圆体形，背面具深皱纹、网状条纹和沟，内果皮骨质。花期4-5月，果期6-9月。生海拔100-1200米的疏林、沟谷灌丛或路边。产中国西南、华南、东南和华中。

Deciduous shrubs. Leaves membranous or thinly papery, margin serrate. Inflorescences cymes, solitary, axillary on scales or leaves. Fruits black, globose, longitudinally ridges and sulcate when dry; pyrenes 4, ellipsoidal, abaxially deeply rugose, reticulately striate and sulcate, endocarps bony. Fl. Apr-May. Fr. Jun-Sep. Sparse forests, shrub forests in valleys or roadsides at 100-1200 m. Distributed in SW, S, SE and C China.

## 紫果冬青

**Ilex tsoi** Merr. et Chun

落叶灌木或小乔木，有显著皮孔。叶纸质，托叶宿存。雄花花萼直径约4毫米，6裂，全缘；退化子房垫状，中央平。果球形，直径6-8毫米，紫黑色，果梗长1-3毫米；分核6。花期5-6月，果期6-8月。生海拔300-2000米的林中、山谷灌丛中或路边。产中国西南、华南、东南、华中和华东。

Deciduous shrubs or small trees, obviously with lenticels. Leaves papery, stipules persistent. Calyx of staminate flowers ca. 4 mm diam, 6-lobed, margin entire; rudimentary ovary pulvinate, flat at center. Fruits globose, 6-8 mm diam, purple-black, fruiting pedicel 1-3 mm long; pyrenes 6. Fl. May-Jun. Fr. Jun-Aug. Forests, thickets in valleys or roadsides at 300-2000 m. Distributed in SW, S, SE, C and E China.

满树星 *Ilex aculeolata*

# 卫矛科 Celastraceae

## 扶芳藤

**Euonymus fortunei** (Turcz.) Hand.-Mazz.

常绿藤本状灌木。枝条圆柱形。叶密集排列于枝上，叶为多变的卵形或卵状椭圆形，边缘具圆齿或细锯齿，薄革质。聚伞花序3-4次分枝。蒴果棕色或红棕色。假种皮红色。花期4-7月，果期9-12月。生海拔3400米以下的林中或灌丛中。产中国大部分地区。南亚、东南亚和东北亚亦有。在非洲、欧洲、美洲和大洋洲亦有栽培。

Evergreen vine-like shrubs. Branches and twigs rounded. Leaves densely arranged on branches, leaves variously ovate or ovate-elliptic, margin crenulate or serrate, thin leathery. Cymes 3-4-branched. Capsules brown or red-brown. Arils red. Fl. Apr-Jul. Fr. Sep-Dec. Forests or scrubs below 3400 m. Distributed in most parts of China. Also in S, SE and NE Asia. Cultivated in Africa, Europe, America and Oceania.

## 南川卫矛

**Euonymus bockii** Loes.

常绿灌木或斜升亚灌木。叶椭圆形或卵状椭圆形，8-16 × 4-8厘米，侧脉6-9对。花梗长3-4厘米，常具数花至有时多于5花；花4数；花瓣绿色。蒴果棕色或棕绿色，密被白点，有时呈白鳞状。假种皮红色。花期4-6月，果期8-12月。生海拔1000-2300米的杂木林。产广西、贵州、四川、云南和重庆。印度和越南亦有。

Evergreen shrubs or ascending subshrubs. Leaves elliptic or ovate-elliptic, 8-16 × 4-8 cm, lateral veins 6-9 pairs. Peduncles 3-4 cm long, usually several-flowered, sometimes more than 5-flowered; flowers 4-merous; petals greenish. Capsules brown or green-brown, densely white spotted, sometimes white scalelike. Arils red. Fl. Apr-Jun. Fr. Aug-Dec. Mixed forests at 1000-2300 m. Distributed in Guangxi, Guizhou, Sichuan, Yunnan and Chongqing. Also in India and Vietnam.

南川卫矛 *Euonymus bockii*

## 曲脉卫矛

**Euonymus venosus** Hemsl.

落叶灌木至小乔木。叶薄革质或革质，侧脉和小脉弯曲，或呈弯扭状。花序梗长2-4厘米，少花；花瓣浅黄色。蒴果球形或近球形，稍具沟，棕粉色至黄棕色，4裂。种子棕黄色；假种皮橘红色。花期5-7月，果期8-9月。生海拔700-2500米的林中、岩石山坡或灌丛中。产四川、湖北、湖南、河南和陕西。

Deciduous shrubs to small trees. Leaves thinly leathery or leathery,

扶芳藤 *Euonymus fortunei*

曲脉卫矛 *Euonymus venosus*

刺果卫矛 *Euonymus acanthocarpus*

lateral veins and veinlets curved or bent, or in tortuous form. Peduncles 2-4 cm long, few-flowered; petals light yellow. Capsules globose or subglobose, slightly grooved, pink-brown to yellow-brown, 4-lobed. Seeds yellow-brown; arils orange-red. Fl. May-Jul. Fr. Aug-Sep. Forests, rock slopes or scrubs at 700-2500 m. Distributed in Sichuan, Hubei, Hunan, Henan and Shaanxi.

## 冬青卫矛

**Euonymus japonicas** Thunb.

常绿直立灌木或小乔木。高至3米，有时矮生。叶革质，卵形、倒卵形、圆卵形或长卵形，有光泽。聚伞花序常腋生，有时顶生，多分枝，具5-13花。蒴果近球形，淡红色。假种皮橘红色。花期6-8月，果期8月至翌年1月。栽培1400米以下处。中国大部分地区有栽培。原产日本，世界各地有栽培。

Evergreen shrubs or small trees, erect. To 3 m tall, sometimes dwarfed. Leaves leathery, ovate, obovate, orbicular-ovate or long ovate, lustrous. Cymes usually axillary, sometimes terminal, many branched with flowers 5-13. Capsules almost orbicular, reddish. Arils orange-red. Fl. Jun-Aug. Fr. Aug to next Jan. Cultivated below 1400 m. Cultivated in most parts of China. Native to Japan, cultivated worldwide.

## 刺果卫矛

**Euonymus acanthocarpus** Franch.

落叶灌木，直立或斜升。叶革质，长圆形、长圆状椭圆形或椭圆形。花序梗常多于三回二叉分枝，多花；花瓣黄绿色。蒴果棕红色，近球形，4裂，密被针刺。假种皮橘黄色。花期5-8月，果期8-11月。生海拔700-2000米的林地。产中国西南、东南、华中和华东。缅甸亦有。

Deciduous shrubs, erect or ascending. Leaves leathery, oblong, oblong-elliptic, or elliptic. Peduncles typically more than 3 × dichotomously branched, many-flowered; petals yellow-green. Capsules brown-red, nearly globose, 4-lobed, densely prickly. Arils orange. Fl. May-Aug. Fr. Aug-Nov. Forests at 700-2000 m. Distributed in SW, SE, C and E China. Also in Myanmar.

冬青卫矛 *Euonymus japonicas*

长刺卫矛 *Euonymus wilsonii*

肉花卫矛 *Euonymus carnosus*

## 长刺卫矛
**Euonymus wilsonii** Sprague

常绿灌木。叶椭圆形、卵状椭圆形或长椭圆形，长10-15厘米，边缘上部2/3具疏锯齿；侧脉6-8对，在正面下凹，在叶背凸起；叶柄长1-1.4厘米。聚伞花序2-3次二歧分枝；花4数；萼片近圆形；花瓣淡绿色，卵形；花丝三角锥状。蒴果近球形，直径1.5-2厘米，密被刺，刺长5-8毫米。种子是橘红色假种皮。花期4-5月，果期7-9月。生海拔1000-2600米的林中和灌丛中。产华中和华南。

Evergreen shrubs. Leaves elliptic, ovate-elliptic, or long elliptic, 10-15 cm long, margin coarsely serrate on distal 2/3; lateral veins 6-8 pairs, impressed adaxially, prominent abaxially; Petioles 1-1.4 cm long. Cymes 2-3 × dichotomously branched; flowers 4-merous; sepals suborbicular; petals greenish, ovate; filaments triangular-coniform. Capsules nearly globose, 1.5-2 cm diam. densely prickly, prickles 5-8 mm long. Seeds with orange-red aril. Fl. Apr-May. Fr. Jul-Sep. Forests or scrub at 1000-2600 m. Distributed in C and S China.

棘刺卫矛 *Euonymus echinatus*

## 棘刺卫矛
**Euonymus echinatus** Wall.

常绿或半常绿攀援灌木。枝和嫩枝细长，有槽，具角。叶厚革质，卵形，边缘有疏锯齿。花梗长2-3厘米，生1-3花；花4基数。蒴果密被细刺，刺长1-2毫米。假种皮亮橙色。花期4-7月，果期9月到翌年1月。生海拔1300-3500米的林中或灌丛中。产中国西南、华南、东南、华中和华东。南亚和日本南部亦有。

Evergreen or semievergreen shrubs, scandent. Branches and twigs slender, striate, angulate. Leaves thickly leathery, ovate, margin sparsely serrate. Peduncles 2-3 cm long, 1-3-flowered; flowers 4-merous. Capsules densely prickly, prickles 1-2 mm long. Arils bright orange. Fl. Apr-Jul. Fr. Sep to next Jan. Forests or scrubs at 1300-3500 m. Distributed in SW, S, SE, C and E China. Also in S Asia and S Japan.

## 肉花卫矛
**Euonymus carnosus** Hemsl.

落叶灌木至小乔木。叶近革质，基部楔形或渐狭，叶缘有细圆齿，先端钝尖或具突尖。聚伞花序疏松；花瓣黄色或绿褐色。蒴果近球状，常具有窄翅棱。花期5-8月，果期8-11月。生海拔2000米以下的林地。产华南、东南、华中和华东。日本亦有。

Deciduous shrubs to small trees. Leaves almost leathery, base

大花卫矛 *Euonymus grandiflorus*

cuneate or attenuate, margin crenulate, apex obtuse or mucronulate. Cymes sparse; petals yellow or brown-green. Capsules almost orbicular, with narrow winged ribs. Fl. May-Aug. Fr. Aug-Nov. Forests below 2000 m. Distributed in S, SE, C and E China. Also in Japan.

### 大花卫矛
**Euonymus grandiflorus** Wall.

落叶灌木至小乔木。叶窄长方形或窄倒卵形，先端钝或急尖。花序梗单生或簇生，1-3个二歧分枝，每枝几朵至多朵黄白色花；花4数，直径17-22毫米。蒴果具4棱，棕色、黄棕色或红棕色。花期3-5月，果期8-11月。生海拔1400-3300米的林中。产中国西南、华中和华西。印度、尼泊尔、不丹、缅甸和越南亦有。

Deciduous shrubs to small trees. Leaves narrowly rectangular or narrowly obovate, apex obtuse or acute. Peduncles single or clustered, 1-3 × dichotomously branched with several to many yellow-white flowers. Flowers 4-merous, 17-22 mm diam. Capsules with 4 ridges, brown, yellow-brown or red-brown. Fl. Mar-May. Fr. Aug-Nov. Forests at 1400-3300 m. Distributed in SW, C and W China. Also in India, Nepal, Bhutan, Myanmar and Vietnam.

### 云南卫矛
**Euonymus yunnanensis** Franch.

常绿灌木。叶革质，多变的条形至椭圆形或倒卵状椭圆形，边缘具疏离圆齿且反卷。聚伞花序具1-3花；花瓣淡绿色。蒴果圆柱状菱形或倒锥状。假种皮橘红色。花期3-4月，果期5-7月。生海拔1700-2400米的林中。产云南、四川和西藏。

Evergreen shrubs. Leaves leathery, variously linear to elliptic or obovate-elliptic, margin remotely crenate and revolute. Cymes 1-3-flowered; petals light green. Capsules cylindric-rhombic or obconic. Arils orange-red. Fl. Mar-Apr. Fr. May-Jul. Forests at 1700-2400 m. Distributed in Yunnan, Sichuan and Xizang.

### 大果卫矛
**Euonymus myrianthus** Hemsl.

常绿灌木至小乔木。叶厚革质。花序梗长2-3.5厘米，一或二回二叉分枝，具数花；花瓣淡黄绿色。蒴果四棱状球形，常具4棱，鲜时粉色或红色，长1.5-1.8 厘米。种子卵形，深棕色；假种皮橙色。花期4-7月，果期8-11月。生海拔1200米以下的林中。产中国西南、华南、华中和东南。

Evergreen shrubs to small trees. Leaves thickly leathery. Peduncles 2-3.5 cm long, 1 or 2 × dichotomously branched, several-flowered; petals greenish yellow. Capsules tetra-globose, usually with 4 ridges, pinkish or reddish when fresh, 1.5-1.8 cm long. Seeds ovoid, dark brown; arils orange. Fl. Apr-Jul. Fr. Aug-Nov. Forests below 1200 m. Distributed in SW, S, C and SE China.

云南卫矛 *Euonymus yunnanensis*

大果卫矛 *Euonymus myrianthus*

白杜 *Euonymus maackii*

## 白杜
**Euonymus maackii** Rupr.

小乔木。叶卵状椭圆形、卵形或窄椭圆形，长4-8厘米，宽2-5厘米。聚伞花序1-2次分枝，有3-7花；花瓣白色。蒴果粉红色，倒圆锥形。假种皮橙色。花期4-7月，果期8-11月。生海拔1000米以下的林缘或林中。产华南以外中国大部分地区。俄罗斯、朝鲜半岛和日本亦有，欧洲和北美洲亦有栽培。

Small trees. Leaves ovate-elliptic, ovate or narrowly elliptic, 4-8 cm long, 2-5 cm wide. Cymes 1-2-branched, 3-7-flowered; petals white. Capsules pink, obconiform. Arils orange. Fl. Apr-Jul. Fr. Aug-Nov. Forest edges or forests below 1000 m. Distributed in most parts of China, except S China. Also in Russia, Korean Peninsula and Japan, cultivated in Europe and North America.

## 西南卫矛
**Euonymus hamiltonianus** Wall.

落叶灌木至小乔木。叶薄革质或厚纸质，边缘具细圆齿，表面粗糙。花序梗长3-4.5厘米，一至三回二叉分枝，具数花；花瓣白色。蒴果菱形，具4棱和深沟。种子椭圆形，深棕色；假种皮橘红色。花期4-7月，果期8-11月。生海拔3000米以下的林地。产中国西南、东南、华中、华西和华东。南亚和东北亚亦有。

Deciduous shrubs to small trees. Leaves thinly leathery or thickly papery, margin finely crenulate, surface rough. Peduncles 3-4.5 cm long, 1-3 × dichotomously branched, several-flowered; petals white. Capsules rhombic, with 4 ridges and deep grooves. Seeds ellipsoid, dark brown; arils orange-red. Fl. Apr-Jul. Fr. Aug-Nov. Forests below 3000 m. Distributed in SW, SE, C, W and E China. Also in S and NE Asia.

## 栓翅卫矛
**Euonymus phellomanus** Loes.

落叶灌木。枝条圆柱形，有木栓翅或细槽。叶厚纸质，卵状椭圆形或长圆状椭圆形。聚伞花序二至三回分枝，具7-15花；花4数；花瓣淡绿白色。蒴果具4棱；假种皮橙色。花期5-7月，果期8-11月。生海拔1000-3000米的林中或干燥山坡。产华中和

西南卫矛 *Euonymus hamiltonianus*

栓翅卫矛 *Euonymus phellomanus*

华西。

Deciduous shrubs. Branches terete, corky winged or striate. Leaves thickly papery, ovate-elliptic or oblong-elliptic. Cymes dichotomous or tricotomous, with flowers 7-15; flowers 4-merous; petals greenish white. Capsules tetragonal; arils orange. Fl. May-Jul. Fr. Aug-Nov. Forests or dry slopes at 1000-3000 m. Distributed in C and W China.

## 小果卫矛

**Euonymus microcarpus** (Oliv. ex Loes.) Sprague

小乔木或灌木。叶薄革质，卵形至卵状椭圆形或卵状条形。花序聚伞状；花瓣亮绿色。蒴果菱形，具4棱。种子卵球形，深棕色，外被橘黄色假种皮。花期4-5月，果期7-10月。生海拔300-2600米的林中或灌丛中。产四川、西藏、湖北、河南和陕西。

Small trees or shrubs. Leaves thinly leathery, ovate to ovate-elliptic or ovate-linear. Inflorescences cymose; petals light green. Capsules rhombic, 4 ridges. Seeds ovoid, dark brown, partially covered by orange arils. Fl. Apr-May. Fr. Jul-Oct. Forests or thickets at 300-2600 m. Distributed in Sichuan, Xizang, Hubei, Henan and Shaanxi.

## 瘤枝卫矛

**Euonymus verrucosus** Scop.

灌木。枝条暗绿色或暗棕色，圆柱形，密被疣状凸起。叶厚纸质，卵状椭圆形或长圆状椭圆形。聚伞花序常1-3花；花瓣淡粉红色至深红色。蒴果倒三角状。假种皮橘黄色。花期5-7月，果期8-11月。生海拔200-1300米的山地林中。产中国西北和东北。中亚和东北亚亦有。

Shrubs. Branches and twigs gray-green to gray-brown, terete, densely verrucate. Leaves thickly papery, ovate-elliptic or oblong-elliptic. Cymes always 1-3-flowered; petals pinkish to dark red. Capsules obo-triangular. Arils orange. Fl. May-Jul. Fr. Aug-Nov. Montane forests at 200-1300 m. Distributed in NW and NE China. Also in C and NE Asia.

小果卫矛 *Euonymus microcarpus*

瘤枝卫矛 *Euonymus verrucosus*

## 中华卫矛
**Euonymus nitidus** Benth.

常绿灌木至小乔木。叶革质至厚纸质，椭圆形或长圆状椭圆形，侧脉7-9对。花梗多数，一至三(四)回二叉分枝，具多花；花4数；花瓣白绿色。蒴果三角卵圆形，具4棱，成熟时粉红色或红色。假种皮橘黄色。花期3-7月，果期7月至翌年1月。生海拔100-1500米的低山林地或山谷。产中国西南、华南、华中和东南。孟加拉国、越南、柬埔寨和日本亦有。

Evergreen shrubs to small trees. Leaves leathery to thickly papery, elliptic or oblong-elliptic, with 7-9 pairs lateral veins. Peduncles typically many, 1-3(-4) × dichotomously branched with many flowers; flowers 4-merous; petals whitish green. Capsules tetraglobose, 4 ridges, pinkish or reddish when fresh. Arils orange. Fl. Mar-Jul. Fr. Jul to next Jan. Forests in lower mountains or valleys at 100-1500 m. Distributed in SW, S, C and SE China. Also in Bangladesh, Vietnam, Cambodia and Japan.

## 疏花卫矛
**Euonymus laxiflorus** Champ. ex Benth.

灌木至小乔木。叶薄革质，椭圆状倒卵形或卵形。聚伞花序一至三回二歧分枝，具5-9花；花5数；花瓣紫色。蒴果倒卵球形，紫红色。假种皮橙黄色。花期3-8月，果期5-11月。生海拔300-2200米的林中或灌丛中。产中国西南、华南、华中和华东。印度、缅甸、柬埔寨和越南亦有。

Shrubs to small trees. Leaves thinly leathery, elliptic-obovate or ovate. Cymes 1-3 × dichotomously branched, 5-9-flowered; flowers 5-merous; petals purple. Capsules obovoid, purplish red. Arils orange. Fl. Mar-Aug. Fr. May-Nov. Forests or thickets at 300-2200 m. Distributed in SW, S, C and E China. Also in India, Myanmar, Cambodia and Vietnam.

疏花卫矛 *Euonymus laxiflorus*

## 卫矛
**Euonymus alatus** (Thunb.) Siebold

落叶灌木。小枝常具2-4列宽阔木栓翅。叶薄革质至纸质，倒卵形或倒卵状椭圆形。聚伞花序具有1-3花；花4数；花瓣绿色、亮黄色或淡黄绿色。蒴果1-4深裂。假种皮鲜红色。花期4-7月，果期7-11月。生海拔2700米以下的山坡或沟谷。产中国大部分地区。俄罗斯、朝鲜半岛和日本亦有，欧洲和北美洲有栽培。

Deciduous shrubs. Young branches usually with 2-4 broad winglike

中华卫矛 *Euonymus nitidus*

卫矛 *Euonymus alatus*

短翅卫矛 *Euonymus rehderianus*

corks. Leaves thinly leathery to papery, obovate or obovate-elliptic. Cymes with 1-3 flowers; flowers 4-merous; petals green, light yellow, or greenish yellow. Capsules 1-4-lobed. Arils bright red. Fl. Apr-Jul. Fr. Jul-Nov. Slopes or valleys below 2700 m. Distributed in most parts of China. Also in Russia, Korean Peninsula and Japan, cultivated in Europe and North America.

## 短翅卫矛

**Euonymus rehderianus** Loes.

落叶灌木至小乔木。叶厚革质，椭圆状卵形或倒卵状椭圆形，全缘。新枝上的花梗细长，2到多次叉状分枝，多花；花5基数；萼片三角形，极小。蒴果鲜时粉色或红色，近扁球形，具5个短而扁的翅。假种皮橘黄色。花期4-5月，果期7-10月。生海拔400-1600米的林中或灌丛中。产云南、四川、贵州和广西。

Deciduous shrubs to small trees. Leaves thickly leathery, elliptic-ovate or obovate-elliptic, margin entire. Peduncles from new branches usually slender, ca. dichotomously branched 2 or moretimes, several-flowered; flowers 5-merous; sepals deltoid, very small. Capsules pink or red when fresh, nearly compressed globose, with 5 short and flat wings. Arils orange. Fl. Apr-May. Fr. Jul-Oct. Forests or scrubs at 400-1600 m. Distributed in Yunnan, Sichuan, Guizhou and Guangxi.

## 垂丝卫矛

**Euonymus oxyphyllus** Miq.

落叶灌木至小乔木。叶厚纸质或薄革质，卵状椭圆形或卵状披针形，基部楔形或渐狭，边缘有细密锯齿，先端渐尖或急尖。花序梗细弱；聚伞花序宽疏，通常7-20花。蒴果近球状，无翅。假种皮亮红色。花期4-6月，果期8-11月。生海拔2300米以下的林中。产华南、华中、华东和东北。朝鲜半岛和日本亦有。

Deciduous shrubs to small trees. Leaves thickly papery or thinly leathery, ovate-elliptic or ovate-lanceolate, base cuneate or attenuate, margin finely serrulate, apex acuminate or acute. Peduncles slender; cymes broad and sparse, usually with flowers 7-20. Capsules almost orbicular, wingless. Arils bright red. Fl. Apr-Jun. Fr. Aug-Nov. Forests below 2300 m. Distributed in S, C, E and NE China. Also in Korean Peninsula and Japan.

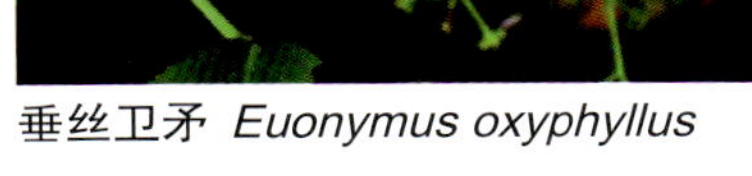
垂丝卫矛 *Euonymus oxyphyllus*

## 岩坡卫矛

**Euonymus clivicolus** W. W. Sm.

常绿灌木。老枝有时具4棱窄栓翅。叶纸质或近膜质，披针形或阔披针形，先端窄缩渐尖；叶柄长2-5毫米。聚伞花序通常3花；花序梗细长；花5数，直径10-12毫米；花瓣半圆形；花盘圆形，边缘5浅裂，雄蕊着生裂片处；子房扁平，柱头圆扁，无花柱。蒴果直径8-10毫米，翅长5-10毫米。生海拔3000米高寒地带的山坡杂木林中。产云南西北部和西藏东南部。

Evergreen shrubs. Branches sometimes with 4-ridged narrow wing bolt. Leaves papery or submembranous, lanceolate or broadly lanceolate; apex narrowly acuminate; Petioles 2-5 mm long. Cymose, 3-flowered; peduncles long; flowers 5-amerous, 10-12 mm diam; petals rounded; disk rounded, margin 5-lobed; stamens at lobes; ovary flat, stigma round flat, without style. Capsules 8-10 mm diam, wings 5-10 mm long. Cold woody zones at 3000 m. Distributed in NW Yunnan and SE Xizang.

岩坡卫矛 *Euonymus clivicolus*

## 冷地卫矛

**Euonymus frigidus** Wall.

落叶灌木至小乔木。叶卵形或阔椭圆形，在中部最宽，基部近圆形或楔形，边缘有细锯齿，先端渐尖或急尖。花深紫色；花序梗细弱，一或二回二歧分枝，具数花。蒴果有4个短而钝的翅。假种皮亮红色。花期5-7月，果期8-11月。生海拔500-4000米的杂木林或灌丛中。产中国西南、华北和华西。印度、尼泊尔、不丹和缅甸亦有。

Deciduous shrubs to small trees. Leaves ovate or broadly elliptic, widest at middle, base subround or cuneate, margin finely crenulate, apex acuminate or acute. Flowers deep purple; peduncles slender, 1 or 2 × dichotomously branched, several-flowered. Capsules with 4 short and obtuse wings. Arils bright red. Fl. May-Jul. Fr. Aug-Nov. Mixed forests or thickets at 500-4000 m. Distributed in SW, N and W China. Also in India, Nepal, Bhutan and Myanmar.

## 角翅卫矛

**Euonymus cornutus** Hemsl.

落叶灌木。叶薄革质或厚纸质，条形、长圆形或竹叶状。聚伞花序常一次分枝，具3花，稀2次分枝，具5-7花。蒴果具4或5翅，近球状。假种皮亮红色。花期4-7月，果期8-11月。生海拔

冷地卫矛 *Euonymus frigidus*

角翅卫矛 *Euonymus cornutus*

2200-4300米的山坡杂木林或灌丛中。产中国西南、华中和华西。印度和缅甸亦有。

Deciduous shrubs. Leaves thinly leathery or thickly papery, linear, oblong, or bamboolike. Cymes usually 1-branched with 3 flowers, rare 2-branched with 5-7 flowers. Capsules with 4 or 5 wings, almost orbicular. Arils bright red. Fl. Apr-Jul. Fr. Aug-Nov. Mixed forests or thickets at 2200-4300 m. Distributed in SW, C and W China. Also in India and Myanmar.

纤齿卫矛 *Euonymus giraldii*

## 纤齿卫矛
**Euonymus giraldii** Loes.

落叶灌木或小乔木。冬芽细长，先端尖。叶椭圆状披针形至卵状椭圆形，长6-8厘米，中部最宽，边缘密具细锯齿；叶柄长3-5毫米。聚伞花序1-2次二歧分枝；花4数；萼片近圆形；花瓣淡绿色，常稍带紫色，卵形。蒴果直径小于1厘米，具4个短且尖的翅。花期4-6月，果期7-10月。生海拔1000-3700米的混交林中。产华中、华北、安徽、青海和云南。

Deciduous shrubs or small trees. Winter buds slender, apex pointed. Leaves elliptic-lanceolate to ovate-elliptic, 6-8 cm long, widest at middle parts, margin finely crenulate; Petioles 3-5 mm long. Cymes 1-2 × dichotomously branched; flowers 4-merous; sepals suborbicular; petals greenish, usually tinged with purple, ovate. Capsules less than 1 cm diam, with 4 short and sharp wings. Fl. Apr-Jun. Fr. Jul-Oct. Mixed forests at 1000-3700 m. Distributed in C and N China, Anhui, Qinghai and Yunnan.

## 陕西卫矛
**Euonymus schensianus** Maxim.

落叶灌木。叶卵状椭圆形、倒卵形或倒卵状椭圆形，长5-8(-10)厘米，先端渐尖或急尖，边缘具细齿；叶柄长3-6毫米。聚伞花序1-2次二歧分枝；花4数；萼片近圆形；花瓣黄绿色，常稍带红色，倒卵形，先端急尖。蒴果直径约1厘米，4个宽且钝的翅。花期4-5月，果期9-11月。生海拔600-1000米的沟边丛林中。产甘肃、宁夏、四川、河南、湖北和陕西。

Deciduous shrubs. Leaves ovate-elliptic, obovate, or obovate-elliptic, 5-8(-10) cm long, apex acuminate or acute, margin finely crenulate; Petioles 3-6 mm long. Cymes 1-2 × dichotomously branched; flowers 4-merous; sepals suborbicular; petals yellow-green, usually slightly tinged with red, obovate, apex acute. Capsules ca. 1 cm diam, with 4 wide and obtuse wings. Fl. Apr-May. Fr. Sep-Nov. Scrubs along ditches at 600-1000 m. Distributed in Gansu, Ningxia, Sichuan, Henan, Hubei and Shaanxi.

陕西卫矛 *Euonymus schensianus*

黄心卫矛 *Euonymus macropterus*

## 黄心卫矛
**Euonymus macropterus** Rupr.

落叶灌木。枝条粗壮有细槽。冬芽长卵球形。小叶纸质，卵状椭圆形、卵形或倒卵形。聚伞花序具3-13花。蒴果绿色至棕色，或干后黄棕色，近球形，具4个长而尖的翅。花期5-7月，果期8-10月。生海拔300-2100米的杂木林或灌丛中。产中国东北。俄罗斯、朝鲜半岛和日本亦有。

Deciduous shrubs. Branches and twigs moderately sturdy, striate. Winter buds long ovoid. Leaves papery, ovate-elliptic, ovate or obovate. Cymes with flowers 3-13. Capsules green to brown, or yellow-brown when dry, nearly globose, with 4 long and sharp wings. Fl. May-Jul. Fr. Aug-Oct. Mixed forests or thickets at 300-2100 m. Distributed in NE China. Also in Russia, Korean Peninsula and Japan.

## 皱叶沟瓣
**Glyptopetalum rhytidophyllum** (Chun et How) C. Y. Cheng.

常绿灌木。幼枝4棱，压扁。叶薄革质，窄长方形或长方阔披针形，边缘具疏短细齿；侧脉8-18对，在叶面下凹较深，在叶背显著突起。聚伞花序1-2次分枝；花序梗长2-4厘米；花淡绿色，4数。蒴果圆球状，灰白色或淡棕色。种子具橙红色假种皮。花期3-4月，果期10-12月。生海拔600-900米的山地密林下或林边。产广西西部和云南东南部。

Evergreen shrubs. Branchlets tetragonous, compressed. Leaves thinly coriaceous, narrowly oblong or oblong-lanceolate, margin sparsely and minutely serrate; lateral veins 8-18 pairs, adaxially depressed, abaxially elevated. Cymes 1-2 times branched; peduncles 2-4 cm long; flowers greenish, 4-merous. Capsules globose, grayish or pale brown. Seeds covered with orange-red aril. Fl. Mar-Apr. Fr. Oct-Dec. Montane thickets or forest edges at 600-900 m. Distributed in W Guangxi and SE Yunnan.

## 永瓣藤
**Monimopetalum chinense** Rehd.

藤状灌木。叶互生，纸质，卵形或窄卵形，先端渐尖，基部圆形，边缘具浅细锯齿；托叶细丝状，宿存。聚伞花序2-3次分枝；花序梗细弱丝状；花小，淡绿色，4数。蒴果4深裂，下有4片增大宿存的花被；花被匙形或长倒卵形，长10-12毫米。花期5-10月，果期6-11月。生山坡、路边或山谷杂林中。产安徽(祁门县)、江西(景德镇)和湖北(通山县)。

Climbing shrubs. Leaves alternate, chartaceous, ovate or narrowly ovate, apex acuminate, base round, margin shallowly and minutely serrate; stipules filamentous, persistent. Cymes 2-3-times branched; peduncles tenuous, filamentous; flowers small, greenish, 4-merous. Capsules 4-parted, base with 4 persistent and swollen petals; petals spatulate or narrowly obovate, 10-12 mm long. Fl. May-Oct. Fr. Jun-Nov. Slopes, roadsides or mixed forests in valleys. Distributed in Anhui (Qimen County), Jiangxi (Jingdezhen) and Hubei (Tongshan County).

皱叶沟瓣 *Glyptopetalum rhytidophyllum*

永瓣藤 *Monimopetalum chinense*

灯油藤 *Celastrus paniculatus*

苦皮藤 *Celastrus angulatus*

## 灯油藤
**Celastrus paniculatus** Willd.

常绿攀援灌木。叶椭圆形、长圆形、矩形、卵形或倒卵形至近圆形，边缘具细锯齿。聚伞圆锥花序顶生，1-2次分枝；花序梗及小花梗偶被短绒毛；花淡绿色；花盘膜质杯状。蒴果压扁，球形，3瓣裂，亮黄色。假种皮橘红色。花期4-6月，果期6-9月。生海拔200-2000米的灌丛或林中。产中国西南和华南。南亚、东南亚、澳大利亚和太平洋岛屿亦有。

Evergreen climbing shrubs. Leaves elliptic, oblong, rectangular, ovate, or obovate to suborbicular, margin serrate. Thyrses terminal, 1- or 2-ramous; rachises and pedicels occasionally with short tomentum; flowers greenish; disc membranous, cup-shaped. Capsules depressed, globose, 3-valved, bright yellow. Arils orange-red. Fl. Apr-Jun. Fr. Jun-Sep. Thickets or forests at 200-2000 m. Distributed in SW and S China. Also in S and SE Asia, Australia and Pacific Islands.

## 苦皮藤
**Celastrus angulatus** Maxim.

落叶攀援灌木。小枝暗棕色，常具4-6纵棱。叶近革质，浅绿色，无毛，基部楔形，边缘有圆锯齿。聚伞圆锥花序顶生；花盘肉质，盘状，5浅裂。蒴果球形，黄色，3瓣裂。假种皮亮红色。花期5-6月，果期8-10月。生海拔1000-2500米的山地丛林和山坡灌丛中。产中国西南、华中、华北、华西和华东。

Deciduous climbing shrubs. Branchlets dark brown, 4-6-angular. Leaves subleathery, light green, glabrous, base cuneate, margin crenate. Paniculate thyrses terminal, apically branched; discs carnose, disciform, 5-lobed. Capsules globose, yellow, 3-valved. Arils bright red. Fl. May-Jun. Fr. Aug-Oct. Thickets or forests on slopes at 1000-2500 m. Distributed in SW, C, N, W and E China.

## 灰叶南蛇藤
**Celastrus glaucophyllus** Rehd. et Wils.

落叶攀援灌木。小枝具稀疏皮孔。叶在果期革质，叶面绿色，叶背灰白色，叶缘疏具锯齿。复总状圆锥花序腋生至顶生，长3-6厘米，腋生花序仅具3-5花；花梗于中部或中部以上具关节。果球形。花期3-6月，果期8-10月。生海拔700-3700米的杂木林中。产中国西南、华中和华西。

Deciduous climbing shrubs. Branchlets with sparse lenticels. Leaves leathery during fruitsing period, adaxially green, abaxially gray-white, margin sparsely finely serrate. Racemose panicles axillary to terminal, 3-6 cm long, axillary inflorescences only 3-5-flowered; pedicels jointed at or above middle. Fruits globose. Fl. Mar-Jun. Fr. Aug-Oct. Mixed forests at 700-3700 m. Distributed in SW, C and W China.

灰叶南蛇藤 *Celastrus glaucophyllus*

长序南蛇藤 *Celastrus vaniotii*

## 长序南蛇藤
**Celastrus vaniotii** (Lévl.) Rehd.

藤状灌木。小枝具稀疏皮孔，腋芽近球形。叶缘具内弯腺齿。顶生单歧聚伞花序长6-18厘米，腋生花序短，长3-4厘米；花梗于中部以下具关节。蒴果球形，果皮内面具棕色斑点。种子椭圆形。花期5-7月，果期8-9月。生海拔500-2200米的杂木林中。产中国西南和华中。

Twining shrubs. Branchlets with sparse lenticels, axillary buds approximately globose. Leaf margin with recurved glandular teeth. Terminal thyrses 6-18 cm long, axillary inflorescences short, 3-4 cm long; pedicels jointed generally below middle. Capsules globose, adaxial side of valves with small brown maculas. Seeds elliptic. Fl. May-Jul. Fr. Aug-Sep. Mixed forests at 500-2200 m. Distributed in SW and C China.

## 大芽南蛇藤
**Celastrus gemmatus** Loes.

藤状灌木。枝具皮孔，冬芽大，长达12毫米。叶缘具浅锯齿。花序顶生及腋生，顶生聚伞花序长3厘米，腋生聚伞花序短，少花；花梗在中部以上具节。蒴果球形。种子宽椭圆形至立方状椭圆形，红棕色，光亮。花期4-9月，果期8-10月。生海拔500-1800米的密林或灌丛中。产中国西南、华南、东南、华中、华北、华西和华东。

Twining shrubs. Branchlets with lenticels, winter buds large, up to 12 mm long. Leaf margin shallowly serrate. Cymes terminal and axillary, terminal inflorescences 3 cm long, axillary ones short and few-flowered; pedicels jointed above middle. Capsules globular. Seeds widely elliptic to rectangular-elliptic, reddish brown, shiny. Fl. Apr-Sep. Fr. Aug-Oct. Dense forests or thickets at 500-1800 m. Distributed in SW, S, SE, C, N, W and E China.

## 短梗南蛇藤
**Celastrus rosthornianus** Loes.

攀援灌木。叶小，纸质，矩状椭圆形至狭矩状椭圆形。花序顶生及腋生，顶生者为总状聚伞花序，腋生者短小，具1至数花，花梗长2-6毫米。蒴果近球形。花期4-5月，果期8-10月。生海拔500-1800米的山坡、林缘、灌丛或路边。产中国西南、东南、华中、华北和华西。

climbing shrubs. Leaves small, papery, rectangular-elliptic to narrowly rectangular-elliptic. Inflorescences terminal and axillary, terminal one cymose-botryose, axillary one short, 1- to several-flowered, pedicels 2-6 mm long. Capsules subglobose. Fl. Apr-May. Fr. Aug-Oct. Slopes, forest edges, thickets or roadsides at 500-1800 m. Distributed in SW, SE, C, N and W China.

## 南蛇藤
**Celastrus orbiculatus** Thunb.

落叶攀援灌木。小枝光滑无毛。叶通常阔倒卵形，近圆形或长方椭圆形。聚伞花序腋生或顶生，小花1-3，偶仅1-2花；花盘厚，肉质。花期5-6月，果期7-10月。生海拔400-2200米的杂木林、林缘或山坡灌丛。产华中、华西、东南、华北和东北。朝鲜半岛和日本亦有。

Deciduous climbing shrubs. Branchlets smooth, glabrous. Leaves generally broadly ovate, suborbicular or rectangular-elliptic. Cymes axillary or terminal, 1-3-flowered, occasionally 1-2; discs thickened, fleshy. Fl. May-Jun. Fr. Jul-Oct. Mixed forests, forest edges or thickets on slopes at 400-2200 m. Distributed in C, W, SE, N and NE China. Also in

大芽南蛇藤 *Celastrus gemmatus*

短梗南蛇藤 *Celastrus rosthornianus*

南蛇藤 *Celastrus orbiculatus*

圆叶南蛇藤 *Celastrus kusanoi*

Korean Peninsula and Japan.

## 圆叶南蛇藤

**Celastrus kusanoi** Hayata

落叶攀援小灌木。小枝常被棕色极短硬毛。叶果期近膜质，阔椭圆形至圆形，先端圆润，具小骤尖。花序腋生和侧生，小，具3-7花。蒴果近球形。花期2-4月。生海拔300-2500米的山地林中。产海南和台湾。

Deciduous climbing shrublets. Branchlets often with very short and brown setae. Leaves approximately membranous during fruiting, broadly elliptic to orbicular, apex widely rounded and mucronate. Inflorescences axillary and lateral, small, 3-7-flowered. Capsules subglobose. Fl. Feb-Apr. Montane forests at 300-2500 m. Distributed in Hainan and Taiwan.

## 刺苞南蛇藤

**Celastrus flagellaris** Rupr.

落叶攀援灌木。小枝光滑，冬芽小，最外一对芽鳞刺状。叶阔椭圆形或阔卵状椭圆形，叶缘具细锯齿至锯齿。聚伞花序腋生，具1-5花或更多。蒴果球形。种子近椭圆形。花期4-5月，果期8-9月。生河畔林中、灌丛、山谷或山地阳坡。产河北、辽宁、吉林和黑龙江。俄罗斯(远东地区)、朝鲜半岛和日本亦有。

Deciduous climbing shrubs. Branchlets glabrous, winter buds small, outer pairs of bud scales thornlike. Leaves broadly elliptic or broadly ovate-elliptic, margin serrulate to serrate. Cymes axillary, 1-5-flowered or more. Capsules globose. Seeds approximately elliptic. Fl. Apr-May. Fr. Aug-Sep. Riverside forests, thickets, valleys or sunny slopes. Distributed in Hebei, Liaoning, Jilin and Heilongjiang. Also in Russia (Far East), Korean Peninsula and Japan.

刺苞南蛇藤 *Celastrus flagellaris*

显柱南蛇藤 *Celastrus stylosus*

过山枫 *Celastrus aculeatus*

## 显柱南蛇藤
**Celastrus stylosus** Wall.

攀援灌木。叶在花期常为膜质，至果期常为近革质。聚伞花序腋生及侧生，具3-7花；果序梗及果梗粗短。果球形。种子一侧突起或稍呈新月状。花期3-5月，果期8-10月。生海拔300-2500米的林地或山坡。产中国西南、华中和华东。印度、尼泊尔、不丹、缅甸和泰国北部亦有。

Climbing shrubs. Leaves membranous in flowering season, almost leathery at fruiting. Cymes axillary and lateral, 3-7-flowered; infrutescences and fruit stalks broad and short. Capsules globose. Seeds convex at one side or slightly crescent. Fl. Mar-May. Fr. Aug-Oct. Forests or slopes at 300-2500 m. Distributed in SW, C and E China. Also in India, Nepal, Bhutan, Myanmar and N Thailand.

## 过山枫
**Celastrus aculeatus** Merr.

攀援灌木。小枝幼时被棕褐色短毛。叶多为椭圆形或矩形，两面光滑无毛。聚伞花序短，腋生或侧生，常具3花。蒴果近球状。种子新月状或弯成半环状，表面密布小疣点。花期3-4月，果期8-9月。生海拔100-1000米的山地灌丛或路边疏林中。产中国西南、华南和东南。

Climbing shrubs. Branchlets with brown short hairs. Leaves mainly elliptic or rectangular, glabrous at both sides. Cymes short, axillary or lateral, generally 3-flowered. Capsules almost orbicular. Seeds crescentiform or semicircular, with tubercles. Fl. Mar-Apr. Fr. Aug-Sep. Mountain thickets or sparse forests by roads at 100-1000 m. Distributed in SW, S and SE China.

## 青江藤
**Celastrus hindsii** Benth.

常绿攀援藤本。小枝圆柱形，灰色或紫色。叶纸质或革质，侧脉5-7对，边缘具疏锯齿。聚伞圆锥花序顶生和腋生，腋生花序近具1-3花；花淡绿色。果近球形。假种皮橙色。花期5-7月，果期7-10月。生海拔300-2500米的山地林中或灌丛。产中国西南、华南和华中。印度、缅甸、越南和马来西亚亦有。

Evergreen climbing shrubs. Branchlets terete, gray or purple. Leaves papery or leathery, lateral veins 5-7 pairs, margin sparsely serrate. Thyrses terminal and axillary, axillary inflorescences 1-3-flowered; flowers lightly green. Fruits approximately globose. Arils orange. Fl. May-Jul. Fr. Jul-Oct. Montane forests or thickets at 300-2500 m. Distributed in SW, S and C China. Also in India, Myanmar, Vietnam and Malaysia.

青江藤 *Celastrus hindsii*

独子藤
*Celastrus monospermus*

## 独子藤

**Celastrus monospermus** Roxb.

常绿攀援藤本。小枝和叶干后深棕色。叶近革质，基部楔形，稀广楔形，边缘具细锯齿。聚伞圆锥花序腋生，有时顶生；花黄绿色或近白色。蒴果阔椭圆形。假种皮紫棕色。花期3-6月，果期6-10月。生海拔300-1500米的林中或灌丛中。产中国西南和华南。南亚亦有。

Evergreen climbing shrubs. Young branches and leaves deep-brown when dry. Leaves subleathery, base cuneate, rarely widely cuneate, margin serrulate. Thyrses axillary, sometimes terminal; flowers yellowish green or whitish. Capsules broadly elliptic. Arils purplish brown. Fl. Mar-Jun. Fr. Jun-Oct. Forests or thickets at 300-1500 m. Distributed in SW and S China. Also in S Asia.

## 变叶裸实

**Gymnosporia diversifolia** Maxim.

灌木或小乔木。嫩枝多刺，棕灰色。叶片倒卵形、阔倒卵形或倒卵状披针形，基部楔形，边缘圆锯齿，先端圆形或微凹。聚伞花序腋生，叉状分枝；花白色或淡黄色。蒴果近倒卵球形，红色或紫色，4室。花期6-9月，果期8-12月。生海拔100米以下的疏林、山坡、海边或路旁。产华南。东南亚和日本南部亦有。

Shrubs or small trees. Twigs spiny, pallid brown. Leaves obovate, broadly obovate, or obovate-lanceolate, base cuneate, margin crenate, apex rounded or emarginated. Cymes axillary, dichotomous; flowers white or light yellow. Capsules subobovoid, red or purple, 4-loculed. Fl. Jun-Sep. Fr. Aug-Dec. Sparse forests, slopes, seashores or roadsides below 100 m. Distributed in S China. Also in SE Asia and S Japan.

变叶裸实 *Gymnosporia diversifolia*

长序美登木 *Maytenus thyrsiflorus*

## 长序美登木

**Maytenus thyrsiflorus** S. J. Pei et Y. H. Li

灌木。叶狭椭圆形、狭倒卵形、倒披针形或椭圆形，纸质或厚纸质。聚伞花序1-6次单歧分枝，花序梗细长；花白色，小，直径2-3毫米。蒴果近球形或倒圆锥形。花期6-9月，果期9-12月。生海拔800-1500米的江边干燥石缝中或疏林下。产云南和广西。

Shrubs. Leaves narrowly elliptic, narrowly obovate, oblanceolate or elliptic, papery or thickly papery. Cymes 1-6-monochotomous-branched; peduncles slender, elongate; flowers white, small, 2-3 mm diam. Capsules subglobose or obconical. Fl. Jun-Sep. Fr. Sep-Dec. Fissures of dry rocks along rivers or open forests at 800-1500 m. Distributed in Yunnan and Guangxi.

滇南美登木 *Maytenus austroyunnanensis*

广西美登木 *Maytenus guangxiensis*

## 滇南美登木
**Maytenus austroyunnanensis** S. J. Pei et Y. H. Li

灌木。嫩枝常无刺，老枝常有刺，刺针状或稍粗壮，直或微下弯。叶倒卵状椭圆形、椭圆形或长圆状披针形，近革质，基部窄楔形。聚伞花序常二或三回二叉分枝，具1-3花。蒴果倒卵球形。假种皮白色，干后淡黄色。花期5-9月，果期9-12月。生海拔500-1100米的山坡路边或江边灌丛中。产云南。

Shrubs. Twigs usually unarmed, old branches typically thorny, thorns needle-like or slightly sturdy, erect or somewhat decurved. Leaves obovate-elliptic, elliptic, or oblong-lanceolate, subleathery, base narrowly cuneate. Cymes usually 2 or 3 × dichotomously branched, 1-3-flowered. Capsules obovoid. Arils white, yellowish when dry. Fl. May-Sep. Fr. Sep-Dec. Roadsides or riverside thickets of mountain slopes at 500-1100 m. Distributed in Yunnan.

## 美登木
**Maytenus hookeri** Loes.

灌木。小枝柔细稍呈藤本状，小枝通常少刺，老枝有明显疏刺。聚伞花序1-6丛生短枝上，花序多2-4(-5)次单歧分枝或第一次二歧分枝；花梗细线状。蒴果2裂，稍扁。假种皮白色。花期12月至翌年6月，果期翌年6-11月。生海拔600-1200米疏林、山坡或山谷中。产云南。不丹和印度亦有。

Shrubs. Twigs slender, sometimes trailing, sparsely thorny or unarmed, older branches typically thorny. Cymes 1-6, fascicled, each cyme 2-4(-5) × monochasially branched or once dichotomously branched; peduncles slender. Capsules 2-lobed, slightly flattened. Arils white. Fl. Dec to next Jun. Fr. next Jun-Nov. Sparse forests, mountain slopes or valleys at 600-1200 m. Distributed in Yunnan. Also in Bhutan and India.

## 广西美登木
**Maytenus guangxiensis** C. Y. Cheng et W. L. Sha

灌木。小枝具刺。叶厚纸质，椭圆形或卵状椭圆形，边缘具圆齿；叶柄长5-13毫米。聚伞花序2-4次分枝，花7-25朵；花序梗短；苞片及小苞片边缘有毛。花白色；萼片卵圆形；花瓣长圆形；花盘厚，5波状浅内凹，雄蕊着生内凹外缘；柱头3裂。蒴果3裂，倒卵状。种子直径约5毫米，基部有白色杯状假种皮。花期11-12月。生石灰岩山地。产广西。

Shrubs. Twigs thorny. Leaves thickly papery, elliptic or ovate-elliptic, margin crenate; Petioles 5-13 mm long. Cymes fascicled, 2-4 × branched, 7-25-flowered; peduncle short; bracts and bractlets pubescence. Flowers white; sepals ovate-rounded; petals oblong; disk thick, 5 shallow concave, anther on the concave outside; stigma 3-lobed. Capsules obovoid, 3-valved. Seeds ca. 5 mm diam, covered by white aril at base. Fl. Nov-Dec. Dry calcareous mountain slopes. Distributed in Guangxi.

美登木 *Maytenus hookeri*

## 密花美登木
**Maytenus confertiflora** J. Y. Luo et X. X. Chen

灌木。枝条具刺，刺粗壮。叶纸质，侧脉细，9-13对。聚伞花序多数簇生于叶腋，约具60花，近球形；花白色；萼片淡紫色。蒴果淡紫褐色，3裂。种子长圆形，基部具白色假种皮。花期11-12月。生石灰质山坡或丛林中。产广西。

密花美登木 *Maytenus confertiflora*

Shrubs. Twigs thorny, thorns sturdy. Leaves papery, lateral veins tenuous, 9-13 pairs. Cymes numerous, fascicled at axils, ca. 60-flowered, subglobose; flowers white; sepals purplish. Capsules purplish brown, 3-valved. Seeds oblong, covered by white arils at base. Fl. Nov-Dec. Dry calcareous mountain slopes or jungles. Distributed in Guangxi.

## 大明假卫矛

**Microtropis thyrsiflora** C. Y. Cheng et T. C. Kao

小乔木。小枝粗壮。叶对生，厚革质，椭圆形或倒卵状椭圆形，侧脉6-8对，网脉不显著。聚伞花序大，2-5次分枝；花序梗粗壮，扁平；花坛状，5数；花瓣椭圆形。蒴果椭圆形或倒卵状椭圆形，长达2厘米。种子橙红色。花果期11-12月。生海拔1000-2300米的密林中。产广西(大明山)。

Small trees. Branchlets stout. Leaves opposite, thickly coriaceous, elliptic or obovate-elliptic, lateral veins 6-8 pairs, reticulate veins obscure. Cymes large, 2-5 times branched; peduncles stout, flat; flowers urceolate, 5-merous; petals elliptic. Capsules elliptic or obovate-elliptic, to 2 cm long. Seeds orange-red. Fl. and fr. Nov-Dec. Thick forests at 1000-2300 m. Distributed in Guangxi (Daming Mountain).

## 麻栗坡假卫矛

**Microtropis malipoensis** Y. M. Shui et W. H. Chen

小乔木。叶大，长可达25厘米，纸质，长圆状椭圆形或长圆形，基部狭楔形，侧脉纤细。聚伞花序疏松，分枝可达4次以上；花序梗纤细。果序开展，下垂；蒴果细圆柱形，弯曲，长2厘米。花期12月至翌年2月，果期11-12月。生海拔约1700米的山沟原始常绿阔叶林或竹林中。产云南(麻栗坡县)。

Small trees. Leaves large, length to 25 cm long, chartaceous, oblong-elliptic or oblong, base narrowly cuneate, lateral veins slender. Cymes very lax, branched 4 times or more; peduncles tenuous. Infructescence lax, pendulous; capsules narrowly cylindrical, falcate, 2 cm long. Fl. Dec to next Feb. Fr. Nov-Dec. Primary evergreen broad-leaved forests in valleys or bamboo groves at ca. 1700 m. Distributed in Yunnan (Malipo County).

大明假卫矛 *Microtropis thyrsiflora*

麻栗坡假卫矛 *Microtropis malipoensis*

大果假卫矛 *Microtropis macrocapa*

## 大果假卫矛

**Microtropis macrocapa** C.Y. cheng et T. C. Kao.

灌木或小乔木。小枝绿色，稍扁。叶对生，薄革质或纸质，椭圆形或阔倒卵形，先端具尾尖，侧脉6-8对，网脉不显著。聚伞花序1-2次分枝；花大，5数，直径达5-6毫米；花瓣近圆形。蒴果大，纺锤状椭圆形，长可达4厘米。花果期8-11月。生海拔1300-1700米的密林中。产云南(屏边县大围山)。

Shrubs or small trees. Branchlets green, somewhat compressed. Leaves opposite, thinly coriaceous or chartaceous, elliptic or broadly obovate, apex caudate, lateral veins 6-8 pairs, reticulate veins obscure. Cymes 1-2 times branched; flowers large, 5-merous, 5-6 mm diam; petals suborbicular. Capsules large, fusoid-elliptoid, to 4 cm long. Fl. and fr. Aug-Nov. Thick forests at 1300-1700 m. Distributed in Yunnan (Dawei Mountain, Pingbian County).

## 塔蕾假卫矛

**Microtropis pyramidalis** C. Y. Cheng et T. C. Kao.

灌木，高0.4-2米。叶薄，近膜质，椭圆形或阔倒卵形，先端具尾尖，侧脉不显著。聚伞花序小，腋上生，1-2次分枝；花小，5数，花蕾尖塔形。蒴果长圆形，长1-2厘米。花期2-4月，果期10-12月。生海拔800-1800米的山谷、溪边、山坡或山顶的原始常绿阔叶林中。产云南(屏边县大围山)。

Shrubs, 0.4-2 m tall. Leaves thin, submembranaceous, elliptic or broadly obovate, apex caudate, lateral veins obscure. Cymes small, extra-axillary, 1-2 times branched; flowers small, 5-merous, flower buds pyramidal. Capsules oblong, 1-2 cm long. Fl. Feb-Apr. Fr. Oct-Dec. Primary evergreen broad-leaved forests in valleys, by streamsides, on slopes or summit at 800-1800 m. Distributed in Yunnan (Dawei Mountain, Pingbian County).

## 广序假卫矛

**Microtropis petelotii** Merr. Freem.

灌木或乔木。小枝淡紫褐色，近四棱形。叶近革质，边缘常为黄白色，基部楔形，主脉细，两面突起。聚伞花序腋生或侧生，疏散。蒴果近圆柱状。花果期6-10月。生海拔1300-2200米的常绿林中。产云南和广西。越南亦有。

Shrubs or trees. Branchlets purplish brown, slightly tetragonal. Leaves subcoriaceous, margin yellowish white, base cuneate, midvein fine, prominent on both surfaces. Cymes axillary or lateral, loose. Capsules nearly terete. Fl. and fr. Jun-Oct. Evergreen forests at 1300-2200 m. Distri-

塔蕾假卫矛 *Microtropis pyramidalis*

广序假卫矛 *Microtropis petelotii*

buted in Yunnan and Guangxi. Also in Vietnam.

## 大围山假卫矛

**Microtropis daweishanensis** Q. W. Lin et Z. X. Zhang

乔木，高8-15米。叶厚革质，椭圆形，侧脉4-6对，疏远。聚伞花序疏松，2-3次分枝；总花序梗粗短，长1-2厘米；花5数，白色，芳香；花瓣长圆形。蒴果椭圆形，长1-1.5厘米。花期9月，果期11月。生海拔1400-2000米的原始常绿阔叶林中。产云南(屏边县大围山)。

Trees, 8-15 m tall. Leaves thickly coriaceous, elliptic, lateral veins 4-6 pairs, distant. Cymes lax, 2-3 times branched; peduncles stout, 1-2 cm long; flowers 5-merous, white, fragrant; petals oblong. Capsules elliptic, 1-1.5 cm long. Fl. Sep. Fr. Nov. Primary evergreen broad-leaved forests at 1400-2000 m. Distributed in Yunnan (Dawei Mountain, Pingbian County).

## 方枝假卫矛

**Microtropis tetragona** Merr. et Freem.

灌木或小乔木。小枝具明显四棱。叶长方椭圆形或狭卵状椭圆形，长8-13厘米，先端渐尖；侧脉6-9对；叶柄长5-10毫米。聚伞花序有花3-7朵，稀稍多；花序梗长5-11毫米；花5数；萼片半圆形；花瓣长方椭圆形或稍宽卵状椭圆形。蒴果近长椭圆状，顶端常具短喙。花期8-10月，果期10-11月。生海拔1000-2100米的原始多苔林中或溪边。产广西、海南和云南。

Shrubs or small trees. Branchlets obviously tetragonal. Leaves rectangular-elliptic or narrowly ovate-elliptic, 8-13 cm long, apex acuminate; lateral veins 6-9 pairs; Petioles 5-10 mm long. Cymes 3-7-flowered, rarely more; peduncle 5-11 mm long; flowers 5-merous; sepals semiorbicular; petals rectangular-elliptic or slightly broadly ovate-elliptic. Capsules nearly oblong, apex often with short rostrum. Fl. Aug-Oct. Fr. Oct-Nov. Mossy evergreen forests or streamsides at 1000-2100 m. Distributed in Guangxi, Hainan and Yunnan.

大围山假卫矛 *Microtropis daweishanensis*

方枝假卫矛 *Microtropis tetragona*

滇东假卫矛 *Microtropis henryi*

## 滇东假卫矛

**Microtropis henryi** Merri. et Freem.

灌木，高2-3米。小枝纤细。叶薄纸质，长圆形或椭圆形，先端具尾尖，侧脉纤细，背面突起。聚伞花序小，腋上生，1-2次分枝；花小，5(-6)数。蒴果近圆形，直径约1厘米。花期6-7月，果期11-12月。生海拔1000-2000米的山坡、山脊及山顶的原始常绿阔叶密林中。产云南(屏边县大围山)。

Shrubs, 2-3 m tall. Branchlets slender. Leaves thinly chartaceous, oblong or elliptic, apex caudate, lateral veins slender, abaxially elevated. Cymes small, extra-axillary, 1-2 times branched; flowers small, 5(-6)-merous. Capsules subglobose, ca. 1 cm diam. Fl. Jun-Jul. Fr. Nov-Dec. Primary evergreen broad-leaved forests on slopes, ridges and summits at 1000-2000 m. Distributed in Yunnan (Dawei Mountain, Pingbian County).

## 三花假卫矛

**Microtropis triflora** Merrill et Freeman

小乔木。叶革质，长圆形或披针形，基部狭楔形，先端长渐尖，侧脉及网脉纤细，两面突起。聚伞花序常具花3朵；总花梗纤细；花小，5数，白色。蒴果椭圆体形，长1-1.5厘米。花期11-12月，果期8-12月。生海拔1100-2500米的原始山地常绿阔叶林。产贵州、湖北、四川和云南。

Small trees. Leaves coriaceous, oblong or lanceolate, base narrowly cuneate, apex long-acuminate, lateral veins and reticulate veins slender, elevated on both surfaces. Cymes usually 3-flowered; flowers small, 5-merous, white. Capsules ellipsoid, 1-1.5 cm long. Fl. Nov-Dec. Fr. Aug-Dec. Primary evergreen broad-leaved forests at 1100-2500 m. Distributed in Guizhou, Hubei, Sichuan and Yunnan.

## 长果假卫矛

**Microtropis longicarpa** Q. W. Lin et Z. X. Zhang

小乔木。叶薄革质，椭圆形或长圆形，基部楔形，先端渐尖。聚伞花序疏松，1-2次分枝；总花序梗细长，长3-4厘米；花4(-5)数，白色，芳香；花瓣长圆形。蒴果细纺锤形，长1.5-2.5厘米。花果期10-12月。生海拔1300-2400米的密林中。产云南东南部。

Small trees. Leaves thinly coriaceous, elliptic or oblong, base cuneate, apex acuminate. Cymes lax, 1-2 times branched; pedun-

三花假卫矛 *Microtropis triflora*

长果假卫矛 *Microtropis longicarpa*

密花假卫矛 *Microtropis gracilipes*

cles slender, 3-4 cm long; flowers 4(-5)-merous, white, fragrant; petals oblong. Capsules narrowly fusiform, 1.5-2.5 cm long. Fl. and fr. Oct-Dec. Thick forests at 1300-2400 m. Distributed in SE Yunnan.

## 密花假卫矛

**Microtropis gracilipes** Merr. et Metc.

灌木。小枝稍具棱。叶阔倒披针形，革质。聚伞花序球形，密集；花瓣紫色，5数，无柄，丛生呈头状。蒴果宽椭圆形。种子椭圆形，深红色。花期4月。生海拔700-1500米的林中、溪边或沼泽。产贵州、广西、广东、福建和湖南。

Shrubs. Branchlets slightly angulate. Leaves broadly oblanceolate, leathery. Cymes glomerate, dense; petals purple, flowers 5-merous, sessile, clustered capitulum-like. Capsules broadly ellipsoid. Seeds elliptic, dark red. Fl. Apr. Forests, streamsides or swamps at 700-1500 m. Distributed in Guizhou, Guangxi, Guangdong, Fujian and Hunan.

## 福建假卫矛

**Microtropis fokienensis** Dunn

灌木。小枝略四棱形。叶近革质。聚伞花序密集成簇，短小，腋生或生枝轴，稀顶生，小花3-9；花具极短梗或无柄，4数或5数。蒴果椭圆状或倒卵球状椭圆体形。花果期10月。生海拔800-2000米的山坡或沟谷林中。产华南、华中和东南。

Shrubs. Branchlets slightly tetragonal. Leaves almost leathery. Cymes glomerate, short and small, axillary or on branched axis, rarely terminal, 3-9-flowered; flowers very shortly pedicellate or sessile, 4-merous or 5-merous. Capsules ellipsoid or obovoid-ellipsoid. Fl. and fr. Oct. Slopes or valley forests at 800-2000 m. Distributed in S, C and SE China.

异色假卫矛 *Microtropis discolor*

## 异色假卫矛

**Microtropis discolor** (Wall.) Wall. ex Meisn.

小乔木。叶薄革质，椭圆形或阔倒卵形，基部楔形，先端急尖，正面深绿色，背面苍白色。聚伞花序腋生，1-3次分枝；总花梗短；花小，5数，白色；花瓣内部中部具龙骨状突起。蒴果卵圆形或椭圆形，长1-1.5厘米。花果期10-12月。生海拔600-2200米的原始常绿阔叶林、山地林、混交林或竹灌丛中。产云南和广西。不丹、印度、孟加拉国、缅甸、泰国、越南、老挝、柬埔寨和马来半岛亦有。

Small trees. Leaves thinly coriaceous, elliptic or broadly obovate, base cuneate, apex acute, adaxially dark green, abaxially pale white. Cymes axillary, 1-3 times branched; peduncles short; flowers small, 5-merous, white; petals distinctly keeled on inner side. Capsules ovoid or elliptic, 1-1.5 cm long. Fl. and fr. Oct-Dec. Primary evergreen broad-leaved forests, montane forests, mixed forests or bamboo groves at 600-2200 m. Distributed in Yunnan and Guangxi. Also in Bhutan, India, Bangladesh, Myanmar, Thailand, Vietnam, Laos, Cambodia and Malay Peninsula.

福建假卫矛 *Microtropis fokienensis*

西藏假卫矛 *Microtropis xizangensis*

吴氏假卫矛 *Microtropis wui*

## 西藏假卫矛

**Microtropis xizangensis** Q. W. Lin et Z. X. Zhang

灌木或小乔木。叶薄革质，椭圆形或阔倒卵形，基部楔形，先端尾尖。聚伞花序极小，腋上生，常具单花或3(-7)花；总花梗极短；花小，5数，白色。蒴果大，圆球形或纺锤形椭圆形，长达3厘米，表面具疣点，先端具显著长喙。花期5-6月，果期8-12月。生海拔1300-2100米的原始常绿阔叶林中。产西藏(墨脱县)。

Shrubs or small trees. Leaves thinly coriaceous, elliptic or broadly obovate, base cuneate, apex caudate. Cymes small, extra-axillary, usually with 1 or 3(-7) flowers; peduncles very short; flowers small, 5-merous, white. Capsules large, globose or fusiform-elliptic, to 3 cm long, surface verrucose, apex distinctly rostrate. Fl. May-Jun. Fr. Aug-Dec. Primary evergreen broad-leaved forests at 1300-2100 m. Distributed in Xizang (Mêdog County).

## 木犀假卫矛

**Microtropis osmanthoides** (Hand.-Mazz.) Hand.-Mazz.

灌木或小乔木。叶厚革质，椭圆形，侧脉及网脉均不明显。聚伞花序腋生或簇生老茎上，无梗；花小，坛形，5数，白色。蒴果椭圆形。种子橙红色。花果期10-12月。生海拔800-1500米的原始常绿阔叶林或石灰岩山地。产广西、广东、香港和海南。

Shrubs or small trees. Leaves thickly coriaceous, elliptic, lateral veins and reticulate veins obscure. Cymes axillary or glomerate on old trunks and branches, sessile; flowers small, ampullaceous, 5-merous, white. Capsules elliptic. Seeds orange-red. Fl. and fr. Oct-Dec. Primary evergreen broad-leaved forests or limestone mountains at 800-1500 m. Distributed in Guangxi, Guangdong, Hong Kong and Hainan.

## 吴氏假卫矛

**Microtropis wui** Y. M. Shui et W. H. Chen

大灌木。叶极大，长可达50厘米以上，倒卵形或菱形，鲜时肉质，基部耳状，侧脉12-15

木犀假卫矛 *Microtropis osmanthoides*

雷公藤 *Tripterygium wilfordii*

对。聚伞花序簇生老茎上，无梗；花多数，坛状，5数，白色。蒴果椭圆形，长1.5-2厘米。花果期11月至翌年1月。生海拔600-800米的石灰岩山地次生林。产云南(河口县)。

Large shrubs. Leaves very large, length to 50 cm long or more, obovate or rhombic, carnose when fresh, base auriculate, lateral veins 12-15 pairs. Cymes glomerate on old trunks and branches, sessile; flowers numerous, ampullaceous, 5-merous, white. Capsules elliptic, 1.5-2 cm long. Fl. and fr. Nov to next Jan. Limestone forests at 600-800 m. Distributed in Yunnan (Hekou County).

## 雷公藤

**Tripterygium wilfordii** Hook. f.

落叶半灌木或攀援、木质藤本。小枝棕红色，有4-6纵棱。叶常卵形或圆卵形，纸质、草质至革质，无毛或疏具秕糠状绒毛与红褐色毛。圆锥聚伞花序较窄小；花白色、微绿色或黄绿色。翅果成熟时常绿色或浅绿褐色。花期5-10月，果期8-11月。生海拔100-3500米的山地林中阴湿处。产中国西南、华南、华中和华东。缅甸东北部、朝鲜半岛和日本亦有。

Deciduous subshrubs, or scandent and scrambling, or sometimes semiwoody vines. Branchlets brownish red, 4-6 thin ribbed. Leaves usually ovate or rounded-ovate, papery, herbaceous to leathery, glabrous or sparsely scurfy tomentose with reddish brown hairs. Paniculate cymes narrow; flowers whitish, greenish, or yellow-green, small. Samaras usually green or greenish brown when mature. Fl. May-Oct. Fr. Aug-Nov. Damp sites in montane forests at 100-3500 m. Distributed in SW, S, C and E China. Also in NE Myanmar, Korean Peninsula and Japan.

## 东北雷公藤

**Tripterygium regelii** Sprague et Takeda

攀援木质藤本。叶椭圆形或长方卵形，纸质，叶背淡绿色，光滑。圆锥聚伞花序顶生，大型；花小，密集；花瓣白色；花盘淡黄色。翅果成熟时淡黄色，果翅边缘波状。花期6-7月，果期9-10月。生海拔1100-2100米的山地路旁林缘。产吉林和辽宁。朝鲜半岛和日本亦有。

Scandent woody vines. Leaves elliptic or oblong-ovate, chartaceous, abaxially greenish, glabrous. Thyrses terminal, large; flowers small, dense; petals white; floral disc yellowish. Samaras usually yellowish at maturity, margin of wings undulate. Fl. Jun-Jul. Fr. Sep-Oct. Forest edges by roadsides in mountain areas at 1100-2100 m. Distributed in Jilin and Liaoning. Also in Korean Peninsula and Japan.

东北雷公藤 *Tripterygium regelii*

核子木 *Perrottetia racemosa*

## 核子木

**Perrottetia racemosa** (Oliv.) Loes.

灌木。枝条和小枝紫褐色，幼时具柔毛，渐无毛。叶片纸质，互生。花多数组成窄总状聚伞花序，花白色，4-5数。浆果近球状，红色或暗红色。每室具1或2粒种子。花期5-9月，果期8-11月。生海拔500-2900米的林中、林缘、山谷或溪边。产中国西南和华中。

Shrubs. Branches and branchlets purplish brown, puberulent when young, glabrescent. Leaves papery, alternate. Narrow cymose-botryose with many flowers, flowers white, 4-5-merous. Berries almost globose, red or dark red. Seeds 1 or 2 per locule. Fl. May-Sep. Fr. Aug-Nov. Forests, forest edges, valleys or riverbanks at 500-2900 m. Distributed in SW and C China.

## 十齿花

**Dipentodon sinicus** Dunn

半常绿灌木或小乔木。叶纸质，披针形或窄椭圆形。聚伞花序近球形，花序梗长2.5-3.5厘米；苞片4或5，生花序梗顶端，膜质。蒴果椭圆状卵球形，密被灰棕色绒毛。种子黑褐色，椭圆体形。花期5-9月，果期8-10月。生海拔900-3200米的山坡、溪边或路旁。产云南、西藏(墨脱县)、贵州和广西。缅甸北部亦有。

Semievergreen shrubs or small trees. Leaves papery, lanceolate or narrowly-elliptic. Cymes subglobose, peduncles 2.5-3.5 cm long; bracts 4 or 5, at apex of peduncle, membranous. Capsules ellipsoid-ovoid, densely gray-brown tomentose. Seeds blackish brown, ellipsoid. Fl. May-Sep. Fr. Aug-Oct. Slopes, streamsides or roadsides at 900-3200 m. Distributed in Yunnan, Xizang (Mêdog County), Guizhou and Guangxi. Also in N Myanmar.

十齿花 *Dipentodon sinicus*

# 翅子藤科
# Hippocrateaceae

多籽五层龙 *Salacia polysperma*

## 多籽五层龙
**Salacia polysperma** Hu

攀援灌木。叶轮生，长圆形或长圆状椭圆形，薄革质，叶缘波状，浅锯齿，中脉凸出。束状花序腋生，多花，花浅绿色；花盘杯状、近五角形；雄蕊3。浆果密被疣状突起或光滑。常有12粒种子。花期8-9月，果期10-11月。生海拔500-1800米的山谷疏林中。产云南南部和广西。

Climbing shrubs. Leaves alternate, oblong or oblong-elliptic, thinly coriaceous, margin sinuate, weakly serrulate, midvein abaxially prominent. Fascicles axillary, many-flowered, flowers light green; disk cupular, subpentagonous; stamens 3. Berries minutely verruciform or smooth. Usually 12-seeded. Fl. Aug-Sep. Fr. Oct-Nov. Sparse forests in montane valleys at 500-1800 m. Distributed in S Yunnan and Guangxi.

## 无柄五层龙
**Salacia sessiliflora** Hand.-Mazz.

灌木。小枝具疣状皮孔。叶薄革质，光亮，网脉横出。束状花序腋生，少花；花淡绿色；雄蕊3；花梗长约1毫米。浆果橘黄色至橘红色，外果皮干时薄革质。具3或4粒种子。花期6月，果期10月。生海拔(200-)600-1600米的山地小树丛中。产云南东南部、贵州、广西、广东和湖南南部。

Shrubs. Branchlets with verruciform lenticels. Leaves thinly coriaceous, shiny, reticulate veins horizontally spreading. Fascicles axillary, few-flowered; flowers light green; stamens 3; pedicel ca. 1 mm long. Berries orange-yellow to orange-red, exocarps thinly coriaceous when dry. 3- or 4-seeded. Fl. Jun. Fr. Oct. Mountain busks at (200-)600-1600 m. Distributed in SE Yunnan, Guizhou, Guangxi, Guangdong and S Hunan.

## 橙果五层龙
**Salacia aurantiaca** C. Y. Wu

攀援灌木。小枝纤细，平滑，皮孔稀疏。叶对生或近对生，纸质，长椭圆形，叶缘具显著圆齿，叶面光亮，网脉横出。浆果橙黄色，直径2-3厘米，光滑，果柄纤细。种子4粒。果期11月。生海拔120-1400米的疏林中。产云南东南部。

climbing shrubs. Branchlets slender, smooth, lenticels sparse. Leaves opposite or subopposite, chartaceous, oblong, margin distinctly crenate, adaxially shiny, reticulate veins horizontally spreading. Berries orange-yellow, 2-3 cm diam, smooth, Fruit stalks tenuous. With 4 seeds. Fr. Nov. Open forests at 120-1400 m. Distributed in SE Yunnan.

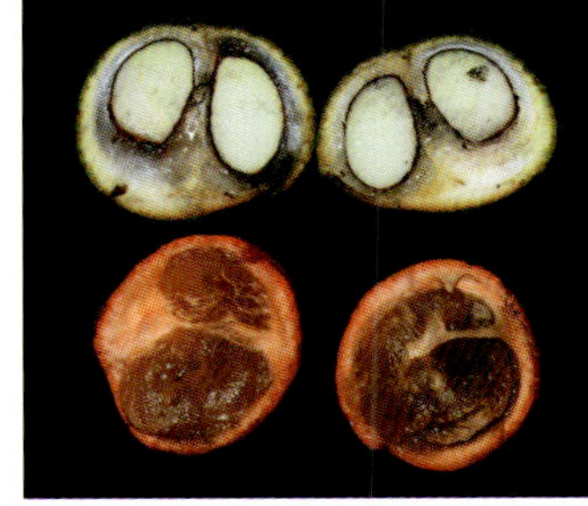

无柄五层龙 *Salacia sessiliflora*

橙果五层龙 *Salacia aurantiaca*

五层龙 *Salacia chinensis*

## 五层龙
**Salacia chinensis** L.

攀援灌木。叶椭圆形，狭卵状圆形或倒卵状椭圆形，边缘具疏牙齿。束状花序腋生，具3-6花；花梗长6-10毫米；花瓣阔卵形，向外开展或弯曲；每室2胚珠。浆果球形或卵形，直径约1厘米，成熟时红色。具1粒种子。花期12月，果期翌年1-2月。生海拔100-700米的林中。产广西和广东。南亚和东南亚亦有。

Climbing shrubs. Leaves elliptic, narrowly ovate-round or obovate-elliptic, margin sparsely denticulate. Fascicles axillary, 3-6-flowered; pedicels 6-10 mm long; petals broadly ovate, broadly spreading or curved outward; ovules 2 per locule. Berries globose or ovate, ca. 1 cm diam, red when mature. 1-seeded. Fl. Dec. Fr. Jan to Feb next year. Forests at 100-700 m. Distributed in Guangxi and Guangdong. Also in S and SE Asia.

## 毛扁蒴藤
**Pristimera setulosa** A. C. Sm.

藤本。幼小枝稍四棱形，密具细刺状腺毛，一年生枝变无毛。叶椭圆形，纸质，边缘疏具牙齿至圆锯齿。聚伞花序单生或两个腋生；花黄白色。果实为1-3个束生蒴果；蒴果长圆状椭圆体形。花期6月至翌年2月，果期9-10月。生海拔600-1600米的石灰岩疏林中。产云南东南部和广西西南部。

Lianas. Young branchlets slightly quadrangular, copiously setulose, annually glabrescent. Leaves elliptic, papery, margin sparsely denticulate to crenulate-serrate. Cymes solitary or binate, axillary; flowers yellow-white. Fruits 1-3-fascicled capsules; capsules oblong-ellipsoid. Fl. Jun to next Feb. Fr. Sep-Oct. Sparse limestone forests at 600-1600 m. Distributed in SE Yunnan and SW Guangxi.

## 二籽扁蒴藤
**Pristimera arborea** (Roxb.) A. C. Sm.

藤本。叶阔卵形或卵状长圆形，8-15 × 5-17厘米，纸质。聚伞花序单生、腋生或顶生；花浅黄色。果实为1-3个束生蒴果；蒴果狭椭圆体形，6.5-8.5(-12) × 2.5-3(-3.8)厘米。具2粒种子。花期6月，果期10月。生海拔300-1100米的山坡灌丛及沟谷中。产云南南部和广西西南部。印度、不丹和缅甸亦有。

Lianas. Leaves broadly ovate or ovate-oblong, 8-15 × 5-17 cm, papery. Cymes solitary, axillary or terminal; flowers light yellow. Fruits 1-3-fascicled capsules; capsules narrowly ellipsoid, 6.5-8.5 (-12) × 2.5-3(-3.8) cm. 2-seeded. Fl. Jun. Fr. Oct. Shrubs of mountainous slopes and valleys at 300-1100 m. Distributed in S Yunnan and SW Guangxi. Also in India, Bhutan and Myanmar.

毛扁蒴藤 *Pristimera setulosa*

二籽扁蒴藤 *Pristimera arborea*

# 刺茉莉科 Salvadoraceae

## 刺茉莉

**Azima sarmentosa** (Blume) Benth. et Hook. f.

直立灌木。腋生刺长2-16毫米，劲直，锐尖。托叶2枚，钻形；叶卵形、倒卵形、椭圆形至近圆形，长2.5-8厘米，先端急尖；中脉在两面突起。苞片狭三角形；花淡绿色；雄花：花萼4深裂，裂片钝，直立；花瓣长圆形；雌花：花冠与雄花的相似，但较短。浆果球形，白色或绿色。花期1-3月。生近海平面海岸林中。产海南。印度和东南亚亦有。

Erect shrubs. Axillary spines 2-16 mm long, strict and acute. Stipules 2, subulate; leaves ovate or obovate, elliptic to subrounded, 2.5-8 cm long, apex acute; midrib prominent on both surfaces. Bracts narrowly triangular; flowers greenish.staminate flowers: calyx 4-parted, lobes obtuse, erect; petals oblong; pistillate flowers: corolla similar to in staminate flower, only shorter. Berries globose, white or green. Fl. Jan-Mar. Coastal forests near sea level. Distributed in Hainan. Also in India and SE Asia.

刺茉莉 *Azima sarmentosa*

# 省沽油科 Staphyleaceae

## 瘿椒树 (银鹊树)

**Tapiscia sinensis** Oliv.

落叶乔木，8-15米高。树皮灰黑色或灰白色。小枝无毛。托叶缺无；奇数羽状复叶，小叶5-9，两面无毛或仅背面脉腋被毛。圆锥花序腋生；花萼花瓣边缘具毛。不开裂的浆果状核果，近球形或椭圆体形。花期3-5月，果期5-6月。生海拔500-2200米的山坡林中、山谷或溪边。产中国西南、华南、华中、东南和华东。

Deciduous trees, 8-15 m tall. Bark gray to grayish black. Twigs glabrous. Stipules absent; leaves odd-pinnate, with 5-9 leaflets; blades glabrous or with tufts of hairs on axils of veins. Panicles axillary; calyx and petals margin ciliate. Indehiscent berry-like drupes almost globose or ellipsoid. Fl. Mar-May. Fr. May-Jun. Forests on mountain slopes, valleys or streamsides at 500-2200 m. Distributed in SW, S, C, SE and E China.

## 云南瘿椒树

**Tapiscia yunnanensis** W. C. Cheng et C. D. Chu

乔木，高20-25米。奇数羽状复叶，叶轴幼时被短绒毛。小叶7-9对，对生，纸质，密具柔毛。花序长达14厘米，具淡黄色长柔毛；花瓣5，白色，花萼与花瓣全缘。浆果倒卵球形或椭圆状倒卵球形，绿色。花期3-4月，果期5月。生海拔1500-2300米的疏林中。产云南、四川和湖北。

Trees, 20-25 m tall. Leaves odd-pinnate, rachises pubescent when young; leaflets 7-9 pairs, opposite, papery, densely pubescent. Inflorescences to 14 cm long, yellowish villous; petals 5, white, margin of calyx and petals entire. Berries obovoid or ellipsoid-obovoid, green. Fl. Mar-Apr. Fr. May. Spares forests at 1500-2300 m. Distributed in Yunnan, Sichuan and Hubei.

瘿椒树（银鹊树） *Tapiscia sinensis*

云南瘿椒树 *Tapiscia yunnanensis*

省沽油 *Staphylea bumalda*

## 省沽油

**Staphylea bumalda** DC.

落叶灌木，树皮紫红色。三小叶复叶；顶生小叶叶柄长达1厘米；小叶椭圆形、卵形或披针状卵形。圆锥花序直立；花瓣白色。蒴果膀胱状，压扁，2室，顶端开裂。花期4-5月，果期8-9月。生疏林或路旁。产华中、华北、华东和东北。朝鲜半岛和日本亦有。

Deciduous shrubs, bark reddish purple. Leaves trifoliolate; petiolules of terminal leaflet up to 1 cm long; leaflets elliptic, ovate, or lanceolate-ovate. Panicles erect; petals white. Capsules bladderlike, compressed, 2-locular, dehiscing at apex. Fl. Apr-May. Fr. Aug-Sep. Sparse forests or roadsides. Distributed in C, N, E and NE China. Also in Korean Peninsula and Japan.

## 膀胱果

**Staphylea holocarpa** Hemsl.

乔木或灌木。三小叶，小叶长圆状披针形至椭圆形，近革质，无毛，正面淡白色。伞房花序扩展；花白色或玫瑰色，叶后开花。蒴果梨形，基部狭，先端截形，3处开口。花期4-5月，果期9月。生海拔1200-2200米的林中。产中国西南、华南、华中和东南。

Trees or shrubs. Leaves trifoliolate; leaflets oblong-lanceolate to elliptic, nearly leathery, glabrous, adaxially pale white. Corymbs broad; flowers white or rosy, flowering after emergence of foliage. Capsules pear-shaped, base narrow, apex truncate with 3 openings. Fl. Apr-May. Fr. Sep. Forests at 1200-2200 m. Distributed in SW, S, C and SE China.

## 野鸦椿

**Euscaphis japonica** (Thunb.) Kanitz

小型落叶乔木或灌木。小叶5-9，揉碎后有难闻气味，纸质。顶生圆锥花序；花瓣黄绿色；萼片基部合生，边缘具缘毛。蓇葖果紫红色。种子近球形，假种皮肉质，黑色。花期

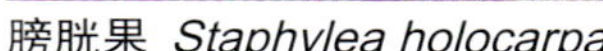

膀胱果 *Staphylea holocarpa*

野鸦椿 *Euscaphis japonica*

4-6月，果期8-11月。生海拔900-1800米的疏林或灌丛中。产中国除西北以外大部分地区。越南、朝鲜半岛和日本亦有。

Small deciduous trees or shrubs. Leaflets 5-9, with unpleasant odor when crushed, papery. Inflorescences a terminal panicle; petals yellowish green; sepals base united, margin ciliate. Follicles purple-red. Seeds subglobose, arils fleshy, black. Fl. Apr-Jun. Fr. Aug-Nov. Sparse forests or thickets at 900-1800 m. Distributed in most parts of China, except NW China. Also in Vietnam, Korean Peninsula and Japan.

## 锐尖山香圆

**Turpinia arguta** (Lindl.) Seem.

灌木。单叶，对生，厚纸质，椭圆形至长圆形，先端渐尖。短圆锥花序顶生；子房及花柱均被柔毛；花白色，较大，直径8-10(-12)毫米。果近球形，幼时绿色，干后黑色。花期3-4月，果期9-10月。生海拔400-700米的林中、灌丛中或路边。产中国西南、华南、东南、华中和华东。

Shrubs. Leaves simple, opposite, thickly papery, elliptic to oblong, apex acuminate. Panicles terminal, short; ovary and styles pubescent; flowers white, large, 8-10(-12) mm diam. Fruits subglobose, green when young, black when dry. Fl. Mar-Apr. Fr. Sep-Oct. Forests, thickets or roadsides at 400-700 m. Distributed in SW, S, SE, C and E China.

## 亮叶山香圆

**Turpinia simplicifolia** Merr.

小乔木或灌木。叶柄基部显著膨大，单叶，薄革质，侧脉12-14对，两面凸起。伞房状圆锥花序长7-12厘米；花萼白色；花瓣白色或黄色；子房3室。浆果浅绿色，近球形，外果皮表面粗糙，顶部具3凹坑。花期3-6月，果期7-9月。生海拔400-1000米的密林或山谷中。产广西和广东。马来西亚、印度尼西亚和菲律宾亦有。

Small trees or shrubs. Petioles base greatly swollen, leaves simple, thinly leathery, with 12-14 lateral veins, prominent on both surfaces. Paniculate cymes 7-12 cm long; sepals white; petals white or yellow; ovary 3-locular. Berries light green, subglobose, surface of exocarps coarse, apex with 3 concave slits. Fl. Mar-Jun. Fr. Jul-Sep. Dense forests or valleys at 400-1000 m. Distributed in Guangxi and Guangdong. Also in Malaysia, Indonesia and the Philippines.

锐尖山香圆 *Turpinia arguta*

亮叶山香圆 *Turpinia simplicifolia*

硬毛山香圆 *Turpinia affinis*

## 硬毛山香圆
**Turpinia affinis** Merr. et Perry

乔木。奇数羽状复叶；小叶2-4，长圆状椭圆形，革质。圆锥花序长约30厘米，开展，具柔毛；花瓣倒卵状椭圆形，内部具绒毛；胚珠6-8。浆果近球形，长1-1.5厘米，花柱宿存，具硬毛。花期3-4月，果期8-11月。生海拔(500-)1000-2000米的山谷或峡谷林中。产云南、四川、贵州和广西。

Trees. Leaves odd-pinnate; leaflets 2-4, oblong-elliptic, leathery. Panicles ca. 30 cm long, spreading, pubescent; petals obovate-elliptic, interior tomentose; ovules 6-8. Berries subglobose, 1- 1.5 cm diam, styles persistent with stiff hairs. Fl. Mar-Apr. Fr. Aug-Nov. Forests in ravines or valleys at (500-)1000-2000 m. Distributed in Yunnan, Sichuan, Guizhou and Guangxi.

大果山香圆 *Turpinia pomifera*

## 大果山香圆
**Turpinia pomifera** (Roxb.) DC.

乔木或灌木。叶薄革质，奇数羽状，有小叶3-9；小叶长圆状椭圆形，网脉在两面明显可见。圆锥花序顶生，长达21厘米，花大，疏生。浆果大，直径2-2.5厘米，外果皮表面幼时粗糙。花期1-4月，果期6-8月。生海拔300-1400米的林缘或路边。产云南南部、广西南部和西部。印度、不丹、尼泊尔、越南北部和马来西亚亦有。

Trees or shrubs. Leaves thinly leathery, imparipinnate, 3-9-foliolate; leaflets oblong-elliptic, with veins distinct on both sides. Panicles terminal, to 21 cm long, with loosely and large flowers. Berries large, 2-2.5 cm diam, surface of exocarp coarse when young. Fl. Jan-Apr. Fr. Jun-Aug. Forest edges or roadsides at 300-1400 m. Distributed in S Yunnan, and S and W Guangxi. Also in India, Bhutan, Nepal, N Vietnam and Malaysia.

## 山香圆
**Turpinia montana** (Blume) Kurz

小乔木。叶纸质，奇数羽状复叶，有小叶5；小叶长圆形至长圆状卵形，(4-)4.5-6 × 1.5-4厘米，网脉在两面几不可见。圆锥花序顶生，花小，疏生。浆果紫红色，球形。花期3-6月，果期8-11月。生山坡林中，密林或湿润林中。产云南、贵州、广西和广东。南亚和东南亚亦有。

Small trees. Leaves papery, odd-pinnate, 5-folioate; leaflets oblong to oblong-oval, (4-)4.5-6 × 1.5-4 cm, veins nearly invisible on both sides. Panicles terminal, with sparse and small flowers. Berries purple-red, globose. Fl. Mar-Jun. Fr. Aug-Nov. Forests on slopes, dense forests or moist forests. Distributed in Yunnan, Guizhou, Guangxi and Guangdong. Also in S and SE Asia.

山香圆 *Turpinia montana*

# 茶茱萸科 Icacinaceae

### 粗丝木
**Gomphandra tetrandra** (Wall.) Sleum.

灌木或小乔木。叶狭披针形，窄椭圆形至阔椭圆形，无毛。聚伞花序与叶对生，有时腋生；雄花黄白色或白绿色，5数；花冠钟形，先端内弯。核果浆果状。花果期全年。生海拔500-2200米的林中、沟谷或路旁灌丛。产云南、贵州、广西、广东和海南。南亚和东南亚亦有。

Shrubs or small trees. Leaves narrowly lanceolate, narrowly elliptic or broadly elliptic, glabrous. Cymes opposite leaves, sometimes axillary; staminate flowers yellow-white or white-green, 5-merous; corollas campanulate, apex incurved. Drupes berry-like. Fl. and fr. all year. Forests, valleys or thickets by roads at 500-2200 m. Distributed in Yunnan, Guizhou, Guangxi, Guangdong and Hainan. Also in S and SE Asia.

粗丝木 *Gomphandra tetrandra*

### 柴龙树
**Apodytes dimidiata** E. Meyer ex Arn.

灌木或乔木。树皮光滑，灰白色。嫩枝密被黄色微柔毛。叶干后黑色。圆锥花序顶生，密被黄色微柔毛；花淡黄色或白色，花梗短，密被黄色微柔毛；花萼黄绿色。核果成熟后红色至深红色，基部具一碟形肉质附属物。花果期全年。生海拔500-1900米的林中、石灰岩山上或村旁灌丛中。产云南南部、广西西部和海南南部。南亚、东南亚、热带和亚热带非洲亦有。

Shrubs or trees. Bark smooth, gray-white. Young branches densely yellow puberulent. Leaves black when dry. Panicles terminal, densely yellow puberulent; flowers light yellow or white, pedicels short, densely yellow puberulent; calyxs yellow-green. Drupes red to black-red when mature, base with a discoid fleshy appendage. Fl. and fr. all year. Forests, limestone hills, or thickets near villages at 500-1900 m. Distributed in S Yunnan, W Guangxi and S Hainan. Also in S and SE Asia, and tropical and subtropical Africa.

柴龙树 *Apodytes dimidiata*

### 假海桐
**Pittosporopsis kerrii** Craib

灌木或小乔木。叶纸质，互生。花瓣5，匙形，内面无毛；花梗黄色，被柔毛；3或4枚小苞片，鳞片状；柱头棒状。核果白绿色或棕色，干后有2棱。种子具淡红棕色、极薄的种皮。花期10月至翌年5月，果期2-10月。生海拔350-1600米的溪边密林中。产云南南部。缅甸、老挝、泰国和越南北部亦有。

Shrubs or small trees. Leaves papery, alternate. Petals 5, spatulate, glabrous inside; pedicels yellow puberulent; bractlets 3 or 4, scalelike; styles clavate. Drupes white-green or brown and 2-ribbed when dry. Seeds with a light red-brown, very thin testa. Fl. Oct to next May. Fr. Feb-Oct. Dense forests by streams at 350-1600 m. Distributed in S Yunnan. Also in Myanmar, Laos, Thailand and N Vietnam.

假海桐 *Pittosporopsis kerrii*

马比木 *Nothapodytes pittosporoides*

## 马比木
**Nothapodytes pittosporoides** (Oliv.) Sleum.

小灌木，稀为乔木。叶薄革质。聚伞花序顶生；花萼绿色，钟状，膜质；花瓣长6.3-7.4毫米；花盘肉质，果时宿存，具不整齐的裂片或深圆齿，里面疏被长硬毛。核果椭圆体形至长圆状卵球形。花期4-6月，果期6-8月。生海拔100-1600(-2500)米的林中。产四川、贵州、广西、广东、湖北、湖南和甘肃。

Shrubs small, rarely trees. Leaves thin coriaceous. Cymes terminal; calyx green, campanulate, membranous; petals 6.3-7.4 mm long; disk persistent in fruit, fleshy, irregularly lobed or deeply crenulate, inside sparsely hirsute. Drupes ellipsoid to oblong-ovoid. Fl. Apr-Jun. Fr. Jun-Aug. Forests at 100-1600(-2500) m. Distributed in Sichuan, Guizhou, Guangxi, Guangdong, Hubei, Hunan and Gansu.

## 定心藤
**Mappianthus iodoides** Hand.-Mazz.

木质藤本。卷须粗壮，与叶轮生。叶狭椭圆形至长圆形，革质。雄花序交替腋生，被黄褐色糙伏毛；小苞片极小；花冠内面有毛。核果从淡绿色或黄绿色至橙黄色或橙红色，椭圆体形。花期4-8月，果期6-12月。生海拔700-1900米的疏林中或灌丛中。产中国西南、华南、华中和东南。越南北部亦有。

Woody lianas. Tendrils hairy, verticillate with leaves. Leaves narrowly elliptic to oblong, coriaceous. Staminate cymes alternate and axillary, yellow-brown strigose; bractlets very small; corolla hairy inside. Drupes from light green or yellow-green to orange-yellow or orange-red, ellipsoid. Fl. Apr-Aug. Fr. Jun-Dec. Sparse woods or thickets at 700-1900 m. Distributed in SW, S, C and SE China. Also in N Vietnam.

## 瘤枝微花藤
**Iodes seguinii** (Lévl.) Rehd.

木质藤本。茎具多数瘤状皮孔。叶卵形或近圆形，背面密具硬糙毛，稀具柔毛。伞房圆锥花序腋生或侧生，密被锈色卷曲柔毛。核果幼时黄绿色，成熟后红色，倒卵球状长圆形。花期1-5月，果期4-6

定心藤 *Mappianthus iodoides*

瘤枝微花藤 *Iodes seguinii*

月。生海拔200-1200米的石灰山地林中。产广西、贵州和云南。

Woody lianas. Stems with many tuberculiform lenticels. Leaves ovate or suborbicular, abaxially densely somewhat rigidly strigose and less puberulent. Corymbose panicles axillary or lateral, densely rust-colored crispate pubescent. Drupes yellow-green when young, red when mature, obovoid-oblong. Fl. Jan-May. Fr. Apr-Jun. Limestone forests at 200-1200 m. Distributed in Guangxi, Guizhou and Yunnan.

## 微花藤

**Iodes cirrhosa** Turcz.

木质藤本。小枝、叶背和花序密被红黄色或锈色柔毛。叶卵形或阔卵形。花序具短柄，密被黄褐色绒毛；雄花为密伞房花序，有时复合成大型圆锥花序；雌花少。核果成熟时红色，卵球形。花期1-4月，果期5-10月。生海拔400-1000(-1300)米的山谷疏林中。产云南南部和广西南部。南亚和东南亚亦有。

Woody lianas. Branchlets, lower surfaces of leaves and inflorescences densely red-yellow or ferruginous-pubescent. Leaves ovate or broadly ovate. Inflorescences shortly pedunculate, densely yellow-brown tomentose; staminate corymbs dense, sometimes combined into large panicles; pistillates few-flowered. Drupes red when mature, ovoid. Fl. Jan-Apr. Fr. May-Oct. Open forests in valleys at 400-1000(-1300) m. Distributed in S Yunnan and S Guangxi. Also in S and SE Asia.

微花藤 *Iodes cirrhosa*

## 小果微花藤

**Iodes vitiginea** (Hance) Hemsl.

木质藤本。小枝压扁，被淡黄色硬伏毛，卷须腋生或生于叶柄的一侧。叶狭卵形至卵形，薄纸质，背面密具白色或浅黄色糙且稍硬的伏毛。伞房圆锥花序腋生。核果卵球形或阔卵球形，1.3-2.2 × 1.2-1.6厘米。花期12月至翌年6月，果期5-8月。生海拔100-1300米的雨林山谷林中或次生灌丛中。产云南、贵州、广西和广东。老挝、泰国和越南亦有。

Woody lianas. Branchlets compressed, light yellow somewhat rigidly strigose, with axillary or lateral tendrils. Leaves narrowly ovate to ovate, thin papery, abaxially densely white or light yellow roughly and somewhat rigidly strigose. Corymbose panicles axillary. Drupes ovoid or wide ovoid, 1.3-2.2 × 1.2-1.6 cm. Fl. Dec to next Jun. Fr. May-Aug. Monsoon valley forests or secondary scrubs at 100-1300 m. Distributed in Yunnan, Guizhou, Guangxi and Guangdong. Also in Laos, Thailand and Vietnam.

小果微花藤 *Iodes vitiginea*

无须藤 *Hosiea sinensis*

## 无须藤

**Hosiea sinensis** (Oliv.) Hemsl. et Wils.

藤状灌木。树皮灰色或灰黄色，光滑，具明显皮孔。叶柄长2-7.5厘米。聚伞花序多花，长2-8厘米；花瓣绿色；柱头4裂。核果幼时绿色，成熟后红色或棕红色，干后表面具多角形孔穴。种子1枚，具胚乳；子叶椭圆形。花期4-5月，果期6-8月。生海拔1200-2100米的林中或缠绕于树上。产四川(峨眉山)、浙江、湖北和湖南。

Scandent shrubs. Bark gray or yellow-gray, smooth, conspicuously lenticellate. Petioles 2-7.5 cm long. Cymes many-flowered, 2-8 cm long; petals green; stigmas 4-lobed. Drupes green when young, red or red-brown when mature, polygonous reticulate-lacunose when dry. Seeds 1, albuminous; cotyledons elliptic. Fl. Apr-May. Fr. Jun-Aug. Forests or twining on trees at 1200-2100 m. Distributed in Sichuan (Emei Mountain), Zhejiang, Hubei and Hunan.

## 心翼果

**Cardiopteris quinqueloba** (Hassk.) Hassk.

草质藤本。叶宽广，叶柄长3-12厘米，叶基5出脉，分叉广。圆锥花序单生花梗，1-3歧分枝，花小，杂性。翅果近圆形，金黄色，被宿存花萼，长2.5-3厘米，宽2-3厘米。花期5-11月，果期10月至翌年3月。生海拔100-1300米的石灰岩山地疏林或灌丛中，或山谷密林中。产云南、广西和海南。南亚和东南亚亦有。

Climbing herbs. Leaves widely spaced, petioles 3-12 cm long, basal veins 5, widely divergent. Panicles solitary on peduncles, 1-3-forked, flowers small, polygamous. Samaras suborbicular, golden, with persistent calyx, 2.5-3 cm long, 2-3 cm wide. Fl. May-Nov. Fr. Oct to next Mar. Sparse forests or thickets in karst mountains, or dense valley forests at 100-1300 m. Distributed in Yunnan, Guangxi and Hainan. Also in S and SE Asia.

心翼果 *Cardiopteris quinqueloba*

# 槭树科
# Aceraceae

金钱槭 *Dipteronia sinensis*

## 金钱槭
**Dipteronia sinensis** Oliv.

落叶乔木。小枝紫色或紫绿色。羽状复叶长20-40厘米；小叶常7-11。圆锥花序顶生或腋生，直立，无毛或具灰色柔毛；花瓣5，白色。翅果具一个阔的圆形的翅。种子近球形。花期4月，果期9月。生海拔1000-2400米的林中或林缘。产中国西南、华中和华西。

Deciduous trees. Branchlets purple or purplish green. Leaves pinnate, 20-40 cm long; leaflets usually 7-11. Panicles terminal or axillary, erect, glabrous or gray puberulent; petals 5, white. Samaras with a broad encircling wing. Seeds nearly globose. Fl. Apr. Fr. Sep. Forests or forest edges at 1000-2400 m. Distributed in SW, C and W China.

## 云南金钱槭
**Dipteronia dyerana** Henry

乔木，高7-13米。叶为奇数羽状复叶，具9-15小叶，长30-40厘米；小叶纸质，披针形或长圆披针形，9-14厘米，先端锐尖或尾状锐尖，边缘具很稀疏粗锯齿；侧脉13-14对，被短柔毛。圆锥花序密被黄绿色的短柔毛。果实压扁状，嫩时绿色；翅圆形，连同小坚果长4.5-6厘米；果梗长2厘米，密被短柔毛。果期9月。生海拔2000-2500米的疏林中。产云南和贵州。

Trees, up to 7-13 m tall. Leaves odd-pinnate, 9-15-foliolate, 30-40 cm long; leaflets papery, lanceolate or oblong-lanceolate, 9-14 cm long, apex acute or caudate-acute, margin sparsely coarsely serrate; lateral veins 13-14 pairs, pubescent. Panicles densely yellow-green pubescent. Fruits compressed, green when young; wing rounded, including nutlet 4.5-6 cm long; fruiting pedicel 2 cm long, densely pubescent. Fr. Sep. Sparse forests at 2000-2500 m. Distributed in Yunnan and Guizhou.

云南金钱槭 *Dipteronia dyerana*

庙台槭 *Acer miaotaiense*

## 庙台槭

**Acer miaotaiense** P. C. Tsoong

落叶乔木，高20-25米。小枝无毛。叶纸质，近阔卵形，长7-9厘米，常3-5裂；裂片卵形，边缘微呈浅波状；叶柄长6-7厘米，无毛。伞房状圆锥花序顶生。果序连同总果梗约长5厘米，无毛；小坚果扁平，被很密的黄色绒毛；翅连小坚果长2-2.5厘米，张开近水平。花期5月，果期9月。生海拔700-1600米的混交林中。产浙江和华中。

Deciduous trees, up to 20-25 m. Branchlets glabrous. Leaves papery, broadly subovate, 7-9 cm long, usually 3-5-lobed; lobes ovate, margin slightly undulate; petioles 6-7 cm long, glabrous. Inflorescences corymbose-paniculate, terminal. Infructescences including fruiting peduncle ca. 5 cm long, glabrous; nutlets flat, densely yellow tomentose; wing including nutlet 2-2.5 cm long, wings spreading nearly horizontally. Fl. May. Fr. Sep. Mixed forests at 700-1600 m. Distributed in Zhejiang and C China.

## 薄叶槭

**Acer tenellum** Pax

落叶乔木，雄花两性花同株。叶柄长3-6厘米。叶近圆形或卵形，常3裂。伞房花序顶生；花5数。坚果压扁，翅开展至水平。花期5月，果期9月。生海拔1200-1900米的杂木林或山谷中。产重庆和湖北西部。

Deciduous trees, andromonoecious. Petioles 3-6 cm long. Leaves suborbicular or ovate, usually 3-lobed. Inflorescences terminal, corymbose; flowers 5-merous. Nutlets compressed, with wings spreading nearly horizontally. Fl. May.

薄叶槭 *Acer tenellum*

元宝槭 *Acer truncatum*

大翅色木槭 *Acer pictum* subsp. *macropterum*

Fr. Sep. Mixed forests or valleys at 1200-1900 m. Distributed in Chongqing and W Hubei.

## 元宝槭

**Acer truncatum** Bunge

落叶乔木，雄花两性花同株。叶下无毛，常5浅裂，有时3或7浅裂，基部常截形，裂片有时具牙齿，纸质，顶端微凹。花序直立，伞房状。小坚果微凹且较厚；翅绿白色，常与小坚果几等长。花期4月，果期8月。生海拔400-1000米的林中。产华北、华西、华东和东北。朝鲜半岛亦有。

Deciduous trees, andromonoecious. Leaves abaxially glabrous, usually 5-lobed, sometimes 3- or 7-lobed, base usually truncate, lobes sometimes dentate, papery, emarginated. Inflorescences erect, corymbose. Nutlets convex and thick; wing greenish white, usually ca. as long as nutlets. Fl. Apr. Fr. Aug. Forests at 400-1000 m. Distributed in N, W, E and NE China. Also in Korean Peninsula.

## 五角槭

**Acer pictum** Thunb. subsp. **mono** (Maxim.) H. Ohashi

落叶乔木，雄花两性花同株。叶5或7裂至中部或浅裂，背面无毛或主脉上具柔毛。花序顶生，圆锥状伞房花序；花瓣5，近白色；子房无毛或渐无毛。小坚果平，压扁；翅开展为不同角度。花期4-5月，果期9月。生海拔3000米以下的阔叶林

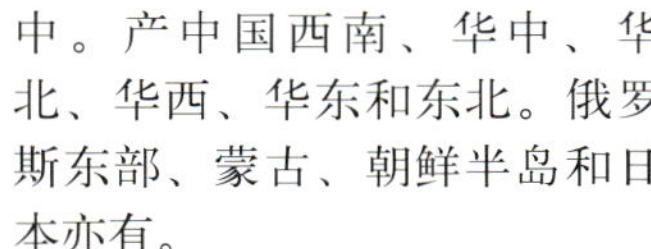

中。产中国西南、华中、华北、华西、华东和东北。俄罗斯东部、蒙古、朝鲜半岛和日本亦有。

Deciduous trees, andromonoecious. Leaves 5- or 7-lobed, to middle or shallowly lobed, abaxially not hairy, or pilose on main veins. Inflorescences terminal, paniculate-corymbose; petals 5, whitish; ovary glabrous or glabrescent. Nutlets flat, compressed; wings spreading variously. Fl. Apr-May. Fr. Sep. Broad-leaved forests below 3000 m. Distributed in SW, C, N, W, E and NE China. Also in E Russia, Mongolia, Korean Peninsula and Japan.

## 大翅色木槭

**Acer pictum** subsp. **macropterum** (W. P. Fang) Ohashi.

落叶乔木。叶较大，常长5-10厘米，基部近心形或截形，常5裂，稀3裂；裂片近卵形，先端锐尖。圆锥状伞房花序具多数花。翅果成熟后黄色，小坚果压扁状，近圆形或卵形，较小，长8-10毫米；翅长圆形，较大，连同小坚果长3.5-4厘米，张开近水平，稀近钝角。花期4-5月，果期9月。生海拔1900-3300米的混交林中。产甘肃、四川、云南和西藏。

Deciduous trees. Leaves large, usually 5-10 cm long, base subcordate or truncate, usually 5-lobed, rarely 3-lobed; lobes subovate, apex acute. Inflorescences terminal, paniculate-corymbose, numerous flowered. Samaras yellowish when mature; nutlets compressed, subglobose or ovoid, small, 8-10 mm long; wing oblong, large, including nutlet 3.5-4 cm long, wings spreading nearly horizontally or rarely nearly obtusely. Fl. Apr-May. Fr. Sep. Mixed forests at 1900-3300 m. Distributed in Gansu, Sichuan, Yunnan and Xizang.

五角槭 *Acer pictum* subsp. *mono*

紫花槭 *Acer pseudosieboldianum*

阔叶槭 *Acer amplum*

## 小叶青皮槭

**Acer cappadocicum** Gled. var. **sinicum** Rehd.

落叶乔木。叶纸质，长5-8厘米，基部近心形或截形，常5裂；裂片短而宽，先端锐尖至尾状锐尖，背面除脉腋其余部分无毛；叶柄细瘦，淡红色。花序伞房状，无毛；花杂性，雄花与两性花同株。翅连同小坚果长2.5-3厘米，稀达3.5厘米，张开常成锐角，淡紫色。果期9月。生海拔1500-2500米的混交林中。产陕西、湖北、四川、贵州、云南和西藏。

Deciduous trees. Leaves papery, 5-8 cm long, base subcordate or truncate, usually 5-lobed; lobes short and broad, apex acuminate to caudate-acute, abaxially glabrous except vein axils; petioles slender, redish. Inflorescences corymbose, glabrous; flowers polygamous, andromonoecious. Wing including nutlet 2.5-3 cm long, rarely to 3.5 cm long, wings spreading usually acutely, purplish. Fr. Sep. Mixed forests at 1500-2500 m. Distributed in Shaanxi, Hubei, Sichuan, Guizhou, Yunnan and Xizang.

## 阔叶槭

**Acer amplum** Rehd.

落叶乔木，雄花两性花同株。叶3或5裂，或不裂。花序顶生，伞状，多花。坚果扁平；翅张开成钝角。花期4月，果期9月。生海拔1000-2000米的林中或山谷中。产中国西南、华南、东南、华中和华东。

Deciduous trees, andromonoecious. Leaves 3- or 5-lobed or unlobed and entire. Inflorescences terminal, corymbose, numerous flowered. Nutlets compressed, flat; wings spreading at obtuse angle. Fl. Apr. Fr. Sep. Forests or valleys at 1000-2000 m. Distributed in SW, S, SE, C and E China.

## 紫花槭

**Acer pseudosieboldianum** (Pax) Kom.

落叶灌木或小乔木。叶近圆形，纸质，背面密具白色绒毛，基部深心形至心形，掌状分裂，裂片9-13，三角状卵形或卵状披针形，边缘具重锯齿。花序顶生，伞房状；花瓣5，白色或黄白色。果翅开展为钝角或近水平。花期5-6月，果期9月。生海拔700-900米的林中。产辽宁东部、吉林东南部和黑龙江。俄罗斯东部和朝鲜半岛亦有。

Deciduous shrubs or small trees. Leaves suborbicular, papery, abaxially densely white pubescent, base deeply cordate to cordate, usually 9-13-lobed; lobes triangular-ovate or ovate-lanceolate, margin doubly serrate. Inflorescences terminal, corymbose; petals 5, white or yellowish white. Fruits purplish yellow, wings spreading obtusely or nearly horizontally. Fl. May-Jun. Fr. Sep. Forests at 700-900 m. Distributed in E Liaoning, SE Jilin and Heilongjiang. Also in E Russia and Korean Peninsula.

## 黄毛槭

**Acer fulvescens** Rehd.

落叶乔木。叶纸质，长7-10厘

小叶青皮槭 *Acer cappadocicum* var. *sinicum*

黄毛槭 *Acer fulvescens*

察隅槭 *Acer tibetense*

米，常3或5裂；裂片三角卵形，先端有长尖尾；下面有黄色或褐色的长柔毛；叶柄长3-9厘米。伞房花序，长8-10厘米，无毛。花杂性，雄花与两性花同株。翅果成熟后紫黄色；小坚果压扁状，长1.3-1.5厘米；翅连同小坚果共长3-3.5厘米，张开几成水平。花期4月，果期8月。生海拔1800-3200米的疏林中。产四川和西藏。

Deciduous trees. Leaves papery, 7-10 cm long, usually 3- or 5-lobed; lobes triangular-ovate, apex long acuminate; abaxially fulvous or brown villose; petioles 3-9 cm long. Inflorescences corymbose, 8-10 cm long, glabrous; flowers polygamous, andromonoecious. Samaras purplish yellow when mature; nutlets compressed, 1.3-1.5 cm long; wing including nutlet 3-3.5 cm long, wings spreading nearly horizontally. Fl. Apr. Fr. Aug. Sparse forests at 1800-3200 m. Distributed in Sichuan and Xizang.

## 陕甘槭

**Acer shenkanense** W. P. Fang ex C. C. Fu

落叶乔木，高常达10米。叶纸质，直径10-19厘米，常3-5裂，先端锐尖或尾状锐尖，嫩时下面密被淡黄绿色短柔毛；叶柄长3-5厘米。果序伞房状，连同总果梗共长4-5厘米，直径5厘米。翅果长3-3.2厘米；翅较窄，长圆形或长圆倒卵形，张开成钝角或近水平。花期4-5月，果期7-9月。生海拔700-3000米的溪边树林、混交林或山谷中。产华中。

Deciduous trees, usually up to 10 m tall. Leaves papery, 10-19 cm diam, usually 3-5-lobed, apex acute or caudate-acute, abaxially densely pale yellowish-green pubescent; petioles 3-5 cm long. Infructescence corymbose, 4-5 cm long including fruiting peduncles, 5 cm diam. Samaras 3-3.2 cm long; wing narrow, oblong or oblong-obovate, wings spreading obtusely or nearly horizontally. Fl. Apr-May. Fr. Jul-Sep. Forests along streams, mixed forests, valleys at 700-3000 m. Distributed in C China.

## 察隅槭

**Acer tibetense** Fang

落叶乔木，高约10米。冬芽卵形，鳞片覆瓦状排列，先端被短柔毛。叶纸质，长6-9厘米，5裂；裂片三角卵形，先端锐尖或尾状锐尖，背面密被黄灰色浅绒毛；叶柄长5-7厘米，顶端被浅绒毛。花序伞房状，长5-6厘米；花瓣5。小坚果压扁状；翅连同小坚果长3.5-4厘米，张开近水平。花期4月，果期9月。生海拔1600-2700米的林中、山谷中。产西藏。

Deciduous trees, ca. 10 m tall. Winter buds ovoid, scales imbricate, apex pubescent. Leaves papery, 6-9 cm long, 5-lobed; lobes triangular-ovate, apex acute or caudate-acute, abaxially densely yellow-gray tomentose; petioles 5-7 cm long, puberulent near apex. Inflorescences corymbose, 5-6 cm long; petals 5. Nutlets compressed; wing 3.5-4 cm long including nutlet, wings spreading nearly horizontally. Fl. Apr. Fr. Sep. Forests, valleys at 1600-2700 m. Distributed in Xizang.

## 梓叶槭

**Acer amplum** Rehd. subsp. **catalpifolium** (Rehd.) Y. S. Chen

本亚种与阔叶槭的区别在于本亚种的叶常全缘不裂，稀有2或3浅裂的叶生同一小枝。翅果长5-5.5厘米。花期4月，果期8-10月。生海拔500-2000米的林中或山谷中。产四川、贵州和广西。

This subspecies differs from the typical subspecies in its leaves usually entire and unlobed, rarely 2- or 3-lobed leaves on same branchlet. Samaras 5-5.5 cm long. Fl. Apr. Fr. Aug-Oct. Forests or valleys at 500-2000 m. Distributed in Sichuan, Guizhou and Guangxi.

陕甘槭 *Acer shenkanense*

梓叶槭 *Acer amplum* subsp. *catalpifolium*

稀花槭 *Acer pauciflorum*

鸡爪槭 *Acer palmatum*

## 稀花槭

**Acer pauciflorum** Fang

落叶乔木。叶膜质，近圆形，直径3-4厘米，5裂；裂片边缘具尖锐的重锯齿；叶柄长约1厘米，被卷曲的长柔毛。伞房花序，花少数，杂性。总果梗长5-10毫米，被长柔毛；果梗长约1厘米，近无毛；小坚果凸起，长6毫米，被稀疏的长柔毛；翅连同小坚果共长1.5厘米，张开成直角。花期5月，果期9月。生海拔500-1000米的疏林中。产浙江和安徽。

Deciduous trees. Leaves membranous, suborbicular, 3-4 cm diam, 5-lobed; lobes margin sharply and doubly serrate; petioles ca. 1 cm long, curly villous. Inflorescences corymbose, few-flowered, andromonoecious. Fruiting peduncles 5-10 mm long, villous; fruiting pedicels ca. 1 cm long, nearly glabrous; nutlets convex, 6 mm long, sparsely villous; wing 1.5 cm long including nutlets, wings spreading erectly. Fl. May. Fr. Sep. Sparse forests at 500-1000 m. Distributed in Zhejiang and Anhui.

## 鸡爪槭

**Acer palmatum** Thunb.

落叶乔木，高达15米，雄花两性花同株。叶膜质至纸质，近圆形，5-9掌状分裂，两面无毛。伞房状圆锥花序，具10-20花，长3-4厘米，半下垂；花瓣淡黄色至粉白色。翅果翅长约1.5厘米；翅开成钝角。花期4-5月，果期9月。生海拔700-1200米的林中。中国广泛栽培。原产朝鲜半岛和日本。

Deciduous trees up to 15 m tall, andromonoecious. Leaves membranous to papery, suborbicular, palmately 5-9-parted, both surfaces of leaves glabrous. Inflorescences corymbose-paniculate, 10-20-flowered, 3-4 cm long, half-pendulous; petals pale yellow to pinkish white. Samaras ca. 1.5 cm long with wing; wings spreading at obtuse angle. Fl. Apr-May. Fr. Sep. Forests at 700-1200 m. Widely cultivated in China. Native to Korean Peninsula and Japan.

## 深灰槭

**Acer caesium** Wall. ex Brandis

落叶乔木，高达25米，雄花两性花同株。叶5裂，稀3裂，裂片三角形，边缘具锯齿，基部心形，叶背被白粉。花瓣5，白色；子房被短柔毛。小坚果强烈压扁。花期5-6月，果期9月。生海拔2000-3700米的高山林中。产中国西南、华中、陕西、甘肃和宁夏。喜马拉雅亦有。

Deciduous trees up to 25 m tall, andromonoecious. Leaves 5-lobed, rarely 3-lobed, lobes deltoid, margin toothed, cordate at base, abaxially with white powder. Petals 5, white; ovary pubescent. Nutlets strongly convex. Fl. May-Jun. Fr. Sep. Alpine forests at 2000-3700 m. Distributed in SW and C China, Shaanxi, Gansu and Ningxia. Also in Himalaya.

茶条槭 *Acer tataricum* subsp. *ginnala*

深灰槭 *Acer caesium*

苦条槭 *Acer tataricum* subsp. *theiferum*

## 茶条槭

**Acer tataricum** L. subsp. **ginnala** (Maxim.) Wesm.

灌木或乔木。叶椭圆状长圆形，3或5裂，纸质，背面近无毛，具锯齿或重锯齿。伞房花序具数花。翅果黄绿色或黄褐色，连同小坚果长 2.5-3厘米，无毛。花期5月，果期10月。生海拔800米以下的疏林中。产华中、华北、华西、华东和东北。俄罗斯(远东地区)、蒙古、朝鲜半岛和日本亦有。

Shrubs or trees. Leaves elliptic-oblong, 3- or 5-lobed, papery, abaxially subglabrous, serrate or doubly serrate. Inflorescences corymbose, several flowered. Samaras yellowish green or yellowish brown, wings including nutlets 2.5-3 cm long, glabrous. Fl. May. Fr. Oct. Sparse forests below 800 m. Distributed in C, N, W, E and NE China. Also in Russia (Far East), Mongolia, Korean Peninsula and Japan.

## 苦条槭

**Acer tataricum** subsp. **theiferum** (W. P. Fang) Y. S. Chen et P. C. de Jong

本亚种与茶条槭的区别在于本亚种的叶卵形或长圆状卵形，不分裂或稍3或5裂，下面具白色柔毛。花期5月，果期9月。生海拔1800米以下的疏林中。产广东北部、浙江、湖北、河南、陕西、江苏和江西。

This subspecies differs from *Acer tataricum* subsp. *ginnala* in its leaves ovate or oblong-ovate, unlobed or slightly 3- or 5-lobed, abaxially white pilose. Fl. May. Fr. Sep. Sparse forests below 1800 m. Distributed in N Guangdong, Zhejiang, Hubei, Henan, Shaanxi, Jiangsu and Jiangxi.

## 天山槭

**Acer tataricum** L. subsp. **semenovii** (Regel et Herder) A. E. Murray

落叶灌木或小乔木。叶卵形，近革质，不分裂或3或5裂，边缘具锯齿或重锯齿。花序伞房状，具腺状柔毛，花多而密集，淡绿色。翅果幼时具柔毛或腺毛，连同小坚果长3-3.5厘米，翅开展成直角。花期5-5月，果期9月。生海拔2000-2200米的河谷或山坡疏林中。产新疆(天山)。西南亚和中亚亦有。

Deciduous shrubs or small trees. Leaves ovate, subleathery, undivided or 3- or 5-lobed, margin serrate or doubly serrate. Inflorescences corymbose, glandular pubescent, flowers many and clustered, light green. Samaras pubescent or glandular pubescent when young, wings including nutlets 3-3.5 cm long, wings spreading at right angle. Fl. May-Jun. Fr. Sep. Valleys or sparse forests on slopes at 2000-2200 m. Distributed in Xinjiang (Tianshan Mountain). Also in SW and C Asia.

天山槭 *Acer tataricum* subsp. *semenovii*

花楷槭 *Acer ukurunduense*

## 长尾槭
**Acer caudatum** Wall.

落叶乔木，雄花两性花同株。小枝粗壮，当年生枝紫色或紫绿色。叶背浅绿色，腹面深绿色，膜质，基部心形，5裂，稀7裂，顶端尾状或渐尖。顶生总状圆锥花序，红色，密集，总状圆锥状，具长柔毛。翅开展成锐角或直角。花期5月，果期9月。生海拔1700-4000米的高山林中。产中国西南、华中和华西。印度、尼泊尔、不丹和缅甸亦有。

Deciduous trees, andromonoecious. Branchlets robust, current year's branches purple or purplish green. Leaves abaxially light green, adaxially dark green, membranous, base cordate, 5-lobed, rarely 7-lobed, apex caudate or acuminate. Racemose panicles terminal, red, compact, racemose-paniculate, villous. Wings spreading acutely or erectly. Fl. May. Fr. Sep. Alpine forests at 1700-4000 m. Distributed in SW, C and W China. Also in India, Nepal, Bhutan and Myanmar.

## 花楷槭
**Acer ukurunduense** Trautv. et Mey.

落叶乔木，雄花两性花同株。树皮坚硬。叶近圆形，膜质或纸质，下面密具浅黄色绒毛，5(-7)裂，边缘具糙齿。总状圆锥花序顶生，紧密，直立；总花梗长(8-)10-12厘米；花梗长(5-)8厘米。花期5月，果期9月。生海拔500-2500米的混生林或山坡灌丛中。产黑龙江、吉林和辽宁。俄罗斯、朝鲜半岛和日本亦有。

Deciduous trees, andromonoecious. Bark rough. Leaves suborbicular, membranous or papery, abaxially densely yellowish tomentose, 5(-7)-lobed, margin coarsely serrate. Inflorescences terminal, racemose-paniculate, compact, erect; peduncles (8-)10-12 cm long; pedicels (5-)8 cm long. Fl. May. Fr. Sep. Mixed forests or thickets on slopes at 500-2500 m. Distributed in Heilongjiang, Jilin and Liaoning. Also in Russia, Korean Peninsula and Japan.

## 扇叶槭
**Acer flabellatum** Rehd.

落叶乔木。叶纸质或膜质，近圆形，直径8-12厘米，常7裂；裂片边缘具不整齐的紧贴的尖锯齿；叶脉上有长柔毛，脉腋有丛毛；叶柄长达7厘米。花杂性，雄花与两性花同株；子房无毛。圆锥果序下垂；小坚果凸起，近卵圆形；翅连同小坚果长3-3.5厘米，张开近水平。花期6月，果期10月。生海拔800-3500米的疏林中。产华中和华南。缅甸和越南亦有。

Deciduous trees. Leaves papery or membranous, suborbicular, 8-12 cm diam, usually 7-lobed; lobes margin with irregularly appressed and sharp serration; villous on veins and tufts of hairs at vein axils; petioles 7 cm long. Flowers polygamous, andromonoecious; ovary glabrous. Paniculate infructescence pendulous; nutlets convex, subglobose; wing 3-3.5 cm long including nutlet, wings spreading nearly horizontally. Fl. Jun. Fr. Oct. Sparse forests at 800-3500 m. Distributed in C and S China. Also in Myanmar and Vietnam.

长尾槭 *Acer caudatum*

扇叶槭 *Acer flabellatum*

## 毛花槭

**Acer erianthum** Schwer.

落叶乔木，雌雄同株。叶纸质，常5裂；裂片卵形或三角卵形，边缘有尖锐而紧贴的锯齿，背面嫩时常被短柔毛；叶柄长5-9厘米。圆锥花序直径1-1.8厘米；花单性；花瓣5或4。小坚果凸起，近球形，脉纹显著，翅连同小坚果长2.5-3厘米，张开近水平或微向外侧反卷。花期5月，果期9月。生海拔1000-2300米的混交林中。产甘肃、华中、云南和广西。

Deciduous trees, monoecious. Leaves papery, usually 5-lobed; lobes ovate or triangular-ovate, margin with sharp and appressed serration, abaxially usually slightly pubescent when young; petioles 5-9 cm long. Panicles 1-1.8 cm diam; flowers unisexual; petals 5 or 4. Nutlets convex, subglobose, veined, wing 2.5-3 cm long including nutlet, spreading nearly horizontally or slightly reflexed outside. Fl. May. Fr. Sep. Mixed forests at 1000-2300 m. Distributed in Gansu, C China, Yunnan and Guangxi.

藏南槭 *Acer campbellii*

## 藏南槭

**Acer campbellii** Hook. f. et Thomson ex Hiern

落叶乔木，雄花两性花同株。叶膜质，(5-)7(-9)裂。圆锥花序顶生；花瓣黄绿色；果实成熟后淡黄色。小坚果凸起；翅镰状，张开近水平。花期5月，果期9月。生海拔2500-3700米的混交林中。产云南西北部和西藏南部。印度北部、尼泊尔、不丹、缅甸和越南亦有。

Deciduous trees, andromonoecious. Leaves membranous, (5-)7(-9)-lobed. Inflorescences terminal, paniculate; petals yellow-green; Fruits yellowish when mature. Nutlets convex; wings falcate, spreading nearly horizontally. Fl. May. Fr. Sep. Mixed forests at 2500-3700 m. Distributed in NW Yunnan and S Xizang. Also in N India, Nepal, Bhutan, Myanmar and Vietnam.

## 国楣槭

**Acer kuomeii** W. P. Fang et M. Y. Fang

落叶乔木，高8-10米，雄花两性花同株。叶7裂，稀5裂，近革质，中脉在腹面明显，背面凸起。圆锥花序顶生，多花；萼片5，紫色；花瓣5，白色。翅果张开成钝角。小坚果球形。花期4-5月，果期7-9月。生海拔1300-2300米的山地林中或山谷中。产云南东南部和广西西部。

Deciduous trees, 8-10 m tall, andromonoecious. Leaves 7-lobed, rarely 5-lobed, sub-leathery, midnerve obvious at adaxial side, convex at abaxial side. Inflorescences terminal, paniculate, many flowered; sepals 5, purple; petals 5, white. Samaras dehiscent to obtuse angle. Nutlets globose. Fl. Apr-May. Fr. Jul-Sep. Montane forests or valleys at 1300-2300 m. Distributed in SE Yunnan and W Guangxi.

毛花槭 *Acer erianthum*

国楣槭 *Acer kuomeii*

中华槭 *Acer sinense*

五裂槭 *Acer oliverianum*

## 中华槭
**Acer sinense** Pax

落叶小乔木，雄花两性花同株。叶近革质，常5裂。顶生圆锥花序下垂，多花；花瓣5，白色；萼片淡绿色；子房被白色柔毛。果实黄色，无毛。小坚果椭圆体形，强烈凸出。花期5月，果期9月。生海拔500-2500米的林中或山谷中。产四川、广东、福建、湖北和河南。

Deciduous small trees, andromonoecious. Leaves subleathery, usually 5-lobed. Inflorescences pendulous, terminal, paniculate, many-flowered; petals 5, white; sepals light green; ovary white pilose. Fruits yellowish, glabrous. Nutlets ellipsoid, strongly convex. Fl. May. Fr. Sep. Forests or valleys at 500-2500 m. Distributed in Sichuan, Guangdong, Fujian, Hubei and Henan.

## 桂林槭
**Acer kweilinense** W. P. Fang et M. Y. Fang

落叶乔木，雄花两性花同株。叶柄具长柔毛；叶纸质，背面主脉明显被长柔毛，叶缘具锯齿。圆锥花序直立；花瓣5，浅绿色；雄蕊8。翅果成熟时棕黄色。小坚果凸起，球形；翅伸展开成钝角。花期4月，果期9月。生海拔1000-1500米的疏林中。产贵州东南部和广西东北部。

Deciduous trees, andromonoecious. Petioles villous; leaves papery, abaxially distinctly villous on primary veins, margin serrate. Inflorescences erect, paniculate; petals 5, whitish green; stamens 8. Samaras yellowish brown when mature. Nutlets convex, globose; wings spreading at obtuse angle. Fl. Apr. Fr. Sep. Sparse forests at 1000-1500 m. Distributed in SE Guizhou and NE Guangxi.

## 五裂槭
**Acer oliverianum** Pax

落叶乔木。高达7米，雄花两性花同株。树皮平滑，常被蜡粉。叶柄长2.5-6厘米，无毛；叶片5裂，裂片上面深绿色或略带黄色。伞房状花序顶生于具2叶的小枝上；萼片淡紫色；花瓣白色。翅果无毛；翅开展几至水平。花期5月，果期9月。生海拔1000-2000米的林中或山谷中。产中国西南、华南、东南、华中、华东和华西。

Deciduous trees, up to 7 m tall, andromonoecious. Bark smooth, usually covered with waxy powders. Petioles 2.5-6 cm long, glabrous; leaves 5-lobed, lobes

桂林槭 *Acer kweilinense*

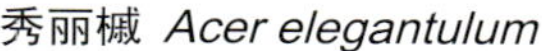
秀丽槭 *Acer elegantulum*

三峡槭 *Acer wilsonii*

adaxially dark green or yellowish. Inflorescences corymbose, terminal on a 2-leaved branchlet; sepals purplish; petals white. Samaras glabrous; wings spreading nearly horizontally. Fl. May. Fr. Sep. Forests or valleys at 1000-2000 m. Distributed in SW, S, SE, C, E and W China.

## 秀丽槭

**Acer elegantulum** Fang et P. L. Chiu

落叶乔木。叶薄纸质或纸质，长5.5-8厘米，通常5裂；裂片卵形或长圆形，边缘具紧贴的细圆齿；叶柄长2.8-6厘米，无毛。花序圆锥状；花杂性，雄花两性花同株；子房密被淡黄色长柔毛。小坚果凸起，近球形；翅连同小坚果共长2-2.3厘米，张开近于水平。花期5月，果期9月。生海拔200-1400米的高山、树林和山谷中。产华东和华南。

Deciduous trees. Leaves thinly papery or papery, 5.5-8 cm long, usually 5-lobed; lobes ovate or oblong, margin with appressed crenulations; petioles 2.8-6 cm long, glabrous. Inflorescences paniculate; flowers polygamous, andromonoecious; ovary densely yellowish villous. Nutlets convex, subglobose; wing 2-2.3 cm long together with nutlets, wings spreading nearly horizontal. Fl. May. Fr. Sep. Mountains, forests, valleys at 200-1400 m. Distributed in E and S China.

## 三峡槭

**Acer wilsonii** Rehd.

落叶乔木，雄花两性花同株。小枝细瘦，无毛。叶卵形，膜质或近革质。圆锥花序，长5-6厘米，无毛；花梗长2-3厘米，纤细；花瓣白色。小坚果卵球形，翅水平开展。花期4月，果期9月。生海拔900-2000米的林中。产中国西南、华南、华中和华东。缅甸、泰国和越南亦有。

Deciduous trees, andromonoecious. Branchlets slender, glabrous. Leaves ovate, membranous or subleathery. Inflorescences paniculate, 5-6 cm long, glabrous; peduncles 2-3 cm long, slender; petals white. Nutlets ovoid, wings spreading horizontally. Fl. Apr. Fr. Sep. Forests at 900-2000 m. Distributed in SW, S, C and E China. Also in Myanmar, Thailand and Vietnam.

## 岭南槭

**Acer tutcheri** Duthie

落叶乔木。树皮棕色或深棕色。叶片通常3裂，稀5裂，基部圆形或圆截形；裂片三角状卵形，稀卵状长圆形，叶缘有细锯齿，裂片尖端锐尖或渐尖，稀长尾尖。花序为短圆锥花序。小坚果凸起；翅幼时常红色，张开成锐角到近水平。花期4月，果期9月。生海拔300-1000米的森林中。产华南和东南。

Deciduous trees. Bark brown or dark brown. Leaves usually 3-lobed, rarely 5-lobed, base rounded or rounded-truncate; lobes triangular-ovate, rarely ovate-oblong, margin serrulate, apex acute or acuminate, rarely caudate-acuminate. Inflorescences shortly paniculate. Nutlets convex; wings usually red when young, spreading acutely to nearly horizontally. Fl. Apr. Fr. Sep. Forests at 300-1000 m. Distributed in S and SE China.

岭南槭 *Acer tutcheri*

细裂槭 *Acer pilosum* var. *stenolobum*

粗柄槭 *Acer tonkinense*

## 细裂槭

**Acer pilosum** Maxim. var. **stenolobum** (Rehd.) W. P. Fang

落叶小乔木。叶纸质，长3-6(-8)厘米，宽 2.5-7(-12)厘米，深3裂。伞房花序生小枝顶端，无毛，具多花；总花梗长1-1.5厘米；萼片5，卵形；花瓣5，线状长圆形。小坚果略凸起。花期4月，果期9月。生海拔1000-1500米较湿润的山坡或山谷中。产内蒙古、陕西、宁夏和甘肃。

Deciduous small trees. Leaves papery, 3-6(-8) cm long, 2.5-7(-12) cm wide, deeply 3-lobed. Corymboses at branch tops, glabrous, numerous flowered; peduncles 1-1.5 cm long; sepals 5, ovate; petals 5, linear-oblong. Nutlets slightly convex. Fl. Apr. Fr. Sep. Damp slopes or valleys at 1000-1500 m. Distributed in Neimenggu, Shaanxi, Ningxia and Gansu.

## 粗柄槭

**Acer tonkinense** H. Lec.

落叶乔木。叶近革质，近椭圆形，长10-15厘米，中段以上3裂；裂片三角形，边缘近全缘，背面的脉腋有丛毛；叶柄粗壮，长2-3.5厘米。花序圆锥状，多花；花瓣5；子房有很密的短柔毛。小坚果近卵圆形；翅镰刀形，连小坚果长1.8-3.5厘米，张开近水平。花期4-5月，果期9月。生海拔300-1800米的混交林中。产广西、贵州、西藏和云南。东南亚亦有。

Deciduous trees. Leaves subleathery, subelliptic, 10-15 cm long, 3-lobed above middle; lobes triangular, margin nearly entire, abaxially with tufts of hairs in axil of veins; petioles stout, 2-3.5 cm long. Inflorescences paniculate, many-flowered; petals 5; ovary densely pubigerous. Nutlets nearly ovoid; wing falcate, 1.8-3.5 cm long including nutlet, wings spreading nearly horizontally. Fl. Apr-May. Fr. Sep. Mixed forests at 300-1800 m. Distributed in Guangxi, Guizhou, Xizang and Yunnan. Also in SE Asia.

## 三角槭

**Acer buergerianum** Miq.

常绿或落叶乔木，高5-20米，雄花两性花同株。叶纸质，背面有白粉，3浅裂，中裂片卵形、三角形或三角状卵形，侧裂片与中裂片近等大。花多数，组成顶生的伞房花序；花瓣5，黄白色。小坚果明显凸起。花期4月，果期8-9月。生海拔1500米以下的混交林中。产中国西南、华南、东南、华中、华西和华东。日本亦有。

Evergreen or deciduous trees, 5-20 m tall, andromonoecious. Leaves papery, with white powder at abaxial side, shallowly 3-lobed, middle lobe ovate, triangular, or triangular-ovate, lateral lobes nearly same size as middle lobe. Flowers many in terminal corymb; petals 5, yellowish white. Nutlets obviously convex. Fl. Apr. Fr. Aug-Sep. Mixed forests below 1500 m. Distributed in SW, S, SE, C, W and E China. Also in Japan.

## 台湾三角槭

**Acer buergerianum** Miq. var. **formosanum** (Hayata ex Koidz.) Sasaki

落叶乔木。冬芽褐色，椭球形，鳞片边缘被长柔毛。叶卵形，长4-10厘米，3浅裂，裂片三角状卵形，正面被白粉，略被毛，背面无毛；叶柄长2.5-5(-8)厘米，无毛。伞房花序顶生，被短柔毛。小坚果强烈凸起，直径6-7毫米；翅与小坚果

三角槭 *Acer buergerianum*

台湾三角槭 *Acer buergerianum* var. *formosanum*

共长2.5-3厘米，张开成钝角或近于水平。花期4月，果期8月。生海拔100米以下的混交林中。产台湾。

Deciduous trees. Winter buds brown, ellipsoidal, scales villous along margin. Leaves ovate, 4-10 cm long, 3-lobed, lobes triangular-ovate, abaxially glaucous, slightly pubescent, adaxially glabrous; Petioles 2.5-5 (-8) cm long, glabrous. Inflorescences terminal, corymbose, pubescent. Nutlets strongly convex, 6-7 mm diam; wings including nutlet 2.5-3 cm long, spreading obtusely or nearly horizontally. Fl. Apr. Fr. Aug. Mixed forests below 100 m. Distributed in Taiwan.

## 金沙槭

**Acer paxii** Franch.

常绿乔木，雄花两性花同株。叶背面灰色，中央裂片顶端渐尖，侧裂片三角形。花序顶生于具叶小枝上，伞房状，具多花；花瓣5，白色，条状披针形；雄蕊8；子房具白绒毛。果实黄绿色。小坚果特别凸起。花期3月，果期8月。生海拔1500-2500米的高山林中。产云南西北部、四川、贵州和广西。

Evergreen trees, andromonoecious. Leaves abaxially glaucous, central lobes apically acuminate, lateral lobes triangular. Inflorescences terminal on leafy branchlets, corymbose, numerous-flowered; petals 5, white, linear-lanceolate; stamens 8; ovary white tomentose. Fruits greenish yellow. Nutlets strongly convex. Fl. Mar. Fr. Aug. Alpine forests at 1500-2500 m. Distributed in NW Yunnan, Sichuan, Guizhou and Guangxi.

## 樟叶槭

**Acer coriaceifolium** Lévl.

常绿乔木。小枝淡紫棕色。叶革质，侧脉5-6对，背面具绒毛，基部宽楔形、楔形或稀钝，叶缘全缘，顶端细尖。花序伞房状，顶生，有黄绿色绒毛；子房有白色柔毛。小坚果浅褐色，凸起。花期5月，果期8月。生海拔1500-2500米的林中。产中国西南、华南、东南、华中和华东。

Evergreen trees. Branchlets purplish brown. Leaves leathery, lateral veins 5-6 pairs, abaxially tomentose, base broadly cuneate, cuneate, or rarely obtuse, margin entire, apex apiculate. Inflorescences terminal, corymbose, yellowish green tomentose; ovary white pubescent. Nutlets light brown, convex. Fl. May. Fr. Aug. Forests at 1500-2500 m. Distributed in SW, S, SE, C and E China.

金沙槭 *Acer paxii*

樟叶槭 *Acer coriaceifolium*

紫果槭 *Acer cordatum*

滨海槭 *Acer sino-oblongum*

## 紫果槭
**Acer cordatum** Pax

常绿乔木。叶卵形或卵状长圆形，纸质，背面无毛且稍具网脉，叶缘近全缘或具稀疏锯齿，顶端短渐尖。伞房花序着生于叶小枝的顶端，具花3-5朵；花瓣浅黄绿色。小坚果凸起，无毛。花期4月，果期9月。生海拔500-1200米的林中或山谷中。产中国西南、华南、华中、东南和华东。

Evergreen trees. Leaves ovate or ovate-oblong, papery, abaxially glabrous and slightly reticulate, margin nearly entire or remotely serrate, apex shortly acuminate. Inflorescences terminal on leafy branchlets, corymbose, 3-5-flowered; petals yellowish green. Nutlets convex, glabrous. Fl. Apr. Fr. Sep. Forests or valleys at 500-1200 m. Distributed in SW, S, C, SE and E China.

## 厚叶槭
**Acer crassum** Hu et W. C. Cheng

常绿乔木，高10-12米。冬芽卵球形或卵状圆锥形，鳞片多数。叶厚革质，长圆椭圆形或椭圆形，稀长圆倒卵形，长8-14厘米，宽3.5-6厘米，全缘；背面灰绿色，微被白粉；侧脉8-10对。伞房花序顶生，长5-6厘米，密被淡黄色长柔毛；花杂性。翅果嫩时紫色；翅连同小坚果长2.8-3.2厘米，张开直立。花期4月，果期9月。生海拔1000米的混交林中。产云南。

Evergreen trees, to 10-12 m tall. Winter buds ovoid or ovate-conic, scales numerous. Leaves thickly leathery, oblong-elliptic or elliptic, rarely oblong-obovate, 8-14 cm long, 3.5-6 cm wide, margin entire, abaxially gray-green, mealy; lateral veins 8-10 pairs. Inflorescences corymbose, terminal, 5-6 cm long, densely yellowish villous; flowers polygamous. Samaras purple when young; wing including nutlet 2.8-3.2 cm long, wings spreading erectly. Fl. Apr. Fr. Sep. Mixed forests at 1000 m. Distributed in Yunnan.

## 滨海槭
**Acer sino-oblongum** Metc.

常绿乔木，雄花两性花同株。树皮粗糙，深灰褐色或深灰色。小枝细瘦，无毛。叶革质，全缘，正面绿色或淡绿色，无毛，背面被白粉，侧脉5-7对。伞房花序顶生。小坚果强烈凸起。花期4月，果期9月。生近海岸常绿林中。产广东。

Evergreen trees, andromonoecious. Bark coarse, dark brownish gray or dark gray. Branchlets slender, glabrous. Leaves leathery, entire, adaxia-

厚叶槭 *Acer crassum*

lly green or pale green, glabrous, abaxially whitish glaucous, lateral veins 5-7 pairs. Corymbs terminal. Nutlets strongly convex. Fl. Apr. Fr. Sep. Evergreen forests near sea shores. Distributed in Guangdong.

光叶槭 *Acer laevigatum*

## 光叶槭

**Acer laevigatum** Wall.

常绿乔木。树皮灰色、灰褐色或深灰色。叶革质，披针形或长圆披针形，幼时背面脉腋具簇毛，后无毛，正面无毛。伞形圆锥花序；花5数，花瓣白色。小坚果强烈凸起。花期4月，果期8-9月。生海拔1000-2000米的山谷林中或溪边。产中国西南、华南、华中和华西。印度北部、不丹、尼泊尔、缅甸和越南亦有。

Evergreen trees. Bark gray, gray-brown or dark gray. Leaves leathery, lanceolate or oblong-lanceolate, abaxially with tufts of hairs at vein axils when young, then glabrous, adaxially glabrous. Inflorescences corymbose-paniculate; flowers 5-merous; petals white. Nutlets strongly convex. Fl. Apr. Fr. Aug-Sep. Forests in valleys or along streams at 1000-2000 m. Distributed in SW, S, C and W China. Also in N India, Bhutan, Nepal, Myanmar and Vietnam.

## 罗浮槭

**Acer fabri** Hance

常绿小乔木。树皮褐灰色。叶柄无毛；叶长圆形、披针形、长圆状倒披针形或椭圆形，革质，背面无毛，不分裂，全缘。花序圆锥状。小坚果凸起，幼翅红色；果梗长1-1.5厘米，细瘦，无毛。花期3-4月，果期9月。生海拔500-2000米的林中或山谷。产中国西南、华南和华中。越南亦有。

Evergreen small trees. Bark brownish gray. Petioles glabrous; leaves oblong, lanceolate, oblong-oblanceolate, or elliptic, leathery, abaxially glabrous, undivided, margin entire. Inflorescences paniculate. Nutlets convex, unmature wings red; fruiting pedicels 1-1.5 cm long, slender, glabrous. Fl. Mar-Apr. Fr. Sep. Forests or valleys at 500-2000 m. Distributed in SW, S and C China. Also in Vietnam.

罗浮槭 *Acer fabri*

疏花槭 *Acer laxiflorum*

## 疏花槭
**Acer laxiflorum** Pax

落叶乔木。小枝树皮有绿色条纹。叶长圆状卵形，纸质，侧裂片常不明显，边缘具糙齿，先端急尖。总状花序顶生，下垂，花不及10朵，后叶开放；花瓣5，绿黄色。翅果的翅基部较宽。花期4月，果期9月。生海拔1800-2500米的混生林中。产云南和四川。

Deciduous trees. Young branch bark green striated. Leaves oblong-ovate, papery, lateral lobes usually inconspicuous, margin coarsely serrate, apex acute. Racemes terminal, pendulous, less than 10-flowered, appearing after leaves; petals 5, greenish yellow. Wings of samaras wide at base. Fl. Apr. Fr. Sep. Mixed forests at 1800-2500 m. Distributed in Yunnan and Sichuan.

## 青榨槭
**Acer davidii** Franch.

落叶乔木。小枝树皮有绿色条纹。叶卵形或卵状长圆形，长8-12厘米，基部近心形或圆形，边缘具锯齿和不等的粗齿，不裂或微3裂，裂片顶端钝。总状花序悬垂，具20-30花；花杂性，黄绿色。翅果绿色至黄棕色。花期4月，果期9月。生海拔1000-3000米的混生林中。产中国西南、华南、华西、华中和华东。缅甸亦有。

Deciduous trees. Young branch bark green striated. Leaves ovate or ovate-oblong, 8-12 cm long, base subcordate or rounded, margin serrate with unequal crenations, not lobed or slightly 3-lobed, lobes apically obtuse. Racemes pendulous, 20-30-flowered; flowers polygamous, yellow-green. Samaras green to yellow-brown. Fl. Apr. Fr. Sep. Mixed forests at 1000-3000 m. Distributed in SW, S, W, C and E China. Also in Myanmar.

## 葛萝槭
**Acer davidii** Franch. subsp. **grosseri** (Pax) P. C. de Jong

本亚种与青榨槭的区别在于本亚种的叶近圆状心形，基部近心形，边缘具重锯齿及平伏的锐齿，5裂，中裂片三角形，先端渐尖；侧裂片和基部裂片顶端锐尖，在老枝上全缘。花期4月，果期9月。生海拔1000-1600米的混生林中。产中国东南、华中和华北。

This subspecies differs from *Acer davidii* in its leaves suborbicular-ovate, base subcordate, margin doubly serrate with appressed acute teeth, 5-lobed, middle lobe

青榨槭 *Acer davidii*

葛萝槭 *Acer davidii* subsp. *grosseri*

triangular, apex acuminate; lateral and basal lobes apically acute, entire on old branches. Fl. Apr. Fr. Sep. Mixed forests at 1000-1600 m. Distributed in SE, C and N China.

## 丽江槭
**Acer forrestii** Diels

落叶乔木，高达17米，雄花两性花同株。小枝树皮有绿色条纹。叶基部心形，叶缘具重锯齿，3裂或不裂，叶背具白粉。花序顶生于多叶的小枝上，总状，无毛，具5-20花。翅果紫色至黄褐色；果梗长6-8毫米，细瘦，无毛。花期5月，果期9月。生海拔3000-3800米的混交林或山谷中。产云南和四川。

Deciduous trees, up to 17 m tall, andromonoecious. Young branch bark green striated. Leaves cordate at base, margin doubly serrulate, 3-lobed or not, abaxially glaucescent. Inflorescences terminal on leafy branchlets, racemose, glabrous, 5-20-flowered. Samaras purple to yellow-brown; fruiting pedicel 6-8 mm long, slender, glabrous. Fl. May. Fr. Sep. Mixed forests or valleys at 3000-3800 m. Distributed in Yunnan and Sichuan.

## 滇藏槭
**Acer wardii** W. W. Sm.

落叶乔木，雌雄异株。叶卵形，长7-9厘米，边缘有紧密的细锯齿，常3裂；裂片先端长尾状锐尖；叶柄长3-5厘米。花序着生于叶的小枝顶端，圆锥状总状；花单性。翅果成熟后紫黄色；小坚果微扁平，长圆形；翅连小坚果长2.2-2.5厘米，张开成钝角；果梗长1-2厘米。花期5月，果期9月。生海拔2400-3600米的高山林中。产云南和西藏。缅甸和印度亦有。

Deciduous trees, dioecious. Leaves ovate, 7-9 cm long, margin densely serrulate, usually 3-lobed; lobes apex elongated caudate-acuminate; petioles 3-5 cm long. Inflorescences terminal on leafy branchlets, paniculate-racemose; flowers unisexual. Samaras purplish yellow when mature; nutlets flat, oblong; wing including nutlet 2.2-2.5 cm long, wings spreading obtusely; fruiting pedicels 1-2 cm long. Fl. May. Fr. Sep. Alpine forests at 2400-3600 m. Distributed in Yunnan and Xizang. Also in Myanmar and India.

丽江槭 *Acer forrestii*

滇藏槭 *Acer wardii*

玉山槭 *Acer morrisonense*

## 玉山槭

**Acer morrisonense** Hayata

落叶乔木。小枝无毛；冬芽椭圆体形，外部鳞片无毛。叶纸质，近圆状卵形，长8-10厘米，基部近截形或近心形，背面无毛，侧脉5-6对，边缘有具粗锐尖齿的重锯齿，5浅裂；中裂片短卵形，顶端渐尖。果序总状；果梗7-10毫米；小坚果近椭圆体形；翅连同翅长1.8-2.3厘米，展开成钝角。花期3-4月，果期10月。生海拔1800-2200米的混交林中。产台湾。

Deciduous trees. Branchlets glabrous; winter buds ellipsoid, outer scales glabrous. Leaves papery, suborbicular-ovate, 8-10 cm long, base nearly truncate or subcordate, abaxially glabrous, lateral veins 5 or 6 pairs, margin doubly serrate with coarse acute teeth, 5-lobed; middle lobe shortly ovate, apex acuminate. Infructescence racemose; fruiting pedicel 7-10 mm long; nutlets subellipsoid; wing including nutlet 1.8-2.3 cm long, wings spreading obtusely. Fl. Mar-Apr. Fr. Oct. Mixed forests at 1800-2200 m. Distributed in Taiwan.

## 五尖槭

**Acer maximowiczii** Pax

落叶乔木，雄花两性花同株。小枝树皮有绿色条纹。叶卵形或三角状卵形，纸质，明显5裂，中裂片与侧裂片顶端渐尖，基部裂片顶端锐尖，边缘具锐锯齿。花序顶生于多叶小枝，下垂，总状，完全于叶后开放。翅果张开成钝角。花期5月，果期9月。生海拔1800-2500米的林缘或疏林中。产中国西南、华中、华北和华西。

Deciduous trees, andromonoecious. Young branch bark green striated. Leaves ovate or triangular-ovate, papery, distinctly 5-lobed, middle and lateral lobes apically acuminate, basal lobes apically acute, margin sharply serrulate. Inflorescences terminal on leafy branchlets, pendulous, racemose, appearing after leaves develop fully. Samaras dehiscent to obtuse angle. Fl. May. Fr. Sep. Forest edges or sparse forests at 1800-2500 m. Distributed in SW, C, N and W China.

## 南岭槭

**Acer metcalfii** Rehd.

落叶乔木，高达10米。叶近卵形，近革质，长10-14厘米，3裂；裂片三角状卵形，先端锐尖，边缘常有具钝齿的稀疏锯齿；叶柄长2-3厘米，无毛。花序总状。6-9枚黄褐色翅果成总状果序；小坚果长8毫米；翅连同小坚果长2.2-2.5厘米，张开成钝角；果梗长5毫米，无毛。果期9月。生海拔800-1500米的混交林或溪边。产湖南、贵州、广东和广西。

Deciduous trees, up to 10 m tall. Leaves subovate, subleathery, 10-14 cm long, 3-lobed; lobes triangular-ovate, apex acuminate, margin usually coarsely serrate with obtuse teeth; petioles 2-3 cm long, glabrous. Inflorescences racemose. Racemose infructe-

五尖槭 *Acer maximowiczii*

南岭槭 *Acer metcalfii*

篦齿槭 *Acer pectinatum*

青楷槭 *Acer tegmentosum*

scences with 6-9 yellowish-brown samaras; nutlets 8 mm long; wing including nutlet 2.2-2.5 cm long, wings spreading obtusely; fruiting pedicels 5 mm long, glabrous. Fr. Sep. Mixed forests or by streamsides at 800-1500 m. Distributed in Hunan, Guizhou, Guangdong and Guangxi.

## 篦齿槭

**Acer pectinatum** Wall. ex Nichols.

落叶乔木，雌雄异株。叶近圆形，长7-10厘米，背面嫩时沿叶脉被棕色长柔毛，3裂或基部的裂片发育而成5裂；叶柄无毛。花单性。果序长6-7厘米；小坚果微扁平；翅镰刀形，连同小坚果长1.6-2.5厘米，张开近水平；果梗长5-7毫米，无毛。花期4月，果期9月。生海拔2300-3700米的混交林中。产四川、西藏和云南。东南亚和印度亦有。

Deciduous trees, dioecious. Leaves suborbicular, 7-10 cm long, abaxially densely rufous pubescent on veins when young, 3-lobed or 5-lobed when basal lobe developed; petioles glabrous. Flowers unisexuail. Infructescence 6-7 cm long; nutlets slightly flat; wing falcate, 1.6-2.5 cm long including nutlets, wings spreading nearly horizontally; fruiting petioles 5-7 mm long, glabrous. Fl. Apr. Fr. Sep. Mixed forests at 2300-3700 m. Distributed in Sichuan, Xizang, Yunnan. Also in SE Asia and India.

## 青楷槭

**Acer tegmentosum** Maxim.

落叶乔木，雄花两性花同株。小枝树皮有绿色条纹。树皮深灰色。叶背面浅绿色，脉腋具黄色簇毛，上面无毛，主脉5条，边缘具重锯齿，叶常3或5裂。总状花序下垂，具15朵花。果翅开展成钝角至近水平。花期4月，果期9月。生海拔500-1000米的针叶林或杂木林中。产辽宁、吉林和黑龙江。俄罗斯东部和朝鲜半岛亦有。

Deciduous trees, andromonoecious. Young branch bark green striated. Leaves abaxially pale green, with yellowish barbed hairs at vein axils, adaxially glabrous, primary veins 5, margin doubly serrate, usually 3- or 5-lobed. Inflorescences pendulous, racemose, 15-flowered. Wings spreading obtusely or nearly horizontally. Fl. Apr. Fr. Sep. Coniferous or mixed forests at 500-1000 m. Distributed in Liaoning, Jilin and Heilongjiang. Also in E Russia and Korean Peninsula.

## 十蕊槭

**Acer laurinum** Hassk.

落叶乔木，雌雄异株。冬芽具7-11对鳞片。叶革质或近革质，全缘，卵状椭圆形或长圆卵形，8-15厘米；侧脉5-6对；叶柄长5-7厘米，无毛。花序聚伞状圆锥状；花单性；子房无毛。小坚果扁平；翅连同小坚果共长6-7厘米，张开近锐角或直角；果梗细瘦，无毛。花期6-9月，果期9-12月。生海拔700-2500米的常绿林中。产华南和西藏。东南亚和印度亦有。

Deciduous trees, dioecious. Winter buds with 7-11-paired scales. Leaves leathery or subleathery, margin entire, ovate-elliptic or oblong-ovate, 8-15 cm long; lateral veins 5-6 pairs; petioles 5-7 cm long, glabrous. Inflorescences cymose-paniculate; flowers unisexual; ovary glabrous. Nutlets flat; wing 6-7 cm long including nutlets, wings spreading nearly acutely or erectly; fruiting pedicels slender, glabrous. Fl. Jun-Sep. Fr. Sep-Dec. Evergreen forests at 700-2500 m. Distributed in S China and Xizang. Also in SE Asia and India.

十蕊槭 *Acer laurinum*

楠叶槭 *Acer pinnatinervium*

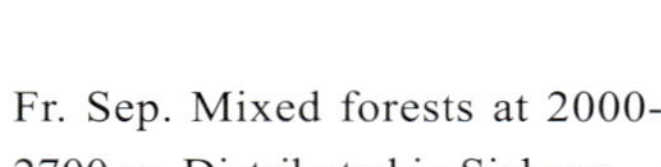

毛叶槭 *Acer stachyophyllum*

## 楠叶槭

**Acer pinnatinervium** Merr.

常绿乔木，高达10米，雄花两性花异株。树皮灰棕色或暗棕色，粗糙。单叶，具羽状脉，侧脉8-12对，背面无毛。伞形圆锥花序；子房淡紫色，被长柔毛。翅连同小坚果长2-4厘米，无毛。花期8-10月，果期12月至翌年2月。生海拔500-2400米的杂木林中或散生。产云南和西藏。印度北部和泰国亦有。

Evergreen trees, up to 10 m tall, androdioecious. Bark gray-brown or dark brown, rough. Leaves simple, leaves pinnati-nerved, lateral veins 8-12 pairs, abaxially glabrous. Inflorescences cymose-paniculate; ovary purplish, villous. Wings including nutlets 2-4 cm long, glabrous. Fl. Aug-Oct. Fr. Dec to next Feb. Mixed forests or scattered at 500-2400 m. Distributed in Yunnan and Xizang. Also in N India and Thailand.

## 毛叶槭

**Acer stachyophyllum** Hiern

落叶乔木，雌雄异株。幼枝树皮青绿色。叶卵形或长圆形，不分裂或3裂，纸质，背面密具灰色或白色柔毛。雄花总状，无毛，具5-8花；花瓣4，黄绿色。果序下垂，翅开展成直角。花期4-5月，果期9-10月。生海拔1400-3500米的高山林中。产中国西南、华中和华西。喜马拉雅亦有。

Deciduous and dioecious trees. Bark of young branchets green. Leaves ovate or oblong, undivided or 3-lobed, papery, abaxially densely pale or white pubescent. Staminate inflorescences racemose, glabrous, 5-8-flowered; petals 4, yellowish green. Infructescences pendulous, wings spreading erectly. Fl. Apr-May. Fr. Sep-Oct. Alpine forests at 1400-3500 m. Distributed in SW, C and W China. Also in Himalaya.

## 雷波槭

**Acer leipoense** Fang et Soong

落叶乔木。冬芽卵圆形，鳞片覆瓦状排列。叶纸质，近圆形卵形，长9-11厘米，上段3裂；裂片三角状卵形，边缘有稀疏的钝齿或稍呈浅波状，背面被白粉及短柔毛。花单性；花瓣5。果序总状，长15-25厘米；果梗无毛，长2.5-4厘米；小坚果强烈凸起；翅镰刀形，连同小坚果长4-4.5厘米，张开成锐角或直角。果期9月。生海拔2000-2700米的混交林中。产四川。

Deciduous trees. Winter buds ovoid, scales imbricate. Leaves papery, suborbicular-ovate, 9-11 cm long, upper parts 3-lobed; lobes triangular-ovate, margin usually sparsely crenate or shallowly undulate, abaxially powdery and pubescent. Flowers unisexual; petals 5. Infructescences racemose, 15-25 cm long; fruiting pedicels glabrous, 2.5-4 cm long; nutlets strongly convex; wing falcate, 4-4.5 cm long including nutlets, wings spreading acutely or erectly. Fr. Sep. Mixed forests at 2000-2700 m. Distributed in Sichuan.

## 苹婆槭

**Acer sterculiaceum** Wall.

落叶乔木，高约20米，雌雄同株。冬芽卵圆形。叶长15-20厘米，常5裂；裂片先端钝尖，边缘具稀疏的粗锯齿。总状花序由小枝旁边生出，被长柔毛；花单性。翅果嫩时淡紫绿色，有长柔毛；小坚果凸起；翅连同小坚果长5-6厘米，张开近直立。花期4-5月，果期9月。生海拔1800-3100米的林中或山谷。产西藏、华中和西南。东南亚和印度亦有。

Deciduous trees, up to ca. 20 m tall, monoecious. Winter buds ovoid. Leaves 15-20 cm long,

雷波槭 *Acer leipoense*

苹婆槭 *Acer sterculiaceum*

房县槭 *Acer sterculiaceum* subsp. *franchetii*

usually 5-lobed; lobes apex acuminate, margin with sparsely crenate. Racemes produced from branchlets side, pilose; flowers unisexual. Samaras pale purplish green when young, villous; nutlets convex; wing 5-6 cm long including nutlets, wings spreading erectly. Fl. Apr-May. Fr. Sep. Forests or valleys at 1800-3100 m. Distributed in Xizang, C and SW China. Also in SE Asia and India.

## 房县槭

**Acer sterculiaceum** Wall. subsp. **franchetii** (Pax) A. E. Murray

落叶乔木。树皮深棕色。叶常3裂，稀5裂，裂片钝。总状花序或圆锥状总状花序侧生；花瓣5，黄绿色。翅连同小坚果长4-6.5厘米；小坚果近球形。花期5月，果期9月。生海拔1800-2500米的山谷林中。产云南、四川、贵州、湖北、河南和陕西。

Deciduous trees. Bark dark brown. Leaves usually 3-lobed, rarely 5-lobed, lobes obtuse. Racemes or paniculate racemes lateral; petals 5, yellowish green. Samaras wings including nutlets 4-6.5 cm long; nutlets subglobose. Fl. May. Fr. Sep. Forests in valleys at 1800-2500 m. Distributed in Yunnan, Sichuan, Guizhou, Hubei, Henan and Shaanxi.

## 贡山槭

**Acer kungshanense** Fang et C. Y. Chang

落叶乔木，高15-20米。叶近革质，长和宽均15-25厘米，3裂；裂片卵形，先端锐尖，边缘具稀疏的圆齿；背面密被宿存的淡黄色短柔毛。果序圆锥状，长7-9厘米，有短柔毛；果梗粗壮，长1.5-3厘米；小坚果凸起，近于球形，有淡黄色疏柔毛；翅镰刀形，连同小坚果长4-4.5厘米，伸展近直立或成锐角。果期9月。生海拔2000-3200米的混交林或山谷中。产云南。

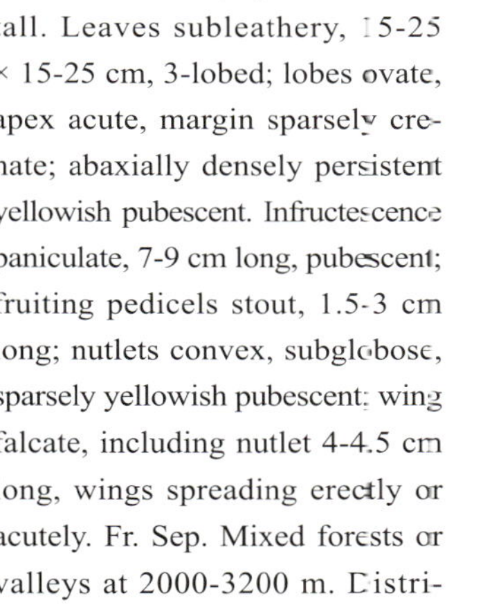

Deciduous trees, up to 15-20 m tall. Leaves subleathery, 15-25 × 15-25 cm, 3-lobed; lobes ovate, apex acute, margin sparsely crenate; abaxially densely persistent yellowish pubescent. Infructescence paniculate, 7-9 cm long, pubescent; fruiting pedicels stout, 1.5-3 cm long; nutlets convex, subglobose, sparsely yellowish pubescent; wing falcate, including nutlet 4-4.5 cm long, wings spreading erectly or acutely. Fr. Sep. Mixed forests or valleys at 2000-3200 m. Distributed in Yunnan.

## 血皮槭

**Acer griseum** (Franch.) Pax

落叶乔木，高达20米。树皮橙棕色或红褐色，撕裂。掌状复叶有3小叶；小叶片椭圆形或椭圆状长圆形，边缘有粗齿。聚伞花序通常有3朵小花；花5数。翅果棕黄色，小坚果球形，密被毛；翅张开成近直角。花期4月，果期9月。生海拔1500-2000米的混交林中。产重庆、湖北、湖南、河南、山西、陕西和甘肃。

Deciduous trees, up to 20 m tall. Bark orange-brown or red-brown, fissured. Leaves with 3 leaflets per petiole; leaflet blades elliptic or elliptic-oblong, margin coarsely dentate. Inflorescences cymose, usually 3-flowered; flowers 5-merous. Fruit yellowish brown; nutlets globose, densely pubescent; wings spreading at 90° or nearly erectly. Fl. Apr. Fr. Sep. Mixed forests at 1500-2000 m. Distributed in Chongqing, Hubei, Hunan, Henan, Shanxi, Shaanxi and Gansu.

贡山槭 *Acer kungshanense*

血皮槭 *Acer griseum*

东北槭 *Acer mandshuricum*

复叶槭 *Acer negundo*

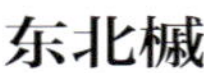

## 东北槭
**Acer mandshuricum** Maxim.

落叶乔木，高常达20米。复叶具3小叶；小叶披针形或长圆披针形，长5-10厘米，先端锐尖，边缘具钝锯齿；叶柄长4-7厘米。伞房花序仅具3-5花；花杂性，雄花两性花异株。小坚果凸起，紫褐色；翅连同小坚果长3-3.5厘米，张开成锐角，钝角或近于直角。花期6月，果期9月。生海拔500-2300米的杂交林中。产中国东北、陕西和甘肃。朝鲜半岛和俄罗斯亦有。

Deciduous trees, usually up to 20 m tall. Compound leaves 3-foliolate; leaflets lanceolate or oblong-lanceolate, 5-10 cm long, apex acute, margin obtusely serrate; petioles 4-7 cm long. Inflorescences corymbose, 3-5-flowered; flowers polygamous, androdioecious. Nutlets convex, purplish brown; wing including nutlet 3-3.5 cm long, wings spreading acutely, obtusely, or nearly erectly. Fl. Jun. Fr. Sep. Mixed forests at 500-2300 m. Distributed in NE China, Shaanxi and Gansu. Also in Korean Peninsula and Russia.

角叶槭 *Acer sycopseoides*

## 角叶槭
**Acer sycopseoides** F. Chun

常绿乔木。小枝、花序、叶背及叶柄密具柔毛。叶柄长约2.5厘米；叶革质，3出脉，边缘强烈反卷，具1或2个退化裂片。果序顶生，聚伞状，密被柔毛，具少量翅果。小坚果强烈凸起。花期4月，果期9月。生海拔600-2100米的林中。产云南、贵州和广西。

Evergreen trees. Branchlets, inflorescences, abaxially of leaves and petioles densely pubescent. Petioles ca. 2.5 cm long; leaves leathery, 3-veined, margin strongly revolute and shortly 1- or 2-lobulate. Infructescences terminal, corymbose, densely pubescent, with few samaras. Nutlets strongly convex. Fl. Apr. Fr. Sep. Forests at 600-2100 m. Distributed in Yunnan, Guizhou and Guangxi.

## 复叶槭
**Acer negundo** L.

落叶乔木，高达20米，雌雄异株。羽状复叶具有3-7(-9)枚小叶；小叶卵形或椭圆状披针形。雄花的花序聚伞状，雄花常4朵；雌花的花序总状，下垂；花4数。瘦果黄褐色；翅开展成锐角或近直角。花期4月，果期9月。中国广泛栽培和归化。原产北美洲。

Deciduous trees, up to 20 m tall, dioecious. Pinnate compound leaves with 3-7(-9) leaflets; leaflets ovate or elliptic-lanceolate.

建始槭 *Acer henryi*

Male inflorescences cymose, staminate flowers usually 4; female inflorescences racemose, pendulous; flowers 4-merous. Samaras brownish yellow; wings spreading acutely or nearly erectly. Fl. Apr. Fr. Sep. Widely cultivated and naturalized in China. Native to North America.

## 建始槭

**Acer henryi** Pax

落叶乔木，高达10米，雌雄异株。掌状复叶有3小叶；小叶边缘全缘或有疏锯齿。总状花序下垂；花4数，黄绿色，花梗极短或无花梗。小坚果强烈凸起，长圆形；翅开展成近直角。花期4月，果期9月。生海拔500-1500米的混交林中。产山西、陕西、西南、东南、华中和华东。

Deciduous trees, up to 10 m tall, dioecious. Leaves 3-foliolate, leaflet margin entire or remotely serrate apically. Inflorescences pendulous, racemose; flowers 4-merous, yellow-green, subsessile or sessile. Nutlets strongly-convex, oblong; wings spreading at 90° or erectly. Fl. Apr. Fr. Sep. Mixed forests at 500-1500 m. Distributed in Shanxi, Shaanxi, SW, SE, C and E China.

## 五小叶槭

**Acer pentaphyllum** Diels

落叶乔木，高达10米。掌状复叶有(4-)5(-7)片小叶；小叶狭披针形或披针形。伞房花序；花5数。小坚果强烈凸起；翅开展成近直角。花期4月，果期9月。生海拔2300-2900米的河谷混交林中。产四川西南部。

Deciduous trees, up to 10 m tall. Leaves with (4-)5(-7) leaflets; leaflet narrowly lanceolate or lanceolate. Inflorescences corymbose; flowers 5-merous. Nutlets strongly convex; wings spreading at 90°. Fl. Apr. Fr. Sep. Mixed forests in valleys at 2300-2900 m. Distributed in SW Sichuan.

五小叶槭 *Acer pentaphyllum*

## 漾濞槭

**Acer yangbiense** Y. S. Chen et Q. E. Yang

落叶乔木，高达20米。叶5浅裂，背面有灰绿色绒毛，基部心形。总状花序下垂，生于2-3年生枝条的侧芽；花5数。小坚果凸起，球形，有长毛。翅开展成锐角或近直角。花期4月，果期9月。生海拔2200-2400米的山谷混交林中。产云南(漾濞县苍山)。

Deciduous trees, up to 20 m tall. Leaves shallowly 5-lobed, abaxially pale green tomentose, base cordate. Racemes pendulous, arising from leafless lateral buds of 2-3-year-old branchlets; flowers 5-merous. Nutlets convex, globose, villous. Wings spreading at acute or nearly right angle. Fl. Apr. Fr. Sep. Mixed forests in valleys at 2200-2400 m. Distributed in Yunnan (Cangshan Mountain, Yangbi County).

漾濞槭 *Acer yangbiense*

# 七叶树科 Hippocastanaceae

## 七叶树

**Aesculus chinensis** Bunge

乔木。叶柄长7-15厘米；小叶5-7(-9)，纸质，背面光滑或幼时沿脉具灰色绒毛，基部楔形或阔楔形。花瓣4，白色，有黄斑，长圆状倒卵形至长圆状倒披针形。蒴果球形或倒卵球形。花期4-5月，果期9-10月。栽培于海拔800米以下地区，特别是庙宇和房舍旁。产华北和华东。

Trees. Petioles 7-15 cm long; leaves 5-7(-9)-foliolate; leaflets papery, abaxially glabrous, or grayish tomentose on veins when young, base cuneate to broadly so. Petals 4, white, with yellow spots, subequal, oblong-obovate to oblong-oblanceolate. Capsules globose or obovoid. Fl. Apr-May. Fr. Sep-Oct. Cultivated, especially by temples and by houses below 800 m. Distributed in N and E China.

七叶树 *Aesculus chinensis*

## 天师栗

**Aesculus chinensis** Bunge var. **wilsonii** (Rehd. ) Turland et N. H. Xia

本变种与七叶树的区别在于本变种的小叶背面多少被一致的灰绒毛或长柔毛，或沿脉具浅灰色的绒毛(有时仅在幼时)，基部阔楔形至圆形或稍心形。花期4-6月，果期9-10月。生海拔600-2000(-2300)米的阔叶林、溪旁或灌丛中。产中国西南、华北、华南和华西。

This variety differs from the typical variety in its leaflets abaxially ± uniformly grayish tomentose or villous, or grayish tomentose on veins (sometimes only when young), base broadly cuneate to rounded or slightly cordate. Fl. Apr-Jun. Fr. Sep-Oct. Broad-leaved forests, by streams or thickets at 600-2000(-2300) m. Distributed in SW, N, S and W China.

天师栗 *Aesculus chinensis* var. *wilsonii*

## 小果七叶树

**Aesculus tsiangii** Hu et Fang

落叶乔木，高达32米。掌状复叶直径约30厘米；小叶5-8枚，近革质，倒披针形或倒卵状长圆形，基部楔形，两面无毛，侧脉21-24对；小叶柄长3-7毫米。花序基部直径达10厘米，连同8厘米长的花序梗总长达35厘米。蒴果卵圆形，无刺，顶端具短尖头。种子近球形，种脐约占种子的1/2。花期4月，果期7月。生海拔350-400米的林中。产贵州和广西。

Deciduous trees, up to 32 m tall. Leaves palmately compound, ca. 30 cm diam; leaflets 5-8, subleathery, oblanceolate or obovate-oblong, cuneate at base, glabrous on both surfaces, with lateral veins 21-24 paired; petiolules 3-7 mm long. Inflorescences up to 10 cm diam at base, 35 cm long including 8 cm long peduncle. Capsules ovate, unarmed, mucronate at apex. Seeds subglobous, hilums about 1/2 as long as seeds. Fl. Apr. Fr. Jul. Forests at 350-400 m. Distributed in Guizhou and Guangxi.

小果七叶树 *Aesculus tsiangii*

## 长柄七叶树

**Aesculus assamica** Griff.

乔木。叶柄长8-30厘米；小叶5-9，近革质，背面无毛，或幼时沿脉具微柔毛或柔毛，基部楔形或宽楔形或圆形。花瓣4，白色或淡黄色，具紫色或褐色斑点，2个匙形至长圆形，2个长圆倒卵形或倒卵形。蒴果卵形至倒卵形，近球形或扁球形。花期2-5月，果期6-11月。生海拔100-2000米的阔叶林中。产广西、贵州、云南和西藏。印度东北部、不丹、孟加拉国、泰国、缅甸北部和越南北部亦有。

Trees. Petioles 8-30 cm; leaves 5-9-foliolate; petiolules 0.3-1.5 cm, 5-9-foliolate; leaflets subleathery, abaxially glabrous, or puberulent/pubescent on veins when young, base cuneate or broadly cuneate or rounded. Petals 4, white or pale yellow with purple or brown spots, 2 spatulate to oblong and 2 oblong-obovate or obovate. Capsules ovate to obovate, nearly spherical or oblate. Fl. Feb-May. Fr. Jun-Nov. Broad-leaved forests at 100-2000 m. Distributed in Guangxi, Guizhou, Yunnan and Xizang. Also in NE India, Bhutan, Bangladesh, Thailand, N Myanmar and N Vietnam.

长柄七叶树 *Aesculus assamica*

# 无患子科 Sapindaceae

## 倒地铃

**Cardiospermum halicacabum** L.

一年生攀援草本。二回三出复叶，轮廓三角形。圆锥花序少花，基部具卷须，与叶近等长或稍长；花瓣乳白色，倒卵形。蒴果膜质，褐色，梨形、螺旋状倒三角形或有时几椭圆体形。花期夏秋，果期秋冬。生海拔约1850米的林缘、灌丛、草地或废弃地。产中国西南、华南和华东。全球热带和亚热带亦有。

Annual climbing herbs. Leaves biternate, triangular in outline. Panicles few-flowered, with basal tendrils, as long as or slightly longer than leaves; petals milky-white, obovate. Capsules membranous, brown, pear-like, turbinate-obtriangular or sometimes nearly ellipsoid. Fl. summer-autumn. Fr. autumn-winter. Forest edges, shrublands, grasslands or wastelands at ca. 1850 m. Distributed in SW, S and E China. Also in tropical and subtropical regions of the world.

无患子 *Sapindus saponaria*

## 无患子

**Sapindus saponaria** L.

落叶乔木。叶及叶柄共长25-45厘米或更长；小叶5-8对，常近对生。花序顶生，直立，常三回分枝；花辐射对称；花瓣5，基部具长爪，上面具2个耳状鳞片。可育分果爿橙色，干后黑色，近球形。花期春季，果期夏秋季。通常栽培于寺庙旁、庭院或村旁。产中国西南、华南、东南、华中和华东。南亚、东南亚和东亚亦有。

Deciduous trees. Leaves with petioles 25-45 cm long or longer; leaflets 5-8 pairs, usually subopposite. Inflorescences terminal, erect, usually 3-branched; flowers actinomorphic; petals 5, basally long clawed, with 2 earlike scales at base adaxially. Fertile schizocarps orange, black when dry, subglobose. Fl. spring. Fr. summer-autumn. Usually cultivated by temples, in gardens, or by villages. Distributed in SW, S, SE, C and E China. Also in S, SE and E Asia.

## 毛瓣无患子

**Sapindus rarak** DC.

落叶乔木，高约20米。小叶7-12对，长圆状披针形；叶轴和小叶下面无毛。花序顶生，直立；花两侧对称，花瓣4，不具爪，上面基部具一个大的鳞片；萼片和花瓣下面密具丝毛；花蕾阔卵球形。花期夏季，果期初秋。生海拔500-2100米的疏林中。产云南和台湾。东南亚和南亚亦有。

Deciduous trees, ca. 20 m tall. Leaflets 7-12 pairs, oblong-lanceolate; leaf axis and abaxial surface of leaflets glabrous. Inflorescences terminal, erect; flowers zygomorphic, petals 4, not clawed, with 1 large scale at base adaxially; sepals and petals abaxially densely sericeous; flower buds broadly ovoid.

倒地铃 *Cardiospermum halicacabum*

毛瓣无患子 *Sapindus rarak*

川滇无患子（皮哨子） *Sapindus delavayi*

赛木患 *Aphania oligophylla*

Fl. summer. Fr. early autumn. Sparse forests at 500-2100 m. Distributed in Yunnan and Taiwan. Also in SE and S Asia.

## 川滇无患子 (皮哨子)

**Sapindus delavayi** (Franch.) Radlk.

落叶乔木。小叶4-7对，卵形至卵圆形，背面被长柔毛或近无毛。花序顶生，直立，常3分枝；花蕾球形，花两侧对称，花萼和花瓣外面被柔毛；花瓣4(5或6)；鳞片大，长为花瓣的近2/3。花期初夏，果期晚秋。生海拔1200-2600米的林中。产云南、四川、贵州、湖北和陕西。

Deciduous trees. Leaflets 4-7 pairs, ovate or ovate-oblong, abaxially villous or subglabrous. Inflorescences terminal, erect, often 3-branched; flower buds globose, flowers zygomorphic, sepals and petals abaxially pilose; petals 4(5 or 6); scales large, nearly as long as 2/3 of petals. Fl. early summer. Fr. late autumn. Forests at 1200-2600 m. Distributed in Yunnan, Sichuan, Guizhou, Hubei and Shaanxi.

## 茎花赤才

**Lepisanthes cauliflora** C. F. Liang et S. L. Mo

灌木至小乔木。羽状复叶互生，连叶柄长25-40厘米；小叶常4对，对生，长圆形或卵状长圆形，纸质。花序总状或聚伞圆锥状，生茎及老枝上。果实球形至扁球形，具3钝棱。种子1-3，近球形或半球形。花期9-10月。生林中。产广西。

Shrubs to small trees. Leaves paripinnate, alternate, 25-40 cm long with petioles; leaflets usually 4 pairs, opposite, oblong or ovate-oblong, papery. Inflorescences racemose or thyrsoid, on stems and old branches. Fruits globose to compressed globose, with 3 obtuse angles. Seeds 1-3, subglobose or semiglobose. Fl. Sep-Oct. Forests. Distributed in Guangxi.

## 赛木患

**Aphania oligophylla** (Merr. et Chun) H. S. Lo

常绿灌木或小乔木。小叶1或2对，薄革质或纸质，长圆形或窄长椭圆形，全缘，上面有光泽。聚伞圆锥花序顶生，主轴粗壮；花梗纤细；花小，单性。果卵圆形，绿色，光滑。花期春季，果期夏季。生密林中。产海南东南部和南部。

Evergreen shrubs or small trees. Leaflets 1 or 2 pairs, thinly coriaceous or chartaceous, oblong or narrowly elliptic, margin entire, adaxially shiny. Cymose panicles terminal, peduncles stout; pedicels slender; flowers small, monoecious. Fruits ovoid, green, smooth. Fl. spring. Fr. summer. Dense forests. Distributed in SE and S Hainan.

茎花赤才 *Lepisanthes cauliflora*

龙眼 *Dimocarpus longan*

荔枝 *Litchi chinensis*

## 龙眼

**Dimocarpus longan** Lour.

常绿乔木。小枝散生苍白色皮孔。小叶(3或)4或5(或6)对，长圆状椭圆形至长圆状披针形，常两侧不对称。花序大，多分枝，密被星状毛；花瓣5。果实常黄褐色或有时灰黄色，近球形，外面具皱及稍具突起小瘤。花期春夏间，果期夏季。原产云南、广西、广东和海南；中国南方广泛栽培。南亚和东南亚亦有。

Evergreen trees. Branchlets sparsely clothed with white lenticels. Leaflets (3 or)4 or 5(or 6) pairs, oblong-elliptic to oblong-lanceolate, often bilaterally asymmetrical. Inflorescences large, many-branched, densely stellate hairy; petals 5. Fruits usually yellowish brown or sometimes grayish yellow, subglobose, abaxially rugose and with slightly prominent tubercles. Fl. spring and summer. Fr. summer. Native to Yunnan, Guangxi, Guangdong and Hainan; widely cultivated in south of China. Also in S and SE Asia.

## 荔枝

**Litchi chinensis** Sonn.

常绿乔木。小叶2或3(或4)对，披针形或卵状披针形，有时椭圆状披针形，薄革质或革质，下面粉绿色，无毛。花序顶生，具绒毛；萼片瓣裂。果卵球形或近球形，外皮红色。种子全为白色、肉质、多汁的假种皮包围。花期春季，果期夏季。生海拔1200米以下的林内，或栽培。原产广东西南部和海南；中国南方栽培。东南亚各国亦有；亚热带有引进。

Evergreen trees. Leaflets 2 or 3(or 4) pairs, lanceolate or ovate-lanceolate, sometimes elliptic-lanceolate, thinly leathery or leathery, abaxially glaucous, glabrous. Inflorescences terminal, tomentose; sepals valvate. Fruits ovoid or subglobose, exterior red. Seeds entirely enveloped by white, fleshy and juicy arils. Fl. spring. Fr. summer. Forests below 1200 m, or culticated. Native to SW Guangdong and Hainan; cultivated in S China. Also in S and SE Asia; introduced to subtropical regions.

## 番龙眼

**Pometia pinnata** J. R. et G. Forst.

常绿乔木。叶连柄长达1.5米；小叶5-9(-15)对，第一对叶圆形，基部心形、托叶状；其余的长圆形或上部的近楔形，边缘具整齐的锯齿。花序分枝粗且挺直，被微柔毛；花梗基部有关节；萼片被微柔毛；花瓣倒卵状三角形，长为萼片的2倍。果椭圆形或近球形，长3厘米，无毛。生雨林中。产台湾和云南。南亚、东南亚和太平洋岛屿亦有。

Evergreen trees. Leaves with Petioles up to 1.5 m long; leaflets 5-9 (-15) pairs, first pair blades orbicular, base cordate, stipulelike;

番龙眼 *Pometia pinnata*

others oblong or upper ones nearly cuneate, margin regularly serrate. Inflorescences branches strong and straight, pilosulose; pedicels jointed at base; sepals pilosulose; petals obovate-triangular, twice as long as sepals. Fruit ellipsoid or subglobose, 3 cm long, glabrous. Rain forests. Distributed in Taiwan and Yunnan. Also in S and SE Asia, and Pacific Islands.

红毛丹 *Nephelium lappaceum*

## 红毛丹
**Nephelium lappaceum** L.

常绿乔木。叶连柄长15-45厘米；小叶常2或3对，椭圆形或倒卵形，长6-18厘米，基部楔形，全缘，两面无毛；侧脉7-9对，仅在背面凸起。花序被锈色短绒毛；花萼革质，长约2毫米，裂片卵形，被绒毛；无花瓣；雄蕊长约3毫米。果阔椭圆形，红黄色，连刺长约5厘米，刺长约1厘米。花期初夏，果期初秋。广东和海南栽培。原产东南亚。

Evergreen trees. Leaves with Petioles 15-45 cm long; leaflets usually 2 or 3 pairs, elliptic or obovate, 6-18 cm long, base cuneate, margin entire, glabrous; lateral veins 7-9 pairs, only prominent abaxially. Inflorescences shortly ferruginous tomentose; calyx leathery, ca. 2 mm long, lobes ovate, tomentose; petals absent; stamens ca. 3 mm long. Fruit broadly ellipsoid, reddish yellow, including spines ca. 5 cm long, spines ca. 1 cm long. Fl. early summer. Fr. early autumn. Cultivated in Guangdong and Hainan. Native to SE Asia.

## 韶子
**Nephelium chryseum** Blume

常绿乔木。小叶常4对，长圆形，下面具柔毛，侧脉9-14对或更多。花序多分枝，雄花序与叶近等长，雌花序较短。果椭圆体形，密被软刺，刺长约1厘米或更长，幼时金黄色，成熟时橙红色，干时转黑褐色。花期春季，果期夏季。生海拔500-1500米的密林中。产云南、广西和广东。越南、加里曼丹岛和菲律宾亦有。

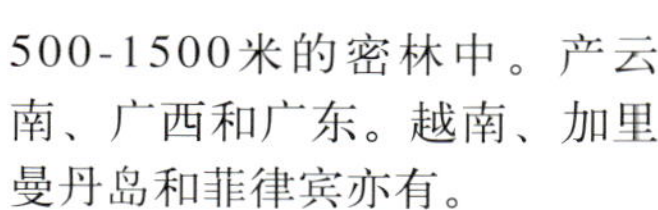

Evergreen trees. Leaflets usually 4 pairs, oblong, abaxially pilose, lateral veins 9-14 pairs or more. Inflorescences many-branched, male ones nearly as long as leaves, female ones shorter. Fruits ellipsoid, densely covered with soft spines, spines ca. 1 cm or longer, golden-yellow when young, orange-red when mature, black-brown when dry. Fl. spring. Fr. summer. Dense forests at 500-1500 m. Distributed in Yunnan, Guangxi and Guangdong. Also in Vietnam, Kalimantan Island and the Philippines.

## 海南韶子
**Nephelium topengii** (Merr.) H. S. Lo

常绿乔木。小叶2-4对，薄革质，长圆形，全缘。果椭圆形，红黄色，具软刺，连刺长约3厘米，宽不超过2厘米，刺长3.5-5毫米。果期5-6月。生林中。产海南。

Evergreen trees. Leaflets 2-4 pairs, thinly coriaceous, oblong, margin entire. Fruits ellipsoid, yellowish-red, softly aculeate, ca. 3 cm long including aculeus, no more than 2 cm width, aculeus 3.5-5 mm long. Fr. May-Jun. Forests. Distributed in Hainan.

韶子 *Nephelium chryseum*

海南韶子 *Nephelium topengii*

滨木患 *Arytera littoralis*

柄果木 *Mischocarpus sundaicus*

## 滨木患
**Arytera littoralis** Blume

小乔木。小叶2或3(或4)对，近对生，长圆状披针形至披针状卵形，下面沿侧脉脉腋具圆形裸露腺点。花序紧缩，多花；花芳香，花瓣5。果实裂为分果爿。种子椭圆体形，假种皮棕色。花期早夏，果期秋季。生海拔500-1200米的林中。产云南、广西、广东和海南。南亚、东南亚至所罗门群岛亦有。

Small trees. Leaflets 2 or 3(or 4) pairs, subopposite, oblong-lanceolate to lanceolate-ovate, abaxially with orbicular naked glands at lateral vein axils. Inflorescences compact, multiflowered; flowers fragrant, petals 5. Fruits parted into schizocarps. Seeds ellipsoid, arils brown. Fl. early summer. Fr. autumn. Forests at 500-1200 m. Distributed in Yunnan, Guangxi, Guangdong and Hainan. Also in S and SE Asia to Solomon Islands.

## 褐叶柄果木
**Mischocarpus pentapetalus** (Roxb.) Radlk.

乔木。小叶常3-5对，极少2对，披针形或长圆状披针形至长圆形，两面色暗，网脉不明显。花序常多分枝，单生及腋生或数个簇生于近枝顶端；花丝和花盘被毛。蒴果较小。着生种子部分宽不超过1厘米。花期春季，果期夏季。生原始或次生林中。产云南、广西和广东。亚洲热带地区亦有。

Trees. Leaflets often 3-5 pairs, rarely 2 pairs, lanceolate or oblong-lanceolate to oblong, both surfaces dull, with obscure reticulate veins. Inflorescences often multibranched, solitary and axillary or several fascicled near branch apices; filaments and disk hairy. Capsules small. Part bearing seed less than 1 cm wide. Fl. spring. Fr. summer. Primary or secondary forests. Distributed in Yunnan, Guangxi and Guangdong. Also in tropical Asia.

## 柄果木
**Mischocarpus sundaicus** Blume

常绿小乔木。小叶常2对，有时1对，叶腹面光亮，卵形或长圆状卵形，革质。复总状花序基部分枝，有时为不分枝总状花序；无花瓣。蒴果梨形，含果柄长8-9毫米，常1室。具1粒种子。花期10-11月，果期春季至夏季。生海拔100-1300米的海边林中。产广西和海南。东南亚亦广布。

Evergreen small trees. Leaflets often 2 pairs, sometimes 1 pairs, adaxially shiny, ovate or oblong-ovate, leathery. Inflorescences compound racemose, branched near base, sometimes racemose and unbranched; petals absent. Capsules pear-shaped, including stalks 8-9 mm long, usually 1-loculed. With 1 seed. Fl. Oct-

褐叶柄果木 *Mischocarpus pentapetalus*

云南檀栗 *Pavieasia yunnanensis*

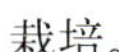
复羽叶栾树 *Koelreuteria bipinnata*

Nov. Fr. spring to summer. Coastal forests at 100-1300 m. Distributed in Guangxi and Hainan. Also widely in SE Asia.

## 云南檀栗

**Pavieasia yunnanensis** H. S. Lo

乔木，高达25米。小枝被粗毛。小叶5-6对，对生或近对生。花序腋生或近顶生；花瓣红色；雄蕊7-8，淡紫红色；子房被柔毛。蒴果短纺锤形，有3棱，两端短尖。花期初夏，果期初秋。生海拔100-900米的林中。产云南。越南北部亦有。

Trees, up to 25 m tall. Branches hirsute. Leaflets 5-6-pairs, opposite or subopposite. Inflorescences axillary or nearly terminal; petals red; stamens 7-8, purplish red; ovary pilose. Capsules shortly fusiform, 3-ridged, base and apex acute. Fl. early summer. Fr. early autumn. Forests at 100-900 m. Distributed in Yunnan. Also in N Vietnam.

## 栾树

**Koelreuteria paniculata** Laxm.

落叶乔木或灌木。小枝具瘤。叶羽状分裂或不完全二回羽状分裂；小叶边缘具不规则的钝锯齿，齿常疏且开裂。聚伞圆锥花序密被微冠毛。蒴果圆锥形，具3棱，顶端渐尖。花期6-8月，果期9-10月。产中国西南、华西、华北、华东和东北。各地广泛栽培。

Deciduous trees or shrubs. Branchlets tuberculate. Leaves pinnate or imperfectly bipinnate; leaflets margin irregularly obtusely serrate, teeth usually sparse and fissured. Racemose cymes densely covered with micropappus. Capsules conical, 3-ridged, apex acuminate. Fl. Jun-Aug. Fr. Sep-Oct. Distributed in SW, W, N, E and NE China. Widely cultivated elsewhere.

## 复羽叶栾树

**Koelreuteria bipinnata** Franch.

乔木，高达20米。叶开展，二回羽状复叶；小叶基部稍倾斜，先端锐尖至具短渐尖，边缘具内卷锯齿或有时全缘。圆锥花序扩展；花瓣4，稀5。蒴果椭圆体形或近球形，具3棱，先端圆形或钝。花期6-9月，果期8-10月。生海拔400-2500米的林中或山坡。产云南、四川、贵州、广西、广东、湖北和湖南。

Trees, up to 20 m tall. Leaves spreading, bipinnate; leaflets basally slightly oblique, apex acute to shortly acuminate, margin incurved serrate or sometimes entire. Panicles spreading, petals 4, rarely 5. Capsules ellipsoid or subglobose, 3-ridged, apex rounded or obtuse. Fl. Jun-Sep. Fr. Aug-Oct. Forests or slopes at 400-2500 m. Distributed in Yunnan, Sichuan, Guizhou, Guangxi, Guangdong, Hubei and Hunan.

栾树 *Koelreuteria paniculata*

车桑子 *Dodonaea viscosa*

## 车桑子

**Dodonaea viscosa** (L.) Jacq.

灌木或小乔木。单叶，形态及大小多变。花序顶生或腋生于近顶端，短于叶，密具花；无花瓣；枝条、叶和花序具黏液。蒴果膜质，边缘延伸成一膜质翅，幼时紫红色，成熟时黄棕色。花期秋末，果期冬末春初。生海拔800-2800米的干燥山坡或海边。产中国西南和华南。全球热带和亚热带地区亦有。

Shrubs or small trees. Leaves simple; variable in shape and size. Inflorescences terminal or axillary near apices, shorter than leaves, densely flowered; petals absent; branches, leaves, and inflorescences with sticky juice. Capsules membranous, margin with a membranous wing, purple-red when young, yellow-brown when old. Fl. late autumn. Fr. late winter to early spring. Dry slopes or seashores at 800-2800 m. Distributed in SW and S China. Also in other tropical and subtropical areas of the world.

## 掌叶木

**Handeliodendron bodinieri** (Lévl.) Rehd.

乔木或灌木，高达15米。叶柄长4-11厘米；小叶片椭圆形至倒披针形。复合圆锥聚伞花序长5-7.5(-12)厘米；萼片下面被白色、球形乳突，边缘密具缘毛；花瓣4或5，黄色至白色。蒴果橙褐色，具斑点。花期3-5月，果期7-8(-10)月。生海拔500-1200米的林中、林缘、山洞或喀斯特岩隙。产贵州南部和广西西北部。

Trees or shrubs, up to 15 m tall. Petioles 4-11 cm long; leaflet blades elliptic to oblanceolate. Compound thyrse 5-7.5(-12) cm long; sepals abaxially covered with whitish, globose papillae, margin densely ciliate; petals 4 or 5, yellow to white. Capsules orange-brown, mottled. Fl. Mar-May. Fr. Jul-Aug(-Oct). Forests, forest edges, caves, or rock crevices in karstic limestone mountain areas at 500-1200 m. Distributed in S Guizhou and NW Guangxi.

## 黄梨木

**Boniodendron minus** (Hemsl.) T. C. Chen

小乔木。树皮深棕色，纵裂。偶数羽状复叶簇生于枝顶；小叶10-20，纸质，基部倾斜，边缘具钝齿。聚伞圆锥花序顶生；分枝开展；花浅黄色至近白色；雄蕊8；子房具3沟槽。蒴果轮廓近球形，具3翅。花期

掌叶木 *Handeliodendron bodinieri*

黄梨木 *Boniodendron minus*

伞花木 *Eurycorymbus cavaleriei*

5-6月，果期7-8月。生石灰岩山地。产云南、贵州、广西、广东和湖南。

Small trees. Bark dark brown, fissured. Leaves fascicled at branch apices, paripinnate; leaflets 10-20, papery, base oblique, margin obtusely serrate. Thyrses terminal; branches spreading; flowers pale yellow to nearly white; stamens 8; ovary 3-furrowed and ridged. Capsules subglobose in outline, 3-winged. Fl. May-Jun. Fr. Jul-Aug. Limestone mountains. Distributed in Yunnan, Guizhou, Guangxi, Guangdong and Hunan.

## 伞花木

**Eurycorymbus cavaleriei** (Lévl.) Rehd. et Hand.-Mazz.

落叶乔木，高达20米。羽状复叶；小叶4-10对，近对生，薄纸质。花序半球形，具密而多的花；花芳香；花瓣长约2毫米。外被长柔毛；子房被绒毛。可育分果爿长约8 × 7毫米，具绒毛。花期5-6月，果期10月。生海拔300-1400米的阔叶林中。产中国西南、华南和华中。

Deciduous trees, up to 20 m tall. Leaves paripinnate; leaflets 4-10 pairs, nearly opposite, thinly papery. Inflorescences hemispheroid, with many dense flowers; flowers fragrant; petals ca. 2 mm long; abaxially villous; ovary tomentose. Fertile schizocarps ca. 8 × 7 mm, tomentose. Fl. May-Jun. Fr. Oct. Broad-leaved forests at 300-1400 m. Distributed in SW, S and C China.

## 山木患 (假山萝)

**Harpullia cupanioides** Roxb.

乔木，高达20米。小叶3-6对，薄革质，两侧不对称。花序疏散，腋生或顶生；花芳香。果实横向椭圆体形或近球形；雄蕊5，子房2室。种子全部为橙色的假种皮所包被。花期春夏间，果期秋末。生海拔500-700米的雨林、村边或路旁。产云南南部、广东和海南。南亚、东南亚和澳大利亚北部亦有。

Trees, up to 20 m tall. Leaflets 3-6 pairs, thinly leathery, asymmetrical. Inflorescences sparse, axillary or terminal; flowers fragrant. Fruits transversely ellipsoid or subglobose; stamens 5; ovary 2-loculed. Seeds entirely enveloped by orange arils. Fl. spring-summer. Fr. late autumn. Rain forests, village or roadside at 500-700 m. Distributed in S Yunnan, Guangdong and Hainan. Also in S and SE Asia, and N Australia.

山木患（假山萝）*Harpullia cupanioides*

茶条木 *Delavaya toxocarpa*

## 茶条木

**Delavaya toxocarpa** Franch.

灌木或小乔木。叶互生，具3小叶；小叶薄革质；中部者椭圆形或卵状椭圆形，有时披针状卵形。花序狭、细弱，疏具花；花瓣白色或粉色。蒴果紫色。种子直径1-1.5厘米。花期4月，果期8月。生海拔500-2000米的密林中。产云南和广西。越南北部亦有。

Shrubs or small trees. Leaves alternate, 3-foliolate; leaflets thinly leathery; middle one elliptic or ovate-elliptic, sometimes lanceolate-ovate. Inflorescences narrow, slender, sparsely flowered; petals white or pink. Capsules purple. Seeds 1-1.5 cm diam. Fl. Apr. Fr. Aug. Dense forests at 500-2000 m. Distributed in Yunnan and Guangxi. Also in N Vietnam.

## 文冠果

**Xanthoceras sorbifolium** Bunge

落叶灌木或小乔木。顶芽和侧芽有覆瓦状排列的芽鳞。小叶4-8对，披针形或近卵形。花序先叶出或与叶同时出；花瓣白色，基部紫红色或黄色。果实无刺，3室。每室具数粒种子。花期春季，果期初秋。生丘陵或山坡。产中国西北和华北。

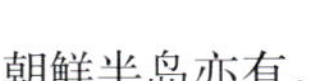
朝鲜半岛亦有。

Deciduous shrubs or small trees. Terminal and lateral buds with imbricate scales. Leaflets 4-8 pairs, lanceolate or subovate. Inflorescences appearing before leaves or simultaneously; petals white, purplish red or yellow at base. Fruits without spines, 3-loculed. With several seeds per locule. Fl. spring. Fr. early autumn. Hills or slopes. Distributed in NW and N China. Also in Korean Peninsula.

文冠果 *Xanthoceras sorbifolium*

# 清风藤科 Sabiaceae

凹萼清风藤 *Sabia emarginata*

## 四川清风藤
**Sabia schumanniana** Diels

木质落叶藤本。叶长圆状卵形，纸质，两面无毛，宽达3.5厘米。聚伞或圆锥花序具1-3(-20)花；花瓣5，浅绿色或绿紫色至浅咖啡色；子房无毛。分果爿倒卵球形或近球形，无毛。花期3-4月，果期6-8月。生海拔600-2600米的林中、山坡、山谷或溪边。产中国西南和华中。

Woody climbers, deciduous. Leaves oblong-ovate, papery, both surfaces glabrous, to 3.5 cm wide. Cymes or panicles 1-3 (-20)-flowered; petals 5, light green or greenish purple to brownish chocolate colored; ovary glabrous. Schizocarps obovoid or subglobose, glabrous. Fl. Mar-Apr. Fr. Jun-Aug. Forests, slopes, valleys or streamsides at 600-2600 m. Distributed in SW and C China.

## 凹萼清风藤
**Sabia emarginata** Lecomte

落叶木质藤本或近直立灌木。叶纸质，两面无毛。伞形花序具2(-3)花；花萼5，稍不等大，最大者顶部明显凹，其他萼片圆形；花盘膨大，具2或3条不明显的肋状突起，其上各具一个小腺体。分果爿近球形，基部具宿存萼片。花期4月，果期6-7月。生海拔400-1500米的林中。产四川、贵州、广西、湖北和湖南。

Woody climbers or suberect shrubs, deciduous. Leaves papery, both surfaces glabrous. Cymes 2(-3)-flowered; sepals 5, slightly unequal, largest one distinctly emarginate at apex, others rounded; disk swollen, with 2 or 3 obscure projecting veins, each vein with a tiny gland. Schizocarps subglobose, with persistent sepals at base. Fl. Apr. Fr. Jun-Jul. Forests at 400-1500 m. Distributed in Sichuan, Guizhou, Guangxi, Hubei and Hunan.

## 清风藤
**Sabia japonica** Maxim.

落叶木质藤本。幼枝具短刺。花先叶开放，单生；花萼5，近圆形或宽卵形；花瓣5，浅黄绿色；花盘杯形，5浅裂。分果爿近圆形或近肾形；内果皮具明显突出中肋及蜂巢状孔穴。花期2-3月，果期4-7月。生海拔800米以下的密林、林缘、山谷或路边。产中国西南、华南、东南、华中和华东。日本亦有。

Woody climbers, deciduous. Young branches armed with short spines. Flowers appearing before leaves, solitary; sepals 5, suborbicular or broadly ovate; petals 5, light yellowish green; disk cup-shaped, shallowly 5-lobed. Schizocarps suborbicular or subreniform; endocarps with prominent midrib, with faveolate cavities. Fl. Feb-Mar. Fr. Apr-Jul. Dense forests, forest edges, valleys or roadsides below 800 m. Distributed in SW, S, SE, C and E China. Also in Japan.

四川清风藤 *Sabia schumanniana*

清风藤 *Sabia japonica*

尖叶清风藤 *Sabia swinhoei*

柠檬清风藤 *Sabia limoniacea*

## 尖叶清风藤
**Sabia swinhoei** Hemsl.

常绿木质藤本。叶纸质或薄革质，背面具短柔毛或仅于中脉具柔毛，先端渐尖至尾尖状锐尖，或钝至近圆形。聚伞花序具(1-)2-7花；子房无毛。分果爿绿色至红色或深蓝色，近圆形或倒卵球形，压扁；内果皮具蜂巢状孔穴。花期(1-)3-4(-6)月，果期7-9(-10)月。生海拔300-2300米的山谷林中或石灰石山麓灌丛中。产中国西南、华南、东南、华中和华东。越南北部亦有。

Evergreen Woody vines. Leaves papery or thinly leathery, abaxially shortly pilose or pilose only on midvein, apex acuminate to caudate-acute, or obtuse to subrounded. Cymes (1-)2-7-flowered; ovary glabrous. Schizo carps green to red or dark blue, subglobose or obovoid, compressed; endocarps with faveolate cavities. Fl. (Jan-)Mar-Apr(-Jun). Fr. Jul-Sep(-Oct). Valley forests or thickets on limestone hills at 300-2300 m. Distributed in SW, S, SE, C and E China. Also in N Vietnam.

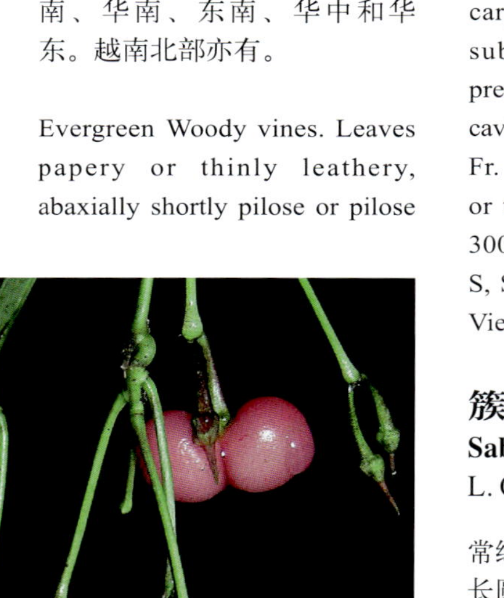

## 簇花清风藤
**Sabia fasciculata** Lecomte ex L. Chen

常绿木质藤本。叶革质，无毛，长圆形至椭圆形，先端尖。聚伞花序有3-4花，再排成伞房花序式，有10-20花；花冠钟状，淡绿色，具紫色斑纹。核果圆球形，红色后变蓝色。花期2-5月，果期5-10月。生海拔600-2000米的山岩、山谷、山坡、林间。产福建、广西、广东和云南。越南和缅甸亦有。

Evergreen woody vines. Leaves coriaceous, glabrous, oblong to elliptic, apex acute. Cymes 3-4-flowered, formed large corymb with 10-20 flowers; corolla campanulate, greenish, purple striate. Drupes globose, red, turned blue later. Fl. Feb-May. Fr. May-Oct. Rocks, valleys, slopes or forests at 600-2000 m. Distributed in Fujian, Guangxi, Guangdong and Yunnan. Also in Vietnam and Myanmar.

## 柠檬清风藤
**Sabia limoniacea** Wall. ex Hook. f. et Thoms.

常绿木质藤本。单叶，革质，无毛，侧脉6或7对。聚伞花序有花2-4朵，排成狭长的圆锥花序，总梗较长；花盘杯状，5浅裂。分果爿粉色至红色或蓝黑

簇花清风藤 *Sabia fasciculata*

小花清风藤 *Sabia parviflora*

色，近球形或近肾形。花期8-11月，果期翌年1-5月。生海拔800-1300米的密林中。产云南、四川、广东、海南和福建。印度北部、孟加拉国、缅甸、泰国、马来西亚和印度尼西亚亦有。

Evergreen woody vines. Leaves simple, leathery, glabrous, lateral veins 6 or 7 pairs. Cymes 2-4-flowered, in a narrow and longer panicle with longer peduncle; disc cup-shaped, 5-lobed. Schizocarps pink to red or bluish black, subglobose or subreniform. Fl. Aug-Nov. Fr. next Jan to May. Dense forests at 800-1300 m. Distributed in Yunnan, Sichuan, Guangdong, Hainan and Fujian. Also in N India, Bangladesh, Myanmar, Thailand, Malaysia and Indonesia.

## 小花清风藤
**Sabia parviflora** Wall.

常绿木质藤本。单叶，纸质或近革质，卵状披针形或狭长圆形，宽1-3厘米，全缘。聚伞花序组成长3-5厘米的圆锥花序；花盘杯状，不整齐5深裂。分果爿近球形、倒卵球形或肾形，直径5-7(-10)毫米。花期3-5月，果期7-9月。生海拔800-2800米的山坡密林或灌木林中。产云南、贵州、四川和广西。印度、尼泊尔、缅甸、泰国、越南、印度尼西亚和菲律宾亦有。

Evergreen woody vines. Leaves simple, papery or subleathery, ovate-lanceolate or narrowly oblong, 1-3 cm wide, margin entire. Cymes in a panicle 3-5 cm long; disk cup-shaped, irregularly 5-parted. Schizocarps subglobose, obovoid, or reniform, 5-7 (-10) mm diam. Fl. Mar-May. Fr. Jul-Sep. Dense forests or shrubs on slopes at 800-2800 m. Distributed in Yunnan, Guizhou, Sichuan and Guangxi. Also in India, Nepal, Myanmar, Thailand, Vietnam, Indonesia and the Philippines.

## 灰背清风藤
**Sabia discolor** Dunn

常绿木质藤本。叶腹面绿色，背面白绿色。聚伞花序呈伞状；花瓣5，黄色，卵形或椭圆状卵形；子房无毛；花盘杯状。分果爿红色或紫红色，内果皮具明显中肋。花期3-4月，果期5-8月。生海拔1000米以下的山地灌木林中。产贵州、广西、广东、福建、浙江和江西。

Evergreen woody vines. Leaves green adaxially, pale green abaxially. Cymes umbellate; petals 5, yellow, ovate or elliptic ovate; ovary glabrous; disk cup-shape. Schizocarps red or purplish red, endocarps with distinct midrib. Fl. Mar-Apr. Fr. May-Aug. Mountain bushes below 1000 m. Distributed in Guizhou, Guangxi, Guangdong, Fujian, Zhejiang and Jiangxi.

灰背清风藤 *Sabia discolor*

泡花树 *Meliosma cuneifolia*

## 泡花树
**Meliosma cuneifolia** Franch.

落叶灌木或乔木。树皮深棕色。叶纸质，倒卵状楔形或狭倒卵状楔形，边缘具锯齿，先端短渐尖。圆锥花序顶生，直立，分枝3(或4)次。核果扁球形，成熟时黑色。花期5-7月，果期9-11月。生海拔500-3300米的林中。产中国西南和华中部分地区。

Deciduous shrubs or trees. Bark dark brown. Leaves papery, obovate-cuneate or narrowly obovate-cuneate, margin serrate, apex shortly acuminate. Panicles terminal, erect, branched 3(or 4) times. Drupes compressed-globose, black when mature. Fl. May-Jul. Fr. Sep-Nov. Forests at 500-3300 m. Distributed in SW and some parts of C China.

## 垂枝泡花树
**Meliosma flexuosa** Pamp.

落叶小乔木。叶柄基部稍扩大；单叶，膜质；两面均疏被柔毛，侧脉12-18对。圆锥花序顶生，下垂；花梗和小枝之字形；花瓣白色，钝角2叉裂至中部；子房无毛。核果倒卵球形；内果皮倾斜，中脉突出。花期5-6月，果期7-9月。生海拔600-2800米的林中。产中国西南、华南、东南、华中、华北、华西和华东。

Deciduous small trees. Petioles base slightly enlarged. Leaves simple, membranous, both surfaces sparsely pubescent, lateral veins 12-18 pairs. Panicles terminal, pendulous; peduncles and branches zigzag; petals white, bifid to half at an obtuse angle; ovary glabrous. Drupes obovoid; endocarps oblique, midrib prominent. Fl. May-Jun. Fr. Jul-Sep. Forests at 600-2800 m. Distributed in SW, S, SE, C, N, W and E China.

## 柔毛泡花树
**Meliosma myriantha** Siebold et Zucc. var. **pilosa** (Lecomte) Y. W. Law

落叶乔木。小枝平，细长，3棱形。单叶膜质或纸质，背面密被柔毛，上面稍具柔毛，侧脉10-20对，从中部至顶端具锯齿。圆锥花序顶生，直立，具短柔毛。核果倒卵球形或球形；内果皮扁球形，中脉稍突出。花期夏季，果期5-9月。生海拔100-2000米的林中溪边或山谷中。产中国西南、东南、华中和华东。

Deciduous trees. Branches flat, long and slender, axis 3-angulate. Leaves simple, membranous or papery, densely pilose abaxially, ± pubescent adaxially, lateral veins 10-20 pairs, margin serrate from middle to apex. Panicles terminal, erect, pubescent. Drupes

垂枝泡花树 *Meliosma flexuosa*

柔毛泡花树 *Meliosma myriantha* var. *pilosa*

obovoid or globose; endocarps compressed-globose, midrib slightly prominent. Fl. summer. Fr. May-Sep. Streamsides in forests or valleys at 100-2000 m. Distributed in SW, SE, C and E China.

## 山様叶泡花树
**Meliosma thorelii** Lecomte

乔木。单叶，革质，倒披针状椭圆形或倒披针形，基部狭楔形，先端渐尖，边缘具锐尖小锯齿；侧脉每边15-22条，两面均凸起。圆锥花序大，顶生，直立；花芳香，外面3片花瓣白色，近圆形，内面2片花瓣狭披针形。核果大，球形，熟时黑色。花期5-7月，果期10-11月。生海拔200-1700米的林中。产福建、广东、广西、贵州和云南。越南和老挝亦有。

Trees. Leaves simple, coriaceous, obovate-elliptic or oblanceolate, base narrowly cuneate, apex acuminate, margin sharply and minutely serrate; lateral veins 15-22 pairs, elevated on both surfaces. Panicles large, terminal, erect; flowers fragrant, outer 3 white, sub-round, inner 2 narrowly lanceolate. Drupes large, globose, black at maturity. Fl. May-Jul. Fr. Oct-Nov. Forests at 200-1700 m. Distributed in Fujian, Guangdong, Guangxi, Guizhou and Yunnan. Also in Vietnam and Laos.

## 腋毛泡花树
**Meliosma rhoifolia** Maxim. var. **barbulata** (Cufod.) Y. W. Law

乔木。小叶下面淡绿色，侧脉6-9对，叶腋被黄色髯毛。圆锥花序顶生或腋生于上部枝条，具锈色柔毛，迅速无毛，分枝3(或4)次；花瓣白色；花盘浅杯状，5齿裂。核果近球形。花期5-6月，果期8-10月。生海拔400-1100米的常绿阔叶林中。产贵州、广西、广东、福建、浙江、湖南和江西。

Trees. Leaflets abaxially pale, lateral veins 6-9 pairs, axils yellowish crinite. Panicles terminal or axillary on upper branches, ferruginous pubescent, soon glabrescent, branched 3(or 4) times; petals white; disk shallowly cup-shaped, 5-toothed. Drupes subglobose. Fl. May-Jun. Fr. Aug-Oct. Evergreen broad-leaved forests at 400-1100 m. Distributed in Guizhou, Guangxi, Guangdong, Fujian, Zhejiang, Hunan and Jiangxi.

山様叶泡花树 *Meliosma thorelii*

腋毛泡花树 *Meliosma rhoifolia* var. *barbulata*

腺毛泡花树 *Meliosma glandulosa*

暖木 *Meliosma veitchiorum*

## 暖木

**Meliosma veitchiorum** Hemsl.

落叶乔木。羽状复叶，轴基部膨大；小叶7-11，纸质，侧脉6-12对，脉腋不具髯毛。圆锥花序顶生，直立，具4(-5)次分枝；花瓣白色，2裂至1/3。核果近球形；内果皮近球形，光滑或具不明显、散生网状条纹，中脉明显突出。花期5月，果期8-9月。生海拔1000-3000米的潮湿林中。产中国西南、东南、华中、华北和华西。

Deciduous trees. Leaves pinnate, axis swollen at base; leaflets 7-11, papery, lateral veins 6-12 pairs, not crinite at vein axils. Panicles terminal, erect, branched 4(-5) times; petals white, bifid to 1/3. Drupes subglobose; endocarps subglobose, smooth or with inconspicuous, scattered netlike strips, midrib distinctly prominent. Fl. May. Fr. Aug-Sep. Humid forests at 1000-3000 m. Distributed in SW, SE, C, N and W China.

## 腺毛泡花树

**Meliosma glandulosa** Cufod.

常绿乔木。奇数羽状复叶；小叶7-9，正面色浅，中部卵形或长圆状卵形，顶端椭圆形，近革质，背面散生棍棒状腺点，脉腋具软毛。圆锥花序顶生，3次分枝；花瓣亮绿色。核果球形。花期夏季，果期8-10月。生海拔400-1400米的山地常绿阔叶林。产贵州、广西东部和广东北部。

Evergreen trees. Leaves odd pinnate; leaflets 7-9, abaxially pale, ovate or oblong-ovate at middle, elliptic terminally, subleathery, abaxially scattered with claviform glands, crinite on vein axils. Panicles terminal, branched 3 times; petals light green. Drupes globose. Fl. summer. Fr. Aug-Oct. Evergreen broad-leaved forests in mountains at 400-1400 m. Distributed in Guizhou, E Guangxi and N Guangdong.

## 红柴枝

**Meliosma oldhamii** Miq. ex Maxim.

落叶乔木。奇数羽状复叶，小叶7-15，薄纸质，稍具柔毛，侧脉7或8对。圆锥花序顶生，直立，具3次分枝。花瓣白色，2裂至中部，有时3裂；子房具黄色柔毛。核果球形；内果皮凸出，具明显散生网状条纹，中肋明显突出。花期5-6月，果期8-9月。生海拔300-1900米的山谷或湿山坡。产中国西南、华南、东南、华中和华东。朝鲜半岛南部和日本亦有。

Deciduous trees. Leaves odd pinnate; leaflets 7-15, thinly papery, ± pubescent, lateral veins 7 or 8 pairs. Panicles terminal, erect, branched 3 times; petals white, bifid to half, sometimes 3-lobed; ovary yellow pubescent. Drupes globose; endocarps convex, with distinct, scattered netlike strips, midrib conspicuously prominent. Fl. May-Jun. Fr. Aug-Sep. Valleys or humid mountain slopes at 300-1900 m. Distributed in SW, S, SE, C and E China. Also in S Korean Peninsula and Japan.

红柴枝 *Meliosma oldhamii*

# 凤仙花科 Balsaminaceae

苏丹凤仙花(非洲凤仙花) *Impatiens walleriana*

## 华凤仙
**Impatiens chinensis** L.

一年生草本。茎无毛，有不定根。叶对生，线形。总花梗簇生于叶腋，1-3花；花紫红色或白色；苞片线形；侧生萼片2，线形；旗瓣圆形；翼瓣无柄，下部裂片近圆形；唇瓣漏斗状，基部成内弯或旋卷的长距；雄蕊5；子房纺锤形。蒴果椭圆形。种子圆球形。花期6-8月，果期9-11月。生海拔100-1200米的池塘、水沟旁、田边或沼泽地。产中国西南、华南、华中和华东。印度和中南半岛亦有。

Annual herbs. Stem glabrous, with adventitious roots. Leaves opposit, linear. Inflorescences 1-flowered, or flowers 2- or 3-fascicled in leaf axils; bracts linear; flowers purple-red or white; lateral sepals 2, linear; upper petal orbicular; lateral united petals not clawed, basal lobes suborbicular; lower sepal funnelform, gradually narrowed into an incurved or involute, slender spur; stamens 5; ovary fusiform. Capsules ellipse. Seeds globose. Fl. Jun-Aug. Fr. Sep-Nov. Often beside ponds, streamsides, field margins or swamps at 100-1200 m. Distributed in SW, S, C and E China. Also in India and Indo-China Peninsula.

## 凤仙花
**Impatiens balsamina** L.

一年生草本。茎直立。叶互生，披针形、狭椭圆形。花单生或2-3朵簇生于叶腋，白色、粉红色或紫色，单瓣或重瓣；苞片线形；侧生萼片2；旗瓣圆形；翼瓣具柄，下部裂片长圆形，上部裂片近圆形；唇瓣深舟状，基部成内弯的距；雄蕊5；子房纺锤形。蒴果宽纺锤形。种子圆球形，黑褐色。花期7-10月。中国广泛栽培。原产印度，世界各地普遍栽培。

Annual herbs. Stem erect. Leaves alternate, lanceolate, narrowly elliptic. Inflorescences 1-flowered, or 2 or 3 flowers fascicled in leaves axils; flowers white, pink, or purple, simple or double petalous; bracts linear; lateral sepals 2; upper petal orbicular; lateral united petals shortly clawed; basal lobes obovate-oblong; distal lobes suborbicular; lower sepal deeply navicular, abruptly narrowed into an incurved spur; stamens 5; ovary fusiform. Capsules broadly fusiform. Seeds globose, black-brown. Fl. Jul-Oct. Widely cultivated in China. Native to India, cultivated worldwide.

## 苏丹凤仙花(非洲凤仙花)
**Impatiens walleriana** Hook. f.

多年生肉质草本。茎直立。叶互生或上部螺旋状排列。总花梗生于叶腋，具2花；苞片线状披针形或钻形；侧生萼片2；旗瓣宽倒心形或倒卵形；翼瓣无柄，基部裂片与上部裂片同形且近等大，倒卵形或倒卵状匙形；唇瓣浅舟状，基部收缩为线状内弯的细距。蒴果纺锤形。花期6-10月。中国各地栽培。原产东非；世界各地广泛栽培。

Perennial herbs, succulent. Stem erect. Leaves alternate or spirally arranged. Inflorescences in leaf axils, usually 2-flowered; upper petals broadly obcordate or obovate; lateral united petals not clawed; basal lobes and distal lobes obovate-oblong; lower sepal shallowly navicular, abruptly narrowed into an incurved spur, slender. Capsules fusiform. Fl. Jun-Oct. Cultivated in most parts of China. Native to E Africa; widely cultivated in many parts of the world.

华凤仙 *Impatiens chinensis*

凤仙花 *Impatiens balsamina*

丰满凤仙花 *Impatiens obesa*

## 丰满凤仙花

**Impatiens obesa** Hook. f.

肉质草本，全株无毛。茎直立肥厚，不分枝或稀中部短分枝。叶互生，卵形或倒披针形。总花梗生于叶腋，极短，单花或2花，花梗细；花粉紫色；侧生萼片4；旗瓣宽倒卵形或楔形；翼瓣无柄；唇瓣短囊状或杯状，基部急狭成内弯的短矩；子房纺锤形。蒴果纺锤形。种子圆形。花期6-7月。生海拔400-750米的山坡林缘、山谷、水旁。产广东和广西。

Herbs succulent, glabrous. Stem erect, thick, simple or rarely shortly branched from middle. Leaves alternate, ovate or oblanceolate. Inflorescences in leaf axils, very short, 1- or 2-flowered; flowers pink-purple; lateral sepals 4; upper petal broadly obovate or cuneate; lateral united petals not clawed; lower sepal shallowly saccate or cup-shaped, abruptly constricted into an incurved, short spur; ovary fusiform. Capsules fusiform. Seeds orbicular. Fl. Jun-Jul. Forest edges on slopes, valleys, watersides at 400-750 m. Distributed in Guangdong and Guangxi.

## 多脉凤仙花

**Impatiens polyneura** K. M. Liu

一年生草本，无毛。茎直立，常具紫色斑点。叶互生，具柄，椭圆形。总花梗具2花；花梗纤细，基部有三角状卵形苞片；花淡紫色；侧生萼片4；旗瓣倒卵状长圆形；翼瓣具柄，2裂；唇瓣宽漏斗状，具紫色斑点；花丝线形；花药卵形；子房纺锤形。蒴果纺锤形。种子被乳头突状微毛。花果期未知。生海拔约400米的山谷溪流边。产湖南(资兴)。

Annual herbs, glabrous. Stem erect, often purple spotted. Leaves alternate, petiolate elliptic. Inflorescences in upper leaf axils, 2-flowered; peduncles slender, bracteate at base; bracts triangular-ovate; flowers purplish; lateral sepals 4; upper petal obovate-oblong; lateral united petals clawed, 2-lobed. lower sepal purple spotted, broadly funnelform; Filaments linear; anthers ovoid; ovary fusiform. Capsules fusiform. Seeds papillose. Fl. and fr. unknown. Streamsides in valleys at ca. 400 m. Distributed in Hunan (Zixing).

## 龙州凤仙花

**Impatiens morsei** Hook. f.

一年生草本。茎肉质。叶互生，叶片卵形、椭圆形或卵状长圆形，侧脉9-12对。无总花梗，花梗单生于叶腋，无苞片；花白色，粉红色或紫色；侧生萼片2；旗瓣圆形；翼瓣具柄，基部裂片镰刀状扇形，上部裂片长圆形；唇瓣檐部舟状或漏斗状，基部成内弯短于檐部的短距；子房纺锤形。花果期6-10月。生海拔400-950米的山谷水旁密林下阴湿处。产广西。

Annual herbs. Stem succulent. Leaves alternate, ovate, elliptic, or

多脉凤仙花 *Impatiens polyneura*

龙州凤仙花 *Impatiens morsei*

海南凤仙花 *Impatiens hainanensis*

ovate-oblong, lateral veins 9-12 pairs. Peduncles absent; pedicels without bracts; flowers solitary in leaf axils, white, pink, or purple; lateral sepals 2; upper petal orbicular; lateral united petals clawed, basal lobes falcate-flabellate, distal lobes oblong; lower sepal navicular or funnelform, abruptly constricted into an incurved spur shorter than limb; ovary fusiform. Fl. and fr. Jun-Oct. Dense forest understories, streamsides in valleys, shaded and humid places at 400-950 m. Distributed in Guangxi.

## 海南凤仙花

**Impatiens hainanensis** Y. L. Chen

多年生草本，全株无毛。茎粗壮，肉质。叶互生，具柄，密集于茎枝上端；叶片纸质，椭圆形。总花梗腋生，花梗细；花较大，黄色或淡黄色；侧生萼片2，淡黄绿色；唇瓣短囊状；旗瓣倒卵状椭圆形；翼瓣具短柄，2裂；花丝线形，花药卵形；子房纺锤状。蒴果棒状。种子长圆形，褐色。花期6-7月。生密林中或石灰岩石缝中。产海南。

Perennial herbs, glabrous. Stem robust, succulent. Leaves alternate, usually crowded at apices of stem and branches; thinly papery, ovate-elliptic. Inflorescences axillary, pedicels slender; flowers large, yellow or yellowish; lateral sepals 2, yellowish green; lower sepal shallowly saccate; upper petal obovate-elliptic; lateral united petals shortly clawed, 2-lobed; filaments linear; anthers ovoid; ovary fusiform. Capsules clavate. Seed oblong, brown. Fl. Jun-Jul. Dense forest understories, crevices and ledges of limestone cliffs. Distributed in Hainan.

## 湖北凤仙花

**Impatiens pritzelii** Hook. f.

多年生草本，全株无毛，具串珠状地下茎。茎肉质，节膨大。叶互生，无柄，披针形。总花梗生于叶腋，总状花序，基部有苞片；花黄色；侧生萼片4；旗瓣椭圆形膜质；翼瓣2裂，背部有小耳；唇瓣囊状，距内弯；花丝线形；花药顶端钝；子房纺锤形。蒴果。花期10月。生海拔400-1800米的山谷林下、沟边及湿润草丛中。产湖北和重庆。

Perennial herbs, glabrous, with a procumbent tortuous subterranean stem. Stem succulent, swollen nodes. Leaves alternate, sessile, leaves lanceolate. Inflorescences in leaf axils, racemes, bracteate at base; flowers yellow; lateral sepals 4; upper petal elliptic, membranous; lateral united petals 2-lobed; auricle inflexed; lower sepal incurved, saccate, involute spur; filaments linear; anthers obtuse; ovary fusiform. Capsule. Fl. Oct. Forest understories, humid grasslands in valleys, streamsides at 400-1800 m. Distributed in Hubei and Chongqing.

湖北凤仙花 *Impatiens pritzelii*

匙叶凤仙花 *Impatiens spathulata*

## 匙叶凤仙花

**Impatiens spathulata** Y. X. Xiong

一年生草本。茎肉质，直立，具根。叶互生，具柄，膜质，倒卵形、倒披针形、匙形或长圆形。总花梗生于上部叶腋，基部具苞片；花总状排列，具2-4花，黄色；旗瓣椭圆形，具窄龙骨状突起；翼瓣无柄；唇瓣囊状；花丝线形；花药钝；子房纺锤形。蒴果未知。花期10月。生海拔391米的洞内瀑布旁湿地。产贵州(赤水)。

Annual herbs. Stem erect, succulent, with roots. Leaves alternate, petiolate, leaf blade obovate, oblanceolate, spatulate, or oblong, membranous. Inflorescences in upper leaf axils, racemose, 2-4-flowered; bracteate at base; flowers yellow; upper petal elliptic, abaxial midvein thickened, narrowly carinate; lateral united petals not clawed; lower sepal saccate; filaments linear; anthers obtuse; ovary fusiform. Capsules unknown. Fl. Oct. Humid places by waterfalls at 391 m. Distributed in Guizhou (Chishui).

## 峨眉凤仙花

**Impatiens omeiana** Hook. f.

直立草本，节膨大。叶互生，披针形或卵状矩圆形。总花梗顶生，花5-8朵排成总状花序；花梗细，基部有1卵状矩圆形苞片；花黄色；萼片4；旗瓣三角状圆形；翼瓣无柄，2裂；唇瓣漏斗状，基部延成卷曲的短距；花药钝；子房纺锤形。花期8-9月。生海拔900-1000米的灌木林下或林缘。产四川。

Herbs erect, nodes turgid. Leaves alternate, leaves lanceolate or ovate-oblong. Inflorescences terminal, racemose, 5-8-flowered; pedicels slender, bracteate at base; bract ovate-oblong; flowers yellow; sepals 4; upper petal triangular-orbicular; lateral united petals not clawed, 2-lobed; lower sepal narrowly funnelform, narrowed into an involute, short spur; anthers obtuse; ovary fusiform. Capsules unknown. Fl. Aug-Sep. Understories of thickets and forests, forest edges at 900-1000 m. Distributed in Sichuan.

峨眉凤仙花 *Impatiens omeiana*

## 白花凤仙花

**Impatiens wilsonii** Hook. f.

直立草本。茎粗壮，节膨大。叶互生，倒卵形或倒披针形；叶柄短或几无柄，有疏腺体。总花梗生于顶叶腋，排成总状花序；花白色；侧生萼片4；旗瓣椭圆形；翼瓣无柄，2裂，基部裂片矩圆形，上部裂片斧形，背部有短耳；唇瓣囊状，基部圆形，有内弯的短距；花药钝；子房纺锤形。花期8-9月。生沟边或林下阴湿处。产四川(峨眉山)。

Herbs erect. Stem robust, nodes turgid. Leaves alternate, leaves obovate or oblanceolate; shortly petiolate or subsessile; Petioles with sparse glands. Inflorescences in terminal leaf axils, racemose; flowers white; lateral sepals 4; upper petal elliptic; lateral united petals not clawed, 2-lobed, basal lobes oblong, distal lobes dolabriform, large, abaxial auricle short; lower sepal saccate, base rounded, with an in curved short spur; anthers obtuse; ovary fusiform. Fl. Aug-Sep. Forest understories, along canals, shaded moist places. Distributed in Sichuan (Emei Mountain).

白花凤仙花 *Impatiens wilsonii*

## 滇南凤仙花

**Impatiens duclouxii** Hook. f.

一年生草本。茎直立。叶互生，叶片薄纸质，卵形。总花梗生于叶腋，总状排列；苞片，宽卵形，宿存，肉质；花黄色；侧生萼片2；旗瓣圆形；翼瓣具宽短柄，基部裂片圆形，上部裂片斧形；唇瓣囊状，基部内弯，顶端钩状卷曲的距；子房纺锤形。蒴果棒状。种子球形，褐色。花期8-9月，果期9-10月。生海拔1500-2500米的林下。产云南。泰国亦有。

Annual herbs. Stem erect. Leaves alternate, thinly papery, ovate. Inflorescences in terminal leaf axils; bracteate persistent, broadly ovate, fleshy; flowers yellow; lateral sepals 2; upper petal orbicular; lateral united petals broadly shortly clawed, basal lobes orbicular, distal lobes dolabriform; lower sepal saccate, narrowed into an incurved, involute, hooked spur; ovary fusiform. Capsules clavate. Seeds globose, brown. Fl. Aug-Sep. Fr. Sep-Oct. Forests at 1500-2500 m. Distributed in Yunnan. Also in Thailand.

滇南凤仙花 *Impatiens duclouxii*

## 湖南凤仙花

**Impatiens hunanensis** Y. L. Chen

一年生草本。茎肉质，节膨大，具纤维状根。叶互生，卵形。总花梗生于叶腋，总状花序；基部具苞片；花黄色；侧生萼片2，卵形；旗瓣圆形；翼瓣具短柄，背部具反折的小耳；唇瓣囊状，距内卷；花丝线形；花药尖；子房纺锤状。蒴果棒状。种子多数。花期6月，果期6月以后。生海拔700-800米的山谷林下河边或岩石上。产广东、广西、湖南和江西。

Annual herbs. Stem succulent, lower nodes turgid, with many fibrous roots. Leaves alternate, ovate. Inflorescences, axially racemose; bracteate at base; flowers yellow; lateral sepals 2, obliquely ovate; upper petal orbicular; lateral united petals shortly stipitate, auricle inflexed; lower sepal saccate, abruptly narrowed into a hooked or incurved spur; filaments linear; anthers acute; ovary fusiform. Capsules clavate. Seeds many. Fl. Jun. Fr. after Jun. Forests, riversides, stony places in valleys at 700-800 m. Distributed in Guangdong, Guangxi, Hunan and Jiangxi.

湖南凤仙花 *Impatiens hunanensis*

贝苞凤仙花 *Impatiens conchibracteata*

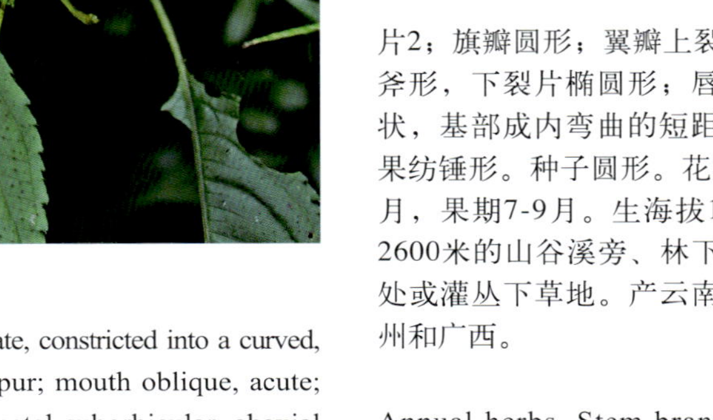

## 贝苞凤仙花

**Impatiens conchibracteata** Y. L. Chen

一年生草本。茎直立。叶互生，叶柄长1-2厘米；叶片椭圆形，上面被毛。总花梗长1-1.5厘米，1-3花；花黄色；侧生萼片2；旗瓣近圆形；翼瓣具柄，基生裂片圆形，上部裂片斧形，背耳反折；唇瓣深囊状；花丝扁平；花药卵形；子房棒状。蒴果棒形，具3-5粒种子。种子黑褐色。花果期8-9月。生海拔1800-2800米的林缘阴湿处。产四川(峨眉山)。

Annual herbs. Stem erect. Leaves alternate; Petioles 1-2 cm long; leaves ovate, abaxially hairy. Inflorescences 1-3-flowered, peduncles 1-1.5 cm long; flowers yellow; lateral sepals 2; lower sepal deeply saccate, constricted into a curved, short spur; mouth oblique, acute; upper petal suborbicular, abaxial midvein rostellate at middle; lateral united petals clawed, 2-lobed; basal lobes orbicular; distal lobes broadly dolabriform, large, auricle inflexed, suborbicular; filaments flattened; anthers ovoid; ovary erect, clavate. Capsules clavate, 3-5-seeded. Seeds brown-black. Fl. and fr. Aug-Sep.Forest understories, shaded moist places at 1800-2800 m. Distributed in Sichuan (Emei Mountain).

## 红纹凤仙花

**Impatiens rubrostriata** Hook. f.

一年生草本。茎有分枝。叶互生，卵形，侧脉5-8对。总花梗腋生，总状；苞片1，披针形；花白色，具红色条纹；侧生萼片2；旗瓣圆形；翼瓣上裂片宽斧形，下裂片椭圆形；唇瓣囊状，基部成内弯曲的短距。蒴果纺锤形。种子圆形。花期6-7月，果期7-9月。生海拔1700-2600米的山谷溪旁、林下潮湿处或灌丛下草地。产云南、贵州和广西。

Annual herbs. Stem branched. Leaves alternate, ovate, lateral veins 5-8 pairs. Inflorescences axillary, racemose; bract 1, lanceolate; flowers white, red striate; lateral sepals 2; upper petals orbicular; lateral united petals broadly hatchet, basal lobes orbicular, distal lobes dolabriform; lower sepals saccate, base narrowed into an incurved short spur. Capsules fusiform. Seeds globose. Fl. Jun-Jul. Fr. Jul-Sep. Watersides in valleys, grasslands, bushes or under sparse forests at 1700-2600 m. Distributed in Yunnan, Guizhou and Guangxi.

## 路南凤仙花

**Impatiens loulanensis** Hook. f.

一年生草本。茎粗壮，直立，有分枝。叶互生，卵状矩圆形或卵状披针形，侧脉6-8对。总花梗腋生，总状；花梗基部有1苞片，卵形；花黄色，侧生萼生2；旗瓣圆形；翼瓣近无柄，基部裂片近圆形，上部裂片狭披针形；唇瓣漏斗状，基部延成内弯的长距。蒴果线形。花期8-9月。生海拔700-2500米的山谷湿地、林下草丛和水沟边。产云南和贵州。

Annual herbs. Stem erect, robust, branched. Leaves alternate, ovate-oblong or ovate-lanceolate, lateral veins 6-8 pairs. Inflorescences axillary, racemose; bracteate at base, 1, ovate; flowers yellow; lateral sepals 2; upper petal orbicular; lateral united petals nearly sessile, basal lobes suborbicular, distal lobes narrowly lanceolate; lower sepal funnelform, narrowed into a curved, long spur. Capsules linear. Fl. Aug-Sep. Forest understories, moist places in valleys, grasslands, along canals at 700-2500 m. Distributed in Yunnan and Guizhou.

红纹凤仙花 *Impatiens rubrostriata*

路南凤仙花 *Impatiens loulanensis*

黄金凤 *Impatiens siculifer*

蓝花凤仙花
*Impatiens cyanantha*

## 蓝花凤仙花
**Impatiens cyanantha** Hook. f.

一年生草本。茎直立，有分枝。叶互生，椭圆形或披针形，侧脉每边5-6条。总花梗细，苞片小；花大，蓝色或紫蓝色；侧生萼片2；旗瓣小，圆形；翼瓣上裂片斧形，下裂片小，圆形；唇瓣囊状，基部下延为细长内弯的长距。蒴果狭纺锤形。种子长圆形。花期7-9月，果期8-10月。生海拔1000-2500米的林下、沟边、路旁等阴湿处。产贵州和云南。

Annual herbs. Stem erect, branched. Leaves alternate, elliptic or lanceolate, lateral veins 5-6 pairs. Bracts small; flowers blue or purple-blue, large; lateral sepals 2; upper petal orbicular, small; lateral united petals, distal lobes dolabriform; basal lobes orbicular, small; lower sepal saccate, narrowed into a slender, incurved spur. Capsules narrowly fusiform. Seeds oblong. Fl. Jul-Sep. Fr. Aug-Oct. Forest understories, along canals, shaded moist places, roadsides at 1000-2500 m. Distributed in Guizhou and Yunnan.

## 黄金凤
**Impatiens siculifer** Hook. f.

一年生草本。茎细弱。叶互生，披针形，侧脉5-11对。总花梗生于叶腋，花5-8，总状；苞片1，披针形，宿存；花黄色；侧生萼片2；旗瓣近圆形；翼瓣无柄，基部裂片近三角形，上部裂片条形；唇瓣狭漏斗状，基部延长成内弯或下弯的长距；花药钝。蒴果棒状。花期7-9月，果期10-11月。生海拔800-2500米的密林中、草丛、沟边等潮湿地。产中国中部和南部。

Annual herbs. Stem slender. Leaves alternate, lanceolate, lateral veins 5-11 pairs. Inflorescences in leaf axils, racemose, 5-8-flowered; bracteate 1, persistent, lanceolate; flowers yellow; lateral sepals 2; upper petal suborbicular; lateral united petals not clawed, basal lobes subtriangular, distal lobes lorate; lower sepal narrowly funnelform, narrowed into an incurved or recurved, long spur. Capsules clavate. Fl. Jul-Sep. Fr. Oct-Nov. Dense forests, grasslands on slopes, along canals, moist places at 800-2500 m. Distributed in C and S China.

## 澜沧凤仙花
**Impatiens principis** Hook. f.

一年生草本，全株无毛。茎不分枝或自下部具短分枝。叶互生，或近轮生，具柄，卵形。总花梗生于上部叶腋，总状，通常具3-5花；花梗细丝状，基部具苞片，脱落；花黄色，卵形；旗瓣倒卵形，翼瓣基部裂片长渐尖，上部裂片线形，背部具狭小耳；唇瓣檐部舟状，口部斜上；花丝线形；花药钝；子房纺锤状。未成熟的蒴果棒状。花期7-8月。生海拔1700-2000米的山坡沟边溪旁。产云南。

Annual herbs, glabrous. Stem simple or shortly branched from lower part. Leaves alternate or subverticillate, petiolate; leaves ovate. Inflorescences in upper leaf axils, racemose, usually 3-5-flowered; pedicels fibrous, bracteate at base, bracts caducous; flowers yellow, ovate; lateral sepals 2, ovate; upper petal obovate; lateral united 2-lobed, basal lobes long acuminate, distal lobes linear, auricle small, narrow; lower sepal navicular, mouth oblique; filaments linear; anthers obtuse; ovary fusiform. Capsules (immature) clavate. Fl. Jul-Aug. Streams and along canals on slopes at 1700-2000 m. Distributed in Yunnan.

澜沧凤仙花 *Impatiens principis*

同距凤仙花 *Impatiens holocentra*

野凤仙花 *Impatiens textorii*

## 同距凤仙花

**Impatiens holocentra** Hand.-Mazz.

一年生草本。叶互生，矩圆状卵形或矩圆状披针形，侧脉7-11对。总花梗腋生，花4-6，总状花序；花梗中部有1条形苞片，早落；花较小，黄色；侧生萼片4；旗瓣倒卵形；翼瓣近无柄，基部裂片三角形，上部裂片披针形；唇瓣狭漏斗状，基部伸长成直而细长的距。蒴果条形。花果期7-10月。生海拔2700-2800米的亚高山山谷溪流或阴湿处。产云南。缅甸亦有。

Annual herbs. Leaves alternate, oblong-ovate or oblong-lanceolate, lateral veins 7-11 pairs. Inflorescences axillary, racemose, 4-6-flowered; bracteate at middle; bracts caducous, linear; flowers yellow, small; lateral sepals 4; upper petal obovate; lateral united petals ± clawed, basal lobes triangular, distal lobes lanceolate; lower sepal narrowly funnelform, narrowed into an erect, slender spur. Capsules linear. Fl. and fr. Jul-Oct. Shaded moist places in subalpine valleys, by streams at 2700-2800 m. Distributed in Yunnan. Also in Myanmar.

## 东北凤仙花

**Impatiens furcillata** Hemsl.

一年生草本。茎直立。叶互生，卵形或披针形。总花梗腋生，花3-9，排成总状花序；花梗基部有1条形苞片；花黄色或淡紫色；侧生萼片2；旗瓣圆形，翼瓣有柄，2裂，基部裂片近卵形，上部裂片斜卵形；唇瓣漏斗状，基部突然延长成长距；花药钝。蒴果近圆柱形，先端具短喙。花果期8-9月。生河边草丛中或林缘。产中国东北、河北和内蒙古。朝鲜半岛和俄罗斯亦有。

Annual herbs. Stem erect. Leaves alternate; leaves ovate or lanceolate. Inflorescences axillary, racemose, 3-9-flowered; pedicels slender, bracteate at base; bracts linear; flowers yellow or pale purple; lateral sepals 2; upper petal orbicular; lateral united petals clawed, 2-lobed; basal lobes subovate, distal lobes obliquely ovate; lower sepal funnelform, abruptly elongated into a long spur; anthers obtuse. Capsules subcylindric. Fl. and fr. Aug-Sep. Forest edges or grasslands in valleys, riversides. Distributed in NE China, Hebei and Neimenggu. Also in Korean Peninsula and Russia (Far East).

## 野凤仙花

**Impatiens textorii** Miq.

一年生草本。茎直立，多分枝。叶互生或近轮生，卵形或披针形，侧脉7-8对。总花梗生于叶腋，具4-10花；苞片披针形；花紫红色；侧生萼片2；旗瓣方形；翼瓣具柄，基部裂片长圆形，上部裂片斧形；唇瓣漏斗形，基部渐狭成向内卷曲的距；子房纺锤形。蒴果纺锤状。种子椭圆形，褐色。花期8-9月。生海拔1050米的山沟溪流旁。产吉林、辽宁和山东。日本、朝鲜半岛和俄罗斯(远东地区)亦有。

Annual herbs. Stem erect, branched. Leaves alternate or subverticillate, ovate or lanceolate, lateral veins 7-8 pairs. Peduncles in axillary, with 4-10 flowers, bracts lanceolate; flowers purple-red; lateral sepals 2; upper petal ovate-square; lateral united petals clawed, basal lobes ovate-oblong, distal lobes oblong-dolabriform; lower sepal infundibuliform, narrowed into an incurved spur. Capsules fusiform. Seeds elliptic, brown. Fl. Aug-Sep. By streams in valleys at 1050 m. Distributed in Jilin, liaoning and Shandong. Also in Japan, Korean Peninsula and Russia (Far East).

## 水凤仙花

**Impatiens aquatilis** Hook. f.

一年生草本。茎直立，具不定根。叶互生，披针形。总花梗生于叶腋，具6-10花；苞片早

东北凤仙花 *Impatiens fucillata*

水凤仙花 *Impatiens aquatilis*

滇水金凤 *Impatiens uliginosa*

落，膜质；花粉紫色；侧生萼片2；旗瓣圆形；翼瓣无柄，基生裂片圆形；唇瓣檐部内弯，短囊状，基部狭成内弯、顶端小棒状的距；子房纺锤状。蒴果线形。种子长圆形，褐色，瘤状突起。花期8-9月，果期10月。生海拔1500-3000米的河边、溪边阴潮处。产云南。

Annual herbs. Stem erect, with few adventitious roots. Leaves alternate, lanceolate. Inflorescences in leaf axils, 6-10-flowered; bracts caducous, membranous; flowers pink-purple; lateral sepals 2; upper petal orbicular; lateral united petals not clawed, basal lobes orbicular; lower sepal incurved, shortly saccate, narrowed into an incurved spur; ovary fusiform. Capsules linear. Seeds oblong, brown, strumae. Fl. Aug-Sep. Fr. Oct. Lakesides, riversides, shaded moist places at 1500-3000 m. Distributed in Yunnan.

## 滇水金凤

**Impatiens uliginosa** Franch.

一年生草本。茎粗壮，有不定根。叶互生，膜质，披针形。总花梗多生于叶腋，近伞房状；苞片早落，卵形；花红色；侧生萼片2；旗瓣圆形；翼瓣无柄，基部裂片圆形，上部裂片半月形；唇瓣檐部漏斗形，基部成内弯的距；子房纺锤形。蒴果近圆柱形。种子长圆形，黑色。花期7-8月，果期9月。生海拔1500-2600米的林下、水沟边、溪边潮湿处。产云南。

Annual herbs. Stem robust, with adventitious roots. Leaves alternate, lanceolate, membranous. Inflorescences often in leaf axils, subcorymbose; bracts caduceus; flowers red; lateral sepals 2; upper petal orbicular; lateral united petals not clawed, basal lobes rounded, distal lobes semilunar; lower sepal funnelform, narrowed into an incurved spur; ovary fusiform. Capsules subcylindric. Seeds black, oblong. Fl. Jul-Aug. Fr. Sep. Forest understories, along canals, by streams or shaded moist places at 1500-2600 m. Distributed in Yunnan.

## 井岗山凤仙花

**Impatiens jinggangshanensis** Y. L. Chen.

一年生草本，高30-90厘米。叶互生，卵状披针形或长圆状披针形，长8-13厘米，宽1.5-2.5厘米。3-8花，花紫色或鲜粉红色；花梗细。蒴果线形，长17-22毫米。种子褐色，倒卵形。生海拔800-1240米河边阴湿处或山谷密林下。花果期8-10月。产湖南和江西。

Annual herbs, 30-90 cm tall. Leaves alternate, ovate-lanceolate or oblong-elliptic lanceolate, 8-13 × 1.5-2.5 cm. 3-8-flowered, purple or shocking pink; pedicels thin. Capsules linear, 17-22 mm long. Seeds brown, obovoid. Wet places along the river or valley forests at 800-1240 m. Fl. and fr. Aug-Oct. Distributed in Hunan and Jiangxi.

井岗山凤仙花 *Impatiens jinggangshanensis*

槽茎凤仙花 *Impatiens sulcata*

## 槽茎凤仙花

**Impatiens sulcata** Wall.

一年生草本。茎直立，具明显的槽沟，节上常具疏腺体。叶对生或上部轮生，卵形或披针形。花较大，排成近伞房状总状花序；花梗上端膨大，基部有苞片；侧生萼片2；旗瓣近圆形；翼瓣宽，无柄；唇瓣囊状，基部骤狭成短距；花药钝。蒴果短棒状。种子倒卵圆形。花果期8-9月。生海拔3000-4000米的冷杉林下、水沟边或阴湿地。产西藏南部。印度、尼泊尔和不丹亦有。

Annual herbs. Stem erect, conspicuously grooved, sparsely glandular at nodes. Leaves opposite or verticillate in upper part of stem; leaves ovate or lanceolate. Inflorescences subcorymbose-racemose, many flowered; pedicels swollen at apex, bracteate at base; flowers purplish pink, large; lateral sepals 2; upper petal suborbicular; lateral united petals not clawed, broad; lower sepal saccate abruptly narrowed into an incurved spur; anthers obtuse. Capsules shortly clavate. Seeds obovoid. Fl. and fr. Aug-Sep. Picea forests, along canals, or shaded moist places at 3000-4000 m. Distributed in S Xizang. Also in India, Nepal and Butan.

## 草莓凤仙花

**Impatiens fragicolor** Marq. et Airy Shaw

一年生草本。茎粗壮，四棱形或圆柱形，无毛。叶具柄，下部对生，上部互生，披针形；叶柄基部具球状腺体。总花梗少数，生于上部叶腋，近伞房状排列，具1-6花；花梗顶端膨大，基部有苞片；花紫色或淡紫色；侧生萼片2；旗瓣宽卵形；翼瓣无柄；唇瓣宽漏斗状，基部有内弯的细距；花药钝。蒴果线形。花期7-8月。生海拔3100-3200米的河边草地、潮湿地或路边。产西藏东部。

Herbs annual. Stem stout, tetragonous or terete, glabrous. Leaves opposite in lower part of stem, alternate in upper part of stem; Petioles with globose glands at base; leaves lanceolate. Inflorescences in upper leaf axils, subcorymbose, 1-6-flowered; pedicels apex often swollen, bracteate at base; flowers purple or pale purple to light yellowish; lateral sepals 2; upper petal broad ovate; lateral united petals not clawed; lower sepal broadly funnelform, narrowed into an incurved, slender spur; anthers obtuse. Capsules oblong-linear, ca. 2 cm long. Fl. Jul-Aug. Grasslands by rivers, moist places, or roadsides at 3100-3200 m. Distributed in E Xizang.

## 抱茎凤仙花

**Impatiens amplexicaulis** Edgew.

抱茎凤仙花 *Impatiens amplexicaulis*

一年生草本。茎四棱形。叶下部对生，上部互生，长圆形。总花梗腋生；花粉色，6-12个，伞形或总状花序；苞片披针形；侧生萼片2；旗瓣近圆形；翼瓣无柄，基部裂片近圆形，上部裂片卵形；唇瓣斜囊状，基部成内弯的短距。蒴果近圆柱形。种子倒卵形，黑褐色。花期7-8月，果期8-9月。生海拔2900-3200米的路边灌丛中。产西藏。喜马拉雅亦有。

Annual herbs. Stem tetragonous. Leaves opposite in lower part of stem, alternate in upper part of stem, oblong. Inflorescences axillary, umbellate or racemose, 6-12-flowered; bracts lanceolate; flowers pink; lateral sepals 2; upper petal suborbicular; lateral united petals not clawed, basal lobes suborbicular, distal lobes ovate; lower sepal obliquely saccate, narrowed into an inflexed, short spur. Capsules subcylindric. Seeds obovoid, black-brown. Fl. Jul-Aug. Fr. Aug-Sep. Thickets along roadsides at 2900-3200 m. Distributed in Xizang. Also in Himalaya.

草莓凤仙花 *Impatiens fragicolor*

双角凤仙花 *Impatiens bicornuta*

## 双角凤仙花

**Impatiens bicornuta** Wall

一年生高大草本，高达1米。茎粗壮，肉质，无毛。叶膜质互生，具柄，椭圆形或椭圆状披针形。总花梗直立，花多数，排成中断的总状花序；花淡蓝紫色；侧生萼片2；旗瓣近圆形；翼瓣无柄；唇瓣宽锥状或弯囊状；花药钝。蒴果圆柱形。种子近圆柱状，有光泽。花期6-8月。生海拔2400-2800米的水边草地或阔叶林和铁杉林下。产西藏南部。印度北部和尼泊尔亦有。

Annual herbs, up to 1 m tall. Stem robust, succulent, glabrous. Leaves alternate, petiole; leaves elliptic or elliptic-lanceolate, membranous. Inflorescences crowded in leaf axils, racemose, many flowered; peduncles erect; flowers pale blue-purple; lateral sepals 2; upper petal suborbicular; lateral united petals not clawed; lower sepal broadly sigmoid-curved-saccate; anthers obtuse. Capsules cylindric. Seeds subcylindric, shiny. Fl. Jun-Aug. Understories of *Tsuga* forests, grasslands by water at 2400-2800 m. Distributed in S Xizang. Also in N India and Nepal.

## 辐射凤仙花

**Impatiens radiata** Hook. f.

一年生草本。茎多分枝。叶互生，长圆状卵形或披针形，侧脉7-9对。总花梗生于上部叶腋，轮生；苞片披针形；花黄白色或浅紫色；侧生萼片2；旗瓣近圆形；翼瓣3裂，下部2裂片近圆形，上部裂片长圆形；唇瓣锥状，基部狭成短直距。蒴果线形。种子倒卵形。花期8-9月。生海拔2100-3500米的山坡湿润草丛中或林下阴湿处。产贵州、四川、西藏和云南。印度东北部、不丹、缅甸和尼泊尔亦有。

Annual herbs. Stem many branched. Leaves alternate, oblong-ovate or lanceolate, lateral veins 7-9 pairs. Inflorescences in upper leaf axils, verticillate; bracts persistent, lanceolate; flowers yellowish white or pale purple; lateral sepals 2; upper petal suborbicular; lateral united petals 3 crack, 2 basal lobes suborbicular, distal lobes oblong; lower sepal navicular, gradually narrowed into an erect spur. Capsules linear. Seeds obovoid. Fl. Aug-Sep. Shaded and moist places of forest understories, humid grasslands at 2100-3500 m. Distributed in Guizhou, Sichuan, Xizang and Yunnan. Also in NE India, Bhutan, Myanmar and Nepal.

## 瘤果凤仙花

**Impatiens tuberculata** Hook. f. et Thoms.

一年生草本。茎直立，无毛。叶互生，具短柄，椭圆形或椭圆状卵形；叶柄基部无腺体。总花梗腋生或顶生，短或长于叶，具4-8花，总状排列；花梗纤细；花小，淡紫色；旗瓣圆形；翼瓣2裂；唇瓣舟状；花药短而宽。蒴果短棒状。种子褐色。花期8-9月。生海拔3800米的冷杉林缘草丛中或水沟边。产西藏西部。印度北部和不丹亦有。

Annual herbs. Stem erect, glabrous throughout. Leaves alternate; Petioles short, without basal glands; leaves elliptic or elliptic-ovate. Inflorescences axillary or terminal, racemose, 4-8-flowered; peduncles shorter or longer than leaves; pedicels slender; flowers purplish, small; upper petal orbicular; lateral united petals 2-lobed; lower sepal navicular; anthers short, broad. Capsules shortly clavate. Seeds few, brown. Fl. Aug-Sep. Margins of *Abies* forests, grasslands or along canals at 3800 m. Distributed in W Xizang. Also in N India and Bhutan.

辐射凤仙花 *Impatiens radiata*

瘤果凤仙花 *Impatiens tuberculata*

无距凤仙花 *Impatiens margaritifera*

锐齿凤仙花 *Impatiens arguta*

## 无距凤仙花

**Impatiens margaritifera** Hook. f.

一年生草本。茎直立，无毛。叶互生，卵形，薄膜质。总花梗细长，腋生，花较小，6-8个排成总状花序；花梗短；花白色；旗瓣椭圆状倒卵形或近圆形；翼瓣少数，具宽柄，基部裂片卵状长圆形；唇瓣舟状，无距；花药钝。蒴果线形。种子倒卵形。花期7-9月。生海拔2600-3800米的河滩湿地或溪边草丛中，或冷杉林下。产四川西南部、西藏和云南西北部。

Annual herbs. Stem erect, glabrous. Leaves alternate; leaves ovate, membranous. Inflorescences axillary, racemose, 6-8-flowered; pedicels short; flowers white, small; upper petal elliptic-obovate or suborbicular; lateral united petals broadly clawed; basal lobes ovate-oblong or suborbicular; anthers obtuse. Capsules linear. lower sepal navicular, spur absent; Seeds few, obovoid. Fl. Jul-Sep. Understories of *Abies* forests, forest edges, grasslands, riverbanks, streamsides, moist places at 2600-3800 m. Distributed in SW Sichuan, Xizang and NW Yunnan.

## 锐齿凤仙花

**Impatiens arguta** Hook. f. et Thomson

多年生草本。茎坚硬，有分枝。叶互生，卵形，侧脉7-8对。总花梗腋生，具1-2花；苞片2，刚毛状；花粉红色或紫红色；侧生萼片4；旗瓣圆形；翼瓣无柄，基部裂片宽长圆形，上部裂片斧形；唇瓣囊状，基部延长成内弯的短距。蒴果纺锤形。种子圆球形。花期7-9月。生海拔1850-3200米的林下潮湿处或水沟边。产四川、西藏和云南。印度北部、不丹、缅甸和尼泊尔亦有。

Herbs perennial. Stem rigid, branched. Leaves alternate, ovate, lateral veins 7-8 pairs. Inflorescences axillary, 1-2-flowered; often with 2 setose bracts; flowers pink or purple-red; lateral sepals 4; upper petal orbicular; lateral united petals not clawed, basal lobes broadly oblong, distal lobes dolabriform; lower sepal saccate, narrowed into an incurved, short spur. Capsules fusiform. Seeds globose. Fl. Jul-Sep. Forest understories, along canals or moist places at 1850-3200 m. Distributed in Sichuan, Xizang and Yunnan. Also in N India, Bhutan, Myanmar and Nepal.

## 镰瓣凤仙花

**Impatiens falcifer** Hook. f.

一年生草本。茎直立。叶互生，卵状长圆形；叶柄基部具2球形腺体。总花梗单生于叶腋，具1或2花；花短，中部有

镰瓣凤仙花 *Impatiens falcifer*

苞片；花黄色，具红色斑点或无斑点；侧生萼片2；旗瓣圆形，盔状；翼瓣具爪，2裂，基部裂片卵形至圆形，上部裂片大；唇瓣檐部漏斗状；花药小。蒴果线形。花期8-9月。生海拔2300-3300米的河边草地或栎林下。产西藏。印度、尼泊尔和不丹亦有。

Annual herbs. Stem erect. Leaves alternate; Petioles with 2 globose basal glands; leaves ovate-oblong. Inflorescences axillary, 1(or 2)-flowered; peduncles short, pedicels bracteate at middle; flowers yellow, red spotted or not; lateral sepals 2; upper petal cucullate, orbicular; lateral united petals clawed, 2-lobed; basal lobes ovate to orbicular; distal lobes large; lower sepal funnelform, narrowed into an incurved or erect spur; filaments linear; anthers small. Capsules linear-cylindric. Fl. Aug-Sep. Understories of *Quercus* forests or grasslands along riverbanks at 2300-3300 m. Distributed in Xizang. Also in India, Nepal and Bhutan.

那坡凤仙花 *Impatiens napoensis*

## 鸭跖草状凤仙花(类鸭跖草凤仙花)

**Impatiens commelinoides** Hand.-Mazz.

一年生草本。茎纤细，平卧。叶互生，卵形。总花梗具1花，苞片1枚，草质，披针形，宿存；花蓝紫色；侧生萼片2；旗瓣圆形；翼瓣具柄，裂片近圆形，上部裂片较大；唇瓣宽漏斗状，基部成内弯或螺旋状卷曲的距；雄蕊5；子房纺锤形。蒴果圆柱形。种子球形，褐色。花期8-10月，果期11月。生海拔300-900米的田边或沟边。产中国西南。

Annual herbs. Stem procumbent, slender. Leaves alternate, ovate. Inflorescences 1-flowered; bracts persistent, lanceolate, herbaceous; flowers blue-purple; lateral sepals 2; upper petal orbicular; lateral united petals clawed, both lobes orbicular, distal lobes larger; lower sepal broadly funnelform, gradually narrowed into an incurved or involute spur; ovary fusiform. Capsules cylindric. Seeds brown, globose. Fl. Aug-Oct. Fr. Nov. Along canals in valleys or margins of fields at 300-900 m. Distributed in SW China.

## 那坡凤仙花

**Impatiens napoensis** Y. L. Chen

一年生草本。茎横走或平卧，有分枝。叶互生，膜质，卵形。总花梗生于小枝顶端叶腋，具1花；花梗具2苞片，宿存；花大，粉红色；侧生萼片2；旗瓣

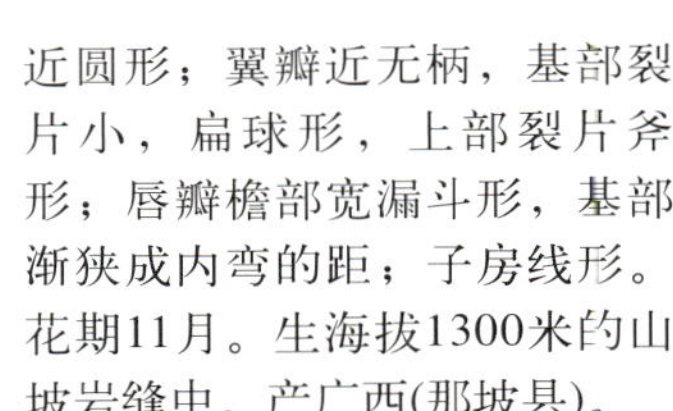

近圆形；翼瓣近无柄，基部裂片小，扁球形，上部裂片斧形；唇瓣檐部宽漏斗形，基部渐狭成内弯的距；子房线形。花期11月。生海拔1300米的山坡岩缝中。产广西(那坡县)。

Annual herbs. Stem prostrate or procumbent, branched. Leaves alternate, ovate, membranous. Inflorescences in leaf axils of branchlets, 1-flowered; pedicel with 2 bracts, bracts persistent; flowers pink, large; lateral sepals 2; upper petal suborbicular; lateral united petals ± clawed, basal lobes oblate, small, distal lobes dolabriform; lower sepal broadly funnelform, gradually narrowed into an incurved spur; ovary linear. Fl. Nov. Stony places on slopes at 1300 m. Distributed in Guangxi (Napo County).

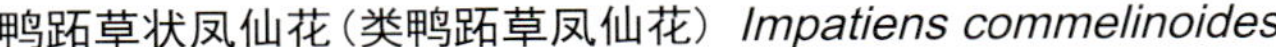

鸭跖草状凤仙花(类鸭跖草凤仙花) *Impatiens commelinoides*

菱叶凤仙花 *Impatiens rhombifolia*

## 菱叶凤仙花

**Impatiens rhombifolia** Y. Q. Lu et Y. L. Chen

一年生草本，光滑无毛，具匍匐生根茎。茎不分枝。叶互生，具短柄；叶片菱形，具有柄小腺，无毛。总花梗光滑单生于上部叶腋内，常2花；花梗细，基生一苞片；花黄色，侧生萼片2，卵圆，黄绿色；旗瓣大，圆形；翼瓣基部裂片圆，具枣红色斑点，上部裂片斧状，背耳狭，反折；唇瓣舟形，具1直而细长的距；花丝细；花药微尖；子房纺锤形。蒴果线形。种子小，卵球形。花期8-9月。生海拔800-1000米的路旁草地。产四川。

Annual herbs, glabrous. Stem prostrate, simple. Leaves alternate, petiolate; leaves rhombic, glabrous, with stipitate glands. Inflorescences in upper leaf axils, often 2-flowered; pedicels slender, bracteate at base; flowers yellow, lateral sepals 2, yellow-green, ovate-orbicular; upper petal orbicular; lateral united petals orbicular, basal lobes red spotted, distal lobes dolabriform, auricle inflexed, narrow; lower sepal navicular, narrowed into an erect spur; spur long, slender; filaments linear; anthers, apex acute; ovary fusiform. Capsules linear. Seeds few, ovoid-globose. Fl. Aug-Sep. Grasslands at roadsides at 800-1000 m. Distributed in Sichuan.

## 浙皖凤仙花

**Impatiens neglecta** Y. L. Xu et Y. L. Chen

一年生草本，无毛。茎直立。叶互生，具柄，膜质，长圆状卵形，顶端渐尖，基部楔形。总花梗粗，直立，生于上部叶腋；具1花，花淡紫色；旗瓣宽卵形；翼瓣具柄，2裂；唇瓣宽漏斗形；花丝线形；花药卵状，顶端尖；子房线形。蒴果线状圆柱形。种子多数，椭圆状。花果期7-10月。生海拔1000-1200米的山坡林下或溪边潮湿处。产浙江和安徽。

Annual herbs, glabrous. Stem erect. Leaves alternate, petiole; leaves oblong-ovate, membranous, base cuneate, apex acuminate. Inflorescences in upper leaf axils, 1-flowered; peduncles erect, stout; flowers pale purple; upper petal broadly ovate; lateral united petals clawed, 2-lobed; lower sepal broadly funnelform; filaments linear; anthers ovoid, apex acute; ovary linear. Capsules linear-cylindric. Seeds many, elliptic. Fl. and fr. Jul-Oct. Forest understories on slopes, moist places by streams at 1000-1200 m. Distributed in Zhejiang and Anhui.

## 华丽凤仙花

**Impatiens faberi** Hook. f.

一年生草本，近无毛。茎有分枝。叶互生，无柄，披针形，无腺体。总花梗生于叶腋，无毛，具2花，具苞片；花大，紫红色；侧生萼片2，绿色，卵形；旗瓣圆形；翼瓣无柄，2裂，背部有小耳；唇瓣角状。花丝短宽；花药尖；子房纺锤形。蒴果线形。种子多数，圆形，褐色。花期8-9月。生海拔1350-2100米的山坡林缘或路边潮湿处。产四川。

Annual herbs, subglabrous. Stem branched. Leaves alternate sessile. without basal glands, leaves lanceolate. Inflorescences in leaf axils, 2-flowered, glabrous, bracteate at base; flowers purple-red, large; lateral sepals 2, green,

浙皖凤仙花 *Impatiens neglecta*

华丽凤仙花 *Impatiens faberi*

ovate; upper petal orbicular; lateral united petals not clawed, 2-lobed, auricle linear; lower sepal cornute; filaments short, broad; anthers acute; ovary fusiform. Capsules linear. Seeds many, brown. Fl. Aug-Sep. Forest edges on slopes, moist places at roadsides at 1350-2100 m. Distributed in Sichuan.

## 侧穗凤仙花

**Impatiens lateristachys** Y. L. Chen et Y. Q. Lu

一年生草本，无毛。叶互生，无柄或具短柄，叶片披针形。总花梗常具5花，在花序轴上呈一侧的总状花序排列；花梗基部生1枚苞片；苞片钻形；花红色、浅红色或白色；侧生萼片2枚；旗瓣基部2裂；翼瓣近无柄，基部裂片圆形，上部裂片斧形，背耳线形，插入距内；唇瓣角状，距直。花丝扁平；花药近尖；子房棒形。种子卵圆形，黑色。花期8-9月。生海拔2000-2500米的山坡林缘中。产四川(峨眉山、瓦山)。

Annual herbs, glabrous. Leaves alternate, sessile or shortly petiolate; leaf blade lanceolate. Inflorescences axillary, racemose, often 5-flowered. Pedicelse-longate at fruiting, bracteate at base; bracts subulate; flowers red, pale red, or white; lateral sepals 2; lower sepal navicular-cornute, narrowed into an erect, stout spur; upper petal suborbicular; lateral united petals ± clawed, 2-lobed, basal lobes orbicular, distal lobes dolabriform, auricle linear, inserted into spur; filaments short, compressed; anthers acute; ovary clavate. Seeds black, ovoid-orbicular. Fl. Aug-Sep. Forest edges on slopes at 2000-2500 m. Distributed in Sichuan (Emei Mountain, Washan Mountain).

## 块节凤仙花

**Impatiens pinfanensis** Hook. f.

一年生草本。茎被白色绒毛，基部匍匐茎节膨大成球状块茎，上着生不定根。单叶互生，卵形。总花梗腋长，仅1花，上部具1小苞片；花红色，中等大；侧生萼片2，椭圆形；旗瓣圆形或倒卵形；翼瓣2裂，上裂片斧形，下裂片圆形；唇瓣漏斗状；花药尖。蒴果线形，具条纹。种子近球形，褐色。花期6-8月，果期7-10月。生海拔900-2000米的林下、沟边等潮湿环境。产贵州和重庆。

Annual herbs. Stem is white hairs, base runner section expands into spherical tubers, born on the adventitious roots. Leaves simple, alternate, ovate. Peduncle axillary long, only a flower, in the upper 1 bract; flowers red, medium large; lateral sepals 2, elliptic; upper petal round or obovate; lateral united petal 2, basal lobes hatchet, distal lobes orbicular; lower sepal funnel-shaped; anther acute. Capsules linear, with stripes. Seeds nearly spherical, brown. Fl. Jun-Aug. Fr. Jul-Oct. Forest understories, moist places at roadsides at 900-2000 m. Distributed in Guizhou and Chongqing.

块节凤仙花 *Impatiens pinfanensis*

侧穗凤仙花 *Impatiens lateristachys*

## 红雉凤仙花

**Impatiens oxyantghera** Hook. f.

一年生草本，全株无毛。茎直立。叶互生，具短柄；叶片卵形，基部无腺体。总花梗生于上部叶腋，具2花；苞片卵形，顶端突尖；花大，红色；侧生萼片2，圆形；旗瓣圆形；翼瓣无柄，2裂，基部裂片圆形，上部裂片斧形，背部具小耳；唇瓣漏斗形；花药卵圆形；子房纺锤状。蒴果线形。花期9月。生海拔1900-2200米的山坡林缘或路旁阴湿处。产四川。

Annual herbs, glabrous.Stems erect. Leaves alternate, shortly stipitate; leaves ovate, base without glands. Peduncles, was born in upper axillary, 2 flowers; bracts ovate, apex acuminate; flowers large, red; lateral sepals 2, rounded; upper petal round; lateral sessile, 2 crack, basal lobes rounded, upper lobes hatchet, back with small auricle; lower sepal funnel-shaped; anthers ovoid; ovary fusiform. Capsules linear. Fl. Sep. Hillside forests or wet places of the road at 1900-2200 m. Distributed in Sichuan.

红雉凤仙花 *Impatiens oxyantghera*

陇南凤仙花 *Impatiens potaninii*

## 陇南凤仙花

**Impatiens potaninii** Maxim.

一年生草本。茎直立。叶互生，卵形或长圆形；叶柄较短，顶生者常近无柄。总花梗腋生，较短，具花2-3朵；花淡黄色；旗瓣圆形或扁圆形；翼瓣2裂；唇瓣漏斗状；花丝扁平；花药尖；子房线形。蒴果狭线形。花期8-10月。喜阴湿环境，生海拔1200-2300米的山谷、林缘或水沟旁湿处。产甘肃、陕西和四川。

Annual herbs. Stem erect. Leaves alternate, shortly petiolate, upper leaves sessile; leaves ovate or oblong. Inflorescences axillary, 2-3-flowered; peduncles short; flowers yellowish; upper petal orbicular or oblate; lateral united petals 2-lobed; lower sepal funnelform; filaments flattened; anthers acute. Capsules linear. Fl. Aug-Oct. Forest edges, shaded moist places in valleys, along canals at 1200-2300 m. Distributed in Gansu, Shaanxi and Sichuan.

山地凤仙花 *Impatiens monticola*

## 山地凤仙花

**Impatiens monticola** Hook. f.

一年生草本，高30-60厘米。叶互生，卵状椭圆形或倒卵形，长5-13厘米，宽3-4.5厘米。具2花；花浅黄色；萼片绿色，草质，披针形或卵状披针形。蒴果狭纺锤形，长2-2.5厘米。生海拔900-1800米的林缘阴湿处或路边石缝中。花期7-9月，果期10月。产四川。

Annual herbs, 30-60 cm tall. Leaves alternate, ovate-elliptic or obovate, 5-13 × 3-4.5 cm. 2-flowered; flowers yellowish; sepals green, herbaceous, lanceolate or ovate-lanceolate. Capsules narrowly fusiform, 2-2.5 cm long. Forest edges, shaded moist places or roadsides at 900-1800 m. Fl. Jul-Sep. Fr. Oct. Distributed in Sichuan.

## 毛萼凤仙花

**Impatiens trichosepala** Y. L. Chen

一年生矮小草本。茎直立，肉质，具根。叶互生，具柄，叶片狭披针形或倒披针形。总花

毛萼凤仙花 *Impatiens trichosepala*

梗单生于上部叶腋；花单生，黄色；旗瓣兜状，圆形；翼瓣具短柄或近无柄，2裂；唇瓣宽漏斗状；花丝线形，上端扩大；花药卵圆形；子房纺锤状。未成熟的蒴果线状圆柱形。花期7-9月。生海拔500-700米的山谷河边或疏林中，或潮湿草丛中。产广西和贵州。

Annual herbs, small. Stem erect, succulent, with roots. Leaves alternate, petiole; leaf narrowly lanceolate or oblanceolate. Inflorescences in upper leaf axils; flower solitary, yellow; upper petal cucullate, orbicular; lateral united petals shortly clawed or ± clawed, 2-lobed; lower sepal broadly funnelform; filaments linear, apex dilated; anthers ovoid; ovary fusiform. Capsules (immature) linear-cylindric. Fl. Jul-Sep. Sparse forests in valleys or moist grasslands or riversides at 500-700 m. Distributed in Guangxi and Guizhou.

## 绿萼凤仙花

**Impatiens chlorosepala** Hand.-Mazz.

一年生草本。茎肉质，直立。叶互生，膜质，卵形。总花梗生于上部叶腋，苞片披针形，宿存；花大，淡红色；侧生萼片2；旗瓣圆形；翼瓣具短柄，基部裂片半圆形，上部裂片长圆形；唇瓣檐部漏斗状基部急狭成内弯的距，具粉红色条纹；子房纺锤形。蒴果披针形。花期10-12月。生海拔300-1300米的山谷水旁阴处或疏林溪旁。产中国南部。

Annual herbs. Stem erect, succulent. Leaves alternate, ovate, membranous. Inflorescences in upper leaf axils; bracts persistent, lanceolate; flowers reddish, large; lateral sepals 2; upper petal orbicular; lateral united petals shortly clawed, basal lobes suborbicular; distal lobes oblong; lower sepal funnelform, abruptly narrowed into an involute or incurved spur; spur pink striate; ovary, fusiform. Capsules lanceolate. Fl. Oct-Dec. Sparse forests in valleys or by water in shaded places at 300-1300 m. Distributed in S China.

## 黄麻叶凤仙花

**Impatiens corchorifolia** Franch.

一年生草本。茎直立。叶互生，卵形或卵状披针形。总花梗细，花2朵；卵形宿存的苞片，或花梗中部有1-2线形苞片；花大，黄色，或有紫斑；侧生萼片4；旗瓣圆形；翼瓣近无柄，基部裂片圆形，上部裂片宽斧形；唇瓣囊状，距极短，内弯，2裂。蒴果条形。花期9-10月。生海拔2100-3500米的杂木林下或山谷林缘阴湿处。产云南和四川。

Annual herbs. Stem erect. Leaves alternate, ovate or ovate-lanceolate, Inflorescences thin, 2-flowered; bracts persistent, ovate, or with 1-2 linear bracts at middle; flowers yellow, sometimes purple spotted, large; lateral sepals 4; upper petal orbicular; lateral united petals ± clawed, basal lobes orbicular, distal lobes broadly dolabriform; lower sepal saccate, abruptly narrowed into an incurved spur, short, 2-fid. Capsules linear. Fl. Sep-Oct. Understories of mixed forests, forest edges, shaded moist places at 2100-3500 m. Distributed in Yunnan and Sichuan.

绿萼凤仙花 *Impatiens chlorosepala*

黄麻叶凤仙花 *Impatiens corchorifolia*

水金凤 *Impatiens noli-tangere*

## 水金凤
**Impatiens noli-tangere** L.

一年生草本。茎直立。叶互生，卵形。总花梗具2-4花，总状排列；苞片1，草质，宿存；花黄色；侧生萼片2；旗瓣圆形；翼瓣无阔柄，下部裂片矩圆形，上部裂片斧形；唇瓣宽漏斗状，基部渐狭成内弯的距；雄蕊5；子房纺锤形。蒴果圆柱形。种子圆球形，褐色。花期7-9月。生海拔900-2400米的山坡林下、林缘草地或沟边。产中国大部分地区。广泛分布于亚洲。

Annual herbs. Stem erect. Leaves alternate, ovate. Inflorescences 2-4-flowered, raceme; bract 1, persistent, herbaceous; flowers yellow; lateral sepals 2; upper petal orbicular; lateral united petals not clawed, basal lobes oblong, upper lobes broadly dolabriform; lower sepal broadly funnelform, gradually narrowed into an incurved spur; stamens 5; ovary fusiform. Capsules cylindric. Seeds globose, brown. Fl. Jul-Sep. Forest understories, forest edges, grasslands, along canals at 900-2400 m. Distributed in most parts of China. Widely distributed in Asia.

## 阔苞凤仙花
**Impatiens latebracteata** Hook. f.

一年生草本。茎纤细，分枝。叶互生，硬质，长圆形。总花梗纤细，具2-5花；苞片圆形，宿存；花黄色；侧生萼片2；旗瓣圆形，具角；翼瓣无柄，基部裂片圆形或宽长圆形，上部裂片斧形；唇瓣漏斗状全部内弯；子房纺锤状。蒴果狭椭圆形。种子长圆形或倒卵形，褐色。花期8月。生海拔约1900米的林缘、山坡、阴湿处。产四川和陕西。

Annual herbs. Stem slender, branched. Leaves alternate, oblong, rigid. Inflorescences thin, 2-5-flowered; bracts persistent, orbicular; flowers yellow; lateral sepals 2; upper petal with angle, orbicular; lateral united petals not clawed, basal lobes orbicular or broadly oblong, distal lobes dolabriform; lower sepal funnelform, wholly curved; ovary fusiform. Capsules

阔苞凤仙花 *Impatiens latebracteata*

narrowly elliptic. Seeds brown, oblong or obovoid. Fl. Aug. Forest edges, slopes, shaded moist places at ca. 1900 m. Distributed in Sichuan and Shaanxi.

## 耳叶凤仙花

**Impatiens delavayi** Franch.

一年生草本。茎细弱。叶互生，卵形，膜质。总花梗生于叶腋，具1-5花；苞片1，卵形，宿存。花紫红色或黄色；侧生萼片2；旗瓣圆形；翼瓣基部裂片近方形，上部裂片斧形；唇瓣囊状，基部成内弯的短距，距端2浅裂。蒴果线形。种子椭圆状长圆形，褐色。花期7-9月。生海拔3400-4200米的山麓、山沟水边或林下。产云南、四川和西藏。

Annual herbs. Stem slender. Leaves alternate, ovate, thinly membranous. Inflorescences in leaf axils, 1-5-flowered; bracts ovate, persistent; flowers purplish or yellow; lateral sepals 2; upper petal, orbicular; lateral united petals, basal lobes subtetragonous, distal lobes dolabriform; lower sepal saccate, abruptly narrowed into an incurved spur, shallowly 2-fid. Capsules linear. Seeds brown, elliptic-oblong. Fl. Jul-Sep. Under forests, by streams, along canals at 3400-4200 m. Distributed in Yunnan, Sichuan and Xizang.

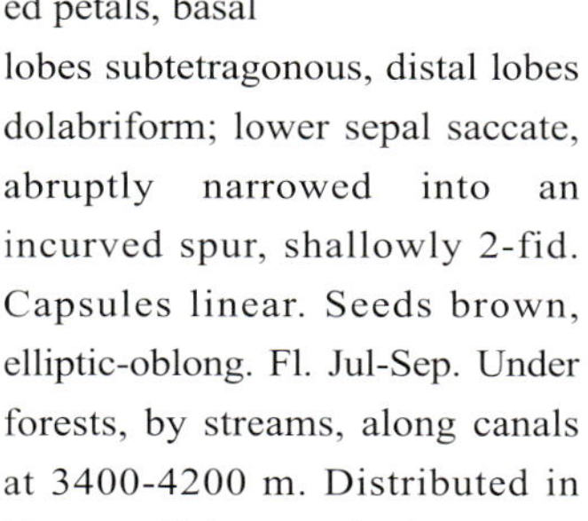

波缘凤仙花 *Impatiens undulata*

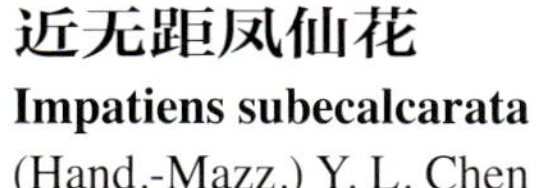

## 近无距凤仙花

**Impatiens subecalcarata** (Hand.-Mazz.) Y. L. Chen

一年生草本，高30-60厘米。叶近对生，卵形或卵状长圆形，长2.5-9厘米，宽1.5-3厘米，膜质。花紫红色，长2.5-3厘米。子房纺锤形，长4-5毫米，喙尖。蒴果线形，长3-3.5厘米。生海拔3500-3700米的亚高山杂木林下或山坡林缘。花期7-8月，果期9月。产云南西北部和四川。

Annual herbs, 30-60 cm tall. Leaves subopposite, ovate or ovate-oblong, 2.5-9 × 1.5-3 cm, membranous. Flowers purple-red, 2.5-3 cm long; ovary fusiform, 4-5 mm long, rostellate. Capsules linear, 3-3.5 cm long. Understories of alpine mixed forests, forest edges on slopes at 3500-3700 m. Fl. Jul-Aug. Fr. Sep. Distributed in NW Yunnan and Sichuan.

## 波缘凤仙花

**Impatiens undulata** Y. L. Chen et Y. Q. Lu

一年生草本，光滑无毛。茎直立。叶互生，在下部有叶柄，在上部无叶柄，叶片卵形。总花梗腋生，1-3花；苞片1，钻状；花黄色，小；侧生萼片2枚；旗瓣近圆形；翼瓣近无柄，背耳狭；唇瓣高脚碟形，口部近水平，距细。花丝短；花药锐尖。蒴果纺锤形，近黑色。花果期8-9月。生海拔1800-2000米的林缘或林间草地。产四川(峨眉山)。

Annual herbs, glabrous. Stem erect. Leaves alternate, petiolate in lower part of stem, sessile in upper part of stem, leaves ovate. inflorescences axillary, 1-3-flowered; bracts 1, subulate; flowers yellow, small; lateral sepals 2; upper petal suborbicular; lateral united petals ± clawed, 2-lobed; auricle narrow; lower sepal salver-shaped, gradually narrowed into slender spur, mouth subvertical, obtuse; Filaments short; anthers acute. Capsules nearly black, fusiform. Fl. and fr. Aug-Sep. Forest edges or grasslands between forests at 1800-2000 m. Distributed in Sichuan (Emei Mountain).

耳叶凤仙花 *Impatiens delavayi*

近无距凤仙花 *Impatiens subecalcarata*

睫毛萼凤仙花 *Impatiens blepharosepala*

微绒毛凤仙花
*Impatiens tomentella*

## 微绒毛凤仙花

**Impatiens tomentella** Hook. f.

一年生草本。茎肉质，密被绣微绒毛。叶互生，披针形。总花梗生于上部叶腋，具2花；苞片披针形，宿存；花橘黄色；侧生萼片2；旗瓣近肾形；翼瓣无柄，基部裂片宽楔形，上部裂片半月形；唇瓣漏斗状，基部渐狭成向上内弯的细距。蒴果线形。种子长圆形，褐色。花期7-8月。生海拔1400-1800米的林下阴湿处。产云南。

Annual herbs. Stem succulent, with young leaves densely covered rusty-tomentose. Leaves alternate, lanceolar. Inflorescences in upper leaf axils, 2-flowered; bract lanceolate, persistent; flowers orange-yellow; lateral sepals sparsely pubescent, 2; upper petal subreniform; lateral united petals not clawed, basal lobes broadly cuneate, distal lobes distal lobes semilunar; lower sepal funnelform, gradually narrowed into an upper incurved slender spur. Capsules linear. Seeds ovate-oblong, brown. Fl. Jul-Aug. Wet places of forests at 1400-1800 m. Distributed in Yunnan.

## 睫毛萼凤仙花

**Impatiens blepharosepala**
Pritz. ex Diels

一年生草本。茎直立，不分枝或基部有分枝。叶互生，矩圆形。总花梗腋生，花1-2朵；苞片条形；花紫色；侧生萼片2，边缘有睫毛，脱落；旗瓣近肾形；翼瓣无柄，基部裂片矩圆形，上部裂片斧形；唇瓣宽漏斗状，基部成内弯的距。蒴果条形。花果期5-11月。生海拔500-1600米的山谷水旁、沟边林缘或山坡阴湿处。产湖南、湖北、江西、贵州、安徽和福建。

Annual herbs. Stem erect, simple or branched from base. Leaves alternate, oblong. Inflorescences axillary, 1-2-flowered; bract linear; flowers purple; lateral sepals 2, margin ciliate; upper petal subreniform; lateral united petals, basal lobes oblong, distal lobes dolabriform; lower sepal broadly funnelform, not saccate, gradually narrowed into an incurved slender spur. Capsules linear. Fl. and fr. May-Nov. Forest edges, shaded moist places on slopes, along watercourses or canals at 500-1600 m. Distributed in Hunan, Hubei, Jiangxi, Guizhou, Anhui and Fujian.

## 大旗瓣凤仙花

**Impatiens macrovexilla** Y. L. Chen

一年生草本。茎肉质，直立，不分枝或中部分枝。叶互生，膜质，长圆形。总花梗具1-2花；苞片披针形，宿存；花紫色；侧生萼片2；旗瓣肾形；翼瓣无柄，基部裂片长圆形，上部裂片斧形；唇瓣窄漏斗形，基部渐狭成内弯的距；子房纺锤状。蒴果长圆形。种子球形，褐色。花期9-10月。生海拔1000-1640米的山谷阴处、林下或路边草地。产广西。

annual herbs. Stem erect, succulent, simple or branched at middle part. Leaves alternate, membranous, oblong. Inflorescences (1 or)2-flowered; bracts persistent, lanceolate; flowers purple; lateral sepals 2; upper petal reniform; lateral united petals not clawed, basal lobes oblong, distal lobes dolabriform; lower sepal narrowly funnelform, gradually narrowed into an incurved spur; ovary fusiform. Capsules oblong. Seeds brown, globose. Fl. Sep-Oct. Forests, grasslands at roadsides, shaded moist places at 1000-1640 m. Distributed in Guangxi.

大旗瓣凤仙花 *Impatiens macrovexilla*

## 浙江凤仙花

**Impatiens chekiangensis** Y. L. Chen

一年生草本。茎直立或上部略弯。叶互生，长圆形，膜质。总花梗单生于叶腋，2-3花；苞片宿存；花粉紫色；侧生萼片2；旗瓣近圆形；翼瓣近无柄，基部裂片卵状长圆形，上部裂片斧形；唇瓣狭漏斗状基部渐狭成内弯的距；子房线形。蒴果纺锤形。种子长圆形，黄褐色。花期7-8月。生海拔400-960米的山谷河边林下或阴湿岩石上。产浙江。

封怀凤仙花 *Impatiens fenghwaiana*

Annual herbs. Stem erect or slightly curved in upper part. Leaves alternate, ovate-oblong, membranous. Inflorescences axillary, 2(or 3)-flowered, bracts persistent, linear-lanceolate or linear, apex shortly acute; flowers pink-purple; lateral sepals 2; upper petal suborbicular; lateral united petals ± clawed, basal lobes ovate-oblong, distal lobes dolabriform; lower sepal narrowly funnelform, gradually narrowed into an incurved spur; ovary linear. Capsules fusiform. Seeds yellow-brown, ovoid-oblong. Fl. Jul-Aug. Forest understories at riversides in valleys or shaded moist places at 400-960 m. Distributed in Zhejiang.

## 封怀凤仙花

**Impatiens fenghwaiana** Y. L. Chen

一年生草本。叶互生或在茎上端密集，卵形或卵状披针形。总花梗单生于上部叶腋，2-3花，近总状排列；花梗细，基部具一粉红色苞片；花粉红色；侧生萼片2；旗瓣近圆形；翼瓣近无柄，2裂；唇瓣狭漏斗状；花丝线形；花药卵圆形；子房纺锤状。蒴果线形。花果期6-8月。生海拔500-1000米的林缘或草地潮湿处。产江西(庐山)。

Annual herbs. Leaves alternate, or crowded in upper part of stem; leaves ovate or ovate-lanceolate. Inflorescences in upper leaf axils, subracemose, (1 or)2- or 3(or 4)-flowered; pedicels slender, bracteate at base; bracts pink; flowers pink; Lateral sepals 2; upper petal suborbicular; lateral united petals ± clawed, 2-lobed; lower sepal narrowly funnelform; filaments linear; anthers ovoid; ovary fusiform. Capsules linear. Fl. and fr. Jun-Aug. Forest edges or grasslands, moist places at 500-1000 m. Distributed in Jiangxi (Lushan Mountain).

## 阔萼凤仙花

**Impatiens platysepala** Y. L. Chen

一年生草本。茎直立或基部平卧。叶互生，叶片膜质，披针形。总花梗生于叶腋，2-3花，近伞状排列；苞片膜质；花粉红色；旗瓣近圆形；翼瓣具柄，基部裂片短圆状倒卵形，上部裂片宽斧形；唇瓣宽漏斗状；子房纺锤形。蒴果线装圆柱形。种子近圆球形，褐色。花期8-10月。生海拔约1000米的山地林下或林缘水沟边。产江西和浙江。

Annual herbs. Stem erect or base prostrate. Leaves alternate, lanceolate, submembranous. Inflorescences axillary, subumbellate, 2-3-flowered; bracts membranous; flowers pink; upper petal suborbicular; lateral united petals clawed, basal lobes oblong-obovate, distal lobes broadly dolabriform; lower sepal broadly funnelform; ovary fusiform. Capsules linear-cylindric. Seeds brown, subglobose. Fl. Aug-Oct. Forest understories or forest edges, along canals at ca. 1000 m. Distributed in Jiangxi and Zhejiang.

浙江凤仙花 *Impatiens chekiangensis*

阔萼凤仙花 *Impatiens platysepala*

淡黄凤仙花 *Impatiens chloroxantha*

## 淡黄凤仙花

**Impatiens chloroxantha** Y. L. Chen

一年生草本。茎直立。叶互生，叶片椭圆形。总花梗具3或2花，排成近伞形花序；花梗基部具披针形苞片；花黄绿色；侧生萼片2；旗瓣阔椭圆形；翼瓣无柄，2裂，基部裂片长圆形，上部裂片圆斧形；唇瓣深黄色，舟状，基部成内弯的距；花丝线形；花药卵圆形；子房纺锤形。花期5月。生海拔500-700米的山地沟谷林中或沟边阴湿处。产浙江(遂昌县)。

Annual herbs. Stem erect. Leaves alternate, oblong or elliptic. Inflorescences subumbellate, 3-flowered, sometimes 2-flowered; pedicels slender, bracteate at base, bracts lanceolate; flowers yellow-green. lateral sepals 2; upper petal broadly elliptic; lateral united petals not clawed, 2-lobed; basal lobes ovate-oblong, small, obtuse; distal lobes orbicular-dolabriform; auricle inflexed; lower sepal deep yellow, navicular, narrowed into an incurved spur; filaments linear; anthers ovoid; ovary fusiform. Fl. May. Forests or shaded moist places along canals at 500-700 m. Distributed in Zhejiang (Suichang County).

## 扭萼凤仙花

**Impatiens tortisepala** Hook. f.

一年生草本，全株无毛。茎直立。叶互生，具柄；叶片近膜质，长圆形或卵状长圆形，无腺体。总花梗具6-8花，花总状排列；花黄色；旗瓣圆肾形；翼瓣具宽柄；唇瓣檐部囊状；花丝短而宽；花药卵圆形，顶端尖。蒴果线形。种子多数，长圆状倒卵形，褐色。花期8-9月。生海拔1500-2900米的山谷阴湿处。产四川。

Annual herbs, glabrous. Stem erect. Leaves alternate; petiole, without glands; leaves oblong or ovate-oblong, submembranous. Inflorescences in upper racemose, 6-8-flowered; flowers yellow; upper petal reniform; lateral united petals broadly clawed; lower sepal saccate; filaments short, broad; anthers ovoid, apex acute. Capsules linear. Seeds many, brown, oblong-obovoid. Fl. Aug-Sep. Shaded moist places in valleys at 1500-2900 m. Distributed in Sichuan.

扭萼凤仙花 *Impatiens tortisepala*

## 高山凤仙花

**Impatiens nubigena** W. W. Sm.

一年生草本，具少数支柱根及须根。茎直立。叶互生，宽卵形。总花梗生于茎枝上部叶腋，下部的总花梗通常具1花，较上部的具2花；花梗线状，具苞片；花极小，白色；侧生萼片2；旗瓣圆形；翼瓣无柄；唇瓣檐部舟状；花丝稍扁；花药钝；子房纺锤状。蒴果线形。花期8月，果期9月。生海拔2700-4000米的高山栎林、冷杉林、山坡草地或水沟边。产云南西北部、四川西南部和西藏东南部。

高山凤仙花 *Impatiens nubigena*

Annual herbs, with few supporting roots and fibrous roots. Stem erect. Leaves alternate; leaves broadly ovate. Inflorescences in upper leaf axils, 1- or 2-flowered; pedicels linear, bracteate below flowers; flowers very small, white; lateral sepals 2; lower sepal navicular; upper petal orbicular; lateral united petals not clawed; filaments slightly compressed; anthers obtuse; ovary fusiform. Capsules linear. Fl. Aug. Fr. Sep. Alpine *Quercus* or *Abies* forests, grasslands on slopes, along canals at 2700-4000 m. Distributed in NW Yunnan, SW Sichuan and SE Xizang.

## 撕裂萼凤仙花

**Impatiens lacinulifera** Y. L. Chen

一年生草本。茎肉质。叶互生，膜质。总花梗生于叶腋，具1花，苞片1枚，宿存；花黄色；侧生萼片2，边缘具不规则撕裂；旗瓣圆形；翼瓣近无柄，基部裂片圆形，上部裂片长圆状斧形；唇瓣檐部深囊状，急狭成钩状内弯、顶端2裂的短距；子房纺锤状。蒴果线形。种子近圆球形，褐色。花期8-9月。生海拔1600米的山谷潮湿处。产四川。

Annual herbs. Stem succulent. Leaves alternate, membranous. Inflorescences in leaf axils, 1-flowered; bracts 1, persistent; flowers yellow; lateral sepals 2, margin irregularly fimbriate-lacerate; upper petal orbicular; lateral united petals notclawed, basal lobes orbicular; distal lobes oblong-dolabriform; lower sepal deeply saccate, abruptly narrowed into a hooked, incurved spur, apex 2-lobed; ovary fusiform. Capsules linear. Seeds brown, subglobose. Fl. Aug-Sep. Shaded moist places in valleys at 1600 m. Distributed in Sichuan.

## 滇西北凤仙花

**Impatiens lecomtei** Hook. f.

一年生草本，无毛。茎直立。叶互生，卵形。总花柄具1花，中部具苞片；苞片卵形，宿存；花粉红色，侧生萼片2；旗瓣圆形；翼瓣近无柄，2裂；基部裂片卵形；上部斧形；唇瓣裂片漏斗形，基部成顶端2裂的细距；花丝线形；花药钝；子房纺锤状。蒴果线形。种子近球形。花期8-9月。生海拔2300-3000米的沟边杂木林或岩石缝中。产云南西北部。

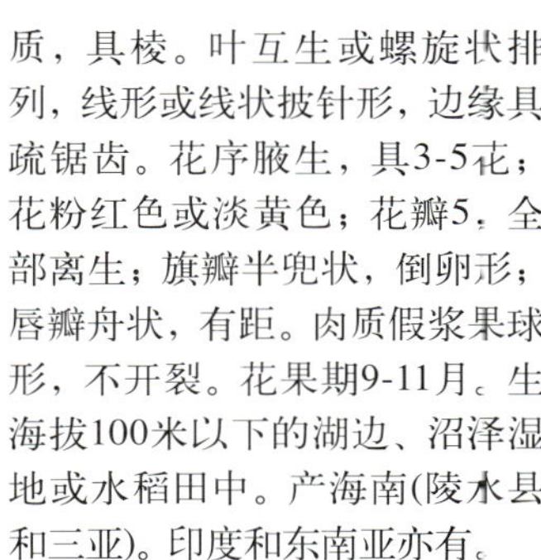

滇西北凤仙花 *Impatiens lecomtei*

Annual herbs, glabrous. Stem erect. Leaves alternate, leaves ovate. Inflorescences 1-flowered; pedicels bracteate at middle; bracts persistent, ovate; flowers pink; lateral sepals 2; upper petal rounded; lateral united petals ± clawed, 2-lobed; basal lobes ovate; distal lobes dolabriform; lower sepal violet striate, broadly funnelform, base gradually narrowed into an incurved spur; spur 2-lobed; filaments linear; anthers obtuse; ovary fusiform. Capsules linear. Seeds subglobose. Fl. Aug-Sep. Mixed forests or stony crevices at 2300-3000 m. Distributed in NW Yunnan.

## 水角

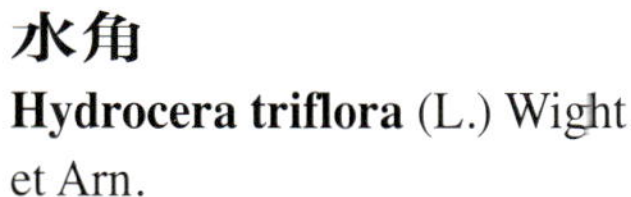

**Hydrocera triflora** (L.) Wight. et Arn.

多年生水生草本。茎直立，肉质，具棱。叶互生或螺旋状排列，线形或线状披针形，边缘具疏锯齿。花序腋生，具3-5花；花粉红色或淡黄色；花瓣5，全部离生；旗瓣半兜状，倒卵形；唇瓣舟状，有距。肉质假浆果球形，不开裂。花果期9-11月。生海拔100米以下的湖边、沼泽湿地或水稻田中。产海南(陵水县和三亚)。印度和东南亚亦有。

Perennial aquatic herbs. Stems erect, fleshy, ridged. Leaves alternate or spirally formed, linear or linear-lanceolate, margin sparsely serrate. Inflorescences axillary, 3-5-flowered; flowers pink or yellowish; petals 5, separate; upper petal subcucullate, obovate; lower 2 longer and narrower sepal navicular, with a spur. pseudoberries fleshy, globose, indehiscent. Fl. and fr. Sep-Nov. Lakesides, wetlands or paddyfields below 100 m. Distributed in Hainan (Lingshui County and Sanya). Also in India and SE Asia.

撕裂萼凤仙花 *Impatiens lacinulifera*

水角 *Hydrocera triflora*

# 鼠李科
# Rhamnaceae

## 枣
**Ziziphus jujuba** Mill.

落叶小乔木，稀灌木，具刺或无刺。叶纸质，卵形、卵状椭圆形或椭圆状长圆形，下面沿主脉多少具柔毛或无毛，基出3脉，基部稍不对称。花单生或2-8个密集成聚伞花序。核果矩圆形或长卵形，成熟时红色，直径1.5-2厘米。花期5-7月，果期8-10月。生海拔1700米以下的山地、丘陵或平原。中国大部分地区及其他国家栽培。

Deciduous small trees, rare shrubs, spinose or unarmed. Leaves papery, ovate, ovate-elliptic, or elliptic-oblong, abaxially ± puberulent on major veins or glabrous, 3-veined from base, base slightly asymmetric. Flowers solitary or 2-8-clustered to cymes. Drupes oblong or long ovoid, red after mature, 1.5-2 cm diam. Fl. May-Jul. Fr. Aug-Oct. Mountains, hills or plains below 1700 m. Cultivated in most parts of China and other countries.

## 酸枣
**Ziziphus jujuba** Mill. var. **spinosa** (Bunge) Hu ex H. F. Chow

本变种与酸枣的区别在于本变种的植株常为灌木。枝直立，不弯曲，具刺。叶小。核果近球形或阔矩圆形，小，直径0.7-1.2厘米；中果皮味酸，薄；果核两端钝。花期6-7月，果期8-9月。生于旱山坡、丘陵或平原。产华北、华西、华东和西北。俄罗斯和朝鲜半岛亦有。

This variety differs from the typical variety in its plants often shrubs. Branches erect, not tortuous, with spines. Leaves small. Drupes subglobose or broadly oblong, small, 0.7-1.2 cm diam; mesocarp sour-tasting, thin; stones obtuse at both ends. Fl. Jun-Jul. Fr. Aug-Sep. Dry slopes, hills or plains.

酸枣 *Ziziphus jujuba* var. *spinosa*

枣 *Ziziphus jujuba*

凹叶雀梅藤 *Sagereta horrida*

Distributed in N, W, E and NW China. Also in Russia and Korean Peninsula.

## 凹叶雀梅藤
**Sageretia horrida** Pax et Hoffm.

直立灌木。叶纸质，倒卵形或长圆形，长0.5-2(-2.5)厘米，宽0.4-1.4厘米，先端圆形，常微凹，边缘全缘或具锯齿；侧脉3-4对；叶柄长1-2毫米，被疏短柔毛。花无梗，排成穗状或稀穗状圆锥花序；花序轴被疏短柔毛；萼片三角状卵形。核果近球形，成熟时黑紫色。花期6-9月，果期翌年4-6月。生海拔1900-3600米的山地林缘或多石山坡。产四川、云南和西藏。

Erect shrubs. Leaves papery, obovate or oblong, 0.5-2(-2.5) cm long, 0.4-1.4 cm wide, apex rounded, often emarginate, margin entire or serrulate; lateral veins 3-4 pairs; Petioles 1-2 mm long, sparsely puberulent. Flowers sessile, in spikes or rarely in spicate panicles; rachis sparsely puberulent; sepals triangular-ovate. Drupe subglobose, black-purple when mature. Fl. Jun-Sep. Fr. next Apr-Jun. Forest edges on mountains or stony slopes at 1900-3600 m. Distributed in Sichuan, Yunnan and Xizang.

雀梅藤 *Sagereta thea*

## 皱叶雀梅藤
**Sageretia rugosa** Hance

藤状或直立灌木。叶卵状长圆形或卵形，纸质或厚纸质，干后常具皱，下面具锈色或灰白色不脱落绒毛。花芳香，常成穗状或穗状圆锥状。核果成熟时红色或紫红色，圆球形，具有2分核。花期7-12月，果期翌年3-4月。生海拔1600米以下的山坡灌丛或林中。产中国西南、华南和华中。

皱叶雀梅藤 *Sageretia rugosa*

Vine-like or erect shrubs. Leaves ovate-oblong or ovate, papery or thickly papery, often rugose when dry, abaxially persistently ferruginous or canescent-tomentose. Flowers fragrant, usually in spikes or spicate panicles. Drupes red or purple-red at maturity, globose, with 2 pyrenes. Fl. Jul-Dec. Fr next Mar-Apr. Bushes or forests on slopes below 1600 m. Distributed in SW, S and C China.

## 雀梅藤
**Sageretia thea** (Osbeck) Johnst.

灌木。小枝具刺。叶纸质，长圆状或卵状椭圆形，2-4.5 × 0.7-2.5厘米，背面无毛或叶脉被毛，腹面光滑，侧脉3-5(-7)对。花黄色，芳香，常2至数朵簇生成疏松穗状或圆锥状花序。核果成熟时黑色或紫黑色，近圆球形至倒卵球形。花期7-9月，果期翌年3-5月。生海拔2100米以下的丘陵、山坡林下或灌丛中。产中国西南、华南、东南、华中和华东。印度、泰国、越南、朝鲜半岛和日本亦有。

Shrubs. Branchlets spiny. Leaves papery, oblong, or ovate-elliptic, 2-4.5 × 0.7-2.5 cm, abaxially glabrous or pubescent on veins, adaxially glabrous, lateral veins 3-5(-7) pairs. Flowers yellow, fragrant, usually 2- to few fascicled lax spikes or paniculate spikes. Drupes black or purple-black at maturity, subglobose to obovoid. Fl. Jul-Sep. Fr. next Mar-May. Hills, under forests of mountain slopes, or thickets below 2100 m. Distributed in SW, S, SE, C and E China. Also in India, Thailand, Vietnam, Korean Peninsula and Japan.

钩刺雀梅藤 *Sageretia hamosa*

长叶冻绿 *Rhamnus crenata*

## 钩刺雀梅藤

**Sageretia hamosa** (Wall.) Brongn.

常绿藤状灌木。小枝常具有钩状的粗刺。叶长圆形或狭椭圆形，革质，下面于脉腋处具髯毛或疏具柔毛，侧脉7-10对；叶柄无毛。花常2或3朵簇生为疏松的穗状圆锥花序。核果成熟后深红色或紫黑色，近球形。花期7-8月，果期8-10月。生海拔1600米以下的山坡灌丛或林中、河道沿岸的密林中。产中国西南、东南和华中。印度、尼泊尔、斯里兰卡、越南和菲律宾亦有。

Evergreen vine-like shrubs. Branchlets with unciform coarse prickles. Leaves oblong or narrowly elliptic, leathery, abaxially barbate at vein axils or sparsely pubescent, glabrescent, lateral veins 7-10 pairs; petioles glabrous. Flowers usually 2- or 3-fascicled in lax spicate panicles. Drupes deep red or purple-black at maturity, subglobose. Fl. Jul-Aug. Fr. Aug-Oct. Forests or thickets on slopes, dense forests along canals below 1600 m. Distributed in SW, SE and C China. Also in India, Nepal, Sri Lanka, Vietnam and the Philippines.

## 梗花雀梅藤

**Sageretia henryi** Merr. et Chun

匍匐灌木。叶互生或近对生，长圆形，狭椭圆形或卵状椭圆形，纸质。花白色或黄色，单生或少数簇生于腋或顶生为疏松总状；花序轴长达15厘米；花梗长1-3毫米；花瓣匙形，稍短于雄蕊。核果成熟紫红色，椭圆形或倒卵球形。花期7-11月，果期翌年3-6月。生海拔400-2500米的密林或山坡灌丛。产中国西南、东南、华中和华西。

Shrubs scandent. Leaves alternate or subopposite, oblong, narrowly elliptic, or ovate-elliptic, papery. Flowers white or yellow, solitary or few fascicled in axillary or terminal lax racemes; rachis to 15 cm long; pedicels 1-3 mm long; petals spatulate, slightly shorter than stamens. Drupes purple-red at maturity, elliptic or obovoid. Fl. Jul-Nov. Fr. next Mar-Jun. Dense forests or mountain thickets at 400-2500 m. Distributed in SW, SE, C and W China.

## 长叶冻绿

**Rhamnus crenata** Siebold et Zucc.

落叶灌木或小乔木。叶倒卵状椭圆形、椭圆形或倒卵形，纸质，侧脉7-12对，下面具柔毛或至少沿脉具柔毛。花少数至10朵簇生成腋生聚伞花序；花柱不分裂。核果成熟后红色、黑色或紫黑色，球形或倒卵球形。花期5-8月，果期8-10月。生海拔2000米以下的山地林下或灌丛中。产中国西南、华南、东南、华中和华东。老挝、泰国、越南、柬埔寨、朝鲜半岛和日本亦有。

梗花雀梅藤 *Sageretia henryi*

尼泊尔鼠李 *Rhamnus napalensis*

川滇鼠李 *Rhamnus gilgiana*

Deciduous shrubs or small trees. Leaves obovate-elliptic, elliptic, or obovate, papery, lateral veins 7-12 pairs, abaxially pubescent or at least ± pubescent on veins. Flowers few or up to 10 crowded in axillary cymes; styles undivided. Drupes red, black, or purple-black at maturity, globose or obovoid. Fl. May-Aug. Fr. Aug-Oct. Under montane forests or among bushes below 2000 m. Distributed in SW, S, SE, C and E China. Also in Laos, Thailand, Vietnam, Cambodia, Korean Peninsula and Japan.

## 尼泊尔鼠李

**Rhamnus napalensis** (Wall.) Laws.

直立或藤状落叶灌木，雌雄异株，无刺。叶小异形，小叶2-5厘米长，大叶6-17(-20)厘米长，下面脉腋具簇生毛。花5数，腋生成聚伞状总状，长达12厘米；花柱3浅裂至中裂。核果幼时紫红色，成熟后紫黑色，倒卵球形。花期5-9月，果期8-11月。生海拔1800米以下的林中或灌丛中。产中国西南、东南和华中。南亚亦有。

Shrubs erect or scandent, deciduous, dioecious, unarmed. Leaves variable in shape and size, small ones 2-5 cm long, large ones 6-17(-20) cm long, abaxially with clustered hairy vein axils. Flowers 5-merous, in axillary cymose racemes, up to 12 cm long; styles 3-fid or cleft to half. Drupes reddish purple when young, purplish black at maturity, obovoid. Fl. May-Sep. Fr. Aug-Nov. Forests or thickets below 1800 m. Distributed in SW, SE and C China. Also in S Asia.

## 川滇鼠李

**Rhamnus gilgiana** Heppel.

多刺灌木。小枝顶端具细刺。叶下常浅绿色，椭圆形或卵状椭圆形，纸质或厚革质，上面于脉腋簇生毛或近无毛。花单性，常3-5朵簇生于短枝叶腋。核果棕色，近球形，基部被浅杯状萼筒包被。花期4-5月，果期6-8月。生海拔2200-2700米的混生林下或灌丛中。产云南和四川。

Spiny shrubs. Top of branchlets with thin prickles. Leaves abaxially pale green, elliptic or ovate-elliptic, papery to thickly papery, abaxially clustered hairy in vein axils or subglabrous. Flowers unisexual, usually 3-5-clustered at axils of short branches. Drupes brown, subglobose, with shallow cup-shaped persistent calyx tube at base. Fl. Apr-May. Fr. Jun-Aug. Mixed forests or shrubs at 2200-2700 m. Distributed in Yunnan and Sichuan.

## 卵叶鼠李

**Rhamnus bungeana** J. Vass.

具刺灌木，雌雄异株。叶卵形、卵状披针形或卵状椭圆形，背面脉上或脉腋处具白色柔毛，侧脉2或3对。花单性，4数。核果成熟时紫色或紫黑色，具2核。种子卵球形，背面具长为种子4/5的纵沟。花期4-5月，果期6-9月。生海拔约1800米的灌丛或开阔山坡。产湖北、河南、河北、山西、山东和吉林。

Shrubs dioecious, spinose. Leaves ovate, ovate-lanceolate or ovate-elliptic, abaxially white pubescent on veins or vein axils, lateral veins 2 or 3 pairs. Flowers unisexual, 4-merous. Drupes purple or black-purple at maturity, with 2 stones. Seeds ovoid, abaxially with margined furrow extending over 4/5 of length. Fl. Apr-May. Fr. Jun-Sep. Thickets or open slopes at ca. 1800 m. Distributed in Hubei, Henan, Hebei, Shanxi, Shandong and Jilin.

卵叶鼠李 *Rhamnus bungeana*

锐齿鼠李 *Rhamnus arguta*

## 锐齿鼠李
**Rhamnus arguta** Maxim.

灌木或小乔木。树皮灰褐色。叶卵状心形或卵状圆形，薄纸质或纸质，边缘密具锐锯齿。花单性，4数；雄花10-20朵簇生，雌花几个簇生。核果成熟时黑色，球形或倒卵球形。花期5-6月，果期6-9月。生海拔2000米以下的林中和灌丛中。产河北、山西、陕西、山东、黑龙江和辽宁。

Shrubs or small trees. Bark grayish brown. Leaves ovate-cordate or ovate-orbicular, thinly papery or papery, margin densely sharply serrate. Flowers unisexual, 4-merous; staminate flowers 10-20-clustered; pistillate flowers several-clustered. Drupes black at maturity, globose or obovoid. Fl. May-Jun. Fr. Jun-Sep. Forests and thickets below 2000 m. Distributed in Hebei, Shanxi, Shaanxi, Shandong, Heilongjiang and Liaoning.

## 薄叶鼠李
**Rhamnus leptophylla** Schneid.

灌木，雌雄异株，具刺。叶纸质，倒卵形或倒卵状椭圆形，宽2-5厘米，先端短渐尖或锐尖，侧脉3-5对。花单性，雌花少数至10朵簇生；花柱2中裂。核果成熟后黑色，球形，具2或3核。花期3-5月，果期5-10月。生海拔1700-2600米的山坡、山谷灌丛中或林缘。产中国西南、东南、华中和华东。

Shrubs, dioecious, spinose. Leaves papery, obovate or obovate-elliptic, 2-5 cm wide, apex shortly acuminate or acute, lateral nerves 3-5 pairs. Flowers unisexual, pistillate flowers few- to 10-fascicled; styles 2-cleft to half. Drupes black at maturity, globose, with 2 or 3 stones. Fl. Mar-May. Fr. May-Oct. Slopes, thickets in valleys or forest edges at 1700-2600 m. Distributed in SW, SE, C and E China.

薄叶鼠李 *Rhamnus leptophylla*

## 甘青鼠李
**Rhamnus tangutica** J. Vass.

落叶灌木，具刺。老枝先端成刺状。叶椭圆形、倒卵状椭圆形或倒卵形，上面沿脉与脉腋疏具柔毛，边缘具钝或圆锯齿。花单性，4数；柱头2浅裂。核果成熟时黑色，具2核。种子背侧面3/4-4/5具纵沟。果实可制染料。花期5-6月，果期6-9月。生海拔1200-3700米的山谷林中或灌丛中。产四川、西藏、河南、陕西、甘肃和青海。

Dioecious shrubs, spinose. Older branches terminating in a spine. Leaves elliptic, obovate-elliptic, or obovate, abaxially sparsely

甘青鼠李 *Rhamnus tangutica*

帚枝鼠李 *Rhamnus virgata*

小叶鼠李 *Rhamnus parvifolia*

puberulent on veins and vein axils, margin obtuse or crenulate. Flowers unisexual, 4-merous; styles 2-fid. Drupes black at maturity, with 2 stones. Seeds abaxially-laterally with margined furrow extending over 3/4-4/5 of length. Fruits for making dye. Fl. May-Jun. Fr. Jun-Sep. Forests or thickets in valleys at 1200-3700 m. Distributed in Sichuan, Xizang, Henan, Shaanxi, Gansu and Qinghai.

## 帚枝鼠李
**Rhamnus virgata** Roxb.

灌木或乔木，雌雄异株，具刺。一年生枝无毛。叶背近无毛或叶脉、脉腋被疏柔毛，叶柄被微绒毛；叶对生或近对生或在短枝上簇生，侧脉4-5对，干后红色。花单性，4数；花柱2裂。核果蓝黑色，近球形，具2核。花期4-5月，果期6-10月。生海拔1200-3800米的山坡灌丛或林中。产云南、四川、西藏和贵州。印度、尼泊尔、不丹和泰国亦有。

Shrubs or trees, dioecious, spinose. Annual branches glabrous. Abaxial leaf surface subglabrous or sparsely pubescent on veins or vein axils, petioles adaxially puberulent; leaves opposite or subopposite, or fascicled on short branch apex, lateral nerves 4-5 pairs, turning red when dry. Flowers unisexual, 4-merous; styles 2-cleft. Drupes bluish black, subglobose, with 2 stones. Fl. Apr-May. Fr. Jun-Oct. Thickets or forests on slopes at 1200-3800 m. Distributed in Yunnan, Sichuan, Xizang and Guizhou. Also in India, Nepal, Bhutan and Thailand.

## 小叶鼠李
**Rhamnus parvifolia** Bunge

灌木，雌雄异株。小枝对生或近对生，紫褐色。叶下绿色，菱状倒卵形或菱状椭圆形，1.2-4 × 0.8-2(-3)厘米。花黄绿色，常数朵簇生于短枝；柱头2中裂。核果成熟时黑色，倒卵球形。花期4-5月，果期6-9月。生海拔400-2300米以下的灌丛、草地、石质山坡或岩石上。产中国东南、华北和东北。俄罗斯(东西伯利亚)、蒙古和朝鲜半岛亦有。

Shrubs, dioecious. Branchlets opposite or subopposite, purplish brown. Leaves abaxially greenish, rhombic-obovate or rhombic-elliptic, 1.2-4 × 0.8-2(-3) cm. Flowers yellowish green, usually few clustered on short shoots; styles 2-cleft to half. Drupes black at maturity, obovoid. Fl. Apr-May. Fr. Jun-Sep. Thickets, grasslands, stony slopes or rocks at 400-2300 m. Distributed in SE, N and NE China. Also in Russia (E Siberia), Mongolia and Korean Peninsula.

## 圆叶鼠李
**Rhamnus globosa** Bunge

灌木，雌雄异株，多刺。小枝灰褐色。叶近圆形，倒卵状圆形或卵状圆形，纸质或薄纸质，下面全部或仅于脉处具柔毛。花常少数至20朵簇生；花柱2-3浅裂或半裂。核果成熟时黑色，球形或倒卵球形。花期4-5月，果期6-10月。生海拔1600米以下的山坡、林下或灌丛中。产华中、华北、华西和华东。

Shrubs, dioecious, spinose. Branchlets gray-brown. Leaves suborbicular, obovate-orbicular, or ovate-orbicular, papery or thinly papery, abaxially throughout or on veins pubescent. Flowers usually few- to 20-fascicled; styles 2-3-fid or cleft to half. Drupes black at maturity, globose or obovoid. Fl. Apr-May. Fr. Jun-Oct. Mountain slopes, under forests or among bushes below 1600 m. Distributed in C, N, W and E China.

圆叶鼠李 *Rhamnus globosa*

鼠李 *Rhamnus davurica*

冻绿 *Rhamnus utilis*

## 鼠李

**Rhamnus davurica** Pall.

小乔木或灌木。小枝粗壮，顶端常有大芽而不形成刺。叶宽椭圆形或卵形，纸质，下面沿脉疏具毛。花黄绿色；雌花1-3朵生叶腋或少数至20朵簇生于短枝；花柱2或3浅裂或裂至中部。核果黑色，球形，具2分核。花期5-6月，果期7-10月。生海拔1800米以下的山坡林下、灌丛、林缘或河渠旁湿处。产河北、山西、黑龙江、吉林和辽宁。俄罗斯(远东地区、西伯利亚)、蒙古和朝鲜半岛亦有。

Small trees or shrubs. Branchlets robust, apex with large buds and without prickle. Leaves broadly elliptic or ovate, papery, abaxially sparsely pilose on veins. Flowers yellowish green; pistillate flowers 1-3 in leaf axils or few- to 20-fascicled on short shoots; styles 2- or 3-fid or cleft to half. Drupes black, globose, with 2 stones. Fl. May-Jun. Fr. Jul-Oct. Under forests of mountain slopes, thickets, forest edges or damp places by canals below 1800 m. Distributed in Hebei, Shanxi. Heilongjiang, Jilin and Liaoning. Also in Russia (Far East, Siberia), Mongolia and Korean Peninsula.

## 冻绿

**Rhamnus utilis** Decne.

灌木或小乔木，雌雄异株。小枝顶端具针刺。叶椭圆形、长圆形或倒卵状椭圆形，纸质至近革质，下面沿脉具金黄色柔毛。花单性，4数；柱头长，2浅裂或中裂。核果成熟后黑色，球形或近球形，具2核。花期4-6月，果期5-8月。生海拔3300米以下的山地、丘陵、山坡草丛、灌丛中或疏林下。产中国西南、东南、华中、华北、华西和华东。朝鲜半岛和日本亦有。

Shrubs or small trees, dioecious. Branchlet apices with prickles. Leaves elliptic, oblong, or obovate-elliptic, papery to subleathery, abaxially golden-yellow pubescent on veins. Flowers unisexual, 4-merous; styles long, 2-fid or cleft to half. Drupes black at maturity, globose or subglobose, with 2 stones. Fl. Apr-Jun. Fr. May-Aug. Mountains, hills, among grasses of mountain slopes, thickets or under sparse forests below 3300 m. Distributed in SW, SE, C, N, W and E China. Also in Korean Peninsula and Japan.

## 柳叶鼠李

**Rhamnus erythroxylum** Pall.

灌木，雌雄异株，多刺。叶狭披针形至狭倒卵形，背面棕色，两面无毛。花黄绿色，单性；子房2或3室；花柱2或3浅裂或裂至中部。核果成熟后黑色，球形，常具2(-3)核。种子浅棕色，背部具长达其4/5的纵沟。花期5月，果期6-7月。生海拔1000-2100米的灌丛、山地、开阔山坡、多石山坡或干沙地。产河北、山西、内蒙古、陕西、甘肃和青海。亚洲西南部、俄罗斯(西伯利亚)和蒙古亦有。

Shrubs, dioecious, very spinose. Leaves narrowly lanceolate to narrowly obovate, abaxially brownish, both surfaces glabrous. Flowers yellow-green, unisexual; ovary 2- or 3-loculed; styles 2(or 3)-fid or cleft to half. Drupes black at maturity, globose, usually 2(-3) stones. Seeds pale brown, abaxially with narrow margined furrow extending over 4/5 of length. Fl. May. Fr. Jun-Jul. Thickets, hills, open slopes, stony and rocky slopes or dry sands at

柳叶鼠李 *Rhamnus erythroxylum*

小冻绿树 *Rhamnus rosthornii*

1000-2100 m. Distributed in Hebei, Shanxi, Neimenggu, Shaanxi, Gansu and Qinghai. Also in SW Asia, Russia (Siberia) and Mongolia.

## 小冻绿树

**Rhamnus rosthornii** E. Pritz.

灌木或小乔木，雌雄异株，具刺。枝互生或近对生，顶端具钝刺。叶匙形、菱状椭圆形或倒卵状椭圆形，下面常于脉腋处具带须毛腺点，侧脉2-4对。花单性。核果成熟时黑色，球形。花期4-5月，果期6-9月。生海拔600-2600米的山坡阳处灌丛或溪边林中。产中国西南、华中和华西。

Shrubs or small trees, dioecious, spinose. Branches alternate or subopposite, terminating in a spine. Leaves spatulate, rhombic-elliptic, or obovate-elliptic, abaxially often with bearded domatia in axils veins, lateral nerves 2-4 pairs. Flowers unisexual. Drupes black at maturity, globose. Fl. Apr-May. Fr. Jun-Sep. Thickets on sunny slopes or forests along streams at 600-2600 m. Distributed in SW, C and W China.

## 山鼠李

**Rhamnus wilsonii** Schneid.

灌木，雌雄异株。叶柄及叶两面均无毛或仅叶背脉处稍被柔毛。花黄绿色，单性，少数至20朵簇生于一年生枝基部或1到少数几朵腋生；柱头（2或）3浅裂或近中裂。核果成熟时紫黑色或黑色。花期4-5月，果期6-10月。生海拔300-1600米的林中、林缘、灌丛、山坡或路边。产中国西南、东南和华东。

Shrubs, dioecious. Petioles and both surfaces of leaves glabrous or leaves abaxially pubescent especially on veins. Flowers yellow-green, unisexual, few- to 20-fascicled at base of annual branches or 1 to few in axils; styles (2 or)3-fid or nearly cleft to half. Drupes purple-black or black at maturity. Fl. Apr-May. Fr. Jun-Oct. Forests, forests edges, thickets, slopes or roadsides at 300-1600 m. Distributed in SW SE and E China.

## 钩齿鼠李

**Rhamnus lamprophylla** Schneid.

灌木或小乔木，雌雄异株。小枝互生，先端刺状。叶狭椭圆形或椭圆形，边缘具钩状内弯圆锯齿。花黄绿色；柱头2或3浅裂至近中裂。核果成熟时黑色，具2(-3)核。种子背面仅下部1/4具短槽。花期4-5月，果期6-9月。生海拔400-1600米的林中或山地灌丛中。产中国西南、华南和华中。

Shrubs or small trees, dioecious. Branchlets alternate, terminating in a spine. Leaves narrowly elliptic or elliptic, margin hooked incurved-crenate. Flowers yellow-green; styles 2- or 3-fid or nearly cleft to half. Drupes black at maturity, with 2(-3) stones. Seeds abaxially with short margined furrow extending over 1/4 of length. Fl. Apr-May. Fr. Jun-Sep. Forests or mountain thickets at 400-1600 m. Distributed in SW, S and C China.

山鼠李 *Rhamnus wilsonii*

钩齿鼠李 *Rhamnus lamprophylla*

湖北鼠李 *Rhamnus hupehensis*

## 湖北鼠李
**Rhamnus hupehensis** Schneid.

无刺灌木。叶椭圆形或长圆状卵形，下面浅绿色，两面无毛。花单性，少数簇生于短枝上。核果常1或2个生于短枝，成熟后黑色，倒卵球形，具2或3核。种子紫黑色，背部具占其5/7的纵沟。花期未知，果期6-10月。生海拔1700-2300米的林中、灌丛或山坡。产湖北西部。

Shrubs, unarmed. Leaves elliptic or oblong-ovate, abaxially pale green, both surfaces glabrous. Flowers unisexual, few fascicled at short shoots. Drupes usually 1 or 2 on short shoots, black at maturity, obovoid, with 2 or 3 stones. Seeds purple-black, abaxially with margined furrow extending over 5/7 of length. Fl. unknown. Fr. Jun-Oct. Forests, thickets or slopes at 1700-2300 m. Distributed in W Hubei.

## 皱叶鼠李
**Rhamnus rugulosa** Hemsl.

灌木，雌雄异株，具刺。叶柄及叶两面均密被白柔毛；侧脉于上面突出。花黄绿色，单性，疏被短毛；柱头长，顶端短2或3浅裂。核果成熟后紫黑色或黑色，具2或3个核。种子背部具约与其等长的纵沟。花期4-5月，果期6-9月。生海拔500-2300米的灌丛林缘、山坡、渠边或路边。产华南、华中、华北、华西和华东。

Shrubs, dioecious, spinescent. Petioles and both surfaces of leaves densely whitish pubescent, lateral veins prominent abaxially. Flowers yellow-green, unisexual, sparsely pilose; styles long, shortly 2- or 3-cleft apically. Drupes purple-black or black at maturity, with 2 or 3 stones. Seeds abaxially with margined furrow ca. as long as seeds. Fl. Apr-May. Fr. Jun-Sep. Thicket edges, slopes, along canals or roadsides at 500-2300 m. Distributed in S, C, N, W and E China.

皱叶鼠李 *Rhamnus rugulosa*

朝鲜鼠李 *Rhamnus koraiensis*

## 朝鲜鼠李
**Rhamnus koraiensis** Schneid.

具刺灌木。小枝互生，灰色。叶宽椭圆形，倒卵状椭圆形或卵形，纸质，两面密具柔毛。花瓣黄绿色，雌花少数至10朵簇生；柱头2深裂。核果倒卵球形，紫黑色，具2核。花期4-5月，果期6-9月。生低海拔的林中或灌丛中。产山东、吉林和辽宁。朝鲜半岛亦有。

Shrubs, spinescent. Branchlets alternate, grayish. Leaves broadly elliptic, obovate-elliptic, or ovate, papery, both surfaces densely pilose. Petals yellowish green, pistillate flowers few- to 10-fascicled; styles deeply 2-cleft. Drupes obovoid, purplish black, with 2 stones. Fl. Apr-May. Fr. Jun-Sep. Forests or thickets at low elevations. Distributed in Shandong, Jilin and Liaoning. Also in Korean Peninsula.

## 山绿柴
**Rhamnus brachypoda** C. Y. Wu ex Y. L. Chen et P. K. Chou

灌木，雌雄异株，具刺。叶背绿色或黄绿色，无毛。花黄绿色，单性，1-3朵生于下部枝腋或生于短枝上；柱头3深裂至半裂。核果成熟时黑色，倒卵球形，具 (2-)3核，基部具浅杯形花萼。花期5-6月，果期7-11月。生海拔500-1700米的山谷

山绿柴 *Rhamnus brachypoda*

疏林、山坡灌丛或路边。产中国西南、华中和东南。

Shrubs, dioecious, spinescent. Leaves abaxially green or yellow-green, glabrous. Flowers yellow-green, unisexual, 1-3 in leaf axils of lower part of branchlets or on short shoots; styles 3-cleft to half. Drupes black at maturity, obovoid, with (2-)3 stones, with shallow cup-shaped calyx tube at base. Fl. May-Jun. Fr. Jul-Nov. Sparse forests in valleys, thickets on slopes or roadsides at 500-1700 m. Distributed in SW, C and SE China.

## 枳椇 (拐枣)

**Hovenia acerba** Lindl.

高大乔木。叶厚纸质或纸质，下面无毛或沿脉或脉腋具柔毛。二歧聚伞圆锥花序顶生或腋生。果实和柱头无毛；果实成熟时黄褐色或褐色，近球形，果梗和花梗膨大，近肉质。花期5-7月，果期8-10月。生海拔2100米以下的开阔地、山坡林缘或疏林中。产中国西南、华南、华中和华东。印度、尼泊尔、不丹和缅甸亦有。

Tall trees. Leaves thick papery or papery, abaxially glabrous or pilose on veins or in vein axils. Dichotomous racemose cymes terminal or axillary. Fruits and styles glabrous; fruits yellow-brown or brown at maturity, subglobose, fruiting peduncles and pedicels dilated and ± fleshy. Fl. May-Jul. Fr. Aug-Oct. Open places, forest edges of mountain slopes or sparse forests below 2100 m. Distributed in SW, S, C and E China. Also in India, Nepal, Bhutan and Myanmar.

枳椇 (拐枣) *Hovenia acerba*

北枳椇（北拐枣） *Hovenia dulcis*

## 北枳椇（北拐枣）

**Hovenia dulcis** Thunb.

高大乔木，稀灌木。叶片纸质至厚膜质，两面无毛或下面沿主脉具柔毛。花黄绿色，排成不对称的聚伞圆锥花序。果实为具3粒种子的坚果，成熟时黑色，近球形，无毛，花梗和小花梗于果实成熟后变为肉质多汁。花期5-7月，果期8-10月。生海拔200-1400米的次生林中，亦在花园中栽培。产中国西南、华北、华西和华东。泰国、朝鲜半岛和日本亦有。

Tall trees, rare shrubs. Leaflets papery to thick membranous, both surfaces glabrous or abaxially pilose on major veins. Flowers yellowish green, arranged to asymmetric racemose cymes. Fruits a 3-seeded nut, black at maturity, subglobose, glabrous, peduncles and pedicels becoming fleshy and juicy at fruit maturity. Fl. May-Jul. Fr. Aug-Oct. Secondary forests, also cultivated in gardens at 200-1400 m. Distributed in SW, N, W and E China. Also in Thailand, Korean Peninsula and Japan.

## 毛果枳椇

**Hovenia trichocarpa** Chun et Tsiang

乔木。小枝淡褐色或黑紫色。叶密被黄褐色或黄灰色绒毛。花序密具锈色毛或黄褐色绒毛；花黄绿色，二歧聚伞花序顶生或兼腋生。果实成熟时黄褐色或褐色，近球形。花期5-6月，果期8-10月。生海拔600-1300米的林中。产贵州、广东、湖南、湖北和江西。日本亦有。

Trees. Branchlets brownish or blackish purple. Leaves densely yellow-brown or yellow-gray tomentose adaxially. Inflorescences densely ferruginous or yellow-brownish tomentose; flowers yellow-green, dichotomous cymes terminal or axillary. Fruits yellow-brown or brown at maturity, subglobose. Fl. May-Jun. Fr. Aug-Oct. Forests at 600-1300 m. Distributed in Guizhou, Guangdong, Hunan, Hubei and Jiangxi. Also in Japan.

## 光叶毛果枳椇

**Hovenia trichocarpa** Chun et Tsiang var. **robusta** (Nakai et Y. Kimura) Y. L. Chen et P. K. Chou

本变种与枳椇的区别在于本变种的小叶两面无毛或背面沿脉被疏柔毛。花期5-6月，果期8-10月。生海拔600-1300米的山坡密林中。产中国西南、华南和东南。日本亦有。

This variety differs from the typical variety in its leaves glabrous on both surfaces or pilose on veins adaxially. Fl. May-Jun. Fr. Aug-Oct. Dense forests of mountain slopes at 600-1300 m. Distributed in SW, S and SE China. Also in Japan.

毛果枳椇 *Hovenia trichocarpa*

光叶毛果枳椇 *Hovenia trichocarpa* var. *robusta*

蛇藤 *Colubrina asiatica*

## 蛇藤
**Colubrina asiatica** (L.) Brongn.

攀援灌木。叶互生，近膜质或薄纸质，卵形，顶端微凹，基部圆形，边缘具粗圆齿。聚伞花序腋生；花黄色，5基数，花萼5裂；花梗长2-3毫米。蒴果状核果圆球形，基部为愈合的萼筒所包围，3室，熟时室背开裂。花期6-9月，果期9-12月。生沿海沙地上的林中或灌丛中。产广东、广西和台湾。热带亚洲、澳大利亚、非洲和太平洋岛屿亦有。

Scandent shrubs. Leaves alternate, submembranous or thinly chartaceous, ovate, apex concave, base round, margin coarsely crenate. Cymes axillary; flowers yellow, 5-merous, calyx 5-lobed; pedicels 2-3 mm long. Drupes capsule-like, globose, base surrounded by connate calyx, 3-loculed, dehiscent at maturity. Fl. Jun-Sep. Fr. Sep-Dec. Forests or scrubs on coastal sandlots. Distributed in Guangdong, Guangxi and Taiwan. Also in tropical Asia, Australia, Africa and Pacific Islands.

## 猫乳
**Rhamnella franguloides** (Maxim.) Weberb.

落叶灌木或小乔木。叶下黄绿色，上面绿色，下面至少沿脉具柔毛，上面无毛。花黄绿色，两性，6-18朵成腋生聚伞状。核果成熟时红色或橘红色，干后变为黑色或紫黑色，圆柱形。花期5-7月，果期7-10月。生海拔1100米以下的山坡、路旁或林中。产华中、华北和华东。朝鲜半岛和日本亦有。

Shrubs or small trees. Leaves abaxially yellow-green, adaxially green, abaxially pubescent at least on veins, adaxially glabrous Flowers yellow-green, bisexual 6-18 in axillary cymes. Drupes red or orange at maturity, turning black or purple-black when dry, cylindric, Fl. May-Jul. Fr. Jul-Oct. Mountain slopes, by the road or in the forests below 1100 m. Distributed in C, N and E China. Also in Korean Peninsula and Japan.

## 西藏猫乳
**Rhamnella gilgitica** Mansf. et Melch.

落叶灌木。叶面灰色，背面深绿色，椭圆形或披针状椭圆形，中部最宽，纸质，两面无毛。花黄绿色，无毛，单生或2-5朵簇生于叶腋，或为有花梗的聚伞花序。核果成熟时橙色，近球形。花期5-7月，果期9月。生海拔2600-2900米的亚高山森林或灌丛中。产云南西北部、四川西南部和西藏东南部。印度北部亦有。

Deciduous shrubs. Leaves abaxially gray, adaxially dark green, elliptic or lanceolate-elliptic, broadest at middle, papery, both surfaces glabrous. Flowers yellow-green, glabrous, solitary or 2-5-fascicled in axils of leaves or in pedunculate cymes. Drupes orange at maturity, subglobose. Fl. May-Jul. Fr. Sep. Subalpine forests or thickets at 2600-2900 m. Distributed in NW Yunnan, SW Sichuan and SE Xizang. Also in N India.

猫乳 *Rhamnella franguloides*

西藏猫乳 *Rhamnella gilgitica*

多脉猫乳 *Rhamnella martinii*

## 多脉猫乳
**Rhamnella martinii** (Lévl.) Schneid.

灌木或小乔木。叶纸质，狭椭圆形、披针状椭圆形或长圆状椭圆形，侧脉6-8对，两面无毛，边缘具细锯齿。聚伞花序，腋生，无毛；花黄绿色。核果成熟时紫黑色，近圆柱形。花期4-6月，果期7-9月。生海拔800-2800米的山坡灌丛中或杂木林中。产云南、四川、西藏、贵州、广东和湖北。尼泊尔亦有。

Shrubs or small trees. Leaves papery, narrowly elliptic, lanceolate-elliptic, or oblong-elliptic, lateral nerves 6-8 pairs, both surfaces glabrous, margins serrulate. Cymes axillary, glabrous; flowers yellow-green. Drupes purple-black at maturity, subcylindric. Fl. Apr-Jun. Fr. Jul-Sep. Thickets or mixed forests on slopes at 800-2800 m. Distributed in Yunnan, Sichuan, Xizang, Guizhou, Guangdong and Hubei. Also in Nepal.

## 苞叶木
**Rhamnella rubrinervis** (Lévl.) Rehd.

常绿灌木或小乔木。叶下深绿色，光亮，上面浅绿色，长圆形或卵状长圆形，革质，下面无毛，上面无毛或沿脉具柔毛。聚伞花序腋生，具苞叶；花盘圆形，稍厚。核果成熟后紫

红色或橘黄色，卵球状圆柱形。花期7-9月，果期8-11月。生海拔1500米以下的林中或灌丛中。产云南、贵州、广西和广东。越南亦有。

Evergreen shrubs or small trees. Leaves abaxially dark green, shiny, adaxially pale green, oblong or ovate-oblong, leathery, abaxially glabrous, adaxially glabrous, or puberulent on veins. Cymes axillary, with bracteal leaves; discs rounded, slightly thick. Drupes purple-red or orange at maturity, ovoid-cylindric. Fl. Jul-Sep. Fr. Aug-Nov. Forests or thickets below 1500 m. Distributed in Yunnan, Guizhou, Guangxi and Guangdong. Also in Vietnam.

## 小勾儿茶
**Berchemiella wilsonii** (Schneid.) Nakai

落叶灌木。叶腹面灰白色，椭圆形，无毛，背面密被柔毛或于叶腋被髯毛，叶基偏斜。花序长约3.5厘米；花绿色，无毛，顶生聚伞总状花序或为少量腋生聚伞花序。果实幼时红色，成熟后近黑色，圆柱形至略倒卵球形。花期7月，果期8-9月。生海拔500-1500米的林中或山谷林中。产浙江、湖北西部和安徽。

Deciduous shrubs. Leaves adaxially gray-white, elliptic, glabrous, abaxially densely pubescent or with barbate vein axils, base oblique. Inflorescences ca. 3.5 cm long; flowers greenish, glabrous, in terminal cymose racemes or few in axillary cymes. Fruits red when young, nearly black at maturity, cylindric to slightly obovoid. Fl. Jul. Fr. Aug-Sep. Forests or valley forests at 500-1500 m. Distributed in Zhejiang, W Hubei and Anhui.

## 细梗勾儿茶
**Berchemia longipedicellata** Y. L. Chen et P. K. Chou

直立矮小灌木，高达1米。叶薄纸质，矩圆形，有大小二型，但长不超过2厘米。花小，黄绿色，单生或2-3个簇生于叶腋。核果近圆柱形，长5-6毫米；果梗纤细。花期夏季，果期7-10月。生海拔2100-3100米的山坡疏林中。产西藏东部和东南部。

Small shrubs, erect, up to 1 m tall. Leaves thinly chartaceous,

苞叶木 *Rhamnella rubrinervis*

小勾儿茶 *Berchemiella wilsonii*

细梗勾儿茶
*Berchemia longipedicellata*

oblong, dichotypic, length of both types less than 2 cm long. Flowers small, greenish-yellow, axillary, solitary or clustered with 2-3 flowers. Drupes sub-cylindrical, 5-6 mm long; stalks tenuous. Fl. summer. Fr. Jul-Oct. Open forests on slopes at 2100-3100 m. Distributed in E and SE Xizang.

## 多叶勾儿茶
**Berchemia polyphylla** Wall. ex M. A. Y. W. Lawson

攀援灌木。小枝黄褐色，具柔毛。叶两面无毛，侧脉7-9对。花浅绿色或白色，常2-10朵簇生成具梗的聚伞总状花序。核果成熟时红色，后为黑色，圆柱形，基部具宿存的花盘和萼筒。花期5-9月，果期7-11月。生海拔100-2100米的林中、山坡或灌丛。产中国西南、华南、华中和华西。印度和缅甸亦有。

Scandent shrubs. Branchlets yellow-brown, pubescent. Leaves glabrous on both surfaces, lateral veins 7-9 pairs. Flowers greenish or white, usually 2-10-fascicled in pedunculate cymose racemes. Drupes red at maturity, turning black, cylindric, base with persistent disk and tube. Fl. May-Sep. Fr. Jul-Nov. Forests, slopes or thickets at 100-2100 m. Distributed in SW, S, C and W China. Also in India and Myanmar.

云南勾儿茶 *Berchemia yunnanensis*

## 云南勾儿茶
**Berchemia yunnanensis** Franch.

攀援灌木。叶纸质，较小，卵状椭圆形、长圆状椭圆形或卵形，长2.5-6厘米，顶端锐尖，干时背面变金黄色或黄色。花黄色，2或3朵簇生成具长梗的顶生长2-5厘米的总状花序。核果幼时红色，成熟后黑色，圆柱形。花期6-8月，果期翌年4-5月。生海拔1500-3900米的灌丛、林中、溪边或山坡。产中国西南和华西。

Scandent shrubs. Leaves papery, small, ovate-elliptic, oblong-elliptic, or ovate, 2.5-6 cm long, apex acute, abaxially golden yellow or yellow when dry. Flowers yellow, 2- or 3-fascicled in long pedunculate, terminal racemes 2-5 cm long. Drupes red when young, turning black at maturity, cylindric. Fl. Jun-Aug. Fr. next Apr-May. Thickets, forests, along streams or slopes at 1500-3900 m. Distributed in SW and W China.

多叶勾儿茶 *Berchemia polyphylla*

牯岭勾儿茶 *Berchemia kulingensis*

## 牯岭勾儿茶

**Berchemia kulingensis** Schneid.

攀援灌木。叶无毛，卵状椭圆形或卵状长圆形。花序长3-5厘米，无毛；花绿色，无毛，常2或3朵簇生成花梗短或几无的疏松聚伞总状花序。核果幼时红色，成熟时紫黑色，狭圆柱形。花期6-7月，果期翌年4-6月。生海拔300-2200米的山谷林中、林缘或灌丛中。产中国西南、东南、华中和华东。

Scandent shrubs. Both surfaces of leaves glabrous, ovate-elliptic or ovate-oblong. Inflorescences 3-5 cm long, glabrous; flowers green, glabrous, usually 2- or 3-fascicled in subsessile or shortly pedunculate lax cymose racemes. Drupes red when young, black-purple at maturity, narrowly cylindric. Fl. Jun-Jul. Fr. next Apr-Jun. Forests in valleys, forest edges or thickets at 300-2200 m. Distributed in SW, SE, C and E China.

## 大叶勾儿茶

**Berchemia huana** Rehd.

藤状灌木。叶长6-10厘米，宽3-6厘米，下面密被黄色柔毛或于叶脉或脉腋处疏被柔毛。宽聚伞状圆锥花序，轴长20厘米，分枝长8厘米，被柔毛；花黄绿色，常顶生。核果成熟后紫红色或紫黑色。花期7-9月，果期翌年5-6月。生海拔1000米以下的山地林中或灌丛中。产中国东南、华中和华东。

Scandent shrubs. Leaves 6-10 cm long, 3-6 cm wide, abaxially densely yellowish pubescent or sparsely pubescent on veins or in vein axils. Broad cymose panicles, rachises to 20 cm long, branches to 8 cm long, pubescent; flowers yellowish green, usually in terminal. Drupes purple-red or purple-black at maturity. Fl. Jul-Sep. Fr. next May-Jun. Forests on slopes or thickets below 1000 m. Distributed in SE, C and E China.

## 峨眉勾儿茶

**Berchemia omeiensis** W. L. Fang ex Y. L. Chen et P. K. Chou

藤状或攀援灌木。叶革质或近革质，常2-5枚簇生于短枝上；托叶宽卵状披针形。花序长16厘米；花黄色或淡绿色，常2-5朵簇生排成具短柄的阔聚伞状圆锥花序。核果红色，成熟时变紫黑色，圆柱状椭球形。花期7-8月，果期翌年5-6月。生海拔400-1700米的山地林中。产四川、贵州和湖北。

Scandent or scandent shrubs. Leaves coriaceous or almost coriaceous, usually 2-5-fascicled on short shoots; stipules wide oval-lanceolate. Inflorescences to 16 cm long; flowers yellow or greenish, usually 2-5 in fascicles, in shortly pedunculate, broad cymose panicles. Drupes red, turning purple-black at maturity, cylindric-ellipsoid. Fl. Jul-Aug. Fr. next May-Jun. Montane forests at 400-1700 m. Distributed in Sichuan, Guizhou and Hubei.

## 黄背勾儿茶

**Berchemia flavescens** (Wall.) Brongn.

攀援灌木。叶较大，纸质，长7-15厘米，背面干时变黄色，侧脉12-18对。花黄绿色，极小，无毛，常1至少数簇生，形成狭聚伞圆锥状花序。核果近圆柱形，成熟时紫红色或紫黑色。花期6-8月，果期翌年5-7月。生海拔1200-4000米的山坡灌丛中或林下。产中国西南、华中和华西。印度、尼泊尔和不丹亦有。

Scandent shrubs. Leaves large, papery, 7-15 cm long, abaxially yellow when dry, lateral nerves 12-18 pairs. Flowers yellow-green, very small, glabrous,

大叶勾儿茶 *Berchemia huana*

峨眉勾儿茶 *Berchemia omeiensis*

黄背勾儿茶
*Berchemia flavescens*

多花勾儿茶 *Berchemia floribunda*

usually 1 to few in fascicles, in narrow cymose panicles. Drupes sub-cylindric, purple-red or purple-black at maturity. Fl. Jun-Aug. Fr. next May-Jul. Thickets or forests on slopes at 1200-4000 m. Distributed in SW, C and W China. Also in India, Nepal and Bhutan.

## 多花勾儿茶

**Berchemia floribunda** (Wall.) Brongn.

藤状或直立灌木。叶干后下面深褐色；托叶宿存。花序长达15厘米；花多数，通常数个簇生成聚伞圆锥花序，或下部有腋生聚伞总状花序。核果红色，成熟时蓝黑色，圆柱状椭球形至卵状长圆球形。花期7-10月，果期翌年4-7月。生海拔2600米以下的山坡、沟谷、林缘、林下或灌丛中。产中国西南、东南、华中、华北和华东。印度、尼泊尔、不丹、泰国、越南和日本亦有。

Vine-like or erect shrubs. Leaves abaxially dark brown when dry; stipules persistent. Inflorescences up to 15 cm long; flowers many, several clustered to racemose cymes or with axillary botry-cymose at lower part. Drupes red, bluish black at maturity, cylindric-elliptic to ovoid-oblong. Fl. Jul-Oct. Fr. next Apr-Jul. Slopes, valleys, forest edges, under forests or bushes below 2600 m. Distributed in SW, SE, C, N and E China. Also in India, Nepal, Bhutan, Thailand, Vietnam and Japan.

## 铜钱树

**Paliurus hemsleyanus** Rehd. ex Schir. et Olabi

常绿灌木或中型乔木。植物无托叶刺。叶宽卵形至宽椭圆形，纸质或厚纸质，两面无毛，基出3脉。花无毛，排成聚伞状或聚伞圆锥状。核果球形，周围具革质宽翅，红棕色或紫棕色。花期4-6月，果期7-9月。生海拔1600米以下的山地林中，庭园中常有栽培。产中国西南、华南、华西、华中和华东。

Evergreen shrubs to medium-sized trees. Plants without spiny stipules. Leaves broadly ovate to broadly elliptic, papery or thickly papery, both surfaces glabrous, 3-veined from base. Flowers glabrous, in cymes or cymose panicles. Drupes orbicular, surrounded by broad coriaceous wings, umber or purple-brown. Fl. Apr-Jun. Fr. Jul-Sep. Montane forests or cultivated in gardens below 1600 m. Distributed in SW, S, W, C and E China.

铜钱树 *Paliurus hemsleyanus*

滇刺枣 *Ziziphus mauritiana*

皱枣 *Ziziphus rugosa*

## 滇刺枣

**Ziziphus mauritiana** Lam.

常绿乔木或灌木。老枝紫红色，有2个托叶刺，1个斜上，另1个钩状下弯。叶卵形至长圆状椭圆形，背面密被黄色或灰白色绒毛。花黄绿色，数个至10个密集成腋生二歧聚伞花序。核果橘黄色或红色，成熟后变黑，长圆球形或球形。花期8-11月，果期9-12月。生海拔1800米以下的山坡、疏林或河岸灌丛中。产云南、四川、广西、广东、台湾和福建。南亚、东南亚、澳大利亚和非洲亦有。

Evergreen trees or shrubs. Old branches purple-red, stipular spines 2, one oblique and another one hook-like recurved. Leaves ovate to oblong-elliptic, abaxially densely yellow or grayish white tomentose. Flowers green-yellow, few to 10 in axillary dichotomous cymes. Drupes orange or red, turning black at maturity, oblong or globose. Fl. Aug-Nov. Fr. Sep-Dec. Slopes, open forests, or thickets along river banks below 1800 m. Distributed in Yunnan, Sichuan, Guangxi, Guangdong, Taiwan and Fujian. Also in S and SE Asia, Australia and Africa.

## 皱枣

**Ziziphus rugosa** Lam.

常绿灌木或小乔木。幼枝被锈色或黄褐色密绒毛，老枝红褐色，粗糙，有条纹，具明显的皮孔。托叶刺1(-2)，内弯，紫红色；叶宽卵形或宽椭圆形，下面密具锈色或黄褐色绒毛。顶生或腋生的大圆锥花序或总状花序。核果倒卵球形或近球形，成熟时黑色。花期3-5月，果期4-6月。生海拔1400米以下的丘陵疏林中或灌丛中或山地向阳地。产云南、广西和海南。南亚亦有。

Evergreen shrubs or small trees. Young branches densely ferruginous or yellow-brown tomentose, old branches red-brown, scabrous, striate, with conspicuous lenticels. Stipular spines 1(-2), recurved, purple-red; leaves broadly ovate or broadly elliptic, abaxially densely ferruginous or yellow-brown tomentose. Big terminal or axillary panicles or racemes. Drupes obovoid or almost orbicular, black at maturity. Fl. Mar-May. Fr. Apr-Jun. Sparse forests and thickets on hills, or sunny places on mountains below 1400 m. Distributed in Yunnan, Guangxi and Hainan. Also in S Asia.

翼核果 *Ventilago leiocarpa*

## 翼核果

**Ventilago leiocarpa** Benth.

藤状灌木。叶卵状长圆形或卵状椭圆形，薄革质。花小，两性，5数，单生或2至少数簇生于叶腋。核果扁球形，长3-5(-6)厘米；果核具翅，被宿存萼筒包被1/4-1/3，具1或2室，含1粒种子。花期3-5月，果期4-7月。生海拔1500米以下的疏林或灌丛中。产中国西南、华中和华南。印度、缅甸、泰国和越南亦有。

Scandent shrubs. Leaves ovate-oblong or ovate-elliptic, thinly coriaceous. Flowers small, bisexual, 5-merous, solitary or 2- to few- fascicled in axils of leaves. Drupes depressed-globose, 3-5 (-6) cm long; stones with wings mucronulate, surrounded by persistent calyx tube at base for 1/4-1/3, 1- or 2-loculed, 1-seeded. Fl. Mar-May. Fr. Apr-Jul. Sparse forests or thickets below 1500 m. Distributed in SW, C and S China. Also in India, Myanmar, Thailand and Vietnam.

### 毛咀签
**Gouania javanica** Miq.

攀援灌木。叶纸质，背面被锈色绒毛或灰色丝状柔毛。花杂性，5数，单生或几个簇生于短枝形成聚伞圆锥花序；花盘五角形，包围子房，每个角都狭缩为一舌状附属物。蒴果成熟时黄色，具3翅。花期7-9月，果期11月至翌年3月。生低、中海拔疏林中或溪边，常攀援树上。产中国西南、华南和东南。南亚和东南亚亦有。

Scandent shrubs. Leaves papery, abaxially ferruginous-tomentose, or gray-sericeous. Flowers polygamous, 5-merous, solitary or few fascicled in shortly pedunculate cymes; disk pentagonous, surrounding ovary, each angle elongated into a liguliform appendage. Capsules yellow at maturity, 3-winged. Fl. Jul-Sep. Fr. Nov to next Mar. Open forests or along streams at low to medium elevations, often scandent on trees. Distributed in SW, S and SE China. Also in S and SE Asia.

## 葡萄科 Vitaceae

### 火筒树
**Leea indica** (Burm. f.) Merr.

直立灌木至小乔木。二(至三)回羽状复叶，小叶13-32厘米长，边缘有不整齐浅锯齿。花序与叶对生，复二歧聚伞状或伞形；花白色或绿白色。浆果直径0.8-1厘米。具4-6粒种子。花期4-7月，果期8-12月。生海拔200-1200米的山坡疏林或灌丛中。产云南、贵州、广西、广东和海南。南亚、东南亚、澳大利亚北部和太平洋岛屿亦有。

Erect shrubs to small trees. Leaves 2(-3)-pinnate, leaflets 13-32 cm long, margins irregularly and shallowly serrate. Inflorescences opposite to leaves, compound dichasial or umbelliform; flowers white or greenish white. Berries 0.8-1 cm diam. 4-6-seeded. Fl. Apr-Jul. Fr. Aug-Dec. Forests or thickets on mountain slopes, open at 200-1200 m. Distributed in Yunnan, Guizhou, Guangxi, Guangdong and Hainan. Also in S Asia, SE Asia, N Australia and Pacific Islands.

毛咀签 *Gouania javanica*

火筒树 *Leea indica*

长柄地锦 *Parthenocissus feddei*

## 长柄地锦

**Parthenocissus feddei** (Lévl.) C. L. Li

木质藤本。卷须具6-11分枝，顶端嫩时稍膨大为拳头状吸盘。叶柄长7.5-15厘米，小叶柄长0.5-2.5厘米；叶小，3小叶。多歧聚伞花序顶生或假顶生，主轴明显。浆果球形。种子1或2粒，倒卵球状椭圆体形。花期4-7月，果期8-10月。生海拔600-1100米的山谷多石地区。产贵州、广东、湖北和湖南。

Woody lianas. Tendrils with 6-11 branches, young apices of tendrils slightly expanded as fist-shaped adhesive disks. Petioles 7.5-15 cm long, petiolules 0.5-2.5 cm long; leaves small, 3-foliolate. Polychasium terminal or pseudoterminal, with a conspicuous, well-developed axis. Berries globose. Seeds 1 or 2, obovoid-ellipsoid. Fl. Apr-Jul. Fr. Aug-Oct. Rocky areas in valleys at 600-1100 m. Distributed in Guizhou, Guangdong, Hubei and Hunan.

## 三叶地锦

**Parthenocissus semicordata** (Wall.) Planch.

木质藤本。小枝、叶柄和叶脉下面疏具柔毛。小枝圆柱形，卷须4-6分裂。叶为3小叶，小叶背面被短柔毛，中央小叶5-13 × 3-6.5厘米，每边具6-11个齿。花瓣5；花盘不明显。花期5-7月，果期8-10月。生海拔500-3800米的山坡林中或灌丛中。产中国西南、华南、华中和华西。南亚和东南亚亦有。

Woody lianas. Branchlets, petioles, and leaflet veins abaxially sparsely pilose. Branchlets terete, tendrils with 4-6 branches. Leaves ternate, leaflets abaxially pubescent, central leaflets 5-13 × 3-6.5 cm, margin with 6-11 teeth on each side. Petals 5; discs inconspicuous. Fl. May-Jul. Fr. Aug-Oct. Forests or thickets on hillsides at 500-3800 m. Distributed in SW, S, C and W China. Also in S and SE Asia.

## 爬山虎(地锦)

**Parthenocissus tricuspidata** (Siebold et Zucc.) Planch.

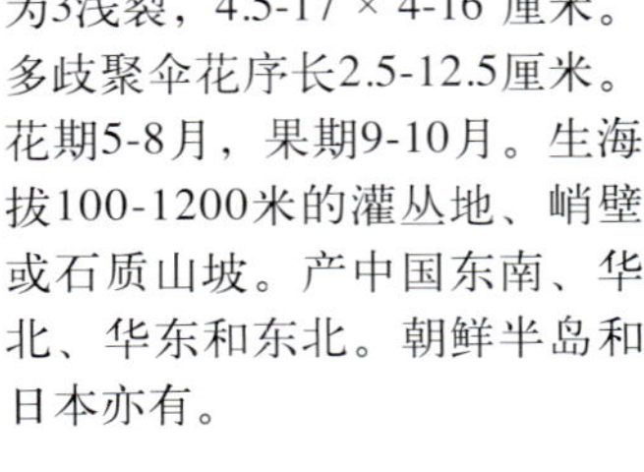

木质藤本。老枝无木质翅，小枝无毛或具稀疏的软毛；卷须5-9分枝。叶柄和叶无毛或仅在下面脉上疏具柔毛；叶单生，在短枝上为3浅裂，4.5-17 × 4-16 厘米。多歧聚伞花序长2.5-12.5厘米。花期5-8月，果期9-10月。生海拔100-1200米的灌丛地、峭壁或石质山坡。产中国东南、华北、华东和东北。朝鲜半岛和日本亦有。

Woody lianas. Old branches without woody wings, branchlets glabrous or with sparse and soft hairs; tendrils 5-9-branched. Petioles and leaves glabrous or only veins abaxially sparsely pubescent; leaves simple, usually 3-lobed on short branches, 4.5-17 × 4-16 cm. Polychasiums 2.5-12.5 cm long. Fl. May-Aug. Fr. Sep-Oct. Shrublands, cliffs or rocky hillsides at 100-1200 m. Distributed in SE, N, E and NE China. Also in Korean Peninsula and Japan.

三叶地锦 *Parthenocissus semicordata*

爬山虎(地锦) *Parthenocissus tricuspidata*

异叶爬山虎（异叶地锦） *Parthenocissus dalzielii*

## 异叶爬山虎(异叶地锦)

**Parthenocissus dalzielii** Gagnep.

木质藤本。卷须总状，5-8分枝，幼卷须顶端膨大成球状吸盘。叶二型，3小叶及单叶。多歧聚伞花序疏松，主轴细弱，假顶生于短枝顶端。浆果成熟时黑紫色。种子1-4粒。花期5-7月，果期7-11月。生海拔200-3800米的峭壁、山坡或山谷林中或灌丛中。产中国西南、华南、东南和华中。

Woody vines. Tendrils racemose, 5-8-branched, young apex of tendril expanded to ball-like adhesive disk. Leaves with 2 types, 3-foliolate and simple. Inflorescences a corymbose polychasium, loose, main axis slender, false terminal at top of short branches. Berries black-purple at maturity. 1-4-seeded. Fl. May-Jul. Fr. Jul-Nov. Forests or bushes on cliffs, slopes or valleys at 200-3800 m. Distributed in SW, S, SE and C China.

## 绿叶地锦

**Parthenocissus laetevirens** Rehd.

木质藤本。小枝具棱，卷须具5-10分枝，幼时顶部膨大成块状。叶为掌状5小叶，小叶腹面明显具乳突。圆锥状多歧聚伞花序假顶生，常具明显主轴和退化小叶。浆果具1-4粒种子。花期7-8月，果期9-11月。生海拔100-1100米的山谷林中或山腰坡灌丛。产华南、东南、华中和华东。

Woody lianas. Branchlets ridged, tendrils 5-10-branched, with young apices expanded as tubercle adhesive disks. Leaves palmately 5-foliolate, leaflets adaxially conspicuously bullate. Paniculate polychasium pseudoterminal, usually with obvious rachises and degenerative leaflets. Berries 1-4-seeded. Fl. Jul-Aug. Fr. Sep-Nov. Forests in valleys or shrubs on hillsides at 100-1100 m. Distributed in S, SE, C and E China.

## 花叶地锦

**Parthenocissus henryana** (Hemsl.) Graebn. ex Diels et Gilg

木质藤本。小枝方形，卷须4-7分枝，嫩时顶部膨大呈块状。叶掌状5裂；小叶表面不具乳突。圆锥状多歧聚伞花序假顶生，常具明显主轴和退化小叶。浆果具1-3粒种子。种子倒卵球形。花期5-7月，果期7-10月。生海拔100-1500米的山谷岩石上或山坡。产中国西南、华中、华北和华西。

Woody lianas. Branchlets quadrangular, tendrils 4-7-branched, with young apices expanded as tubercle adhesive disks. Leaves palmately 5-foliolate; leaflet surface not bullate. Paniculate polychasium pseudoterminal, usually with a conspicuous rachises and degenerative leaflets. Berries 1-3-seeded. Seeds obovoid. Fl. May-Jul. Fr. Jul-Oct. Rocks in valleys or hillsides at 100-1500 m. Distributed in SW, C, N and W China.

绿叶地锦 *Parthenocissus laetevirens*

花叶地锦 *Parthenocissus henryana*

俞藤 *Yua thomsonii*

## 俞藤
**Yua thomsonii** (Laws.) C. L. Li

木质藤本。小枝棕色，圆柱形。卷须二叉分枝；小枝、叶柄、叶腹面、花序轴、花柄和花瓣无毛。小叶纸质，边缘具细锐齿，小脉明显但不凸起，背面无毛或沿叶脉疏生短柔毛。复二歧聚伞花序与叶对生。浆果黑紫色，直径1-1.3厘米。花期5-6月，果期7-9月。生海拔200-1300米的山腰林中。产华南、华东和华中。印度和尼泊尔亦有。

Woody, lianas. Branchlets brown, terete. Tendrils bifurcate; branchlets, petioles, adaxial leaflet surface, peduncles, pedicels, and petals glabrous. Leaflets papery, margin with fine, sharp teeth, veinlets conspicuous but not raised, abaxial surface glabrous or sparsely pubescent on veins. Compound dichasium leaf-opposed. Berries black-purple, 1-1.3 cm diam. Fl. May-Jun. Fr. Jul-Sep. Forests on hillsides at 200-1300 m. Distributed in S, E and C China. Also in India and Nepal.

## 蓝果蛇葡萄
**Ampelopsis bodinieri** (Lévl. et Vaniot) Rehd.

木质藤本。卷须二叉分枝。叶单生；小叶卵形或卵状椭圆形，不分裂或稍3浅裂，下面密具浅灰色毛，基部心形或稍心形。复二歧聚伞花序疏松；花梗长2.5-3毫米；花盘5裂。浆果具3-4粒种子。花期4-6月，果期7-8月。生海拔200-3000米的山谷林中或阴丘陵灌丛中。产中国西南、华南和华中。

Woody lianas. Tendrils bifurcate. Leaves simple; leaflets ovate or ovate-elliptic, undivided or slightly 3-lobed, abaxially with dense grayish hairs, base cordate or slightly so. Compound dichasium cymes loose; pedicels 2.5-3 mm long; disk 5-lobed. Berries 3-4-seeded. Fl. Apr-Jun. Fr. Jul-Aug. Forests in valleys or shrubs on shaded hillsides at 200-3000 m. Distributed in SW, S and C China.

## 葎叶蛇葡萄
**Ampelopsis humulifolia** Bunge

木质藤本。小枝圆柱形，卷须二叉分枝。叶心形或略心形至具5角，具3-5个宽裂(裂至中部)，基部缺凹成圆形，边缘具大的锐齿。多歧聚伞花序与叶对生。浆果球形，具2-4粒种子。花期5-7月，果期5-9月。生海拔400-1100米的林中或灌丛中。产中国西北、东北和华北。

Woody vines. Branchlets terete, tendrils bifurcate. Leaves cordate or roughly so with 5 angles, with 3-5 broad lobes (lobed up to middle), notch rounded, margin with large, sharp teeth. Polychasiums opposite to leaves. Berries globose, 2-4-seeded. Fl. May-Jul. Fr. May-Sep. Forests or thickets at 400-1100 m. Distributed in NW, NE and N China.

蓝果蛇葡萄 *Ampelopsis bodinieri*

葎叶蛇葡萄 *Ampelopsis humulifolia*

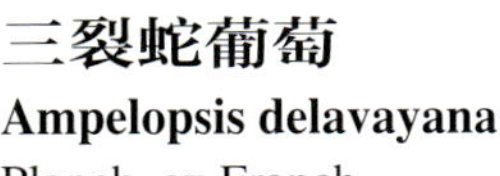
三裂蛇葡萄 *Ampelopsis delavayana*

白蔹 *Ampelopsis japonica*

## 三裂蛇葡萄

**Ampelopsis delavayana** Planch. ex Franch.

木质藤本。小枝、叶柄或叶下面疏具柔毛。小枝圆柱形，卷须2-3叉，疏生短柔毛。叶具3小叶，小叶不分裂或仅基部侧生小叶分裂。多歧聚伞花序与叶对生；花盘明显。浆果球形。花期5-8月，果期7-11月。生海拔100-2200米的林中或灌丛中。产中国西南、华南和华中。

Woody vines. Branchlets, petioles, or leaves abaxially sparsely pubescent. Branchlets cylindric, tendrils 2-3-branched, sparsely pubescent. Leaves 3-foliolate, leaflets undivided, or only base of lateral leaflets divided. Polychasiums opposite to leaves; flower disc obvious. Berries globose. Fl. May-Aug. Fr. Jul-Nov. Forests or shrubs at 100-2200 m. Distributed in SW, S and C China.

## 白蔹

**Ampelopsis japonica** (Thunb.) Makino

木质藤本。小枝、叶柄和叶下面无毛。卷须不分枝或在近顶部具短分枝。叶为掌状3-5小叶，小叶羽状深裂，下半部渐狭成翅状。聚伞花序通常聚生于花序梗顶端。浆果球形。花期5-6月，果期7-9月。生海拔100-900米的山坡、灌丛或草地。产中国西南、华南、华中、华北、华东和东北。日本有栽培。

Woody vines. Branchlets, petioles, and leaves abaxially glabrous. Tendrils unbranched or with short branches near tip. Palmate leaves with 3-5 leaflets, leaflets pinnatipartite, lower half parts attenuate into wings. Cymes usually clustered at top of peduncles. Berries globose. Fl. May-Jun. Fr. Jul-Sep. Slopes, bushes or grasslands at 100-900 m. Distributed in SW, S, C, N, E and NE China. Cultivated in Japan.

## 掌裂草葡萄

**Ampelopsis aconitifolia** Bunge var. **palmiloba** (Carr.) Rehd.

木质藤本。小枝圆柱形，具纵脊；卷须2或3叉。小叶大多不分裂，边缘具大齿，或混生稍浅裂的叶，无毛或叶下面疏被微柔毛。花小，黄绿色，花瓣5。浆果球形，种子2或3粒。花期5-8月，果期7-9月。生海拔200-2200米的灌丛或山谷。产中国西南、华北和东北。

Woody vines. Branchlets terete, with longitudinal ridges; tendrils 2- or 3-branched. Leaflets mostly undivided, margin with large teeth, or mixed with slightly lobed leaves, glabrous or abaxially sparsely pilose. Flowers small, yellowish green, with 5 petals. Berries globose, 2- or 3-seeded. Fl. May-Aug. Fr. Jul-Sep. Thickets or valleys at 200-2200 m. Distributed in SW, N and NE China.

掌裂草葡萄 *Ampelopsis aconitifolia* var. *palmiloba*

大叶蛇葡萄 *Ampelopsis megalophylla*

羽叶蛇葡萄 *Ampelopsis chaffanjonii*

## 大叶蛇葡萄

**Ampelopsis megalophylla**
Diels et Gilg

木质藤本。卷须三分叉。羽状复叶，基部一对小叶常为3小叶；小叶长圆形或卵状椭圆形，长4-12厘米。花序伞房状或为宽复二歧聚伞状，顶生或与叶对生。浆果稍倒卵球形，具1-4粒种子。花期5-8月，果期7-10月。生海拔1600-2000米的林中、山坡或山谷。产中国西南和华西。

Woody lianas. Tendrils trifurcate. Leaves bipinnate, with basal pinnae usually 3-foliolate; leaflets oblong or ovate-elliptic, 4-12 cm long. Inflorescences corymbose or a broad compound dichasium, terminal or leaf-opposed. Berries slightly obovoid, 1-4-seeded. Fl. May-Aug. Fr. Jul-Oct. Forests, mountain slopes or valleys at 1600-2000 m. Distributed in SW and W China.

## 广东蛇葡萄

**Ampelopsis cantoniensis**
(Hook. et Arn.) K. Koch.

木质藤本。小枝圆柱形，被灰色短柔毛。叶为(1-)2回羽状复叶；小叶边缘近波状，中央小叶倒卵形或卵形，背面脉基部疏生短柔毛，后变无毛，末回细脉明显。聚伞花序与叶对生；花瓣5；花盘明显。花期4-7月，果期8-11月。生海拔100-850米的林中或灌丛中。产中国西南、华南、中南和东南。泰国、越南、马来西亚和日本亦有。

Woody vines. Branchlets terete with gray-pubescent. Leaves (1-) 2-pinnate; leaflets margin ± undulate, central leaflets obovate or ovate, abaxially sparse pubescent at base of nerves, later glabrescent, ultimate veinlets conspicuous. Cymes opposite to leaves; petals 5; discs conspicuous. Fl. Apr-Jul. Fr. Aug-Nov. Forests or thickets at 100-850 m. Distributed in SW, S, SC and SE China. Also in Thailand, Vietnam, Malaysia and Japan.

广东蛇葡萄 *Ampelopsis cantoniensis*

## 羽叶蛇葡萄

**Ampelopsis chaffanjonii** (Lévl. et Vaniot) Rehd.

木质藤本。卷须二叉分枝。羽状复叶具2或3对小叶；小叶长圆形或卵状椭圆形。多歧聚伞花序顶生或与叶对生；子房下部贴生于花盘。浆果球形，具2或3粒种子。花期5-7月，果期7-9月。生海拔500-2000米的山谷林中或灌丛中。产中国西南和华中。

Woody lianas. Tendrils bifurcate. Leaves pinnate, usually with 2 or 3 pairs of leaflets; leaflets oblong or ovate-elliptic. Corymbose polychasium terminal or leaf-opposed; lower part of ovary adnate to disk. Berries globose, 2- or 3-seeded. Fl. May-Jul. Fr. Jul-Sep. Forests or shrubs in valleys at 500-2000 m. Distributed in SW and C China.

## 显齿蛇葡萄

**Ampelopsis grossedentata**
(Hand.-Mazz.) W. T. Wang

木质藤本，植株全部无毛。卷须二叉分枝。叶1至2回羽状，基部小叶具3小叶；小叶薄纸质，2-5 × 1-2.5厘米，每边具2-5齿。聚伞花序有梗。浆果近球形，具2-4粒种子。种子倒卵球形。花期5-8月，果期8-12月。生海拔200-1500米的山谷林中或山坡灌丛中。产中国西南、华南、东南和华中。越南亦有。

Woody vines, whole plant glabrous. Tendrils bifurcate. Leaves

显齿蛇葡萄 *Ampelopsis grossedentata*

四棱白粉藤 *Cissus subtetragona*

1-2-pinnate, basal pinnae 3-foliolate; leaflets thin papery, 2-5 × 1-2.5 cm, margin with 2-5 teeth on each side. Cymes petiolate. Berries almost globose, 2-4-seeded. Seeds obovoid. Fl. May-Aug. Fr. Aug-Dec. Valley forests or shrubs on mountain slopes at 200-1500 m. Distributed in SW, S, SE and C China. Also in Vietnam.

## 四棱白粉藤

**Cissus subtetragona** Planch

木质藤本。卷须不分枝。叶长圆形或三角状长圆形，6-19 × 2-7厘米，基部近截形。复二歧聚伞花序顶生或与叶对生；花蕾圆锥椭圆形，3-4毫米，花萼全缘；花瓣三角状长圆形，无毛；子房贴生于盘下部；雄蕊4；花盘4裂；花柱钻形。果实近球形。种子1枚，表面平滑。花期9-10月，果期10-12月。生海拔100-1300米的森林和灌丛中。产广东、广西、海南和云南。

Lianas, woody. Tendrils unbranched. Leaves blade oblong or triangular-oblong, 6-19 × 2-7 cm, base nearly truncate. Compound dichasium terminal or leaf-opposed; buds conical-elliptic, 3-4 mm; calyx entire; petals triangular-oblong, glabrous; lower part of ovary adnate to disk; anthers 4; disk 4-lobed; style conical. Berry global. 1-seeded, seed surface smooth. Fl. Sep-Oct. Fr. Oct-Dec. Forests and shrubs at 100-1300 m. Distributed in Guangdong, Guangxi, Hainan and Yunnan.

## 掌叶白粉藤

**Cissus triloba** (Lour.) Merr.

草质藤本。小枝圆柱形，有纵棱，无毛，常被白粉；卷须不分枝。叶多3-5深裂或混生不裂叶，每侧边缘具20-30齿，齿锐尖。复多歧聚伞花序顶生或与叶对生。种子表面光滑。花期6-10月，果期8-11月。生海拔900-1400米的河边林中或山坡。产云南。越南亦有。

Vines, herbaceous. Branchlets terete, with longitudinal ridges, glabrous, usually glaucous; tendrils unbranched. Leaves mostly with 3-5 deep lobes or mixed with undivided ones, margin 20-30-toothed on each side, teeth sharp. Compound polychasium terminal or leaf-opposed. Seeds surface smooth. Fl. Jun-Oct. Fr. Aug-Nov. Forests by rivers or hillsides at 900-1400 m. Distributed in Yunnan. Also in Vietnam.

掌叶白粉藤 *Cissus triloba*

翼茎白粉藤 *Cissus pteroclada*

## 翼茎白粉藤
**Cissus pteroclada** Hayata

草质藤本。小枝4棱，棱有翅，无毛。卷须二叉分枝。叶卵圆形，基部心形，无毛；托叶卵圆形。伞形花序；花序梗长1-2厘米，被毛；花萼杯形，全缘，无毛；花瓣4；花药卵圆形；花盘4裂；花柱钻形。浆果倒卵椭圆形。种子表面棱纹尖锐，种脊突出。花期6-8月，果期8-12月。生海拔300-2100米的山谷树上或灌丛中。产台湾、福建、广东、广西、海南和云南。

Lianas, herbaceous. Branchlets with 4 wings, glabrous. Tendrils bifurcate. Leaves oval, base cordate, glabrous; stipule oval. Inflorescences umbelliform; peduncle 1-2 cm long, pubescent; calyx cup-shaped, entire, glabrous; petals 4; anthers oval; disk 4-lobed; stigma conical. Berry fruit, obovate-elliptic. Seed surface with sharp ridges, raphe raised. Fl. Jun-Aug. Fr. Aug-Dec. On trees or shrubs in valleys at 300-2100 m. Distributed in Taiwan, Fujian, Guangdong, Guangxi, Hainan and Yunnan.

## 青紫葛
**Cissus javana** DC.

草质藤本。小枝具4纵棱；卷须二叉分枝。单叶，戟形或卵状戟形，长多于宽的2倍；托叶广椭圆形或卵状椭圆形；叶戟形或卵状戟形。伞形花序。种子表面具钝脊。花期6-10月，果期11-12月。生海拔600-2000米的林中或灌丛中。产云南和四川。南亚和东南亚亦有。

Herbaceous lianas. Branchlets longitudinally nearly 4-ridged; tendrils bifurcate. Leaves simple, hastate or ovate-hastate, length more than 2 × width; stipules oval or ovate-elliptic. Inflorescences umbelliform. Seed surface with obtuse ridges. Fl. Jun-Oct. Fr. Nov-Dec. Forests or shrubs at 600-2000 m. Distributed in Yunnan and Sichuan. Also in S and SE Asia.

## 白粉藤
**Cissus repens** Lam.

草质藤本。小枝圆柱状，常被白粉，无毛；卷须二叉分枝。叶单生，心状椭圆形，无毛，侧脉3或4对。花序伞形，顶生或与叶对生。浆果具1粒种子。种子表面光滑。花期7-10月，果期11月至翌年5月。生海拔100-1800米的沟谷林中或山坡灌丛中。产中国西南和华南。南亚、东南亚和澳大利亚亦有。

Vines, herbaceous. Branchlets terete, usually glaucous, glabrous; tendrils bifurcate. Leaves simple, cordate-oval, glabrous, lateral veins 3 or 4 pairs. Inflorescences umbelliform, terminal or leaf-opposed. Berries 1-seeded. Seeds surface smooth. Fl. Jul-Oct. Fr. Nov to next May. Forests in valleys or shrubs on hillsides at 100-1800 m. Distributed in SW and S China. Also in S and SE Asia, and Australia.

## 苦郎藤
**Cissus assamica** (M. A. Lawson) Craib

木质藤本。卷须粗壮。单叶，中脉在下面稍具长柔毛及“T”字形毛；托叶广椭圆形。花序伞形，与叶对生；花瓣光滑。果实倒卵球形，成熟时紫黑

青紫葛 *Cissus javana*

白粉藤 *Cissus repens*

苦郎藤 *Cissus assamica*

色，有种子1粒。种子表面具锐及凸起的脊。花期5-6月，果期7-10月。生海拔200-1600米的林中或溪边灌丛中。产中国西南、华南和东南。南亚亦有。

Woody vines. Tendrils robust. Leaves simple, midvein on abaxial ± villous with T-shaped hairs; stipules broadly ovate. Inflorescences umbelliform, opposite to leaves; petals glabrous. Fruits obovoid, purple-black at maturity, 1-seeded. Seed surface with sharp and raised ridges. Fl. May-Jun. Fr. Jul-Oct. Forests or shrubs by streams at 200-1600 m. Distributed in SW, S and SE China. Also in S Asia.

## 三叶乌蔹莓

**Cayratia trifolia** (L.) Domin

木质藤本。卷须3-5分叉。叶为

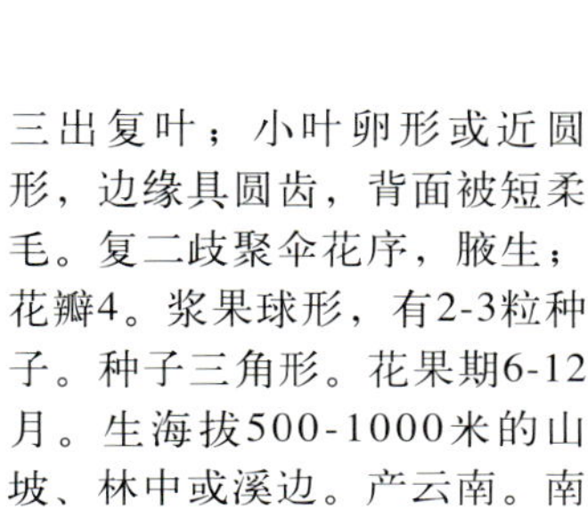

三出复叶；小叶卵形或近圆形，边缘具圆齿，背面被短柔毛。复二歧聚伞花序，腋生；花瓣4。浆果球形，有2-3粒种子。种子三角形。花果期6-12月。生海拔500-1000米的山坡、林中或溪边。产云南。南亚和东南亚亦有。

Woody vines. Tendrils 3-5-branched. Leaves ternately compound; leaflets ovate or nearly orbicular, margin crenate, abaxially pubescent. Compound dichasium axillary; petals 4. Berries globose, 2-3-seeded. Seeds triangular. Fl. and fr. Jun-Dec. Slopes, forests or by streams at 500-1000 m. Distributed in Yunnan. Also in S and SE Asia.

三叶乌蔹莓 *Cayratia trifolia*

## 乌蔹莓

**Cayratia japonica** (Thunb.) Gagnep.

草质藤本。小枝、叶柄和小叶背面脉上具柔毛。卷须二叉分枝。叶鸟足状5裂；小叶长圆形或卵形，4-8 × 2-4厘米。复二歧聚伞花序，腋生。浆果球形，具2-4粒种子。种子三角状倒卵球形。花期3-8月，果期7月至翌年1月。生海拔300-2500米的林中、灌丛或草地。产中国西南、华南、华中、华北、华西和华东。南亚、东南亚、东北亚和澳大利亚亦有。

Vines, herbaceous. Branchlets, petioles and veins on abaxial surface of leaflets pilose. Tendrils bifurcate. Leaves pedately 5-foliolate; leaflets oblong or ovate, 4-8 × 2-4 cm. Compound dichasium axillary. Berries globose, 2-4-seeded. Seeds triangular-obovoid. Fl. Mar-Aug. Fr. Jul to next Jan. Forests, shrubs or grassy areas at 300-2500 m. Distributed in SW, S, C, N, W and E China. Also in S, SE and NE Asia, and Australia.

乌蔹莓 *Cayratia japonica*

白毛乌蔹莓 *Cayratia albifolia*

## 白毛乌蔹莓
**Cayratia albifolia** C. L. Li

半木质或草质攀援藤本。卷须三叉分枝。叶为鸟足状5小叶；小叶下面浅灰色，上面绿色，具短的灰色柔毛。伞房状多歧聚伞花序，腋生；花瓣具乳突；子房下部贴生于花盘。浆果球形，具2-4粒种子。种子倒卵球状椭圆体形。花期5-7月，果期7-9月。生海拔300-2000米的山谷林中、灌丛、岩石上或山腰。产中国西南、华南、华中、东南和华东。

Climbers, semiwoody or herbaceous. Tendrils trifurcate. Leaves pedately 5-foliolate; leaflets abaxially light glaucous, adaxially green, pubescent with short grayish hairs. Corymbose polychasium axillary; petals with papillose; lower part of ovary adnate to disk. Berries globose, 2-4-seeded. Seeds obovoid-ellipsoid. Fl. May-Jul. Fr. Jul-Sep. Forests in valleys, Shrubs. on rocks or cliffs at 300-2000 m. Distributed in SW, S, C, SE and E China.

## 十字崖爬藤
**Tetrastigma cruciatum** Craib et Gagnep.

木质藤本。小枝有瘤状突起。叶具3小叶；小叶椭圆形、椭圆状披针形或倒卵状椭圆形，下面不具白粉；中央小叶常倒卵状卵形。花序团伞形；花瓣具角，无毛；花柱不明显。花期4-8月，果期7-11月。生海拔600-1600米的溪边林下或山坡灌丛中。产云南南部。泰国和越南亦有。

Woody lianas. Branchlets verrucose. Leaves 3-foliolate; leaflets elliptic, elliptic-lanceolate, or obovate-elliptic, abaxially not glaucous; central leaflet usually obovate-oval. Inflorescences glomerate; petals corniculate, glabrous; styles indistinct. Fl. Apr-Aug. Fr. Jul-Nov. Forests by streams or montane shrubs at 600-1600 m. Distributed in S Yunnan. Also in Thailand and Vietnam.

## 茎花崖爬藤
**Tetrastigma cauliflorum** Merr.

木质藤本。小枝稍扁平，具纵脊，无毛；卷须不分枝。叶具掌状5小叶，叶下无毛；伞形或复合二歧状，着生于老茎上，长9-11厘米，基部有节，节上有苞片。浆果1.5-2 × 1.2-2厘米。花期4月，果期6-12月。生海拔100-1100米的山谷林中。产云南、广西、广东和海南。老挝和越南亦有。

十字崖爬藤 *Tetrastigma cruciatum*

茎花崖爬藤 *Tetrastigma cauliflorum*

Woody lianas. Branchlets slightly flat, with longitudinal ridges, glabrous; tendrils unbranched. Leaves palmately 5-foliolate, abaxially glabrous. Inflorescences umbelliform or a compound dichasium, cauliflorous, 9-11 cm long, with nodes and bracts. Berries 1.5-2 × 1.2-2 cm. Fl. Apr. Fr. Jun-Dec. Forests in valleys at 100-1100 m. Distributed in Yunnan, Guangxi, Guangdong and Hainan. Also in Laos and Vietnam.

扁担藤 *Tetrastigma planicaule*

## 扁担藤

**Tetrastigma planicaule** (Hook. f.) Gagnep.

大木质藤本。茎扁压，卷须不分枝。叶为掌状5小叶，中央小叶披针形、狭披针形或卵状披针形。花序伞形，腋生；花柱不明显；花瓣4；柱头4裂。浆果球形，具1-2(-3)粒种子。花期4-6月，果期8-12月。生海拔100-2100米的山地林中或岩石。产云南、西藏、贵州、广西、广东和福建。印度、斯里兰卡、老挝和越南亦有。

Large woody vines. Stems compressed, tendrils not branched. Leaves palmately 5-foliolate, central leaflets lanceolate, narrowly lanceolate, or ovate-lanceolate. Inflorescences umbelliform, axillary; styles inconspicuous; petals 4; stigmas 4-lobed. Berries globose, 1-2(-3)-seeded. Fl. Apr-Jun. Fr. Aug-Dec. Montane forests or rocks at 100-2100 m. Distributed in Yunnan, Xizang, Guizhou, Guangxi, Guangdong and Fujian. Also in India, Sri Lanka, Laos and Vietnam.

## 三叶崖爬藤

**Tetrastigma hemsleyanum** Diels et Gilg

细长藤蔓。卷须不分枝。叶具三小叶，小叶披针形。花序为小伞形二歧状，腋生，长1-5厘米；花瓣4，卵形，无毛；柱头4裂。浆果球形或倒卵球状球形，具1粒种子。花期4-6月，果期8-11月。生海拔300-1300米的山谷。产中国西南、华南、东南、华中和华东。印度亦有。

Slender lianas. Tendrils unbranched. Leaves ternate; leaflets lanceolate. Inflorescences small umbelliform dichasium, axillary, 1-5 cm long; petals 4, ovate, glabrous; stigmas 4-lobed. Berries globose or obovoid-spheroid, 1-seeded. Fl. Apr-Jun. Fr. Aug-Nov. Valleys at 300-1300 m. Distributed in SW, S, SE, C and E China. Also in India.

## 狭叶崖爬藤

**Tetrastigma serrulatum** (Roxb.) Planch.

细弱藤本。老枝绿褐色或淡紫色，不具瘤。卷须二叉分枝或有时不分枝。叶鸟足状5小叶；小叶卵状披针形。花序伞形，腋生，长1-8厘米，具节和苞片。浆果成熟时紫黑色，球形，具2粒种子。花期3-6月，果期7-10月。生海拔500-2900米的山谷林中、山腰灌丛或石缝中。产中国西南、华南和华中。印度、尼泊尔、不丹、缅甸和泰国亦有。

Lianas, slender. Old branches green-brownish or purplish, not tuberculate. Tendrils biforked or sometimes unbranched. Leaves pedately 5-foliolate; leaflets ovate-lanceolate. Inflorescences umbelliform, axillary, 1-8 cm long, with nodes and bracts. Berries purple-black at maturity, spheroid, 2-seeded. Fl. Mar-Jun. Fr. Jul-Oct. Forests in valleys, shrubs on hillsides or rocky gaps at 500-2900 m. Distributed in SW, S and C China. Also in India, Nepal, Bhutan, Myanmar and Thailand.

三叶崖爬藤 *Tetrastigma hemsleyanum*

狭叶崖爬藤 *Tetrastigma serrulatum*

叉须崖爬藤 *Tetrastigma hypoglaucum*

## 叉须崖爬藤
**Tetrastigma hypoglaucum** Planch. ex Franch.

木质藤本。小枝圆柱形，直径1.5-2毫米，卷须二分枝。叶为掌状5小叶，中央小叶披针形；托叶棕色。花序单伞形，花瓣4；柱头4裂。浆果球形，具1-3粒种子。花期6月，果期8-9月。生海拔2300-2500米的山谷林中或灌丛中。产云南和四川。

Woody vines. Branchlets terete, 1.5-2 mm diam, tendrils 2-branched. Leaves palmately 5-foliolate; central leaflets lanceolate; stipules brown. Inflorescences umbellate; petals 4; stigmas 4-lobed. Berries spheroid, 1-3-seeded. Fl. Jun. Fr. Aug-Sep. Forests or thickets in valleys at 2300-2500 m. Distributed in Yunnan and Sichuan.

## 崖爬藤
**Tetrastigma obtectum** (Wall. ex M. A. Lawson) Planch. ex Franch.

半木质藤本。卷须掌状4-7分枝。叶为掌状5小叶；小叶菱状椭圆形或椭圆状披针形。聚伞花序常单生，顶生或在短枝上与1或2片叶假顶生；子房圆锥形。浆果球形，具1粒种子。花期4-6月，果期8-11月。生海拔200-2400米的林中或岩石上。产中国西南、华南、华中和华西。尼泊尔、不丹和越南亦有。

Lianas, semi-woody. Tendrils palmately 4-7-branched. Leaves palmately 5-foliolate; leaflets rhombic-elliptic or elliptic-lanceolate. Umbels usually simple, terminal or pseudoterminal with 1 or 2 leaves on short branches; ovary coniform. Berries spheroid, 1-seeded. Fl. Apr-Jun. Fr. Aug-Nov. Forests or rocks at 200-2400 m. Distributed in SW, S, C and W China. Also in Nepal, Bhutan and Vietnam.

崖爬藤 *Tetrastigma obtectum*

酸蔹藤 *Ampelocissus artemisiifolia*

## 酸蔹藤
**Ampelocissus artemisiifolia** Planch.

木质藤本。卷须二叉分枝。3小叶复叶；小叶下面密被白色蛛丝状绒毛，上面疏被蛛丝状绒毛。复二歧聚伞花序与叶对生，基部具一条卷须；花柱呈锥形，具约10条纵棱。浆果球形，具2或3粒种子。花期6月，果期8月。生海拔1600-1800米的林中或灌丛中。产云南和四川。

Woody lianas. Tendrils bifurcate. Leaves 3-foliolate; leaflets abaxially with dense white and arachnoid tomentum, adaxially with sparse arachnoid tomentum. Compound dichasium leaf-opposed, with a tendril at base; styles conical, ca. 10 ridges. Berries globose, 2- or 3-seeded. Fl. Jun. Fr. Aug. Forests or shrubs at 1600-1800 m. Distributed in Yunnan and Sichuan.

## 刺葡萄
**Vitis davidii** (Rom. Caill.) Föex

木质藤本。小枝具刺，老枝上变为瘤状，渐无毛；卷须二叉分枝。叶卵形或卵状椭圆形，不分裂或3浅裂，每侧具12-33个锐齿。圆锥花序与叶对生。浆果成熟时紫色，球形。花期4-6月，果期7-10月。生海拔500-2300米的林中、灌丛中或山谷。产中国西南、华南、华中、华西和华东。

刺葡萄 *Vitis davidii*

Woody vines. Branchlets prickly, becoming tuberculate on old branches, glabrescent; tendrils bifurcate. Leaves oval or oval-elliptic, undivided or shallowly 3-lobed, margin with 12-33 teeth on each side, teeth sharp. Panicles opposite to leaves. Berries purple at maturity, globose. Fl. Apr-Jun. Fr. Jul-Oct. Forests, thickets or valleys at 500-2300 m. Distributed in SW, S, C, W and E China.

## 桦叶葡萄
**Vitis betulifolia** Diels et Gilg

木质藤本。小枝幼时疏具毡绒毛，渐无毛；卷须二叉分枝。叶单生，椭圆形或卵状椭圆形，不分裂或3裂，下面幼时密被绒毛，成熟后近无毛或仅叶腋具柔毛。圆锥花序与叶对生，疏松。浆果成熟时紫黑色，球形。花期3-6月，果期6-11月。生海拔600-3600米的山谷林中、灌丛地或山腰。产中国西南、华中和华西。

Woody lianas. Branchlets sparse lanate tomentum when young, then becoming glabrescent; tendrils bifurcate. Leaves simple, oval or ovate-elliptic, undivided or 3-lobed, abaxially densely tomentose when young, then hairs falling off and only veins pubescent or nearly glabrous. Panicles leaf-opposed, loose. Berries purple-black at maturity, globose. Fl. Mar-Jun. Fr. Jun-Nov. Forests in valleys, shrublands or hillsides at 600-3600 m. Distributed in SW, C and W China.

## 东南葡萄
**Vitis chunganensis** Hu

木质藤本。老枝具明显的纵脊，无毛；卷须二叉分枝。叶卵形至卵状椭圆形，下面具白粉，稀苍白色。圆锥花序疏散，与叶对生；子房椭圆形；花柱短，纤细；柱头膨大。浆果成熟时紫黑色。花期4-6月，果期6-8月。生海拔500-1400米的山坡灌丛或山谷林中。产中国西南、华南、东南和华东。

Woody vines. Old branches with conspicuous longitudinal ridges, glabrous; tendrils bifurcate. Leaves ovate or ovate-elliptic, abaxially glaucous, rarely glaucescent. Panicles sparse, opposite to leaves; ovary oval; styles short, slender; stigmas expanded. Berries purple-black when mature. Fl. Apr-Jun. Fr. Jun-Aug. Bushes on slopes or valley forests at 500-1400 m. Distributed in SW, S, SE and E China.

桦叶葡萄 *Vitis betulifolia*

东南葡萄 *Vitis chunganensis*

变叶葡萄 *Vitis piasezkii*

## 变叶葡萄
**Vitis piasezkii** Maxim.

木质藤本。小枝和叶片无毛或近无毛；卷须二叉分枝。叶形高度多变，具3-5小叶，或混生单叶及不裂叶，或单叶与不同程度裂叶混生。圆锥花序与叶对生，疏散。浆果球形，黑紫色。花期6月，果期7-9月。生海拔900-2100米的林中、林缘、灌丛中、山谷或河边。产中国西南、东南、华北、华中和华西。

Woody vines. Branchlets and leaflets glabrous or subglabrous; tendrils bifurcate. Leaves highly variable in shape, 3-5-foliolate, or mixed with simple and unlobed, or simple and variously lobed leaves. Panicles opposite to leaves, loose. Berries globose, blackish purple. Fl. Jun. Fr. Jul-Sep. Forests, forest edges, thickets, valleys or riversides at 900-2100 m. Distributed in SW, SE, N, C and W China.

## 网脉葡萄
**Vitis wilsonae** H. J. Veitch

木质藤本。小枝疏具淡褐色蛛丝状绒毛；卷须二叉分枝，与叶对生。单叶心形至卵状椭圆形，网脉凸起，背面叶脉上被棕色蛛丝状毛。圆锥花序与叶对生，疏松，长4-16厘米，基部分枝发达。浆果球形。花期5-7月，果期6月至翌年1月。生海拔400-2000米的林中、灌丛地、山谷或溪边。产中国西南、东南、华中、华西和东南。

Woody lianas. Branchlets with sparse, brownish arachnoid tomentum; tendrils bifurcate, leaf-opposed. Leaves simple, cordate to ovate-elliptic, veinlets raised, abaxially with brownish arachnoid tomentum on veins. Panicles leaf-opposed, loose, 4-16 cm long, with well-developed basal branches. Berries globose. Fl. May-Jul. Fr. Jun to next Jan. Forests, shrublands, valleys or streamsides at 400-2000 m. Distributed in SW, SE, C, W and SE China.

网脉葡萄 *Vitis wilsonae*

## 葛藟葡萄
**Vitis flexuosa** Thunb.

木质藤本。嫩枝疏被蛛丝状毛；卷须二叉分枝。叶卵形、三角状卵形、卵圆形或卵状椭圆形，下面幼时疏具蛛丝状绒毛。圆锥花序与叶对生，疏松，长4-12厘米。浆果球形。花期3-5月，果期7-11月。生海拔100-2300米的林中、灌丛、山谷、草甸或田野。产中国西南、华南、华中、华西、华北和华东。南亚、东南亚和日本亦有。

Woody vines. Branchlets with sparse arachnoid tomentum when young; tendrils bifurcate. Leaves ovate, triangular-ovate, oval, or ovate-elliptic, abaxially with sparse arachnoid tomentum when young. Panicles opposite to leaves, loose, 4-12 cm long. Berries globose. Fl. Mar-May. Fr. Jul-Nov. Forests, thickets, valleys, meadows or fields at 100-2300 m. Distributed in SW, S, C, W, N and E China. Also in S and SE Asia, and Japan.

## 华东葡萄
**Vitis pseudoreticulata** W. T. Wang

木质藤本，杂性异株。嫩枝密被蛛丝状绒毛；卷须二叉分枝。叶卵圆形或长圆形，两面幼时具长柔毛状绒毛，后渐稀疏。圆锥花序疏散，与叶对生，长5-11厘米。浆果球形，成熟时紫黑色。花期4-6月，果期6-10月。生海拔100-300米的河边、山坡、草丛、灌丛或林中。产华南、东南、华中和华东。朝鲜半岛亦有。

Woody vines, polygamo-dioecious. Young branches densely clothed with arachnoid tomentum; tendrils bifurcate. Leaves oval or oblong, both surfaces with villous tomentum when young, then becoming sparsely so. Panicles

葛藟葡萄 *Vitis flexuosa*

华东葡萄 *Vitis pseudoreticulata*

狭叶葡萄 *Vitis tsoi*

sparse, opposite to leaves, 5-11 cm long. Berries globose, black-purple when mature. Fl. Apr-Jun. Fr. Jun-Oct. By rivers, slopes, grasslands, bushes or forests at 100-300 m. Distributed in S, SE, C and E China. Also in Korean Peninsula.

## 狭叶葡萄
**Vitis tsoi** Merr.

木质藤本。小枝密具柔毛；卷须不分枝。叶卵状披针形或三角状卵形，下面沿脉具柔毛。圆锥花序狭窄，长2-6厘米；子房圆锥状；花柱短。浆果成熟时紫黑色，球形，直径5-8毫米。花期4-5月，果期6-9月。生海拔300-700米的山坡林中或灌丛中。产广西、广东和福建。

Woody vines. Branchlets densely pubescent; tendrils unbranched. Leaves ovate-lanceolate or triangular-ovate, abaxially pubescent on veins. Panicles narrow, 2-6 cm long; ovary conical; styles short. Berries purple-black at maturity, globose, 5-8 mm diam. Fl. Apr-May. Fr. Jun-Sep. Forests on slopes or bushes at 300-700 m. Distributed in Guangxi, Guangdong and Fujian.

## 山葡萄
**Vitis amurensis** Rupr.

木质藤本。小枝圆柱状，光滑；卷须2-3分枝。叶阔卵圆形，上面幼时疏具蛛丝状绒毛，渐无毛。圆锥花序与叶对生，疏散，长5-13厘米；子房圆锥形；花柱明显，基部稍加厚。花期5-6月，果期7-9月。生海拔100-2100米的林中、灌丛、丘陵或山谷。产华北、华东和东北。

Woody vines. Branchlets terete, glabrous; tendrils 2-3-branched. Leaves broadly oval, adaxially with sparse arachnoid tomentum when young, then glabrescent. Panicles opposite to leaves, loose, 5-13 cm long; ovary conical; styles obvious, slightly thick at base. Fl. May-Jun. Fr. Jul-Sep. Forests, thickets, hills or valleys at 100-2100 m. Distributed in N, E and NE China.

山葡萄 *Vitis amurensis*

葡萄 *Vitis vinifera*

## 葡萄

**Vitis vinifera** L.

木质藤本。小枝圆柱形，有纵棱纹；卷须二叉分枝，每隔2节间断与叶对生。叶卵圆形，显著3-5浅裂或中裂，基部深心形，两侧常靠合，边缘有22-27个锯齿，齿深而粗大，不整齐，下面淡绿色，无毛或被疏柔毛。圆锥花序密集或疏散，多花，与叶对生。花期4-5月，果期8-9月。中国各地栽培，为著名水果并可酿酒。原产亚洲西南部和欧洲东南部。

Woody vines. Branchlets terete, with longitudinal ridges; tendrils bifurcate, interruptedly opposite to leaves every 3 nodes. Leaves ovate, conspicuously 3-5-lobed or cleft, base deeply cordate, 2 sides usually overlapping to nearly so, margin 22-27-toothed on each side, teeth deep and large, irregular, abaxially greenish, glabrous or sparsely pilose. Panicles dense or sparse, with numerous flowers, opposite to leaves. Fl. Apr-May. Fr. Aug-Sep. Widly cultivated in China for grapes and wine-making. Native to SW Asia and SE Europe.

## 毛葡萄

**Vitis heyneana** Roem. et Schult.

木质藤本，杂性异株。小枝圆柱形，具灰色或褐色绵毛；卷须二叉分枝。单叶，有时3浅裂，卵圆形、卵状长圆形或卵状五方形，背面密被灰色或棕色绒毛，渐脱落。圆锥花序与叶对生，长4-14厘米。浆果球形。花期4-6月，果期6-10月。生海拔100-3200米的林中、灌丛地，山腰或山谷中。产中国西南、华南、东南、华中、华北、华西和华东。印度、尼泊尔和不丹亦有。

Woody lianas, polygamo-dioecious. Branchlets terete, gray or brown lanate; tendrils bifurcate. Leaves simple, sometimes 3-lobed, oval, ovate-oblong, or ovate-quinquangular, abaxially densely grayish or brown tomentose, gradually becoming less so. Panicles leaf-opposed, 4-14 cm long. Berries globose. Fl. Apr-Jun. Fr. Jun-Oct. Forests, shrublands, hillsides or valleys at 100-3200 m. Distributed in SW, S, SE, C, N, W and E China. Also in India, Nepal and Bhutan.

毛葡萄 *Vitis heyneana*

## 桑叶葡萄

**Vitis heyneana** Roem. et Schult. subsp. **ficifolia** (Bunge) C. L. Li

本亚种与毛葡萄的区别在于本亚种的叶通常3浅裂至中裂。花期5-7月，果期7-9月。生海拔100-1300米的林中、灌丛或山谷。产河南、河北、山西、陕西、江苏和山东。

This subspecies differs from the typical subspecies in its leaves usually 3-lobed to

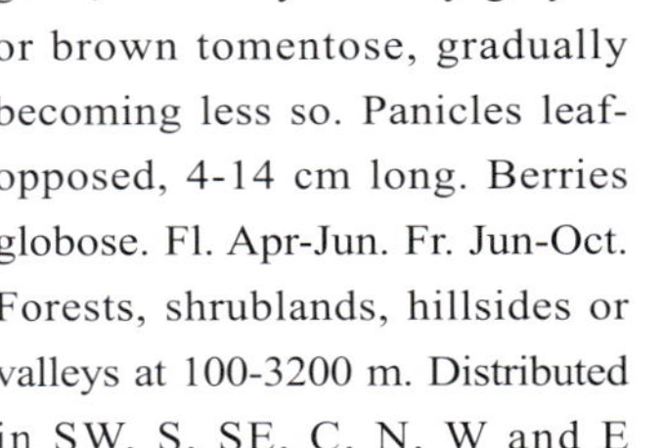

桑叶葡萄 *Vitis heyneana* subsp. *ficifolia*

cleft. Fl. May-Jul. Fr. Jul-Sep. Forests, thickets or valleys at 100-1300 m. Distributed in Henan, Hebei, Shanxi, Shaanxi, Jiangsu and Shandong.

## 蘡薁
**Vitis bryoniifolia** Bunge

木质藤本，杂性异株。小枝圆柱形，幼时密具绵毛，渐稀疏；卷须二叉分枝。叶卵形，均深裂至少至中部，有时深裂的裂片羽状半裂。圆锥花序与叶对生。浆果成熟时紫红色，直径5-8毫米。花期4-8月，果期6-10月。生海拔100-2500米的林中、灌丛中、溪边或田野中。产中国西南、东南、华中、华东和华北。

Woody vines, polygamo-dioecious. Branchlets terete, densely lanate when young, becoming sparsely so; tendrils bifurcate. Leaves ovate, all divided deeply or at least to middle, sometimes deep lobes pinnatifid. Panicles opposite to leaves. Berries purple-red at maturity, 5-8 mm diam. Fl. Apr-Aug. Fr. Jun-Oct. Forests, thickets, streamsides or fields at 100-2500 m. Distributed in SW, SE, C, E and N China.

蘡薁 *Vitis bryoniifolia*

## 小叶葡萄
**Vitis sinocinerea** W. T. Wang

木质藤本。小枝疏具柔毛及蛛丝毛状绒毛；卷须不分枝或二叉分枝。单叶，卵形，长3-8厘米，宽 3-6厘米，背面密被棕色和蛛丝状绒毛。圆锥花序与叶对生，较小且细弱，长3-6厘米。浆果成熟时紫黑色。花期5-6月，果期7-10月。生海拔200-2800米的林中、灌丛地或山腰。产中国西南、东南、华中和华东。

Woody lianas. Branchlets sparsely pubescent with arachnoid tomentum; tendrils unbranched or bifurcate. Leaves simple, oval, 3-8 mm long, 3-6 cm wide, abaxially with dense brown and arachnoid tomentum. Panicles leaf-opposed, small, 3-6 cm long, slender. Berries purple-black at maturity. Fl. May-Jun. Fr. Jul-Oct. Forests, shrublands or hillsides at 200-2800 m. Distributed in SW, SE, C and E China.

## 鸡足葡萄
**Vitis lanceolatifoliosa** C. L. Li

木质藤本。小枝密具绵毛；卷须二叉分枝。叶为掌状3-5小叶，背面密被褐色蛛丝状绒毛，中部小叶披针形。圆锥花序与叶对生，疏松。浆果球形。种子倒卵球形，种脐在背面中部椭圆体形，脊凸起。花期5月，果期8-9月。生海拔600-800米的林中、溪边灌丛或山腰。产广东、湖南和江西。

Woody lianas. Branchlets densely rubiginous lanate; tendrils bifurcate. Leaves palmate, 3-5-foliolate, abaxially with dense brown arachnoid tomentum, central leaflets lanceolate. Panicles leaf-opposed, loose. Berries globose. Seeds obovoid, chalazal knot ellipsoid, raphe raised. Fl. May. Fr. Aug-Sep. Forests or shrublands by streams or hillsides at 600-800 m. Distributed in Guangdong, Hunan and Jiangxi.

小叶葡萄 *Vitis sinocinerea*

鸡足葡萄 *Vitis lanceolatifoliosa*

# 杜英科 Elaeocarpaceae

## 水石榕
**Elaeocarpus hainanensis** Oliv.

灌木或小乔木。叶革质，狭倒披针形，纸质，下面具柔毛，侧脉14-30对。总状花序生当年生叶腋，上部具2-6花；花瓣白色，外面有柔毛，先端撕裂；雄蕊63-75；花药行具约4毫米长的芒；花盘多裂，具绒毛；子房2室。核果纺锤形。花期6-7月，果期7-9月。生海拔200-500米的湿处或山谷水边。产云南东南部、广西、广东和海南。缅甸、泰国和越南亦有。

Shrubs or small trees. Leaves leathery, narrowly oblanceolate, papery, abaxially pubescent, lateral veins 14-30 pairs. Racemes in axils of current leaves, 2-6-flowered in upper part; petals white, abaxially pubescent, lacerate at apex; stamens 63-75; anthers with awn ca. 4 mm long; disk multi-lobed, tomentose; ovary 2-loculed. Drupes fusiform Fl. Jun-Jul. Fr. Jul-Sep. Wet places or river banks in valleys at 200-500 m. Distributed in SE Yunnan, Guangxi, Guangdong and Hainan. Also in Myanmar, Thailand and Vietnam.

## 假樱叶杜英
**Elaeocarpus prunifolioides** Hu

乔木。叶长圆形、卵状长圆形或椭圆形，薄革质或纸质，无毛，边缘有钝齿，侧脉8-12对。总状花序具8-15花；花瓣5，下面具柔毛，上部1/5撕裂状；雄蕊20-30；花药具长1毫米的芒；花盘10裂；子房3室。核果椭圆体形，长约2厘米。花期1-2月，果期2-5月。生海拔600-1700米的常绿林中。产云南南部。

Trees. Leaves oblong, ovate-oblong, or elliptic, thinly leathery or papery, glabrous, at magin obtuse-serrate, lateral veins 8-12 pairs. Racemes 8-15-flowered; petals 5, abaxially pubescent, upper 1/5 laciniate; stamens 20-30; anthers with awn ca. 1 mm long; disk 10-lobed; ovary 3-loculed. Drupes ellipsoid, ca. 2 cm long. Fl. Jan-Feb. Fr. Feb-May. Evergreen forests at 600-1700 m. Distributed in S Yunnan.

假樱叶杜英 *Elaeocarpus prunifolioides*

## 老挝杜英
**Elaeocarpus laoticus** Gagnep.

乔木。叶倒披针形，薄革质，侧脉约10对。总状花序具7-10花；花瓣上半部撕裂，裂片17-20条，内侧有毛；雄蕊24-28；花药具长1毫米的芒；花盘10裂，腺体球形；子房3室，被毛。核果椭圆体形。花期5-7月，果期12月。生海拔1300-1500米的常绿林中。产云南东南部。老挝

水石榕 *Elaeocarpus hainanensis*

老挝杜英 *Elaeocarpus laoticus*

显脉杜英 *Elaeocarpus dubius*

薯豆 *Elaeocarpus japonicus*

亦有。

Trees. Leaves oblanceolate, thin-leathery, lateral veins ca. 10-paired. Flowers 7-10 in racemes; petals above lacerate, lobes 17-20, abaxially hairy; stamens 24-28; anthers with awn 1 mm long; disk 10-lobed, glands globose; ovary 3-locular, hairy. Drupes ellipsoid. Fl. May-Jul. Fr. Dec. Evergreen forests at 1300-1500 m. Distributed in SE Yunnan. Also in Laos.

## 显脉杜英

**Elaeocarpus dubius** A. DC.

常绿乔木。叶簇生于枝顶；叶长圆形或披针形，长4-8厘米，薄革质，无毛，侧脉明显在两面凸起。总状花序生落叶或当年生叶腋，具4-7花；花瓣5，长圆形，上部1/3流苏状；裂片9-11；雄蕊20-23；子房3室。核果椭圆体形。花期3-4月，果期4-6月。生海拔600-700米的低海拔林中。产云南、贵州、广西、广东和海南。越南亦有。

Evergreen trees. Leaves crowded at twig apices; leaves oblong or lanceolate, 4-8 cm long, thinly leathery, glabrous, lateral veins conspicuously raised on both surfaces. Racemes in axils of fallen and current leaves, 4-7-flowered; petals 5, oblong, upper 1/3 laciniate; segments 9-11; stamens 20-23; ovary 3-loculed. Drupes ellipsoid. Fl. Mar-Apr. Fr. Apr-Jun. Low-elevation forests at 600-700 m. Distributed in Yunnan, Guizhou, Guangxi, Guangdong and Hainan. Also in Vietnam.

## 薯豆

**Elaeocarpus japonicus** Siebold et Zucc.

乔木。叶卵形至狭长圆形，革质，密具银灰色绢毛，迅速无毛，下面散生细小黑色腺点。总状花序长3-6厘米；雌花花瓣边缘全缘或具锯刻；雄蕊15；花药顶端不具芒；花盘5或10裂，环形；子房具柔毛，3室。核果光亮，椭圆体形。花期4-5月，果期5-7月。生海拔400-2800米的常绿林中。产中国西南、华南、东南、华中和华东。越南和日本亦有。

Trees. Leaves ovate to narrowly oblong, leathery, densely silvery-gray sericeous at first, soon glabrescent, abaxially sparsely black-glandular-punctate. Racemes 3-6 cm long; female flower petals margin entire or incised; stamens 15; anthers not awned at apices; disk 5- or 10-lobed, in circle; ovary pubescent, 3-loculed. Drupes shiny, ellipsoid. Fl. Apr-May. Fr. May-Jul. Evergreen forests at 400-2800 m. Distributed in SW, S, SE, C and E China. Also in Vietnam and Japan.

## 华杜英

**Elaeocarpus chinensis** (Gardn. et Chanp.) Hook. f. ex Benth.

常绿乔木。叶柄长1.5-2厘米；叶卵状披针形或披针形，宽2-3厘米，纸质，幼时具柔毛，迅速无毛，下面具黑色腺点，侧脉4-6对，下面稍凸起。总状花序生落叶叶腋；花瓣4，边缘近全缘；雄蕊8或9；子房2室。核果椭圆体形。花期5-6月，果期6-9月。生海拔300-900米的常绿林中。产贵州、广西、广东、福建、浙江和江西。越南亦有。

Evergreen trees. Petioles 1.5-2 cm long; leaves ovate-lanceolate or lanceolate, 2-3 cm wide, papery, pubescent when young, soon glabrescent, abaxially black glandular punctate, lateral veins 4-6 per side, slightly raised abaxially. Racemes in axils of fallen leaves; petals 4, margin nearly entire; stamens 8 or 9; ovary 2-loculed. Drupes ellipsoid. Fl. May-Jun. Fr. Jun-Sep. Evergreen forests at 300-900 m. Distributed in Guizhou, Guangxi, Guangdong, Fujian, Zhejiang and Jiangxi. Also in Vietnam.

华杜英 *Elaeocarpus chinensis*

山杜英 *Elaeocarpus sylvestris*

## 山杜英

**Elaeocarpus sylvestris** (Lour.) Poir.

乔木。叶倒卵形或倒披针形，纸质，两面无毛，侧脉常4或5对。总状花序；花瓣5，上部1/2撕裂；雄蕊15；花药顶端不具芒；花盘5浅裂，球形，完全分离；子房具柔毛，2或3室。核果椭圆体形。花期4-5月，果期5-8月。生海拔300-2000米的常绿林中。产中国西南、华南和东南。越南亦有。

Trees. Leaves obovate or oblanceolate, papery, both surfaces glabrous, lateral veins usually 4 or 5 per side. Racemes; petals 5, laciniate in upper 1/2; stamens 15; anthers not awned at apices; disk 5-lobed, globose, completely separate; ovary pubescent, 2- or 3-loculed. Drupes ellipsoid. Fl. Apr-May. Fr. May-Aug. Evergreen forests at 300-2000 m.

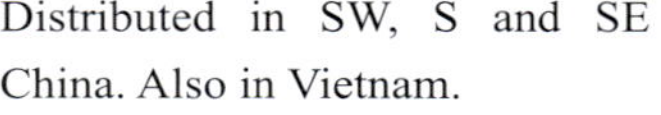

Distributed in SW, S and SE China. Also in Vietnam.

## 秃瓣杜英

**Elaeocarpus glabripetalus** Merr.

乔木。叶倒披针形，纸质或膜质，无毛，下面无腺点，侧脉7-9对；叶柄几无至长4(-10)毫米。总状花序生第二年生枝腋；花序梗具柔毛；花瓣5，白色，14-18裂；雄蕊20-30；子房2或3室，具柔毛。核果椭圆体形。花期7月，果期8-10月。生海拔300-1500米的常绿林中。产中国西南、华南、东南、华中和华东。

秃瓣杜英 *Elaeocarpus glabripetalus*

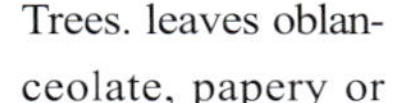

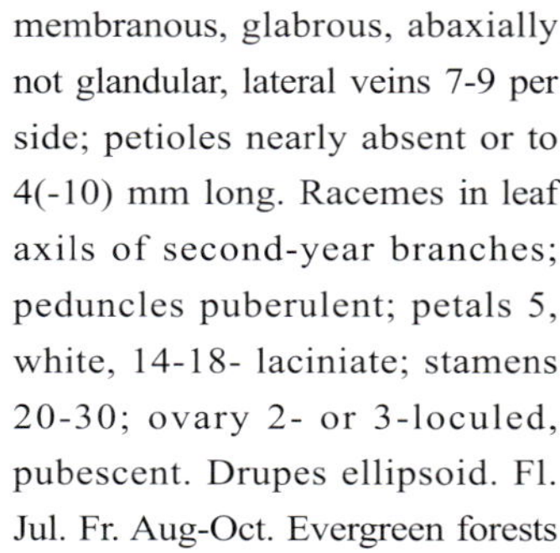

Trees. leaves oblanceolate, papery or membranous, glabrous, abaxially not glandular, lateral veins 7-9 per side; petioles nearly absent or to 4(-10) mm long. Racemes in leaf axils of second-year branches; peduncles puberulent; petals 5, white, 14-18- laciniate; stamens 20-30; ovary 2- or 3-loculed, pubescent. Drupes ellipsoid. Fl. Jul. Fr. Aug-Oct. Evergreen forests at 300-1500 m. Distributed in SW, S, SE, C and E China.

## 多沟杜英

**Elaeocarpus lacunosus** Wall. ex Kurz

乔木。叶片长圆状倒披针形或长圆形至椭圆形，革质，下面无毛或仅沿中脉具柔毛，侧脉7-8对。总状花序多花；花瓣5，上半部撕裂，边缘有纤毛；雄

多沟杜英 *Elaeocarpus lacunosus*

灰毛杜英 *Elaeocarpus limitaneus*

蕊30-40；花药先端不具芒，具柔毛；花盘5裂，完全离生。核果卵球形；内果皮多沟。花期4-5月，果期5-6月。生海拔1400-2600米的林中。产云南西部。南亚和东南亚亦有。

Trees. Leaves oblong-oblanceolate or oblong to elliptic, leathery, abaxially glabrous or puberulent only on midvein, lateral veins 7-8 pairs. Racemes multi-flowered; petals 5, above lacerate, margin ciliate; stamens 30-40; anthers not awned but pubescent at apices; disk 5-lobed, completely separated. Drupes ovoid; endocarp lacunose. Fl. Apr-May. Fr. May-Jun. Forests at 1400-2600 m. Distributed in W Yunnan. Also in S and SE Asia.

## 灰毛杜英

**Elaeocarpus limitaneus** Hand. - Mazz.

常绿乔木。叶上干后光亮，椭圆形或倒卵形，革质，下面被银灰色平伏绒毛。总状花序生当年或落叶叶腋；花两性；花瓣5，白色，上部1/2撕裂；裂片11-16。核果椭圆状卵球形。花期7月，果期8-9月。生海拔1000-1700米的林中。产云南东南部、广西、广东、海南和福建。越南亦有。

Evergreen trees. Leaves adaxially shiny when dry, elliptic or obovate, leathery, abaxially silvery-gray appressed tomentose. Racemes in axils of current and fallen leaves; flowers bisexual; petals 5, white, laciniate in upper 1/2; segments 11-16. Drupes elliptic-ovoid. Fl. Jul. Fr. Aug-Sep. Forests at 1000-1700 m Distributed in SE Yunnan, Guangxi, Guangdong, Hainan and Fujian. Also in Vietnam.

## 滇藏杜英

**Elaeocarpus braceanus** Watt ex C. B. Clarke

乔木。小枝上具有灰白色皮孔。叶长圆形或椭圆形，薄革质，下面具锈褐色柔毛。总状花序腋生；小苞片近肾形；雄蕊40-50；花药先端不具芒，具柔毛；花盘5浅裂，具绒毛；子房3室。核果椭圆体形。花期10-11月，果期11月至翌年1月。生海拔800-3000米的林中。产云南南部和西藏东南部。印度、缅甸和泰国亦有。

Trees. Branchlets with gray-white lenticels. Leaves oblong or elliptic, thinly leathery, abaxially rust-brown puberulent. Racemes axillary; bracteoles nearly reniform; stamens 40-50; anthers not awned at apices, puberulent; disk 5-lobed, tomentose; ovary 3-loculed. Drupes ellipsoid. Fl. Oct-Nov. Fr. Nov to next Jan. Forests at 800-3000 m. Distributed in S Yunnan and SE Xizang. Also in India, Myanmar and Thailand.

## 杜英

**Elaeocarpus decipiens** Hemsl

常绿乔木。叶革质，披针形或倒披针形，革质或厚纸质，无毛，侧脉7-9对。总状花序生落叶的叶腋，长5-10厘米；花瓣5片，白色；雄蕊20或25-32；子房3室。核果椭圆体形；内果皮明显具疣。花期6-7月，果期11月至翌年1月。生海拔400-2400米的林中。产中国西南、华南和东南。越南和日本亦有。

Evergreen trees. Leaves leathery, lanceolate or oblanceolate, leathery or thickly papery, glabrous, lateral veins 7-9 pairs. Racemes in axils of fallen leaves, 5-10 cm long; petals 5, white; stamens 20 or 25-32; ovary 3-locular. Drupes ellipsoid; endocarp prominently verrucose. Fl. Jun-Jul. Fr. Nov to next Jan. Forests at 400-2400 m. Distributed in SW, S and SE China. Also in Vietnam and Japan.

滇藏杜英 *Elaeocarpus braceanus*

杜英 *Elaeocarpus decipiens*

毛果猴欢喜 *Sloanea dasycarpa*

## 毛果猴欢喜

**Sloanea dasycarpa** (Benth.) Hemsl.

常绿乔木。叶长圆形或倒卵状长圆形，宽4-7厘米，膜质或纸质，基部渐狭且钝或圆形。花腋生于上部小枝；花瓣不等长，先端具齿；子房具糙柔毛。蒴果球形，3或4瓣裂；针刺褐色，长6-8毫米。花期4-5月，果期9-10月。生海拔1400-2100米的林中。产云南、西藏、海南、台湾和福建。印度东部、不丹、缅甸和越南亦有。

Evergreen trees. Leaves oblong or obovate-oblong, 4-7 cm wide, membranous or papery, tapered into narrow and obtuse or rounded base. Flowers axillary in upper part of twigs; petals unequal in length, incised at apices; ovary roughly pubescent. Capsules globose, 3- or 4-valved; prickles brown, 6-8 mm long. Fl. Apr-May. Fr. Sep-Oct. Forests at 1400-2100 m. Distributed in Yunnan, Xizang, Hainan, Taiwan and Fujian. Also in E India, Bhutan, Myanmar and Vietnam.

## 全缘叶猴欢喜

**Sloanea integrifolia** Chun et F. C. How

乔木。枝和叶柄密具深褐色绒毛。叶革质，长圆形，基部楔形，近圆形或稍心形，全缘。蒴果腋生，单生，卵球形；分果爿上面紫色，密具长达2.5毫米的针刺。花期5月，果期9-10月。产广西、广东和海南。越南亦有。

Trees. Branches and petioles densely dark brown tomentose. Leaves leathery, oblong, base cuneate, nearly rounded, or slightly cordate, margin entire. Capsules axillary, solitary, ovoid; valves adaxially purple, with dense prickles to 2.5 mm long. Fl. May. Fr. Sep-Oct. Distributed in Guangxi, Guangdong and Hainan. Also in Vietnam.

## 西畴猴欢喜

**Sloanea xichouensis** Feng ex Y. Fang et Y. C. Hsu

乔木。叶倒卵形至椭圆形，宽5-6厘米，纸质，下面具柔毛，基部楔形，边缘疏具牙齿，侧脉6对；叶柄长2-4.5厘米，密被黄褐色绒毛。蒴果近球形，直径约2厘米，4瓣裂，外面密被刚毛状刺，刺线形，长5-7毫米。花期未知，果期8-9月。生海拔约1300米的常绿林中。产云南东南部。

Trees. Leaves obovate to elliptic, 5-6 cm wide, papery, abaxially pubescent, base cuneate, margin sparsely dentate, lateral veins 6 pairs; petioles 2-4.5 cm long, densely yellow-brown tomentose. Capsules subglobose, ca. 2 cm diam, 4-valved, abaxially densely bristly-spiny, spines filiform, 5-7 mm long. Fl. unknown. Fr. Aug-Sep. Evergreen forests at ca. 1300 m. Distributed in SE Yunnan.

## 滇越猴欢喜

**Sloanea mollis** Gagnep.

乔木。叶卵形或狭椭圆形，宽6-12厘米，薄革质，基部阔圆形或稍心形，边缘稍具锯齿或小圆齿，侧脉8-10对；叶柄圆柱形，3-6厘米或更长。萼片4-5，两面被绒毛；花瓣顶端具大牙齿。蒴果4裂，针刺通常黄色，长1-1.5厘米。花期4-5月，果期10-12月。生海拔1200-1400米的常绿林中。产云南东南部和广西南部。越南北部亦有。

Trees. Leaves ovate or narrowly elliptic, 6-12 cm wide, thin-leathery, base broad and rounded, or slightly

全缘叶猴欢喜 *Sloanea integrifolia*

西畴猴欢喜 *Sloanea xichouensis*

滇越猴欢喜 *Sloanea mollis*

贡山猴欢喜(苹婆猴欢喜) *Sloanea sterculiacea*

cordate, margin minutely denticulate or minutely crenulate, lateral veins 8-10 pairs; petioles terete, 3-6 cm or longer. Sepals 4-5, tomentose on both surfaces; petals large toothed at apices. Capsules 4-lobed, prickles usually yellow, 1-1.5 cm long. Fl. Apr-May. Fr. Oct-Dec. Evergreen forests at 1200-1400 m. Distributed in SE Yunnan and S Guangxi. Also in N Vietnam.

## 贡山猴欢喜(苹婆猴欢喜)

**Sloanea sterculiacea** (Benth.) Rehd. et Wils.

乔木。叶倒卵状长圆形，中部以上最宽，薄革质，下面被褐色绒毛，侧脉7-9对，基部极狭，圆形，边缘稍具锯齿。花单生；萼片4；花盘厚；子房被绒毛，4室；花柱长8-10毫米。蒴果针刺长1.5-2厘米；假种皮长为种子的1/2。花期4月，果期9-11月。生海拔1400-2500米的林中。产云南西北部和西藏东南部。东喜马拉雅、印度和缅甸亦有。

Trees. Leaves obovate-oblong, broadest from middle upward, thinly leathery, abaxially brown puberulent, lateral veins 7-9 pairs, base very narrow, rounded, margin minutely serrulate. Flowers solitary; sepals 4; disk thick, ovary velutinous, 4-locular; styles 8-10 mm long. Bristles of capsules 1.5-2 cm long; arils covering 1/2 of seeds. Fl. Apr. Fr. Sep-Nov. Forests at 1400-2500 m. Distributed in NW Yunnan and SE Xizang. Also in E Himalaya, India and Myanmar.

## 猴欢喜

**Sloanea sinensis** (Hance) Hemsl.

乔木。叶常长圆形或狭倒卵形，宽3-5厘米，薄革质，无毛，侧脉在上面不明显，基部楔形或圆形，边缘常全缘。花多数簇生于枝顶；花瓣4，白色，先端撕裂并有缺刻；花柱连合。蒴果大小不一，3-7裂，针刺通常黄色，长1-1.5厘米。花期6-10月，果期8-10月。生海拔700-1000米的常绿林中。产中国西南、华南和东南。缅甸、老挝、泰国、越南和柬埔寨亦有。

Trees. Leaves usually oblong or narrowly obovate, 3-5 cm wide, thinly leathery, glabrous, lateral veins adaxially inconspicuous, base cuneate, or rounded, margin usually entire. Flowers numerous, clustered at top of branchlets; petals 4, white, apex lacerate and incised; styles connate. Capsules variable in size, 3-7-valved, prickles usually yellow, 1-1.5 cm long. Fl. Jun-Oct. Fr. Aug-Oct. Evergreen forests at 700-1000 m. Distributed in SW, S and SE China. Also in Myanmar, Laos, Thailand, Vietnam and Cambodia.

## 仿栗

**Sloanea hemsleyana** (Ito) Rehd. et Wils.

乔木。叶簇生枝顶，常狭倒卵形或倒披针形，薄革质，宽3-5(-7)厘米，侧脉7-9对。花生于枝顶；萼片4；花瓣白色，先端有撕裂状齿刺；子房被锈色绒毛。蒴果大小不一，(3-)4-5(-6)裂；针刺长1-2厘米。花期7月，果期未知。生海拔1100-1400米的常绿林里。产云南、四川、贵州、广西、湖北和湖南。越南亦有。

Trees. Leaves fascicled at branch apices, usually narrowly obovate or oblanceolate, 3-5(-7) cm wide, thin-leathery, lateral veins 7-9 pairs. Flowers at branch apices; sepals 4; petals white, apex lacerate-pectinate; ovary rusty-tomentose. Capsules unequal in size, (3-)4-5(-6)-lobed; prickles 1-2 cm long. Fl. Jul. Fr. unknown. Evergreen forests at 1100-1400 m. Distributed in Yunnan, Sichuan, Guizhou, Guangxi, Hubei and Hunan. Also in Vietnam.

猴欢喜 *Sloanea sinensis*

仿栗 *Sloanea hemsleyana*

# 椴树科
# Tiliaceae

## 糠椴

**Tilia mandshurica** Rupr. et Maxim.

乔木。小枝幼时具灰白色星状绒毛。叶卵状圆形，侧脉5-7对，边缘齿芒状，下面密被星状毛。聚伞花序，有6-12花，长6-9厘米；退化雄蕊花瓣状；子房被星状毛。果实球形，卵球形或倒卵球形，具5棱，有时具瘤；外果皮木质，不开裂。花期7月，果期9月。产华北、华东和东北。俄罗斯(南西伯利亚)、朝鲜半岛和日本亦有。

Trees. Branchlets gray-white stellate tomentose when young. Leaves ovate-orbicular, lateral veins 5-7 per side, marginal teeth awnlike, abaxially densely stellate-hairy. Cymes 6-12-flowered, 6-9 cm long; staminodes petaloid; ovary stellate-hairy. Fruits globose, ovoid or obovoid, 5-angled, sometimes tuberculate; exocarps woody, indehiscent. Fl. Jul. Fr. Sep. Distributed in N, E and NE China. Also in Russia (S Siberia), Korean Peninsula and Japan.

## 毛糯米椴

**Tilia henryana** Szyszy.

乔木。小枝和芽具黄色星状绒毛或无毛。叶圆形，下面被黄色星状绒毛；侧脉4-6对。聚伞花序长10-12厘米，具30-100花；花梗有星状柔毛；子房有毛。果实倒卵球形，具5棱，具星状毛；外果皮木质，质硬，不开裂。花期6月。生山坡杂木林中。产华中和华东。

毛糯米椴 *Tilia henryana*

Trees. Branchlets and buds yellow stellate tomentose or glabrous. Leaves orbicular, abaxially yellowish stellate-tomentose; lateral veins 4-6 pairs. Cymes 10-12 cm long, 30-100-flowered; pedicels stellate-pubescent; ovary hairy. Fruits obovoid, 5-angled, stellate hairy; exocarps woody, hard, indehiscent. Fl. Jun. Mixed forests on slopes. Distributed in C and E China.

## 华椴

**Tilia chinensis** Maxim.

乔木。小枝无毛。叶阔椭圆形或圆形至卵状圆形，纸质，侧脉7-9对，下面密具星状绒毛。聚伞花序具1-3花；苞片长条形；子房具灰黄色绒毛。果实椭圆体形或球形，具突出5棱及灰黄色绒毛；外果皮木质，质硬，不开裂。花期6-7月，果期8-10月。生海拔1800-3900米的林中。产中国西南、华中和华西。

Trees. Branchlets glabrous. Leaves broadly elliptic or orbicular to ovate-orbicular, papery, lateral veins 7-9 pairs, abaxially densely stellate tomentose. Cymes 1-3-flowered; bracts long band-shaped; ovary gray-yellow tomentose. Fruits ellipsoid or globose, prominently 5-angled, gray-yellow tomentose; exocarps woody, hard, indehiscent. Fl. Jun-Jul. Fr. Aug-Oct. Forests at 1800-3900 m. Distributed in SW, C and W China.

## 紫椴

**Tilia amurensis** Rupr.

乔木，高达25米。小枝具白色或浅红色星状毛，渐无毛。叶阔卵形或卵圆形。聚伞花序纤细，长3-5厘米，具3-20花；萼片阔披针形；子房有毛。果实卵球形，具5棱或不明显具棱；外果皮厚革质，脆弱，不开裂。花期7月。生海拔1300-1400米的林中。产黑龙江、吉林和辽宁。俄罗斯和朝鲜半岛亦有。

Trees, to 25 m tall. Branchlets white or reddish stellate pubescent, glabrescent. Leaves broadly ovate or ovate-orbicular. Cymes slender, 3-5 cm long, 3-20-flowered; sepals

糠椴 *Tilia mandshurica*

华椴 *Tilia chinensis*

紫椴 *Tilia amurensis*

broadly lanceolate; ovary hairy. Fruits ovoid, 5-angled or obscurely angled; exocarps thickly leathery, fragile, indehiscent. Fl. Jul. Forests at 1300-1400 m. Distributed in Heilongjiang, Jilin and Liaoning. Also in Russia and Korean Peninsula.

## 椴树

**Tilia tuan** Szyszyl.

乔木。小枝无毛或具绒毛。叶卵圆形，心形，全缘或近顶端具少量小齿，或具明显牙齿。聚伞花序长8-13厘米；苞片窄披针形，长10-16厘米，下面有星状毛。果实球形，不具脊，被褐色或灰色毛，具疣；外果皮木质，质硬，不开裂。花期6-7月，果期7-11月。生海拔1200-2400米的林中。产中国西南、华中和华东。

Trees. Branchlets glabrous or tomentose. Leaves ovate-orbicular, cordate, margin entire or with a few minute teeth near apex or prominently dentate. Cymes 8-13 cm long; bracts narrowly lanceolate, 10-16 cm long, abaxially stellate-hairy; sepals hairy. Fruits globose, not ridged, brown or gray hairy, verrucose; exocarps woody, hard, indehiscent. Fl. Jun-Jul. Fr. Jul-Nov. Forests at 1200-2400 m. Distributed in SW, C and E China.

## 华东椴

**Tilia japonica** (Miq.) Simonk.

乔木。小枝具长柔毛，很快秃净。叶圆形或近圆形，革质，侧脉6-7对。聚伞花序长5-7厘米；苞片狭倒披针形或狭长圆形；退化雄蕊花瓣状；萼片外面被疏星状毛。果卵球状圆形，不具棱，有星状毛；外果皮厚革质，易碎，不开裂。花期5-6月，果期7-8月。生海拔1100-1700米的山坡杂木林中。产浙江、安徽、江苏和山东。日本亦有。

Trees. Branchlets villous, soon glabrescent. Leaves orbicular or nearly so, leathery, lateral veins 6-7 pairs. Cymes 5-7 cm long; bracts narrowly oblanceolate or narrowly oblong; staminodes petaloid; sepals abaxially sparsely stellate-hairy. Fruits ovoid-orbicular, not angled, stellate-hairy; exocarps thickly leathery, fragile, indehiscent. Fl. May-Jun. Fr. Jul-Aug. Mixed forests on slopes at 1100-1700 m. Distributed in Zhejiang, Anhui, Jiangsu and Shandong. Also in Japan.

椴树 *Tilia tuan*

华东椴 *Tilia japonica*

少脉椴 *Tilia paucicostata*

## 少脉椴
**Tilia paucicostata** Maxim.

乔木。小枝细弱，无毛。叶薄革质，卵形，边有细锯齿，两面无毛或下面于脉腋疏具绒毛，侧脉5-6对。聚伞花序无毛，具3-10花；子房被星状绒毛。果倒卵球形，先端常具喙；外果皮厚革质，易碎，不开裂。花期6-8月，果期9-10月。生海拔1300-2400米的林中。产中国西南、华中、华北和华西。

Trees. Branchlets slender, glabrous. Leaves thinly leathery, ovate, margin serrulate, both surfaces glabrous or abaxially sparsely tomentose in vein axils, lateral veins 5-6 pairs. Cymes glabrous, 3-10-flowered; ovary stellate-tomentose. Fruits obovoid, apex usually beaked; exocarps thickly leathery, fragile, indehiscent. Fl. Jun-Aug. Fr. Sep-Oct. Forests at 1300-2400 m. Distributed in SW, C, N and W China.

## 南京椴
**Tilia miqueliana** Maxim.

乔木。小枝具黄褐色绒毛。叶卵圆形，下面被灰色或黄色星状绒毛。聚伞花序长6-8厘米，具3-12花；苞片窄倒披针形；退化雄蕊花瓣状。果实球形，不具棱，被星状柔毛，具疣；外果皮木质，质硬，不开裂。花期7月。生海拔1000米以下的山坡林中。产华南和华东。日本亦有。

Trees. Branchlets yellow-brown tomentose. Leaves ovate-orbicular, abaxially gray- or yellow-stellate-hairy. Cymes 6-8 cm long, 3-12-flowered; bracts narrowly oblanceolate; staminodes petaloid. Fruits globose, not angled, stellate puberulent, verrucose; exocarps woody, hard, indehiscent. Fl. Jul. Forests on slopes below 1000 m. Distributed in S and E China. Also in Japan.

南京椴 *Tilia miqueliana*

## 甜麻
**Corchorus aestuans** L.

一年生草本。叶卵形或阔卵形，基脉5-7，基部一对齿常伸长为线形或尾尖状附属物。花单生或数个组成聚伞状；萼片5，外面紫红色；花瓣5，黄色；子房长圆柱形，被柔毛。蒴果长筒形，具角，3-5裂。花期夏秋季。生荒地、旷野或村边；广栽培。产长江以南各省区。热带亚洲、澳大利亚、热带非洲和中美洲亦有。

Annual herbs. Leaves ovate or broadly ovate, basal veins 5-7, basal pair of teeth usually elongating into filiform or caudate appendages. Flowers solitary or several together in cymes; sepals 5, abaxially purple-red; petals 5, yellow; ovary long terete, pubescent. Capsules long cylindric, angled, 3-5-valved. Fl. summer-autumn. Wastelands, wilderness or villages; widely cultivated. Distributed throughout the provinces in the south of Yangtze River. Also in tropical Asia, Australia, tropical Africa and Central America.

## 长蒴黄麻
**Corchorus olitorius** L.

草本带木质。叶纸质，长圆披针形，无毛，基出脉5条。花单生或几朵成聚伞花序；花萼顶端具长芒；花瓣长圆形，与花萼等长或稍短，基部具柄。蒴果稍反折，具10棱，粗壮，5-6瓣裂，长3-8厘米。花期夏至秋

甜麻 *Corchorus aestuans*

长蒴黄麻 *Corchorus olitorius*

季。生受干扰区域的杂草中。中国西南、华南和华东栽培。热带地区广泛栽培。

Herbs woody. Leaves papery, oblong-lanceolate, glabrous, basal nerves 5. Flowers solitary or several in a cyme; sepals apex long awned; petals oblong, as long as or slightly shorter than sepals, stalked at base. Capsules slightly curved, 10-angled, robust, 5-6-valved, 3-8 cm long. Fl. summer to autumn. Weed of disturbed areas. Cultivated in SW, S and E China. Also widely cultivated in tropical areas.

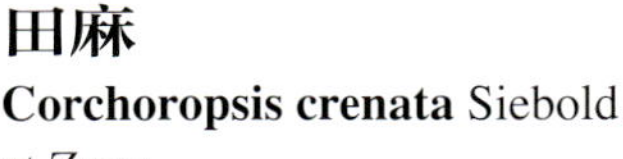

## 田麻

**Corchoropsis crenata** Siebold. et Zucc.

一年生草本。叶卵形至窄卵形，纸质或膜质，3出脉。花单生叶腋，具细花梗；花瓣黄色，倒卵形；雄蕊(5-)10-15，5组，长为花瓣的1/2，反折；退化雄蕊5，匙状条形。蒴果窄圆柱形。花期4-6月，果期秋季。产中国大部分地区。朝鲜半岛和日本亦有。

Annual herbs. Leaves ovate to narrowly ovate, papery or membranous, ternate. Flowers solitary, axillary, with slender pedicels; petals yellow, obovate; stamens (5-)10-15, in 5 groups, 1/2 as long as petals, reflexed; staminodes 5, spatulate-linear. Capsules narrowly cylindric. Fl. Apr-Jun. Fr. autumn. Distributed in most parts of China. Also in Korean Peninsula and Japan.

田麻 *Corchoropsis crenata*

一担柴 *Colona floribunda*

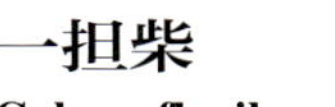

## 一担柴

**Colona floribunda** (Wall. ex Kurz) Craib

乔木。叶片阔倒卵状圆形或近圆形，两面均被粗糙灰褐色星状毛，背面较密，基部稍心形，边缘具细齿，有时3-5浅裂。花序顶生，长达27厘米；花萼披针形；花瓣黄色，匙形；雄蕊约40枚，5束；子房3-5室。蒴果具3-5翅。花期6月，果期11月。生海拔300-2000米的林中。产云南南部。印度、老挝、泰国、越南和柬埔寨亦有。

Trees. Leaves broadly obovate-round or suborbicular, both surfaces scattered grayish brown stellate-hairy, more dense abaxially, base slightly cordate, margin serrulate, sometimes 3-5-lobed. Inflorescences terminal, to 27 cm long; sepals lanceolate; petals yellow, spatulate; stamens ca. 40, in 5 bundles; ovary 3-5-loculed. Capsules 3-5-winged. Fl. Jun. Fr. Nov. Forests at 300-2000 m. Distributed in S Yunnan. Also in India, Laos, Thailand, Vietnam and Cambodia.

破布叶 *Microcos paniculata*

## 破布叶
**Microcos paniculata** L.

灌木或小乔木。叶卵形至长卵形，薄革质，起初两面疏具星状毛，后渐无毛。圆锥花序顶生，被星状毛，基部圆形，边缘密具圆齿；花瓣长圆形，腺体长约2毫米；子房无毛。核果倒卵球形至近球形。花期5-7月，果期9-11月。生海拔500-1400米的疏林中。产云南、广西、广东和海南。南亚和东南亚亦有。

Shrubs or small trees. Leaves ovate to oblong, thinly leathery, very sparsely stellate at first and glabrescent both abaxially and adaxially, base rounded, margin finely crenate. Panicles terminal, stellate-hairy; petals oblong, glands ca. 2 mm long; ovary glabrous. Drupes obovoid to subglobose. Fl. May-Jul. Fr. Sep-Nov. Forests at 500-1400 m. Distributed in Yunnan, Guangxi, Guangdong and Hainan. Also in S and SE Asia.

## 海南破布叶
**Microcos chungii** (Merr.) Chun

乔木。叶具柄，近革质，长圆形或披针形，顶端长渐尖，基部圆钝，全缘。花序顶生或腋生；花淡黄色；子房密被长柔毛。核果梨形，被毛。花期5-6月，果期7-8月。生山地疏林或密林中。产海南。

Trees. Leaves petiolate, sub-coriaceous, oblong or lanceolate, apex acuminate, base round or obtuse, margin entire. Inflorescences terminal or axillary; flowers yellowish; ovaries densely villous. Drupes pyriform, pubescent. Fl. May-Jun. Fr. Jul-Aug. Montane forests or thickets. Distributed in Hainan.

## 毛果扁担杆
**Grewia eriocarpa** A. Juss.

灌木或小乔木。叶卵形或卵状长圆形，纸质，下面具柔软灰色星状绒毛，基部斜圆形或截形。聚伞花序1-3，腋生；花萼狭长圆形；花瓣无腺；雌雄蕊柄缺无。核果近球状，具沟槽，被星状毛。花期夏季，果期秋冬季。生山坡灌丛中。产云南、贵州、广西、广东和台湾。南亚和东南亚亦有。

Shrubs or small trees. Leaves ovate or ovate-oblong, papery, softly gray stellate tomentose abaxially, base obliquely rounded or truncate. Cymes 1-3, axillary; sepals narrowly oblong; petals eglandular; androgynophores absent. Drupes subglobose, furrowed, stellate hairy. Fl. summer. Fr. autumn-winter. Bushes on

海南破布叶 *Microcos chungii*

毛果扁担杆 *Grewia eriocarpa*

扁担杆 *Grewia biloba*

Slopes. Distributed in Yunnan, Guizhou, Guangxi, Guangdong and Taiwan. Also in S and SE Asia.

## 扁担杆

**Grewia biloba** G. Don

灌木或小乔木。叶椭圆形或倒卵状椭圆形，有时浅3裂，薄革质，下面疏具星状毛，侧脉3-5对。聚伞花序腋生，多花；苞片钻形；萼片窄长圆形，外面被毛；子房有毛；柱头扩大，盘状，有浅裂。核果红色，有2-4分核。花期5-7月，果期9-12月。生海拔1000米以下的山坡灌丛中。产中国西南、华南、华中、华北和华东。朝鲜半岛亦有。

Shrubs or small trees. Leaves elliptic or obovate-elliptic, sometimes shallowly 3-lobed, thinly leathery, abaxially sparsely stellate hairy, lateral veins 3-5 paired. Cymes axillary, many-flowered; bracts subulate; sepals narrowly oblong, abaxially hairy; ovary hairy; styles inflated, disc, lobed. Drupes red, with 2-4 pyrenes. Fl. May-Jul. Fr. Sep-Dec. Bushes on slopes below 1000 m. Distributed in SW, S, C, N and E China. Also in Korean Peninsula.

## 小花扁担杆

**Grewia biloba** G. Don var. **parviflora** (Bunge) Hand.-Mazz.

灌木或小乔木。叶卵状圆形或倒卵状椭圆形，有时浅3裂，薄革质，3出脉，背面被锈黄色绒毛。聚伞花序腋生，具多花；苞片钻形；花较小；雌雄蕊柄被毛。核果红色，2-4裂。花期5-7月，果期9-12月。生海拔1000米以下的山坡灌丛中。产中国西南、华南、东南、华北和华东。朝鲜半岛亦有。

Shrubs or small trees. Leaves ovate-orbicular or obovate-elliptic, sometimes shallowly 3-lobed, thinly leathery, ternate, abaxially yellowish-rusty-tomentose. Cymes axillary, many flowered; bracts subulate; flowers smaller; androgynophores hairy. Drupes red, 2-4-lobed. Fl. May-Jul. Fr. Sep-Dec. Bushes on slopes below 1000 m. Distributed in SW, S, SE, N and E China. Also in Korean Peninsula.

小花扁担杆 *Grewia biloba* var. *parviflora*

苘麻叶扁担杆 *Grewia abutilifolia*

## 苘麻叶扁担杆

**Grewia abutilifolia** Vent ex Juss.

灌木或小乔木。叶阔卵形至近圆形，常上部浅裂，纸质，下面密具黄色及褐色粗糙星状绒毛，边缘具细齿。花序多枝，腋生，花序柄长3-6毫米；花萼白色，狭长圆形；子房具长柔毛。核果不明显2或4裂，具绒毛。花期4-8月，果期9-10月。生海拔160-1600米的灌丛、草坡或林中。产中国西南和华南。南亚亦有。

Shrubs or small trees. Leaves broadly ovate to nearly orbicular, usually lobed distally, papery, densely yellow and brown, coarsely stellate tomentose abaxially, margin serrulate. Inflorescences ramoses, axillary, peduncles 3-6 mm long; sepals white, narrowly oblong; ovary villous. Drupes obscurely 2- or 4-lobed, tomentose. Fl. Apr-Aug. Fr. Sep-Oct. Scrubs, slopes on grasslands or forests at 160-1600 m. Distributed in SW and S China. Also in S Asia.

## 黄麻叶扁担杆

**Grewia henryi** Burret

灌木或小乔木。具糙星状毛；叶阔长圆形，薄革质，下面淡绿色或具糙黄绿色星状毛，渐无毛，基部阔楔形；叶柄长7-9毫米。叶腋具1或2聚伞花序，具3-4花；总花梗长1-2.5厘米；花梗长5-11厘米。核果4裂。花果期夏秋季。生山坡灌丛和疏林中。产云南、贵州、广西、广东、福建和江西。

Shrubs or small trees. Petioles 7-9 mm long, coarsely stellate hairy; leaves broadly oblong, thinly leathery, abaxially greenish or coarsely yellow-green stellate, glabrescent, base broadly cuneate. Cymes 1 or 2 per leaf axil, 3-4-flowered; peduncles 1-2.5 cm long; pedicels 5-11 cm long. Drupe 4-lobed. Fl. and fr. summer-autumn. Bushes or sparse forests on slopes. Distributed in Yunnan, Guizhou, Guangxi, Guangdong, Fujian and Jiangxi.

黄麻叶扁担杆 *Grewia henryi*

## 毛刺蒴麻

**Triumfetta cana** Blume

木质化草本或亚灌木。叶卵形或卵状披针形，长4-8厘米，宽约2.4厘米，下面密具星状绒毛，先端渐尖。每叶腋具聚伞花序1至数个；花蕾密被紧贴灰白绒毛。蒴果圆球形，4瓣开裂；针刺细弱，具柔毛，弯曲，先端直。花期夏秋季，果期11-12月。生海拔120-1700米的灌丛或废弃地。产中国西南和华南。南亚和东南亚亦有。

Herbs woody or subshrubs. Leaves ovate or ovate-lanceolate, 4-8 cm long, ca. 2.4 cm wide, abaxially densely stellate tomentose, apex acuminate. Cymes 1 to several per axil; flowers buds densely appressed gray-white puberulent. Capsules globose, dehiscent into 4 valves; spines slender, puberulent, curved, tips straight. Fl. summer-autumn. Fr. Nov-Dec. Thickets or wastelands

毛刺蒴麻 *Triumfetta cana*

单毛刺蒴麻 *Triumfetta annua*

indehiscent. Fl. winter-spring. Sandy coasts or coastal wastelands. Distributed in Guangdong and Hainan. Also in Thailand, Vietnam, Cambodia and Malaysia.

at 120-1700 m. Distributed in SW and S China. Also in S and SE Asia.

## 单毛刺蒴麻

**Triumfetta annua** L.

一年生草本或亚灌木。叶卵形或卵状披针形，5-13 × 3-7厘米，纸质，两面具单柔毛；先端尾尖或渐尖。蒴果扁球形，具刺，开裂，3或4瓣裂。针刺无毛或仅基部有毛，先端具钩。花期5-10月，果期10-12月。生海拔450-2100米的开阔地、路边、草地或废弃地。产中国西南、东南和华中。热带亚洲至非洲亦有。

Annual herbs, or subshrubs. Leaves ovate or ovate-lanceolate, 5-13 × 3-7 cm, papery, both surfaces simple-pilose, apex caudate or acuminate. Capsules depressed-globose, spiny, dehiscent, 3- or 4-valved. Spines glabrous or hairy at base, tips hooked. Fl. May-Oct. Fr. Oct-Dec. Open areas, roadsides, grasslands or wastelands at 450-2100 m. Distributed in SW, SE and C China. Also in tropical Asia to Africa.

## 粗齿刺蒴麻

**Triumfetta grandidens** Hance

木质草本，多分枝。叶柄长5-10毫米，多毛；下部叶菱形，3-5裂；上部叶长圆形，1-2.5 × 0.7-1.5厘米，边缘具锯齿。聚伞花序长10-20毫米；总花梗长5-7毫米；花梗长2-3毫米。蒴果球形，具刺，不开裂。花期冬季至春季。产海岸沙地或海岸湿地。产广东和海南。泰国、越南、柬埔寨和马来西亚亦有。

Herbs woody, many branched. Petioles 5-10 mm long, hairy; lower Leaves rhomboid, 3-5-lobed; upper ones oblong, 1-2.5 × 0.7-1.5 cm, margin serrate. Cymes 10-20 mm long; peduncles 5-7 mm long; pedicels 2-3 mm long. Capsules globose, spiny,

## 刺蒴麻

**Triumfetta rhomboidea** Jacq.

亚灌木或草本。下部叶阔卵状圆形、菱形或阔卵形，3浅裂，3-9.5 × 2-8.5厘米，下面具星状柔毛。聚伞花序；花瓣黄色；雄蕊10；子房有刺毛。果实球形，不开裂；刺具灰黄色柔毛，先端具弯钩。花期夏秋季。生海拔100-1500米的田间、荒地或开阔的山丘。产云南、广西、广东、台湾和福建。热带亚洲和非洲亦有。

Subshrubs or herbs. Lower leaves broadly ovate-orbicular, rhomboid, or broadly ovate, 3-lobed, 3-9.5 × 2-8.5 cm, abaxially stellate pilose. Cymes; petals yellow; stamens 10; ovary spinecent-hairy. Fruits globose, indehiscent; spines gray-yellow puberulent, tips hooked. Fl. summer to autumn. Fields, wastelands or open hills at 100-1500 m. Distributed in Yunnan, Guangxi, Guangdong, Taiwan and Fujian. Also in tropical Asia and Africa.

粗齿刺蒴麻 *Triumfetta grandidens*

刺蒴麻 *Triumfetta rhomboidea*

# 锦葵科 Malvaceae

柄翅果 *Burretiodendron esquirolii*

## 柄翅果

**Burretiodendron esquirolii** (Lévl.) Rehd.

落叶乔木。叶纸质，具星状柔毛，稍偏斜。聚伞花序具3花；花萼长圆形，上面基部具腺体，腺体长为花萼的1/3；花瓣阔倒卵形；雄蕊约30枚；子房柄长3-4毫米。蒴果椭圆体形，基部圆形。花期6-7月，果期8-9月。生海拔100-700米的常绿阔叶林中。产云南、贵州和广西。缅甸和泰国亦有。

Deciduous trees. Leaves papery, stellate puberulent, slightly oblique. Cymes 3-flowered; sepals oblong, glandular at base adaxially, glands as long as 1/3 sepals; petals broadly obovate; stamens ca. 30; ovary petioles 3-4 mm long. Capsules ellipsoid, base round. Fl. Jun-Jul. Fr. Aug-Sep. Evergreen forests at 100-700 m. Distributed in Yunnan, Guizhou and Guangxi. Also in Myanmar and Thailand.

## 海南椴

**Diplodiscus trichosperma** (Merr.) Y. Tang, M. G. Gilbert et Dorr

乔木。树皮灰白色，老枝深褐色，无毛；小枝密具灰褐色绒毛。叶6-14 × 4-10厘米，下面密具平伏灰黄色星状柔毛，基部近心形或截形；叶柄长2.5-5.5厘米，多毛。圆锥花序长达26厘米；总花梗密具灰黄色星状柔毛；花瓣黄色或白色。蒴果长2-2.5厘米。花期秋季，果期冬季。生海拔200-300米的开阔林中。产广西和海南。

Trees. Bark gray-white, old branches dark brown, glabrous; branchlets densely gray-brown tomentose. Leaves 6-14 × 4-10 cm, abaxially densely appressed gray-yellow stellate puberulent, base subcordate or truncate; petioles 2.5-5.5 cm long, hairy. Panicles to 26 cm long; peduncles densely gray-yellow stellate puberulent; petals yellow or white. Capsules 2-2.5 cm long. Fl. autumn. Fr. winter. Open forests at 200-300 m. Distributed in Guangxi and Hainan.

## 锦葵

**Malva cathayensis** M. G. Gilbert, Y. Tang et Dorr

二年或多年生草本，具硬毛。叶心形或肾形。花3-11腋部簇生；小苞片3，长圆形；花萼杯形，阔三角形；花冠紫红色或白色，直径3-5厘米。蒴果扁球形；分果爿9-11。花期5-10月。中国各地区常见栽培。原产印度。

Herbs biennial or perennial, strigose. Leaves cordate or reniform. Flowers 3-11-fascicled, axillary; epicalyx lobes 3, oblong; calyx cup-shaped, broadly triangular; corolla purplish red or white, 3-5 cm diam. Capsules flat globose; mericarps 9-11. Fl. May-Oct. Cultivated throughout China. Native to India.

海南椴 *Diplodiscus trichosperma*

锦葵 *Malva cathayensis*

圆叶锦葵 *Malva pusilla*

## 圆叶锦葵
**Malva pusilla** Sm.

多年生草本，高20-50厘米，分枝多而常匍生，被粗毛。叶肾形，长1-3厘米，边缘具细圆齿，上面疏被长柔毛，下面疏被星状柔毛。花梗不等长，长2-5厘米；小苞片披针形，被星状柔毛；花白色至浅粉红色，直径10-12毫米。果实扁球形，直径5-6毫米，分果爿12-15，不为网状，被短柔毛。花期夏季。生草坡和荒野。几遍全中国。欧洲和亚洲亦有。

Perennial herbs, 20-50 cm tall, much branched and usually procumbent, scabrous. Leaves reniform, 1-3 cm long, margin minutely denticulate, adaxially sparsely velutinous, abaxially sparsely stellate puberulent. Pedicels unequal, 2-5 cm long; bractlets lanceolate, stellate puberulent; corolla white to pinkish, 10-12 mm diam. Fruits flat globose, 5-6 mm diam, mericarps 12-15, not reticulate, puberulent. Fl. summer. Grassy slopes and open areas. Distributed almost throughout China. Also in Europe and Asia.

## 野葵
**Malva verticillata** L.

二年生草本，高达1米。茎疏具星状丝毛。叶肾形或近圆形，5-7浅裂。花3至数朵簇生，腋生；小苞片裂片线状披针形；花萼杯形；花冠白色至浅红色。裂果扁球形；分果爿10-12。花期3-11月。生丘陵或平原。产中国各省区。印度、缅甸、朝鲜半岛、埃及、埃塞俄比亚和欧洲亦有。

Herbs biennial, up to 1 m tall. Stems sparsely stellate velutinous. Leaves reniform or suborbicular, 5-7-lobed. Flowers 3- to many-fascicled, axillary; epicalyx lobes filiform-lanceolate; calyx cup-shaped; corolla whitish to reddish. Schizocarps flat-globose; mericarps 10-12. Fl. Mar-Nov. Hills or plains. Distributed throughout China. Also in India, Myanmar, Korean Peninsula, Egypt, Ethiopia and Europe.

## 蜀葵
**Alcea rosea** L.

二年生草本。枝被刺毛。叶近圆形，掌状5-7浅裂或具波状棱角。花单生或簇生，聚集为一顶生的穗状花序；小苞片杯形，常6或7裂，裂片卵状披针形；花大，直径5-10厘米，红色、紫色、白色、粉红色、黄色或黑紫色。裂果碟形，分果爿多数。花期2-8月。产中国西南，全国各地栽培。世界各国广栽培。

Biennial herbs. Branches setose. Leaves nearly orbicular, palmately 5-7-lobed or crenate-angled. Flowers solitary or fascicled, aggregated into a terminal, spike-like inflorescence; epicalyx cup-shaped, usually 6- or 7-lobed, lobes ovate-lanceolate; flowers large, 6-10 cm diam, red, purple, white, pink, yellow or black-purple. Schizocarps disk-shaped, mericarps many. Fl. Feb-Aug. Distributed in SW China, cultivated throughout China. Cultivated worldwide.

野葵 *Malva verticillata*

蜀葵 *Alcea rosea*

## 赛葵

**Malvastrum coromandelianum** (L.) Garcke

亚灌木，可达1(-1.5)米高，大部分具柔毛和平伏星状柔毛。叶卵状披针形或卵形，下面疏被柔毛和星状柔毛。花单生于叶腋；小苞片线形；花萼浅杯状；花冠杏黄色。裂果分果爿8-12(-14)。花期夏秋。生海拔500米以下的干热草坡或路边。产云南、广西、广东、台湾和福建。泛热带分布，可能原产美洲。

Subshrubs, up to 1(-1.5) m tall, most parts pilose and appressed stellate pubescent. Leaves ovate-lanceolate or ovate, abaxially pilose and stellate pilose. Flowers solitary, axillary; epicalyx lobes filiform; calyx shallowly cup-shaped; corolla apricot-yellow. Schizocarps mericarps 8-12(-14). Fl. summer and autumn. Dry and hot grassy slopes or roadsides below 500 m. Distributed in Yunnan, Guangxi, Guangdong, Taiwan and Fujian. Pantropical, probably native to America.

## 黄花稔

**Sida acuta** Burm. f.

亚灌木或直立草本。叶线状披针形，基部圆形或钝形。花单生或成对，腋生，有时聚集于茎顶；花萼浅杯状，合生至1/2；花冠黄色，稀白色或橘黄色。裂果近球形；分果爿(4-)6(-9)。花期冬春季。生海拔200-1400米的灌丛、山坡或路旁。产中国西南和华南。南亚亦有。

Subshrubs or herbs erect. Leaves linear-lanceolate, base round or obtuse. Flowers solitary or paired, axillary, sometimes congested at stem apex; calyx shallowly cup-shaped, connate in basal 1/2; corolla yellow, less often white or yellow-orange. Schizocarps nearly globose; mericarps (4-)6(-9). Fl. winter to next spring. Thickets, slopes or roadsides at 200-1400 m. Distributed in SW and S China. Also in S Asia.

榛叶黄花稔 *Sida subcordata*

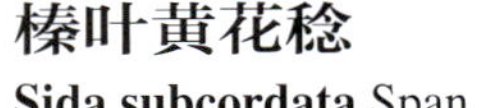

## 榛叶黄花稔

**Sida subcordata** Span.

直立亚灌木。叶卵形至长圆形，基部圆形，边缘稍具圆齿，两面疏被星状毛，或下面无毛。花序多近顶生，形成伞房状顶生成簇，常生于退化枝条腋处；花冠黄色，花瓣5。裂果近球形；分果爿8或9，顶端具2长芒。花期冬春季。生海拔500-1400米的山谷疏林、草地或路旁。产云南、广西、广东和海南。亚洲热带地区亦有。

Subshrubs erect. Leaves ovate to oblong, base rounded, margin minutely crenate, both surfaces densely stellate-hairy, sometimes glabrescent abaxially. Flowers mostly subterminal, in umbel-like terminal clusters, often on reduced shoots axillary; corolla yellow; petals 5. Schizocarps nearly globose; mericarps 8 or 9, with 2 long awns at apex. Fl. winter to next spring. Open forests in valleys, grasslands or roadsides at 500-1400 m. Distributed in Yunnan, Guangxi, Guangdong and Hainan. Also in tropical Asia.

赛葵 *Malvastrum coromandelianum*

黄花稔 *Sida acuta*

## 心叶黄花稔

**Sida cordifolia** L.

直立亚灌木。小枝、托叶、叶柄和叶密具星状糙毛。叶卵形，基部微心形或圆形。花单生或簇生于叶腋或枝端；花萼杯形；花冠黄色。裂果分果爿10，具纵沟，先端具2芒。花期全年。生山坡灌丛中或路边草丛中。产中国西南和华南。泛热带分布。

Erect subshrubs. Branchlets, stipules, petioles, and leaves densely stellate strigose. Leaves ovate, base minutely cordate or rounded. Flowers solitary or clustered in leaf axils or branch tops; calyx cup-shaped; corolla yellow. Schizocarps mericarps 10, with vertical grooves, apex 2-awned.

心叶黄花稔 *Sida cordifolia*

Fl. all year. Bushes on mountain slopes or grasses by roads. Distributed in SW and S China. Also in antropical.

## 滇西苘麻

**Abutilon gebauerianum** Hand.-Mazz.

灌木。小枝具灰色星状绒毛。叶卵心形，具不规则圆齿，两面密被毡毛。单花腋生，较大；花萼杯形，裂片5，披针形；花橘黄色；子房8-10室。裂果分果爿8-10，具绒毛，先端凸尖，外果皮革质。花期1-2月。生海拔700-2500米的干热河谷灌丛中。产云南西南部。

Shrubs. Branchlets gray stellate tomentose. Leaves ovate-cordate, margin irregularly crenate, both surfaces densely pannose. Flowers solitary, axillary, large; sepals cup-shaped, lobes 5, lanceolate; corolla orange; ovary 8-10-loculed. Schizocarps mericarps 8-10, tomentose, apex pointed, pericarps leathery. Fl. Jan-Feb. Thickets in dry and hot valleys at 700-2500 m. Distributed in SW Yunnan.

## 金铃花

**Abutilon pictum** (Gillies ex Hooker) Walp.

常绿灌木。叶掌状，3-5深裂，直径5-8厘米。花单生于叶腋；花橙黄色，直径约3厘米；雄蕊柱长约5.5厘米；花药黄色，多数，集生于雄蕊柱的顶端；花柱10裂。花期5-7月。中国广泛栽培。原产南美洲；世界各国广泛栽培。

Evergreen shrubs. Leaves palmately 3-5-parted, 5-8 cm diam. Flowers solitary, axillary; flowers orange-yellow, ca. 3 cm diam; phalanx ca. 5.5 cm long; anther yellow, numerous, aggregated at top of phalanx; styles 10-lobed. Fl. May- Jul. Commonly cultivated in China. Native to South America; cultivated worldwide.

## 苘麻

**Abutilon theophrasti** Medik.

一年生亚灌木状草本。叶圆心形，边缘具微圆齿。花单生于叶腋；花萼杯形，裂片5，卵形；花冠纯黄色。蒴果半球形，分果爿15-20，先端具2芒，芒开展，长3-5毫米。花期7-8月。生路边、荒地或田野中。产中国大部分地区。南亚、东南亚、澳大利亚、非洲、欧洲和北美洲亦有。

Herbs subshrublike, annual. Leaves orbicular-cordate, margin minutely crenate. Flowers solitary, axillary; calyx cup-shaped, lobes 5, ovate; corolla uniformly yellow. Capsules semiglobose, mericarps 15-20, apex 2-awned, awns spreading, 3-5 mm long. Fl. Jul-Aug. Roadsides, wastelands or fields. Distributed in most parts of China. Also in S and SE Asia, Australia, Africa, Europe and North America.

滇西苘麻 *Abutilon gebauerianum*

金铃花 *Abutilon pictum*

苘麻 *Abutilon theophrasti*

磨盘草 *Abutilon indicum*

## 磨盘草
**Abutilon indicum** (L.) Sweet

一年生或多年生近灌木状草本，全株具灰色柔毛。叶卵状圆形或近圆形，边缘具不规则锯齿。花单生叶腋；花萼碟形，裂片5，阔卵形；花冠纯黄色；子房15-20室。果实黑色，平顶，分果爿15-20，稍具芒。花期7-10月。生海拔800米以下的平原、海边、沙地、旷野、山坡、河边和路边。产中国西南和华南。南亚和东南亚亦有。

Herbs subshrublike, annual or perennial, entire plant gray puberulent. Leaves ovate-orbicular or nearly orbicular, margin irregularly serrate. Flowers solitary, axillary; calyx disk-shaped, lobes 5, broadly ovate; corolla uniformly yellow; ovary 15-20-loculed. Fruits black, flat topped, mericarps 15-20, slightly awned. Fl. Jul-Oct. Plains, seashores, sandy places, open lands, slopes, river banks or roadsides below 800 m. Distributed in SW and S China. Also in S and SE Asia.

翅果麻 *Kydia calycina*

## 翅果麻
**Kydia calycina** Roxb.

乔木。小枝圆柱形，密具浅褐色星状柔毛。叶近圆形，掌状3-5浅裂，下面被星状绵毛；圆锥花序；小苞片4(或6)，长圆形；花萼浅杯状，1/2合生；花瓣浅红色，倒心形，先端具腺状流苏。蒴果球形。花期9-11月。生海拔500-1600米的山谷疏林中。产云南南部。南亚亦有。

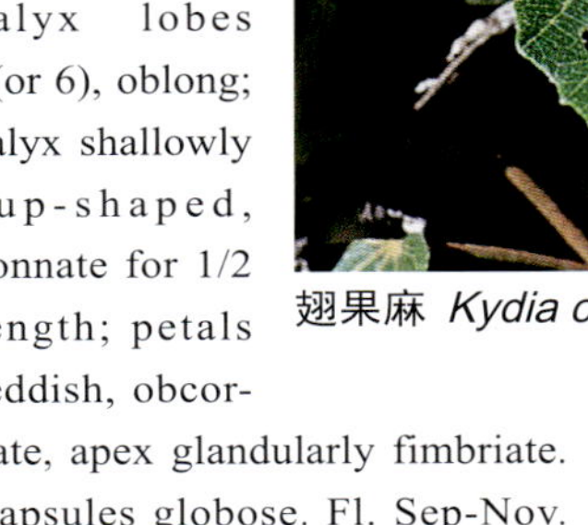

Trees. Branchlets terete, densely brownish stellate pubescent. Leaves suborbicular, palmately 3-5-lobed, abaxially stellate-lanate. Inflorescences paniculate; epicalyx lobes 4(or 6), oblong; calyx shallowly cup-shaped, connate for 1/2 length; petals reddish, obcordate, apex glandularly fimbriate. Capsules globose. Fl. Sep-Nov. Open forests in valleys at 500-1600 m. Distributed in S Yunnan. Also in S Asia.

## 地桃花
**Urena lobata** L.

直立亚灌木状草本。近轴叶轮廓近圆形，下部叶卵形且稍3浅裂，下面具绒毛和粗毛。小苞片裂片线形，稍长于花萼，密具黄色绵毛；花冠淡红色，直径约15毫米。果实扁球形，分果爿具星状柔毛和钩状刺。花期7-10月。生海拔500-2200米的干热空旷地、荒地、路边或疏林下。产中国西南、华南、东南、华中和华东。南亚、东南

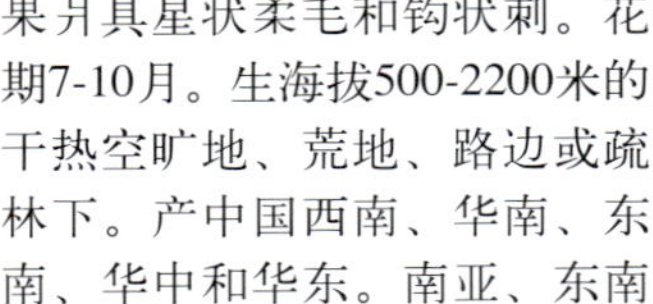

地桃花 *Urena lobata*

粗叶地桃花
*Urena lobata* var. *glauca*

波叶梵天花 *Urena repanda*

亚和日本亦有；泛热带分布。

Subshrublike herbs, erect. Leaves on proximal part of stem suborbicular in outline, lower leaves ovate and slightly 3-lobed, abaxially tomentose and hirsute. Epicalyx lobes filiform, slightly longer than sepals, densely yellow woolly; corolla reddish, ca. 15 mm diam. Fruits flattened globose; mericarps stellate puberulent and spiny with hooked spines. Fl. Jul-Oct. Dry and hot fields, wastelands, roadsides or open forests at 500-2200 m. Distributed in SW, S, SE, C and E China. Also in S and SE Asia, and Japan; pantropical.

## 粗叶地桃花

**Urena lobata** L. var. **glauca** (Blume) Borss. Waalk.

本变种与地桃花的主要区别为其叶密具糙柔毛和绵毛；下部叶稍宽且稀裂，基部近心形，先端常3浅裂，下部叶卵形或近圆形，边缘具锯齿。花萼裂片线形，稍长于花萼，密具黄色绵毛；花瓣长1-1.3厘米。花期7-10月。生海拔500-1500米的草坡、灌丛或路边。产云南、四川、贵州、广东和福建。印度、孟加拉国、缅甸和印度尼西亚亦有。

This variety differs from *Urena lobata* var. *lobata* in its leaves densely rough puberulent and woolly; leaves on proximal part of stem wider and rarely lobed, base subcordate, apex usually 3-lobed, those on distal part of stem ovate or suborbicular, margin serrate. Epicalyx lobes filiform, slightly longer than sepals, densely yellow woolly; petals 1-1.3 cm long. Fl. Jul-Oct. Grassy slopes, thickets or roadsides at 500-1500 m. Distributed in Yunnan, Sichuan, Guizhou, Guangdong and Fujian. Also in India, Bangladesh, Myanmar and Indonesia.

## 梵天花

**Urena procumbens** L.

亚灌木。小枝被星状绒毛。茎下部叶掌状3-5深裂至1/2；花单生或近簇生；小苞片约1/3合生；萼被星状毛；花瓣粉色或白色。果实球形，具刺和长硬毛，刺端有倒钩。花期6-9月。生海拔500米的山坡。产华南和东南。

Subshrubs. Branchlets stellate-tomentose. Leaves on proximal part of stem palmately 3-5-lobed to 1/2 blade; flowers solitary, or nearly clustered; epicalyx connate for ca. 1/3 length; calyx stellate-hairy; petals pink or white. Fruits globose, spinecent and hispid, spines at apex hooked. Fl. Jun-Sep. Mountain slopes at 500 m. Distributed in S and SE China.

## 波叶梵天花

**Urena repanda** Roxb. ex Sm.

多年生草本。小枝圆柱形，密具星状柔毛。叶卵形，茎下部者常3浅裂，边缘具锯齿。花集生于小枝顶端，近总状花序；小苞片钟形，合生至约1/2；花冠粉红色，花瓣5。果实近球形；分果爿倒三角形。花期8-11月。生海拔300-1600米的山坡灌丛中。产云南、贵州和广西。南亚亦有。

Perennial herbs. Branches terete, densely stellate puberulent. Leaves ovate, on proximal part of stem usually 3-lobed, margin serrate. Flowers aggregated at tops of branchlets, subracemose; epicalyx campanulate, connate for ca. 1/2 length; corolla pink, petals 5. Fruits nearly globose; mericarps obtriangular. Fl. Aug-Nov. Thickets on slopes at 300-1600 m. Distributed in Yunnan, Guizhou and Guangxi. Also in S Asia.

梵天花 *Urena procumbens*

黄蜀葵 *Abelmoschus manihot*

## 黄蜀葵

**Abelmoschus manihot** (L.) Medic.

一年生或多年生草本，大部分均被长单硬毛及微单至少辐柔毛。茎无刺毛。叶掌状5-9裂。花单生，近顶生且形成顶生总状；小苞片裂片4或5，卵状披针形；花冠淡黄色，中央紫色。蒴果卵球状椭圆体形。花期8-10月。生海拔1000-2100米的山谷、草丛中或农田边，或栽培。产中国东南、华南和华中。印度和尼泊尔亦有。

Annual herbs or perennial, most parts long simple-hispid and minutely simple- or few-rayed pubescent. Stems without prickly hairs. Leaves palmately 5-9-lobed. Flowers solitary, subapical and forming terminal raceme; epicalyx lobes 4 or 5, ovate-lanceolate; corolla yellowish with purple center. Capsules ovoid-ellipsoid. Fl. Aug-Oct. Valleys, grassy places or margins of farms at 1000-2100 m, or cultivated. Distributed in SW, S and C China. Also in India and Nepal.

## 刚毛黄蜀葵

**Abelmoschus manihot** (L.) Medic. (Roxb.) Hochr.

此变种与黄蜀葵的主要区别为包括外萼片边缘在内的全株具明显的黄色刺毛。花期6-8月，果期9-11月。生海拔1000-2100米的山谷、草丛中或农田边，或栽培。产中国西南和华南。印度、尼泊尔、泰国和菲律宾亦有。

This variety differs from the typical variety in plants with conspicuous yellow prickly hairs throughout including margins of epicalyx lobes. Fl. Jun-Aug. Fr. Sep-Nov. Valleys, grassy places or margins of farms at 1000-2100 m, or cultivated. Distributed in SW and S China. Also in India, Nepal, Thailand and the Philippines.

## 咖啡黄葵 (秋葵)

**Abelmoschus esculentus** (L.) Moench

一年生草本。植物体大部分疏被刺毛。茎通常中空。叶掌状3-7裂。花单生叶腋；小苞片裂片7-10(-12)，线形；花冠黄色或白色，中央紫色。蒴果圆柱状至塔形。花期5-9月。产中国西南、华南、华中、华北和华东栽培。原产印度。

Annual herbs. most parts with very sparse prickly hairs. Stems often hollow. Leaves palmately 3-7-lobed. Flowers solitary, axillary; epicalyx lobes 7-10(-12), filiform; corolla yellow or white, with purple center. Capsules cylindric to tower-shaped. Fl. May-Sep. Cultivated in SW, S, C, N and E China. Native to India.

刚毛黄蜀葵 *Abelmoschus manihot* var. *pungens*

咖啡黄葵（秋葵） *Abelmoschus esculentus*

黄葵 *Abelmoschus moschatus*

## 黄葵

**Abelmoschus moschatus** Medic.

一年生或多年生草本。大部分具单一的黄色硬毛或糙毛。叶常掌状(3-)5-7裂。花单生叶腋；小苞片裂片6-10，线形至狭长圆形，内折；花萼佛焰苞状；花黄色，中央暗紫色。蒴果长圆形。花期6-10月。生平原、山谷、溪边或山坡灌丛中。产中国西南和华南。印度、老挝、泰国、越南和柬埔寨亦有；热带地区广泛栽培。

Annual herbs or perennial, most parts uniformly yellow hispid/setose. Leaves usually palmately (3-)5-7-lobed. Flowers solitary, axillary; epicalyx lobes 6-10, linear to narrowly oblong, incurved; calyx spatulate; corolla yellow with dark purple center. Capsules oblong. Fl. Jun-Oct. Plains, valleys, by streams or thickets on mountain slopes. Distributed in SW and S China. Also in India, Laos, Thailand, Vietnam and Cambodia; widely cultivated in tropical regions.

## 大叶木槿

**Hibiscus macrophyllus** Roxb. ex Hornem.

乔木。叶近圆形，直径20-36厘米，两面密被星状长柔毛。花序顶生，多花聚伞状；小苞片裂片10-12，线形，基部贴生；花萼钟形；花冠黄色，中央紫色。蒴果长圆形，长2.5-3厘米，密被星状长柔毛。花期3-5月，果期7月。生海拔400-1000米的常绿阔叶林或村庄附近。产云南南部。南亚和东南亚亦有。

Trees. Leaves orbicular, 20-36 cm diam, densely stellate-villose on both surfaces. Inflorescences terminal, multi-flowered cymes; epicalyx lobes 10-12, filiform, connate at base; calyx campanulate; corolla yellow with purple center. Capsules oblong, 2.5-3 cm long, densely stellate-pubescent. Fl. Mar-May. Fr. Jul. Evergreen broad-leaved forests or near villages at 400-1000 m. Distributed in S Yunnan. Also in S and SE Asia.

## 黄槿

**Hibiscus tiliaceus** L.

常绿灌木或乔木。小枝无毛或近无毛。叶近圆形至阔卵形，革质。花序具1至数花，总状聚伞状，顶生或腋生；小苞片7-10，1/3-1/2合生，三角状渐尖；花瓣黄色，外密被黄色星状毛；花柱5。蒴果近球形至卵球形，木质。花期6-8月。生海拔300米以下的海边、溪边或沙土。产广东、海南、台湾和福建。南亚和东南亚亦有。

Evergreen shrubs or trees. Branchlets glabrous or nearly so. Leaves nearly orbicular to broadly ovate, leathery. Inflorescences a 1- to few-flowered racemelike cyme, terminal or axillary; epicalyx lobes 7-10, joined for 1/3-1/2 of length, triangular-acuminate; flowers yellow, abaxially densely yellow-stellate-hairy; styles 5. Capsules subglobose to ovoid, woody. Fl. Jun-Aug. Seashores, along streams or sandy soil below 300 m. Distributed in Guangdong, Hainan, Taiwan and Fujian. Also in S and SE Asia.

大叶木槿 *Hibiscus macrophyllus*

黄槿 *Hibiscus tiliaceus*

朱槿（扶桑） *Hibiscus rosa-sinensis*

## 朱槿（扶桑）
**Hibiscus rosa-sinensis** L.

常绿灌木。叶阔或狭卵形，不裂，边缘具牙齿或浅裂。花单生，腋生于上部枝，常下垂，单瓣或重瓣；小苞片裂片6-7，线形，基部贴生；花萼钟状；花冠漏斗状，玫瑰红、淡红或淡黄色；花瓣不开裂或有微缺刻。蒴果卵球形，无毛。花果期全年。栽培于中国西南和华南。

Evergreen shrubs. Leaves broadly or narrowly ovate, not lobed, margin dentate or lobed. Flowers solitary, axillary on upper branches, usually pendulous, simple or double; epicalyx lobes 6-7, filiform, connate at base; calyx campanulate; corolla funnel form, rose red, pale red or pale yellow; petals not divided or nicked. Capsules ovoid, glabrous. Fl. and fr. all year. Cultivated in SW and S China.

## 吊灯扶桑
**Hibiscus schizopetalus** (Dyer ex Mast.) Hook. f.

常绿灌木。叶椭圆形或长圆形，边缘缺刻状齿，两面无毛。花单生，腋生于上部小枝腋处，下垂；小苞片5，披针形，具缘毛；花萼管状；花瓣红色，5，深羽状裂，反折。蒴果圆柱形，长约4厘米。花期全年。栽培于中国西南和华南。原产东非，现热带地区普遍栽培。

Evergreen shrubs. Leaves elliptic or oblong, margin incised, glabrous on both surfaces. Flowers solitary, axillary on upper branchlets, pendulous; epicalyx lobes 5, lanceolate, ciliate; calyx tubular; petals red, 5, deeply pinnatifid, reflexed. Capsules terete, ca. 4 cm long. Fl. all year. Cultivated in SW and S China. Native to E Africa, now cultivated throughout tropical regions.

## 美丽芙蓉
**Hibiscus indicus** (Burm. f.) Hochr.

落叶灌木。叶心形，长8-12厘米，两面密被星状短柔毛。花单生于上部小枝腋处；花粉红至白色，直径约10厘米；外萼片4或5，卵形，基部合生。蒴果近球形，被硬毛；分果爿5-6。花期7-12月。生海拔700-2000米的山谷灌丛或珊瑚灰岩中。产中国西南和华南。南亚和东南亚亦有。

Deciduous shrubs. Leaves cordate, 8-12 cm long, densely stellate-pubescent on both surfaces. Flowers solitary, axillary on upper branchlets; flowers pink to white, ca. 10 cm diam; epicalyx lobes 4 or 5, ovate, connate at base. Capsules subglobose, hirsute, mericarps 5-6. Fl. Jul-Dec. Thickets in valleys or coral limestone at 700-2000 m. Distributed in SW and S China. Also in S and SE Asia.

吊灯扶桑 *Hibiscus schizopetalus*

美丽芙蓉 *Hibiscus indicus*

木芙蓉 *Hibiscus mutabilis*

## 木芙蓉
**Hibiscus mutabilis** L.

灌木或小乔木。叶5-7裂，纸质。花单生，腋生于上部枝；小苞片裂片8，线形，基部贴生；花初开时白色或淡红色，后变成深红色。蒴果扁球形，具浅黄色硬毛和绵毛；分果爿5。花期8-10月。生溪边灌丛。原产云南、广东、台湾、福建和湖南；中国大部分地区广泛栽培。东南亚和日本亦有。

Shrubs or small trees. Leaves 5-7-lobed, papery. Flowers solitary, axillary on upper branches; epicalyx lobes 8, filiform, connate at base; flowers white or pale red when young, later becoming deeply red. Capsules flattened globose, yellowish hispid and woolly; mericarps 5. Fl. Aug-Oct. Thickets along streams. Native to Yunnan, Guangdong, Taiwan, Fujian and Hunan; commonly planted in most parts of China. Also in SE Asia and Japan.

## 木槿
**Hibiscus syriacus** L.

落叶灌木。小枝具黄色星状柔毛。叶菱形至三角状卵形，三裂或不裂，基部楔形。小苞片6-8；花冠蓝紫色、紫罗兰色、白色、粉色或淡红色，有时中央色深，钟形，有时重瓣；花瓣5，倒卵形。蒴果卵球形。花期7-10月。生山腰、溪边、路边，海拔1200米以下亦广泛栽培。产中国大部分地区，东北和西北地区除外。世界广泛栽培。

Deciduous shrubs. Branchlets yellow stellate puberulent. Leaves rhomboid to triangular-ovate, variously 3-lobed or entire, base cuneate. Bracteoles 6-8; corolla blue-purple, violet, white, pink, or reddish, sometimes with darker center, campanulate, sometimes double; petals 5, obovate. Capsules ovoid. Fl. Jul-Oct. Hillsides, along streams, roadsides, also extensively cultivated below 1200 m. Distributed in most parts of China, except NE and NW parts. Cultivated worldwide.

## 刺芙蓉
**Hibiscus surattensis** L.

一年生草本，近灌木状，常匍匐，有时攀援，大部分疏具长柔毛和刺，刺反向弯曲。叶掌状，3-5深裂；托叶耳状。花单生叶腋；小苞片8-10，线状披针形至线形；花萼果期变为淡红色，浅杯状；花冠浅黄色，中央深红色。蒴果卵球形。花期9月至翌年3月。生海拔300-1200米的山谷、溪边、山坡或林缘。产云南、海南和香港。南亚、东南亚、大洋洲和热带非洲亦有。

Annual herbs, subshrublike, usually procumbent, sometimes scandent, most parts sparsely villous and aculeate, prickles retrorsely curved. Leaves palmately 3-5-parted; stipules auriculate. Flowers solitary, axillary; epicalyx lobes 8-10, linear-lanceolate to filiform; calyx turning reddish in fruit, shallowly cup-shaped; corolla pale yellow with dark red center. Capsules ovoid. Fl. Sep to next Mar. Valleys, streamsides, slopes or forest edges at 300-1200 m. Distributed in Yunnan, Hainan and Hong Kong. Also in S and SE Asia, Oceania and tropical Africa.

木槿 *Hibiscus syriacus*

刺芙蓉 *Hibiscus surattensis*

野西瓜苗 *Hibiscus trionum*

## 野西瓜苗
**Hibiscus trionum** L.

一年生平卧或直立草本。茎细弱，具白色星状毛。叶二型，下部的圆形不裂，上部的3-5深裂。花单生叶腋；小苞片12，线形，基部合生；花萼钟形，膨大，膜质；花冠极浅黄色，中央紫色。蒴果长圆状球形。花期7-10月。生田间。中国各地均有。中亚和热带地区亦有。原产非洲中部。

Annual herbs, prostrate or erect. Stems slender, white stellate hirsute. Leaves dimorphic, lower ones orbicular, not lobed, upper ones 3-5-parted. Flowers solitary, axillary; epicalyx lobes 12, filiform, connate at base; calyx campanulate, swollen, membranous; corolla very pale yellow with purple center. Capsules oblong-globose. Fl. Jul-Oct. Weeds in fields. Distributed throughout China. Also in C Asia and tropical regions. Native to C Africa.

## 玫瑰茄 (洛神花)
**Hibiscus sabdariffa** L.

一年生草本。叶二型，下部的卵形，不裂，上部的掌状3深裂。花单生叶腋；小苞片裂片8-12，红色，披针形，基部合生，近顶端具刺状附属物；花瓣黄色，基部深红色，直径6-7厘米。蒴果卵球形。花期夏秋间。栽培于云南、广东、海南、台湾和福建。可能起源自非洲，现热带地区普遍栽培。

Annual herbs. Leaves dimorphic, lower ones ovate, not lobed, upper ones palmately 3-parted. Flowers solitary, axillary; epicalyx lobes 8-12, red, lanceolate, connate at base, with spiny appendix near apex; petals yellow, base deeply red, 6-7 cm diam. Capsules ovoid. Fl. summer to autumn. Cultivated in Yunnan, Guangdong, Hainan, Taiwan and Fujian. Probably originating in Africa, now cultivated throughout tropical regions.

## 桐棉
**Thespesia populnea** (L.) Solander ex Corrêa

常绿乔木或灌木。小枝具微小褐色盾状鳞片，有时较密。叶

桐棉 *Thespesia populnea*

玫瑰茄(洛神花) *Hibiscus sabdariffa*

陆地棉 *Gossypium hirsutum*

卵状心形至三角形，下面具鳞片，基部截形至心形，全缘。单花腋生；花冠黄色，钟形。蒴果球形至梨形。种子具褐色毛或光滑。花期全年。生海平面附近的海岸或开阔地。产广东、海南和台湾。印度、泰国、越南、柬埔寨、菲律宾和日本亦有。

Evergreen trees or shrubs. Branchlets with minute brown peltate scales, sometimes dense. Leaves ovate-cordate to triangular, abaxially with scales, base truncate to cordate, margin entire. Flowers solitary, axillary; corolla yellow, campanulate. Capsules globose to pyriform. Seeds brown hairy or glabrous. Fl. year-round. Sea coasts or open situations at near sea level. Distributed in Guangdong, Hainan and Taiwan. Also in India, Thailand, Vietnam, Cambodia, the Philippines and Japan.

## 陆地棉

**Gossypium hirsutum** L.

一年生草本。叶阔卵形，3(-5)裂。花单生于叶腋，白色或淡黄色，后变淡红色或紫色；小苞片裂片3，离生。蒴果卵球形，具喙。种子分离，卵圆形，具白色长绵毛和灰白色不易剥离的短柔毛。花期夏秋季。中国各地棉区广泛栽培。原产中美洲。

Annual herbs. Leaves broadly ovate, 3(-5)-lobed. Flowers solitary, axillary, white or pale yellow, later pale red or purple; epicalyx lobes 3, free. Capsules ovoid, apex beaked. Seeds free, ovoid, with white long cottony and gray-white short not peel off cottony hairs. Fl. summer-autumn. Widely cultivated in cotton-growing regions in China. Native to Central America.

## 大萼葵

**Cenocentrum tonkinense** Gagnep.

落叶灌木。叶掌状脉5-9，裂片阔三角形。花单生；小苞片裂片卵形，边缘具白色刺毛；花萼膨大；花冠黄色，中央紫色，直径约10厘米。蒴果近球形，10瓣裂。花期9-11月。生海拔700-1600米的山谷、林缘或草地上。产云南南部和贵州西南部。老挝、泰国和越南亦有。

Deciduous shrubs. Leaves palmate veins 5-9, lobes broadly triangular. Flowers solitary; epicalyx lobes ovate, margin white spiny hairy; calyx swollen; corolla yellow, purple in center, ca. 10 cm diam. Capsules nearly globose, 10-valved. Fl. Sep-Nov. Valleys, forest edges or grasslands at 700-1600 m. Distributed in S Yunnan and SW Guizhou. Also in Laos, Thailand and Vietnam.

大萼葵 *Cenocentrum tonkinense*

# 木棉科 Bombacaceae

## 木棉 (攀枝花)
**Bombax ceiba** L.

乔木。树干具板根，幼树常多刺。小叶5-7，长圆形至长圆状披针形。花单生枝顶叶腋，直径约10厘米，肉质；花萼杯状，长2-3(-4.5)厘米，内面被黄色短绢毛；花瓣红色或橙红色。蒴果长10-15毫米，密具灰白色长绒毛及星状柔毛。花期3-4月，果期5-7月。生海拔1700米以下的干热河谷或稀树草原。产中国西南和华南。亚洲热带地区和澳大利亚亦有。

Trees. Trunk buttressed, usually very spiny on young trees. Leaflets 5-7, oblong to oblong-lanceolate. Flowers solitary, terminal, ca. 10 cm diam, fleshy; calyx cup-shaped, 2-3(-4.5) cm long, yellow short-sericeous inside; petals red or orange-red. Capsules 10-15 mm long, densely gray-white villous and stellate puberulent. Fl. Mar-Apr. Fr. May-Jul. Hot and dry valleys or savanna below 1700 m. Distributed in SW and S China. Also in tropical Asia and Australia.

木棉（攀枝花） *Bombax ceiba*

## 马拉巴栗
**Pachira glabra** Pasq.

小乔木。树皮光滑，绿色。掌状复叶互生；小叶5-11，长圆形至倒卵状长圆形。花单生近枝顶叶腋；花萼杯状，近革质；花瓣淡黄绿色，狭披针形至线形，上半部反卷；雄蕊多数，花丝下部黄色，向上变红色。蒴果近梨形，长9-10厘米。花果期5-11月。福建、广东、广西、海南和云南栽培。原产中美洲。

Small trees. Bark smooth, green. Palmately compound leaves alternate; leaflets 5-11, oblong to obovate-oblong. Flowers solitary, subterminal; calyx cuplike, subcoriaceous; petals yellowish-green, narrowly lanceolate to linear, upper part revolute; stamens numerous, lower part of filaments yellow, upper part tinted with red. Capsules nearly pyriform, 9-10 cm long. Fl. and

马拉巴栗 *Pachira glabra*

fr. May-Nov. Cultivated in Fujian, Guangdong, Guangxi, Hainan and Yunnan. Native to Central America.

吉贝 *Ceiba pentandra*

## 吉贝
**Ceiba pentandra** (L.) Gaertn.

大乔木。板根小或无，树干常疏具刺。主枝轮生。小叶5-9，长圆形至披针形。花近顶生，单个或数朵组成密伞花序；花瓣粉红色或白色。蒴果长圆形，近顶端渐狭，内果皮革质，光滑。花期3-4月。云南、广西和广东栽培。原产热带美洲。

Large trees. Buttresses small or absent, trunk often sparsely spiny. Main branches verticillate; leaflets 5-9, oblong to lanceolate. Flowers subterminal, solitary or several clustered to fascicles; petals pink or white. Capsules oblong, tapering toward tips, endocarps leathery, smooth. Fl. Mar-Apr. Cultivated in Yunnan, Guangxi and Guangdong. Native to tropical America.

## 美丽异木棉
**Ceiba speciosa** (A. St.-Hil.) Ravenna

落叶乔木。树皮绿色，光滑，具刺。掌状复叶互生，小叶常5片。花大，1-3朵腋生或数朵聚生枝端；花瓣5，粉红色或红色，带紫斑。蒴果纺锤形，内有绵毛。花期10-12月，果期翌年5月。中国南方城市广泛栽培。原产南美洲；世界热带地区栽培。

Deciduous trees. Bark green, smooth, aculeate. Compound leaves alternate, palmate, with 5 leaflets. Flowers large, solitary or fascicled, axillary or terminal; petals 5, pink or red, purple striate. Capsules fusiform, with silky wool inside. Fl. Oct-Dec. Fr. May next year. Widely cultivated in southern cities of China. Native to South America; cultivated in tropical regions of the world.

美丽异木棉 *Ceiba speciosa*

# 梧桐科 Sterculiaceae

## 滇桐

**Craigia yunnanensis** W. W. Sm. et W. E. Evans

乔木。叶纸质，椭圆形，无毛，基部心形，基三出脉。聚伞花序腋生，具2-5花；花瓣缺；萼片紫红色，长圆形；内轮雄蕊短于萼片；子房无毛。果实椭圆体形，翅具5棱。花期8-9月，果期10-12月。生海拔500-1600米的路旁、河边或山坡混交林中。产云南、西藏、贵州和广西。越南北部亦有。

Trees. Leaves papery, elliptic, glabrous, base cordate, trinerved. Cymes axillary, 2-5-flowered; petals absent; sepals purplish pink, oblong; stamens of inner series shorter than sepals; ovary glabrous. Fruits ellipsoid, wings 5-angled. Fl. Aug-Sep. Fr. Oct-Dec. Roadsides, river banks or mixed forests on slopes at 500-1600 m. Distributed in Yunnan, Xizang, Guizhou and Guangxi. Also in N Vietnam.

## 香苹婆

**Sterculia foetida** L.

乔木。小枝轮生，伸展。叶簇生在小枝顶端，具掌状7-9小叶；托叶剑状，早落；小叶椭圆状披针形，起初具柔毛，成熟后无毛。花萼红紫色，5深裂。蓇葖果椭圆体形及船形，木质。花期4-5月。广西、广东和海南栽培。原产印度；栽培于南亚、东南亚、澳大利亚、热带非洲和南美洲。

Trees. Branchlets whorled, spreading. Leaves clustered at apices of branchlets, palmately 7-9-foliolate; stipules ensiform, caducous; leaflets elliptic-lanceolate, at first pilose, glabrescent when mature. Calyx red-purple, 5-parted. Follicles ellipsoid and boat-shaped, woody. Fl. Apr-May. Cultivated in Guangxi, Guangdong and Hainan. Native to India; cultivated in S and SE Asia, Australia, tropical Africa and South America.

## 家麻树

**Sterculia pexa** Pierre

乔木。掌状复叶；小叶7-9，倒卵形至披针形或狭椭圆形，下面密具星状柔毛，侧脉22-44，平行。花序簇生于小枝顶，总状或圆锥状，长达20厘米；花萼白色，钟形，裂至中部。蓇葖果红褐色，密被绒毛和长刚毛。花期10月。生坡地或村旁。产云南和广西。老挝、泰国、越南和柬埔寨亦有。

Trees. Palmate compound leaves; leaflets 7-9, obovate-lanceolate or narrowly elliptic, abaxially densely stellate pubescent, lateral veins 22-44, parallel. Inflorescences clustered at branchlet tips, race-

滇桐 *Craigia yunnanensis*

香苹婆 *Sterculia foetida*

家麻树 *Sterculia pexa*

mose or paniculate, up to 20 cm long; calyx white, campanulate, divided to 1/2 length. Follicles red-brown, densely downy and long setose. Fl. Oct. Slopes or near villages. Distributed in Yunnan and Guangxi. Also in Laos, Thailand, Vietnam and Cambodia.

## 苹婆

**Sterculia monosperma** Vent.

乔木。叶单生，长圆形或椭圆形，薄革质，无毛。圆锥花序多花，腋生，疏散，下垂；花萼钟状；萼片条形，几与萼筒等长；雌花少数；子房被毛；花柱弯曲。蓇葖果革质，成熟时暗红色，被短绒毛。花期4-5月。生密林中。产云南、广西、广东、台湾和福建(常栽培)。印度、泰国、越南、马来西亚和印度尼西亚(苏门答腊)亦有。

Trees. Leaves simple, oblong or elliptic, thinly leathery, glabrous. Panicles many-flowered, axillary, sparse, pendulous; calyx campanulate; sepals linear, nearly as long as tube; pistillate flowers few; ovary hairy; styles curved. Follicles leathery, dark red when mature, velutinous. Fl. Apr-May. Dense forests. Distributed in Yunnan, Guangxi, Guangdong, Taiwan and Fujian (often cultivated). Also in India, Thailand, Vietnam, Malaysia and Indonesia (Pulau Sumatera).

## 北越苹婆

**Sterculia tonkinensis** A. DC.

灌木或小乔木。叶椭圆形，背面密被淡黄褐色绒毛，全缘。圆锥花序于顶端簇生；花星状；外萼片三角形；花萼红色。蓇葖果下垂，红色，纺锤形，具3-6粒种子，下面密具长柔毛和黄褐色毛。花期4月。生海拔100-1500米的次生林中。产云南东南部。越南亦有。

苹婆 *Sterculia monosperma*

Shrubs or small trees. Leaves elliptic, abaxially densely pale-fulvous tomentose, entire. Inflorescences paniculate, apically clustered; flowers stellate; epicalyx lobes triangular; calyx red. Follicles pendulous, red, fusiform, 3-6-seeded, densely villous abaxially with yellowish-brown hairs. Fl. Apr. Secondary forests at 100-1500 m. Distributed in SE Yunnan. Also in Vietnam.

北越苹婆 *Sterculia tonkinensis*

蒙自苹婆 *Sterculia henryi*

## 蒙自苹婆
**Sterculia henryi** Hemsl.

灌木或小乔木。小枝幼时密被黄褐色柔毛。叶纸质，侧脉15对。总状花序腋生，长5-15厘米；萼片红色，裂至近基部；小苞片条状披针形，几乎等长于花梗；花柱弯曲；子房卵球形，密被毛。花期3-4月。生海拔800-1500米的林中。产云南(蒙自)。越南亦有。

Shrubs or small trees. Branchlets when young densely yellowish-brown hairy. Leaves papery, lateral veins 15 per side. Racemes axillary, 5-15 cm long; calyx red, divided to near base; bracteoles linear-lanceolate, nearly as long as pedicels; styles curved; ovary ovoid, densely hairy. Fl. Mar-Apr. Forests at 800-1500 m. Distributed in Yunnan (Mengzi). Also in Vietnam.

## 短柄苹婆
**Sterculia brevissima** Hsue

灌木或小乔木。叶在顶端簇生。叶几无柄，倒披针形，集生于小枝顶端。花序细弱，总状或圆锥状，腋生，下垂；外萼片线状披针形；花萼粉色，中部以下紫色。蓇葖果红褐色，椭圆体形，下面密具柔毛，两面渐狭。花期4月。生海拔500-1300米的山谷或山坡的混交林或雨林中。产云南南部。

Shrubs or small trees. Leaves apically clustered. Leaves subsessile, oblanceolate, congregated at apices of branchlets. Inflorescences slender, racemose or paniculate, axillary, pendulous; epicalyx lobes linear-lanceolate; calyx pink, purple below middle. Follicles red-brown, ellipsoid, abaxially densely puberulent, both ends attenuate. Fl. Apr. Mixed or rain forests on slopes or in valleys at 500-1300 m. Distributed in S Yunnan.

短柄苹婆 *Sterculia brevissima*

## 假苹婆
**Sterculia lanceolata** Cav.

乔木。叶椭圆形、披针形或椭圆状披针形，下面近无毛，侧脉7-9对，向上弯曲，边缘处联结。花序圆锥状，密具多分枝；花萼淡红色，裂至近基部；花无花瓣。蓇葖果鲜红色，先端具喙。花期4-6月。生海拔800-1500米的山坡次生林中或路边。产云南、四川、贵州、广西和广东。缅甸、老挝、泰国和越南亦有。

Trees. Leaves elliptic, lanceolate, or elliptic-lanceolate, abaxially nearly glabrous, lateral veins 7-9 per side, curved upward, connected near margin. Inflorescences paniculate, densely many-branched; calyx reddish, divided almost to base; flowers without petals. Follicles scarlet, apex with beak. Fl. Apr-Jun. Secondary forests on slopes or roadsides at 800-1500 m. Distributed in Yunnan, Sichuan, Guizhou, Guangxi and Guangdong. Also in Myanmar, Laos, Thailand and Vietnam.

假苹婆 *Sterculia lanceolata*

## 梧桐
**Firmiana simplex** (L.) W. F.

落叶乔木。叶心形，掌状3-5裂，两面无毛或稍具柔毛。花

梧桐 *Firmiana simplex*

单性或杂性；圆锥花序顶生，长20-50厘米；花萼黄绿色，裂至近基部，裂片条形，向外扭转。蓇葖果有2-4粒种子；果皮膜质。花期6月。产中国西南、华南、东南、华中至华西(大多为栽培)。日本亦有。欧洲和北美洲(美国)有栽培。

Deciduous trees. Leaves cordate, palmately 3-5-lobed, both surfaces glabrous or minutely puberulent. Flowers unisexual or polygamous; panicles terminal, 20-50 cm long; calyx yellowish green, divided nearly to base, lobes linear, twisted outward. Follicles 2-4-seeded; pericarps membranous. Fl. Jun. Distributed in SW, S, SE, C to W China (mostly cultivated). Also in Japan. Cultivated in Europe and North America (USA) .

## 海南梧桐

**Firmiana hainanensis** Kosterm.

乔木。树皮灰白色。叶卵形，全缘，基部截形或略成浅心形，下面密被灰白色星状短柔毛，基生脉5条。圆锥花序顶生或腋生；花黄白色。蓇葖果卵形。花期4月。生沙质土上。产海南中部和西部。

Trees. Bark offwhite. Leaves ovate, margin entire, base cuneate or slightly and shallowly cordate; lower surfaces densely covered with offwhite short stellate hairs; basal veins 5. Panicles terminal or axillary; flowers yellowish-white. Follicles ovoid. Fl. Apr. Sandy soils. Distributed in C and W Hainan.

## 火桐

**Firmiana colorata** (Roxb.) R. Br.

落叶乔木。叶阔心形，薄革质，两面疏具淡黄色星状柔毛，基脉5-7，先端3-5浅裂。花先叶开放，组成聚伞花序，密被红色星状短柔毛；花单性，无花瓣。蓇葖果舌状，红色。花期3-4月。生海拔700-1000米的山坡密林中。产云南。南亚和东南亚亦有。

Deciduous trees. Leaves broadly cordate, thinly leathery, both surfaces sparsely yellowish stellate puberulent, basal veins 5-7, apex 3-5-lobed. Flowering before appearance of leaves, in a cyme, densely red stellate pubescent. Flowers unisexual, without petals. Follicles ligulate, red. Fl. Mar-Apr. Dense forests on slopes at 700-1000 m. Distributed in Yunnan. Also in S and SE Asia.

海南梧桐 *Firmiana hainanensis*

火桐 *Firmiana colorata*

美丽火桐 *Firmiana pulcherrima*

广西火桐 *Erythropsis kwangsiensis*

## 美丽火桐

**Firmiana pulcherrima** Hsue

落叶乔木。叶异形，薄纸质，掌状3-5裂或全缘，裂片长9-14厘米，顶端尾状渐尖。聚伞圆锥花序顶生；萼近钟形，长16毫米，密被棕红褐色星状短柔毛。花期4-5月。生森林中和山谷溪旁。产海南东部。

Deciduous trees. Leaves heterotypic, thinly chartaceous, palmately 3-5-lobed or entire, lobes 9-14 cm long, apex caudate-acuminate. Cymose panicles terminal; calyx sub-campanulate, 16 mm long, densely covered with reddish-brown short stellate hairs. Fl. Apr-May. Forests or streamsides in valleys. Distributed in E Hainan.

## 广西火桐

**Erythropsis kwangsiensis** (Hsue) Hsue

落叶乔木。树皮灰白色。叶阔卵形或近圆形，纸质，两面疏具柔毛，基出脉5-7。花序为聚伞状总状花序；花萼圆锥形，先端5裂，下面被金色至红褐色星状毛；花药15，呈头状簇生。花期6月。生海拔900-1000米的山谷灌丛中。产广西。

Deciduous trees. Bark grayish white. Leaves broadly ovate or nearly round, papery, both surfaces sparsely puberulent, basal veins 5-7. Inflorescences cymose-racemose; calyx cylindric, apically 5-lobed, abaxially golden to red-brown stellate hairy; anthers 15, in capitate cluster. Fl. Jun. Thickets in valleys at 900-1000 m. Distributed in Guangxi.

## 银叶树

**Heritiera littoralis** Aiton

常绿乔木。树皮灰棕色。叶革质，背面密被银白色鳞屑片。圆锥花序腋生，密具星状毛或鳞片；花萼红褐色，钟形。果实木质，干后黄棕色，近椭圆体形，约6 × 3.5厘米。花期夏季。生红树林。产广西、广东、海南和台湾。南亚、东南亚、澳大利亚和东非亦有。

Evergreen trees. Bark gray-brown. Leaves leathery, abaxially densely silver-white scurfy scaly. Inflorescences paniculate, axillary, densely stellate hairy or with scales; calyx red-brown, campanulate. Fruits woody, drying yellow-brown, nearly ellipsoid, ca. 6 × 3.5 cm. Fl. summer. Mangrove forests. Distributed in Guangxi, Guangdong, Hainan and Taiwan. Also in S and SE Asia, Australia and E Africa.

银叶树 *Heritiera littoralis*

## 粗齿梭罗
**Reevesia rotundifolia** Chun

乔木。小枝密具黄褐色星状柔毛。叶圆形或倒卵状圆形，薄革质，上部边缘具2-3齿。花序平顶，密集；花萼漏斗形；花瓣白色。蒴果倒卵球状长圆形，5棱，长3-4厘米，具黄色柔毛或灰色鳞片。花期5月。生海拔约1000米的山地。产广西南部和广东。

Trees. Branchlets densely yellowish brown stellate puberulent. Leaves orbicular or obovate-orbicular, thinly leathery, with 2-3 teeth on upper margin. Inflorescences flat-topped, dense; calyx funnel-shaped; petals white. Capsules obovoid-oblong, 5 angled, 3-4 cm long, yellowish puberulent or gray scaly. Fl. May. Mountains at ca. 1000 m. Distributed in S Guangxi and Guangdong.

粗齿梭罗 *Reevesia rotundifolia*

## 梭罗树
**Reevesia pubescens** Mast.

乔木。叶薄革质或纸质，下面密具星状绒毛、柔毛或无毛。花序伞房状、聚伞状；花瓣5，白或粉红色；子房球形，5室，密具毛。蒴果梨形或长圆状梨形，具5棱。花期5-6月。生海拔500-2500米的山坡、灌丛或山谷疏林。产云南、四川、贵州、广西、海南和湖南。印度、不丹、缅甸、老挝、泰国和越南亦有。

Trees. Leaves thinly leathery or papery, abaxially densely stellate tomentose, pilose, or glabrous. Inflorescences cymose, corymbose; petals 5, white or pink; ovary globose, 5-locular, densely hairy. Capsules pyriform or oblong-pyriform, 5-ribbed. Fl. May-Jun. Slopes, thickets or open forests in valleys at 500-2500 m. Distributed in Yunnan, Sichuan, Guizhou, Guangxi, Hainan and Hunan. Also in India, Bhutan, Myanmar, Laos, Thailand and Vietnam.

梭罗树 *Reevesia pubescens*

两广梭罗 *Reevesia thyrsoidea*

## 两广梭罗
**Reevesia thyrsoidea** Lindl.

常绿乔木。小枝干后黑褐色，疏具星状柔毛。叶革质，两面无毛。聚伞状伞房花序，花密集，被毛；花萼钟状；花瓣5，白色；子房5室，被毛。蒴果长圆状梨形。种子含翅长约2厘米。花期3-4月。生海拔500-1500米的山坡或山谷溪边。产广西、广东和海南。越南和柬埔寨亦有。

Evergreen trees. Branchlets brownish black when dried, sparsely stellate puberulent. Leaves leathery, glabrous in both surfaces. Inflorescences cymose-corymbose, densely flowered, puberulent; calyx campanulate; petals 5, white; ovary 5-locular, pilose. Capsules oblong-pyriform. Seeds ca. 2 cm long including wings. Fl. Mar-Apr. Mountain slopes or by streams in valleys at 500-1500 m. Distributed in Guangxi, Guangdong and Hainan. Also in Vietnam and Cambodia.

山芝麻 *Helicteres angustifolia*

## 山芝麻
**Helicteres angustifolia** L.

小灌木，高达1米。小枝具灰绿色柔毛。叶狭长圆形或条状披针形。花序聚伞状，具2或多花；花瓣5，不等长，浅蓝色、粉色或淡紫色，冠檐基部色深，花期反折；每室约有10胚珠。蒴果卵球状长圆形，密具鳞片，疏具毛。花期几近全年。生斜坡草地。产中国西南、华南和东南。南亚、东南亚、日本和澳大利亚亦有。

Small shrubs, up to 1 m tall. Branchlets gray greenish puberulent. Leaves narrowly oblong or linear-lanceolate. Inflorescences cymose, 2- or many-flowered; petals 5, unequal in length, pale blue, pink, or purplish, darker at base of limb, reflexed at anthesis; ovules ca. 10 in each locule. Capsules ovoid-oblong, densely setose, sparsely hairy. Fl. almost throughout year. Sloping grasslands. Distributed in SW, S and SE China. Also in S and SE Asia, Japan and Australia.

## 细齿山芝麻
**Helicteres glabriuscula** Wall. ex Mast.

灌木。叶斜披针形，疏被星状绒毛，边缘具小齿。花序总状，腋生，具2-3花；花萼管状，5裂；花瓣5，紫色或蓝紫色；子房5室，具毛。蒴果长圆柱形，密被长柔毛。花果期几全年。生草坡上或灌丛中。产云南、贵州和广西。缅甸亦有。

Shrubs. Leaves obliquely lanceolate, sparsely stellate puberulent, margin denticulate. Inflorescences cymose, axillary, 2-3-flowered; calyx tube-shaped, 5-lobed; petals 5, purple or blue-purple; ovary 5-locular, hairy. Capsules long terete, densely villous. Fl. and fr. almost all year. Grassy slopes or thickets. Distributed in Yunnan, Guizhou and Guangxi. Also in Myanmar.

黏毛山芝麻 *Helicteres viscida*

## 黏毛山芝麻
**Helicteres viscida** Blume

灌木。叶卵形或近圆形，下面密具白色星状绒毛，边缘有不规则锯齿。花单生或聚伞状，腋生；花瓣5，白色。蒴果圆柱形，先端锐尖，密具长至4毫米的星状长柔毛。花期5-6月。生海拔300-900米的山坡上。产云南和海南。缅甸、老挝、泰国、越南、马来西亚和印度尼西亚亦有。

细齿山芝麻 *Helicteres glabriuscula*

雁婆麻 *Helicteres hirsuta*

Shrubs. Leaves ovate or suborbicular, abaxially densely whitish stellate tomentose, margin irregular-serrate. Flowers solitary or cymose, axillary; petals 5, white. Capsules cylindric, apex acute, densely stellate villous and up to 4 mm long hairs. Fl. May-Jun. Slopes at 300-900 m. Distributed in Yunnan and Hainan. Also in Myanmar, Laos, Thailand, Vietnam, Malaysia and Indonesia.

## 雁婆麻

**Helicteres hirsuta** Lour.

灌木。小枝具星状毛。叶卵形或卵状长圆形，两面密具星状柔毛，边缘具不规则牙齿。花序腋生，伸长，穗状，一侧生，近长于叶的1/2，具多花；花瓣红色或紫红色；雄蕊10。蒴果圆柱形，密具长柔毛及疣。花期4-9月。生开阔林中或灌丛。产广西南部、广东和海南。南亚和东南亚亦有。

Shrubs. Branchlets stellate hairy. Leaves ovate or ovate-oblong, both surfaces densely stellate puberulent, margin irregularly dentate. Inflorescences axillary, elongate, spikelike, one-sided, less than 1/2 as long as Leaves, several-flowered; petals red or red-purple; stamens 10. Capsules cylindric, densely villous and verrucose. Fl. Apr-Sep. Open forests or thickets. Distributed in S Guangxi, Guangdong and Hainan. Also in S and SE Asia.

## 桂火绳

**Eriolaena kwangsiensis** Hand.-Mazz.

乔木或灌木。叶圆形或阔心形，下面密具星状柔毛，基部心形。花序聚伞总状，腋生，具数花；花瓣(4或)5，倒卵状匙形；子房卵球形，具黄色星状绒毛。蒴果轮廓椭圆状披针形，先端具长喙。种子具翅。花期6-8月。生海拔800-1200米的沟谷密林或灌木中。产云南南部和广西。

Trees or shrubs. Leaves orbicular or broadly cordate, abaxially densely stellate puberulent, base cordate. Inflorescences cymelike racemose, axillary, with several flowers; petals (4 or)5, obovate-spatulate; ovary ovoid, yellowish stellate velutinous. Capsules elliptic-lanceolate in outline, apex long beaked. Seeds winged. Fl. Jun-Aug. Dense valley forests or scrubs at 800-1200 m. Distributed in S Yunnan and Guangxi.

桂火绳 *Eriolaena kwangsiensis*

马松子 *Melochia corchorifolia*

蛇婆子 *Waltheria indica*

## 马松子

**Melochia corchorifolia** L.

草本或半灌木。叶卵形、长圆状卵形或披针形，薄革质。花序为密聚伞状或团伞状，顶生或腋生；外花萼裂片4，条形，具毛；花萼钟形，5裂；花瓣5，白色，干后变淡红色。蒴果圆球形，具5棱。种子卵形，黑褐色。花期夏秋。生田野或低山丘陵。产中国长江以南各省区。日本和亚洲热带地区亦有。

Herbs or subshrubs. Leaves ovate, oblong-ovate, or lanceolate, thinly papery. Inflorescences a dense cyme or glomerule, terminal or axillary; epicalyx lobes 4, linear, hairy; calyx campanulate, 5-lobed; petals 5, white, drying reddish. Capsules globose, 5-ribbed. Seeds ovoid, black- brown. Fl. summer to autumn. Fields or low hills. Distributed throughout the provinces to the south of Yangtze River, China. Also in Japan and tropical Asia.

## 蛇婆子

**Waltheria indica** L.

半灌木，密被短柔毛。叶有短柄，叶片卵形或长椭圆状卵形，顶端钝，基部圆形，边缘有小齿。聚伞花序腋生，头状；萼筒状；花瓣5，淡黄色。蒴果小，二瓣裂。花果期夏秋季。生向阳草坡、海边或丘陵地。产福建、广东、广西、海南、台湾和云南。世界热带地区广布。

Semishrubs, densely pubescent. Leaves shortly petiolate, Leaves ovate or narrowly ellipticovate, apex obtuse, base round, margin serrulate. Cymes axillary, capitate; calyx tubiform; petals 5, yellowish. Capsules small, 2-valved. Fl. and fr. Summer-Autumn. Sunny grassy slopes, seacoasts or hills. Distributed in Fujian, Guangdong, Guangxi, Hainan, Taiwan and Yunnan. Also widely in tropical areas of the world.

## 可可

**Theobroma cacao** L.

常绿乔木。叶大，有短柄，狭卵状长椭圆形，顶端长渐尖。聚伞花序小，簇生老茎上；花淡黄色。核果大，椭圆形，长15-20厘米，具纵沟。花果期全年。海南和云南栽培。原产美洲，世界热带地区栽培。

可可 *Theobroma cacao*

午时花 *Pentapetes phoenicea*

Evergreen trees. Leaves very large, shortly petiolate, narrowly ovate-elliptic, apex narrowly acuminate. Cymes small, fasciated on old trunks; flowers yellowish. Drupes large, ellipsoid, 15-20 cm long, longitudinally sulcate. Fl. and fr. all year. Cultivated in Hainan and Yunnan. Native to America, cultivated in tropical areas worldwide.

## 午时花

**Pentapetes phoenicea** L.

一年生草本。叶条状披针形，边缘有钝锯齿。花1或2朵生于叶腋，开于午间，闭于明晨；萼片5枚，披针形；花瓣5片，红色，广倒卵形。蒴果近圆球形。花期夏秋季。广东、广西和云南栽培。原产印度；亚洲热带地区栽培。

Annual herbs. Leaves linear-lanceolate, margin obtusely serrate. Flowers 1 or 2, axillary, opening at midday, withered at morning of next day; sepals 5, lanceolate; petals 5, red, broadly obovate. Capsules subglobose. Fl. summer-autumn. Cultivated in Guangdong, Guangxi and Yunnan. Native to India; cultivated in tropical areas of Asia.

翻白叶树 *Pterospermum heterophyllum*

## 翻白叶树

**Pterospermum heterophyllum** Hance

乔木。叶盾形，掌状5裂，下面密具黄褐色星状柔毛。花单生或2-4朵组成腋生聚伞花序；外萼片与花萼紧密贴生，鳞片状；花瓣青白色。蒴果木质，长圆状卵形，被黄褐色绒毛。种子具膜质翅。花期秋季。生山地或丘陵林中。产广西、广东、海南和福建。

Trees. Leaves peltate, palmately 5-lobed, abaxially densely yellow-brown stellate pubescent. Flowers solitary or 2-4 in cymes, axillary; epicalyx lobes closely adnate to calyx, scalelike; petals greenish white. Capsules woody, oblong-ovoid, yellow-brown-tomentose. Seeds with membranous wings. Fl. autumn. Forests on mountains or hills. Distributed in Guangxi, Guangdong, Hainan and Fujian.

## 窄叶半枫荷

**Pterospermum lanceifolium** Roxb.

乔木。小枝幼时具黄褐色绒毛。叶披针形或长圆披针形，背面被黄棕色或黄白色绒毛，腹面光滑。花单生；花瓣白色，披针形。蒴果木质，圆柱状卵球形，具黄褐色绒毛。花期春夏季。生海拔800-900米的林中、山坡或山谷。产云南、广西和广东。印度、缅甸、越南和马来西亚亦有。

Trees. Branchlets yellow-brown velutinous when young. Leaves lanceolate or oblong-lanceolate, abaxially densely yellow-brown or yellow-white velutinous, adaxially glabrous. Flowers solitary; petals white, lanceolate. Capsules woody, cylindrical-ovoid, yellow-brown tomentose. Fl. spring-summer. Forests, slopes or valleys at 800-900 m. Distributed in Yunnan, Guangxi and Guangdong. Also in India, Myanmar, Vietnam and Malaysia.

窄叶半枫荷 *Pterospermum lanceifolium*

刺果藤 *Byttneria grandifolia*

昂天莲 *Ambroma augusta*

## 昂天莲

**Ambroma augusta** (L.) L. f.

灌木。叶片心形或卵状心形，有时为3-5浅裂。花序聚伞状，具1-5花；花下垂；萼片披针形；花瓣深紫红色。蒴果倒圆锥形，具5纵翅，边缘有长柔毛。种子长圆形，黑色。花期春夏季。生海拔200-1200米的山谷沟边或林缘。产云南、贵州、广西和广东。东亚、东南亚和太平洋岛屿亦有。

Shrubs. Leaves cordate or ovate-cordate, sometimes 3-5-lobed. Inflorescences cymose, 1-5-flowered; flowers pendulous; sepals lanceolate; petals dark reddish purple. Capsules obconical, with 5 longitudinal wings, margin villous. Seeds oblong, black. Fl. spring to summer. Valleys by streams or forest edges at 200-1200 m. Distributed in Yunnan, Guizhou, Guangxi and Guangdong. Also in S and SE Asia, and Pacific Islands.

## 梅蓝

**Melhania hamiltoniana** Wall.

小灌木。小枝密被淡黄褐色短柔毛。叶阔卵形或椭圆状卵形，两面密具柔毛。花序腋生，短总状，常具3-5花；外萼片2或3；花瓣黄色。蒴果卵形。种子黑褐色。花期11月，果期翌年5月。生海拔400-500米的多石草坡。产云南。印度和缅甸亦有。

Small shrubs. Branchlets with densely pale fulvous pubescences. Leaves broadly ovate or ellipticovate, both surfaces densely puberulent. Inflorescences axillary, shortly racemose, usually 3-5-flowered; epicalyx lobes 2 or 3; petals yellow. Capsules ovoid. Seeds black-brown. Fl. Nov. Fr. May next year. Stony grassy slopes at 400-500 m. Distributed in Yunnan. Also in India and Myanmar.

## 刺果藤

**Byttneria grandifolia** DC.

木质大藤本。叶宽卵形、心形或近圆形，长7-23厘米，下面被白色星状毛。花瓣黄白色，上面紫红色，先端2裂，具长舌状附属物，与花萼近等长。蒴果球形或卵球形，具刺，刺短且粗壮，具柔毛。花期春季至夏季。生海拔200-300米的疏林下、山谷或路旁。产云南、广西、广东和海南。南亚和东南亚亦有。

Woody, large lianas. Leaves broadly ovate, cordate, or suborbicular, 7-23 cm long, abaxially white-stellate-hairy. Petals yellowish white, and purple-red adaxially, apex 2-lobed, with long ligulate appendix, nearly as long as sepals. Capsules globose or ovoid, spiny, spines short and robust, puberulent. Fl. spring to summer. Sparse forests, valleys or roadsides at 200-300 m. Distributed in Yunnan, Guangxi, Guangdong and Hainan. Also in S and SE Asia.

梅蓝 *Melhania hamiltoniana*

# 五桠果科 Dilleniaceae

锡叶藤 *Tetracera sarmentosa*

## 锡叶藤
**Tetracera sarmentosa** (L.) Vahl

攀援木质藤本，长达20厘米。叶片圆形，革质，次生脉10-15。圆锥花序顶生；萼片5，宿存；花瓣3，白色；雄蕊多数，短于花萼；心皮1(或2)，无毛；花柱长于雄蕊。蓇葖果橘黄色，外果皮薄革质，干后稍亮。假种皮红色或紫色。花期4-5月。生疏林或灌丛中。产云南、广西、广东和海南。南亚和东南亚亦有。

Climbing woody vines, up to 20 cm long. Leaves orbicular, leathery, secondary veins 10-15. Panicles terminal; sepals 5, persistent; petals 3, white; stamens numerous, shorter than sepals; carpels 1(or 2), glabrous; styles longer than stamens. Follicles orange, pericarps thinly leathery and slightly bright when dry. Arils red or purple. Fl. Apr-May. Sparse forests or thickets. Distributed in Yunnan, Guangxi, Guangdong and Hainan. Also in S and SE Asia.

## 五桠果
**Dillenia indica** L.

常绿乔木，高达30米。叶长圆形，侧脉25-50对。花单生；萼片5，近圆形，厚肉质；花瓣白色，倒卵形；雄蕊成明显2组；心皮16-20；花柱开展。聚合果球形，不开裂。花期7-10月，果期10-12月。生海拔200-300米的山谷或溪边。产云南南部和广西。南亚和东南亚亦有。

Evergreen trees, up to 30 m tall. Leaves oblong, lateral nerves 25-50 pairs. Flowers solitary; sepals 5, approximately rounded, thickly fleshy; petals white, obovate; stamens in 2 distinct groups; carpels 16-20; stylodia spreading. Aggregates fruits globose, indehiscent. Fl. Jul-Oct. Fr. Oct-Dec. Valleys or by streams at 200-800 m. Distributed in S Yunnan and Guangxi. Also in S and SE Asia.

五桠果 *Dillenia indica*

### 大花五桠果
**Dillenia turbinata** Finet et Gagnep.

常绿乔木，高达30米。叶倒卵形，侧脉15-25对。总状花序顶生，具3-5花；花2-7，芳香；花萼卵形，厚肉质；花瓣黄色、黄白色或淡红色；雄蕊明显2组；心皮8或9。假果皮不开裂，淡红色至暗红色。花期4-5月。生海拔700-1000米的林中。产云南南部、广西南部和海南。越南亦有。

Evergreen trees, up to 30 m tall. Leaves obovate, lateral nerves 15-25 pairs. Racemes terminal, 3-5-flowered; flowers 2-7, fragrant; sepals oval, thickly fleshy; petals yellow, yellowish white, or reddish; stamens in 2 distinct groups; carpels 8 or 9. Pseudocarps indehiscent, reddish to dull red. Fl. Apr-May. Forests at 700-1000 m. Distributed in S Yunnan, S Guangxi and Hainan. Also in Vietnam.

大花五桠果 *Dillenia turbinata*

# 猕猴桃科 Actinidiaceae

### 软枣猕猴桃
**Actinidia arguta** (Siebold. et Zucc.) Planch.

大型落叶攀援灌木。髓片层状。叶常卵形至阔卵形至近圆形，膜质至纸质，下面无毛至具锈色绒毛至具硬毛。花绿黄色或白色；花瓣4-6；花药暗紫色；子房瓶状，无毛。果实绿黄色，成熟时偶尔紫红色，球形至长圆形，长2-3厘米，顶端有钝喙。花期4月，果期8-10月。生海拔700-3600米的山林中、灌丛或潮湿地，亦广泛栽培。产中国西南、东南、华中、华北、华西、华东和东北。朝鲜半岛和日本亦有。

Large deciduous climbing shrubs. Pith lamellate. Leaves usually ovate to broadly ovate to suborbicular, membranous to papery, abaxially glabrous to rusty tomentose to strigillose. Flowers greenish yellow or white; petals 4-6; anthers dark-purple; ovaries bottle-shaped, glabrous. Fruits greenish yellow or purple-red when mature, globose to oblong, 2-3 cm long, apex obtuse-beaked. Fl. Apr. Fr. Aug-Oct. Montane forests, thickets or moist places at 700-3600 m, also widely cultivated. Distributed in SW, SE, C, N, W, E and NE China. Also in Korean Peninsula and Japan.

软枣猕猴桃 *Actinidia arguta*

### 黑蕊猕猴桃
**Actinidia melanandra** Franch.

中型落叶攀援藤本。髓片层状。叶片卵状披针形，卵状长圆形，长圆形或卵形，纸质至革质，无毛，叶背面有白粉。聚伞花序，具1-7花；花绿色至黄白色或白色；花药黑色；子房瓶状，无毛。果实球形至长圆形，长2-4.5厘米，无毛，顶部具喙。花期5-6月，果期9月。生海拔1000-1600米的阔叶林、山地林中或潮湿处，亦广泛栽培。产中国西南、东南、

黑蕊猕猴桃 *Actinidia melanandra*

四萼猕猴桃 *Actinidia tetramera*

华中、华西和华东。

Mid-sized deciduous climbing shrubs. Pith lamellate. Leaves ovate-lanceolate, ovate-oblong, oblong or ovate, papery to leathery, glabrous, glaucous abaxially. Panicles cymose, 1-7-flowered; flowers greenish to yellowish white or white; anthers black; ovary bottle-shaped, glabrous. Fruits globose to oblong, 2-4.5 cm long, glabrous, rostrate at apex. Fl. May-Jun. Fr. Sep. Broad-leaved forests, montane forests or moist places at 1000-1600 m, also widely cultivated. Distributed in SW, SE, C, W and E China.

## 狗枣猕猴桃

**Actinidia kolomikta** (Maxim. et Rupr.) Maxim.

大型落叶攀援灌木。髓片层状。叶阔卵形至长圆状卵形或长圆状倒卵形，膜质或薄纸质，基部偏斜，上面散生短刺毛。花粉红色或白色；花瓣5或6。果实成熟时淡橙色或紫红色，卵圆形，长2-2.5厘米，无毛。花期4-7月，果期9-10月。生海拔1600-2900米的山地混交林中，亦广泛栽培。产中国西南、华北、华西和东北。俄罗斯(远东地区)、朝鲜半岛和日本亦有。

Large deciduous climbing shrubs. Pith lamellate. Leaves broadly ovate to oblong-ovate or oblongobovate, membranous or thinly papery, base oblique, adaxially sparsely shortly spinecent. Flowers pink or white; petals 5 or 6. Fruits pale orange or purple when mature, ovoid, 2-2.5 cm long, glabrous. Fl. Apr-Jul. Fr. Sep-Oct. Mixed forests on mountains at 1600-2900 m, also widely cultivated. Distributed in SW, N, W and NE China. Also in Russia (Far East), Korean Peninsula and Japan.

## 四萼猕猴桃

**Actinidia tetramera** Maxim.

中型落叶攀援灌木。髓片层状。叶长圆状卵形至长圆状椭圆形或椭圆状披针形，薄纸质，下面无毛至沿脉稍具硬毛，背面侧脉脉腋通常有极显著的髯毛。萼片4(-5)；花稍粉白色，花瓣4(-5)片。果实成熟时橙色，长1.5-2厘米，无毛。花期5-6月，果期8-9月。生海拔1100-2700米的林中、灌丛或湿润处。产中国西南、华北和华西。

Mid-sized deciduous climbing shrubs. Pith lamellate. Leaves oblong-ovate to oblong-elliptic or elliptic-lanceolate, thinly papery, abaxially glabrous to slightly strigillose on midvein, always very conspicuously white barbate at axils of lateral veins. Sepals 4 (-5); flowers somewhat pinkish white, petals 4(-5). Fruits orange when mature, 1.5-2 cm long, glabrous. Fl. May-Jun. Fr. Aug-Sep. Forests, thickets or moist places at 1100-2700 m. Distributed in SW, N and W China.

狗枣猕猴桃 *Actinidia kolomikta*

葛枣猕猴桃 *Actinidia polygama*

## 葛枣猕猴桃

**Actinidia polygama** (Siebold. et Zucc.) Maxim.

大型落叶攀援藤本。髓白色，实心。叶片卵形至长圆状卵形，膜质至薄纸质，叶背常无毛，叶面疏被硬毛。花白色，花瓣5。果实成熟后橘黄色，卵形至圆柱状卵形至长圆状卵形，无毛，顶部具喙。花期6-7月，果期9-10月。生海拔500-1900米的山地林中，亦广泛栽培。产中国西南、华中、华北、华西、华东和东北。俄罗斯、朝鲜半岛和日本亦有。

Large deciduous climbing shrubs. Pith white, solid. Leaves ovate to oblong-ovate, membranous to thinly papery, abaxially usually glabrous, adaxially sparsely strigillose. Flowers white, petals 5. Fruits orange when mature, ovoid to cylindric-ovoid to oblong-ovoid, glabrous, rostrate at apex. Fl. Jun-Jul. Fr. Sep-Oct. Montane forests at 500-1900 m, also widely cultivated. Distributed in SW, C, N, W, E and NE China. Also in Russia, Korean Peninsula and Japan.

## 大籽猕猴桃

**Actinidia macrosperma** C. F. Liang

大型落叶攀援灌木。髓白色，实心。叶卵形至椭圆形，成熟时革质，下面无毛至沿脉具糙毛。花常单生，白色；萼片2-3；花瓣5-12；子房瓶状。果成熟时橘黄色，无毛，无皮孔。种子直径约3毫米，长4-5毫米。花期4-5月，果期9-10月。生丘陵或低山地的丛林中或林缘。产广东、浙江、湖北、安徽、江苏和江西。

Large deciduous climbing shrubs. Pith white, solid. Leaves ovate to elliptic, leathery when mature, abaxially glabrous to strigillose on midvein. Flowers usually solitary, white; sepals 2-3; petals 5-12; ovary bottle-shaped. Fruits orange when ripe, glabrous, lenticels absent. Seeds ca. 3 mm diam, 4-5 mm long. Fl. Apr-May. Fr. Sep-Oct. Hills or low mountain places, thickets or forest edges. Distributed in Guangdong, Zhejiang, Hubei, Anhui, Jiangsu and Jiangxi.

大籽猕猴桃 *Actinidia macrosperma*

## 对萼猕猴桃

**Actinidia valvata** Dunn

中型落叶攀援灌木。髓白色，实心。叶阔卵形至狭卵形，膜质至纸质，两面无毛，侧脉5-6对。花序2-3花或单生；萼片2-3，花白色，花瓣7-9。果实卵球形至倒卵球形，不具皮孔，成熟时橙色，长2-2.5厘米，无毛，顶端有尖喙。花期5月，果期8月。生海拔约1000米的疏林、灌丛或山谷中。产中国东南、华中和华东。

Mid-sized, deciduous climbing shrubs. Pith white, solid. Leaves broadly ovate to narrowly ovate, membranous to papery, glabrous on both surfaces, lateral veins 5-6 pairs. Panicles 2-3-flowered or solitary; sepals 2-3; flowers white, petals 7-9. Fruits ovoid to obovoid, not lenticellate, orange when mature, 2-2.5 cm long, glabrous, rostrate at apex. Fl. May. Fr. Aug. Sparse forests, thickets or valleys at ca. 1000 m. Distributed in SE, C and E China.

## 革叶猕猴桃

**Actinidia rubricaulis** Dunn var. **coriacea** (Fin. et Gagn.) C. F. Liang

半常绿攀援灌木。髓实心。叶片长圆状披针形至椭圆形至倒披针形，厚革质，无毛。花序

对萼猕猴桃 *Actinidia valvata*

革叶猕猴桃 *Actinidia rubricaulis* var. *conacea*

常仅具1花；花红色，花瓣5。果实深绿色，卵形至球形，长1.5-2厘米，成熟后无毛，皮孔棕色。花期4-5月，果期9-11月。生海拔1000米以上的阔叶林中。产云南、四川、贵州、重庆、广西、湖北和湖南。泰国亦有。

Semi-evergreen climbing shrubs. Pith solid. Leaves oblong-lanceolate to elliptic to oblanceolate, thickly leathery, glabrous. Panicles often 1-flowered; flowers reddish, petals 5. Fruits dark green, ovoid to globose, 1.5-2 cm long, glabrous when mature, lenticels brown. Fl. Apr-May. Fr. Sep-Nov. Broad-leaved forests above 1000 m. Distributed in Yunnan, Sichuan, Guizhou, Chongqing, Guangxi, Hubei and Hunan. Also in Thailand.

## 山梨猕猴桃

**Actinidia rufa** (Siebold et Zucc.) Planch. ex Miq.

大型半落叶藤本。花枝被棕色绒毛，髓棕色，片层状。叶片阔卵圆形或圆形，纸质，背腹两面光滑，背面叶脉明显，边缘具锯齿状短尖头。聚伞花序腋生，被棕色毛；雄花序小花多数；雌花序较少；花白色，花瓣倒卵形。果实长圆形至卵形，光滑，皮孔模糊。花期5月，果期10-11月。生海拔1000-2000米的山地疏林中。产台湾。日本和朝鲜半岛亦有。

Climbing shrubs, large, semideciduous. Floral branchlets brownish puberulent; pith brown, small, lamellate. Leaves broadly ovate or orbicular, papery, both surfaces glabrous, midvein and lateral veins conspicuous abaxially, margin shallowly mucronate-serrate. Inflorescences cymose, axillary, brownish velutinous; male inflorescences many flowered; female inflorescences few flowered; flowers white, petals obovate. Fruit oblong to ovoid, glabrous, lenticels obscure. Fl. May. Fr. Oct-Nov. Montane forests at 1000-2000 m. Distributed in Taiwan. Also in Japan and Korean Peninsula.

## 硬齿猕猴桃

**Actinidia callosa** Lindl.

大型落叶攀援灌木。髓片层状或有时实心，花枝无毛，稀被粗毛，皮孔显著。叶卵形或卵状长圆形，锯齿端常骨质，下面除脉腋具簇毛外常常无毛。花白色，花瓣5；花药黄色。果灰绿色，近球形至卵球形或乳头状，长1.5-5厘米，无毛，皮孔明显。花期4-6月，果期9-10月。生海拔400-2600米的林中、灌丛、溪边或湿润处，亦广泛栽培。产中国西南、华南、华西和华中。印度、尼泊尔和不丹亦有。

Large deciduous climbing shrubs. Floral branchlets glabrous, rarely strigose, lenticels very conspicuous; pith brown, lamellate, or sometimes solid. Leaves ovate to ovate-oblong, tips of serrations usually callose, abaxially glabrous except for barbate vein axils. Flowers white, petals 5; anthers yellow. Fruits grayish green, subglobose to ovoid or mammilliform, 1.5-5 cm long, glabrous, lenticels conspicuous. Fl. Apr-Jun. Fr. Sep-Oct. Forests, thickets, streamsides or moist places at 400-2600 m, also widely cultivated. Distributed in SW, S, W and C China. Also in India, Nepal and Bhutan.

山梨猕猴桃 *Actinidia rufa*

硬齿猕猴桃 *Actinidia callosa*

京梨猕猴桃 *Actinidia callosa* var. *henryi*

显脉猕猴桃 *Actinidia venosa*

## 京梨猕猴桃

**Actinidia callosa** Lindl. var. **henryi** Maxim.

落叶木质藤本。小枝坚硬，干后土黄色，洁净无毛。叶纸质，卵形或卵状椭圆形至倒卵形，长8-10厘米，宽4-5.5厘米，先端具短尖，边缘锯齿细小，背面脉腋上有髯毛。萼片无毛。果乳头状至长圆圆柱状，长可达5厘米，有显著的淡褐色圆形斑点，具反折的宿存萼片。花期4-6月，果期9-10月。生海拔500-2600米的林中、灌丛和山谷。产华东、四川、湖北、湖南、甘肃和陕西。

Deciduous woody vines. Branchlets hard, khaki after dry, glabrous. Leaves papery, ovate or ovate-elliptic to obovate, 8-10 cm long, 4-5.5 cm wide, apex macronate, margin minutely serrate, abaxially often with barbate lateral vein axils. Calyx glabrous. Fruits mammilliform to cylindric, to 5 cm long, with conspicuous brownish rounded lenticels, with reflexed and persistent sepals. Fl. Apr-Jun. Fr. Sep-Oct. Forests, thickets and valleys at 500-2600 m. Distributed in E China, Sichuan, Hubei, Hunan, Gansu and Shaanxi.

## 显脉猕猴桃

**Actinidia venosa** Rehd.

大型落叶攀援灌木。髓白色，片层状。叶卵状长圆形至卵形或卵状圆形，中脉与侧脉在下面明显凸出，叶背通常有蛛网状柔毛。聚伞花序腋生，具1-7花；花黄色；花瓣5。果实圆柱状至近球形，长1.5-2.5厘米，成熟后无毛，皮孔小。花期5-7月，果期9月。生海拔1200-2400米的疏林或灌丛中。产云南、四川和西藏。印度亦有。

Large deciduous climbing shrubs. Pith white, lamellate. Leaves ovate-oblong to ovate or ovate-orbicular, midvein and lateral veins conspicuously raised abaxially, abaxially white arachnoid-pubescent throughout or only on midvein and lateral veins. Panicles cymose, axillary, 1-7-flowered; flowers yellow, petals 5. Fruits cylindric to subglobose, 1.5-2.5 cm long, glabrous when mature, lenticels small. Fl. May-Jul. Fr. Sep. Open forests or thickets at 1200-2400 m. Distributed in Yunnan, Sichuan and Xizang. Also in India.

## 柱果猕猴桃

**Actinidia cylindrica** C. F. Liang

半常绿藤本。髓白色，片层状。芽体锥形，不被毛。叶椭圆形至矩圆披针形，顶端急尖至渐尖，厚膜质；隔年老叶呈革质，均洁净无毛。花序聚伞状；花序柄和花柄均略被微绒毛。果成熟时变黄绿色，圆柱形，幼时可见残存细绒毛，老时完全秃净，有枯褐色的稍突起的斑点；宿存萼片无毛，不反折。花期5月。果期10月。生海拔400-800米的低山森林中。产广西。

Climbing shrubs, semi-evergreen. Pith white or brown, lamellate. Buds conical, glabrous. Leaves elliptic to oblong, apex acute to acuminate, firmly membranous; but leathery on two-year-old leaves, glabrous. Inflorescences cymose; pedicels slightly tomentose. Fruit yellowish green when mature, cylindric, sparsely and finely tomentose when young, glabrous when mature, lenticels brownish, raised; persistent sepals reflexed or not, glabrous. Fl. May. Fr. Oct. Low montane forests, thickets at 400-800 m. Distributed in Guangxi.

柱果猕猴桃 *Actinidia cylindrica*

滑叶猕猴桃 *Actinidia laevissima*

## 滑叶猕猴桃
**Actinidia laevissima** C. F. Liang

中型落叶藤本。枝条皮孔极显著；芽体被紧密锈色绒毛。髓白色，片层状。叶膜质，卵形至长卵形或矩状卵形，边缘有芒尖状的斜举或开展的小锯齿，背面无毛，叶脉很不发达。花单生，粉红色，花瓣倒卵形。果实暗绿色，秃净，黄褐色斑点显著，突起。花期5-6月。生海拔800-2000米的山地上的灌丛中或疏林中。产贵州和湖北。

Mid-sized climbing shrubs deciduous. Branchlets lenticels conspicuously raised; buds densely rusty strigose. Pith white, lamellate. Leaves ovate to narrowly ovate or oblong-ovate, thickly leathery, abaxially glabrous, midvein and lateral veins unconspicuous abaxially, margin usually setose-serrulate, occasionally coarsely dentate. Inflorescences 1-flowered; flowers pink, petals obovate. Fruit dark green, glabrous, lenticels yellowish brown. Fl. May-Jun. Sparse montane forests or thickets at 800-2000 m. Distributed in Guizhou and Hubei.

金花猕猴桃 *Actinidia chrysantha*

## 金花猕猴桃
**Actinidia chrysantha** C. F. Liang

大型落叶藤本。小枝、花序和萼片被有茶褐色粉末状绒毛。髓茶褐色，片层状。叶软纸质，阔卵形或卵形至披针状长卵形，叶缘具圆锯齿；腹面草绿色，洁净无毛。花金黄色。果实成熟时栗褐色或绿褐色，秃净，具枯黄色斑点，圆球形或卵珠形。花期5月，果期11月。生海拔900-1300米的疏林、灌丛或山林等阳坡。产江西、广西、广东和湖南。

Climbing shrubs, large, deciduous. Branchlets, inflorescences and sepals brown tomentose. Pith brown, lamellate. Leaves adaxially green, broadly ovate to lanceolate-ovate, soft papery, margin conspicuously crenate, abaxially green, glabrous. Flowers yellow. Fruit maroon-brown to greenish brown, ovoid to globose, glabrous, lenticels yellowish; persistent sepals reflexed. Fl. May. Fr. Nov. Sparse forests, thickets or open sunny places in forests at 900-1300 m. Distributed in Jiangxi, Guangxi, Guangdong and Hunan.

## 粉叶猕猴桃
**Actinidia glaucocallosa** C. Y. Wu

大型落叶攀援灌木。髓片层状或实心。叶纸质，长卵形或卵状披针形，叶背有白粉，侧脉7-9对。花绿黄色；花瓣5-7，卵形至倒卵形。果实球形，直径约3厘米，幼时密具绒毛，成熟后无毛。花期5-6月，果期7-10月。生海拔2300-2800米的沟谷或常绿阔叶林中。产云南。

Large deciduous climbing shrubs. Pith lamellate or solid. Leaves papery, long ovate or ovate-lanceolate, abaxially glaucous, lateral nerves 7-9 pairs. Flowers greenish yellow; petals 5-7, ovate to obovate. Fruits globose, ca. 3 cm diam, densely tomentose when young, glabrous when mature. Fl. May-Jun. Fr. Jul-Oct. Valleys or evergreen broad-leaved forests at 2300-2800 m. Distributed in Yunnan.

粉叶猕猴桃 *Actinidia glaucocallosa*

糙叶猕猴桃 *Actinidia rudis*

## 糙叶猕猴桃
**Actinidia rudis** Dunn

大型半落叶藤本。枝条和叶柄密被粗硬毛；皮孔少，且不显著；髓白色，片层状。叶长卵形，顶端渐尖，背腹两面均被硬毛。聚伞花序近无柄，密被黄褐色绒毛；花白色；萼片无毛；花瓣矩卵形。果圆柱状，幼时被绒毛，成熟时秃净，有斑点。花期5-6月，果期7-9月。生海拔1200-2300米的疏林、灌丛、溪边、湿润处或路边。产云南(屏边县和蒙自)。

Climbing shrubs, large, deciduous. Branchlets and petioles densely yellowish or brownish rigidly strigose; pith white, lamellate. Leaves narrowly ovate, apex acute to acuminate; both surfaces sparsely to densely strigose, Inflorescences densely fasciculate, densely ferruginous tomentose; flowers white; sepals glabrous; petals oblong-ovate. Fruit cylindric, tomentose when young, glabrous when mature, lenticellate. Fl. May-Jun. Fr. Jul-Sep. Sparse forests, thickets, streamsides, moist places, roadsides at 1200-2300 m. Distributed in Yunnan (Pingbian County, Mengzi).

葡萄叶猕猴桃 *Actinidia vitifolia*

## 葡萄叶猕猴桃
**Actinidia vitifolia** C. Y. Wu

大型落叶藤本。叶膜质或软纸质，两面无毛，圆卵形，偏斜，顶端急尖，基部钝圆或浅心形。花大，白色或粉色，单生，花瓣8片，匙状倒卵形，大小不等；子房圆球形至球状圆柱状，被黄褐色绒毛。果实球形，具密集黄褐色斑点。花期5-6月，果期9-10月。生海拔1600-1900米的阔叶林下、林缘和石灰岩山地。产四川(马边县、峨边县、雷波县)和云南(彝良县)。

Climbing shrubs, large, deciduous. Leaves membranous to soft papery, both surface glabrous, round-ovate, oblique, apex acute, base rounded to shallow cordate. Flowers white to pink, solitary; petals 8, speculate-obovate, unequal; ovary globose to globose-cylindric, densely tomentose. Fruit globose, with brownish lenticels. Fl. May-Jun. Fr. Sep-Oct. Broad-leaved forests, forest edges, limestone mountains at 1600-1900 m. Distributed in Sichuan (Mabian County, Ebian County, Leibo County) and Yunnan (Yiliang County).

条叶猕猴桃 *Actinidia fortunatii*

美丽猕猴桃 *Actinidia melliana*

## 条叶猕猴桃

**Actinidia fortunatii** Fin. et Gagn.

落叶攀援藤本。髓白色，片层状。叶片披针形至长圆状椭圆形至倒卵状披针形，常无毛并明显倾斜，叶背有白粉。聚伞花序，具1-3花；花红色。果实灰绿色，圆柱形或卵状圆柱形，长1.5-1.8厘米。花期4-6月，果期11月。生海拔400-1000米的低山树木、灌丛、山坡或沟谷中。产贵州、广西、广东和湖南。

Deciduous climbing shrubs. Pith white, lamellate. Leaves lanceolate to oblong-elliptic to obovate-lanceolate, usually glabrous and distinctly oblique, glaucous abaxially. Panicles cymose, 1-3-flowered; flowers reddish. Fruits grayish green, cylindric or ovoid-cylindric, 1.5-1.8 cm long. Fl. Apr-Jun. Fr. Nov. Low montane forests, thickets, slopes or valleys at 400-1000 m. Distributed in Guizhou, Guangxi, Guangdong and Hunan.

## 美丽猕猴桃

**Actinidia melliana** Hand.-Mazz.

中型半常绿藤本。髓白色，片层状。叶片长圆状椭圆形至长圆状披针形或长圆状倒卵形，膜质至纸质，两面全部或仅中脉和侧脉密被锈色硬毛，叶背面有白粉。花序多达10花；花白色或粉红色；花药黄色。果实圆柱形，长1.6-2.2厘米，无毛。花期5-7月，果期5-11月。生海拔200-1300米的山地林中或灌丛中。产广西、广东、海南、湖南和江西。

Mid-sized semi-evergreen climbing shrubs. Pith white, lamellate. Leaves oblong-elliptic to oblong-lanceolate or oblong-obovate, membranous to papery, both surfaces densely rusty hispid throughout or only on midvein and lateral veins, glaucous abaxially. Panicles up to 10-flowered; flowers white or pink; anthers yellow. Fruits cylindric, 1.6-2.2 cm long, glabrous. Fl. May-Jun. Fr. May-Nov. Montane forests or thickets at 200-1300 m. Distributed in Guangxi, Guangdong, Hainan, Hunan and Jiangxi.

## 蒙自猕猴桃

**Actinidia henryi** Dunn

中型至大型半常绿藤本。小枝被红褐色长绒毛。叶纸质，长方长卵形至长方披针形，腹面深绿色，背面淡绿色；老叶两面完全无毛，边缘小齿很不发达。花白色，花瓣条状椭圆形至近圆形。果卵状圆柱形，具斑点，成熟时秃净。花期5-6月，果期10月。生海拔1400-2500米的林中或灌丛。产广东、广西、贵州、湖南和云南。

Climbing shrubs, mid-sized to large, semi-evergreen. Young branchlets reddish brown tomentose. Leaves papery, rarrowly oblong-ovate to oblong-lanceolate, adaxially dark green, abaxially pale green, totally glabrout on both sides of mature leaves, margin mucronate-serrulate, undeveloped. Flowers white, petals linear-elliptic to suborbicular. Fruits ovate-cylindric, with lenticels, glabrout at maturity. Fl. May-Jun. Fr. Oct. Montane forests or thickets at 1400-2500 m. Distributed in Guangdong, Guangxi, Guizhou, Hunan and Yunnan.

蒙自猕猴桃 *Actinidia henryi*

粉毛猕猴桃 *Actinidia farinosa*

## 粉毛猕猴桃
**Actinidia farinosa** C. F. Liang

中型半常绿藤本。叶柄和花枝被绵毛。髓污白色，片层状。叶阔卵形或卵状近圆形，腹面绿色，散被刚伏毛；背面苍绿色，密被褐色绒毛，毛被容易脱落。聚伞花序1-3花；花序柄很短；花瓣小，粉色，倒卵形至长方倒卵形。果卵珠状圆柱形，毛被逐渐脱落，具斑点。花期6月。生海拔1000-1200米的阳坡、路边。产广西(田林县)。

Climbing shrubs, mid-sized, deciduous. Branchlets and petioles densely lanate. Pith dirty white, lamellate. Leaves broadly ovate to ovate-orbicular, adaxially green, sparsely strigillose; abaxially pale green, thickly brownish cottony-tomentose, hairs easily caducous. Inflorescences cymose, 1-3-flowered; peduncles very short; flowers pink, petals small, obovate to oblong-obovate. Fruit ovoid-cylindric, glabrescent, lenticellate. Fl. Jun. Sunny places, roadsides at 1000-1200 m. Distributed in Guangxi (Tianlin County).

## 黄毛猕猴桃
**Actinidia fulvicoma** Hance

中型半常绿藤本。髓白色，片层状。叶纸质至亚革质，卵状矩圆形至卵状披针形，背面横脉显著，腹面被短糙伏毛或硬伏毛。聚伞花序密被黄褐色绵毛；苞片钻形；花白色，半开展，花瓣无毛，倒卵形至倒长卵形；宿存萼片反折。花期5-6月，果期10月。生海拔130-1800米的山地疏林中或灌丛中。产福建、广东、广西、云南、贵州、湖南和江西。

黄毛猕猴桃 *Actinidia fulvicoma*

Climbing shrubs, mid-sized, semi-evergreen. Pith white, lamellate. Leaves papery to leathery, ovate-oblong to ovate-lanceolate, adaxially scabrous to densely softly strigose. Inflorescences cymose, densely brownish villous-pubescent; bracts subulate, 2-6 mm; flowers white, petals obovate to narrowly obovate; persistent sepals reflexed. Fl. May-Jun. Fr. Oct. Montane forests, sparse forests or thickets, mountain slopes at 130-1800 m. Distributed in Fujian, Guangdong, Guangxi, Yunnan, Guizhou, Hunan and Jiangxi.

## 栓叶猕猴桃
**Actinidia suberifolia** C. Y. Wu

大型粗壮落叶藤本。髓实心，淡褐色。叶厚纸质，长方椭圆形，背面主脉和侧脉密被茶褐色绒毛，叶脉发达。雌雄花序互异，雄花花序为腋生总状花序；雌花花序1-2回聚伞花序；花柄粗壮；花橘黄色，花瓣倒卵形。果实近球形，幼时密被茶褐色绒毛，具斑点。花期4月。生海拔900-1000米的干燥灌丛中。产云南(屏边县、蒙自）。

Climbing shrubs, large, deciduous. Pith brownish, solid. Leaves thickly papery, narrowly elliptic, abaxially very densely tomentose with cinnamon-colored hairs on midvein and lateral veins, veins conspicuous. Male inflore-

栓叶猕猴桃 *Actinidia suberifolia*

阔叶猕猴桃 *Actinidia latifolia*

scences racemiform; female inflorescences 1-2-branched; pedicels robust; flowers orange; petals obovate. Fruit subglobose, densely ferruginous tomentose when young, lenticellate. Fl. Apr. Dry thickets at 900-1000 m. Distributed in Yunnan (Mengzi, Pingbian County).

## 阔叶猕猴桃

**Actinidia latifolia** (Gardn. et Champ.) Merr.

大型落叶攀援灌木。髓白色，片层状、实心或中空。叶常阔卵形至阔倒卵形，下面渐无毛至密具平伏星状绒毛，坚纸质。花序大型，3-4歧分枝；花瓣5-8，上部及边缘白色，下部中间橙色。果实暗绿色，圆柱形或卵状圆柱形，成熟后无毛或仅基部和顶端具柔毛。花期5-6月，果期10-11月。生海拔400-1700米的林中、山谷、溪边或灌丛中。产中国西南、华南、东南和华东。老挝、泰国、越南、柬埔寨和马来西亚亦有。

Large deciduous climbing shrubs. Pith white, lamellate, solid, or hollow. Leaves usually broadly ovate to broadly obovate, abaxially glabrescent to densely appressed stellate tomentose, hard-papery. Panicles large, 3-4-branched; petals 5-8, white on upper part and margins, orange on middle of lower parts. Fruits dark-green, terete or ovate-terete, glabrous when mature or only pubescent at base and apex. Fl. May-Jun. Fr. Oct-Nov. Forests, valleys, stream banks or thickets at 400-1700 m. Distributed in SW, S, SE and E China. Also in Laos, Thailand, Vietnam, Cambodia and Malaysia.

## 小叶猕猴桃

**Actinidia lanceolata** Dunn

落叶藤本。髓褐色，片层状。小枝、叶柄和花序的毛锈褐色。叶纸质，卵状椭圆形至椭圆披针形，顶端短尖至渐尖，背面密被灰色星状绒毛。花淡绿色。果小，卵形，有显著的浅褐色斑点。花期5-6月，果期11月。生海拔200-800米的疏林中、林缘、山地上的高草灌丛中。产安徽、浙江、江苏、江西、福建、湖南和广东。

Climbing shrubs, deciduous. Pith brown, lamellate. Branchlets, petioles, and inflorescences ferruginous velutinous. Leaves papery, elliptico-vate to elliptic-lanceolate, apex acuminate to shortly acuminate, abaxially densely appressed stellate with grayish hairs. Flowers greenish. Fruit ovoid, small, glabrous, lenticels pale brown. Fl. May-Jun. Fr. Nov. Sparse forests, forest edges, tall grassy thickets at 200-800 m. Distributed in Anhui, Zhejiang, Jiangsu, Jiangxi, Fujian, Hunan and Guangdong.

小叶猕猴桃 *Actinidia lanceolata*

## 毛花猕猴桃
**Actinidia eriantha** Benth.

大型落叶攀援灌木。髓白色，片层状。叶卵形至阔卵形，下面密具白色星状绒毛，纸质。聚伞花序1-3花；花玫瑰粉红色；花瓣5，倒卵形。果实柱状卵球形，直径2.5-3厘米，密被白色绒毛；宿存萼片反折。花期5-6月，果期11月。生海拔200-1000米的林中或低山灌丛中。产中国西南、华南和东南。

Large deciduous climbing shrubs. Pith white, lamellate. Leaves ovate to broadly ovate, abaxially densely white stellate tomentose, papery. Panicles cymose, 1-3-flowered; flowers rose-pink, petals 5, obovate. Fruits terete-ovoid, 2.5-3 cm diam, densely white-tomentose; persistent sepals reflexed. Fl. May-Jun. Fr. Nov. Forests or thickets on low mountains at 200-1000 m. Distributed in SW, S and SE China.

毛花猕猴桃 *Actinidia eriantha*

## 两广猕猴桃
**Actinidia liangguangensis** C. F. Liang

大型常绿藤本。髓白色，片层状。叶软革质，卵形，基部心形，背面淡绿色，密被淡黄褐色压紧的、星状绒毛，侧脉网状。花白色，花瓣瓢状倒卵形。果实卵珠状至柱状长圆形，密被黄褐色绒毛。花期4-5月，果期11月。生海拔200-1000米的山地山谷灌丛中或林中向阳处。产广东、广西和湖南。

Climbing shrubs, large, evergreen. Pith white, lamellate. Leaves softly leathery, ovate, base cordatulate, pale green abaxially, densely brownish appressed stellate tomentose, veinlets reticulate. Flowers white, petals spatulate-obovate. Fruit ovoid to cylindric-oblong, densely brownish tomentose. Fl. Apr-May. Fr. Nov. Forests in sunny places, thickets, mountain valleys at 200-1000 m. Distributed in Guangdong, Guangxi and Hunan.

两广猕猴桃 *Actinidia liangguangensis*

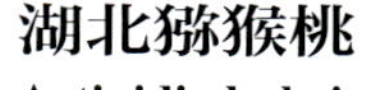

## 湖北猕猴桃
**Actinidia hubeiensis** H. M. Sun et R. H. Huang

落叶藤本。髓褐色，片层状。叶纸质，狭卵形至阔卵形，基部近心形，边缘具睫毛状锯齿，叶背疏被星状柔毛，易脱落；腹面光滑。雌花单生，具绒毛，花瓣5或6，倒卵形，白色至紫色，后期基部变黄。果实深棕色，卵形至圆锥状至近球形，成熟时光滑无毛，皮孔密集，棕色；宿存花萼反折。花期4-5月，果期10月。生山地灌丛中。产湖北(宜昌)。

Climbing shrubs, deciduous. Pith brownish, lamellate. Leaves narrowly ovate to broadly ovate, papery, base mostly cordate, margin ciliate-serrulate, abaxially very sparsely stellate tomentose with hairs easily rubbing off; adaxially glabrous. Female flowers solitary, puberulent; petals 5 or 6, obovate, white to purplish, and then yellowish at base. Fruit dark brown, ovoid to conical to subglobose, glabrescent when mature, densely lenticellate, lenticels brownish; persistent sepals reflexed. Fl. Apr-May. Fr. Oct. Mountain bushes. Distributed in Hubei (Yichang).

湖北猕猴桃 *Actinidia hubeiensis*

## 漓江猕猴桃

**Actinidia lijiangensis** C. F. Liang et Y. X. Lu

大型落叶藤本。小枝密被棕色绒毛；髓棕色，片层状。叶阔卵形或阔倒卵形至倒卵形，顶端锐尖至渐尖至尾尖，纸质，背面疏被星状毛，易脱落。聚伞花序1-3小花，被棕色绒毛；花白色，花瓣倒卵形。果实狭长圆柱形，光滑，密被棕色斑点，宿存萼片

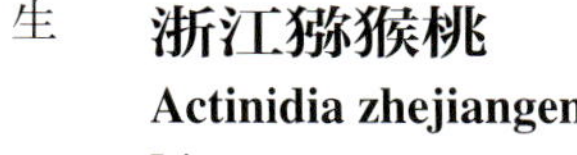

反折。花期4-5月，果期10月。生山坡灌丛中。产广西东北部。

Climbing shrubs, large, deciduous. Young branchlets densely brownish velutinous; pith brown, lamellate. Leaves broadly ovate or broadly obovate to obovate, apex acute to abruptly acuminate to cuspidate, papery, abaxially sparsely stellate tomentose, soon glabrous. Inflorescences cymose, 1-3-flowered, brownish velutinous; flowers white; petals obovate. Fruit narrowly cylindric, glabrous, lenticels brownish, dense; persistent sepals reflexed. Fl. Apr-May. Fr. Oct. Bushes on slopes. Distributed in NE Guangxi.

漓江猕猴桃 *Actinidia lijiangensis*

## 浙江猕猴桃

**Actinidia zhejiangensis** C. F. Liang

大型落叶或半常绿藤本。髓白色或棕色，片层状。叶卵形至披针形，基部耳状心形，纸质，背面白色，密被棕色绒毛及不完全的星状毛，叶脉明显。花序密被棕黄色毛；苞片钻形；花红色或粉色，花瓣倒卵圆形至狭倒卵圆形。果实圆柱形至长圆形，密被白色长绒毛。花期5月，果期10月。生山坡灌丛中。产福建和浙江。

Climbing shrubs, large, deciduous or semi-evergreen. Pith whitish or brownish, lamellate. Leaves ovate-lanceolate, base auriculate-cordatulate, papery, abaxially whitish and densely brownish tomentose with imperfectly stellate hairs, veins conspicuous abaxially. Inflorescences densely yellowish brownish tomentose; bracts subulate; flowers rose-pink; petals obovate to narrowly obovate. Fruit cylindric-oblong, densely long whitish tomentose. Fl. May. Fr. Oct. Bushes on slopes. Distributed in Fujian and Zhejiang.

浙江猕猴桃 *Actinidia zhejiangensis*

中华猕猴桃 *Actinidia chinensis*

## 中华猕猴桃

**Actinidia chinensis** Planch.

大型落叶攀援灌木。髓片层状。叶纸质，长6-17厘米，下面被星状绒毛。花白色或橙黄色；花瓣(3-)5(-8)。果实近球形至圆柱形至倒卵球形或椭圆体形，长4-6厘米，被易脱落的绒毛或被长硬毛或刺毛。花期4-5月，果期9月。生海拔200-2600米的山地林中、灌丛或次生疏林中，亦广泛栽培。产中国西南、华南、东南、华中、华西和华东。

Large deciduous climbing shrubs. Pith lamellate. Leaves papery, 6-17 cm long, abaxially densely stellate-tomentose. Flowers white or orange-yellow; petals (3-)5 (-8). Fruits subglobose to cylindric to obovoid or ellipsoidal, 4-6 cm long, pubescent, hirsute or setiform. Fl. Apr-May. Fr. Sep. Montane forests, thickets or secondary sparse forests at 200-2600 m, also widely cultivated. Distributed in SW, S, SE, C, W and E China.

## 贡山猕猴桃

**Actinidia pilosula** (Fin. et Gagn.) Stapf ex Hand.-Mazz.

中型落叶攀援灌木。髓片层状。叶卵状长圆形至卵状圆形，薄纸质，下面全部或仅沿中脉和侧脉密具白色柔毛，叶背横脉突出。花序聚伞状，腋生，具(1-)5-7花；花淡黄色，花瓣5。果实球形，直径约2.3厘米，无毛，具白色圆皮孔。花期5-6月，果期9-10月。生海拔约2000米的山地林中。产云南(贡山县)。

Mid-sized deciduous climbing shrubs. Pith lamellate. Leaves ovate-oblong to ovate-orbicular, thinly papery, abaxially densely white pubescent throughout or ± so on midvein and lateral veins, midvein and lateral veins conspicuous abaxially with raised parallel cross-bars. Panicles cymose, axillary, (1-)5-7-flowered; flowers yellowish, petals 5. Fruits globose, ca. 2.3 cm diam, glabrous, with whitish and rounded lenticels. Fl. May-Jun. Fr. Sep-Oct. Montane forests at ca. 2000 m. Distributed in Yunnan (Gongshan County).

贡山猕猴桃 *Actinidia pilosula*

## 大花猕猴桃

**Actinidia grandiflora** C. F. Liang

大型落叶藤本。髓褐色，片层状。叶纸质，倒卵形，两侧对称，基部圆形至微心形，叶背薄被淡黄色的不分枝或分枝乃至近星状的柔毛，叶脉相当发达，在叶背隆起呈圆线形。花序一回分歧，一序3花；苞片线形；花黄色，花瓣瓢状倒卵形，退化子房圆球形。花期5-6月。生海拔1800米的山地疏林中。产四川(天全二郎山)。

Climbing shrubs, large, deciduous. Pith brown, lamellate. Leaves, obovate, papery, abaxially sparsely yellowish and long pubescent, hairs simple to furcate to substellate, midvein and lateral veins abaxially conspicuous, and raised parallel cross-bars abaxially. Inflorescences cymose, 1-3-flowered; bracts linear; flowers yellowish, petals spatulate-obovate; sterile ovary globose. Fl. May-Jun. Montane forests at 1800 m. Distributed in Sichuan (Erlang Mountain, Tianquan).

大花猕猴桃 *Actinidia grandiflora*

刚毛藤山柳 *Clematoclethra scandens*

尼泊尔水东哥(锥序水东哥) *Saurauia napaulensis*

## 刚毛藤山柳

**Clematoclethra scandens** Maxim.

落叶木质藤本。小枝褐色，被刚毛。叶纸质至革质，卵形、长圆形、披针形或倒卵形，边缘不反卷，有胼胝质睫状小锯齿，腹面叶脉上有刚毛；背面全部被或厚或薄的细绒毛；叶柄只被刚毛。花序被细绒毛或兼被刚毛；花白色；萼片矩卵形；花瓣瓢状倒矩卵形。果实蒴果，具5棱。花期6-7月，果期7-9月。生海拔1000-3900米的山林中。产四川。

Climbing shrubs, deciduous. Branchlets brown, strigillose. Leaves papery to leathery, highly variable, usually ovate, elliptic, broadly ovate, lanceolate, or oblong-obovate, margin not reflexed, finely bristle-toothed, veins strigose adaxially; densely or sparsely hairy abaxially; petioles only strigose. Cymes hairy or setose; flowers white; sepals oblong-ovate; petals broadly ovate to broadly elliptic. Capsule, 5-ridged. Fl. Jun-Jul. Fr. Jul-Sep. Forests, thickets, mountain slopes, valleys at 1000-3900 m. Distributed in Sichuan.

## 朱毛水东哥

**Saurauia miniata** C. F. Liang et Y. S. Wang

灌木或小乔木。小枝密具锈色绒毛，混生有爪甲状鳞片。叶长圆状椭圆形，具23-30对侧脉。聚伞花序3-4簇生于老枝叶腋；萼片无毛；花粉红色。果实绿色至白色，扁球状。花期5-6月，果期10月。生海拔500-1500米的林中、灌丛中、河岸或山谷。产广西和云南。

Shrubs or small trees. Branchlets densely ferruginous tomentose, hairs intermixed with unguiculate scales. Leaves oblong-elliptic, with 23-30 pairs of lateral veins. Cymes 3-4-fascicled, axillary on old branches; sepals glabrous; flowers pink. Fruits green to white, depressed-globose. Fl. May-Jun. Fr. Oct. Forests, thickets, riverbanks or valleys at 500-1500 m. Distributed in Guangxi and Yunnan.

朱毛水东哥 *Saurauia miniata*

## 尼泊尔水东哥(锥序水东哥)

**Saurauia napaulensis** DC.

乔木。小枝具褐色柔毛至渐无毛。叶狭椭圆形至长圆状倒卵形，具28-40(-46)对侧脉。圆锥花序单个腋生，圆锥状，长12-33厘米；花粉色至淡紫色。果实绿色至淡黄色，球状至扁球形。花果期6-12月。生海拔400-3200米的林中、灌丛或山谷。产贵州、四川、广西和云南。南亚和东南亚亦有。

Trees. Branchlets brown pubescent to glabrescent. Leaves narrowly elliptic to oblong-obovate, with 28-40(-46) pairs of lateral veins. Panicles axillary, solitary, paniculate, 12-33 cm long; flowers pink to purplish. Fruits green to yellowish, globose to depressed-globose. Fl. and fr. Jun-Dec. Forests, thickets or valleys at 400-3200 m. Distributed in Guizhou, Sichuan, Guangxi and Yunnan. Also in S and SE Asia.

红果水东哥 *Saurauia erythrocarpa*

## 红果水东哥

**Saurauia erythrocarpa** C. F. Liang et Y. S. Wang

乔木。小枝疏具褐色或锈色丝毛。叶狭椭圆形，边缘为重锯齿，齿尖有刺状尖头；侧脉密，22-28对。花粉红色；雄蕊70-80；柱头中部以上4或5裂。果粉红色，扁球形或近球形。花期9-10月，果期10-11月。生海拔1400-1500米的山地林中或山谷中。产云南和西藏。

Trees. Branchlets sparsely brown or ferruginous velutinous. Leaves narrowly elliptic, margin double serrate, teeth apex spiny-mucronulate, lateral nerves dense, 22-28 pairs. Flowers pink; stamens 70-80; styles 4- or 5-fid above middle. Fruits pink, depressed-globose or subglobose. Fl. Sep-Oct. Fr. Oct-Nov. Montane forests or valleys at 1400-1500 m. Distributed in Yunnan and Xizang.

## 大花水东哥

**Saurauia punduana** wall.

小乔木。小枝被爪甲状鳞片，无毛。叶革质，狭椭圆形至狭倒卵形，长33厘米，先端渐尖，侧脉约34对，背面密被尘埃状褐色小柔毛。花序聚伞式，1-3枚簇生于叶腋，长约5厘米，有花2-3朵；花粉白色，大，直径约2厘米；苞片阔椭圆形，长4毫米；花柱中部以上5裂。花果期全年。生海拔700-1700米的山地阔叶林中。产西藏。不丹、印度和缅甸亦有。

Small trees. Branchlets with unguicular scales, glabrous. Leaves leathery, narrowly elliptic to narrowly obovate, 33 cm long, apex acuminate, lateral veins ca. 34 pairs, abaxially densely brown scurfy-puberulent. Inflorescences cymose, 1-3-fascicled in leaf axils, ca. 5 cm long, 2-3-flowered; flowers pinkish white, large, ca. 2 cm diam; bracts broadly elliptic, ca. 4 mm long; styles 5 lobed above middle. Fl. and fr. all year. Broad-leaved forests on mountains at 700-1700 m. Distributed in Xizang. Also in Bhutan, India and Myanmar.

## 水东哥

**Saurauia tristyla** DC.

乔木。叶倒卵形至阔椭圆状倒卵形，10-28 × 4-11厘米，纸质。聚伞花序生落叶老枝，腋生，长1-5厘米，具1-3花；花粉红色至白色。果绿白色至淡黄色，球状。花期3-7月，果期8-12月。生海拔 100-1700米的阔叶林、灌丛或山谷。产中国西南和华南。印度、尼泊尔、

大花水东哥 *Saurauia punduana*

水东哥 *Saurauia tristyla*

多脉水东哥 *Saurauia polyneura*

泰国和马来西亚亦有。

Trees. Leaves obovate to broadly elliptic-obovate, 10-28 × 4-11 cm, papery. Cymose inflorescences in axils of fallen leaves on old branches, 1-5 cm long, 1-3-flowered; flowers pink to white. Fruits greenish white to pale yellow, globose. Fl. Mar-Jul. Fr. Aug-Dec. Broad-leaved forests, thickets or valleys at 100-1700 m. Distributed in SW and S China. Also in India, Nepal, Thailand and Malaysia.

## 多脉水东哥

**Saurauia polyneura** C. F. Liang et Y. S. Wang

乔木。叶膜质或薄革质，边缘具细锯齿，侧脉28-40对或更多，两面无毛。圆锥花序单生叶腋，长7-33厘米；花粉色至白色，直径6-10毫米；子房近球形，绿色或淡绿色；雄蕊50。花期7-9月，果期9-11月。生海拔 1200-3200米的山地林中或山谷。产西藏和云南。

Trees. Leaves membranous or thinly coriaceous, margin denticulate, lateral veins 28-40 pairs or more, glabrous on both surfaces. Inflorescences solitary, paniculate, 7-33 cm long, axillary; flowers pink to white, 6-10 mm diam; ovary subglobose, green or yellowish-green; stamens 50. Fl. Jul-Sep. Fr. Sep-Nov. Montane forests or valleys at 1200-3200 m. Distributed in Xizang and Yunnan.

## 长毛水东哥

**Saurauia macrotricha** Kurz ex Dyer

乔木。植物体遍被褐色长硬刺毛，毛长5-6毫米。叶狭披针形，纸质，侧脉17对。花序长约2.5厘米，腋生，2或3个束生，具1-3花，生幼小枝上；花粉红色。花期6月。生海拔900-1400米的林中或山谷。产云南。印度和缅甸亦有。

Trees. Whole plant with brown and long setose hairs, hairs 5-6 mm long. Leaves narrowly lanceolate, papery, lateral veins 17 pairs. Inflorescences ca. 2.5 cm long, axillary, 2- or 3-fascicled, 1-3-flowered, on young branchlets; flowers pink. Fl. Jun. Forests or valleys at 900-1400 m. Distributed in Yunnan. Also in India and Myanmar.

长毛水东哥 *Saurauia macrotricha*

# 金莲木科 Ochnaceae

金莲木 *Ochna integerrima*

## 金莲木
**Ochna integerrima** (Lour.) Merr.

灌木或小乔木。叶椭圆形、倒卵状长圆形或倒卵状披针形，纸质。花序近伞房状；萼片长圆形，花期反折，果期红色；雄蕊3轮；子房10-12室；花柱圆柱状；柱头盘状，5-6裂。花期3-4月，果期5-6月。生海拔300-1400米的山谷石旁或溪边湿润的空旷地。产广西、广东西南部和海南。南亚和东南亚亦有。

Shrubs or small trees. Leaves elliptical, obovate-oblong, or obovate-lanceolate, papery. Inflorescences subcymose; sepals oblong, reflexed during anthesis, red in fruit; stamens 3-whorled; ovary 10-12-locular; styles terete; stigmas disc, 5-6-lobed. Fl. Mar-Apr. Fr. May-Jun. By rocks in valleys or wet open places by rivers at 300-1400 m. Distributed in Guangxi, SW Guangdong and Hainan. Also in S and SE Asia.

## 赛金莲木
**Campylospermum striatum** (Tiegh.) M. C. E. Amaral

常绿灌木。叶近革质，窄长圆形至披针形，边近全缘或有不明显疏细齿。圆锥花序短，腋生或顶生；花萼5片，红色，结果时增大，宿存；花瓣5，长圆状披针形，黄色。核果近肾形。花期4-11月，果期8-12月。生海拔60-700米的热带雨林中。产海南西南部和中部。越南亦有。

Evergreen shrubs. Leaves subcoriaceous, narrowly oblong to lanceolate, margin entire or inconspicuous and sparsely serrulate. Panicles short, axillary or terminal; sepals 5, red, swollen at fruit, persistent; petals 5, oblong-lanceolate, yellow. Drupes sub-reniform. Fl. Apr-Nov. Fr. Aug-Dec. Tropical rain forests at 60-700 m. Distributed in SW and C Hainan. Also in Vietnam.

## 合柱金莲木
**Sauvagesia rhodoleuca** (Diels) M. C. E. Amaral

直立小灌木。叶纸质，狭披针形或狭椭圆形，侧脉多数，近平行。花药箭头形，纵向开裂；退化雄蕊两轮，白色；外轮多数，阔匙形；中轮10枚，花瓣状，长圆形。蒴果卵球形，成熟时3瓣裂。花期4-5月，果期6-7月。生海拔1000米的山谷水旁或密林中。产广西和广东。

Small erect shrubs. Leaves papery, narrowly lanceolate or narrowly ellipsoid, lateral veins numerous, nearly parallel. Anthers sagittate, longitudinally dehiscent; staminodes in 2 whorls, white; in outer whorl numerous, broadly spatulate; in middle whorl 10, petaloid, oblong. Capsules ovoid, 3-valved when ripe. Fl. Apr-May. Fr. Jun-Jul. River banks in valleys or dense forests at 1000 m. Distributed in Guangxi and Guangdong.

赛金莲木 *Campylospermum striataum*

合柱金莲木 *Sauvagesia rhodoleuca*

# 山茶科
# Theaceae

油茶 *Camellia oleifera*

## 大苞白山茶
**Camellia granthamiana** Sealy

灌木或乔木。叶革质，侧脉6-7对。花单生，白色，直径10-14厘米；苞萼片12-17；花瓣8-10；雄蕊多轮；子房有毛；花柱5，离生但是基部贴合。蒴果圆球形，5裂。花期12月至翌年1月，果期翌年8-9月。生海拔100-300米的常绿阔叶林中。产广东东部。

Shrubs or trees. Leaves coriaceous, lateral veins 6-7-paired. Flowers solitary, white, 10-14 cm diam; bracteoles and sepals 12-17; petals 8-10; stamens numerous, many-verticillate; ovary hairy; styles 5, distinct but basally connivent.Capsules globose, 5-valved. Fl. Dec to next Jan. Fr. next Aug-Sep. Evergreen broad-leaved forests at 100-300 m. Distributed in E Guangdong.

## 猴子木
**Camellia yunnanensis** (Pitard ex Diels) Cohen-Stuart

灌木至小乔木。叶先端短渐尖或短尖。花单生；苞萼片9-11；花瓣7-12，白色；子房碟形，无毛或稍具柔毛和绒毛，(3-)5室；花柱(3-)5，离生。蒴果紫红色，球形至扁球形。花期11月至翌年3月，果期翌年8-10月。生海拔(800-)1100-3200米的疏林或灌丛中。产云南、四川和贵州西南部。

Shrubs to small trees. Leaves apex shortly acuminate or mucronate. Flowers solitary; bracteoles and sepals 9-11; petals 7-12, white; ovary discoid, glabrous or ± pilose to tomentose, (3-)5-loculed; styles (3-)5, distinct. Capsules purplish red, globose to oblate. Fl. Nov to next Mar. Fr. Aug-Oct next year. Open forests or thickets at (800-)1100-3200 m. Distributed in Yunnan, Sichuan and SW Guizhou.

## 油茶
**Camellia oleifera** C. Abel

灌木或乔木。叶革质至硬革质。花腋生或近顶生，单生或成对，近无柄；苞萼片8-11；花瓣5-7，白色；子房球形，具白色绒毛，3室。蒴果球形至椭圆体形。花期12月至翌年1月，果期9-10月。生海拔(200-)500-1800米的林中或灌丛中。产中国西南、华南、东南、华中和华东。缅甸北部、老挝北部和越南北部亦有。世界广泛栽培。

Shrubs or trees. Leaves leathery to rigidly leathery. Flowers axillary or subterminal, solitary or paired, subsessile; bracteoles and sepals 8-11; petals 5-7, white; ovary globose, white tomentose, 3-loculed. Capsules globose to ellipsoid. Fl. Dec to next Jan. Fr. Sep-Oct. Forests or thickets at (200-)500-1800 m. Distributed in SW, S, SE, C and E China. Also in N Myanmar, N Laos and N Vietnam. Cultivated. worldwide

大苞白山茶 *Camellia granthamiana*

猴子木 *Camellia yunnanensis*

红皮糙果茶 *Camellia crapnelliana*

## 红皮糙果茶
**Camellia crapnelliana** Tutch.

灌木或小乔木。树皮红色。叶硬革质，椭圆形至长圆椭圆形，侧脉6对。花单生或成对，顶生或腋生；花瓣6-8，白色；子房密具绒毛，3-5室；花柱3-5，离生。蒴果灰褐色，近球形，果皮厚1-2厘米，被糠秕状物。花期12月至翌年1月，果期9-10月。生海拔100-800米的林中。产广东、广西、福建、浙江和江西。

Shrubs or small trees. Bark red. Leaves hard coriaceous, elliptic to oblong-elliptic, lateral veins 6 pairs. Flowers solitary or paired, terminal or axillary; petals 6-8, white; ovary densely tomentose, 3-5-loculed; styles 3-5, distinct. Capsules grayish brown, subglobose, pericarp 1-2 cm thick, furfuraceous. Fl. Dec to next Jan. Fr. Sep-Oct. Forests at 100-800 m. Distributed in Guangdong, Guangxi, Fujian, Zhejiang and Jiangxi.

## 长瓣短柱茶
**Camellia grijsii** Hance

灌木。叶革质，侧脉6-7对。花腋生或近顶生，芳香，白色，直径3-5厘米；苞片及萼片7或8，半圆形至近圆形；花瓣5-6；子房有黄色长粗毛；花柱无毛，先端3浅裂。蒴果球形，直径2-2.5厘米。花期1-3月，果期9-10月。生海拔100-1500米的常绿林中。产中国西南、东南和华中。

Shrubs . Leaves coriaceous, lateral veins 6-7 pairs. Flowers axillary or subterminal, fragrant, white, 3-5 cm diam; bracteoles and sepals 7 or 8, semiorbicular to suborbicular; petals 5-6; ovary yellow-hirsute; styles glabrous, apex 3-lobed. Capsules globose, 2-2.5 cm diam. Fl. Jan-Mar. Fr. Sep-Oct. Forests at 100-1500 m. Distributed in SW, SE and C China.

长瓣短柱茶 *Camellia grijsii*

落瓣油茶 *Camellia kissii*

## 落瓣油茶
**Camellia kissii** Wall.

灌木至小乔木。叶革质，侧脉7对。花腋生；苞萼片7-9；花瓣白色，5-8朵，近离生；子房有白色长绒毛，3室；花柱无毛或基部具绒毛，3全裂或3深裂。蒴果梨形或球形。花期11-12月，果期9-10月。生海拔300-2000(-3100)米的林下或灌丛中。产云南、广西、广东和海南。南亚和东南亚亦有。

Shrubs to small trees. Leaves coriaceous, lateral veins 7-paired. Flowers axillary; bracteoles and sepals 7-9; petals white, 5-8, nearly distinct; ovary white tomentose, 3-loculed; styles glabrous or basally tomentose, 3-divided or 3-parted. Capsules pyriform or globose. Fl. Nov-Dec. Fr. Sep-Oct. Forests or thickets at 300-2000(-3100) m. Distributed in Yunnan, Guangxi, Guangdong and Hainan. Also in S and SE Asia.

## 杜鹃叶山茶
**Camellia azalea** C. F. Wei

常绿灌木。嫩枝灰褐色，当年枝淡红棕色，无毛。叶革质，边全缘，略外卷。花顶生，近无柄；苞片及萼片8-11枚，外轮3-6枚，内轮5枚具缘毛；花瓣6-9，玫瑰色；雄蕊无毛；子房无毛，3或4室；花柱无毛。蒴果卵形，3室，每室2粒种子，果壳3爿裂开。花期10-12月，果期8-9月。生海拔100-500米的丘陵地区河边林中。产广东(阳春)。

Shrubs, evergreen. Young branches grayish brown; current year branchlets reddish brown, glabrous. Leaves leathery, margin entire and slightly revolute.

杜鹃叶山茶 *Camellia azalea*

Flowers subterminal, subsessile; bracteoles and sepals 8-11, outer 3-6, inner 5 margin ciliolate; petals 6-9, roseate; stamens glabrous; ovary glabrous, 3(or 4)-loculed; style glabrous. Capsules ovoid, 3-loculed with 2 seeds per locule; pericarp splitting into 3 valves. Fl. Oct-Dec. Fr. Aug-Sep. Forests in hilly areas, among boulders by rivers at 100-500 m. Distributed in Guangdong (Yangchun).

## 多齿红山茶
**Camellia polyodonta** F. I. How ex Hu

灌木或乔木。叶缘具篦齿状细齿。花单个近顶生或腋生；苞片及萼片约15；花瓣6或7，玫瑰红色或稀白色；子房球形，3室。蒴果球状，外果皮木质，粗糙。花期1-2月，果期9-10月。生海拔(100-)500-1000米的林中。产广西、广东和湖南。

Shrubs or trees. Leaves margin pectinately serrulate. Flowers subterminal or axillary, solitary; bracteoles and sepals ca. 15; petals 6 or 7, rose or rarely white; ovary globose, 3-loculed. Capsules globose, pericarps woody, furfuraceous. Fl. Jan-Feb. Fr. Sep-Oct. Forests at (100-)500-1000 m. Distributed in Guangxi, Guangdong and Hunan.

## 滇山茶
**Camellia reticulata** Lindl.

灌木至乔木。叶革质，边缘具细齿。花腋生或近顶生，单生至3朵一簇；苞片及萼片9-11；花瓣5-7(栽培者更多)，玫瑰红色至粉色；子房球形，具绒毛，3(-5)室；花柱联合。蒴果球形至扁球形。花期1-3月，果期9-10月。生海拔1000-3200米的阔叶林或混交林中。产云南西南部、四川和贵州西部。

Shrubs to trees. Leaves leathery, margin serrulate. Flowers axillary or subterminal, solitary or to 3 in a cluster; bracteoles and sepals 9-11; petals 5-7 (often more for some cultivars), rose to pink; ovary globose, tomentose, 3(-5)-loculed; styles confluent. Capsules globose to oblate. Fl. Jan-Mar. Fr. Sep-Oct. Broad-leaved forests or mixed forests at 1000-3200 m. Distributed in Yunnan, SW Sichuan and W Guizhou.

多齿红山茶 *Camellia polyodonta*

滇山茶 *Camellia reticulata*

西南山茶 *Camellia pitardii*

## 西南山茶
**Camellia pitardii** Cohen-Stuart

灌木至小乔木。叶缘具细锯齿或尖锯齿，先端渐尖至尾尖。花近顶生，单生，近无柄；苞片及萼片9-10；花瓣5或6，玫瑰红色、粉色或白色；子房球形，具绒毛，3室。蒴果扁球形或近球形，2.5-3.5(-7) × 3-5(-8)厘米。花期12月至翌年3月，果期8-10月。生海拔500-2500(-2700)米的阔叶林或灌丛中。产云南、四川、贵州、广西、湖北和湖南。

Shrubs to small trees. Leaves margin serrulate or apiculately serrulate, apex acuminate to caudate. Flowers subterminal, solitary, subsessile; bracteoles and sepals 9-10; petals 5 or 6, rose, pink, or white; ovary globose, tomentose, 3-loculed. Capsules depressed globose or subglobose, 2.5-3.5(-7) × 3-5(-8) cm. Fl. Dec to next Mar. Fr. Aug-Oct. Broad-leaved forests or thickets at 500-2500(-2700) m. Distributed in Yunnan, Sichuan, Guizhou, Guangxi, Hubei and Hunan.

## 怒江山茶
**Camellia saluenensis** Stapf ex Bean

灌木至小乔木。叶片长圆形，硬革质，先端急尖或钝。花单生或成对，近无柄；苞片及萼片8-10；花瓣5-7，玫瑰红色、粉色或白色；子房具白色绒毛，3室。蒴果近球形。花期1-3月，果期10月。生海拔1200-2800(-3200)米的密林下或灌丛中。产云南中部及西部、四川西南部和贵州西部。

Shrubs to small trees. Leaves oblong, rigidly coriaceous, apex acute or obtuse. Flowers solitary or paired, subsessile; bracteoles and sepals 8-10; petals 5-7, rose, pink, or white; ovary white tomentose, 3-loculed. Capsules subglobose. Fl. Jan-Mar. Fr. Oct. Dense forests or thickets at 1200-2800(-3200) m. Distributed in C and W Yunnan, SW Sichuan and W Guizhou.

怒江山茶 *Camellia saluenensis*

## 东南山茶
**Camellia edithae** Hance

灌木或小乔木。幼枝灰棕色，光滑。叶披针形，侧脉7-9对，基部心形，被毛。花瓣5-6，红色；苞片及萼片卵形；子房具绒毛，3室；花柱无毛，顶端3浅裂。蒴果棕色，扁球形至近球形。花期1-3月，果期10月。生海拔200-1000米的林中。产广东东北部、福建和江

东南山茶 *Camellia edithae*

山茶 *Camellia japonica*

西东南部。

Shrubs or small trees. Young branches grayish brown, glabrous. Leaves lanceolate, lateral veins 7-9-paired, base cordate, hairy. Petals 5-6, red; bracteoles and sepals ovate; ovary tomentose, 3-loculed; styles glabrous, apically 3-lobed. Capsules brown, oblate to subglobose. Fl. Jan-Mar. Fr. Oct. Forests at 200-1000 m. Distributed in NE Guangdong, Fujian and SE Jiangxi.

## 山茶

**Camellia japonica** L.

灌木或乔木。叶革质，叶缘具锯齿。花腋生或近顶生，单生或成对，近无柄；苞片及萼片约9；花瓣6或7，栽培品种更多，玫瑰红或白色；子房卵球形，3室；花柱顶端3裂。蒴果球形，直径3-5厘米。花期1-3月，果期9-10月。生海拔300-1100米的林中。产台湾、浙江东部和山东。朝鲜半岛南部和日本南部亦有。

Shrubs or trees. Leaves coriaceous, margin serrate. Flowers axillary or subterminal, solitary or paired, subsessile; bracteoles and sepals ca. 9; petals 6 or 7 but often more for cultivars, rose or white; ovary ovoid, 3-loculed; styles apically 3-lobed. Capsules globose, 3-5 cm diam. Fl. Jan-Mar. Fr. Sep-Oct. Forests at 300-1100 m. Distributed in Taiwan, E Zhejiang and Shandong. Also in S Korean Peninsula and S Japan.

## 抱茎短蕊茶

**Camellia amplexifolia** Merr. et Chun

灌木至小乔木。叶革质，长圆形或长圆状披针形，基部心形，抱茎，无柄；叶柄长1-2毫米。花顶生或腋生；花瓣黄白色，6片；雄蕊极短；子房无毛。花期5-6月。生海拔约1100米的山谷疏林中。产海南(保亭县)。

Shrubs to small trees. Leaves coriaceous, oblong or oblong-lanceolate, base cordate, amplexicaul, sessile; petioles 1-2 mm long. Flowers terminal or axillary; petals yellowish-white, 6; stamens very short; ovary glabrous. Fl. May-Jun. Open forests in valleys ca. 1100 m. Distributed in Hainan (Baoting County).

抱茎短蕊茶 *Camellia amplexifolia*

金花茶 *Camellia petelotii*

显脉金花茶 *Camellia euphlebia*

## 金花茶

**Camellia petelotii** (Merr.) Sealy

灌木或乔木。叶革质，两面无毛。花腋生或近顶生，单生或成对；小苞片(6-)8-10，开展；萼片5，革质；花瓣10-14，金黄色，肉质；子房球形，无毛，3室；花柱3，离生。蒴果扁球形。花期11月至翌年2月，果期9-11月，生海拔100-900米的溪谷或溪边林中。产广西南部。越南北部亦有。

Shrubs or trees. Leaves leathery, both surfaces glabrous. Flowers axillary or subterminal, solitary or paired; bracteoles (6-)8-10, spreading; sepals 5, leathery; petals 10-14, golden yellow, fleshy; ovary globose, glabrous, 3-loculed; styles 3, distinct. Capsules oblate. Fl. Nov to next Feb. Fr. Sep-Nov. Forests in river valleys or along streams at 100-900 m. Distributed in S Guangxi. Also in N Vietnam.

## 淡黄金花茶

**Camellia flavida** Hung T. Chang

常绿灌木。嫩枝灰褐色，当年枝淡紫红色，无毛。叶柄具凹槽，叶薄革质，边缘有细锯齿。单花腋生，花柄长3-5毫米；苞片5-6，萼片5；花瓣7-13，淡黄色；雄蕊无毛；子房无毛；花柱3，完全分离。蒴果扁球形。种子棕色，具红棕色绒毛。花期11月至翌年1月，果期翌年9-10月。生海拔100-500米的石灰质丘陵常绿阔叶林中。产广西南部和西南部。

Shrubs, evergreen. Young branches grayish brown; current year branchlets purplish red, glabrous. Petioles adaxially grooved; leaves thinly leathery, margin serrulate. Flowers axillary, solitary, pedicel 3-5 mm long; bracteoles 5-6, sepals 5; petals 7-13, pale yellow; stamens glabrous; ovary glabrous; styles 3, distinct. Capsules oblate. Seeds brown, reddish brown villous. Fl. Nov to next Jan. Fr. next Sep-Oct. Evergreen broad-leaved forests on calcareous hills at 100-500 m. Distributed in S and SW Guangxi.

## 显脉金花茶

**Camellia euphlebia** Merr. ex Sealy

灌木或小乔木。叶革质，中脉在下面突起，侧脉11-13对。花单生于叶腋；苞片8片；花瓣8-9，金黄色；子房无毛，3室；花柱3，离生。蒴果扁球形。花期2月，果期10月。生海拔100-500米的溪边常绿阔叶林下。产广西(防城县)。越南北部亦有。

Shrubs or small trees. Leaves coriaceous, midvein abaxially elevated, lateral veins 11-13-paired. Flowers solitary, axillary; bracteoles 8; petals 8-9, golden-yellow; ovary glabrous, 3-loculed; styles 3, free. Capsules oblate. Fl. Feb. Fr. Oct. Evergreen broad-leaved forests along streams at 100-500 m. Distributed in Guangxi (Fangcheng County). Also in N Vietnam.

## 凹脉金花茶

**Camellia impressinervis** H. T. Chang et S. Ye Liang

灌木或乔木。叶革质，下面具红褐色腺点，中脉和侧脉在上面极凹陷。花单个腋生；小苞片5或6；花瓣11或12，黄色；

淡黄金花茶 *Camellia flavida*

凹脉金花茶 *Camellia impressinervis*

子房卵球形，无毛，3室；花柱3，离生。蒴果扁球形。花期1-3月，果期10月。生海拔100-500米的石灰岩山地常绿阔叶林中。产广西(龙州县)。

Shrubs or trees. Leaves leathery, abaxially reddish brown glandular punctate, midvein and secondary veins adaxially clearly impressed. Flowers solitary, axillary; bracteoles 5 or 6; petals 11 or 12, yellow; ovary ovoid, glabrous, 3-loculed; styles 3, distinct. Capsules oblate. Fl. Jan-Mar. Fr. Oct. Evergreen broad-leaved forests on calcareous hills at 100-500 m. Distributed in S Guangxi (Longzhou County).

## 顶生金花茶

**Camellia pingguoensis** D. Fang var. **terminalis** (J. Y. Liang et Z. M. Su) T. L.Ming et W. J. Zhang

灌木。叶薄革质，下面具褐色腺点；花近顶生，直径3.5-4.5厘米；小苞片5或6(或7)，萼片绿色，革质；花瓣7或8，淡黄色；子房近球形，无毛，3室；花柱3，离生。蒴果球形至近球形。花期12月至翌年1月。生海拔100-500米的石灰岩山丘林中。产广西(天等县)。

Shrubs. Leaves thinly leathery, abaxially brown glandular punctate. Flowers subterminal, 3.5-4.5 cm diam; bracteoles 5 or 6(or 7), sepals green, coriaceous; petals 7 or 8, pale yellow; ovary subglobose, glabrous, 3-loculed; styles 3, distinct. Capsules globose to subglobose. Fl. Dec to next Jan. Forests on calcareous hills at 100-500 m. Distributed in Guangxi (Tiandeng County).

## 毛瓣金花茶

**Camellia pubipetala** Y. Wan et S. Z. Huang

灌木或乔木。叶革质，边缘具细齿。花单个腋生或近顶生，近无柄；小苞片(4-)6-8；萼片5或6；花瓣9-13，柠檬黄色，外面具灰色柔毛；子房具黄色绒毛，3或(4室)。蒴果扁球形，具3条纵沟。花期1-2月，果期10月。生海拔200-400米的石灰岩山丘林中。产广西(大新县、隆安县)。

Shrubs or trees. Leaves leathery, margin serrulate. Flowers axillary or subterminal, solitary, subsessile; bracteoles (4-)6-8; sepals 5 or 6; petals 9-13, lemon yellow, outside gray puberulent; ovary yellowish tomentose, 3(or 4)-loculed. Capsules oblate with 3 longitudinal grooves. Fl. Jan-Feb. Fr. Oct. Forests on calcareous hill at 200-400 m. Distributed in Guangxi (Daxin County, Long'an County).

毛瓣金花茶 *Camellia pubipetala*

顶生金花茶 *Camellia pingguoensis* var. *terminalis*

茶 *Camellia sinensis*

### 茶

**Camellia sinensis** (L.) Kuntze

灌木或小乔木。叶革质，下面无毛或幼时疏具柔毛，先端钝尖；长4-12厘米，侧脉5-7对。花1-3朵腋生，白色；小苞片2；花萼外无毛；子房密生白色毛。蒴果三角状球形。花期10-12月，果期9-10月。生海拔100-2200米的阔叶林或灌丛中。产中国西南、华南、东南、华中和华东。南亚和东南亚亦有。广泛栽培。

Shrubs or small trees. Leaves coriaceous, abaxially glabrous or sparsely pubescent only when young, apex bluntly acute, 4-12 cm long, lateral veins 5-7-paired. Flowers 1-3, axillary, white; bracteoles 2; sepals outside glabrous; ovary densely white hairy. Capsules trigonous-globose. Fl. Oct-Dec. Fr. Sep-Oct. Broad-leaved forests or thickets at 100-2200 m. Distributed SW, S, SE, C and E China. Also in S and SE Asia. Widely cultivated.

### 普洱茶

**Camellia sinensis** (L.) Kuntze var. **assamica** (Mast.) Kitamura

本变种与原变种的区别在于：叶椭圆形，长8-14厘米，先端渐尖或急尖，下面沿中脉有疏柔毛。子房顶端无毛。花期12月至翌年2月，果期8-10月。生海拔(100-)500-1500(-1900)米的常绿阔叶林中。产云南南部、广西南部、广东南部和海南。老挝、缅甸、泰国和越南亦有。

This rarietes differs from *Camellia sinensis* in its: leaves elliptic, 8-14 cm long, apex acuminate or acute, abaxially pilose along costa. ovary apically glabrous. Fl. Dec to next Feb. Fr. Aug-Oct. Evergreen broad-leaved forests at (100-)500-1500(-1900) m. Distributed in S Yunnan, S Guangxi, S Guangdong and Hainan. Also in Laos, Myanmar, Thailand and Vietnam.

普洱茶 *Camellia sinensis* var. *assamica*

### 连蕊茶

**Camellia cuspidata** (Kochs) H. J. Veitch

灌木。叶革质，长圆状椭圆形、椭圆形或长圆状披针形，先端尾尖。花单生；萼片5；花瓣5-7，白色或淡红色；子房卵球形，无毛，3室。蒴果球形。花期12月至翌年4月，果期8-10月。生海拔(100-)500-1500(-2200)米的林中或灌丛。产中国西南、东南、华中和华东。

Shrubs. Leaves coriaceous, oblong-elliptic, elliptic, or oblong-lanceolate, apex caudate. Flowers solitary; sepals 5; petals 5-7, white or pale red; ovary ovoid, glabrous, 3-loculed. Capsules globose. Fl. Dec to next Apr. Fr. Aug-Oct. Forests or thickets along river at (100-)500-1500(-2200) m. Distributed in SW, SE, C and E China.

连蕊茶 *Camellia cuspidata*

### 粗梗连蕊茶

**Camellia crassipes** Sealy

灌木或乔木。叶革质，椭圆卵圆形至椭圆形。花梗粗壮，先端加厚；小苞片2片；花萼深杯状；长6-7毫米，萼片5，不等大；花冠白色，花瓣6，最外面2片小，内侧4，宽卵形；子房无毛，先端3裂。蒴果近球形。花期2-3月，果期10月。生海拔900-2500米的林中或灌丛中。产云南中部、东北部和东南部。

Shrubs or trees. Leaves coriaceous, elliptic-ovate to elliptic. Pedicels stout, thickened toward

粗梗连蕊茶 *Camellia crassipes*

云南连蕊茶 *Camellia forrestii*

apex; bracteoles 2; sepals deeply cup-shaped, 6-7 mm long; sepals 5, unequal; corolla white; petals 6, outer 2 small, inner 4, broadly ovate; ovary glabrous, apex 3-lobed. Capsules subglobose. Fl. Feb-Mar. Fr. Oct. Forests or thickets at 900-2500 m. Distributed in C, NE and SE Yunnan.

## 云南连蕊茶

**Camellia forrestii** (Diels) Cohen-Stuart

灌木至小乔木。叶中脉于两面凸起。花腋生，单花或成对；萼片半月形至近圆形，先端圆；花瓣5-6，白色，有时花苞粉红色或紫红色；子房卵球形，无毛，3室。蒴果卵球形。花期11月至翌年5月，果期9-10月。生海拔(1200-)1600-2900(-3200)米的密林或灌丛中。产云南。越南北部亦有。

Shrubs to small trees. Leaves midvein raised on both surfaces. Flowers axillary, solitary or paired; sepals semilunar to suborbicular, apex rounded; petals 5-6, white, but sometimes pink or purplish red in bud; ovary ovoid, glabrous, 3-loculed. Capsules ovoid. Fl. Nov to next May. Fr. Sep-Oct. Dense forests or thickets at (1200-)1600-2900(-3200) m. Distributed in Yunnan. Also in N Vietnam.

## 心叶毛蕊茶

**Camellia cordifolia** (F. P. Metcalf.) Nakai

灌木或乔木。叶革质，长圆状披针形或长卵形，侧脉6-7对。花单生或成对；小苞片4-5，半圆形或宽卵形，外面有毛；萼片5，卵形或近圆形，外面密被柔毛；花瓣5-7，白色；子房有丝状毛；花柱多毛。蒴果近球形。花期11月至翌年1月，果期9-10月。生海拔(200-)300-1700(-2000)米的林中或灌丛中。产云南、贵州、广西、广东、福建、湖南和江西。

Shrubs or trees. Leaves coriaceous, oblong-lanceolate or long-ovate, lateral veins 6-7 pairs. Flowers solitary or in pairs; bracteoles 4-5, semiorbicular or broadly ovate, abaxially hairy; sepals 5, ovate to suborbicular, outside appressed pubescent; petals 5-7, white; ovary sericeous; styles hairy. Capsules subglobose. Fl. Nov to next Jan. Fr. Sep-Oct. Forests or thickets at (200-)300-1700(-2000) m. Distributed in Yunnan, Guizhou, Guangxi, Guangdong, Fujian, Hunan and Jiangxi.

心叶毛蕊茶 *Camellia cordifolia*

长尾毛蕊茶 *Camellia caudata*

## 长尾毛蕊茶
**Camellia caudata** Wall.

灌木至小乔木。叶先端长尾尖至尾尖。萼片5，宿存；花腋生，单朵至3朵成一簇；花瓣5-7，白色；子房密具白色绒毛；花柱顶端3浅裂。蒴果椭圆状球形。花期12月至翌年1月，果期9-10月。生海拔(200-)400-1200(-2200)米的常绿阔叶林或灌丛中。产中国西南、华南和华中。印度东北部、尼泊尔、缅甸北部和越南北部亦有。

Shrubs to small trees. Leaves apex long caudate to caudate. Sepals 5, persistent; flowers axillary, solitary or to 3 in a cluster; petals 5-7, white; ovary densely white tomentose; styles apically 3-parted. Capsules ellipsoid-globose. Fl. Dec to next Jan. Fr. Sep-Oct. Evergreen broad-leaved forests or thickets at (200-)400-1200(-2200) m. Distributed in SW, S and C China. Also in NE India, Nepal, N Myanmar and N Vietnam.

## 多瓣核果茶
**Pyrenaria jonquieriana** Pierre ex Laness. subsp. **multisepala** (Merr. et Chun) S. X. Yang

乔木。叶革质，长圆形或倒披针形，边缘有锯齿。花无柄，直径3.5-4.5厘米，黄色；萼片厚革质，8-10片，花后脱落；花瓣8-10，黄色；花柱合生。核果不开裂，狭椭圆形。花期5-6月，果期10-12月。生海拔800米的常绿林中。产海南。

Trees. Leaves coriaceous, oblong or oblanceolate, margin serrate. Flowers sessile, 3.5-4.5 cm diam, yellow; calyx lobes thickly coriaceous, 8-10, shedding after blossom; petals 8-10, yellow; styles connate. Drupes indehiscent, narrowly ellipsoid. Fl. May-Jun. Fr. Oct-Dec. Evergreen forests at 800 m. Distributed in Hainan.

## 小果核果茶(小果石笔木)
**Pyrenaria microcarpa** (Dunn) H. Keng

灌木或乔木。叶卵形、矩圆状卵形或倒披针形。花瓣5-7，白色或淡黄色；子房圆锥形，3室，密具绢毛；花柱无毛。蒴果三角状球形、卵球形、椭圆体形或倒卵球形，1.5-2 × 1-1.5 厘米。花期4-7月，果期8-11月。生海拔100-1000米的山谷林中或溪边。产中国西南、华南和东南。越南和琉球群岛亦有。

Shrubs or trees. Leaves elliptic, oblong-elliptic or oblanceolate. Petals 5-7, white or pale yellow; ovary conic, 3-loculed, densely sericeous; styles glabrous. Capsules triangular-globose, ovoid, ellipsoid or obovoid, 1.5-2 × 1-1.5 cm. Fl. Apr-Jul. Fr. Aug-Nov. Forests in mountain valleys or along streams at 100-1000 m. Distributed in SW, S and SE China. Also in Vietnam and Ryukyu Islands.

## 大头茶
**Polyspora axillaris** (Roxb. ex Ker Gawl.) Sweet

灌木或乔木。叶厚革质，倒披针形，全缘。花生于枝顶叶腋，单

多瓣核果茶 *Pyrenaria jonquieriana* subsp. *multisepala*

小果核果茶(小果石笔木) *Pyrenaria microcarpa*

大头茶 *Polyspora axillaris*

黄药大头茶 *Polyspora chrysandra*

生或成对；萼片宿存；花瓣5，白色；子房具绒毛，5室；花柱长2厘米，有绢毛。蒴果圆柱形，5瓣裂。花期9-10月，果期11-12月。生海拔100-800(-2300)米的林中或灌丛中。产广东、广西、海南和台湾。越南亦有。

Shrubs or trees. Leaves thickly coriaceous, oblanceolate, entire. Flowers axillary at apex of branchlets, solitary or paired; sepals persistent; petals 5, white; ovary tomentose, 5-loculed; styles 2 cm long, sericeous. Capsules cylindric, 5-valved. Fl. Sep-Oct. Fr. Nov-Dec. Forests or thickets at 100-800 (-2300) m. Distributed in Guangdong, Guangxi, Hainan and Taiwan. Also in Vietnam.

## 黄药大头茶

**Polyspora chrysandra** (Cowan) Hu ex B. M. Barthol. et T. L. Ming

灌木或小乔木。叶薄革质，狭倒卵形。花腋生，单生或3-5朵组成短总状，直径4-5厘米，淡黄色，有香气；花瓣5；子房卵球形，具白色绒毛，5室；花柱基部具白色绒毛。蒴果长圆状圆柱形，先端锐尖。花期11-12月，果期翌年8-9月。生海拔1100-2400米的阔叶林或常绿灌丛中。产云南、四川东南部和贵州北部。缅甸北部亦有。

Shrubs or small trees. Leaves thinly coriaceous, narrowly obovate. Flowers axillary, solitary or 3-5 in a short raceme, 4-5 cm diam, yellowish, fragrant; petals 5; ovary ovoid, white tomentose, 5-loculed; styles basally white tomentose. Capsules oblong-cylindric, apex acute. Fl. Nov-Dec. Fr. Aug-Sep next year. Broad-leaved forests or evergreen thickets at 1100-2400 m. Distributed in Yunnan, SE Sichuan and N Guizhou. Also in N Myanmar.

## 长果大头茶

**Polyspora longicarpa** (H. T. Chang) C. X. Ye ex B. M. Barthol. et T. L. Ming

乔木。花单生叶腋，白色；小苞片5；萼片5，外面密生黄色绢毛；花丝基部被短柔毛；子房卵球形，具白色绒毛；花柱具白色绢毛；柱头5，头状。蒴果长圆状圆柱形。花期10-11月，果期翌年9-10月。生海拔(1000-)1700-2500米的沟谷或山坡的常绿阔叶林中。产云南东南部和西南部。缅甸北部、泰国东北部和越南北部亦有。

Trees. Flowers solitary, axillary, white; bracteoles 5; sepals 5, abaxially densely yellow sericeous, filaments pubescent at base; ovary ovoid, white tomentose; styles white velutinous; stigma 5, capitate. Capsules oblong-cylindric. Fl. Oct-Nov. Fr. Sep-Oct next year. Evergreen broad-leaved forests on slopes or in valleys at (1000-)1700-2500 m. Distributed in SE and SW Yunnan. Also in N Myanmar, NE Thailand and N Vietnam.

长果大头茶 *Polyspora longicarpa*

海南大头茶 *Polyspora hainanensis*

## 海南大头茶

**Polyspora hainanensis** (Hung T. chang) C. X. Ye ex B. M. Barthol. et T. L. Ming

乔木。叶革质，狭长圆形或倒披针形，边缘有钝锯齿。花直径4厘米，白色；苞片3，早落；花瓣5。蒴果长筒形，5爿裂开。花期11-12月，果期翌年10月。生海拔300-1500米的常绿林中。产海南。

Tree. Leaves coriaceous, narrowly oblong or oblanceolate, margin obtusely serrate. Flowers 4 cm diam, white; bracts 3, caduceus; petals 5. Capsules oblong, 5-valved. Fl. Nov-Dec. Fr. Oct next year. Evergreen forests at 350-1500 m. Distributed in Hainan.

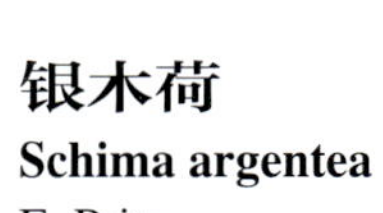

## 银木荷

**Schima argentea** E. Pritz.

乔木，高6-15米。嫩枝有毛。叶薄长圆形至披针形，革质，背面有白粉。花单生或3-8朵排成聚伞状总状花序；

小苞片2；花萼圆形；花瓣白色；子房具白色绒毛，顶端无毛。蒴果球形，开裂成5瓣。花期7-9月，果期12月。生海拔1600-2800(-3200)米的林中。产云南、四川、广西和江西。缅甸和越南亦有。

Trees, 6-15 m tall. Branchlets pubescent. Leaves thinly oblong to lanceolate, leathery, abaxial with white powders. Flowers solitary or 3-8 arranged in a corymbose raceme; bracteoles 2; sepals orbicular; petals white; ovary white tomentose but apically glabrous. Capsules globose, splitting into 5 valves. Fl. Jul-Sep. Fr. Dec. Forests at 1600-2800(-3200) m. Distributed in Yunnan, Sichuan, Guangxi and Jiangxi. Also in Myanmar and Vietnam.

## 短梗木荷

**Schima brevipedicellata** Hung. T. Chang

乔木，高7-25米。叶革质，倒卵形，长5-8厘米，侧脉8-11对。花单生或3-5朵排成聚伞状总状花序；小苞片2；花萼宿存，近圆形；花瓣白色；子房具绒毛，5室。蒴果扁球形，开裂成5瓣。花期7月，果期10-12月。生海拔500-1900米的常绿阔叶林中。产中国西南和华南。越南北部亦有。

Trees, 7-25 m tall. Leaves coriaceous, obovate, 5-8 cm long, lateral veins 8-11 pairs. Flowers solitary or in a corymbose raceme of 3-5; bracteoles 2; sepals persistent, suborbicular; petals white; ovary tomentose, 5-loculed. Capsules compressed globose, splitting into 5 valves. Fl. Jul. Fr. Oct-Dec. Evergreen broad-leaved forests at 500-1900 m. Distributed in SW and S China. Also in N Vietnam.

## 红木荷

**Schima wallichii** (DC.) Korth.

乔木。叶椭圆形至宽椭圆形，革质。花单朵腋生或3朵簇生，白色，芳香；小苞片2；花萼新月形至半圆形；子房球形，具黄色绒毛，顶端无毛，5室；花柱无毛。蒴果棕色，近球形。花期4-5月，果期11-12月。生海拔(300-)800-1800(-2700)米的林中。产云南、西藏、贵州和广西。南亚和东南亚亦有。

Trees. Leaves elliptic to broadly elliptic, leathery. Flowers axil-

银木荷 *Schima argentea*

短梗木荷 *Schima brevipedicellata*

红木荷 *Schima wallichii*

lary, solitary or to 3 in a cluster, white, fragrant; bracteoles 2; sepals lunate to semiorbicular; ovary globose, yellowish tomentose but apically glabrous, 5-loculed; styles glabrous. Capsules brown, subglobose. Fl. Apr-May. Fr. Nov-Dec. Forests at (300-) 800-1800(-2700) m. Distributed in Yunnan, Xizang, Guizhou and Guangxi. Also in S and SE Asia.

## 华木荷

**Schima sinensis** (Hemsl. et E. H. Wils.) Airy Shaw

乔木。叶革质，长圆状椭圆形至长圆形，侧脉12-14对；叶柄紫红色。花单生，直径4-5厘米；小苞片2；花萼近圆形；子房球形，具绒毛，顶端无毛，5室；花柱无毛。蒴果近球形。花期7-8月，果期10-11月。生海拔1400-2200米的林中。产云南、四川、贵州、广西、湖北和湖南。

Trees. Leaves coriaceous, oblong-elliptic to oblong, lateral veins 12-14 pairs; petioles purplish red; Flowers solitary, 4-5 cm diam; bracteoles 2; sepals suborbicular; ovary globose, tomentose but apically glabrous, 5-loculed; styles glabrous. Capsules subglobose. Fl. Jul-Aug. Fr. Oct-Nov. Forests at 1400-2200 m. Distributed in Yunnan, Sichuan, Guizhou, Guangxi, Hubei and Hunan.

## 印度木荷

**Schima khasiana** Dyer

乔木。叶革质，长10-16厘米；侧脉10-12对。花单生枝顶或叶腋，直径约6厘米；小苞片2，卵形；花瓣倒卵形，白色；子房具黄色绒毛；花柱无毛。蒴果球形，直径3-3.5厘米。花期6-7月，果期10-12月。生海拔900-1800(-2800)米的林中。产云南东南部和西部、西藏东南部。印度东北部、不丹、缅甸北部和越南北部亦有。

Trees. Leaves coriaceous, 10-16 cm long, lateral veins 10-12-paired. Flowers solitary at top of branchlets or at leaf axils, ca. 6 cm diam; bracteoles 2, ovate; petals obovate, white; ovary yellowish tomentose; styles glabrous. Capsules globose, 3-3.5 cm diam; Fl. Jun-Jul. Fr. Oct-Dec. Forests at 900-1800 (-2800) m. Distributed in SE and W Yunnan, and SE Xizang. Also in NE India, Bhutan, N Myanmar and N Vietnam.

华木荷 *Schima sinensis*

印度木荷 *Schima khasiana*

木荷 *Schima superba*

## 木荷

**Schima superba** Gardn. et Champ.

乔木。叶革质或薄革质，侧脉7-9对。4-8花组成总状花序，花直径2-3厘米；小苞片2；萼片半圆形；花瓣白色；子房具绒毛。蒴果近球形，直径1-2厘米。花期6-8月，果期10-12月。生海拔100-800(-1600)米的林中。产中国西南、华南、东南、华中和华东。琉球群岛亦有。

Trees. Leaves coriaceous or thinly coriaceous, lateral veins 7-9 pairs. Flowers 4-8 in a raceme, 2-3 cm diam; bracteoles 2; sepals semiorbicular; petals white; ovary tomentose. Capsules subglobose, 1-2 cm diam. Fl. Jun-Aug. Fr. Oct-Dec. Forests 100-800(-1600) m. Distributed in SW, S, SE, C and E China. Also in Ryukyu Islands.

## 云南紫茎

**Stewartia calcicola** T. L. Ming et J. Li

常绿乔木。叶长圆状椭圆形，纸质至薄革质；叶柄具翅。花单生；小苞片2；花萼圆形，先端圆；花瓣白色；子房圆锥形，无毛。蒴果圆锥形。花期5-6月，果期10-12月。生海拔900-1700米的林中。产云南东南部和广西西南部。

Evergreen trees. Leaves oblong-elliptic, papery to thinly leathery; petioles winged. Flowers solitary; bracteoles 2; sepals orbicular, apex rounded; petals white; ovary conic, glabrous. Capsules conic. Fl. May-Jun. Fr. Oct-Dec. Forests at 900-1700 m. Distributed in SE Yunnan and SW Guangxi.

翅柄紫茎 *Stewartia pteropetiolata*

云南紫茎 *Stewartia calcicola*

## 翅柄紫茎

**Stewartia pteropetiolata** W. C. Cheng

乔木。幼枝、叶柄和花梗被平伏柔毛；叶片边缘具锯齿；叶柄具翅。花单生叶腋，白色；花萼紫红色，长卵形，叶状，先端锐尖；子房卵球形，无毛。蒴果长卵球形。花期4-5月，果期9-11月。生海拔1200-2600米的林中。产云南南部和西部。

Trees. Young branches, petioles and pedicels depressed pubescent; leaves serrate at margin; petioles winged. Flowers solitary, axillary, white; sepals purplish red, long ovate, leaflike, apex acute; ovary ovoid, glabrous. Capsules long ovoid. Fl. Apr-May. Fr. Sep-Nov. Forests at 1200-2600 m. Distributed in S and W Yunnan.

## 心叶紫茎

**Stewartia cordifolia** (H. L. Li) J. Li et T. L. Ming

常绿乔木。嫩枝无毛，当年枝被短柔毛。叶长卵形，厚革质，边缘具稀锯齿，基部圆形至心形；叶柄被短柔毛。花单生，花梗4-6毫米，被短柔毛；小苞片披针形，被短柔毛；萼片被短柔毛；花瓣白色，椭圆

心叶紫茎 *Stewartia cordifolia*

形；子房无毛。蒴果圆锥形，5室，每室种子5-6粒。种子阔卵形，具窄翅。花期6-7月，果期9-10月。生海拔400-1300米的林中。产广西东北部、贵州东南部和湖南南部。

Evergreen trees. Young branches glabrous; current year branchlets pubescent. Leaves long ovate, thickly leathery , margin sparsely serrate; petioles pubescent. Flowers solitary, base rounded to cordate; pedicel 4-6 mm long, pubescent; bracteoles lanceolate, pubescent; sepals outside pubescent; petals white, elliptic; ovary glabrous. Capsules conic, 5-loculed with 5-6 seeds per locule. Seeds broadly ovate, narrowly winged. Fl. Jun-Jul. Fr. Sep-Oct. Forests at 400-1300 m. Distributed in NE Guangxi, SE Guizhou and S Hunan.

## 紫茎

**Stewartia sinensis** Rehd. et Wils.

落叶灌木或乔木。当年生小枝紫红色，无毛或具柔毛。叶柄具狭翅；叶缘具疏锯齿。花单生；花瓣白色；花萼叶状，先端急尖；子房圆锥形，具绒毛；花柱无毛。蒴果圆锥状。花期5-7月，果期9-11月。生海拔500-2200米的山上的林中或灌丛中。产中国西南、东南、华中和华东。

Shrubs or trees, deciduous. Current year branchlets purplish red, glabrous or villous. Petioles narrowly winged; leaves margin sparsely serrate. Flowers solitary; petals white; sepals leaflike, apex abruptly acute; ovary conical, tomentose; styles glabrous. Capsules conical. Fl. May-Jul. Fr. Sep-Nov. Forests or dense thickets on mountains at 500-2200 m. Distributed in SW, SE, C and E China.

紫茎 *Stewartia sinensis*

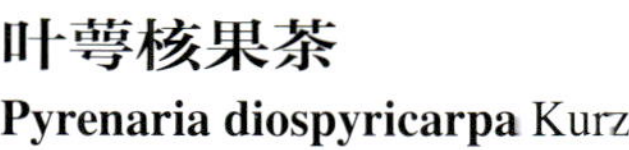

## 叶萼核果茶

**Pyrenaria diospyricarpa** Kurz

乔木，高5-15米。叶片纸质，基部楔形。花腋生，单生；花萼5，绿色，叶状；花瓣5或6，白色；雄蕊多数；子房球形，密具黄色绒毛，5室；花柱5，离生。果实长倒卵形。花期5月，果期9-11月。生海拔1000-2000米的山坡常绿阔叶林中或沟谷。产云南南部。缅甸、泰国和越南亦有。

Trees, 5-15 m tall. Leaves papery, base cuneate. Flowers axillary, solitary; sepals 5, green, leaflike; petals 5 or 6, white; stamens numerous; ovary globose, densely flavescent velutinous, 5-loculed; styles 5, distinct. Fruits long obovoid. Fl. May. Fr. Sep-Nov. Evergreen broad-leaved forests on slopes or in valleys at 1000-2000 m. Distributed in S Yunnan. Also in Myanmar, Thailand and Vietnam.

叶萼核果茶 *Pyrenaria diospyricarpa*

勐腊核果茶 *Pyrenaria menglaensis*

## 勐腊核果茶
**Pyrenaria menglaensis** G. D. Tao

乔木，高至15米。叶倒卵形，薄革质，长20-30厘米。花腋生，单生或3朵成一簇；花萼5(或6)，外面具黄色绢毛；花瓣5-7(-9)，白色；子房球形，具黄色绒毛，5室；花柱5(或6)，离生。果实核果状，球形，直径5-8厘米。花期5月，果期10-11月。生海拔600-700米的石灰岩山地林下。产云南(勐腊县)。

Trees to 15 m tall. Leaves obovate, thinly leathery, 20-30 cm long. Flowers axillary, solitary or to 3 in a cluster; sepals 5(or 6), outside yellow sericeous; petals 5-7(-9), white; ovary globose, yellow velutinous, 5-loculed; styles 5(or 6), distinct. Fruits drupaceous, globose, 5-8 cm diam. Fl. May. Fr. Oct-Nov. Under forests on limestone hills at 600-700 m. Distributed in Yunnan (Mengla County).

## 大果核果茶(石笔木)
**Pyrenaria spectabilis** (Champ.) C. Y. Wu et S. X. Yang

乔木。叶长圆形，革质。花单朵腋生或近顶生，直径4-10厘米；花萼9-11；子房密具黄色绒毛，3-6室；花柱3-6，于近顶部贴生。蒴果球形至扁球形，直径2-8厘米。花期5-6月，果期8-10月。生海拔300-1500米的常绿阔叶林中。产广西、广东、福建、湖南和江西。越南北部亦有。

Trees. Leaves oblong, leathery. Flowers axillary or subterminal, solitary, 4-10 cm diam; sepals 9-11; ovary densely yellowish tomentose, 3-6-loculed; styles 3-6, connate to near apex. Capsules globose to oblate, 2-8 cm diam. Fl. May-Jun. Fr. Aug-Oct. Evergreen broad-leaved forests at 300-1500 m. Distributed in Guangxi, Guangdong, Fujian, Hunan and Jiangxi. Also in N Vietnam.

## 浙江山茶
**Camellia chekiangoleosa** Hu

灌木或乔木。叶革质，侧脉约8对。花单生或成对，近顶生或腋生，直径7-10厘米；花瓣红色或粉红色；苞萼片9-10，宿存；子房卵球形，无毛，3室；花柱无毛，顶端3裂。蒴果球形，3-5室。花期11-12月，果期翌年10月。生海拔500-1300米的林中或灌丛中。产福建、浙

大果核果茶(石笔木) *Pyrenaria spectabilis*

浙江山茶 *Camellia chekiangoleosa*

江、湖南、安徽和江西。

Shrubs or trees. Leaves coriaceous, lateral veins ca. 8 paired. Flowers solitary or paired, subterminal or axillary, 7-10 cm diam; petals red or pink; bracts and sepals 9-10, persistent; ovary ovoid, glabrous, 3-loculed; styles glabrous, apically 3-lobed. Capsules globose, 3-5-loculed. Fl. Nov-Dec. Fr. Oct. next yaer Forests or thickets at 500-1300 m. Distributed in Fujian, Zhejiang, Hunan, Anhui and Jiangxi.

## 厚叶厚皮香(广东厚皮香)

**Ternstroemia kwangtungensis** Merr.

灌木或乔木。嫩枝有毛。叶厚革质，椭圆形，全缘，侧脉9-10对。花3-5朵生于枝顶；小苞片2；花萼卵形至近圆形，先端圆形；花瓣白色。果实扁球形，3-4(-5)室。花期5-6月，果期10-11月。生海拔700-1700米的山坡林中或溪边灌丛中。产广西南部、广东、福建和江西。越南亦有。

Shrubs or trees. Branches hairy when young. Leaves thickly coriaceous, elliptic, margin entire, lateral veins 9-10-paired. Flowers 3-5 at top of branchlets; bracteoles 2; sepals ovate to suborbicular, apex rounded; petals white. Fruits compressed-globose, 3-4(-5)-loculed. Fl. May-Jun. Fr. Oct-Nov. Forests on mountain slopes or thickets along streams at 700-1700 m. Distributed in S Guangxi, Guangdong, Fujian and Jiangxi. Also in Vietnam.

厚叶厚皮香(广东厚皮香) *Ternstroemia kwangtungensis*

## 大果厚皮香

**Ternstroemia insignis** Y. C. Wu

乔木高9-15米。叶较大，坚革质，全缘。花腋生，单生或数朵簇生于无叶小枝上；花萼圆形至卵形，先端圆形至钝；花瓣白色；子房卵球形，4室；花柱顶端4裂。果实圆球形或扁球形，2-2.5 × 1.6-3厘米。花期6-7月，果期8-10月。生海拔800-2600米的山地林中或林缘。产云南东南部、贵州西南部和广西北部。

Trees, 9-15 m tall. Leaves rather large, coriaceous, margin entire. Flowers axillary, solitary or several clustered on leafless branchlets; sepals orbicular to ovate, apex rounded to retuse; petals white; ovary ovoid, 4-loculed; styles apically 4-lobed. Fruits globose to compressed globose, 2-2.5 × 1.6-3 cm. Fl. Jun-Jul. Fr. Aug-Oct. Montane forests or forest edges at 800-2600 m. Distributed in SE Yunnan, SW Guizhou and N Guangxi.

大果厚皮香 *Ternstroemia insignis*

厚皮香 *Ternstroemia gymnanthera*

## 厚皮香
**Ternstroemia gymnanthera** (Wight et Arn.) Bedd.

乔木。叶革质或薄革质，上半部边缘具胼胝质圆齿。花腋生，单生或数朵簇生于无叶小枝，直径1-1.8厘米；花萼卵形至长卵形，先端圆形；花瓣浅黄色；子房卵球形，2室；花柱顶端2裂。果实成熟后紫红色，球形，直径1-1.5厘米。花期5-7月，果期9-11月。生海拔200-2800米的山地林中或灌丛中。产中国西南、东南、华中和华东。南亚和东南亚亦有。

Trees. Leaves coriaceous or thinly coriaceous, margin callose-crenate on upper part. Flowers axillary, solitary or several clustered on leafless branchlets, 1-1.8 cm diam; sepals ovate to long ovate, apex rounded; petals pale yellow; ovary ovoid, 2-loculed; styles apically 2-lobed. Fruits purplish red when mature, globose, 1-1.5 cm diam. Fl. May-Jun. Fr. Sep-Nov. Forests or thickets on mountains at 200-2800 m. Distributed in SW, SE, C and E China. Also in S and SE Asia.

全缘叶杨桐 *Adinandra integerrima*

## 全缘叶杨桐
**Adinandra integerrima** T. Anderson ex Dyer

乔木。叶长圆状椭圆形至长圆状卵形，革质，全缘或顶部具小波状缺刻。花单朵腋生；花梗反曲；花瓣白色；子房扁球形，具黄褐色绢毛，5室；柱头单生。果实成熟时红色，卵球形至球形。花期4-5月，果期9-10月。生海拔700-1900米的林中。产云南南部和东南部。缅甸、泰国、越南、柬埔寨和马来西亚亦有。

Trees. Leaves oblong-elliptic to oblong-ovate, leathery, margin entire or apical portion undulately denticulate. Flowers axillary, solitary; pedicels recurved; petals white; ovary depressed globose, yellowish brown sericeous, 5-loculed; styles simple. Fruits red when mature, ovoid to globose. Fl. Apr-May. Fr. Sep-Oct. Forests at 700-1900 m. Distributed in S and SE Yunnan. Also in Myanmar, Thailand, Vietnam, Cambodia and Malaysia.

杨桐（黄瑞木） *Adinandra millettii*

## 杨桐（黄瑞木）
**Adinandra millettii** (Hook. et Arn.) Benth. et Hook. f. ex Hance

灌木或小乔木。叶革质，长圆状椭圆形，全缘或顶部疏具细齿。花单生于叶腋，白色；子房球形，具柔毛，3室；花柱完全合生。果实成熟时黑色，球形。花期5-7月，果期8-10月。生海拔100-1300(-1800)米的山坡灌丛或林中。产中国西南、东南、华中和华东。越南亦有。

Shrubs or small trees. Leaves coriaceous, oblong-elliptic, margin entire or apically sparsely serrate. Flowers axillary, solitary, white; ovary globose, pubescent, 3-loculed; styles completely united. Fruits black when mature, globose. Fl. May-Jul. Fr. Aug-

红淡比 *Cleyera japonica*

华南毛柃 *Eurya ciliata*

Oct. Thickets or forests on mountain slopes at 100-1300 (-1800) m. Distributed in SW, SE, C and E China. Also in Vietnam.

## 红淡比

**Cleyera japonica** Thunb.

灌木或乔木。叶革质，全缘；顶芽长圆锥形，无毛。花3-5朵簇生；花瓣7-12，白色；子房球形，无毛，2室；花柱顶端2裂。果球形，成熟时紫黑色。花期5-6月，果期10-12月。生海拔200-2000米的山坡或山谷的林中或灌丛中。产中国西南、华南、东南、华中和华东。印度北部、缅甸、尼泊尔和日本亦有。

Shrubs or trees. Leaves leathery, margin entire; terminal buds long conic, glabrous. Flowers 3-5-clustered; petals 7-12, white; ovary globose, glabrous, 2-loculed; styles apically 2-lobed. Fruits globose, purplish black when mature. Fl. May-Jun. Fr. Oct-Dec. Forests or thickets on slopes or in valleys at 200-2000 m. Distributed in SW, S, SE, C and E China. Also in N India, Myanmar, Nepal and Japan.

厚叶红淡比 *Cleyera pachyphylla*

## 厚叶红淡比

**Cleyera pachyphylla** Chun ex Hung T. Chang

灌木或小乔木。叶厚革质，侧脉20-28对。花1-3朵簇生；花瓣白色；子房球形，无毛，2或3室；花柱长约9毫米，先端2或3裂。果成熟时黑色。花期6-7月，果期10-11月。生海拔 300-1800米的山坡或山顶林中。产广西、广东西部、福建、浙江南部、湖南和江西。

Shrubs or small trees. Leaves thickly coriaceous; lateral veins 20-28 pairs. Flowers solitary or to 3 in a cluster; petals white; ovary globose, glabrous, 2- or 3-loculed; styles ca. 9 mm long, apex 2- or 3-lobed. Fruits black when ripe. Fl. Jun-Jul. Fr. Oct-Nov. Forests on mountain slopes or tops at 300-1800 m. Distributed in Guangxi, W Guangdong, Fujian, S Zhejiang, Hunan and Jiangxi.

## 华南毛柃

**Eurya ciliata** Merr.

灌木或乔木。当年生小枝密具黄褐色开展柔毛。叶纸质。花1-3朵簇生；子房球形，密具柔毛，5室；花柱4或5，离生。果实球形。花期10-11月，果期翌年4-5月。生海拔100-1300米的林中或溪边。产云南东南部、广西南部、广东东南部和海南。

Shrubs or trees. Current year branchlets densely yellowish brown spreading villous. Leaves papery. Flowers axillary, solitary or to 3 in a cluster; ovary globose, densely pubescent, 5-loculed; styles 4 or 5, distinct. Fruits globose. Fl. Oct-Nov. Fr. Apr-May next year. Forests or streamsides at 100-1300 m. Distributed in SE Yunnan, S Guangxi, SE Guangdong and Hainan.

尖萼毛柃 *Eurya acutisepala*

## 尖萼毛柃

**Eurya acutisepala** Hu et L. K. Ling

灌木或乔木。叶薄革质，下面浅绿色，疏被柔毛。花2-3朵腋生；花萼卵形至长卵形，先端锐尖；子房卵球形，密具绒毛，3室；花柱顶端3裂。果实卵状椭圆体形至椭圆状球形，成熟时紫黑色。花期10-11月，果期翌年6-8月。生海拔500-2000米的林中或溪边。产中国西南和东南。

Shrubs or trees. Leaves thinly leathery, abaxially pale green and sparsely pubescent. Flowers 2-3 axillary; sepals ovate to long ovate, apex acute; ovary ovoid, densely pubescent, 3-loculed; styles apically 3-lobed. Fruits ovate-ellipsoid to elliptic-globose, purplish black when mature. Fl. Oct-Nov. Fr. Jun-Aug next year. Forests or streamsides at 500-2000 m. Distributed in SW and SE China.

格药柃 *Eurya muricata*

## 格药柃

**Eurya muricata** Dunn

灌木或乔木。叶长圆状椭圆形至椭圆形，侧脉9-11对。花1-5朵簇生；花药具多格；子房球形，无毛，3室；花柱顶部3瓣裂。果实成熟后紫黑色，球形。花期9-12月，果期翌年6-9月。生海拔300-1300米的林下或灌丛中。产中国西南、东南、华中和华东。

Shrubs or trees. Leaves oblong-elliptic to elliptic, secondary veins 9-11 on each side of midvein. Flowers axillary, solitary or to 5 in a cluster; anthers multi-locellate; ovary globose, glabrous, 3-loculed; styles apically 3-lobed. Fruits purplish black when mature, globose. Fl. Sep-Dec. Fr. Jun-Sep next year. Forest or thickets at 300-1300 m. Distributed in SW, SE, C and E China.

滨柃 *Eurya emarginata*

## 滨柃

**Eurya emarginata** (Thunb.) Makino

灌木。当年生小枝红褐色，密具黄褐色柔毛。叶厚革质，两面无毛。花腋生，单生或成对；子房球形，无毛，3室；花柱顶端3裂。果实成熟后黑色，球形。花期10-11月，果期翌年6-8月。生灌丛中或沿海岩石缝中。产台湾、福建东部和浙江东部。朝鲜半岛和日本亦有。

Shrubs. Current year branchlets reddish brown, densely yellowish brown pubescent. Leaves thickly leathery, both surfaces glabrous. Flowers axillary, solitary or paired; ovary globose, glabrous, 3-loculed; styles apically 3-lobed. Fruits black when mature, globose. Fl. Oct-Nov. Fr. Jun-Aug next year. Thickets or rock crevices along seacoasts. Distributed in Taiwan, E

岗柃 *Eurya groffii*

细枝柃 *Eurya loquaiana*

Fujian and E Zhejiang. Also in Korean Peninsula and Japan.

## 岗柃

**Eurya groffii** Merr.

灌木或乔木。叶薄或薄革质，下面密被贴生短柔毛；侧脉10-14对。花腋生，1-9朵簇生；花萼革质，外面被黄褐色柔毛；花瓣5，白色；子房无毛，3室，花柱3深裂近基部。果实球形，成熟后紫黑色。花期9-11月，果期12月至翌年4月。生海拔300-2700米的山地林中或灌丛中。产中国西南和华南。缅甸和越南北部亦有。

Shrubs or trees. Leave thin or thin-coriaceous, abaxially densely adnate-pubescent; lateral veins 10-14 pairs. Flowers axillary, solitary or to 9 in a cluster; sepals leathery, outside yellowish brown pubescent; petals 5, white; ovary glabrous, 3-loculed; styles 3-parted near to base. Fruits globose, purplish black when mature. Fl. Sep-Nov. Fr. Dec to next Apr. Forests or thickets on mountains at 300-2700 m. Distributed in SW and S China. Also in Myanmar and N Vietnam.

## 细枝柃

**Eurya loquaiana** Dunn

灌木或乔木。当年生小枝黄绿色，纤细，具柔毛。侧脉(8-)10-20对，叶缘具细锯齿。花1-4朵簇生；雄蕊10-15；花柱顶端3瓣裂。果实成熟后黑色，球形。花期10-12月，果期翌年7-9月。生海拔400-2000米的山坡或沟谷的林下或灌丛中。产中国西南、华南、东南、华中和华东。

Shrubs or trees. Current year branchlets yellowish green, slender, puberulent. Leaves secondary veins (8-)10-20 on each side of midvein, margin serrulate. Flowers axillary, solitary or to 4 in a cluster; stamens 10-15; styles apically 3-lobed. Fruits black when mature, globose. Fl. Oct-Dec. Fr. Jul-Sep next year. Forests or thickets on mountain slopes or in valleys at 400-2000 m. Distributed in SW, S, SE, C and E China.

## 细齿叶柃

**Eurya nitida** Korth.

灌木或乔木。叶边缘具紧密细锯齿、细圆齿或近全缘，侧脉9-12对。花1-4朵聚为一簇；雄蕊14-20；花药不分格；子房卵球形，无毛，3室；花柱顶端3瓣裂。果实成熟时蓝黑色，球形。花期11月至翌年1月，果期7-9月。生海拔500-1500(-2600)米的林下或灌丛中。产中国西南、华南、东南、华中和华东。南亚和东南亚亦有。

Shrubs or trees. Leaves margin closely serrulate, crenulate, or subentire, secondary veins 9-12 on each side of midvein. Flowers axillary, solitary or to 4 in a cluster; stamens 14-20; anthers not locellate; ovary ovoid, glabrous, 3-loculed; styles apically 3-lobed. Fruits bluish black when mature, globose. Fl. Nov to next Jan. Fr. Jul-Sep. Forests or thickets at 500-1500(-2600) m. Distributed in SW, S, SE, C and E China. Also in S and SE Asia.

细齿叶柃 *Eurya nitida*

柃木 *Eurya japonica*

## 柃木

**Eurya japonica** Thunb.

灌木。叶厚革质，两面无毛。花1-3朵腋生；花梗短，长约2毫米；雄蕊12-15；花药不分室；子房球形，无毛，3室；花柱顶端3裂。果圆球形。花期2-3月，果期9-10月。生海拔300-2500米的山坡或山谷灌丛中。产浙江东部和安徽。朝鲜半岛和日本亦有。

Shrubs. Leaves thickly coriaceous, both surfaces glabrous. Flowers 1-3 axillary; pedicels short, ca. 2 mm long; stamens 12-15; anthers not locellate; ovary globose, glabrous, 3-loculed; styles apically 3-lobed. Fruits globose. Fl. Feb-Mar. Fr. Sep-Oct. Thickets on mountain slopes or in valleys at 300-2500 m. Distributed in E Zhejiang and Anhui. Also in Korean Peninsula and Japan.

## 黑柃

**Eurya macartneyi** Champ.

灌木或乔木。树皮黑褐色，光滑。叶革质，侧脉12-14对。花1-4朵簇生；雄花有2小苞片；萼片革质；雌花花瓣5；子房卵球形，无毛，3室；花柱3，离生。果实成熟时紫黑色，球形。花期11月至翌年1月，果期6-8月。生海拔200-1000米的山坡或山谷灌丛中。产广西、广东、海南、福建、湖南和江西。

Shrubs or trees. Bark blackish brown, smooth. Leaves coriaceous, lateral veins 12-14 per side. Flowers axillary, solitary or to 4 in a cluster; male flowers with 2 bracteoles; sepals coriaceous; female flowers with 5 petals; ovary ovoid, glabrous, 3-loculed; styles 3, distinct. Fruits purplish black when mature, globose. Fl. Nov to next Jan. Fr. Jun-Aug. Thickets on mountain slopes or in valleys at 200-1000 m. Distributed in Guangxi, Guangdong, Hainan, Fujian, Hunan and Jiangxi.

## 单耳柃

**Eurya weissiae** Chun

灌木。叶侧脉9-11对，基部倾斜耳状抱茎。单花腋生或3朵聚成一簇，由一小叶状总苞包被；雄蕊约10；子房卵球形，无毛，3室；花柱顶部3瓣裂。果实成熟时蓝黑色，球形。花期9-10月，果期11月至翌年1月。生海拔300-1200米的山地林中。产贵州、广东、福建、浙江、湖南和江西。

Shrubs. Leaves secondary veins 9-11 on each side of midvein, base obliquely auriculate and amplexicaul. Flowers axillary, solitary or to 3 in a cluster, wrapped by a small leaflike involucre bract; stamens ca. 10; ovary ovoid, glabrous, 3-loculed; styles apically 3-lobed. Fruits bluish black when mature, globose. Fl. Sep-Oct. Fr. Nov to next Jan. Forests on mountain slopes at 300-1200 m. Distributed in Guizhou, Guangdong, Fujian, Zhejiang, Hunan and Jiangxi.

## 微毛柃

**Eurya hebeclados** Y. Ling

灌木或乔木。幼枝灰褐色，无毛或渐无毛。叶侧脉8-10对，两面无毛。花腋生，4-7朵成一簇；雄蕊约15；子房卵球形，无毛，3室；花柱顶端3裂。果实成熟后蓝黑色，球形。花期12月至翌年1月，果期8-10月。生海拔200-1700米的山坡林下或灌丛中。产中国西南、东南、华中和华东。

黑柃 *Eurya macartneyi*

单耳柃 *Eurya weissiae*

微毛柃 *Eurya hebeclados*

Shrubs or trees. Young branches grayish brown, glabrous or glabrescent. Leaves secondary veins 8-10 on each side of midvein, both surfaces glabrous. Flowers axillary, 4-7 in a cluster; stamens ca. 15; ovary ovoid, glabrous, 3-loculed; styles apically 3-parted. Fruits bluish black when mature, globose. Fl. Dec to next Jan. Fr. Aug-Oct. Forests or thickets on mountain slopes at 200-1700 m. Distributed in SW, SE, C and E China.

## 翅柃

**Eurya alata** Kobuski

灌木。幼枝灰褐色，4棱。叶长圆形至椭圆形，革质，两面无毛。花1-3朵簇生；子房球形，无毛，3室；花柱顶端3裂。果实球状，成熟时蓝黑色。花期10-11月，果期翌年6-8月。生海拔300-1600米的林中或山谷。产中国西南、东南、华中和华东。

Shrubs. Young branches grayish brown, 4-ribbed. Leaves oblong to elliptic, leathery, both surfaces glabrous. Flowers axillary, solitary or to 3-clustered; ovary globose, glabrous, 3-loculed; styles apically 3-lobed. Fruits globose, bluish black when mature. Fl. Oct-Nov. Fr. Jun-Aug next year. Forests or valleys at 300-1600 m. Distributed in SW, SE, C and E China.

钝叶柃 *Eurya obtusifolia*

## 钝叶柃

**Eurya obtusifolia** Hung T. Chang

灌木。叶侧脉5-7对，基部楔形，顶部钝或近圆形。花1-4朵簇生；雄蕊约10；子房球形，无毛，3室；花柱顶部3瓣裂。果实成熟时蓝黑色，球形。花期11月至翌年3月，果期8-10月。生海拔400-2600米的林下或灌丛中。产云南、四川、贵州、广西、湖北、湖南和陕西。

Shrubs. Leaves secondary veins 5-7 on each side of midvein, base cuneate, apex obtuse or subrounded. Flowers axillary, solitary or to 4 in a cluster; stamens ca. 10; ovary globose, glabrous, 3-loculed; styles apically 3-lobed. Fruits bluish black when mature, globose. Fl. Nov-next Mar. Fr. Aug-Oct. Forests or thickets at 400-2600 m. Distributed in Yunnan, Sichuan, Guizhou, Guangxi, Hubei, Hunan and Shaanxi.

翅柃 *Eurya alata*

岩柃 *Eurya saxicola*

## 岩柃
**Eurya saxicola** Hung T. Chang

灌木。叶厚革质，侧脉5-7对。花1-4朵簇生于叶腋；雄蕊5(或6)；花药不具分格；子房球形，无毛，3室；花柱顶端3裂。果实球形，成熟时紫黑色。花期9-10月，果期翌年6-8月。生海拔1500-2100米的林中、灌丛中或山崖。产中国西南、东南、华中和华东。

Shrubs. Leaves thickly coriaceous, lateral veins 5-7 per side. Flowers axillary, solitary or to 4-clustered; stamens 5(or 6); anthers not locellate; ovary globose, glabrous, 3-loculed; styles apically 3-lobed. Fruits purplish black when mature, globose. Fl. Sep-Oct. Fr. Jun-Aug next year. Forests, thickets or cliffs at 1500-2100 m. Distributed in SW, SE, C and E China.

## 云南凹脉柃
**Eurya cavinervis** Vesque

灌木或乔木。叶厚革质，侧脉8-10对，中脉上面凹陷，两面无毛。萼片边缘有黑色腺点；花腋生，单生或3朵成一簇；子房球形，无毛；花柱顶端3裂。果实球形。花期11月至翌年1月，果期7-9月。生海拔600-3500米的山坡林中或灌丛中。产广西、云南和西藏。印度(锡金)、尼泊尔、不丹和缅甸北部亦有。

Shrubs or trees. Leaves thickly coriaceous, lateral veins 8-10 pairs, midvein adaxially impressed, both surfaces glabrous. Sepals with black glands at margin; flowers axillary, solitary or to 3 in a cluster; ovary globose, glabrous; styles apically 3-lobed. Fruits globose. Fl. Nov to next Jan. Fr. Jul-Sep. Forests or thickets on mountain slopes at 600-3500 m. Distributed in Guangxi, Yunnan and Xizang. Also in India (Sikkim), Nepal, Bhutan and N Myanmar.

云南凹脉柃 *Eurya cavinervis*

## 茶梨
**Anneslea fragrans** Wall.

灌木或乔木。叶长圆形或长圆状披针形，背面具褐色腺点。花腋生，数朵至多于10朵形成聚伞状；花萼红色；花瓣浅黄色；子房半下位，无毛，2-或3(-5)室；花柱顶端2或3(-5)裂。果实球形至椭圆体形。花期10月至翌年3月，果期7-9月。生海拔300-2700米的林中或灌丛中。产中国西南、华南、东南和华中。缅甸、老挝、泰国、越南、柬埔寨和马来西亚亦有。

Shrubs or trees. Leaves oblong or oblong-lanceolate, abaxially brown glandular punctate. Flowers axillary, several to more than 10 in a corymb; sepals reddish; petals pale yellow; ovary half inferior, glabrous, 2- or 3(-5)-loculed; styles apically 2- or 3(-5)-lobed. Fruits globose to ellipsoid. Fl. Oct to next Mar. Fr. Jul-Sep. Forests or thickets at 300-2700 m. Distributed in SW, S, SE and C China. Also in Myanmar, Laos, Thailand, Vietnam, Cambodia and Malaysia.

茶梨 *Anneslea fragrans*

# 藤黄科
# Guttiferae

## 铁力木
**Mesua ferrea** L.

常绿乔木，高20-30米。叶革质，下垂，幼时淡红色，后为深绿色，革质，侧脉极多数。花两性，单朵腋生；雄蕊离生；花萼圆形，凸起；花瓣白色。果实宽卵球形或侧扁球形，常开裂成2瓣。花期3-5月，果期8-10月。生海拔500-600米的低丘陵坡地。产云南、广西和广东，通常栽培。其他亚洲热带地区亦有。

红厚壳（琼崖海棠） *Calophyllum inophyllum*

Evergreen trees, 20-30 m tall. Leaves coriaceous, pendulous, reddish when young, becoming dark green, leathery, with very numerous lateral nerves. Flowers bisexual, solitary, axillary; stamens free; sepals orbicular, convex; petals white. Fruits broadly ovoid or laterally depressed globose, usually dehiscent by 2 valves. Fl. Mar-May. Fr. Aug-Oct. Slopes of low hills at 500-600 m. Distributed in Yunnan, Guangxi and Guangdong, usually cultivated. Also in other tropical areas of Asia.

## 红厚壳 (琼崖海棠)
**Calophyllum inophyllum** L.

乔木。叶片两面光亮，阔椭圆形或倒卵状椭圆形，厚革质。聚伞圆锥花序生上部腋处，具7-11花；花有香气，白色，花瓣4。果实球形，成熟时黄色，直径约2.5厘米。花期3-6月，果期9-11月。生海拔100(-200)米的丘陵、荒地或海岸。产海南和台湾。南亚、东南亚、马达加斯加、澳大利亚、印度洋岛屿和太平洋岛屿亦有。

Trees. Leaves shiny on both surfaces, broadly elliptic or obovate-elliptic, thick coriaceous. Thyrses in upper axils, 7-11-flowered; flowers scented, white, petals 4. Fruits globose, yellow when mature, ca. 2.5 cm diam. Fl. Mar-Jun. Fr. Sep-Nov. Hills, wastelands or seashores at 100 (-200) m. Distributed in Hainan and Taiwan. Also in S and SE Asia, Madagascar, Australia, Indian Ocean Islands and Pacific Islands.

铁力木 *Mesua ferrea*

滇南红厚壳 *Calophyllum polyanthum*

## 大苞藤黄
**Garcinia bracteata** C. Y. Wu ex Y. H. Li

乔木，高约8米，雌雄异株。叶下绿色，卵形、卵状椭圆形或长圆形，革质，边缘骨质，反卷。花2-7朵排成伞形；花序梗先端具2叶状苞片。果序常具1果实；果实卵球形，先端常偏斜。花期4-5月，果期11-12月。生海拔400-1300(-1800)米的石灰岩山林中。产云南南部及东南部和广西南部。

Trees, ca. 8 m tall, dioecious. Leaves abaxially greenish, ovate, ovate-elliptic or oblong, leathery, margin cartilaginous, involute. Flowers in 2-7-flowered umbels; peduncles with 2 foliaceous bracts at apex. Infructescences usually 1-fruited; fruits ovoid, usually oblique at apex. Fl. Apr-May. Fr. Nov-Dec. Forests on limestone hills at 400-1300 (-1800) m. Distributed in S and SE Yunnan and S Guangxi.

大苞藤黄 *Garcinia bracteata*

## 滇南红厚壳
**Calophyllum polyanthum** Wall. ex Choisy

乔木，高约25米。幼枝密被灰色微柔毛。叶片长圆状椭圆形或卵状长圆形。聚伞圆锥花序顶生，常短于叶；花梗密具锈色柔毛；花白色；花瓣缺无。果序常具1或2个果实；果实球形，直径2-2.5厘米。花期4-5月，果期9-10月。生海拔1100-1800米的山谷密林中。产云南(景洪、澜沧县)。南亚和东南亚亦有。

Trees, ca. 25 m tall. Branchlets densely gray-puberulous. Leaves oblong-elliptic or ovate-oblong. Thyrses terminal, always shorter than leaves; pedicels densely rusty-puberulous; flowers white; petals absent. Infructescences usually with 1 or 2 fruit; fruits globose, 2-2.5 cm diam. Fl. Apr-May. Fr. Sep-Oct. Dense forests in valleys at 1100-1800 m. Distributed in Yunnan (Jinghong, Lancang County). Also in S and SE Asia.

## 大叶藤黄
**Garcinia xanthochymus** Hook. f. et T. Anderson.

乔木，高8-10米。叶厚革质，椭圆形或长圆形至长圆状披针形，(14-)20-34 × (4-)6-12 厘米；侧脉密，35-40对。伞房状聚伞花序；花丝基部连合成5束。浆果成熟后黄色，球形或卵球形，直径3-5厘米。花期3-5月，果期8-11月。生海拔(100-)600-1000(-1400)米的沟谷、丘陵或潮湿林中。产云南南部、西南部及西部、广西西南部和广东(栽培)。南亚亦有。

Trees, 8-10 m tall. Leaves thick coriaceous, elliptic or oblong to oblong-lanceolate, (14-)20-34 × (4-)6-12 cm, lateral veins dense, 35-40 pairs. Corymbose cymes; filaments connate in 5-fascided at base. Mature berries yellow, globose or ovoid, 3-5 cm diam. Fl. Mar-May. Fr. Aug-Nov. Valleys, hills or wet forests at (100-) 600-1000 (-1400) m. Distributed in S, SW and W Yunnan, SW Guangxi and Guangdong (cultivated). Also in S Asia.

大叶藤黄 *Garcinia xanthochymus*

木竹子 *Garcinia multiflora*

## 木竹子
**Garcinia multiflora** Champ. ex Benth.

乔木，雌雄同株。叶卵形、长圆状卵形或长圆状倒卵形，薄革质。雌花1-5；花直径2-3厘米，花瓣橙色；萼片2大2小，子房2室，果卵形至倒卵形，黄色。花期6-8月，果期11-12月。生海拔(100-)400-1200(-1900)米的次生林中或灌丛中。产中国西南、华南和东南。越南北部亦有。

Trees, monoecious. Leaves ovate, oblong-ovate or oblong-obovate, thinly leathery. Female flowers 1-5; flowers 2-3 cm diam; petals orange; sepals 2 large and 2 small; ovary 2-locular. Fruits ovoid to obovoid, yellow. Fl. Jun-Aug. Fr. Nov-Dec. Secondary forests or thickets at (100-) 400-1200 (-1900) m. Distributed in SW, S and SE China. Also in N Vietnam.

## 版纳藤黄
**Garcinia xishuanbannaensis** Y. H. Li

乔木，高6-15米，雌雄同株。叶椭圆形、椭圆状披针形或卵状披针形，纸质。花排成疏松圆锥聚伞状；雄花具退化雄蕊；能育雄蕊联合成束；萼片2大2小。成熟果实黄色，球形，直径4-5厘米。花期1-2月，果期4-5月。生海拔约600米的山谷林中。产云南(勐腊县)。

Trees, 6-15 m tall, monoecious. Leaves elliptic, elliptic-lanceolate, or ovate-lanceolate, papery. Flowers arranged in a lax paniculiform cyme; male flowers with staminodes; fertile stamens united into 1 fascicle; sepals 2 large and 2 small. Mature fruits yellow, globose, 4-5 cm diam. Fl. Jan-Feb. Fr. Apr-May. Forests in valleys at ca. 600 m. Distributed in Yunnan (Mengla County).

版纳藤黄 *Garcinia xishuanbannaensis*

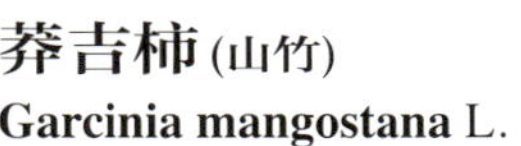

## 莽吉柿(山竹)
**Garcinia mangostana** L.

小乔木。叶光亮，椭圆形或椭圆状长圆形，厚革质，次级脉密生，达40-50对。雄花稀，2-9；雄蕊4束；雌花单生或成对。成熟果实紫红色，有时具黄褐色斑点，球形，直径5-8厘米。果肉白色，多汁，肉质。花期9-10月，果期11-12月。云南、广东、海南、台湾和福建栽培。原产印度尼西亚。

Trees small. Leaves shiny, elliptic or elliptic-oblong, thickly leathery, secondary veins dense, up to 40-50 pairs. Male flowers rare, 2-9; stamen fascicles 4; female flowers solitary or paired. Mature fruits purple-red, sometimes yellow-brown spotted, globose, 5-8 cm diam; pulp white, juicy, fleshy. Fl. Sep-Oct. Fr. Nov-Dec. Cultivated in Yunnan, Guangdong, Hainan, Taiwan and Fujian. Native to Indonesia.

莽吉柿(山竹) *Garcinia mangostana*

金丝李 *Garcinia paucinervis*

## 金丝李
**Garcinia paucinervis** Chun et F. C. How

乔木，雌雄同株。叶幼时紫红色，椭圆形、椭圆状长圆形或卵状长圆形，幼时膜质，后近革质。雌花常单花腋生；花序梗顶端无叶状苞片；柱头全缘。成熟果实椭球形或卵状椭球形。花期6-7月，果期11-12月。生海拔300-800米的石灰山林中。产云南(麻栗坡县)和广西西南部及西部。

Trees, monoecious. Leaves purple-red when young, elliptic, elliptic-oblong, or ovate-oblong, membranous when young, becoming subleathery. Female flowers usually solitary and axillary; peduncles apex without foliaceous bracts; stigmas entire. Mature fruits ellipsoid or ovoid-ellipsoid. Fl. Jun-Jul. Fr. Nov-Dec. Forests on limestone hills at 300-800 m. Distributed in Yunnan (Malipo County) and SW and W Guangxi.

## 云树
**Garcinia cowa** Roxb.

乔木，雌雄异株。叶披针形或长圆状披针形，纸质。雄花3-8，排成伞形，基部具4苞片；花瓣黄色；雄蕊4束。成熟果实暗黄褐色，卵状球形，有棱，沟槽4-8。花期5月，果期7-10月。生海拔(100-)400-900(-1300)米的山坡或沟谷湿润杂木林。产云南南部和西部。印度、孟加拉国东部、老挝、中南半岛、越南、柬埔寨和马来西亚亦有。

Trees, dioecious. Leaves lanceolate or oblong-lanceolate, papery. Male flowers 3-8, in an umbel, 4-bracteate at base; petals yellow; stamen fascicles 4. Mature fruits opaquely yellow-brown, ovoid-globose, angulate, 4-8-sulcate. Fl. May. Fr. Jul-Oct. Humid mixed forests on hills or in valleys at (100-)400-900(-1300) m. Distributed in S and W Yunnan. Also in India, E Bangladesh, Laos, Indo-China Peninsula, Vietnam, Cambodia and Malaysia.

云树 *Garcinia cowa*

## 岭南山竹子
**Garcinia oblongifolia** Champ. ex Benth.

乔木或灌木。叶长圆形、倒卵状长圆形至倒披针形，近革质，侧脉10-18对。花单生或为伞状聚伞花序；雄花花瓣橙色或淡黄色；花药聚集为头状。果实卵球形或球形，长2-4厘米。种子1。花期4-5月，果期10-12月。生海拔200-400(-1200)米的平原密或疏林、山坡或沟谷。产广西、广东和海南。

Trees or shrubs. Leaves oblong, obovate-oblong to oblanceolate, subleathery, secondary veins 10-18 pairs. Flowers solitary or in an umbel-like cyme; male flowers: petals orange or yellowish; anthers aggregated into a head. Fruits ovoid or globose, 2-4 cm long. Seeds 1. Fl. Apr-May. Fr. Oct-Dec. Dense or sparse forests on plains, hills, or in valleys at 200-400(-1200) m. Distributed in Guangxi, Guangdong and Hainan.

## 无柄金丝桃
**Hypericum augustinii** N. Robson

灌木。叶全无柄或下部具长1.5毫米的扁平叶柄。花序具(1-)3-13花，近伞房状；雄蕊5束；萼片直立，宽长圆形至宽椭圆

岭南山竹子 *Garcinia oblongifolia*

无柄金丝桃 *Hypericum augustinii*

尖萼金丝桃 *Hypericum acmosepalum*

形或卵形；花瓣亮金黄色；花药金黄色；柱头头状。蒴果宽卵球形。花期9-10月，果期11月。生海拔1200-1700米的溪岸、山坡或路边。产云南(景洪，石屏县)和贵州(安龙县)。

Shrubs. Leaves all sessile or lower with flat petioles to 1.5 mm long. Inflorescences (1-)3-13-flowered, subcorymboses; stamens 5-fascicled; sepals erect, broadly oblong to broadly elliptic or ovate; petals light golden yellow; anthers golden yellow; stigmas capitate. Capsules broadly ovoid. Fl. Sep-Oct. Fr. Nov. River banks, slopes or roadsides at 1200-1700 m. Distributed in Yunnan (Jinghong, Shiping County) and Guizhou (Anlong County).

## 金丝桃

**Hypericum monogynum** L.

灌木。叶片腺体小而点状。花序具1-15(-30)花；花萼稍开展，宽至狭椭圆形或倒披针形；花瓣金黄色至柠檬黄色；雄蕊每簇具25-35雄蕊，约与花瓣等长。蒴果宽卵球形。花期5-8月，果期8-9月。生海拔0-200(-1500)米的山坡、路旁或灌丛中。产中国西南、华南、东南、华中和华东。在日本归化。世界广泛栽培。

Shrubs. Leaves with minute and dotlike glands. Inflorescences 1-15(-30)-flowered; sepals ± spreading, broadly to narrowly elliptic or oblanceolate; petals golden yellow to lemon yellow; stamens fascicles each with 25-35 stamens, ca. as long as petals. Capsules broadly ovoid. Fl. May-Aug. Fr. Aug-Sep. Slopes, roadsides or thickets at 0-200(-1500) m. Distributed in SW, S, SE, C and E China. Naturalized in Japan. World widely cultivated.

## 尖萼金丝桃

**Hypericum acmosepalum** N. Robson

灌木。叶长圆形或椭圆状长圆形至狭椭圆形。花序具1-3(-6)花，近伞房状；花直径3-5厘米，星状；花萼卵形至狭披针形；苞片叶状至披针形，宿存；雄蕊5束，每束有雄蕊40-60。蒴果卵球形至狭卵状圆锥形。花期5-7月，果期8-9月。生海拔900-2700米的山坡路旁、灌丛、疏林或荒地上。产云南、四川、广西和贵州。

Shrubs. Leaves oblong or elliptic-oblong to narrowly elliptic. Flowers 1-3(-6) in corymbs, 3-5 cm diam, stellate; sepals ovate to narrowly lanceolate; bracts foliaceous to lanceolate, persistent; stamens 5-fascicled, per fascicle 40-60. Capsules ovoid to narrowly ovoid-conic. Fl. May-Jul. Fr. Aug-Sep. Roadsides on slopes, scrubs, open forests or wastelands at 900-2700 m. Distributed in Yunnan, Sichuan, Guangxi and Guizhou.

金丝桃 *Hypericum monogynum*

碟花金丝桃 *Hypericum addingtonii*

## 碟花金丝桃

**Hypericum addingtonii** N. Robson

灌木。叶片坚纸质，腺体点状或线条状。花序具1-3(-5) 花，花直径(3-)5-6.5厘米；苞片披针形，宿存；花瓣金黄色；雄蕊5束。蒴果卵形至圆柱状卵形。花期4-7月，果期10月。生海拔1800-3400米的竹丛中、灌丛、草坡或铁杉林缘。产云南西部和西北部。

Trees. Leaves hard-papery, with dotted or linear glands. Inflorescences 1-3(-5) -flowered, flowers (3-)5-6.5 cm diam; bracts lanceolate, persistent; petals golden yellow; stamens 5-fascicled. Capsules ovoid to cylindric-ovoid. Fl. Apr-Jul. Fr. Oct. Bamboo forests, scrubs, grassy slopes or *Tsuga* forest edges at 1800-3400 m. Distributed in W and NW Yunnan.

## 金丝梅

**Hypericum patulum** Thunb.

灌木。叶下较苍白色。花序具1-15花；花萼直立，常红色，宽卵形至宽椭圆形或近圆形至长圆状椭圆形或倒卵状匙形；花瓣金黄色；雄蕊每束具50-70。蒴果宽卵球形。花期5-9月，果期7-10月。生海拔(300-)450-2400米的山坡、灌丛、悬崖、开阔林地或路边。产中国西南、东南、华中和华东。世界广泛栽培。

Shrubs. Leaves abaxially rather glaucous. Inflorescences 1-15-flowered; sepals erect, often reddish, broadly ovate to broadly elliptic or subcircular to oblong-elliptic or obovate-spatulate; petals golden yellow; stamens fascicles each with 50-70 stamens. Capsules broadly ovoid. Fl. May-Sep. Fr. Jul-Oct. Slopes, thickets, cliffs, open forests or roadsides at (300-) 450-2400 m. Distributed in SW, SE, C and E China. World widely cultivated.

金丝梅 *Hypericum patulum*

## 匙萼金丝桃

**Hypericum uralum** Buch.-Ham. ex D. Don

灌木。叶片腺体呈线状或点状。花序具1-3(-10)花；花萼直立，长圆形或椭圆形至长圆状匙形，先端圆形；花瓣金色至深黄色；雄蕊每束具40-60。蒴果近球形。花期7-9月，果期8-10月。生海拔1500-3600米的草坡、石坡疏林下或草地。产云南西北部和西藏(达旺、察隅县)。印度东北部、尼泊尔、不丹、巴基斯坦和缅甸亦有。

Shrubs. Leaves with linear or punctiform glands. Inflorescences 1-3(-10)-flowered; sepals erect, oblong or elliptic to oblong-spatulate, apex rounded; petals golden to deep yellow; stamens fascicles each with 40-60 stamens. Capsules subglobose. Fl. Jul-Sep. Fr. Aug-Oct. Grassy slopes, open forests on rocky slopes, or grassy places at 1500-3600 m. Distributed in NW Yunnan and Xizang (Dawang, Zayü County). Also in NE India, Nepal, Bhutan, Pakistan and Myanmar.

## 宽萼金丝桃

**Hypericum latisepalum** (N. Robson) N. Robson

灌木。叶片腺点点状或短条状。花序具1-14花，自顶部节间生出，近平顶；萼片直立，卵形至宽椭圆形；花瓣金黄色；雄蕊5束，每束具26-65。蒴果宽卵圆形。花期6-10月，果期8-11月。生海拔2500-2900(-3700)米的草坡、林缘、疏林下或灌丛中。产云南西部和西北部及西藏(察隅县)。印度东北部和缅甸北部亦有。

Shrubs. Glands of leaves dot-shaped or shortly linear. Inflorescences 1-14-flowered, from apical node, nearly flat-topped; sepals erect, ovate to broadly elliptic; petals golden yellow; stamens 5-fascicled, with 26-65 per fascicle. Capsules broadly ovoid. Fl. Jun-Oct. Fr.

匙萼金丝桃 *Hypericum uralum*

宽萼金丝桃 *Hypericum latisepalum*

Aug-Nov. Grassy slopes, forest edges, sparse forests or thickets at 2500-2900(-3700) m. Distributed in W and NW Yunnan, and SE Xizang (Zayü County). Also in NE India and N Myanmar.

## 川滇金丝桃

**Hypericum forrestii** (Chitt.) N. Robson

灌木。叶披针形或三角状卵形至宽卵形，厚纸质。花序具1-20花，着生在1-2节；萼片直立，卵形或稍宽椭圆形至近圆形；花瓣金黄色，内折。蒴果近宽卵球形。花期5-9月，果期8-10月。生海拔1500-3000(-4000)米的山坡或林下。产云南北部及西南部和四川(康定、天全县)。缅甸东北部亦有。

Shrubs. Leaves lanceolate or triangular-ovate to broadly ovate, thickly papery. Inflorescences 1-20-flowered, from 1-2 nodes; sepals erect, ovate or ± broadly elliptic to subcircular; petals golden yellow, incurved. Capsules ± broadly ovoid. Fl. May-Sep. Fr. Aug-Oct. Slopes or under forests at 1500-3000(-4000) m. Distributed in N and SW Yunnan and Sichuan (Kangding, Tianquan County). Also in NE Myanmar.

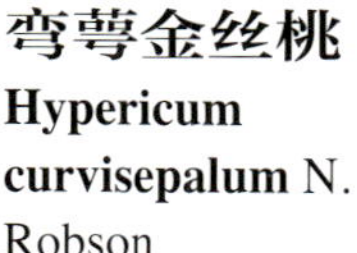

## 弯萼金丝桃

**Hypericum curvisepalum** N. Robson

川滇金丝桃 *Hypericum forrestii*

灌木。叶片腺体条状或点状。花序具1(-3)花；花瓣深黄色；苞片线形或常为叶状；萼片分离，外弯或开展；雄蕊5束，每束有雄蕊60；花药深黄色。蒴果卵状圆锥形至宽卵形。花期5-6月，果期7-9月。生海拔1800-3000米的干燥或多石的山坡，或开阔的林地。产云南、四川和贵州。

Shrubs. Glands of leaves linear or dot-like. Inflorescences 1(-3)-flowered; petals deeply yellow; bracts linear or foliaceous; sepals free, recurved or spreading; stamens 5-fascicled, 60 each fascicle; anthers deeply yellow. Capsules ovoid-conic to broadly ovoid. Fl. May-Jun. Fr. Jul-Sep. Dry or rocky slopes, or open forests at 1800-3000 m. Distributed in Yunnan, Sichuan and Guizhou.

弯萼金丝桃 *Hypericum curvisepalum*

## 北栽秧花

**Hypericum pseudohenryi** N. Robson

灌木。叶常披针形或披针状长圆形至卵状长圆形。花序1-7(至约25)花；花萼直立至外折，宽至狭卵状长圆形；花瓣金黄色；雄蕊每束约具40枚雄蕊；花柱长于子房。蒴果卵状圆锥形至卵球形。花期6-9月，果期9-11月。生海拔1400-3800米的高山松林下、灌丛、草坡或岩坡。产云南东北部及西北部和四川西南部及西部。

北栽秧花 *Hypericum pseudohenryi*

Shrubs. Leaves usually lanceolate or lanceolate-oblong to ovate-oblong. Inflorescences 1-7(to ca. 25)-flowered; sepals erect to outcurved, broadly to narrowly ovate-oblong; petals golden yellow; stamens fascicles each with ca. 40 stamens; styles longer than ovary. Capsules ovoid-conic to ovoid. Fl. Jun-Sep. Fr. Sep-Nov. Pine forests, thickets, grassy slopes or rocky slopes at 1400-3800 m. Distributed in NE and NW Yunnan and SW and W Sichuan.

栽秧花 *Hypericum beanii*

## 栽秧花

**Hypericum beanii** N. Robson

灌木。茎幼时具4纵棱。叶柄长0.5-2(-2.5)毫米。花序具1-11(-14)花，自第1(或2)节间生出，近平顶；萼片先端锐尖，边缘透明，全缘或上方有细小腺齿；花瓣金黄色；花柱与子房等长或短于子房。花期3-7月，果期7-10月。生海拔1500-2100米的溪边、山坡、疏林或灌丛中。产云南、四川和贵州。

Shrubs. Stems 4-angulate when young. Leaves with petioles 0.5-2 (-2.5) mm long. Inflorescences 1-11 (-14)-flowered, from 1 (or 2) nodes, nearly flat-topped; sepal apices acute, margin hyaline, entire or upwards glandular-denticulate; petals golden yellow; styles equal to ovary or shorter. Fl. Mar-Jul. Fr. Jul-Oct. Streamsides, slopes, open forests or thickets at 1500-2100 m. Distributed in Yunnan, Sichuan and Guizhou.

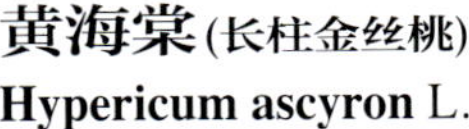

## 黄海棠(长柱金丝桃)

**Hypericum ascyron** L.

多年生草本。叶无柄。花序1至约35朵生1-5节间，整体近平顶至狭金字塔形；花瓣金黄色；雄蕊极多数，5束，每束有雄蕊约30枚；花柱5。蒴果宽至狭卵球形或卵状金字塔形。花期7-8月，果期8-9月。生海拔2800(-3600)米及以下的山坡林、林缘、灌丛、草丛或河边。产除西藏以外全国各地。越南北部、俄罗斯、蒙古、朝鲜半岛、日本和北美洲亦有。

黄海棠(长柱金丝桃) *Hypericum ascyron*

Prennial herbs. Leaves sessile. Inflorescences 1- to ca. 35-flowered from 1-5 nodes, the whole nearly flat-topped to narrowly pyramidal; petals golden yellow; stamens numerous, 5-fascicled, per fascicle ca. 30; styles 5. Capsules broadly to narrowly ovoid or ovoid-pyramidal. Fl. Jul-Aug. Fr. Aug-Sep. Forests on slopes, forest edges, thickets, grasslands or by streams below 2800 (-3600) m. Almost throughout China, except Xizang. Also in N Vietnam, Russia, Mongolia, Korean Peninsula, Japan and North America.

## 地耳草

**Hypericum japonicum** Thunb.

一年生草本。叶通常卵形或卵状三角形至长圆形或椭圆形。花序具1至约30花；花萼离生，直立，狭长圆形至椭圆形；花瓣浅至亮黄色或橘黄色；雄蕊5-30；花柱3。蒴果圆柱状至球形。花期3-10月，果期4-11月。生海拔3000米以下的溪边、田间、草地或废弃地。产中国西南、华南、东南、华中和华东。南亚、东南亚、东北亚、澳大利亚和太平洋岛屿亦有。

Annual herbs. Leaves usually ovate, ovate-deltoid to oblong or elliptic. Inflorescences 1- to ca. 30-flowered; sepals free, erect, narrowly oblong to elliptic; petals pale to bright yellow or orange; stamens 5-30; styles 3. Capsules cylindric to globose. Fl. Mar-Oct. Fr. Apr-Nov. By streams, fields, grasslands or wastelands below 3000 m. Distributed in SW, S, SE, C and E China. Also in S, SE and NE Asia, Australia, and Pacific Islands.

地耳草 *Hypericum japonicum*

## 遍地金

**Hypericum wightianum** Wall. ex Wight et Arn.

一年生或多年生草本。叶上密生腺点或腺条。花序具3-50花，着生在1-3节；花萼基部联合，直立，狭至宽长圆形或椭圆形；花瓣亮黄色；雄蕊7-11，明显3束。蒴果宽卵状至近球状。花期4-7月，果期7-9月。生海拔700-3300米的草坡、开阔林地、溪边或路边。产云南、四川、西藏、贵州和广西。印度、不丹、斯里兰卡、缅甸、老挝和泰国亦有。

Annual or perennial herbs. Leaves with dots to streaks, dense. Inflorescences 3-50-flowered, from 1-3 nodes; sepals basally united, erect, narrowly to broadly oblong or elliptic; petals bright yellow; stamens 7-11, apparently 3-fascicled. Capsules broadly ovoid to subglobose. Fl. Apr-Jul. Fr. Jul-Sep. Grassy slopes, open woodlands, streamsides or roadsides at 700-3300 m. Distributed in Yunnan, Sichuan, Xizang, Guizhou and Guangxi. Also in India, Bhutan, Sri Lanka, Myanmar, Laos and Thailand.

遍地金 *Hypericum wightianum*

## 挺茎遍地金

**Hypericum elodeoides** Choisy

多年生直立草本。叶无柄。花序具(1-)5至约30花，自第1(-4)节间生出，伞房状至圆柱状；花瓣金黄色；苞片及小苞片边缘有小刺齿，齿端有黑色腺体；雄蕊3束，每束约有20。蒴果卵球形。花期7-8月，果期8-10月。生海拔2100-3200米的山坡草地、灌丛、林下及田埂上。产中国西南、东南和华中。印度、尼泊尔、克什米尔地区和缅甸亦有。

Perennial herbs, erect. Leaves sessile. Inflorescences (1-)5- to ca. 30-flowered, from 1(-4) nodes, corymbiform to cylindric; petals golden yellow; bracts and bracteoles spinecent-serrate at margin, teeth apex with black glands; stamens 3-fascicled, 20 each fascicle. Capsules ovoid. Fl. Jul-Aug. Fr. Aug-Oct. Grassy slopes, scrubs, forests or ridges of fields at 2100-3200 m. Distributed in SW, SE and C China. Also in India, Nepal, Kashmir and Myanmar.

挺茎遍地金 *Hypericum elodeoides*

元宝草 *Hypericum sampsonii*

贯叶连翘 *Hypericum perforatum*

## 元宝草
**Hypericum sampsonii** Hance

多年生草本。叶贯穿对生，厚纸质，侧脉4对。花序具20-40花，着生于1-2节上；花萼离生，直立，长圆形至长圆状匙形或线状长圆形；花瓣亮黄色；雄蕊30-42，明显3束。蒴果宽卵球形至宽或狭卵状金字塔形。花期5-7月，果期6-10月。生海拔100-1700米的山坡、路边、草地、灌丛或田边。产中国西南、华南、东南、华中和华东。缅甸、越南北部和日本南部亦有。

Perennial herbs. Leaves in perfoliate pairs; thickly papery, lateral veins 4 per side. Inflorescences 20-40-flowered, from 1-2 nodes; sepals free, erect, oblong to oblong-spatulate or linear-oblong; petals bright yellow; stamens 30-42, apparently 3-fascicled. Capsules broadly ovoid to broadly or narrowly ovoid-pyramidal. Fl. May-Jul. Fr. Jun-Oct. Slopes, roadsides, grasslands, thickets or field sides at 100-1700 m. Distributed in SW, S, SE, C and E China. Also in Myanmar, N Vietnam and S Japan.

## 密腺小连翘
**Hypericum seniavinii** Maxim.

多年生草本。茎直立，圆柱形，无毛。叶厚纸质，侧脉约3对，腺点密集，极大。花序具5-50花，圆柱形；花萼离生，直立，长圆状披针形至线状披针形；花瓣亮黄色；雄蕊24至约55，明显3束。蒴果卵球形。花期7-9月，果期8-11月。生海拔(100-)500-1600(-2000)米的山坡、路边、草地或田埂。产中国西南、东南、华中和华东。越南北部亦有。

Perennial herbs. Stems erect, terete, glabrous. Leaves thickly papery, lateral veins ca. 3 per side, gland dots dense, rather large. Inflorescences 5-50-flowered, cylindric; sepals free, erect, oblong-lanceolate to linear-lanceolate; petals bright yellow; stamens 24 to ca. 55, apparently 3-fascicled. Capsules ovoid. Fl. Jul-Sep. Fr. Aug-Nov. Slopes, roadsides, grasslands or field ridges at (100-)500-1600(-2000) m. Distributed in SW, SE, C and E China. Also in N Vietnam.

密腺小连翘 *Hypericum seniavinii*

## 贯叶连翘
**Hypericum perforatum** L.

多年生草本。叶基部近心状抱茎至极狭楔形。花序具3至多数花，着生于1-3节上；花萼离生，在花蕾时直立，果期反折，狭长圆形或披针形至线形；花瓣金黄色。蒴果卵状圆锥形至卵球形。花期6-9月，果期7-10月。生海拔100-2800米的草甸、林中、草地、山坡、河岸或路边。产中国西南、华中、华北、华西和华东。西南亚、中亚、非洲西北部和欧洲亦有。

Perennial herbs. Leaves base subcordate-amplexicaul to rather narrowly cuneate. Inflorescences 3- to numerous-flowered, from 1-3 nodes; sepals free, erect in bud, recurved in fruit, narrowly oblong or lanceolate to linear; petals golden yellow. Capsules ovoid-conic to ovoid. Fl. Jun-Sep. Fr. Jul-Oct. Meadows, forests, grasslands, slopes, riverbanks or roadsides at 100-2800 m. Distributed in SW, C, N, W and E China. Also in SW and C Asia, NW Africa and Europe.

红芽木 *Cratoxylum formosum* subsp. *pruniflorum*

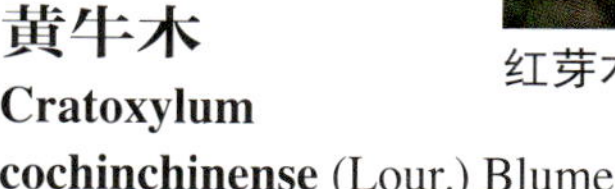

## 黄牛木

**Cratoxylum cochinchinense** (Lour.) Blume

落叶灌木或乔木。树干下部簇生长刺。叶下灰绿色，椭圆形至长圆形或披针形，纸质，两面无毛，下面具透明或深色腺点。聚伞花序具(1或)2或3花；花瓣深绯红色至粉色或粉黄色，基部无鳞片；雄蕊3束，短。蒴果棕色，椭球形。花期4-5月，果期6月以后。生海拔1200米以下的丘陵、干燥山坡、灌丛中或次生林。产云南南部、广西南部和广东南部。缅甸、泰国、越南、马来西亚、印度尼西亚和菲律宾亦有。

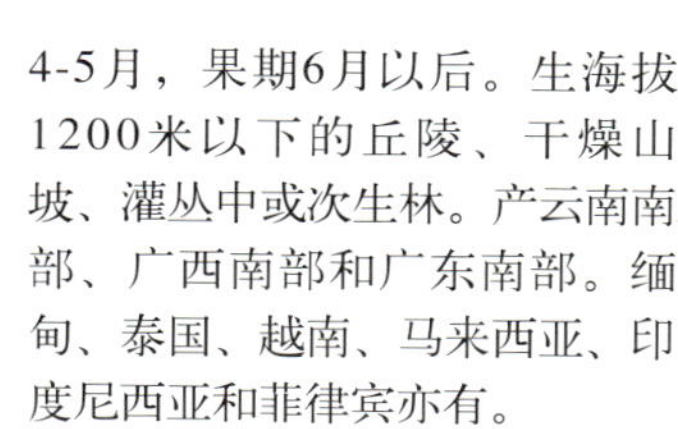

Shrubs or trees, deciduous. Trunk with clusters of long thorns on lower part. Leaves abaxially gray-green, elliptic to oblong or lanceolate, papery, both surfaces glabrous, abaxially with pellucid or dark glands. Cymes (1 or)2 or 3-flowered; petals deep crimson to pink or pinkish yellow, base without scales; stamens 3-fascicled, short. Capsules brown, ellipsoid. Fl. Apr-May. Fr. after Jun. Hills, dry slopes, thickets or secondary forests below 1200 m. Distributed in S Yunnan, S Guangxi and S Guangdong. Also in Myanmar, Thailand, Vietnam, Malaysia, Indonesia and the Philippines.

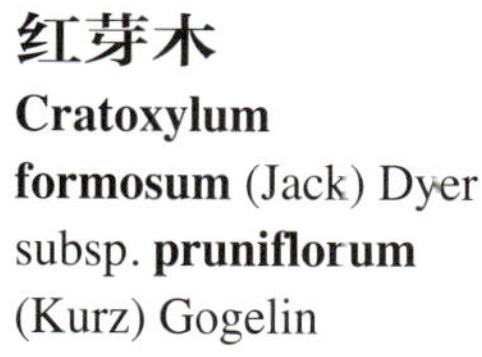

## 红芽木

**Cratoxylum formosum** (Jack) Dyer subsp. **pruniflorum** (Kurz) Gogelin

灌木或乔木。小枝、叶、花梗和花萼密具柔毛。叶椭圆形至长圆形，下面具透明腺点。聚伞花序，5-8花；花瓣基部有鳞片；雄蕊束纤细，20-30。蒴果深褐色，长圆球形，至少1/2包被于宿存花萼。花期3-4月，果期5月以后。生海拔1000米以下的山地次生疏林或灌丛中。产云南南部和广西南部。缅甸、泰国、越南和柬埔寨亦有。

Shrubs or trees. Young twigs, leaves, pedicels and sepals densely villous. Leaves elliptic to oblong, abaxially with pellucid glands. Cymes 5-8-flowered; petal base with scales; stamen fascicles slender, with 20-30 stamens. Capsules dark brown, oblong, up to 1/2 enclosed by persistent calyx. Fl. Mar-Apr. Fr. after May. Secondary open forests or thickets below 1000 m. Distributed in S Yunnan and S Guangxi. Also in Myanmar, Thailand, Vietnam and Cambodia.

黄牛木 *Cratoxylum cochinchinense*

# 龙脑香科
# Dipterocarpaceae

## 东京龙脑香
**Dipterocarpus retusus** Blume

乔木，高达45米。叶阔卵形，下面疏或密具暗金黄色星状柔毛，基部圆形或有时心形。总状花序腋生，2-5花；花瓣淡红色，具甜味，密具鳞片状毛；花萼管直径达3.5厘米；翅状花萼裂片红色，具3-5脉。坚果卵球形。花期5-6月，果期12月至翌年1月。生海拔1000米以下的山谷或雨林中。产云南东南部及南部和西藏东南部。印度、老挝、泰国、越南、柬埔寨、马来西亚和印度尼西亚亦有。

Trees, up to 45 m tall. Leaves broadly ovate, abaxially sparsely or densely golden-buff stellate pubescent, base rounded or somewhat cordate. Racemes axillary, 2-5-flowered; petals reddish, sweetly scented, with dense squamate hairs; calyx tubes to 3.5 cm diam; winglike calyx segments red, 3-5-veined. Nuts ovoid. Fl. May-Jun. Fr. Dec to next Jan. Valleys or rain forests below 1000 m. Distributed in SE and S Yunnan and SE Xizang. Also in India, Laos, Thailand, Vietnam, Cambodia, Malaysia and Indonesia.

东京龙脑香 *Dipterocarpus retusus*

## 羯布罗香
**Dipterocarpus turbinatus** Gaertn. f.

乔木，高达35米。叶卵状长圆形，无毛或疏具星状柔毛。总状花序腋生，3-6花；花萼裂片直径达2.8厘米；翅状花萼裂片条状披针形，近多分枝的单脉微具乳突。坚果卵球形或狭卵球形。花期3-4月，果期6-7月。云南南部及北部有长栽培历史。原产印度、缅甸、泰国和柬埔寨、孟加拉国、老挝和越南。

Trees, up to 35 m tall. Leaves ovate-oblong, glabrous or sparsely stellate pubescent. Racemes axillary, 3-6-flowered; calyx tubes to 2.8 cm diam; winglike calyx segments linear-lanceolate, minutely papillate near much-ramified solitary midvein. Nuts ovoid or narrowly ovoid. Fl. Mar-Apr. Fr. Jun-Jul. Cultivated in S and W Yunnan, for a long time. Native to India, Myanmar, Thailand, Cambodia, Bangladesh, Laos and Vietnam.

羯布罗香 *Dipterocarpus turbinatus*

## 狭叶坡垒
**Hopea chinensis** (Merr.) Hand.-Mazz.

乔木，高15-20米。叶长圆形至长圆状披针形，下面疏具柔毛或无毛，侧脉8-12对。圆锥花序腋生，具少花；花萼覆瓦状，无毛或具柔毛；花瓣淡红色；翅状花萼裂片长圆状披针形或长圆形，具12纵脉。果实深褐色，卵球形。花期6-7月，果期10-12月。生海拔300-600米的密林或山丘。产云南(江城、绿春县、屏边县)和广西南部及西南部。越南北部亦有。

Trees, 15-20 m tall. Leaves oblong to oblong-lanceolate, abaxially sparsely pubescent or glabrous, lateral veins 8-12 pairs. Panicles axillary, few flowered; sepals imbricate, glabrous or pubescent; petals reddish; winglike calyx segments oblong-lanceolate or oblong, longitudinally 12-veined. Fruits dark brown, ovoid. Fl. Jun-Jul. Fr. Oct-Dec. Dense forests or hills at 300-600 m. Distributed in Yunnan (Jiangcheng, Lüchun, County, Pingbian County), and S and SW Guangxi. Also in N Vietnam.

狭叶坡垒 *Hopea chinensis*

坡垒 *Hopea hainanensis*

## 坡垒

**Hopea hainanensis** Merr. et Chun

乔木，高约20米。叶近革质，长圆形至长圆状卵形，两面无毛或具粉屑，侧脉9-12对。单侧总状花序排成疏松的圆锥花序；萼片卵形，外部2个全部具柔毛；翅状花萼裂片长圆形或倒披针形，具9-11纵脉。果实卵球形，蜡质。花期6-7月，果期11-12月。生海拔约700米的密林中。产海南。越南北部亦有。

Trees, ca. 20 m tall. Leaves subleathery, oblong to oblong-ovate, both surfaces glabrous or farinose-scurfy, lateral veins 9-12 pairs. Flowers arranged in lax panicle of unilateral racemes; sepals ovate, outer 2 pubescent on entire surface; winglike calyx segments oblong or oblanceolate, longitudinally 9-11-veined. Fruits ovoid, waxy. Fl. Jun-Jul. Fr. Nov-Dec. Dense forests at ca. 700 m. Distributed in Hainan. Also in N Vietnam.

## 铁凌

**Hopea reticulata** Taidieu

乔木，高约15米。枝条密被灰黄色的绒毛，后变疏被毛。叶革质，全缘，卵形至卵状披针形，基出脉5-6条，侧脉3-5对，叶柄具灰色绒毛。圆锥花序腋生或顶生，纤细，少花，被疏毛；花萼近圆形，无毛，花瓣粉红色，倒卵状椭圆形，外面被绒毛；果花萼裂片均不增大为翅状。实卵圆形，壳薄，无毛。花期3-4月，果期5-6月。生海拔400米左右的坡地。产海南。老挝和越南亦有。

Trees, ca. 15 m tall. Branchlets initially densely grayish yellow tomentose, soon sparsely so. Leaves leathery, leaves ovate to ovate-lanceolate, margin entire, basal veins 5-6, lateral veins 3-5 pairs, petioles with gray tomentose. Panicles axillary or terminal, slender, few flowered, sparsely pubescent; calyx suborbicular, subglabrous; petals reddish, obovate-elliptic, outside tomentose; calyx not becoming winglike. Fruit ovoid, pericarp thin, glabrous. Fl. Mar-Apr. Fr. May-Jun. Forests on mountain at ca. 400 m. Distributed in Hainan. Also in Laos and Vietnam.

云南娑罗双 *Shorea assamica*

铁凌 *Hopea reticulata*

## 云南娑罗双

**Shorea assamica** Dyer

常绿乔木，高达50米。叶卵状椭圆形，薄革质，背面具星状柔毛。伞房状圆锥花序腋生或顶生；花瓣黄白色，具11脉；雄蕊15；花药椭圆形；药隔附属物长，针状；花萼裂片翅状，不等大，具10-14纵脉。花期6-7月，果期12月至翌年1月。生海拔1000米以下的热带山谷林中。产云南和西藏。印度、缅甸、泰国、马来西亚、印度尼西亚和菲律宾亦有。

Evergreen trees, up to 50 m tall. Leaves ovate-elliptic, thinly leathery, abaxially stellate pubescent. Flowers in axillary or terminal cymose panicles; petals yellowish-white, 11-nerved; stamens 15; anthers ellipsoid; connective appendages long, acicular; sepal lobes winglike, unequal, longitudinally 10-14-veined. Fl. Jun-Jul. Fr. Dec to next Jan. Forests in tropical valleys below 1000 m. Distributed in Yunnan and Xizang. Also in India, Myanmar, Thailand, Malaysia, Indonesia and the Philippines.

望天树 *Parashorea chinensis*

## 望天树

**Parashorea chinensis** H. Wang

常绿乔木，高40(-60)米。叶互生，革质，长圆状披针形或披针形，侧脉14-19对，两面具屑状柔毛或绒毛。聚伞状圆锥花序腋生或顶生；花瓣黄白色；花萼裂片近等大，翅状，具5-7纵脉，常开裂。果实椭圆体形，密具银色丝状柔毛。花期5-6月，果期8-9月。生海拔300-1100米的山谷、山坡或丘陵地。产云南南部及东南部和广西西部。越南北部亦有。

Evergreen trees, 40(-60) m tall. Leaves alternate, leathery, oblong-lanceolate or lanceolate, lateral veins 14-19 pairs, both surfaces scurfy pubescent or tomentose. Inflorescences axillary or terminal cymose panicles; petals yellowish-white; calyx segments 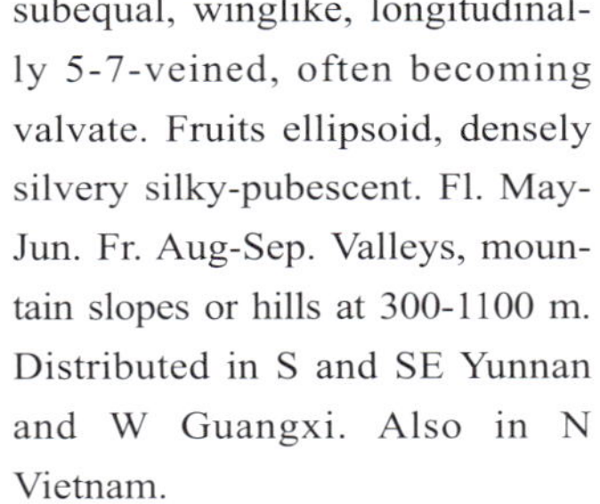 subequal, winglike, longitudinally 5-7-veined, often becoming valvate. Fruits ellipsoid, densely silvery silky-pubescent. Fl. May-Jun. Fr. Aug-Sep. Valleys, mountain slopes or hills at 300-1100 m. Distributed in S and SE Yunnan and W Guangxi. Also in N Vietnam.

## 广西青梅

**Vatica guangxiensis** X. L. Mo

常绿乔木，高达35-40米。叶狭椭圆形至椭圆状披针形，薄革质，两面具灰黄色星状毛，侧脉12-18(-20)对。圆锥花序顶生或腋生，粗壮；花瓣白色或淡红色；花萼无基部杯状贴生于子房；果萼2稍长者长圆状椭圆形。坚果近球形。花期4-5月，果期7-8月。生海拔800-1000米的山坡或山谷。产云南(勐腊县)和广西(那坡县)。越南亦有。

Evergreen trees, up to 35-40 m tall. Leaves narrowly elliptic to elliptic-lanceolate, thinly leathery, both surfaces with grayish yellow stellate hairs, lateral veins 12-18 (-20) pairs. Panicles terminal or axillary, robust; petals white or reddish; calyx without basal cup adnate to ovary; fruits sepals: 2 longer oblong-elliptic. Nuts subglobose. Fl. Apr-May. Fr. Jul-Aug. Mountain slopes or valleys at 800-1000 m. Distributed in Yunnan (Mengla County) and Guangxi (Napo County). Also in Vietnam.

## 青梅

**Vatica mangachapoi** Blanco

乔木。叶革质，长圆形至长圆披针形，全缘，两面无毛或疏具柔毛。圆锥花序顶生或腋生，细弱；花白色，有时淡黄色或淡红色，芳香；2个萼裂片条形，稍长，具5纵脉；3个较短裂片披针形。果实球形。花期5-6月，果期8-9月。生海拔700米以下的山地林中或丘陵。产海南。泰国、越南、马来西亚、印度尼西亚和菲律宾亦有。

Trees. Leaves leathery, oblong to oblong-lanceolate, entire, both surfaces glabrous or sparsely pubescent. Panicles terminal or axillary, slender; flowers white, sometimes pale yellow or pale red, aromatic; 2 calyx segments lorate, longer, longitudinally 5-veined. 3 shorter segments lanceolate. Fruits globose. Fl. May-Jun. Fr. Aug-Sep. Montane forests or hills below 700 m. Distributed in Hainan. Also in Thailand, Vietnam, Malaysia, Indonesia and the Philippines.

广西青梅 *Vatica guangxiensis*

青梅 *Vatica mangachapoi*

# 沟繁缕科 Elatinaceae

### 大叶田繁缕
**Bergia capensis** L.

一年生湿生草本，高15-30厘米。茎肉质。叶对生，纸质，披针形或倒卵形，长1-4厘米，边缘具微细锯齿。聚伞花序小型，腋生；花极小，近无梗，粉红色。蒴果近球形，5瓣裂。花果期10-12月生水田及水沟边。产广东。马来西亚、斯里兰卡、印度、伊朗、高加索地区和埃及亦有。

三蕊沟繁缕 *Elatine triandra*

Annual hygrocolous herbs, 15-30 cm tall. Stems fleshy. Leaves opposite, chartaceous, lanceolate or obovate, 1-4 cm long, margin minutely serrulate. Cymes small, axillary; flowers minute, subsessile, pinkish. Capsules subglobose, 5-valved. Paddy fields or ditch sides. Fl. and fr. Oct-Dec. Distributed in Guangdong. Also in Malaysia, Sri Lanka, India, Iran, Caucasus and Egypt.

### 三蕊沟繁缕
**Elatine triandra** Schkuhr

一年生水生草本。叶对生，卵状长圆形或披针形至条状披针形，近膜质。单花腋生，无柄、近无柄或具短柄，花梗于果期长0.3-0.4毫米；花瓣3，白色或淡红色；雄蕊3，短于花瓣。蒴果扁球形。花果期7-8月。生池塘、溪流、麦田或沼泽地。产广东、台湾、吉林和黑龙江。南亚、东南亚、欧洲、大洋洲和美洲亦有。

Annual herbs, aquatic. Leaves opposite, ovate-oblong or lanceolate to linear-lanceolate, submembranous. Flowers axillary, solitary, sessile, subsessile, or shortly pedicellate and then pedicel 0.3-0.4 mm long at fruiting stage; petals 3, white or reddish; stamens 3, shorter than petals. Capsules compressed globose. Fl. and fr. Jul-Aug. Pools, streams, paddy fields or marshy places. Distributed in Guangdong, Taiwan, Jilin and Heilongjiang. Also in S and SE Asia, Europe, Oceania and America.

大叶田繁缕 *Bergia capensis*

# 瓣鳞花科 Frankeniaceae

### 瓣鳞花
**Frankenia pulverulenta** L.

一年生草本，高6-16厘米，多分枝，有紧贴的白色微柔毛。叶窄倒卵形或倒卵形。花小，常生于茎上部分枝分叉处。萼筒有5个钻形齿；花瓣5，粉红至紫色，有爪和舌状附属物；雄蕊6；子房3室，胚珠多数。蒴果包藏在宿存的萼筒内。花果期5-8月。生海拔1200-1500米的河滩、湖边等盐化草甸中。产内蒙古(额济纳)、甘肃(民勤县)和新疆(新源)。亚洲、欧洲和非洲亦有。

Annual herbs, 6-16 cm long, richly branched from base, sparsely white puberulous. Leaves narrowly obovate or elliptic. Flowers small, borne in terminal or axillary dichasia; calyx tube with 5 subulate teeth; petals 5, pink to violet, attenuate below middle, apex erose denticulate; stamens 6; ovary with numerous ovules on 3 parietal placentas. Capsules oblong-ovoid, ca. 2 × 1 mm long. Fl. and fr. May-Aug. High-salinity grasslands of floodlands, by lakes at 1200-1500 m. Distributed in Neimenggu (Ejina), Gansu (Minqin County) and Xinjiang (Xinyuan). Also in Asia, Europe and Africa.

瓣鳞花 *Frankenia pulverulenta*

# 柽柳科 Tamaricaceae

## 红砂
**Reaumuria soongarica** (Pall.) Maxim.

小灌木，高10-30厘米。小枝常淡红色，多拐曲。叶片常4-6聚集在短枝上，肉质，短圆柱状。花单生于叶腋；花瓣白色，带淡红色；花柱3。蒴果狭椭圆状、纺锤体形或三棱锥体形。花期7-8月，果期8-9月。生荒漠或低地边缘。产内蒙古、甘肃、宁夏、青海和新疆。蒙古亦有。

Dwarf shrubs, 10-30 cm tall. Branchlets often reddish, usually zigzag. Leaves often 4-6 clustered on shortened branches, fleshy, shortly terete. Flowers solitary in axillae; petals white, tinged reddish; styles 3. Capsules narrowly ellipsoid, fusiform, or triqueter-conic. Fl. Jul-Aug. Fr. Aug-Sep. Deserts or margins of lowlands. Distributed in Neimenggu, Gansu, Ningxia, Qinghai and Xinjiang. Also in Mongolia.

## 黄花红砂
**Reaumuria trigyna** Maxim.

小半灌木。多分枝。叶肉质，2-5枚簇生，半圆柱状线形。花单生于叶腋，5数；苞片基部扩展，宽卵形；花瓣黄色，长圆状倒卵形；花柱3。蒴果长圆形，3瓣裂。花期6-7月，果期7-8月。生荒漠沙砾地或石质旱山坡。产内蒙古、宁夏北部和甘肃北部。

Small semishrubs. Multi-branched. Leaves carnose, clustered in 2-5, semicylindrical-linear. Flowers solitary, axillary, 5-merous; bracts swollen at base, broadly ovate; petals yellow, oblong-obovate; styles 3. Capsules oblong, 3-valved. Fl. Jun-Jul. Fr. Jul-Aug. Sandy grounds in deserts or dry stony slopes. Distributed in Neimenggu, N Ningxia and N Gansu.

## 五柱红砂
**Reaumuria kaschgarica** Rupr.

小灌木，达20厘米。基生叶鳞片状，上部叶渐大，线形或近圆柱形，常稍弯曲，肉质。花单生于小枝顶端，近无柄；花瓣粉红色，椭圆形；花柱5。蒴果矩圆状球形。花期7-8月，果期8月。生盐土荒漠、草原或砾质山坡。产西藏、甘肃、青海和新疆。中亚亦有。

Dwarf subshrubs, to 20 cm tall. Basal leaves scale-like, those in upper part becoming longer, linear

黄花红砂 *Reaumuria trigyna*

红砂 *Reaumuria soongarica*

五柱红砂 *Reaumuria kaschgarica*

短穗柽柳 *Tamarix laxa*

or subterete, often slightly curved, fleshy. Flowers solitary at apices of branchlets, subsessile; petals pink, elliptic; styles 5. Capsules oblong-ovoid. Fl. Jul-Aug. Fr. Aug. Salty deserts, steppes or rocky mountain slopes. Distributed in Xizang, Gansu, Qinghai and Xinjiang. Also in C Asia.

## 短穗柽柳

**Tamarix laxa** Willd.

灌木，高1.5(-3)米。老枝灰褐色，幼枝淡紫灰色。营养枝上，叶卵状披针形，先端具尖头。花序早春侧生于二年生枝上，短而粗，长1-3厘米；花瓣4，粉红或紫红色；花柱3，短。蒴果圆锥体形。花期3-5月。生荒漠河流阶地、湖盆、沙丘边缘或盐土上。产中国西北。俄罗斯、蒙古、中亚、伊朗和亦有。

Shrubs, 1.5(-3) m tall. Old branchlets gray-brown, young branches purplish gray. Leaves of vegetative branches ovate-lanceolate, apex acuminate. Racemes lateral in old branches of previous year, blooming in early spring, short and thick, to 1-3 cm long; petals 4, red or red-purple; styles 3, short. Capsules conic. Fl. Mar-May. River banks, lake basins, margins of sand dunes or salty soils. Distributed in NW China. Also in Russia, Mongolia, C Asia and Iran.

## 柽柳

**Tamarix chinensis** Lour.

乔木或灌木，高常3-6米。老枝直立，暗褐红色；嫩枝纤细，悬垂。叶鲜绿色，先端渐尖而内弯。总状花序轴和花梗柔软下垂；萼片5；花瓣5，粉红色；花盘5裂；花柱棍棒状。蒴果圆锥体形，室背开裂。花期4-9月。生河流冲积平原、海滨、盐碱地或潮湿荒地。产华北和华东部分地区。华东和西南栽培。日本和美国亦有栽培。

Trees or shrubs. 3-6 m tall. Old branches straight, dark brown-red; young branches slender, pendulous. Leaves bright green, apex acute and curved. Pedicles and axis of receme pendulous; sepals 5, petals 5, pink; disk 5-lobed; styles clavate. Capsules conic, loculicidal. Fl. Apr-Sep. Alluvial plains along rivers, seashores, saline-alkali of soil or wet sandy wastelands. Distributed in N and part of E China. Cultivated in E and SW China. Also cultivated in Japan and USA.

柽柳 *Tamarix chinensis*

多枝柽柳 *Tamarix ramosissima*

## 多枝柽柳

**Tamarix ramosissima** Ledeb.

灌木或小乔木，高常1-3米。老枝暗灰色，当年生枝淡红或橙黄色。营养枝的叶宽卵状圆形或三角状心形，近抱茎。总状花序聚集呈顶生圆锥花序；花瓣粉红色或紫色，形成闭合的酒杯状；花柱3，棍棒状。蒴果三棱锥体形。花期5-9月。生河漫滩，盐渍化沙土或沙丘。产中国西南和西北。蒙古、中亚、西南亚和欧洲东部亦有。

Shrubs or small trees, 1-3 m tall. Old branches dark gray; growing branches of current year reddish or orange-yellow. Leaves of vegetative branches broadly ovate-orbicular or triangular-cordate, nearly amplexicaul. Racemes clustered into terminal panicles; petals pink or purple, contacting each other and forming a cup-shaped corolla; styles 3, clavate. Capsules triqueter-conic. Fl. May-Sep. Riversides, salty sandy or dunes, or sand dunes. Distributed in SW and NW China. Also in Mongolia, C and SW Asia, and E Europe.

## 甘蒙柽柳

**Tamarix austromongolica** Nakai

灌木或乔木，高常1.5-4米。老枝直立，紫红色；嫩枝质硬直伸。叶灰绿色，生下部枝者宽卵形，上部卵状披针形。总状花序侧生于去年木质化枝条上，总状花序轴质硬而直伸，花梗极短；花瓣5，紫红色；花柱3，反折。蒴果狭圆锥体形。花期5-9月。生盐渍化河漫滩、平原或盐化沙地。产华北和华西。

Shrubs or trees, 1.5-4 m tall. Old branches straight, red-purple; young branches straight, ascending, hard in texture. Leaves gray-green, those in lower part of growing branches broadly ovate, those in upper part ovate-lanceolate. Racemes lateral in ligneous branches of previous year, Inflorescences axis straight, ascending, hard in texture, peduncles short; petals 5, purplish red; styles 3, recurved. Capsules narrowly conic. Fl. May-Sep. Salty riversides, plains or salty sandy fields. Distributed in N and W China.

甘蒙柽柳 *Tamarix austromongolica*

细穗柽柳 *Tamarix leptostachya*

## 细穗柽柳

**Tamarix leptostachya** Bunge

灌木，高1-3米。营养枝上叶狭卵形或卵状披针形，基部下延。总状花序长纤细，组成紧密的圆锥花序；花瓣5，紫红色或粉红色，花后花瓣全部脱落；花柱3。蒴果小，圆锥体形。花期6-7月。生荒漠地区盐化土壤、河边、湖边或丘间低地。产内蒙古、甘肃、宁夏、青海和新疆。蒙古和中亚亦有。

Shrubs, 1-3 m tall. Leaves of vegetative branches narrowly ovate or ovate-lanceolate, base decurrent. Racemes long, slender, clustered into dense panicles; petals 5, purplish red or pink, caducous; styles 3. Capsules small, conic. Fl. Jun-Jul. Salty soils, riversides, lakesides or lowlands in desert regions. Distributed in Neimenggu, Gansu, Ningxia, Qinghai and Xinjiang. Also in Mongolia and C Asia.

## 匍匐水柏枝

**Myricaria prostrata** Hook. f. et Thomson.

矮小匍匐灌木，高5-14厘米。去年枝红褐色，常生根。叶生于当年生枝上，长圆形、狭椭圆形或卵形。总状花序球状，侧生于去年生枝条上，密集，常由1-3花组成；花瓣紫色至粉色。蒴果圆锥体形。花果期6-8月。生海拔4000-5200米的河滩、砾质山坡或溪边。产西藏、甘肃、青海和新疆。印度、巴基斯坦和中亚亦有。

Shrubs prostrate, dwarf, 5-14 cm tall. Branches of previous year red-brown, often rooting. Leaves clustered in branches of current year, oblong, narrowly elliptic, or ovate. Racemes globose, lateral on branches of previous year, dense, often consisting of 1-3 flowers; petals purplish to pink. Capsules conic. Fl. and fr. Jun-Aug. Riversides, rocky slopes or streamsides at 4000-5200 m. Distributed in Xizang, Gansu, Qinghai and Xinjiang. Also in India, Pakistan and C Asia.

匍匐水柏枝 *Myricaria prostrata*

具鳞水柏枝 *Myricaria squamosa*

## 具鳞水柏枝
**Myricaria squamosa** Desv.

灌木，高1-5米。茎直立，上部多分枝。叶披针形、卵状披针形或长圆形。花序侧生或数个花序簇生于枝腋，基部有多数鳞片；花瓣紫红色或粉红色。蒴果圆锥体形。花果期5-8月。生海拔2400-4600米的山地河滩或湖边沙地。产四川、西藏、甘肃、青海和新疆。印度、尼泊尔、巴基斯坦和中亚亦有。

Shrubs, 1-5 m tall. Stem erect, much branched in upper part. Leaves lanceolate, ovate-lanceolate, or oblong. Inflorescences lateral or several ones clustered in axils of branches, with persistent scales at base; petals purplish red or pink. Capsules conic. Fl. and fr. May-Aug. Riversides in mountains or sandy places at lakesides at 2400-4600 m. Distributed in Sichuan, Xizang, Gansu, Qinghai and Xinjiang. Also in India, Nepal, Pakistan and C Asia.

## 三春水柏枝
**Myricaria paniculata** P. Y. Zhang et Y. J. Zhang

灌木，高1-3米。老枝深褐色，红褐色或灰褐色。叶无柄，披针形、卵状披针形或长圆形，边缘狭膜质。春季总状花序侧生，夏秋季形成顶生圆锥花序，花序基部有宿存鳞片；花瓣紫红色或粉红色。蒴果狭圆锥体形。花期3-9月，果期5-10月。生海拔1000-2800米的山地河滩或河谷山坡。产中国西南、华北和华西。

Shrubs, 1-3 m tall. Old branches deep brown, red-brown, or gray-brown. Leaves sessile, lanceolate, or oblong, margin narrowly membranous. Flowers in a lateral raceme in spring, in a terminal panicle in summer and autumn, inflorescences with persistent scales at base; petals purplish red or pink. Capsules narrowly conic. Fl. Mar-Sep. Fr. May-Oct. Riversides in mountains or mountain slopes at 1000-2800 m. Distributed in SW, N and W China.

三春水柏枝 *Myricaria paniculata*

### 宽苞水柏枝
**Myricaria bracteata** Royle

灌木，多分枝。叶小，密生，卵形至线状披针形。总状花序顶生，密集呈穗状；苞片宽卵形或椭圆形；花小，粉红色或淡紫色。蒴果狭圆锥形。花期6-7月，果期8-9月。生海拔1100-3300米的河滩、沙地或戈壁。产华北和西北。中亚和东北亚亦有。

Shrubs, multi-branched. Leaves tiny, dense, ovate to linear-lanceolate. Racemes terminal, dense, spike-like; bracts broadly ovate or elliptic; flowers small, pink or purplish. Capsules narrowly conical. Fl. Jun-Jul. Fr. Aug-Sep. Flood plains, sandy grounds or Gobi at 1100-3300 m. Distributed in N and NW China. Also in C and NE Asia.

# 半日花科 Cistaceae

### 半日花
**Helianthemum songaricum** Schrenk

矮小灌木。小枝对生，老枝先端刺状。单叶对生，革质，披针形或狭卵形，边缘全缘，常反卷，两面被白色短柔毛；托叶钻形或线形披针形；叶柄极短或近无柄。花单生枝顶；萼片5，不等大；花瓣5，黄色；雄蕊多数。蒴果卵形。花果期5-9月。生海拔1000-1400米的干草原荒漠区低山石质坡地。产甘肃、内蒙古和新疆。中亚亦有。

半日花 *Helianthemum songaricum*

宽苞水柏枝 *Myricaria bracteata*

Dwarf shrubs. Branchlets opposite, apex of old branches spinose. Leaves simple, opposite, coriaceous, lanceolate or narrowly ovate, margin entire, often revolute, both surfaces white puberulous; stipules subulate or linear-lanceolate; petioles very short to nearly absent. Flowers solitary, terminal; sepals 5, unequal; petals 5, yellow; stamens numerous. Capsules ovoid. Fl. and fr. May-Sep. Rocky hills and slopes in steppe-desert regions at 1000-1400 m. Distributed in Gansu, Neimenggu and Xinjiang. Also in C Asia.

# 红木科 Bixaceae

## 红木
**Bixa orellana** L.

常绿灌木或小乔木。枝条灰褐色至褐色，密被红褐色腺毛。叶心状卵形或三角状卵形，长10-25厘米，宽6-16厘米，背面具树脂状腺点，正面无毛，基出脉5。圆锥花序顶生，长5-10厘米；萼片5，倒卵形，长0.8-1厘米；花瓣5，倒卵形，长1.5-3厘米，白色或粉红色。蒴果卵形或近球形，长2.5-4厘米，密生长刺。种子多数，倒卵球形，长4-5毫米，暗红色。云南、广东和台湾有栽培。原产热带美洲。泛热带地区常见栽培。

Evergreen shrubs or small trees. Branches grey-brown to brown, densely red-brown glandular hairy. Leaves cordate-ovate to triangular-ovate, 10-25 cm long, 6-16 cm wide, abaxially with resinlike gland dots, adaxially glabrous, basinerved 5. Panicle terminal, 5-10 cm long; sepals 5, obovate, 0.8-1 cm long; petals 5, obovate, 1.5-3 cm long, white or pink. Capsules ovoid or subglobose, 2.5-4 cm long, densely spiny. Seeds numerous, obovoid, 4-5 mm long, dark red. Cultivated in Yunnan, Guangdong and Taiwan. Native to tropical America. Cultivated pantropically.

红木 *Bixa orellana*

# 堇菜科 Violaceae

## 三角车
**Rinorea bengalensis** (Wall.) Kuntze.

灌木或小乔木。幼枝有明显的叶痕。叶互生，椭圆状披针形或椭圆形；叶脉两面凸起；叶柄长5-12毫米，无毛；托叶披针形，早落，留有环状托叶痕。花梗长可达1厘米，略被黄色绒毛；药隔顶部附属物为广卵状。种子有褐色斑点。花期4-5月和10月，果期9月。生海拔600米以下的灌丛或密林中。产广西西南部和海南。东南亚、南亚和澳大利亚亦有。

Shrubs or small trees. Young branches with conspicuous leaf scars. Leaves alternate, elliptic-lanceolate or elliptic; veins raised on both sides; petioles 5-12 mm long, glabrous; stipules lanceolate, caducous, scars circular. Pedicels to 1 cm long, slightly yellow tomentose; appendages at top of connectives broadly ovate. Seeds with brown dots. Fl. Apr-May and Oct. Fr. Sep. Thickets or dense forests below 600 m. Distributed in SW Guangxi and Hainan. Also in SE and S Asia, and Australia.

三角车 *Rinorea bengalensis*

球果堇菜 *Viola collina*

## 球果堇菜
**Viola collina** Bess.

多年生草本。根状茎直立或斜升，黄褐色。叶阔卵形或近圆形，果期明显增大，两面疏被柔毛；托叶披针形，膜质，基部与叶柄合生，边缘有流苏。花淡紫色，具长柄；花瓣基部白色，前方花瓣距白色。蒴果球状，密被柔毛。花期3-5月，果期6-8月。生海拔2800米以下的林中、林缘、灌丛中、草地、草坡或山谷。产中国西南、华中、华北、华西、华东和东北。塔吉克斯坦、俄罗斯、蒙古、朝鲜半岛、日本和欧洲亦有。

Perennial herbs. Rhizomes erect or oblique, yellow-brown. Leaves broadly ovate or suborbicular, conspicuously accrescent at fruiting, both surfaces sparsely puberulous; stipules lanceolate, membranous, base adnate to petioles, margin remotely fimbriate-denticulate. Flowers purplish, long pedicellate; petals whitish at base, spurs of anterior one white. Capsules globose, densely white puberulous. Fl. Mar-May. Fr. Jun-Aug. Forests, forest edges, thickets, grasslands, grassy slopes or valleys below 2800 m. Distributed in SW, C, N, W, E and NE China. Also in Tajikistan, Russia,

Mongolia, Korean Peninsula, Japan and Europe.

## 紫花堇菜
**Viola grypoceras** A. Gray

多年生草本。根茎直立，强壮，节密集。地上茎明显，高5-30厘米。基生叶心形或阔心形，两面无毛或近无毛，密具褐色腺点；托叶披针形，羽状深裂呈流苏状。花淡紫色至白色，芳香；花瓣具褐色腺点。蒴果椭圆体形。花期3-5月，果期5-8月。生海拔2400米以下的草坡或灌丛中。产中国西南、华南、东南、华中、华西和华东。朝鲜半岛和日本亦有。

Perennial herbs. Rhizomes erect, robust, densely nodded. Stem several, distinct, 5-30 cm tall. Basal leaves cordate or broadly cordate, both surfaces glabrous or subglabrous, with dense brown glandular dots; stipules narrowly lanceolate, usually pinnately parted, fimbricate. Flowers purplish to white, fragrant; petals with brown glandular dots. Capsules ellipsoid. Fl. Mar-May. Fr. May-Aug. Grassy slopes or thickets below 2400 m. Distributed in SW, S, SE, C, W and E China. Also in Korean Peninsula and Japan.

## 庐山堇菜
**Viola stewardiana** W. Beck.

多年生草本。茎生叶长卵形、菱形或三角状卵形，基部楔形，叶柄具狭翅。花瓣先端具明显微缺，侧瓣里面基部无须毛，下方花瓣倒长卵形，连距长约1.4厘米；距长约6毫米；下方2枚雄蕊无距；花柱向上方逐渐增粗，具钩状短喙，顶端具较大的柱头孔。花期4-5月，果期5-9月。生海拔400-1500米的山沟溪边或石缝中。产华南、东南、华中和华东。

Perennial herbs. Cauline leaves long-ovate, rhombic or triangular-ovate, base cuneate, petioles narrowly winged. Petals conspicuously emarginate at apex, lateral ones glabrous at base inside, anterior one narrowly obovate, ca. 1.4 cm long (spur included); spur ca. 6 mm long; 2 anterior stamens spurless; styles gradually thickened upward, beaks short and hooked, with a larger stigma hole at tip. Fl. Apr-May. Fr. May-Sep. Streamsides in mountain valleys or rock crevices at 400-1500 m. Distributed in S, SE, C and E China.

紫花堇菜 *Viola grypoceras*

庐山堇菜 *Viola stewardiana*

## 库页堇菜
**Viola sacchalinensis** H. de Boiss.

多年生草本。地上茎逐渐抽出。叶心形、卵状心形或肾形，边缘具钝锯齿，两面具褐色腺点。侧瓣长圆状，里面基部有须毛，下瓣连距长1.3-1.6厘米；花柱由顶部至喙端有乳头状附属物；喙呈钩状，喙端具向上倾斜且较大的柱头孔。花果期4-9月。生海拔400-2400米的稀疏落叶林下、林缘或高山草地。产内蒙古、黑龙江和吉林。俄罗斯、蒙古、朝鲜半岛和日本亦有。

Perennial herbs. Gradually caulescent. Leaves cordate, ovate-cordate or reniform, margin obtusely serrate, both surfaces with brown glandular dots. Lateral petals oblong, bearded at base inside, anterior one, 1.3-1.6 cm long (spur included); styles with papillose appendages from apex to beak; beak hooked, with a larger stigma hole curved upward. Fl. and fr. Apr-Sep. Sparse deciduous forests, forest edges, alpine grasslands at 400-2400 m. Distributed in Neimenggu, Heilongjiang and Jilin. Also in Russia, Mongolia, Korean Peninsula and Japan.

## 石生堇菜
**Viola rupestris** F. W. Schmidt

多年生草本，有地上茎，全体被短柔毛。茎生叶的托叶离生，常卵形或宽披针形，长3-5毫米，边缘具流苏状锯齿或疏生缺刻状锯齿；下方花瓣连距长1.3-1.4厘米；花柱基部稍膝曲，顶部向前弯曲呈喙状，喙平伸，具较粗的柱头孔。蒴果长圆状。花果期4-6月。生海拔1000-2500米的山地林下、草坡或河谷草甸、沙质草地。产华中和新疆。亚洲和欧洲亦有。

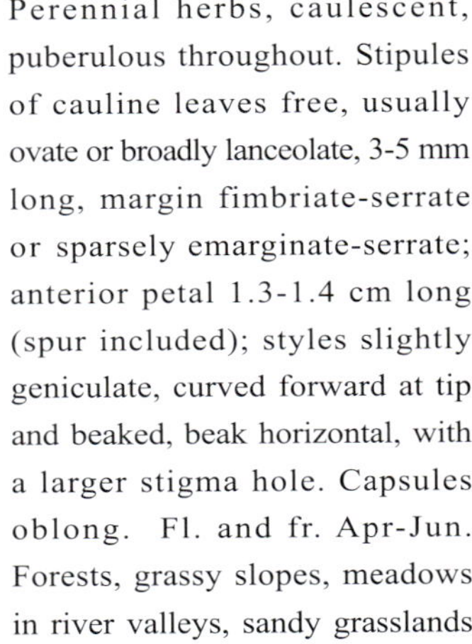

鸡腿堇菜 *Viola acuminata*

Perennial herbs, caulescent, puberulous throughout. Stipules of cauline leaves free, usually ovate or broadly lanceolate, 3-5 mm long, margin fimbriate-serrate or sparsely emarginate-serrate; anterior petal 1.3-1.4 cm long (spur included); styles slightly geniculate, curved forward at tip and beaked, beak horizontal, with a larger stigma hole. Capsules oblong. Fl. and fr. Apr-Jun. Forests, grassy slopes, meadows in river valleys, sandy grasslands at 1000-2500 m. Distributed in C China and Xinjiang. Also in Asia and Europe.

## 鸡腿堇菜
**Viola acuminata** Ledeb.

多年生草本。茎常2-4个簇生，直立。叶片心形；托叶草质，叶状，长1-3.5厘米，通常羽状深裂呈流苏状。花淡紫色或白色；花瓣具点状褐色腺点。蒴果椭圆体形。花期5-6月，果期6-9月。生海拔400-2500米的混交林下、林缘、灌丛、山坡草地或山谷阴湿处。产华中、华北、华东、华西和东北。俄罗斯(西伯利亚、远东地区)、蒙古、朝鲜半岛和日本亦有。

Perennial herbs. Stem usually 2-4 fasciculate, erect. Leaves cordate; stipules herbercous, foliaceous, 1-3.5 cm long, usually pinnately

库页堇菜 *Viola sacchalinensis*

石生堇菜 *Viola rupestris*

深山堇菜 *Viola selkirkii*

紫背堇菜 *Viola violacea*

parted, fimbricate. Flowers pale purple or white; petals dotlike brown glandular. Capsules ellipsoid. Fl. May-Jun. Fr. Jun-Sep. Mixed forests, forest edges, thickets, grassy lands on slopes or wet places in valleys at 400-2500 m. Distributed in C, N, E, W and NE China. Also in Russia (Siberia, Far East), Mongolia, Korean Peninsula and Japan.

## 深山堇菜
**Viola selkirkii** Pursh ex Gold

多年生草本，无地上茎。叶边缘具钝齿；托叶1/2与叶柄合生。侧方花瓣无须毛，下方花瓣连距长1.5-2厘米；距较粗，长5-7毫米；花柱上部明显增粗，柱头顶部平坦，两侧具窄缘边，前方具明显短喙，喙端具向上柱头孔。花果期4-9月。生海拔400-1500米的针阔混交林、落叶阔叶林及灌丛和溪谷。产华北和东北。东北亚、欧洲和北美洲亦有。

Perennial herbs, acaulescent. Leaves margin obtusely dentate; stipules 1/2 adnate to petiole. Lateral petals beardless, anterior one 1.5-2 cm long (spur included); spur thick, 5-7 mm long; styles thickened in upper part, stigmas flat at top, narrowly margined on lateral sides, shortly beaked in front, beak with an upward stigma hole at tip. Fl. and fr. Apr-Sep. Mixed needle-leaved and broad-leaved forests, deciduous broad-leaved forests, thickets and stream valleys at 400-1500 m. Distributed in N and NE China. Also in NE Asia, Europe and North America.

## 紫背堇菜
**Viola violacea** Makino

多年生草本，无茎。叶片狭三角状卵形，基部心形，背面常紫色；托叶基部与叶柄合生。花红紫色；萼片宽披针形，附属物长约1毫米；花瓣8-12毫米，侧花瓣无须毛或稀有稀疏的须毛；距长5-7毫米。花期3-4月。生海拔1000米以下的山谷林下。产华东。日本和朝鲜半岛南部亦有。

Perennial herbs, acaulescent. Leaves narrowly triangular-ovate, base cordate, abaxially usually purplish stipules base adnate to petioles. Flowers red-purple. sepals broadly lanceolate, appendages ca. 1 mm long; petals 8-12 mm long; lateral ones glabrous or rarely sparsely bearded; spur 5-7 mm long. Fl. Mar-Apr. Forest understories in valleys below 1000 m. Distributed in E China. Also in Japan and S Korean Peninsula.

## 心叶堇菜
**Viola yunnanfuensis** W. Beck.

多年生无茎草本。叶卵形、阔卵形或三角状卵形，背面疏具柔毛或渐无毛，基部深或阔心形，边缘具深圆锯齿。花白色或蓝紫色。蒴果椭圆体形。花期2-4月，果期5-10月。生海拔3500米以下的林下、林缘、山坡草丛或山谷。产云南、四川、西藏南部、贵州和广西。不丹亦有。

Perennial herbs, acaulescent. Leaves ovate, broadly ovate, or triangular-ovate, abaxially sparsely puberulous or glabrescent, base deeply or broadly cordate, margin deeply crenate. Flowers white or violet. Capsules ellipsoid. Fl. Feb-Apr. Fr. May-Oct. Under forests, forest edges, grassy slopes or valleys below 3500 m. Distributed in Yunnan, Sichuan, S Xizang, Guizhou and Guangxi. Also in Bhutan.

心叶堇菜 *Viola yunnanfuensis*

西山堇菜 *Viola hancockii*

斑叶堇菜 *Viola variegata*

## 西山堇菜

**Viola hancockii** W. Beck.

多年生草本。根状茎节密。叶多数，基生；叶下粉色至粉绿色，卵状心形，下面基部疏具柔毛或近无毛，基部深心形。花近白色，长达2厘米。蒴果矩圆状，无毛。花期4-5月，果期6-9月。生海拔200-1800米的林中、林缘或溪边。产华北、华西和华东。

Perennial herbs. Rhizomes densely nodded. Leaves numerous, basal; leaves abaxially pink to pinkish green, ovate-cordate, abaxially sparsely puberulous at base or subglabrous, base deeply cordate. Flowers whitish, to 2 cm long. Capsules oblong, glabrous. Fl. Apr-May. Fr. Jun-Sep. Forests, forest edges or streamsides at 200-1800 m. Distributed in N, W and E China.

## 斑叶堇菜

**Viola variegata** Fisch. ex Link

多年生草本。根状茎具几个白色或淡褐色长根。叶基生，丛生；叶下常淡紫红色，上面深绿色或绿色，沿脉具白色斑纹（点至条纹），两面常具短硬毛，边缘具钝圆齿。花红紫色或深紫色。蒴果椭圆体形。花期4-5月，果期6-9月。生海拔300-1700米的草坡、林中、灌丛或岩石缝。产河北、山西、内蒙古、辽宁、吉林和黑龙江。俄罗斯、蒙古、朝鲜半岛和日本亦有。

Perennial herbs. Rhizomes with several whitish or brownish long roots. Leaves basal, rosulate; leaves abaxially usually purplish red, adaxially dark green or green, white variegated (punctate-striate) along veins, both surfaces often covered with short stiff hairs, margin obtusely dentate. Flowers red-purple or dark purple. Capsules ellipsoid. Fl. Apr-May. Fr. Jun-Sep. Grassy slopes, forests, thickets or rock crevices at 300-1700 m. Distributed in Hebei, Shanxi, Neimenggu, Liaoning, Jilin and Heilongjiang. Also in Russia, Mongolia, Korean Peninsula and Japan.

## 犁头草

**Viola japonica** Langsd.

多年生草本，无茎。根状茎短，节密集。叶基生，多数；叶卵形、阔卵形或三角状卵形，两面疏具柔毛，基部深或阔心形，边缘密具圆齿。花淡紫色。蒴果椭圆体形。花期3月及10月，果期5-10月。生海拔1100米以下的阳处或半阴处。产中国西南、东南、华中和华东。朝鲜半岛和日本亦有。

Perennial herbs, acaulescent. Rhizomes short, densely nodded. Leaves numerous, basal; leaves ovate, broadly ovate, or triangular-ovate, both surfaces sparsely puberulous, base deeply or broadly cordate, margin densely crenate. Flowers light purplish. Capsules ellipsoid. Fl. Mar and Oct. Fr. May-Oct. Sunny or half shaded

犁头草 *Viola japonica*

places below 1100 m. Distributed in SW, SE, C and E China. Also in Korean Peninsula and Japan.

## 极细堇菜

**Viola perpusilla** H. Boiss.

多年生矮小草本。无地上茎。叶基生，莲座状；托叶白色，膜质1/3-1/2叶柄合生；叶片卵形或长圆形，基部圆形或楔形，稍下延。花堇紫色；花瓣长圆状倒卵形长5-7毫米，侧花瓣无毛；距长约2毫米。花期6-10月。生海拔约2800米的草坡。产云南。

Perennial herbs, dwarf, acaulescent. Leaves basal, rosulate; stipules whitish, membranous, 1/3-1/2 adnate to petioles; leaves ovate or oblong, base rounded or cuneate, slightly decurrent. Flowers violet; petals oblong-obovate, 5-7 mm long, lateral ones glabrous; spur ca. 2 mm long. Fl. Jun-Oct. Grassy slopes at ca. 2800 m. Distributed in Yunnan.

## 细距堇菜

**Viola tenuicornis** W. Beck.

多年生草本。无地上茎。根状茎通常直立，短，节密。叶2至多数，基生；叶下深紫色至绿色，卵形或阔卵形，无毛或沿脉或边缘具柔毛，边缘具浅圆齿。花紫色；距圆柱形，长5-7(-9)毫米。蒴果椭圆体形。花期4-5月，果期6-9月。生海拔200-2300米的林中、林缘、灌丛中、山坡或开阔地较干处。产华北、华西、华东和东北。俄罗斯和朝鲜半岛亦有。

细距堇菜 *Viola tenuicornis*

Perennial herbs. Acaulescent. Rhizomes usually erect, short, densely nodded. Leaves 2 to numerous, basal; leaves abaxially dark purple to green, ovate or broadly ovate, glabrous or puberulous along veins or margin, margin shallowly crenate. Flowers purple; spurs cylindric, 5-7(-9) mm long. Capsules ellipsoid. Fl. Apr-May. Fr. Jun-Sep. Forests, forest edges, thickets, slopes or dry places in open fields at 200-2300 m. Distributed in N, W, E and NE China. Also in Russia and Korean Peninsula.

极细堇菜 *Viola perpusilla*

## 北京堇菜

**Viola pekinensis** (Regel) W. Beck.

多年生草本，无茎。根状茎具多数白色小根。叶数个，基生；叶卵状心形、心形或椭圆状心形，两面疏具柔毛，边缘具钝锯齿。花白色或稀为淡玫瑰色。蒴果椭圆体形或卵球形。花期4-5月，果期6-8月。生海拔500-1900米的林中、林缘、砾石地或岩缝中。产华北和东北。

Perennial herbs, acaulescent. Rhizomes with numerous white rootlets. Leaves several, basal, leaves ovate-cordate, cordate, or elliptic-cordate, both surfaces sparsely puberulous, margin obtusely serrate. Flowers white or rarely light rose. Capsules ellipsoid or ovoid. Fl. Apr-May. Fr. Jun-Aug. Forests, forest edges, stony places or rock crevices at 500-1900 m. Distributed in N and NE China.

北京堇菜 *Viola pekinensis*

圆叶堇菜 *Viola striatella*

## 圆叶堇菜

**Viola striatella** H. Boiss.

多年生小草本，无地上茎。叶边缘具细锯齿；托叶干膜质，基部与叶柄合生，上部的托叶线状披针形或线形。花深紫色；侧瓣里面有白色须毛，下方花瓣具短距；距长约3毫米；柱头顶端扁平，两侧具狭缘边，前方具明显的短喙，喙端具较细的柱头孔。花期5月，果期6-9月。生海拔1200-3400米的山坡草地或岩石上阴湿处。产中国西南、华中和华西。

Perennial herbs, acaulescent. Leaves margin serrulate; stipules dry membranous, adnate to petioles at base, upper part linear-lanceolate or linear. Flowers deep purple; lateral petals white bearded inside, anterior one with short spur; spur ca. 3 mm long; stigmas slightly flat at top, narrowly margined on lateral sides, shortly beaked in front, with a smaller stigma hole at tip. Fl. May. Fr. Jun-Sep. Grasslands on mountain slopes or moist places on rocks at 1200-3400 m. Distributed in SW, C and W China.

## 长萼堇菜

**Viola inconspicua** Blume

多年生无茎草本。叶基生，丛生；叶片三角状卵形；叶柄具狭翅，常无毛；花淡紫色，具深色条纹；萼片卵状披针形，附属物长约3毫米。蒴果矩圆形。花期11月至翌年4月，果期6-11月。生海拔1600(-2400)米的草地、田边或林下。产中国西南、华南、东南、华中和华东。南亚、东南亚、日本和新几内亚岛亦有。

Perennial herbs, acaulescent. Leaves basal, rosulate; leaves triangular-ovate; petioles narrowly winged, usually glabrous. Flowers pale purple, with darker striations; sepals ovate-lanceolate, appendages ca. 3 mm long. Capsules oblong. Fl. Nov to next Apr. Fr. Jun-Nov. Grasslands, field sides, path edges or forest edges below 1600 (-2400) m. Distributed in SW, S, SE, C and E China. Also in S and SE Asia, Japan and New Guinea.

## 茜堇菜

**Viola phalacrocarpa** Maxim.

多年生草本，无地上茎。最基部叶常呈圆形，其余呈卵形或卵圆形，边缘具低而平的圆齿；叶柄长4-13厘米；托叶1/2以上与叶柄合生。花梗和子房常被短毛；柱头两侧及背部具明显增厚的缘边，前方具较粗的短喙，喙尖端柱头孔较粗。花果期4-9月。生海拔100-600米的向阳山坡草地、灌丛及林缘。产黑龙江、吉林和辽宁。朝鲜半岛、日本和俄罗斯亦有。

Perennial herbs, acaulescent. Leaves of lowest leaves often orbicular, those of others ovate or ovate-orbicular, margin low and flat crenate; petioles 4-13 cm long; stipules over 1/2 adnate to petioles. Pedicels and ovary usually puberulous; stigmas conspicuously margined on lateral sides and abaxially, shortly beaked in front, with a large stigma hole at tip of beak. Fl. and fr. Apr-Sep. Grasslands on sunny mountain slopes, thickets, forest edges at 100-600 m. Distributed in Heilongjiang, Jinlin and Liaoning. Also in Korean Peninsula, Japan and Russia.

## 早开堇菜

**Viola prionantha** Bunge

多年生无茎草本。叶多数，基生；叶在花期长圆状卵形、卵状披针形或狭卵形，边缘密具圆齿。花紫堇色或淡紫色，上方花瓣倒卵形，下方花瓣连距长14-21毫米；花柱棍棒状。蒴果狭椭圆体形。花期3-4月和10月，果期5-9月。生海拔2800米

长萼堇菜 *Viola inconspicua*

茜堇菜 *Viola phalacrocarpa*

旱开堇菜 *Viola prionantha*

以下的山坡草地、沟边或空旷处。产中国西南、华中、华北、华西、西北和东北。俄罗斯(远东地区)和朝鲜半岛亦有。

Perennial herbs, acaulescent. Leaves numerous, basal; leaves at anthesis oblong-ovate, ovate-lanceolate or narrowly ovate, margin densely crenulate. Flowers purple-violet or pale purple; upper petals obovate, lower petals corvered with spurs 14-21 mm long; styles clavate. Capsules narrowly ellipsoid. Fl. Mar-Apr and Oct. Fr. May-Sep. Grassy slopes, by streams or open places below 2800 m. Distributed in SW, C, N, W, NW and NE China. Also in Russia (Far East) and Korean Peninsula.

## 紫花地丁

**Viola philippica** Cav.

多年生无茎草本。叶多数，基生，丛生；叶通常三角状卵形或狭卵形，两面无毛或被细毛。花淡紫色，喉部色浅，具紫色条纹，中等大小。蒴果椭圆体形。花期4-5月，果期5-9月。生海拔1700米以下的田里、荒地、山坡、林下或灌丛。产中国大部分地区。南亚、东南亚和东北亚亦有。

Perennial herbs, acaulescent. Leaves numerous, basal, rosulate; leaves triangular-ovate or narrowly ovate, glabrous or minutely hairy on both surfaces. Flowers purplish, light colored and purple-striate at throat, medium-sized. Capsules ellipsoid. Fl. Apr-May. Fr. May-Sep. Fields, wastelands, slopes, forests or thickets below 1700 m. Distributed in most parts of China. Also in S, SE and NE Asia.

## 白花地丁

**Viola patrinii** DC. ex Ging.

多年生草本，无地上茎。叶疏生波状浅圆齿或近全缘；叶柄常比叶长2-3倍；托叶约2/3与叶柄合生。下方花瓣连距长约1.3厘米；距短而粗，浅囊状；柱头顶部平坦呈三角形，两侧具较狭的缘边，前方具斜升的短喙。花果期5-9月。生海拔200-1700米的沼泽化草甸、河岸湿地、灌丛及林缘阴湿地带。产东北和内蒙古。俄罗斯、蒙古、朝鲜半岛和日本亦有。

Perennial herbs, acaulescent. Leaves margin remotely denticulate or nearly entire; petioles usually 2-3 × exceeding blades; stipules ca. 2/3 adnate to petioles. Anterior petal ca. 1.3 cm long (spur included); spur short, robust, shallowly saccate; stigmas flat at top and triangular, narrowly margined on lateral sides, shortly beaked in front, beak obliquely ascending. Fl. and fr. May-Sep. Marshy meadows, moist places along riversides, thickets, moist and shaded places at forest edges at 200-1700 m. Distributed in NE China and Neimenggu. Also in Russia, Mongolia, Korean Peninsula and Japan.

紫花地丁 *Viola philippica*

白花地丁 *Viola patrinii*

总裂叶堇菜 *Viola dissecta* var. *incisa*

## 裂叶堇菜
**Viola dissecta** Ledeb.

多年生草本。根状茎节密。基部叶圆形、肾形或阔卵形，两面具白色柔毛，后渐无毛或仅上面疏具柔毛，常3深裂，稀5全裂，裂片条形，矩圆形或狭卵状披针形，全缘或疏具不规则牙齿，或近羽状裂。花淡紫色或紫堇色。蒴果矩圆状或椭圆体形。花期4-9月，果期5-10月。生海拔3000米以下的山坡、林缘、灌丛、田野或路边。产中国东南、华北、华西和东北。俄罗斯、蒙古和朝鲜半岛亦有。

Perennial herbs. Rhizomes densely nodded. Basal leaves orbicular, reniform, or broadly ovate, both surfaces white pubescent, later glabrescent or sparsely pubescent only adaxially, usually 3-, rarely 5-sect, segments linear, oblong, or narrowly ovate-lanceolate, margin entire or sparsely irregularly dentate, or subpinnatilobed. Flowers purplish or purple-violet, large. Capsules oblong or ellipsoid. Fl. Apr-Sep. Fr. May-Oct. Slopes, forest edges, thickets, fields or roadsides below 3000 m. Distributed in SW, N, W and NE China. Also in Russia, Mongolia and Korean Peninsula.

裂叶堇菜 *Viola dissecta*

## 总裂叶堇菜
**Viola dissecta** Ledeb. var. **incisa** (Turcz.) Y. S. Chen

本变种与原变种区别在于叶基部阔楔形，边缘具不规则缺刻状浅裂至中裂。花期4-5月，果期6-10月。生海拔1300米以下的林缘或草地。产河北、山西、内蒙古、陕西、辽宁和吉林。俄罗斯和蒙古亦有。

This variety differs from *Viola dissecta* in its leaves ovate, base broadly cuneate, margin irregularly emarginate-lobed to emarginate-fid. Fl. Apr-May. Fr. Jun-Oct. Forest edges or grasslands below 1300 m. Distributed in Hebei, Shanxi, Neimenggu, Shaanxi, Liaoning and Jilin. Also in Russia and Mongolia.

## 辽宁堇菜
**Viola rossii** Hemsl.

多年生草本，无地上茎。叶宽卵形或近肾形，边缘密具细锯齿；托叶离生，干后近膜质。花较大，下方花瓣连距长1.8-2厘米；距囊状，长3-4毫米；子房无毛，花柱顶部两侧具稍肥厚的缘边，中央略平，前方伸长成短喙。花果期4-8月。生海拔100-1300米的针阔混交林或阔叶林中或林缘、灌丛、山坡草地。产湖南、江西、安徽、山东、浙江和辽宁。朝鲜半岛和日本亦有。

Perennial herbs, acaulescent. Leaves broadly ovate or subreniform, margin densely serrulate; stipules free, submembranous when dry. Flowers large, anterior petal 1.8-2 cm long (spur included); spur saccate, 3-4 mm long; ovary glabrous, styles slightly thickly margined on lateral sides, slightly flat in central part, shortly beaked in front. Fl. and fr. Apr-Aug. Mixed needle-leaved and broad-leaved forests, broad-leaved forests or forest edges, thickets or grasslands on mountain slopes at 100-1300 m. Distributed in Hunan, Jiangxi, Anhui, Shandong, Zhejiang and Liaoning. Also in Korean Peninsula and Japan.

辽宁堇菜 *Viola rossii*

荁 *Viola moupinensis*

匍匐堇菜 *Viola pilosa*

## 荁
**Viola moupinensis** Franch.

多年生草本，无地上茎。根状茎粗而较长。托叶离生，长1-1.8厘米。花较大，下方花瓣连距长约1.5厘米；距囊状，较粗；花柱上部增粗，柱头平截，两侧及后方具肥厚的缘边，前方具平伸的短喙。蒴果长约1.5厘米。花期3-5月，果期5-7月。生海拔600-3600米的林缘开旷地或灌丛中、溪旁及草坡。产除华北、东北和西北外中国各地。不丹、尼泊尔和印度亦有。

Perennial herbs, acaulescent. Rhizomes thick and long. Stipules free, 1-1.8 cm long. Flowers large, anterior petal ca. 1.5 cm long (spur included); spur saccate, stout; styles thickened in upper part, stigmas truncate, thickly margined on lateral sides and abaxially, shortly beaked in front, beak horizontal. Capsules ca. 1.5 cm long. Fl. Mar-May. Fr. May-Jul. Open places at forest edges, thickets, streamsides, grassy slopes at 600-3600 m. Distributed throughout China, except N, NE and NW China. Also in Bhutan, Nepal and India.

## 匍匐堇菜
**Viola pilosa** Blume

多年生草本。匍匐枝细长，茎散生叶片。叶片近基生；托叶大部离生，披针形，边缘有流苏齿；叶片卵形或狭卵形，疏被白色硬毛，基部狭且深波状。花淡紫色或白色；花瓣长匝状倒卵形，侧花瓣有须毛，距长2-2.5毫米。蒴果近球形。花期2-4月。生海拔800-3000米的山地林下、山坡或路边。产中国西南。南亚和东南亚亦有。

Herbs perennial. Stolons elongated, with evenly scattered leaves. Leaves nearly basal; stipules mostly free, lanceolate, margin fimbriate-dentate; leaves ovate or narrowly ovate, sparsely white stiffly hairy, base narrowly and deeply sinuate. Flowers purplish or white; petals oblong-obovate, lateral ones bearded, spur 2-2.5 mm long. Capsules subglobose. Fl. Feb-Apr. Montane forests, grasslands or roadsides at 800-3000 m. Distributed in SW China. Also in S and SE Asia.

## 柔毛堇菜
**Viola fargesii** H. Boiss.

多年生草本，全体密被开展的白色柔毛。叶近基生或互生于匍匐茎；叶卵形或阔卵形，有时近圆形，背面具柔毛，沿脉更密，边缘具钝圆齿。花白色，花梗通常高于叶。蒴果窄球形。花期3-6月，果期6-9月。生海拔600-3800米的林下、林缘、草地、山谷或路边。产中国西南、华南、东南、华中和华东。

Perennial herbs. spreading white pubescent throughout. Leaves nearly basal or alternate on stolon; leaves ovate or broadly ovate, sometimes suborbicular, abaxially puberulous, more densely so along veins, margin obtusely dentate. Flowers white, pedicels usually longer than leaves. Capsules narrowly globose. Fl. Mar-Jun. Fr. Jun-Sep. Under forests, forest edges, grasslands, valleys or roadsides at 600-3800 m. Distributed in SW, S, SE, C and E China.

柔毛堇菜 *Viola fargesii*

紫点堇菜 *Viola duclouxii*

## 紫点堇菜

**Viola duclouxii** W. Beck.

多年生草本。无地上茎。匍匐茎长达20厘米，节上生根。叶基生；托叶褐色，披针形，基部与叶柄合生，边缘有长的或疏生流苏状齿；叶片宽卵形或近圆形，具有褐色腺点，基部深心形。花白色；上花瓣和侧花瓣匙形；距长约1毫米。蒴果卵球形。花期3-4月。生海拔1600-2700米的林下。产云南。

Herbs perennial. Acaulescent. Stolons to 20 cm long, producing rootlets at nodes. Leaves basal; stipules brown, lanceolate, margin long or sparsely fimbriate; leaves broadly ovate or suborbicular, dotlike brown glandular, base deeply cordate. Flowers white; upper and lateral petals spatulate; spur ca. 1 mm long. Capsules ovoid. Fl. Mar-Apr. Forests at 1600-2700 m. Distributed in Yunnan.

深圆齿堇菜 *Viola davidii*

小尖堇菜 *Viola mucronulifera*

## 深圆齿堇菜

**Viola davidii** Franch.

多年生草本。有匍匐茎。叶圆形或肾形，边缘疏生深圆齿，背面常具白粉，两面无毛，边缘每侧具6-8个圆齿。花白或淡紫色；花梗超出或近等长于叶。蒴果长圆形。花期3-5月和9月，果期5-10月。生海拔1200-2800米的水边草地、岩石上、林下、草坡或路边。产中国西南、华南、东南和华中。

Perennial herbs. With stolons. Leaves circular or reniform, margin sparsely profoundly crenate, abaxially often glaucous, both surfaces glabrous, margin shallowly 6-8-crenate on each side. Flowers white or purplish; pedicels exceeding or subequaling leaves. Capsules oblong. Fl. Mar-May and Sep. Fr. May-Oct. Grasslands by waters, rocks, forests, grassy slopes or roadsides at 1200-2800 m. Distributed in SW, S, SE and C China.

## 小尖堇菜

**Viola mucronulifera** Hand.-Mazz.

多年生草本，近无茎。匍匐枝长达15厘米，顶端生出新植株。叶基生或聚生于短茎；托叶与叶柄仅在基部合生，离生部分披针形，边缘有流苏齿；叶片卵状心形或窄卵形，边缘有近圆齿，齿端有尖头。花白色或淡紫色；花瓣倒卵形，侧花瓣无须毛；距长2-2.5毫米。蒴果椭球形。花期3-4月，果期5-6月。生海拔1300-1900米的山地林下、林缘或草地。产云南、四川、贵州和广西。

Herbs perennial, nearly acaulescent. Stolons to 15 cm long, top developing into a new plant. Leaves nearly basal or clustered on shortened stem; stipules adnate to petioles only at base, free part lanceolate, margin long fimbriate-dentate; leaves ovate-cordate or narrowly cordate, margin serrate, teeth suborbicular, conspicuously spinose at apices. Flowers white or purplish; petals obovate, lateral ones glabrous; spur 2-2.5 mm long. Capsules ellipsoid. Fl. Mar-Apr. Fr. May-Jun. Montane forests, forest edges or grasslands at 1300-1900 m. Distributed in Yunnan, Sichuan, Guizhou and Guangxi.

毛堇菜 *Viola thomsonii*

## 毛堇菜

**Viola thomsonii** Oudem.

多年生草本，无直立茎。匍匐茎长达15厘米。托叶披针形或宽披针形，边缘有流苏状齿；下部叶心形，较小，上部叶片较大，卵状心形，基部深心形。花白色或浅紫色；花瓣长圆状倒卵形，侧花瓣有短须毛，距短于2毫米。蒴果椭球形。花期4-6月。生海拔800-2400米的山地林下、林缘或山谷溪边。产云南西北部及南部和西藏东南部。喜马拉雅亦有。

Herbs perennial, acaulescent. Stolons to 15 cm long. Stipules lanceolate or broadly lanceolate, margin fimbriate-dentate; lower leaves cordate, smaller, upper ones larger, ovate-cordate, base deeply cordate. Flowers white or light violet; petals oblong-obovate, lateral ones shortly bearded; spur less than 2 mm long. Capsules ellipsoid. Fl. Apr-Jun. Montane forests, forest edges or stream valleys at 800-2400 m. Distributed in NW and S Yunnan and SE Xizang. Also in Himalaya.

## 猫儿山堇菜

**Viola maoershanensis** Y. S. Chen et Q. E. Yang

多年生草本，无直立茎。匍匐茎细长，节间生根，节上常发出新的莲座状叶丛。叶多数基生，或散生于匍匐枝上，长圆状卵形或三角状卵形；托叶绿色基部与叶柄合生。花蓝堇色或粉白色，距长约1毫米。花期3-4月。生海拔约650米的山谷竹林下。产广西(猫儿山)。

Perennial herbs, acaulescent. Stolons elongated, slender, rooting at nodes and producing rosettes of leaves and flowers at the apex. Leaves nearly basal or alternate on stolons; blade oblong-ovate or triangular-ovate; stipules green, adnate to petioles at base. Flowers bluish-violet or pinkish-white, spur ca.1 mm long. Fl. Mar-Apr. Bamboo forests in a montane valley at ca. 650 m. Distributed in Guangxi (Maoer Mountain).

## 华南堇菜

**Viola austrosinensis** Y. S. Chen et Q. E. Yang

多年生草本，无直立茎。匍匐茎细长，节间生根，节上常发出新的莲座状叶丛。叶革质，近基生，或互生于匍匐枝上；托叶边缘有长流苏状齿；叶片卵形或宽卵形，基部心形，边缘有明显圆齿。花粉白色。萼片披针形，6-7毫米；距2-2.5毫米。花期3-4月。生海拔500-1800米的阴湿混交林下。产华南。

Perennial herbs. Acaulescent. Stolons elongated, slender, rooting at nodes and producing rosettes of leaves and flowers at the apex. Leaves coriaceous, nearly basal, alternate on stolons; stipules adnate to petioles only at base, margin long fimbriate-dentate; leaves ovate or broadly ovate, base cordate, margin distinctly crenate. Flowers pinkish white; sepals lanceolate, 6-7 mm long; spur 2-2.5 mm long. Fl. Mar-Apr. Humus under shady mixed forests at 500-1800 m. Distributed in S China.

猫儿山堇菜 *Viola maoershanensis*

华南堇菜 *Viola austrosinensis*

七星莲 *Viola diffusa*

悬果堇菜 *Viola pendulicarpa*

## 七星莲
**Viola diffusa** Ging.

一年生草本。有匍匐茎，全株具硬毛或白色柔毛或近无毛。基生叶丛生；托叶基部贴生到叶柄上；叶卵形或卵状长圆形，边缘具钝齿和缘毛。花小，淡紫色；花柱棍棒状。蒴果长圆形。花期3-5月，果期5-10月。生海拔2000米以下的山地林下、林缘、草坡、河边或岩隙。产中国西南、华南、东南、华中、华西和华东。南亚、东南亚、日本和巴布亚新几内亚亦有。

Annual herbs. With stolons, stiffly hairy or white puberulous throughout or subglabrous. Basal leaves rosulate; stipules adnate to petioles; leaves ovate or ovate-oblong, margin obtusely dentate and ciliate. Flowers small, pale purple; styles clavate. Capsules oblong. Fl. Mar-May. Fr. May-Oct. Forests on mountains, forest edges, grassy slopes, by streams or rock crevices below 2000 m. Distributed in SW, S, SE, C, W and E China. Also in S and SE Asia, Japan and Papua New Guinea.

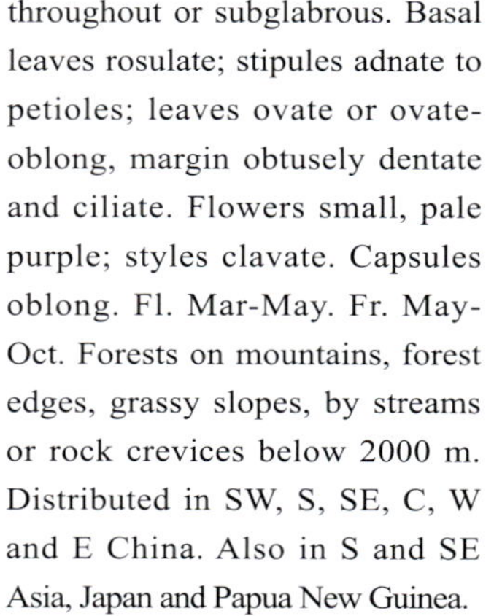

## 悬果堇菜
**Viola pendulicarpa** W. Beck.

多年生草本，有茎或近无茎。叶多基生，莲座状；托叶离生，边缘有长流苏；叶片三角状卵形、卵形或卵状心形，基部截形或浅心形。花白色或紫色；侧花瓣有须毛；距长约2.5毫米。蒴果长球形，有棕色腺点。花期3-5月，果期5-9月。生海拔300-2400(-3500)米的山地林下、林缘或河边。产云南西北部、四川、湖北西部和陕西南部。

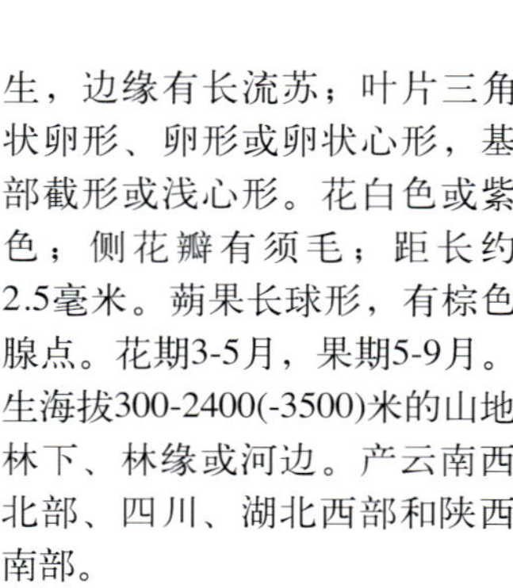

Perennial herbs, caulescent or sometimes nearly acaulescent. Leaves mostly basal, rosulate; stipules free, margin long fimbriate; leaves triangular-ovate, ovate, or ovate-cordate, base truncate or shallowly cordate. Flowers white or purplish; lateral petals bearded; spur ca. 2.5 mm long. Capsules nearly oblong, with sparse brown glandular dots. Fl. Mar-May. Fr. May-Sep. Montane forests, forest edges or riversides at 300-2400(-3500) m. Distributed in NW Yunnan Sichuan, W Hubei and S Shaanxi.

## 三角叶堇菜
**Viola triangulifolia** W. Beck.

多年生草本。茎单一或几个丛生。叶卵状三角形至狭三角形，两面无毛，边缘具浅锯齿。花单生于茎叶叶腋，白色，有紫色条纹。蒴果椭圆体形。花期4-5月，果期5-9月。生海拔200-1800米的山谷、林缘或路边。产中国西南、东南、华中和华东。

Perennial herbs. Stems solitary or several fasciculate. Leaves ovate-triangular to narrowly triangular, both surfaces glabrous, margin shallowly serrate. Flowers solitary in axils of cauline leaves, white, purple striate. Capsules ellipsoid. Fl. Apr-May. Fr. May-Sep. Valleys, forest edges or roadsides at 200-1800 m. Distributed in SW, SE, C and E China.

三角叶堇菜 *Viola triangulifolia*

灰叶堇菜 *Viola delavayi*

## 灰叶堇菜

**Viola delavayi** Franch.

多年生草本。地上茎直立或斜升。叶纸质，卵形，背面呈苍白色，基部疏具柔毛，边缘具波状齿。花黄色；具长梗；花梗超出叶。蒴果卵球形或近长圆形，较小。花期5-8月，果期7-10月。生海拔1800-2800米的林下、草地或山谷。产云南、四川和贵州。

Perennial herbs. Stem erect or ascending. Leaves chartaceous, ovate, abaxially glaucous, sparsely villous at base, margin repand-serrate. Flowers yellow, long pedicellate; pedicels exceeding leaves. Capsules ovoid or nearly oblong, small. Fl. May-Aug. Fr. Jul-Oct. Forests, grasslands or valleys at 1800-2800 m. Distributed in Yunnan, Sichuan and Guizhou.

## 粗齿堇菜

**Viola urophylla** Franch.

多年生草本。地上茎细弱，高达22厘米，无毛。基生叶深心形，长5-9厘米，边缘具内弯的粗锯齿，无毛；托叶长1.5-2.5厘米，边缘疏生粗齿或3浅裂。花黄色；花下有不发育的线形小苞片2；下方花瓣长约8毫米，基部有紫色条纹；距短，长0.5-0.9毫米，先端钝圆；柱头2裂。花期6-8月。生海拔1600-3600米的山地林缘、草甸及溪边。产云南西部和四川。

Perennial herbs. Stems slender, up to 22 cm tall, glabrous. Basal leaves deeply cordate, 5-9 cm long, margin with coarsely inflected serration, glabrous; stipules 1.5-2.5 cm long, margin remotely crenate or shallowly 3-lobed. Flowers yellow; 2-undeveloped-bracteolate below flower; anterior petal ca. 8 mm long, base purple striate; spur short, 0.5-0.9 mm long, apex obtuse; stigmas 2-lobed. Fl. Jun-Aug. Montane forest edges, meadows or streamsides at 1600-3600 m. Distributed in W Yunnan and Sichuan.

粗齿堇菜 *Viola urophylla*

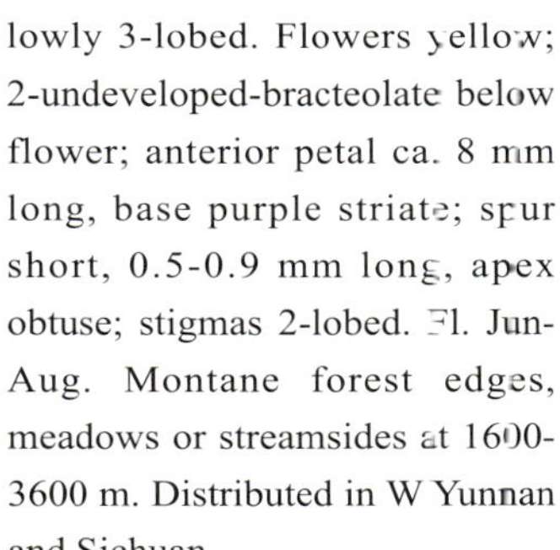

## 木里堇菜

**Viola muliensis** Y. S. Chen et Q. E. Yang

多年生草本。根状茎密具节，生多数深褐色纤维状须根。茎高达28厘米。叶片宽卵形或三角形，稀圆形，三回鸟足状分裂，裂片2-3裂；托叶2，叶状，离生。花腋生，黄色；萼片绿色，线形；花瓣无须毛，距长约0.5毫米。蒴果卵球形，无毛。花期7月，果期9月。生海拔约2500米的林缘苔藓覆盖的岩石缝。产四川(木里县)。

Perennial herbs. Rootstock densely articulate, with numerous dark brown fibrous roots. Stems to 28 cm tall. Leaves broadly ovate or triangular, very rarely orbicular, 2-3-pedatipartite; stipules 2, leaf-like, free. Flowers axillary, yellow; petals glabrous; spur ca. 0.5 mm long. Capsules ovoid or globose, glabrous. Fl. Jul. Fr. Sep. Moss-covered rock crevices in forest edges at ca. 2500 m. Distributed in Sichuan (Muli County).

木里堇菜 *Viola muliensis,*

四川堇菜 *Viola szetschwanensis*

圆叶小堇菜 *Viola biflora* var. *rockiana*

双花堇菜 *Viola biflora*

## 四川堇菜

**Viola szetschwanensis** W. Beck. et H. Boiss.

多年生草本。茎直立，强壮，具3-4节。叶卵状心形或阔卵形，背面疏具柔毛，边缘具浅圆齿。花单生于上部叶腋，黄色。蒴果矩圆状。花期5-9月，果期7-10月。生海拔2400-4000米的林中、林缘、草坡或灌丛中。产云南北部、四川西部和西藏。尼泊尔亦有。

Perennial herbs. Stems erect, robust, 3-4-noded. Leaves ovate-cordate or broadly ovate, abaxially sparsely puberulous, margin shallowly crenate. Flowers solitary in leaf axils in upper part, yellow. Capsules oblong. Fl. May-Sep. Fr. Jul-Oct. Forests, forest edges, grassy slopes or thickets at 2400-4000 m. Distributed in N Yunnan, W Sichuan and Xizang. Also in Nepal.

## 圆叶小堇菜

**Viola biflora** L. var. **rockiana** (W. Beck.) Y. S. Chen

本变种与原变种的区别在于本变种的茎常2(-3)，细弱，高达10厘米，具2节，无毛，仅下部具叶；叶圆形至卵圆形，基部浅心形或近截形，边缘具浅残波状圆齿；距浅囊状，长1-1.5毫米。花期6-7月，果期7-8月。生海拔2500-4300米的灌丛、高山或亚高山草坡或岩石上。产云南、四川、西藏东部、甘肃和青海。

This variety differs from the typical variety in its stems usually 2(-3), slender, to 10 cm tall, 2-noded, glabrous, with leaves only in lower part; leaves orbicular or ovate-orbicular, base shallowly cordate or subtruncate, margin shallowly repand-crenate; spurs shallowly saccate, 1-1.5 mm long. Fl. Jun-Jul. Fr. Jul-Aug. Thickets, alpine or subalpine grassy slopes or rocks at 2500-4300 m. Distributed in Yunnan, Sichuan, E Xizang, Gansu and Qinghai.

## 双花堇菜

**Viola biflora** L.

多年生草本。根状茎垂直，具结节。茎2或数个簇生，具3-5节。叶肾形、阔卵形或近圆形至圆形，背面无毛，边缘具钝圆齿或浅波齿。花黄色，有紫色条纹；蒴果长圆状卵球形。花期6-7月，果期7-8月。生海拔2500-4300米的草坡、林下或灌丛中。产中国大部分高山和亚高山地区。南亚、东南亚、东北亚、欧洲和北美洲亦有。

Perennial herbs. Rhizomes erect, nodded. Stems 2 or several fasciculate, 3-5-noded. Leaves reniform, broadly ovate, or suborbicular to orbicular, abaxially glabrous, margin obtusely dentate or shallowly repand-crenate. Flowers yellow, with purple stripes. Capsules oblong-ovoid. Fl. Jun-Jul. Fr. Jul-Aug. Grassy slopes, under forests or thickets at 2500-4300 m. Distributed in alpine and subalpine places in most parts of China. Also in S, SE and NE Asia, Europe and North America.

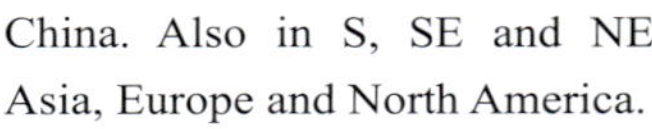

## 纤茎堇菜

**Viola tenuissima** Chang

多年生小草本。茎生叶心形，边缘疏生粗锯齿。叶柄长达5厘米。花黄色；花梗线状，长2-3.5厘米；侧方花瓣无须毛，下方花瓣长约1厘米，基部有紫色条纹；距长4-5毫米，末端变细且钝圆；子房无毛，有红色斑点，花柱先端2裂，裂片钝圆。蒴果表面有红色斑点。花果期6-9月。生海拔2300-3300米的山地林下或阴湿处岩缝中。产贵州和四川。

纤茎堇菜 *Viola tenuissima*

肾叶堇菜 *Viola schulzeana*

Perennial small herbs. Cauline leaves cordate, margin remotely and coarsely serrate. Petiolules up to 5 cm long. Flowers yellow; pedicels filiform, 2-3.5 cm long; lateral petals not barbate, anterior one ca. 1 cm long, base purple striate; spur 4-5 mm long, apex small and obtuse; ovary glabrous, red punctate, styles apex 2-lobed, lobes obtuse. Capsules red punctate. Fl. and fr. Jun-Sep. Montane forests, rock crevices in shaded and moist places at 2300-3300 m. Distributed in Guizhou and Sichuan.

## 肾叶堇菜
**Viola schulzeana** W. Beck.

多年生草本。叶片肾形，长1-1.5厘米，基部浅心形或深心形，边缘具疏锯齿；基生叶叶柄长达8厘米；托叶离生，狭卵形，全缘。花黄色；花梗长约2厘米；萼片基部有极短的截形附属物；侧方花瓣内无须毛，下方花瓣倒卵形，长约1厘米，具长约4毫米的距；花柱2裂。花果期5-6月。生海拔2600-4000米的山地林下、山坡草地。产四川、云南和西藏。

Perennial herbs. Leaves reniform, 1-1.5 cm long, base shallowly or deeply cordate, margin sparsely serrate; petioles of basal leaves up to 8 cm long; stipules free, narrowly ovate, margin entire. Flowers yellow; pedicels ca. 2 cm long; sepals with shortly truncate appendages at base; lateral petals without beards inside, anterior one obovate, ca. 1 cm long, with ca. 4 mm long spur; style 2-lobed. Fl. and fr. May-Jun. Montane forests, slope grasslands at 2600-4000 m. Distributed in Sichuan, Yunnan and Xizang.

## 东方堇菜
**Viola orientalis** (Maxim.) W. Beck.

多年生草本。根状茎通常直立，具密节。叶卵形，阔卵形或椭圆形，下面具柔毛，边缘具钝锯齿；茎生叶3（或4）；托叶小，仅基部与叶柄合生。花黄色。蒴果淡绿色，常有紫黑色斑点，椭圆体形或矩圆形。花期4-5月，果期5-7月。生海拔100-1100米的疏林、林缘、灌丛或草坡。产山东东部、辽宁、吉林东部和黑龙江。俄罗斯(远东地区)、朝鲜半岛和日本亦有。

Perennial herbs. Rhizomes usually erect, densely nodded. Leaves ovate, broadly so or elliptic, abaxially puberulous, margin obtusely serrate; cauline leaves 3(or 4); stipules small, adnate to petiote only base. Flowers yellow. Capsules greenish, often purple-black-punctate, ellipsoid or oblong. Fl. Apr-May. Fr. Jun-Jul. Sparse forests, forest edges, thickets or grassy slopes at 100-1100 m. Distributed in E Shandong, Liaoning, E Jilin and Heilongjiang. Also in Russia (Far East), Korean Peninsula and Japan.

东方堇菜 *Viola orientalis*

# 大风子科 Flacourtiaceae

## 球花脚骨脆
**Casearia glomerata** Roxb.

乔木或灌木。叶形状多变，椭圆形、披针形或长圆形，厚纸质，幼时疏具柔毛，后两面迅速无毛。团伞花序腋生，具花10-15或更多；花淡黄色而小型。蒴果椭圆体形或矩圆状椭圆体形。花期8-12月，果期10月至翌年春季。生低海拔林中。产中国西南和华南。印度、尼泊尔、不丹和越南亦有。

Trees or shrubs. Leaves variable in shape, elliptic, lanceolate, or oblong, thickly papery, sparsely puberulous when young, soon glabrous on both surfaces. Glomerules axillary, with flowers 10-15 or more; flowers yellowish and small. Capsules ellipsoid or oblong-ellipsoid. Fl. Aug-Dec. Fr. Oct to next spring. Forests at low elevations. Distributed in SW and S China. Also in India, Nepal, Bhutan and Vietnam.

球花脚骨脆 *Casearia glomerata*

毛叶脚骨脆 *Casearia velutina*

## 毛叶脚骨脆
**Casearia velutina** Blume

乔木或灌木。小枝密被短柔毛。叶形状和大小多变，椭圆形至长圆形，厚纸质，起初两面具柔毛，下面尤密，两面后疏具毛或中脉及主脉外无毛。花(1至)数朵至多数形成腋生的具柄或无柄的团伞状。蒴果阔卵球形。花期2-12月，果期翌年4-6月。生海拔100-1800米的常绿阔叶林。产中国西南和华南。东南亚亦有。

Trees or shrubs. Branchlets densely pubescent. Leaves variable in shape and size, elliptic to oblong, thickly papery, initially pubescent on both sides, very densely so beneath, both sides becoming more sparsely hairy or glabrous except for midvein and main veins. Flowers (1 to) few to many in axillary sessile or subsessile glomerules. Capsules broadly ellipsoid. Fl. Feb-Dec. Fr. next Apr-Jun. Evergreen broad-leaved forests at 100-1800 m. Distributed in SW and S China. Also in SE Asia.

## 大果刺篱木
**Flacourtia ramontchi** L' Her.

乔木。花和果枝常不具刺。叶阔椭圆形、椭圆形或椭圆状披针形，纸质。花序顶生或腋生，总状。花黄绿色。果实球形，不具纵棱，具宿存花柱。种子4-6。花期4-5月，果期6-10月。生海拔200-1700米的灌丛或疏林中。产云南、贵州和广西。热带亚洲和非洲亦有。

Trees. Flowering and fruiting branches usually not spiny. Leaves broadly elliptic, elliptic, or elliptic-lanceolate, papery. Inflorescences terminal or axillary, racemose. Flowers yellowish green. Fruits globose, not longitudinally angled, with persistent styles. Seeds 4-6. Fl. Apr-May. Fr. Jun-Oct. Thickets or open forests at 200-1700 m. Distributed in Yunnan, Guizhou and Guangxi. Also in tropical Asia and Africa.

## 大叶刺篱木
**Flacourtia rukam** Zoll. et Mor.

乔木，幼时具长达10厘米的单或分枝刺。幼叶松软，下垂，玫红色至褐色；成熟叶近革质。花序腋生，总状；花淡黄绿色，无味；花盘橘红色至淡黄色。果浅绿色、粉色、淡紫色或深红色，球形，干后具4-7棱。种子约12。花期4-5月，果期6-10月。生海拔2000米以下的常绿阔叶林。产云南、广西、广东、海南和台湾。南亚和东南亚亦有。

Trees. With simple or branched thorns to 10 cm long when young. Young leaves flaccid, drooping, rose-red to brown; mature leaves subleathery. Inflorescences axillary, racemose; flowers yellowish green, scentless; disk orange-red to yellowish. Fruits light green, pink, purplish, or dark red, globose, 4-7-angled in dried state. Seeds ca. 12. Fl. Apr-May. Fr. Jun-Oct. Evergreen broad-leaved forests below 2000 m. Distributed in Yunnan, Guangxi, Guangdong, Hainan and Taiwan. Also in S and SE Asia.

## 马蛋果
**Gynocardia odorata** R. Br.

常绿乔木。叶长圆状椭圆形，革质。雄花直径3-4厘米，芳香，花瓣黄绿色，无毛；退化雄蕊10-15，具长柔毛。浆果黄褐色，球形，直径(5-)8-12厘米，生树干上。花期1-2月，果期6-8月。生海拔800-1000米的山谷林中。产云南东南部和西藏东南部。印度、尼泊尔、不丹、孟加拉国和缅甸亦有。

Evergreen trees. Leaves oblong-elliptic, leathery. Staminate flowers 3-4 cm diam, fragrant; petals yellowish green, glabrous; staminodes 10-15, villous. Berries yellowish brown, globose, (5-)8-12 cm diam, on trunk. Fl. Jan-Feb. Fr. Jun-Aug. Forests in valleys at 800-1000 m. Distributed in SE Yunnan and SE Xizang. Also in India, Nepal, Bhutan, Bangladesh and Myanmar.

大果刺篱木 *Flacourtia ramontchi*

大叶刺篱木 *Flacourtia rukam*

马蛋果 *Gynocardia odorata*

天料木 *Homalium cochinchinense*

## 天料木

**Homalium cochinchinense** (Lour.) Druce

灌木或小乔木。小枝顶端和叶柄具毛。叶厚纸质，阔椭圆形、椭圆状长圆形或倒卵形，侧脉7-9对。总状花序，长(5-)8-15厘米，被淡黄色短柔毛；花多数，白色，7或8数；子房被毛；花柱通常3。蒴果倒圆锥状。花期2-11月，果期9-12月。生海拔400-1200米的山地阔叶林中。产中国西南和华南。越南亦有。

Shrubs or small trees. Twig tips and petioles hairy. Leaves thickly papery, broadly elliptic, elliptic-oblong or obovate, lateral veins 7-9 pairs. Racemes (5-)8-15 cm long, yellowish- tomentose; flowers numerous, whitish, 7- or 8-merous; ovary hairy; styles usually 3. Capsules obconical. Fl. Feb-Nov. Fr. Sep-Dec. Broad-leaved forests on mountains at 400-1200 m. Distributed in SW and S China. Also in Vietnam.

## 斯里兰卡天料木

**Homalium ceylanicum** (Gardn.) Benth.

乔木，具板根，除花序被短柔毛外，植物全体无毛；叶形状和大小多变，椭圆形至长圆形，薄革质至厚纸质。花序腋生，总状，下垂；花多数，3至近20簇生4-6数，芳香；花瓣白色或粉红色，两面密具白色柔毛。花期1-11月，果期2-12月。生海拔400-1200米的山谷林中、林缘或溪旁。产中国西南和华南。南亚亦有。

Trees, buttressed, whole plant glabrous except inflorescences pubescent. Leaves variable in shape and size, elliptic to oblong, thinly leathery to thickly papery. Inflorescences axillary, racemose, pendulous; flowers numerous, in fascicles of 3 to ca. 20, 4-6-merous, fragrant; petals whitish or pinkish, both surfaces densely appressed whitish pubescent. Fl. Jan-Nov. Fr. Feb-Dec. Forests in valleys, forest edges or along streams at 400-1200 m. Distributed in SW and S China. Also in S Asia.

斯里兰卡天料木 *Homalium ceylanicum*

窄叶天料木 *Homalium sabiifolium*

## 窄叶天料木

**Homalium sabiifolium** F. C. How et W. C. Ko

灌木。叶柄短，长约2毫米，具柔毛，膨大；叶狭椭圆形，稀狭椭圆状长圆形，长为宽的3倍，革质，两面常无毛，侧脉6-8对。花序顶生或腋生，总状，长4-6厘米；花多数；胎座3，每个具3或4胚珠。花期10月至翌年2月，果期3-11月。生海拔约500米的山谷疏林。产广西。

Shrubs. Petioles short, ca. 2 mm long, pubescent, swollen; leaves narrowly elliptic, rarely narrowly elliptic-oblong, 3 × as long as wide, leathery, both surfaces usually glabrous, lateral veins 6-8 pairs. Inflorescences terminal or axillary, racemose, 4-6 cm long; flowers numerous; placentas 3, each with 3 or 4 ovules. Fl. Oct to next Feb. Fr. Mar-Nov. Sparse forests of mountain valleys at ca. 500 m. Distributed in Guangxi.

大叶龙角 *Hydnocarpus annamensis*

## 大叶龙角

**Hydnocarpus annamensis** (Gagnep.) Lescot et Sleum.

常绿乔木。叶片椭圆状长圆形或长圆状披针形。花序腋生；花单生，或2或3朵合生成聚伞状；花瓣8；子房卵球状圆形，稍8棱，密具柔毛。浆果近球形，具淡红色或淡褐色绒毛，以及长硬刚毛。花期4-6月，果期全年。生海拔200-600米的湿润山坡或溪边灌丛中。产云南南部和广西南部。越南亦有。

Evergreen trees. Leaves elliptic-oblong or oblong-lanceolate. Inflorescences axillary; flowers solitary, or 2 or 3 together in cymes; petals 8; ovary ovoid-orbicular, slightly 8-angled, densely pubescent. Berries subglobose, reddish or brownish tomentose interspersed with longer stiffer bristles. Fl. Apr-Jun. Fr. all year. Moist mountain slopes or thickets along streams at 200-600 m. Distributed in S Yunnan and S Guangxi. Also in Vietnam.

## 泰国大风子

**Hydnocarpus anthelminthicus** Pierre

乔木。叶鲜时绿色，干后红褐色，披针形、卵状披针形或长圆形。花2-3朵组成缩短的假聚伞花序或总状花序，或花单生；花萼5，基部连合；花瓣5，后反折，黄粉色。浆果球状。花期9月，果期11月至翌年6月。生海拔300-1300米的热带雨林或常绿阔叶林中。产云南和广西；广西、海南和台湾有栽培。泰国、越南和柬埔寨亦有。

泰国大风子 *Hydnocarpus anthelminthicus*

Trees. Leaves green when fresh, often drying reddish brown, lanceolate, ovate-lanceolate or oblong. Flowers 2 or 3 in often abbreviated false cymes or racemes, or flowers solitary; sepals 5, united at base; petals 5, becoming reflexed, yellowish pink. Berries globose. Fl. Sep. Fr. Nov to next Jun. Rain forests or evergreen broad-leaved forests at 300-1300 m. Distributed in Yunnan and Guangxi; cultivated in Guangxi, Hainan and Taiwan. Also in Thailand, Vietnam and Cambodia.

## 海南大风子

**Hydnocarpus hainanensis** (Merr.) Sleum.

常绿乔木。叶常长圆形，薄革质；侧脉7-8对，两面无毛。花5-20朵，呈总状花序，腋生或近顶生；萼片4；花瓣4。浆果球形，密生锈色绒毛；果皮革质，外果皮不纤维状。花期4-5月，果期6-8月。生海拔300-1800米的常绿阔叶林中。产云南南部、贵州、广西和海南。越南亦有。

Evergreen trees. Leaves usually oblong, thinly leathery, lateral veins 7-8 pairs, both surfaces glabrous. Racemes, 5-20-flowered, axillary or subterminal; sepals 4; petals 4. Berries globose, densely rusty-tomentose; pericarps leathery, exocarp not fibrous. Fl. Apr-May. Fr. Jun-Aug. Evergreen broad-leaved forests at 300-1800 m. Distributed in S Yunnan, Guizhou, Guangxi and Hainan. Also in Vietnam.

海南大风子 *Hydnocarpus hainanensis*

## 山桂花

**Bennettiodendron leprosipes** (Clos) Merr.

常绿灌木或小乔木。叶多狭至阔椭圆形、椭圆状长圆形或倒卵形，纸质或薄纸质。花序顶生，圆锥形，长6-12(-20)厘米，多花(至少20-30，常更多)；花污白色或绿黄色，具香味；雄花花丝具柔毛；花盘腺体小，截形。浆果成熟时红色，干后黑色，球形。花期3-4月，果期5-11月。生海拔400-1800米的常绿阔叶林。产中国西南和华南。印度、孟加拉国、缅甸、泰国和印度尼西亚亦有。

Shrubs or small trees, evergreen. Leaves mostly narrowly to broadly elliptic, elliptic-oblong, or obovate, papery or thinly papery. Inflorescences terminal, paniculate, 6-12 (-20) cm long, many flowered (at least 20-30, usually more); flowers sordid-white or greenish yellow, scented; staminate flowers filaments pubescent; disk glands small, truncate. Berries red when mature, drying black, globose. Fl. Mar-Apr. Fr. May-Nov. Evergreen broad-leaved forests at 400-1800 m. Distributed in SW and S China. Also in India, Bangladesh, Myanmar, Thailand and Indonesia.

山桂花 *Bennettiodendron leprosipes*

## 山桐子
**Idesia polycarpa** Maxim.

乔木。叶革质，卵形或心状卵形，基部心形。圆锥花序长(13-)20-30厘米；花黄绿色，芳香；花瓣缺；萼片(3-)6。浆果成熟时紫红色，扁圆形；果皮薄，干后脆。花期4-5月，果期10-11月。生海拔400-3000米的针叶林或阔叶林中。产中国西南、华南、东南、华中和华东。朝鲜半岛和日本亦有。

Trees. Leaves leathery, ovate or cordate-ovate, base cordate. Panicles (13-)20-30 cm long; flowers yellowish green, fragrant; petals absent; sepals (3-)6. Berries purple-red at maturity, flat-orbicular; pericarps thin, brittle when dry. Fl. Apr-May. Fr. Oct-Nov. Coniferous or broad-leaved forests at 400-3000 m. Distributed in SW, S, SE, C and E China. Also in Korean Peninsula and Japan.

## 栀子皮 (伊桐)
**Itoa orientalis** Hemsl.

乔木，8-12米高。树皮灰色。叶背面被黄色短柔毛，大，薄革质。花序圆锥状或总状；花单性；萼片3或4；雄花排成直立顶生具柔毛圆锥状；雄蕊120-160；雌花单生于枝顶。蒴果卵球形，密具橘黄色或淡红色绒毛，渐无毛。花期5-6月，果期9-10月。生海拔500-1700米的林中。产云南、四川、贵州、广西和海南。越南亦有。

Trees, 8-12 m tall. Bark gray. Leaves abaxially densely yellow pubescent, large, thinly leathery. Inflorescences paniculate or racemose; flowers unisexual; sepals 3 or 4; staminate flowers arranged in erect terminal pubescent panicles; stamens 120-160; pistillate flowers, solitary at apices of branches. Capsules ovoid, densely orange-yellow or reddish

山桐子 *Idesia polycarpa*

栀子皮(伊桐) *Itoa orientalis*

tomentose, glabrescent. Fl. May-Jun. Fr. Sep-Oct. Forests at 500-1700 m. Distributed in Yunnan, Sichuan, Guizhou, Guangxi and Hainan. Also in Vietnam.

## 箣柊

**Scolopia chinensis** (Lour.) Clos

常绿灌木或小乔木。枝条和小枝常具刺。叶椭圆形至长圆状椭圆形，革质，三出脉。总状花序腋生或顶生；花小，多数，淡黄色；萼片4-5；花盘具10腺体。浆果圆球形。花期6-9月，果期10月至翌年4月。生低丘陵疏林或灌丛中。产广西、广东、海南和福建。南亚亦有。

Shrubs or small trees, evergreen. Branches and branchlets often spiny. Leaves elliptic to oblong-elliptic, leathery, trinerved. Racemes axillary or terminal; flowers small, numerous, pale yellow; sepals 4-5; disk glands 10. Berries globose. Fl. Jun-Sep. Fr. Oct to next Apr. Sparse forests or thickets in hilly regions. Distributed in Guangxi, Guangdong, Hainan and Fujian. Also in S Asia.

箣柊 *Scolopia chinensis*

## 台湾箣柊(鲁花树)

**Scolopia oldhamii** Hance

常绿灌木或小乔木。枝幼时具刺，老时无刺。叶片革质至近革质，侧脉4-6对。总状花序腋生或顶生，具少花；花淡黄色至白色；花盘具10-15腺体。浆果球形，成熟时绿色至黑绿色。花期8-9月，果期11月至翌年5月。生海拔400米以下的山地、平原、向阳山坡、路边、灌丛中或海边。产台湾和福建。琉球群岛亦有。

Shrubs or small trees, evergreen. Branches spiny when young, unarmed when old. Leaves leathery or subleathery, lateral veins 4-6 per side. Racemes axillary or terminal, few flowered; flowers yellowish to white; disk glands 10-15. Berries globose, green to blackish green when mature. Fl. Aug-Sep. Fr. Nov to next May. Mountains, plains, sunny slopes, roadsides, thickets or seaside below 400 m. Distributed in Taiwan and Fujian. Also in Ryukyu Islands.

台湾箣柊(鲁花树) *Scolopia oldhamii*

### 柞木
**Xylosma congesta** (Lour.) Merr.

常绿灌木或小乔木。幼枝有刺。叶革质，两面光滑，广卵圆形至卵圆状椭圆形，侧脉3-4(-5)对。花序腋生，总状，较短；花淡黄色。浆果球形，暗红至黑色。花期7-11月，果期8-12月。生海拔500-1100米的山地林中、丘陵灌丛、平原或村庄周围。产中国西南、华南、东南、华中和华东。印度、朝鲜半岛和日本亦有。

Shrubs or small trees, evergreen, Branches spiny when young. Leaves leathery, glabrous on both surfaces, broadly ovate to ovate-elliptic, lateral veins 3-4(-5) per side. Inflorescences axillary, racemose, short; flowers yellowish. Berries globose, dark red to black. Fl. Jul-Nov. Fr. Aug-Dec. Montane forests, thickets on hills, plains or surrounding villages at 500-1100 m. Distributed in SW, S, SE, C and E China. Also in India, Korean Peninsula and Japan.

# 旌节花科 Stachyuraceae

### 云南旌节花
**Stachyurus yunnanensis** Franch.

常绿灌木。树皮深灰色。叶柄劲直，叶椭圆状长圆形至长圆状披针形，革质或薄革质，两面无毛，边缘均具细锯齿。穗状花序具12-22花；花瓣黄色至白色；子房瓶形，无毛。果实球形。花期3-4月，果期6-9月。生海拔800-1800米的山坡常绿阔叶林中或林缘灌丛中。产中国西南、华南和华中。越南北部亦有。

Evergreen shrubs. Bark dark gray. Petioles stout, leaves elliptic-oblong to oblong-lanceolate, leathery or thinly leathery, both surfaces glabrous, margin serrulate throughout. Spikes 12-22-flowered; petals yellow to white; ovary bottle-shaped, glabrous. Fruits globose. Fl. Mar-Apr. Fr. Jun-Sep. Evergreen broad-leaved forests on mountain slopes, or thickets at forest edges at 800-1800 m. Distributed in SW, S and C China. Also in N Vietnam.

云南旌节花 *Stachyurus yunnanensis*

柞木 *Xylosma congesta*

### 倒卵叶旌节花
**Stachyurus obovatus** (Rehd.) Hand.-Mazz.

常绿灌木或小乔木。叶倒卵形或倒卵状椭圆形，革质或近革质，中部以上具锯齿，先端长尾尖状渐尖。穗状花序具5-8花，长1-2厘米；总花梗长3-5毫米；花黄绿色。果实球形。花期4-5月，果期8月。生海拔500-2000米的山坡常绿阔叶林或林缘。产云南东北部、四川、贵州北部和重庆。

Shrubs or small trees, evergreen. Leaves obovate or obovate-elliptic, leathery or subleathery, margin serrate above middle, apex long caudate-acuminate. Spikes 5-8-flowered, 1-2 cm long; peduncles 3-5 mm long; flowers yellowish green. Fruits globose. Fl.

倒卵叶旋节花 *Stachyurus obovatus*

凹叶旋节花 *Stachyurus retusus*

Apr-May. Fr. Aug. Evergreen broad-leaved forests on mountain slopes or forest edges at 500-2000 m. Distributed in NE Yunnan, Sichuan, N Guizhou and Chongqing.

## 凹叶旋节花
**Stachyurus retusus** Y. C. Yang

落叶灌木。树皮黑褐色，具有稀疏的白色卵状皮孔。叶坚纸质，近圆形，下面具白色柔毛至无毛，先端钝或凹缺。穗状花序下垂，长2-5厘米；花梗近无。荚果近球形，近无梗。花期5月，果期7月。生海拔1600-2500米的山坡林中。产云南(彝良县)和四川。

Deciduous shrubs. Bark dark brown, with sparse white ovate lenticels. Leaves hard papery, suborbicular, abaxially white puberulous to glabrous, apex obtuse or emarginate. Spikes nodding, 2-5 cm long; peduncles nearly absent. Pods almost globose, subsessile. Fl. May. Fr. Jul. Forests of mountain slopes at 1600-2500 m. Distributed in Yunnan (Yiliang County) and Sichuan.

## 中国旋节花
**Stachyurus chinensis** Franch.

落叶灌木。树皮光滑，紫褐色或深褐色。叶卵形或长圆状卵形至长圆状椭圆形或近圆形，纸质至膜质，下面无毛或沿中脉及侧脉疏具柔毛。穗状花序腋生，花先叶开放。果实球形。花期3-4月，果期5-7月。生海拔400-3000米的山谷林中或林缘。产中国西南、华南和华中。

中国旋节花 *Stachyurus chinensis*

Deciduous shrubs. Bark glabrous, purple brown or dark brown. Leaves ovate or oblong-ovate to oblong-elliptic, or suborbicular, papery to membranous, abaxially glabrous or sparsely pubescent along midvein and lateral veins. Spikes axillary, flowers precocious. Fruits globose. Fl. Mar-Apr. Fr. May-Jul. Forests or forest edges in valleys at 400-3000 m. Distributed in SW, S and C China.

## 西域旋节花
**Stachyurus himalaicus** Hook. f. et Thomson ex Benth.

灌木或小乔木。叶长圆形至长圆状披针形，纸质至薄革质，8-13 × 3.5-5.5厘米，边缘具密而锐尖的细锯齿。穗状花序常下垂，长5-13厘米；花黄色。果实近球形，无毛。花期3-4月，果期5-8月。生海拔400-3000米的山坡林中。产云南和西藏。印度北部、尼泊尔、不丹和缅甸北部亦有。

Shrubs or small trees. Leaves oblong to oblong-lanceolate, papery to thinly leathery, 8-13 × 3.5-5.5 cm, margins densely serrulate, serrulations acute. Spikes usually nodding, 5-13 cm long; flowers yellow. Fruits subglobose, glabrous. Fl. Mar-Apr. Fr. May-Aug. Forests on slopes at 400-3000 m. Distributed in Yunnan and Xizang. Also in N India, Nepal, Bhutan and N Myanmar.

西域旋节花 *Stachyurus himalaicus*

# 时钟花科 Turneraceae

### 白时钟花
**Turnera subulata** Sm.

多年生草本。叶互生，叶片羽状脉，宽披针形，长6-9厘米。花生于叶腋，5数，花瓣白色，基部带深色斑。蒴果有毛。花果期全年。中国南部各省引种栽培。原产热带美洲。

Perennial herbs. Leaves alternative, leaves pinnately veined, broad lanceolate, 6-9 cm long, Flowers axillary, 5-merous, petals white with darker bases. Capsules hairy. Fl. and fr. all year. Cultivated in Southern China. Native to tropical America.

时钟花 *Turnera ulmifolia*

### 时钟花
**Turnera ulmifolia** L.

多年生草本。叶互生，叶片具羽状脉，披针形，长6.5-10厘米。花生于叶腋，5数，花瓣黄色。蒴果卵球形。花果期全年。中国南部各省引种栽培。原产墨西哥和西印度群岛。

Perennial herbs. Leaves alternative, leaves pinnately veined, lanceolate, 6.5-10 cm long, Flowers axillary, 5-merous, petals yellow. Capsules ovoid. Fl. and fr. all year. Cultivated in Southern China. Native to Mexico and the West Indies.

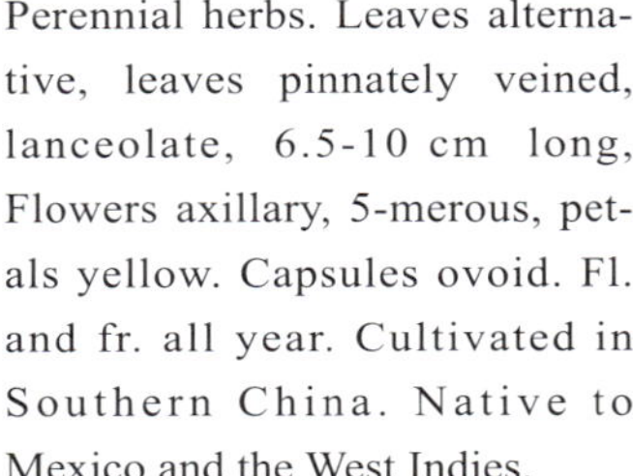

白时钟花 *Turnera subulata*

蛇王藤 *Passiflora cochinchinensis*

# 西番莲科 Passifloraceae

### 蛇王藤
**Passiflora cochinchinensis** Spreng.

草质藤本。叶披针形至椭圆形，革质，下面灰白色，下面具4-6个片状蜜腺，基部近心形。花序无柄，具2-12花；花粉白色；副花冠2层。浆果球形，成熟时蓝色。花果期3-5月。生海拔100-1000米的山谷。产广西、广东和海南。老挝、越南和马来西亚亦有。

Herbaceous vines. Leaves lanceolate to elliptic, leathery, abaxially canescent, abaxially with 4-6 laminar nectaries, base subcordate. Inflorescences sessile, 2-12-flowered; flowers pinkish white; coronas in 2 series. Berries globose, blue when mature. Fl. and fr. Mar-May. Valleys at 100-1000 m. Distributed in Guangxi, Guangdong and Hainan. Also in Laos, Vietnam and Malaysia.

### 圆叶西番莲
**Passiflora henryi** Hemsl.

草质藤本。叶近圆形至阔圆形，革质，下面具4-6个腺体，基部圆形或近心形，边缘全缘，先端截形。聚伞花序2-8花；花黄绿色；副花冠花丝2层，呈丝状。果实成熟后蓝色，球形，无毛。花期6-7月，果期8-10月。生海拔400-1600米的山坡灌丛或山谷。产云南。

Herbaceous vines. Leaves suborbicular to broadly orbicular, leathery, abaxially with 4-6 nectaries, base rounded or subcordate, margin entire, apex truncate. Inflorescences 2-8-flowered; flowers greenish yellow;

圆叶西番莲 *Passiflora henryi*

杯叶西番莲 *Passiflora cupiformis*

coronal filaments in 2 filamentous series. Fruits blue at maturity, globose, glabrous. Fl. Jun-Jul. Fr. Aug-Oct. Thickets on slopes or valleys at 400-1600 m. Distributed in Yunnan.

## 杯叶西番莲
**Passiflora cupiformis** Mast.

缠绕藤本。叶革质，长6-12(-15)厘米，先端截形至2深裂，下面具6-25枚腺体，裂片先端圆形或钝。花序近无梗，有(1-)5-20朵花；花乳白色；萼片顶端通常具顶生附属物，无毛。浆果球形，直径1-1.6厘米，熟时蓝色，无毛。花期4月，果期9月。生海拔1000-2000米的山坡草丛和灌丛、沟谷或路边。产云南、四川、广西、广东和湖北。越南亦有。

Climbing vines. Leaves leathery, 6-12(-15) cm long, apex truncate to 2-parted, abaxially with 6-25 nectaries, lobes apically rounded or obtuse. Inflorescences subsessile, (1-)5-20-flowered; flowers creamy-white; sepals often with apical appendage, glabrescent. Berries globose, 1-1.6 cm diam, blue at maturity, glabrous. Fl. Apr. Fr. Sep. Grass and thickets on slopes, valleys or roadsides at 1000-2000 m. Distributed in Yunnan, Sichuan, Guangxi, Guangdong and Hubei. Also in Vietnam.

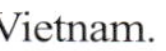

## 镰叶西番莲
**Passiflora wilsonii** Hemsl.

草质藤本。植株无毛或近无毛。叶片膜质，具1(或2)对腺体，基部阔圆形至近心形，先端强烈截形，2(或3)浅裂。花序(近)无柄，具2-15花；花白色；萼片紫褐色；花瓣白色；副花冠单层。果实成熟后蓝色，近球形。果期7月。生海拔1300-2500米的山坡灌丛、林缘或山谷。产云南、西藏和贵州。老挝、泰国和越南亦有。

Herbaceous vines. Plants glabrous or subglabrous. Leaves membranous, with 1(or 2) pairs of glands, base broadly rounded to subcordate, apex strongly truncate, 2(or 3)-lobed. Inflorescences (sub) sessile, 2-15-flowered; flowers white; sepals purplish brown; petals whitish; coronas in a single series. Fruits blue at maturity, subglobose. Fr. Jul. Thickets on slopes, forest edges or valleys at 1300-2500 m. Distributed in Yunnan, Xizang and Guizhou. Also in Laos, Thailand and Vietnam.

## 月叶西番莲
**Passiflora altebilobata** Hemsl.

草质藤本，约2米长。叶下面近中脉顶端具4个小蜜腺，深2裂。花序具2-16花；花萼绿色；花瓣白色；副花冠花丝排成两层；花瓣近白色。浆果球形，具白色脉纹。花期4-5月。生海拔600-1500米的疏林下或水边。产云南(思茅、西双版纳)。

Herbaceous vines, ca. 2 m long. Leaves abaxially with 4 small nectaries near apex of midvein, deeply 2-lobed. Inflorescences 2-16-flowered; sepals green; petals whitish; coronal filaments in two series. Berries globose, with white stripes. Fl. Apr-May. Open forests or watersides at 600-1500 m. Distributed in Yunnan (Simao, Sipsongpanna).

镰叶西番莲 *Passiflora wilsonii*

月叶西番莲 *Passiflora altebilobata*

龙珠果 *Passiflora foetida*

## 龙珠果
**Passiflora foetida** L.

草质藤本，有臭味。卷须腋生。叶膜质，宽卵形至长圆状卵形，顶端3浅裂。苞片3，一至三回羽状分裂，裂片条形，顶端具腺体；花白色或粉色。浆果卵球形，无毛。花期7-8月，果期翌年4-5月。生海拔100-1200米的草坡、林缘或路边。云南、广西、广东、海南和台湾栽培或逸生。原产西印度群岛和美洲。

Herbaceous vines, foetid. Tendrils axillary. Leaves membranous, broadly ovate to oblong-ovate, apex 3-lobed. Bracts 3, 1-3- pinnatifid, lobes linear, apex glandular hairy; flowers white or pink. Berries ovoid, glabrous. Fl. Jul-Aug. Fr. Apr-May next year. Grassy slopes, forest edges or roadsides at 100-1200 m. Cultivated or naturalized in Yunnan, Guangxi, Guangdong, Hainan and Taiwan. Native to the West Indies and America.

## 鸡蛋果(百香果)
**Passiflora edulis** Sims

草质藤本。茎无毛。叶纸质，黄绿色，掌状3深裂。花序为一退化聚伞状，中央花不发育，一侧枝特化为卷须，花与卷须对生；副花冠4或5层。果熟时紫色，卵球形。花期6月，果期11月。云南、广东、台湾和福建栽培或逸生。原产南美洲(可能起源于巴西南部)。

Herbaceous vines. Stems glabrous. Leaves papery, yellow-green, palmately 3-parted. Inflorescences a reduced cyme, central flower not developed, one lateral branch converted to a tendril, flower opposite tendril; coronas in 4 or 5 series. Fruits purple at maturity, ovoid. Fl. Jun. Fr. Nov. Cultivated or escaped Yunnan, Guangdong, Taiwan and Fujian. Native to South America (probably originally from S Brazil).

鸡蛋果(百香果) *Passiflora edulis*

## 三开瓢
**Adenia cardiophylla** (Mast.) Engl.

木质大藤本。叶心形，纸质，侧脉4-5对。花序自中部生卷须，花梗长达18厘米，雄花多至30朵，雌花1-3朵；无副花冠。蒴果纺锤形，长6-8厘米，深红色，成熟时室背开裂。花期4-9月，果期6-11月。生海拔800-2000米的山坡密林中。产云南。南亚和东南亚亦有。

Woody lianas, large. Leaves cordate, papery, lateral veins 4-5 pairs. Inflorescences with tendril emerging from center, peduncles to 18 cm long, to 30-flowered in males, 1-3-flowered in females; coronas absent. Capsules fusiform, 6-8 cm long, deeply red, loculicidal at maturity. Fl. Apr-Sep. Fr. Jun-Nov. Dense forests on slopes at 800-2000 m. Distributed in Yunnan. Also in S and SE Asia.

三开瓢 *Adenia cardiophylla*

# 番木瓜科 Caricaceae

## 番木瓜
**Carica papaya** L.

小乔木或灌木。茎单生，具螺旋状排列的托叶痕。叶近盾形，常5-9掌状裂，裂片羽状分裂。花单性或两性；雄花具10雄蕊；雌花无雄蕊。果实成熟后橘黄色或黄色，圆柱形，卵状圆柱形或近球形；果实肉质软，具温和愉悦的味道。花果期全年。栽培于华南，或逸生。原产中美洲。热带和亚热带地区亦广泛栽培。

Small trees or shrubs. Stems simple, with stipulate scars helically arranged. Leaves subpeltate, usually 5-9 palmatifid, lobes pinnatifid. Flowers unisexual or bisexual; staminate flowers stamens 10; pistillate flowers stamens absent. Fruits orange-yellow or yellow at maturity, cylindric, ovoid-cylindric or subglobose; sarcocarps soft with a mild, pleasant flavor. Fl. and fr. all year. Cultivated in S China, or escaped. Native to Central America. Also in other tropical and subtropical areas.

番木瓜 *Carica papaya*

# 四数木科 Tetramelaceae

## 四数木
**Tetrameles nudiflora** R. Br.

乔木，高25-45米，雌雄异株，具板状根。叶心形、心状卵形或近圆形，幼时稍2或3浅裂。花先叶开放；花萼稍4棱。蒴果成熟后黄褐色，球状坛形，外面具8-10脉，疏具褐色腺点。花期3-4月，果期5月。生海拔500-700米的山谷雨林或石灰岩山丘。产云南南部。南亚、东南亚和澳大利亚(昆士兰)亦有。

Trees, 25-45 m tall, dioecious, with buttresses. Leaves cordate, cordate-ovate, or suborbicular, slightly 2- or 3-lobed when young. Flowers blossom before leaves appearing; calyx slightly 4-angled. Capsules brown-yellow at maturity, globose-urceolate, 8-10-veined outside, sparsely brown glandular punctate. Fl. Mar-Apr. Fr. May. Rain forests in valleys or limestone hills at 500-700 m. Distributed in S Yunnan. Also in S and SE Asia, and Australia (Queensland).

四数木 *Tetrameles nudiflora*

# 秋海棠科 Begoniaceae

## 刺盾叶秋海棠

**Begonia setulosopeltata** C. Y. Wu

草本，具根茎。叶全基生，盾形，具长柄；叶片卵形或阔卵形，上面疏具硬毛或糙毛。花数朵，浅紫色，二歧聚伞状；雄花花被片4；雌花花被片4。蒴果下垂。花期2-4月，果期4-5月。生海拔约300米的阴湿沟谷或石灰岩洞中。产广西(东兰县、河池)。

Herbs, rhizomatous. Leaves all basal, peltate, long petiolate; blade ovate or broadly ovate, adaxially sparsely hispidulous or setulose. Flowers several, pale purple, in dichasial cyme; staminate flower with tepals 4; pistillate flower with tepals 4. Capsules nodding. Fl. Feb-Apr. Fr. Apr-May. Shaded places of valleys or limestone caves at ca. 300 m. Distributed in Guangxi (Donglan County, Hechi).

刺盾叶秋海棠 *Begonia setulosopeltata*

## 乌叶秋海棠

**Begonia ornithophylla** Irmsch.

无茎草本。叶全基生，斜卵形至卵状披针形，近革质，基部浅心形。花序腋生；花粉色；雄花花被片4；雌花花被片3；子房1室，侧膜胎座。蒴果具3个近等大翅。花期1-5月，果期3-6月。生海拔100-600米的林下石灰岩上。产广西。

Herbs acaulous. Leaves all basal, obliquely ovate to ovate-lanceolate, subleathery, base shallowly cordate. Inflorescences axillary; flowers pink; staminate flowers, tepals 4; pistillate flowers, tepals 3; ovary 1-loculed, placentae parietal. Capsules with 3 subequal wings. Fl. Jan-May. Fr. Mar-Jun. Limestone rocks under forests at 100-600 m. Distributed in Guangxi.

## 伞叶秋海棠

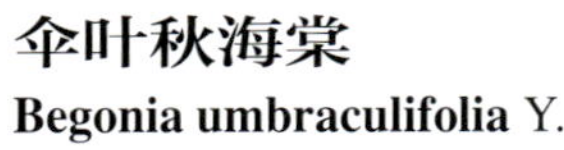

**Begonia umbraculifolia** Y. Wan et B. N. Chang

多年生草本。具根状茎。叶基生，盾状。花序腋生，二歧聚伞状；雄花花被片4，外轮2枚，宽卵形，内轮2枚，椭圆形；雌花花被片3，外轮2枚近圆形或宽卵形，内轮1片椭圆形，子房具不等3翅，侧膜胎座，花柱3，基部合生。蒴果下垂。花期10-11月。生海拔200-500米的林下、山谷或石灰岩上。产广西。

Perennial herbs. Rhizomatous. Leaves all basal, peltate. Inflore-

乌叶秋海棠 *Begonia ornithophylla*

伞叶秋海棠 *Begonia umbraculifolia*

蛛网脉秋海棠 *Begonia arachnoidea*

scences axillary, dichasial cymose. Staminate flowers, tepals 4, outer 2 broadly ovate, inner 2 elliptic; pistillate flowers, tepals 3, outer 2 suborbicular or broadly ovate, inner 1 elliptic; ovary unequally 3 winged, parietal placentation; styles 3, fused at base. Capsules nodding. Fl. Oct-Nov. Forest understories, valleys or limestone at 200-500 m. Distributed in Guangxi.

## 蛛网脉秋海棠

**Begonia arachnoidea** C. I Peng, Yan Liu et S. M. Ku

多年生草本。叶纸质，近圆形或宽卵形，长12-26(-35)厘米，上面沿主脉具白色或灰白色带，密被短刚毛和鬃毛状细刚毛，下面沿脉密被细硬毛状柔毛；三级支脉蛛网状。总花梗长9-31厘米，被鬃毛状绒毛；雄花花被片4，粉色；雌花花梗长4-6厘米；子房长圆形，具不等3翅。花期9-10月，果期10-12月。生石灰岩山上。产广西西南部。

Perennial herbs. Leaves papery, suborbicular or broadly ovate, 12-26(-35) cm long, adaxially with white or pale band along major veins, densely shortly setose and hispid-setulose, abaxially densely hispidulous-pilose on all veins; tertiary veins spiderweb-like. Peduncle 9-31 cm long, hispid-villous; staminate flower with tepals 4, pink; carpellate flower with pedicel 4-6 cm long; ovary oblong, unequally 3-winged. Fl Sep-Oct. Fr. Oct-Dec. Limestone hill. Distributed in SW Guangxi.

## 假厚叶秋海棠

**Begonia pseudodryadis** C. Y. Wu

无茎草本，细弱。根状茎伸长，匍匐。叶全基生，下面红褐色，近革质，两面无毛，基部深心形。二歧聚伞花序；花粉色；雄花花被片4；雌花花被片5；子房无毛，具不等3翅，侧膜胎座。花期5-9月，果期7-11月。生海拔800-1500米的山谷中覆苔藓的石灰岩上。产云南(河口县、马关县、屏边县)。

Herbs acaule, gracilis. Rhizomes lengthened, creeping. Leaves all basal, abaxially reddish brown, subleathery, both surfaces glabrous, base deeply cordate. Dichotomous cymes; flowers pink; staminate flowers with 4 tepals; pistillate flowers with 5 tepals; ovary glabrous, unequally 3-winged, placentae parietal. Fl. May-Sep. Fr. Jul-Nov. Mossy limestone rocks in valleys at 800-1500 m. Distributed in Yunnan (Hekou County, Maguan County, Pingbian County).

假厚叶秋海棠 *Begonia pseudodryadis*

假癞叶秋海棠 *Begonia pseudoleprosa*

## 假癞叶秋海棠

**Begonia pseudoleprosa** C. I Peng, Yan Liu et S. M. Ku

草本，具根茎。叶全基生；叶片全绿色，阔卵形至近圆形，近革质，上面近无毛。花序为二歧聚伞状，具5-15花；雌花花被片3；子房倒卵球状椭圆体形，具3翅。蒴果下垂。花期11-12月。生石灰山。产广西(大新县)。

Herbs, rhizomatous. Leaves all basal; blade all green, broadly ovate to suborbicular, subleathery, adaxially subglabrous. Inflorescences a dichasial cyme with 5-15 flowers; pistillate flowers, tepals 3; ovary obovoid-ellipsoid, 3-winged. Capsules nodding. Fl. Nov-Dec. Limestone hills. Distributed in Guangxi (Daxin County).

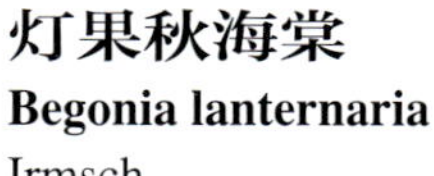

## 灯果秋海棠

**Begonia lanternaria** Irmsch.

草本。根茎伸长。叶基生；叶柄疏具长柔毛；叶片下面浅褐绿色，上面深褐绿色，卵形至阔卵形，不对称，无毛。花序无毛；雌花花被片3，粉色；子房无毛，1室。蒴果下垂，长圆状卵球形。花期8月，果期9月。生林缘石灰岩上。产广西(龙州县)。越南北部亦有。

Herbs. Rhizomes elongate. Leaves basal; petioles sparsely villous; blade abaxially pale brown-green, adaxially deep brown-green, ovate to broadly ovate, asymmetric, glabrous. Inflorescences glabrous; pistillate flowers, tepals 3, pinkish; ovary glabrous, 1-loculed. Capsules nodding, oblong-ovoid. Fl. Aug. Fr. Sep. limestone rocks at forest margins. Distributed in Guangxi (Longzhou County). Also in N Vietnam.

## 蛮耗秋海棠

**Begonia manhaoensis** S. H. Huang et Y. M. Shui

草本。叶片暗绿色，广卵圆形，不对称，草质，叶脉型掌裂，基部歪斜，心形，边缘浅裂，先端急尖至渐尖。雄花花被片4，无毛；雌花花被片5，无毛；子房2室。蒴果弯头，具不等3翅。花期未知。生海拔300-800米的林下叶层。产云南(龙陵县、屏边县)。

Herbs. Leaves dark green, broadly ovate, asymmetric, herbaceous, venation palmate, base oblique, cordate, margin shallowly lobed, apex acute to acuminate. Staminate flowers with 4 glabrous tepals; pistillate flowers with 5 glabrous tepals; ovary 2-loculed. Capsules nodding, unequally 3-winged. Fl. unknown. Forest understories at 300-800 m. Distributed in Yunnan (Longling County, Pingbian County).

## 广西秋海棠

**Begonia guangxiensis** C. Y. Wu

草本，具根茎。叶全基生；叶柄具长柔毛；叶片轮廓斜阔卵形或近圆形，10-33 × 8-24厘

灯果秋海棠 *Begonia lanternaria*

蛮耗秋海棠 *Begonia manhaoensis*

广西秋海棠
*Begonia guangxiensis*

米，纸质，两面具长柔毛状粗毛。花序腋生；花5-20朵组成二歧状聚伞花序；雌花花被片3；子房具长柔毛。蒴果下垂，卵球状长圆形。花期1-3月，果期3-5月。生海拔200-300米的石灰山、湿润多石山坡、林下或岩洞。产广西(都安县、宜山)。

Herbs. Rhizomatous. Leaves all basal; Petioles villous; blade obliquely broadly ovate or suborbicular in outline, 10-33 × 8-24 cm, papery, both surfaces villous-hirsute. Inflorescences axillary; flowers 5-20 in a dichasial cyme; pistillate flowers, tepals 3; ovary villous. Capsules nodding, ovoid-oblong. Fl. Jan-Mar. Fr. Mar-May. Limestone hills, on moist rocky slopes, forest understories or in caverns at 200-300 m. Distributed in Guangxi (Du'an County, Yishan).

## 铁十字秋海棠 (铁甲秋海棠)

**Begonia masoniana** Irmsch. ex Ziesenh.

草本。叶片轮廓宽卵形或近圆形，基部深心形，不对称，下面沿脉被硬毛。花20-200朵组成二歧聚伞花序；雄花花被片4；雌花花被片3；子房疏被紫红色长毛。蒴果弯头，长圆形至椭圆体形，淡红色，具腺状硬毛，具不等大3翅。花期5-9月，果期6-9月。生海拔100-300米的山坡石灰岩石上、密林湿土石穴上或灌丛中。产广西。越南亦有。

Herbs. Leaves broadly ovate or suborbicular in outline, asymmetrical and profoundly cordate at base, abaxially hirsute on nerves. Flowers 20-200 in a dichasial cyme; staminate flowers with tepals 4; pistillate flower with tepals 3; ovary sparsely purplish red villose. Capsules nodding, oblong to ellipsoid, reddish, glandular hispid, unequally 3-winged. Fl. May-Sep. Fr. Jun-Sep. Limestone rocks of slopes, wet caves under dense forests or shrubbery at 100-300 m. Distributed in Guangxi. Also in Vietnam.

铁十字秋海棠(铁甲秋海棠)
*Begonia masoniana*

## 不显秋海棠 (侧膜秋海棠)

**Begonia obsolescens** Irmsch.

草本。叶基生，卵形或长圆状卵形，不对称，疏具硬毛，基部倾斜，心形。花白色至粉色；雄花花被片4；雌花花被片5；子房具柔毛，3室。蒴果弯头，具近等大3翅。花期6月，果期翌年5月。生海拔500-1600米的林下土坡。产广西和云南东南部。

Herbs. Leaves basal, ovate or oblong-ovate, asymmetric, sparsely hispidulous, base oblique, cordate. Flowers white to pink; staminate flowers with tepals 4; pistillate flowers with tepals 5; ovary pubescent, 3-loculed. Capsules nodding, subequally 3-winged. Fl. Jun. Fr. next May. Earth slope in forests at 500-1600 m. Distributed in Guangxi and SE Yunnan.

不显秋海棠(侧膜秋海棠) *Begonia obsolescens*

罗甸秋海棠 *Begonia porteri*

## 罗甸秋海棠

**Begonia porteri** Lévl. et Vaniot

草本。具根茎。叶全基生；叶片下面浅绿色或淡红色，斜卵形，阔卵形或近圆形，纸质，上面具柔毛状绒毛至疏具长柔毛，上面具柔毛状绒毛。花序腋生，花3-17朵组成二歧聚伞花序；雌花花被片3；子房长圆形。蒴果长圆形。花期6-11月，果期7-12月。生海拔100-400米的半阴稍湿润的石灰岩表面及洞穴。产贵州(罗甸县)和广西(罗城县、屏边县、宜山)。

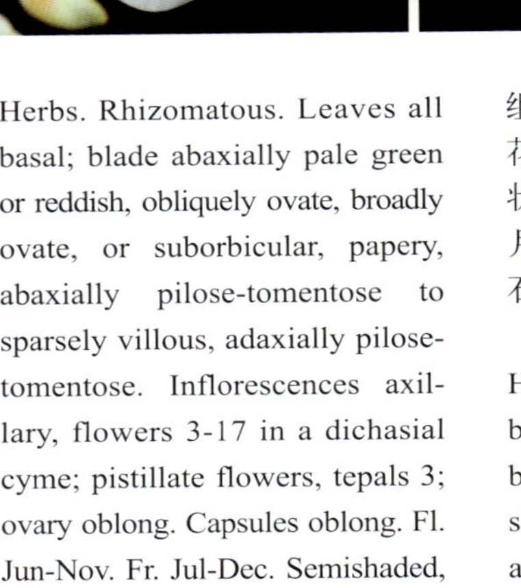

Herbs. Rhizomatous. Leaves all basal; blade abaxially pale green or reddish, obliquely ovate, broadly ovate, or suborbicular, papery, abaxially pilose-tomentose to sparsely villous, adaxially pilose-tomentose. Inflorescences axillary, flowers 3-17 in a dichasial cyme; pistillate flowers, tepals 3; ovary oblong. Capsules oblong. Fl. Jun-Nov. Fr. Jul-Dec. Semishaded, slightly moist limestone surfaces and caves at 100-400 m. Distributed in Guizhou (Luodian County) and Guangxi (Luocheng County, Pingbian County, Yishan).

## 假大新秋海棠

**Begonia pseudodaxinensis** S. M. Ku, Yan Liu et C. I Peng

草本。具根茎。叶全基生；叶片全绿色，阔卵形或近圆形，稍具皱。花序腋生，花25-57朵组成二歧聚伞状；雌花下垂；花被片3；子房椭圆体形或卵球状椭圆体形。蒴果下垂。花期12月至翌年3月。生海拔约400米的石灰山。产广西(大新县)。

Herbs. Rhizomatous. Leaves all basal; blade green throughout, broadly ovate or suborbicular, slightly rugose. Inflorescences axillary, flowers 25-57 in a dichasial cyme; pistillate flowers with pendent; tepals 3; ovary ellipsoid or ovoid-ellipsoid. Capsules nodding. Fl. Dec to next Mar. Limestone hills at ca. 400 m. Distributed in Guangxi (Daxin County).

## 大新秋海棠

**Begonia daxinensis** T. C. Ku

草本。具根茎。叶全基生；叶片上面绿色，具一白色环状带，斜卵形至近圆形，草质至纸质，下

假大新秋海棠 *Begonia pseudodaxinensis*

大新秋海棠 *Begonia daxinensis*

德保秋海棠 *Begonia debaoensis*

面近无毛或粗糙，上面无毛。花序腋生；花序白色或淡粉色，4-10花3-4次二歧状分枝组成聚伞花序。蒴果下垂，卵球状长圆形。花期3-6月，果期4-6月。生石灰山或阴湿多石区。产广西(大新县、隆安县)。

Herbs. Rhizomatous. Leaves all basal; blade adaxially green, with a whitish ring-shaped belt, obliquely ovate to suborbicular, herbaceous or papery, abaxially subglabrous or scaberulous, adaxially glabrous. Inflorescences axillary; flowers white or pinkish, 4-10 in 3-4 times branched dichasial cyme. Capsules nodding, ovoid-oblong. Fl. Mar-Jun. Fr. Apr-Jun. Limestone hills or shaded rocky places. Distributed in Guangxi (Daxin County, Long'an County).

## 德保秋海棠

**Begonia debaoensis** C. I Peng, Yan Liu et S. M. Ku

草本。具根茎。叶基生；叶片上面沿脉绿色或深绿色，阔卵形或近圆形。花序腋生，4-7花组成二歧状聚伞花序；雌花花被片3；子房椭圆体形。蒴果未发育。花期8月至翌年1月。生陡峭石灰山岩壁上。产广西(德保县)。

Herbs. Rhizomatous. Leaves basal; blade adaxially green or dark green on veins, broadly ovate or suborbicular. Inflorescences axillary, flowers 4-7 in a dichasial cyme; pistillate flowers, tepals 3; ovary ellipsoid. Capsules not developed. Fl. Aug to next Jan. Walls of caves at bases of steep limestone hills. Distributed in Guangxi (Debao County).

## 癞叶秋海棠

**Begonia leprosa** Hance

蔓性草本。叶基生，有时极浅盾形、近圆形、倒卵形或阔卵形，对称，下面无毛至沿脉具褐色长柔毛，上面疏刚毛。花序常缩短；雌花花被片4；子房3室。蒴果下垂，棒状，无翅。花期5-9月，果期6-10月。生海拔100-800米的林下或石灰岩灌丛中。产广西和广东。

Herbs creeping. Leaves basal, sometimes very shallowly peltate, suborbicular, obovate, or broadly ovate, asymmetric, abaxially glabrous to nerves brown villous, adaxially laxly setose. Inflorescences usually reduced; pistillate flowers with 4 tepals; ovary 3-locular. Capsules pendulous, clavate, wingless. Fl. May-Sep. Fr. Jun-Oct. Forests or scrubby vegetation, on limestone rocks at 100-800 m. Distributed in Guangxi and Guangdong.

癞叶秋海棠 *Begonia leprosa*

粗喙秋海棠 *Begonia longifolia*

## 盾叶秋海棠
**Begonia peltatifolia** H. L. Li

草本。叶基生，肉质，盾形，阔卵形至圆形，近对称，近无毛至无毛。花葶高达20厘米，无毛；花浅粉色；雄花花被片4；雌花花被片2；子房倒卵状长圆形，3室，具不等3翅。蒴果下垂或弯头，无毛。花期6-7月，果期7月。生海拔约900米的阔叶林、半阴而干的石灰岩上。产海南(白沙县、昌江县)。

Herbs. Leaves basal, succulent, peltate, broadly ovate to orbicular, nearly symmetric, subglabrous to glabrous. Scapes up to 20 cm tall, glabrous; flowers pinkish; staminate flowers with 4 tepals; pistillate flowers with 2 tepals; ovary obovoid-oblong, 3-celled, unequally 3-winged. Capsules pendulous or nodding, glabrous. Fl. Jun-Jul. Fr. Jul. Broad-leaved forests, on limestone rocks in semishaded but dry environments at ca. 900 m. Distributed in Hainan (Baisha County, Changjiang County).

## 粗喙秋海棠
**Begonia longifolia** Blume

多年生草本。叶茎生，披针形至阔披针形，对称，无毛或近无毛，基部倾斜，心形。花白色，具2-4花，腋生，二歧聚伞状；子房无毛，3室。蒴果近球状，下垂，无翅或具3角。花期4-5月，果期7月。生海拔200-2200米的林中、阴湿环境、山坡、山谷或河边。产中国西南和华南。南亚和东南亚亦有。

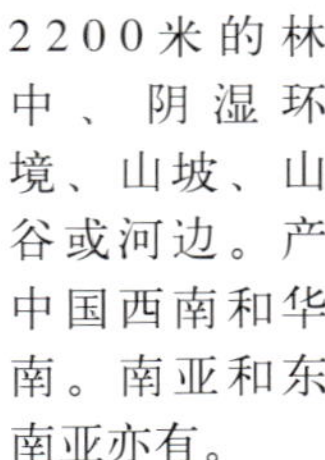

Perennial herbs. Leaves cauline, lanceolate to broadly lanceolate, asymmetric, glabrous or subglabrous, base oblique, cordate. Flowers white, 2-4 in axillary, dichotomous cymose; ovary glabrous, 3-loculed. Capsules subglobose, pendulous, wingless or 3-horned. Fl. Apr-May. Fr. Jul. Forests, shaded moist environments, slopes, valleys or riversides at 200-2200 m. Distributed in SW and S China. Also in S and SE Asia.

盾叶秋海棠 *Begonia peltatifolia*

昌感秋海棠 *Begonia cavaleriei*

## 昌感秋海棠

**Begonia cavaleriei** Lévl.

多年生草本。叶基生，叶片盾形，近对称，卵圆形或广椭圆形，厚纸质，无毛。花淡粉色，呈聚伞状；雄花花被片4；雌花花被片3。蒴果弯头，长圆状，具不等3翅。花期5-7月，果期7月。生海拔700-1300(-1800)米的林下、石灰岩上。产云南、贵州和广西。越南亦有。

Perennial herbs. Leaves basal, blades peltate, nearly symmetric, ovate or broadly elliptic, thickly papery, glabrous. Flowers light pink, cymoses; staminate flowers with 4 tepals; pistillate flowers with 3 tepals. Capsules nodding, oblong, unequally 3-winged. Fl. May-Jul. Fr. Jul. Under forests, on limestone rocks at 700-1300(-1800) m. Distributed in Yunnan, Guizhou and Guangxi. Also in Vietnam.

## 少瓣秋海棠

**Begonia wangii** T. T. Yü

草本。叶盾形，基生，具长柄。花葶高12-17厘米；花数朵呈二歧聚伞状；雄花花被片2，基部心形，先端锐尖；雌花花被片2，圆卵形；子房3室。蒴果弯头，翅小。花期未知。生海拔600-1000米的石灰岩上或灌丛植被。产云南东南部和广西西部。

Herbs. Leaves peltate, basal, with long petioles. Scapes up to 12-17 cm tall. flowers several in dichotomous cymes; staminate flowers 2-tepaled, tepals base cordate, apex acute; pistillate flowers with 2 tepals, orbicular-ovate; ovary 3-locular. Capsules nodding, wings small. Limestone rocks or scrubby vegetation at 600-1000 m. Distributed in SE Yunnan and W Guangxi.

少瓣秋海棠 *Begonia wangii*

海南秋海棠 *Begonia hainanensis*

膜果秋海棠 *Begonia hymenocarpa*

## 海南秋海棠

**Begonia hainanensis** Chun et F. Chun

匍匐草本。叶茎生，下面沿脉红色，椭圆状长圆形，稍不对称，下面疏具褐色粗毛和灰色乳突，基部稍斜，圆形。花序顶生及腋生；花白色，具淡红色脉纹。蒴果卵状长圆形，3室，具近等大3翅，翅斜三角形，无毛。花期4月，果期5月开始。生海拔约1000米的林下溪边石上。产海南(保亭县、陵水县)。

Herbs, prostrate. Leaves cauline, abaxially red on veins, elliptic-oblong, slightly asymmetric, abaxially sparsely brown hirsute and pale papillose, base slightly oblique, rounded. Inflorescences terminal and axillary; flowers white, with reddish veins. Capsules ovoid-oblong, 3-loculed, with 3 subequal wings, wings obliquely triangular, glabrous. Fl. Apr. Fr. since May. Forests, on rocks along streams at ca. 1000 m. Distributed in Hainan (Baoting County, Lingshui County).

## 膜果秋海棠

**Begonia hymenocarpa** C. Y. Wu

细弱草本。叶茎生，狭卵形，不对称，膜质。花序顶生及腋生，无毛；花粉红色；雄花花被片4；雌花花被片5；子房3室，花柱3，离生。蒴果下垂，具不等大3翅。花期7-8月，果期9月。生海拔500-700米的水边或山谷疏林湿地。产广西(大苗山、大瑶山)。

Herbs, slender. Leaves cauline, narrowly ovate, asymmetric, membranous. Inflorescences terminal or axillary, glabrous; flowers pink; staminate flowers with 4 tepals; pistillate flowes with 5 tepals; ovary 3-loculed; styles 3, free. Capsules pendulous, with 3 unequal wings. Fl. Jul-Aug. Fr. Sep. River banks or wet places of sparse forests in valleys at 500-700 m. Distributed in Guangxi (Damiao Mountain, Dayao Mountain).

## 秋海棠

**Begonia grandis** Dryand.

落叶草本。茎粗壮。茎生叶互生，不对称，下面红色或至少脉上红色，卵形至阔卵形，基部不对称，斜心形。顶生花序基部总状，腋生花序伞房状；花白色至粉色；雄花花被片4；雌花花被片3；子房3室。蒴果具明显的狭三角状翅。花期7月开始，果期8月开始。生海拔100-1100米的林下、溪边、石灰岩湿润缝隙、山谷潮湿石壁上或山谷灌丛下。产中国西南、东南、华中、华北、华西和华东。印度、马来西亚、印度尼西亚(爪哇)和日本亦有。

秋海棠 *Begonia grandis*

Herbs, deciduous. Stem stout. Cauline leaves alternate, abaxially red or at least red on veins, ovate to broadly so, asymmetrical and obliquely cordate at base. Terminal Inflorescences racemose

中华秋海棠 *Begonia grandis* subsp. *sinensis*

at base, axillary Inflorescences cymose; flowers white to pink; staminate flowers with tepals 4; pistillate flowers with tepals 3; ovary 3-loculed. Capsules with evident, narrow, triangular wings. Fl. since Jul. Fr. since Aug. Forests, streamsides, wet fissures of limestone rocks, moist rocky walls in valleys, or under shrubbery in valleys at 100-1100 m. Distributed in SW, SE, C, N, W and E China. Also in India, Malaysia, Indonesia (Java) and Japan.

## 中华秋海棠

**Begonia grandis** Dryand. subsp. **sinensis** (A. DC) Irmsch.

本亚种与原亚种区别在于：茎细弱，通常不分枝。叶片下面色浅，偶淡红色，椭圆状卵形或三角状卵形，花丝合生，少于2毫米。花期7月，果期8月。生海拔300-3400米的林中、山坡或沟谷中。产中国西南、东南、华北、华西和华东。

This subspecies differs from the *Begonia grandis* in its stems weak, usually not branched. Leaves abaxially pale, occasionally reddish elliptic-ovate or triangular-ovate, filaments connate for less than 2 mm long Fl. Jul. Fr. Aug. Forests, slopes or valleys at 300-3400 m. Distributed in SW, SE, N, W and E China.

## 阳春秋海棠

**Begonia coptidifolia** H. G. Ye, F. G. Wang, Y. S. Ye et C. I Peng

直立草本。叶基生与茎生，卵形至近圆形，近对称，疏具刚毛，基部心形，边缘具牙齿，掌状3深裂至基部，羽片二回羽状分裂。花白色；雄花花被片4；雌花花被片5；子房无毛，2室。蒴果弯头，具不等3翅。花期7-9月，果期10-12月。生海拔约600米的阔叶林山谷溪沟阴湿石上。产广东(阳春)。

Herbs, erect. Leaves basal and cauline, ovate to suborbicular, nearly symmetric, sparsely setulose, base cordate, margin dentate, palmately 3-cleft to base, pinna bipinnatifid. Flowers white; staminate flowers with tepals 4; pistillate flowers with tepals 5; ovary glabrous, 2-loculed. Capsules nodding, unequally 3-winged. Fl. Jul-Sep. Fr. Oct-Dec. Broad-leaved forests, on mossy rocks along shaded stream banks at ca. 600 m. Distributed in Guangdong (Yangchun).

## 橙花秋海棠

**Begonia crocea** C. I Peng

无茎草本。叶阔卵形，对称，下面疏具柔毛，基部倾斜，深心形，边缘疏微小锯齿。花橘红色；雄花花被片4；雌花花被片5；子房无毛，2室。蒴果弯头，具不等大的3翅。花期未知。生海拔约1200米的林中。产云南(江城县)。

Herbs, acaulescent. Leaves broadly ovate, asymmetric, abaxially sparsely pilose, base oblique, deeply cordate, margin remotely minutely serrulate. Flowers orange-red; staminate flowers with tepals 4; pistillate flowers with tepals 5; ovary glabrous, 2-loculed. Capsules nodding, unequally 3-winged. Fl. unknown. Forests at ca. 1200 m. Distributed in Yunnan (Jiangcheng County).

阳春秋海棠 *Begonia coptidifolia*

橙花秋海棠 *Begonia crocea*

古林箐秋海棠 *Begonia gulinqingensis*

## 古林箐秋海棠

**Begonia gulinqingensis** S. H. Huang et Y. M. Shui

草本。根状茎匍匐，具红色毛。叶均基生，叶片轮廓卵形或近圆形，下面疏具粗毛，基部深心形，略不对称，顶端圆。花粉色；雄花花被片4；雌花花被片5；子房3室。果实具3翅，3翅近等大。花期6月，果期7月。生海拔1600-1900米的林中或近山顶。产云南(马关县)。

Herbs. Rhizomes repent, red hairy. Leaves all basal, leaves ovate or suborbicular in outline, abaxially sparsely hirsute, profoundly cordate, slightly asymmetrical at base, rounded at apex. Flowers pink; staminate flowers with tepals 4; pistillate flowers with tepals 5; ovary 3-loculed. Fruits 3-winged, wings almost equal. Fl. Jun. Fr. Jul. Forests or near summits at 1600-1900 m. Distributed in Yunnan (Maguan County).

## 香花秋海棠

**Begonia handelii** Irmsch.

草本，雌雄异株。叶生根茎，阔卵形至卵形，不对称，两面无毛，近无毛或具红色柔毛，基部斜心形。花序缩短；花芳香，白色至粉色，雄花花被片4；雌花花被片4；子房倒卵球形，具毛或无毛，4(-8)室。果实下垂，无翅或具4(-8)角。花期1月。生海拔100-1500米的密林中潮湿处。产云南、广西、广东和海南。缅甸、老挝、泰国和越南亦有。

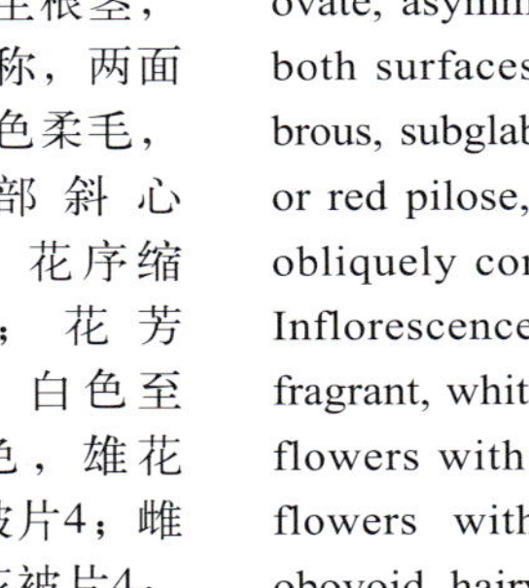

Herbs, dioecious. Leaves on rhizome, broadly ovate to ovate, asymmetric, both surfaces glabrous, subglabrous, or red pilose, base obliquely cordate. Inflorescences reduced; flowers fragrant, white to pink; staminate flowers with tepals 4; pistillate flowers with tepals 4; ovary obovoid, hairy or glabrous, 4(-8)-loculed. Fruits pendulous, wingless or 4(-8)-horned. Fl. Jan. Wet places under dense forests at 100-1500 m. Distributed in Yunnan, Guangxi, Guangdong and Hainan. Also in Myanmar, Laos, Thailand and Vietnam.

中越秋海棠 *Begonia sinovietnamica*

香花秋海棠 *Begonia handelii*

## 中越秋海棠

**Begonia sinovietnamica** C. Y. Wu

草本。根茎伸长。叶基生；叶卵形或阔卵形，不对称，下面密具粗毛，上面具粗毛。花序高9-12厘米，具硬毛或近无毛；雌花花被片5，白色；子房无毛或疏具短柔毛。蒴果下垂，卵球形。花期7月，果期8月。生阔叶林下、山坡或溪边。产广西(东兴、桂平)。

Herbs. Rhizomes elongate.

刚毛秋海棠 *Begonia setifolia*

Leaves basal; blade ovate or broadly ovate, asymmetric, abaxially densely hirsute, adaxially hirsute. Inflorescences 9-12 cm tall, hispidulous or subglabrous; pistillate flowers with tepals 5, white; ovary glabrous or sparsely short-pilose. Capsules nodding, ovoid. Fl. Jul. Fr. Aug. Understories of broad-leaved forests, on slopes or by streams. Distributed in Guangxi (Dongxing, Guiping).

## 刚毛秋海棠

**Begonia setifolia** Irmsch.

草本。叶片广卵圆形至近圆形，上面被疏长刚毛，基部心形，边缘具不明显细锯齿。花序具褐色长柔毛；雄花花被片4，粉色，雄蕊多数；雌花花被片5。蒴果下垂，卵球形，具不等3翅。花果期5月。生海拔1300-2100米的阔叶林下阴湿环境。产云南东南部。

Herbs. Leaves broadly ovate to suborbicular, asymmetric, adaxially sparsely long-setose, base cordate, margin indistinctly serrulate. Inflorescences brown villous; staminate flowers with tepals 4, pink, stamens numerous; pistillate flowers with tepals 5. Capsules nodding, ovoid, unequally 3-winged. Fl. and fr. May. Broad-leaved forests, in shaded moist environments at 1300-2100 m. Distributed in SE Yunnan.

澜沧秋海棠 *Begonia lancangensis*

## 澜沧秋海棠

**Begonia lancangensis** S. H. Huang

草本，雌雄异株。叶卵圆形或卵状矩圆形，基部不对称，掌状6-7条脉。雄花花被片4；雌花2-3朵呈聚伞状，腋生；花被片4，白色或带粉红色；子房倒卵球形，4室，每室2胎座裂片。果实浆果状，具4角。花期2-3月，果期4-7月。生海拔约1600米的常绿阔叶林中。产云南(澜沧县)。

Herbs, dioecious. Leaves ovate or ovate-oblong, base asymmetrical, palmately 6-7-veined. Staminate flowers with 4 tepals; pistillate flowers 2-3 in cymes axillary; tepals 4, white or pinkish; ovary obovoid, 4-loculed, each locule with 2 placental branches. Fruits baccate, 4-horned. Fl. Feb-Mar. Fr. Apr-Jul. Evergreen broad-leaved forests at ca. 1600 m. Distributed in Yunnan (Lancang County).

## 长翅秋海棠

**Begonia longialata** K. Y. Guan et D. K. Tian

草本，全株无毛。叶片轮廓长圆形，掌状3深裂，基部深心形，略不对称；茎生叶小。花序具11-23花；花白色至粉色；雄花花被片4；雌花花被片5；子房常红褐色，2室。果实有3翅，背翅特长。花期8-12月。生海拔约900米的岩坡。产云南(耿马县)。

Herbs, glabrous throughout. Leaves oblong in outline, palmately 3-parted, profoundly cordate, slightly asymmetrical at base; stem leaves small. Inflorescences 11-23-flowered; flowers white to pink; staminate tepals 4, pistillate tepals 5; ovary often brownish red, 2-loculed. Fruits 3-winged, abaxial wing markedly protruded. Fl. Aug-Dec. Rocky slopes at ca. 900 m. Distributed in Yunnan (Gengma County).

长翅秋海棠 *Begonia longialata*

心叶秋海棠 *Begonia labordei*

### 心叶秋海棠
**Begonia labordei** Lévl.

落叶草本。块茎球形。叶均基生，卵形，稍不对称，基部略偏斜心形，近无毛至下面疏具粗毛。花序顶生，基部总状；花粉色；雄花花被片4；雌花花被片3(或4)；子房3室。蒴果下垂，长圆状倒卵球形，具不等大3翅。花期8月，果期9月。生海拔800-3300米的山坡阴湿处岩石上或沟边杂木林中。产云南、四川和贵州。

Herbs, deciduous. Tubers globose. Leaves all basal, ovate, slightly asymmetric, slightly obliquely cordate at base, subglabrous to abaxially sparsely hirsute. Inflorescences terminal, racemose at base; flowers pink; staminate flowers with tepals 4; pistillate flowers with tepals 3(or 4); ovary 3-loculed. Capsules pendulous, oblong-obovoid, unequally 3-winged. Fl. Aug. Fr. Sep. Shady moist rocky slopes or mixed forests by streams at 800-3300 m. Distributed in Yunnan, Sichuan and Guizhou.

### 紫背天葵
**Begonia fimbristipula** Hance

落叶草本。叶单生，阔卵形，近对称，具皱，下面具毛，基部稍斜，心形或深心形，边缘具小齿。聚伞花序顶生；花浅粉色至紫红色；雄花花被片4；雌花花被片3。蒴果下垂，卵状长圆形，3室，无毛，具不等大3翅。花期5月，果期6月。生海拔700-1100米的山坡疏林下石上或山坡。产华南至中国东南。

Herbs, deciduous. Leaf solitary, broadly ovate, nearly symmetric, rugulose, abaxially hairy, base slightly oblique, cordate or deeply cordate, margin denticulate. Inflorescences terminal, cymose; flowers pale pink to purplish pink; staminate flowers with 4 tepals; pistillate flowers with 3 tepals. Capsules pendulous, ovoid-oblong, 3-loculed, glabrous, with 3 unequal wings. Fl. May. Fr. Jun. Rocks under sparse forests or slopes at 700-1100 m. Distributed in S to SE China.

紫背天葵 *Begonia fimbristipula*

### 木里秋海棠
**Begonia muliensis** T. T. Yu

草本。叶1(或2)，阔心状卵形，不对称，下面疏具硬毛，基部心形。花浅粉色；雄花花被片4；雌花花被片3；子房倒卵形，无毛；有不等3翅，3室，每室胎座具2裂片。花期8月，果期9月。生海拔1800-2600米的河边岩石上、山沟岩石下或密林中潮湿处。产云南(香格里拉)和四川西南部。

Herbs. Leaves 1(or 2), broadly cordate-ovate, asymmetric, abaxially sparsely hispidulous, base cordate. Flowers pinkish; staminate flowers with tepals 4; pistillate flowers with tepals 3; ovary obovate, glabrous, unequally 3-winged, 3-loculed, each locule with 2 placental branches. Fl. Aug. Fr. Sep. Rocks by streams,

under rocks by streams, or moist places of dense forests at 1800-2600 m. Distributed in Yunnan (Shangri-La) and SW Sichuan.

木里秋海棠 *Begonia muliensis*

刘演秋海棠 *Begonia liuyanii*

## 刘演秋海棠

**Begonia liuyanii** C. I Peng, S. M. Ku et W. C. Leong

草本。具根茎。叶均基生；叶片下面色淡，上面绿色或深绿色，斜阔卵形或近圆形，23-50 × 16-40厘米，近革质，上面疏具糙毛。花序铺散状聚伞圆锥状，具13-100花；子房椭圆体形。蒴果下垂。花期4-9月，果期6月至翌年2月。生海拔约200米的阔叶林、阴或多石石灰山坡。产广西(龙州县)。

Herbs. Rhizomatous. Leaves all basal; blade abaxially pale, adaxially green or dark green, obliquely broadly ovate or suborbicular, 23-50 × 16-40 cm, subleathery, adaxially sparsely setose. Inflorescences diffusely thyrsoid; flowers 13-100 in a thyrsoid cyme; ovary ellipsoid. Capsules nodding. Fl. Apr-Sep. Fr. Jun to next Feb. Broad-leaved forests, shaded or rocky limestone slopes at ca. 200 m. Distributed in Guangxi (Longzhou County).

## 近革叶秋海棠

**Begonia subcoriacea** C. I Peng, Yan Liu et S. M. Ku

多年生草本。叶近革质，宽卵形或近圆形，长(10-)12-20(-22)厘米，基部斜心形，上面无毛，具光泽，下面沿主脉被绒毛，支脉疏被长柔毛。花序扩散聚伞状；总花梗被淡红色具腺的长柔毛；雌花：子房三角状椭圆体形，红色，无毛，具3个不等的、黄绿色翅。蒴果开裂，长6.5-11毫米。花期3-6月，果期5至翌年3月。生石灰岩山上。产广西西南部。

Perennial herbs. Leaves subcoriaceous, widely ovate or suborbicular, (10-)12-20(-22) cm long, base oblique cordate, adaxially glabrous, nitid, abaxially tomentose on major veins, sparsely pilose on tertiary venation. Inflorescences diffusely cymose; peduncle covered with reddish glandulose-pilose trichomes; pistillate flowers: ovary trigonous-ellipsoid, red, glabrous, with 3 unequally greenish-yellow wings. Capsules dehiscent, 6.5-11 mm long. Fl. Mar-Jun. Fr. May to next Mar. Limestone hill. Distributed in SW Guangxi.

近革叶秋海棠 *Begonia subcoriacea*

靖西秋海棠 *Begonia jingxiensis*

## 靖西秋海棠

**Begonia jingxiensis** D. Fang et Y. G. Wei

草本。具根茎。叶全基生；叶片斜阔卵形或近圆形，草质，下面具纤毛，脉上具长柔毛，毛起初白色，后为锈色。花序腋生，3-40朵花组成二歧聚伞花序；雄花花被片2(-4)，雌花花被片2。蒴果下垂，卵球形。花期6-12月，果期8-12月。生海拔100-600米的石灰山及石灰洞入口或林中。产广西。

Herbs. Rhizomatous. Leaves all basal; blade obliquely broadly ovate or suborbicular, herbaceous, abaxially fibrillose, villous on veins, hairs initially whitish, later rusty. Inflorescences axillary; flowers 3-40 in a dichasial cyme; staminate flowers with tepals 2(-4), pistillate flowers with tepals 2. Capsules nodding, ovoid. Fl. Jun-Dec. Fr. Aug-Dec. Limestone hills and entrances to limestone caves or forests at 100-600 m. Distributed in Guangxi.

## 桂西秋海棠

**Begonia guixiensis** Yan Liu, S.M. Ku et C. I Peng

多年生草本。叶盾状着生，近革质，宽卵形，长6.5-15厘米，上面无毛或近无毛，下面被细硬毛；基出脉6-7条。总花梗短于叶，密被细硬毛；雄花花被片4；雄蕊22-36，花丝基部融合；雌花花被片3；花柱3，基部1/4-1/3融合。果蒴果状，不裂，三角状椭圆体形，长1-1.2厘米，具3翅。花期7-10月，果期8-11月。生石灰岩山上。产广西西南部。

Perennial herbs. Leaves peltate, subcoriaceous, broadly ovate, 6.5-15 cm long, adaxially glabrous or nearly so, abaxially minutely hispidulous; basally 6-7-veined. Peduncle shorter than leaves, densely and minutely hispidulous; staminate flowers with tepals 4; stamens 22-36, filaments fused at base; pistillate flowers with tepals 3; styles 3, fused 1/4 to 1/3 at base. Fruit capsule-like but indehiscent, trigonous-ellipsoid, 1-1.2 cm long, 3-winged. Fl. Jul-Oct. Fr. Aug-Nov. Limestone hills. Distributed in SW Guangxi.

## 一点血

**Begonia wilsonii** Gagnep.

草本。叶常1(或2)，有时下面紫色，阔卵形，稍不对称，近无毛。花粉红色；雄花花被片4；雌花花被片3；子房纺锤形，无毛，3室。蒴果下垂，棍棒状，无翅。花期8月，果期9月。生海拔700-1950米的山坡密林下、沟边或阴湿环境石壁上或山坡阴处岩石上。产四川(洪溪县、峨眉山)和重庆(南川区)。

Herbs. Leaves usually 1(or 2), sometimes purple abaxially, broadly ovate, slightly asymmetric, subglabrous. Flowers pink; staminate flowers with 4 tepals; pistillate flowers with 3 tepals; ovary fusiform, glabrous, 3-celled. Capsules pendulous, clavate, not winged. Fl. Aug. Fr. Sep. Dense forests on slopes, rocky walls by streams or shaded moist environments at 750-1950 m. Distributed

桂西秋海棠 *Begonia guixiensis*

一点血 *Begonia wilsonii*

巴马秋海棠 *Begonia bamaensis*

in Sichuan (Hongxi County, Emei Mountain) and Chongqing (Nanchuan District).

## 巴马秋海棠

**Begonia bamaensis** Yan Liu et C. I Peng

多年生草本。叶不对称，宽卵形或近圆形，长(7-)10-25(-32)厘米，基部深心形，上面密被短刚毛或糙硬毛。二歧聚伞花序具8-36花；雄花花被片4，外轮2枚长7-14.5毫米，外面被长柔毛；花丝基部部分融合；雌花花被片3；子房1室，具3翅。蒴果三角状椭圆体形，长7-12毫米。花期5-12月，果期6至翌年3月。生石灰岩山上。产广西西部。

Perennial herbs. Leaves asymmetric, broadly ovate or suborbicular, (7-)10-25(-32) cm long, base deeply cordate, adaxially densely setulose to hispidulous. Flowers 8-36 in a dichasial cyme; staminate flowers with tepals 4, 2 outer tepals 7-14.5 mm long, abaxially pilose; filaments partly fused at base; carpellate flowers with tepals 3; ovary1-locular, 3-winged. Capsules trigonous-ellipsoid, 7-12 mm long. Fl. May-Dec. Fr. Jun to next Mar. Limestone hill. Distributed in W Guangxi.

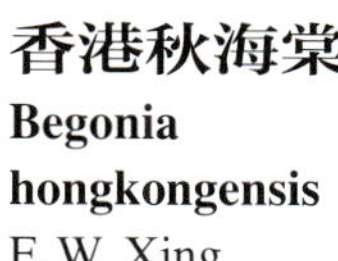

## 香港秋海棠

**Begonia hongkongensis** F. W. Xing

草本。叶基生及茎生；叶片长圆状卵形至菱状卵形，稍不对称，7-13.6 × 2.3-4.5厘米，下面疏具短毛，上面无毛。雄花花被片4，白色；雌花花被片5，柱头2，离生。蒴果下垂。花期7-9月，果期10-12月。生海拔100-400米的常绿次生林或沟谷溪边潮湿岩石上。产香港。

Herbs. Leaves basal and cauline; blade oblong-ovate to rhombic-ovate, slightly asymmetric, 7-13.6 × 2.3-4.5 cm, abaxially sparsely shortly hairy, adaxially glabrous. Staminate flowers with tepals 4, white; pistillate flowers with tepals 5; styles 2, free. Capsules nodding. Fl. Jul-Sep. Fr. Oct-Dec. Evergreen secondary forests or moist rocks by streams in ravines at 100-400 m. Distributed in Hong Kong.

小叶秋海棠 *Begonia parvula*

## 小叶秋海棠

**Begonia parvula** Lévl. et Vaniot

落叶草本。叶常2或3，基生；叶片阔卵形，近对称，2-3.4(-4) × 2-4厘米，具柔毛或近无毛。花序顶生，具2或3花；雄花花被片4，淡粉色；雌花花被片5；胎座不分裂。蒴果下垂。花期9月。生海拔200-1300米的岩石上。产云南和贵州。

Herbs, deciduous. Leaves usually 2 or 3, basal; blade broadly ovate, nearly symmetric, 2-3.4(-4) × 2-4 cm, pilose or subglabrous. Inflorescences terminal, with 2 or 3 flowers; staminate flowers with tepals 4, pinkish; pistillate flowers with tepals 5; placentae undivided. Capsules pendulous. Fl. Sep. Rocks at 200-1300 m. Distributed in Yunnan and Guizhou.

香港秋海棠 *Begonia hongkongensis*

凤山秋海棠 *Begonia chingii*

## 凤山秋海棠
**Begonia chingii** Irmsch.

落叶草本。叶通常1(-3)，基生，卵形至阔卵形，近对称，具长柔毛，基部心形，边缘具细锯齿。花序顶生，聚伞状；花粉红色；雄花花被片4；雌花花被片3。蒴果下垂，长圆形，3室，具不等3翅。花期7月。生海拔200-800米的石灰岩山丘或水流旁洞穴。产广西。

Herbs, deciduous. Leaves usually 1(-3), basal, ovate to broadly ovate, nearly symmetric, villous, base cordate, margin serrulate. Inflorescences terminal, cymose; flowers pink; staminate flowers with 4 tepals; pistillate flowers with 3 tepals. Capsules pendulous, oblong, 3-celled, with 3 unequal wings. Fl. Jul. Limestone hills or caves by streams at 200-800 m. Distributed in Guangxi.

## 丝形秋海棠
**Begonia filiformis** Irmsch.

草本。根茎匍匐，弯曲。叶全基生；叶片上面绿色，脉间具白色斑点，斜卵形至近圆形，纸质，下面具绵毛状长柔毛。花序腋生；花序梗具腺状硬毛；雄花花被片4；雌花花被片3。蒴果下垂。花期(3-)4-6月，果期5月。生石灰石上或林下潮湿岩洞。产广西(隆安县、龙州县)。

Herbs. Rhizomes creeping, tortuous. Leaves all basal; blade adaxially green, with white spot between veins, obliquely ovate to suborbicular, papery, abaxially woolly-villous. Inflorescences axillary; peduncles glandular hispid; staminate flowers with tepals 4; pistillate flowers with tepals 3. Capsules nodding. Fl. (Mar-) Apr-Jun. Fr. May. limestone rocks or in moist rocky caves of forest understories. Distributed in Guangxi (Long'an County, Longzhou County).

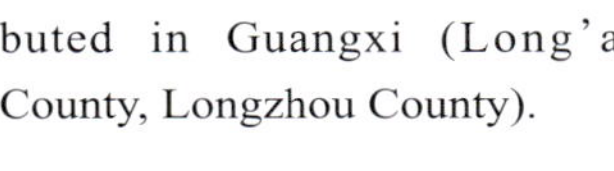

无翅秋海棠 *Begonia acetosella*

丝形秋海棠 *Begonia filiformi*

## 无翅秋海棠
**Begonia acetosella** Craib

直立草本，雌雄异株。叶茎生；叶长圆形、卵状披针形或长圆状披针形。花序极度缩短；子房倒卵球形，无毛，(3或)4室；胎座腋生，2裂。果实浆果状，下垂，具(3或)4(或5)角。花期4-5月，果期5月。生海拔500-1800米的常绿阔叶林湿润环境。产云南南部和西藏东南部。缅甸、老挝、泰国和越南亦有。

Herbs, dioecious, erect. Leaves cauline; leaves oblong, ovate-lanceolate, or oblong-lanceolate. Inflorescences much reduced; ovary obovoid, glabrous, (3 or) 4-loculed; placentae axile, bilamellate. Fruit berrylike, pendulous, (3 or 4)5-horned. Fl. Apr-May. Fr. May. Moist shaded environments in evergreen broad-leaved forests at 500-1800 m. Distributed in S Yunnan and SE Xizang. Also in Myanmar, Laos, Thailand and Vietnam.

## 角果秋海棠
**Begonia ceratocarpa** S. H. Huang et Y. M. Shui

无茎草本。叶片纸质，轮廓卵状长圆形，极不对称，基部斜心形，近全缘。花序腋生，基部总状；花小，粉色；雄花花被片4；雌花花被片5。果浆果状，菱形，不开裂，具3个角

角果秋海棠 *Begonia ceratocarpa*

状突起。花果期8-10月。生海拔300-400米的溪边阔叶林或阴湿处石上。产云南(河口县)。越南亦有。

Herbs, acaulescent. Leaves papery, ovate-oblong in outline, very asymmetrical and obliquely cordate at base, margin subentire. Inflorescences axillary, racemose at base; flowers small, pink; staminate flowers with tepals 4; pistillate flowers with tepals 5. Fruits baccate, rhomboid, indehiscent, 3-horned. Fl. and fr. Aug-Oct. Broad-leaved forests by streams or shady moist rocks at 300-400 m. Distributed in Yunnan (Hekou County). Also in Vietnam.

## 厚壁秋海棠

**Begonia silletensis** (A. DC.) C. B. Clarke subsp. **mengyangensis** Tebbitt et K. Y. Guan

草本。叶基部，卵形或阔卵形，不对称，下面稍具毛，基部倾斜，深心形。花序腋生及顶生；花芳香；花白色；雄花花被片4；雌花花被片4-6；子房4室。果实浆果状，椭圆体形，不开裂，下垂。叶片宽大，圆心形，直径达20厘米以上。花期3-5月，果期4-5月。生海拔500-800(-1200)米的林中潮湿处。产云南南部和西南部。印度亦有。

Herbs. Leaves basal, ovate or broadly ovate, asymmetric, abaxially minutely hairy, base oblique, deeply cordate. Inflorescences axillary and terminal; flowers fragrant; flowers white; staminate flowers with tepals 4; pistillate flowers with tepals 4-6; ovary 4-loculed. Fruits baccate, ellipsoidal, indehiscent, pendulous. Fl. Mar-May. Fr. Apr-May. Damp places in forests at 500-800(-1200) m Distributed in S and SW Yunnan. Also in India.

## 多花秋海棠

**Begonia sinofloribunda** Dorr

草本。叶极浅盾状，长圆状披针形或卵状披针形，稍不对称，无毛。花紫色，多数，呈二歧聚伞状；雌花花被片2；子房长圆形，具3翅，3室。蒴果下垂或弯头，矩圆状倒卵球形，无毛。花期7月，果期8月。生海拔约200米的林地、石灰岩或石山林阴下。产广西(龙州县)。

Herbs. Leaves very shallowly peltate, oblong-lanceolate or ovate-lanceolate, slightly asymmetric, glabrous. Flowers purple, numerous, in dichasial cymes; pistillate flowers with 2 tepals; ovary oblong, 3-winged, 3-loculed. Capsules pendulous or nodding, oblong-obovoid, glabrous. Fl. Jul. Fr. Aug. Forests, on limestone rocks or shady places of stony hills at ca. 200 m. Distributed in Guangxi (Longzhou County).

厚壁秋海棠 *Begonia silletensis* subsp. *mengyangensis*

多花秋海棠 *Begonia sinofloribunda*

光滑秋海棠 *Begonia psilophylla*

## 光滑秋海棠

**Begonia psilophylla** Irmsch.

草本。茎直立。叶基生及茎生，卵状心形，基部略偏斜，无毛。花粉色至深红褐色；雄花花被片4；雌花花被片5；子房无毛，2室。蒴果弯头，椭圆体形，具不等3翅。花期2月。生海拔100-700米的林下、阴湿处石灰岩石上。产云南(河口县)。

Herbs. Stems erect. Leaves basal and cauline, ovate-cordate, slightly oblique at base, glabrous. Flowers pink to dark red-brown; staminate flowers with 4 tepals; pistillate flowers with 5 tepals; ovary glabrous, 2-loculed. Capsules nodding, ellipsoid, unequally 3-winged. Fl. Feb. Under forests, shady moist limestone rocks at 100-700 m. Distributed in Yunnan (Hekou County).

## 肿柄秋海棠

**Begonia pulvinifera** C. I Peng et Yan Liu

草本。根茎伸长、匍匐。叶自根茎伸出；托叶三角形，先端渐尖或具尖头；叶片绿色、盾形、卵形，稍不对称；叶柄膨大。花序无毛；花被粉色，无毛。果序长17-45厘米；蒴果下垂。花期未知。生海拔约300米的林中或石灰岩上。产广西(靖西)。

Herbs. Rhizomes elongate, creeping. Leaves arising from rhizome; stipules triangular, apex acuminate or cuspidate; blade green, peltate, ovate, slightly asymmetric; petioles swollen; Inflorescences glabrous; tepals pinkish, glabrous. Infructescences 17-45 cm long; capsules nodding. Fl. unknown. Forests or limestone rocks at ca. 300 m. Distributed in Guangxi(Jingxi).

## 黄氏秋海棠

**Begonia huangii** Y. M. Shui et W. H. Chen

草本。具根茎。叶全基生；叶片上面脉间具白色斑块，倾斜，近圆形或极阔卵形，纸

肿柄秋海棠 *Begonia pulvinifera*

黄氏秋海棠 *Begonia huangii*

细茎秋海棠 *Begonia discrepans*

质，下面密具柔毛，上面稍具柔毛。花序具6或7花；雄花花被片4，白色或红色；雌花花被片3；子房无毛。蒴果下垂，椭圆体形，无毛。花期8-11月，果期10月。生海拔(300-)700-1000米的石灰岩林中岩石上。产云南(个旧、屏边县)。

Herbs. Rhizomatous. Leaves all basal; blade adaxially with white patches between veins, oblique, suborbicular or very broadly ovate, papery, abaxially densely pilose, adaxially moderately pilose. Inflorescences 6- or 7-flowered; staminate flowers with tepals 4, white or red; pistillate flowers with tepals 3; ovary glabrous. Capsules nodding, ellipsoid, glabrous. Fl. Aug-Nov. Fr. Oct. On rocks in limestone forests at (300-)700-1000 m. Distributed in Yunnan(Gejiu, Pingbian County).

## 细茎秋海棠

**Begonia discrepans** Irmsch.

草本。茎高约40厘米，直径4-5毫米，具褐色毛。叶卵形，不对称，上面疏具粗毛。花序具柔毛；雄花花被片4；雌花花被片5；子房长圆形，近等长3翅，2室。花期、生境未知。产云南(腾冲)。

Herbs. Stem ca. 40 cm tall, 4-5 mm diam, brown hairy. Leaves ovate, asymmetric, adaxially sparsely hirsute. Inflorescences pilose; staminate flowers with tepals 4; pistillate flowers with tepals 5; ovary oblong, subequally 3-winged, 2-loculed. Fl. and habitat unknown. Distributed in Yunnan (Tengchong).

## 滇缅秋海棠

**Begonia rockii** Irmsch.

草本。基生叶2-4，具长柄。叶片宽卵形，下面淡红色，基部极不对称。花少数，白色，呈二歧聚伞状；雌花花被片5；子房椭圆形，疏被卷曲毛，2室，具不等3翅。蒴果弯头。花期11月。生海拔700-800米的林中、岩石上。产云南(盈江县)。缅甸亦有。

Herbs. Basal leaves 2-4, with long petioles. Leaves broadly ovate, abaxially reddish, base strongly asymmetrical. Flowers few, white, in dichasial cymes; pistillate flowers with 5 tepals; ovary elliptic, sparsely floccose-villose, 2-loculed, unequally 3-winged. Capsules nodding. Fl. Nov. Forests, on rocks at 700-800 m. Distributed in Yunnan (Yingjiang County). Also in Myanmar.

滇缅秋海棠 *Begonia rockii*

多毛秋海棠 *Begonia polytricha*

## 多毛秋海棠

**Begonia polytricha** C. Y. Wu

草本。根茎伸长。茎密具褐色粗毛。叶基生及茎生；叶片卵形，不对称，下面疏具硬毛，上面具长硬毛，6-7 × 4-5厘米；托叶及苞片具缘毛。花序密具褐色粗毛；雄花花被片4；雌花花被片5。花期10月。生海拔1800-2200米的阔叶林或山坡阴湿环境。产云南(绿春县、元阳县)。

Herbs. Rhizomes elongate. Stems densely brown hirsute. Leaves basal and cauline; blade ovate, asymmetric, abaxially sparsely hispidulous, adaxially long hirsute, 6-7 × 4-5 cm; stipules and bracts ciliate. Inflorescences densely brown hirsute; staminate flowers with tepals 4; pistillate flowers with tepals 5. Fl. Oct. Broad-leaved forests or shaded moist environments on slopes at 1800-2200 m. Distributed in Yunnan (Lüchun County, Yuanyang County).

## 宁明秋海棠

**Begonia ningmingensis** D. Fang, Y. G. Wei et C. I Peng

草本。具根茎。叶下常淡红色，上面具沿主脉具白斑，阔卵形、近圆形或肾形，纸质或薄纸质，具皱，上面具柔毛状刚毛。花序腋生，5-21朵花组成二歧状聚伞花序；雌花花被片3。蒴果下垂，椭圆体形，先端具宿存花柱。花期6-8月。生海拔100-400米的石灰山阔叶林中。产广西。

Herbs. Rhizomatous. Leaves abaxially usually reddish, adaxially often with white spots along major veins, broadly ovate, suborbicular, or reniform, papery or thinly so, rugose, adaxially pilose-setose. Inflorescences axillary, flowers 5-21 in a dichasial cyme; pistillate flowers with tepals 3. Capsules nodding, ellipsoid, apex with persistent styles. Fl. Jun-Aug. Broad-leaved forests on limestone hills at 100-400 m. Distributed in Guangxi.

## 花叶秋海棠

**Begonia cathayana** Hemsl.

直立草本。茎生叶互生，基部极不对称，上面带红褐色并具有白色带状纹，并密被硬毛，下面紫红色。花序下垂；花粉色或橘黄色；雄花花被片4；雌花花被片5；子房具柔毛，2室。蒴果弯头，倒卵球状长圆形，具不等3翅。花期8月，果期9月开始。生海拔800-1500米的混交林下或溪边山谷阴处。产云南东南部和广西南部。越南亦有。

Erect herbs. Cauline leaves alternate, strongly asymmetrical at base, adaxially with whitish ring-shaped belt at middle and densely hispidulous, abaxially pur-

宁明秋海棠 *Begonia ningmingensis*

花叶秋海棠 *Begonia cathayana*

ple-red. Inflorescences puberulous; flowers pink or orangish; staminate flowers with tepals 4; pistillate flowers with tepals 5, pink or orangish; ovary pubescent, 2-loculed. Capsules nodding, obovoid-oblong, unequally 3-winged. Fl. Aug. Fr. since Sep. Mixed forests or shady places of valleys or by streams at 800-1500 m. Distributed in SE Yunnan and S Guangxi. Also in Vietnam.

## 山地秋海棠

**Begonia oreodoxa** Chun et F. Chun ex C. Y. Wu et Ku

草本。叶基生，卵状近圆形或阔卵形，不对称，下面具红褐色粗毛。花葶高7-9厘米，密被褐色卷曲长毛；雄花花被片4，雌花花被片(3-)4，外面被长毛；子房具长柔毛，基部2室。蒴果弯头，具不等大3翅。花期4月。生海拔100-1200米的山坡阴而湿的密林下。产云南东南部。越南亦有。

Herbs. Leaves basal, ovate-suborbicular or broadly ovate, asymmetric, abaxially reddish brown hirsute. Scapes 7-9 cm tall, densely brown-floccose-villose; staminate flowers with 4 tepals, pistillate flowers with (3-)4 tepals, outside villose; ovary villous, 2-loculed at base. Capsules nodding, unequally 3-winged. Fl. Apr. Shady moist dense forests on mountain slopes at 100-1200 m. Distributed in SE Yunnan. Also in Vietnam.

山地秋海棠 *Begonia oreodoxa*

## 耳托秋海棠

**Begonia auritistipula** Y. M. Shui et W. H. Chen

藤状草本。根茎长匍匐状，之字形。叶全基生；托叶基部斜耳状；叶片卵形，纸质，具皱，下面沿脉具粗柔毛，上面具刚毛。花序腋生，8-13花组成2-4次分枝的二歧聚伞花序；雌花花被片3；子房疏具刚毛状粗毛。花期5-11月。广西药用植物园栽培。

Herbs vinelike. Rhizomes long creeping, zigzag. Leaves all basal; stipules base obliquely auriculate; blade ovate, papery, rugose, abaxially hirsute-pilose on veins, adaxially setulose. Inflorescences axillary, flowers 8-13 in 2-4 times branched dichasial cyme; pistillate flowers with tepals 3; ovary sparsely strigose-hirsute. Fl. May-Nov. Cultivated in Guangxi Medicinal Botanical Garden.

耳托秋海棠 *Begonia auritistipula*

红孩儿 *Begonia palmata* var. *bowringiana*

### 红孩儿

**Begonia palmata** D. Don var. **bowringiana** (Champ. ex Benth.) J. Golding et C. Kareg.

草本。茎和叶柄密被褐色绵状绒毛。叶基生及茎生，不对称，卵形或卵状圆形，下面沿脉被褐色绒毛。花白色至粉色，花被片外面密被毛；雄花花被片4；雌花花被片5(-7)；子房2室。蒴果弯头，倒卵球形，具不等大3翅。花期6月，果期7月。生海拔100-2500米的常绿阔叶林下、山谷河边阴处湿地岩石上。产中国西南和华南。

Herbs. Stems and petioles densely brown woolly-tomentose. Leaves basal and cauline, asymmetric, ovate or oblate-orbicular, abaxially brown tomentose along veins. Flowers white to pink, tepals abaxially densely hairy; staminate flowers with tepals 4; pistillate flowers with tepals 5(-7); ovary 2-loculed. Capsules nodding, obovoid, unequally 3-winged. Fl. Jun. Fr. Jul. Evergreen broad-leaved forests, on rocks in shaded moist environments by streams in valleys at 100-2500 m. Distributed in SW and S China.

### 食用秋海棠

**Begonia edulis** Lévl.

多年生肉质草本。叶基生及茎生。叶片圆形或扁圆形，略不对称，叶脉型掌裂，基部心形，边缘具疏锯齿，浅裂至叶长的近1/3。雄花花被片4，无毛；雌花花被片5-6，无毛。蒴

果弯头，具不等大3翅。花期6-9月，果期8月。生海拔500-1500米的林下阴湿岩上。产云南、广西和广东。越南亦有。

Perennial herbs, fleshy. Leaves basal and cauline. Leaves orbicular or oblate-orbicular, slightly asymmetric, venation palmate, base cordate, margin remotely denticulate, shallowly lobed, divided to ca. 1/3 of leaf length. Staminate flowers with 4 glabrous tepals; pistillate flowers with 5-6 glabrous tepals. Capsules nodding, oblong, unequally 3-winged. Fl. Jun-Sep. Fr. Aug. Under forests, on shaded moist rocks at 500-1500 m. Distributed in Yunnan, Guangxi and Guangdong. Also in Vietnam.

### 大围山秋海棠

**Begonia daweishanensis** S. H. Huang et Y. M. Shui

匍匐型草本。根状茎伸长。叶多基生叶，卵形，不对称，下面具柔

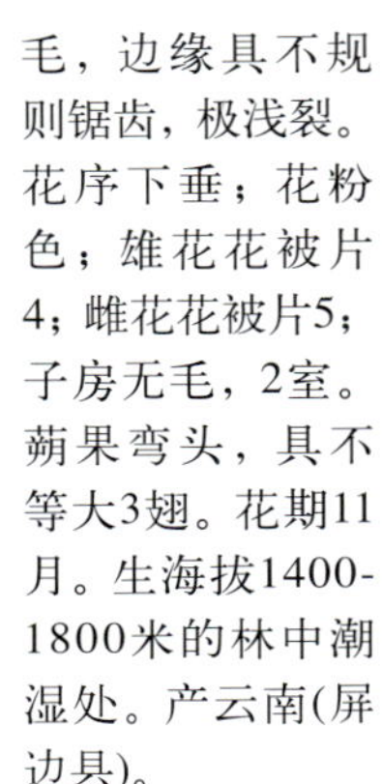

毛，边缘具不规则锯齿，极浅裂。花序下垂；花粉色；雄花花被片4；雌花花被片5；子房无毛，2室。蒴果弯头，具不等大3翅。花期11月。生海拔1400-1800米的林中潮湿处。产云南(屏边县)。

Creeping herbs. Rhizomes elongate. Leaves mostly basa, lovate, asymmetric, abaxially puberulous, margin irregularly serrulate, very shallowly lobed. Inflorescences puberulous; flowers pink; staminate flowers with tepals 4; pistillate flowers with tepals 5; ovary puberulous, 2-loculed. Capsules nodding, unequally 3-winged Fl. Nov. Moist places in forests at 1400-1800 m. Distributed in Yunnan (Pingbian County).

### 突脉秋海棠

**Begonia retinervia** D. Fang, D. H. Qin et C. I Peng

草本。具根茎。根茎匍匐。叶

食用秋海棠 *Begonia edulis*

大围山秋海棠 *Begonia daweishanensis*

突脉秋海棠 *Begonia retinervia*

变色秋海棠 *Begonia versicolor*

全基生；叶片沿主脉具白色斑点，近圆形，不开裂，厚纸质，具皱。5-40朵花组成二歧聚伞花序；雄花花被片4；雌花花被片3；子房绿白色或红白色，阔卵球形至倒卵球形，具长柔毛。花期8-12月，果期11月至翌年3月。生海拔200-600米的石灰岩山坡或潮湿洞穴。产广西(都安县)。

Herbs. Rhizomatous. Rhizome creeping. Leaves all basal; blade with white spots along major veins, suborbicular, unlobed, thickly papery, rugose. Inflorescences with 5-40 flowers in a dichasial cyme; staminate flowers with tepals 4; pistillate flowers with tepals 3; ovary greenish white or reddish white, broadly ovoid to obovoid, villous. Fl. Aug-Dec. Fr. Nov to next Mar. Limestone slopes or moist caves at 200-600 m. Distributed in Guangxi (Du'an County).

## 厚叶秋海棠

**Begonia dryadis** Irmsch.

草本。叶茎生，托叶早落，叶上面近无毛；叶柄被短小卷毛；叶卵形至阔卵形，不对称，下面疏具柔毛，基部斜心形，边缘疏具不明显的小牙齿。花序幼时下垂；子房被毛，2室。蒴果弯头，椭圆体形，具不等大3翅。花期11-12月，果期12月开始。生海拔600-1200米的山谷林下阴湿处。产云南南部。

Herbs. Leaves cauline, stipules caducous, abaxially subglabrous; petioles with short crisped hairs; leaves ovate or broadly ovate, asymmetric, abaxially sparsely puberulous, base obliquely cordate, margin remotely and indistinctly denticulate. Inflorescences puberulous when young; ovary puberulous, 2-loculed. Capsules nodding, ellipsoid, unequally 3-winged. Fl. Nov-Dec. Fr. since Dec. Moist places in forests of valleys at 600-1200 m. Distributed in S Yunnan.

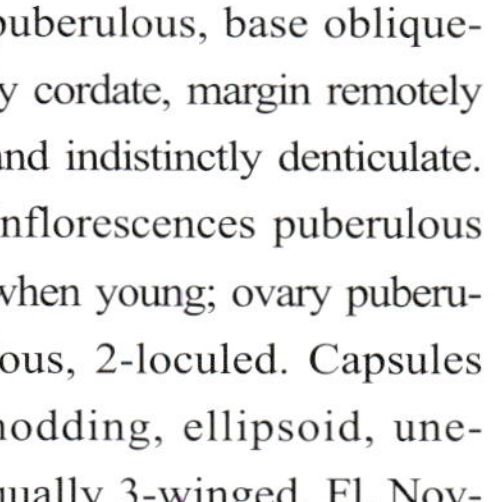

厚叶秋海棠 *Begonia dryadis*

## 变色秋海棠

**Begonia versicolor** Irmsch.

草本。叶均基生，极不对称，基部极偏斜深心形，两面被粗绒毛。雄花花被片4(或5)，粉色；雌花花被片5；子房矩圆体，被褐色卷曲毛，2室。蒴果弯头，倒卵球状长圆形，具不等大3翅。花期6-9月，果期7月。生海拔1800-2100米的密林中潮湿处、混交林下或沟谷石壁上。产云南东南部。

Herbs. Leaves all basal, strongly asymmetrical, base strongly oblique, deeply cordate, hirsute-tomentose. Staminate flowers with tepals 4(or 5), pink; pistillate flowers with tepals 5; ovary oblong, brown-floccose-villose, 2-locular. Capsules nodding, obovoid-oblong, unequally 3-winged. Fl. Jun-Sep. Fr. Jul. Shady moist places of dense forests, under mixed forests, or rocks by streams at 1800-2100 m. Distributed in SE Yunnan.

大叶秋海棠 *Begonia megalophyllaria*

歪叶秋海棠 *Begonia augustinei*

## 大叶秋海棠

**Begonia megalophyllaria** C. Y. Wu

草本。叶片广卵圆形，不对称，无毛，基部歪斜，心形，边缘不规则疏锯齿，先端圆形或急尖。雄花花被片4，白色；雌花花被片5，白色；子房2室。蒴果弯头，具不等3翅。果期10月。生海拔800-1000米的阴湿陡坡森林。产云南(屏边县)。

Herbs. Leaves broadly ovate, asymmetric, glabrous, base oblique, cordate, margin irregularly remotely serrulate, apex rounded or acute. Staminate flowers with 4 white tepals; pistillate flowers with 5 white tepals; ovary 2-loculed. Capsules nodding, unequally 3-winged. Fr. Oct. Shaded moist forests, on steep slopes at 800-1000 m. Distributed in Yunnan (Pingbian County).

## 长柱秋海棠

**Begonia longistyla** Y. M. Shui et W. H. Chen

草本。具根茎。叶全基生；叶片卵形，纸质，具皱，下面沿脉密具毛，上面密具瘤基柔毛至刚毛。花序具20-40朵花组成二歧聚伞状；雌花花被片3，无毛；子房圆锥形。蒴果卵球形。花期2-6月，果期4-6月。生海拔200-300米的溪边密阔叶林石灰山坡。产云南(个旧、河口县)。

Herbs. Rhizomatous. Leaves all basal; blades ovate, papery, rugose, abaxially densely hairy on veins, adaxially densely tuberculate based pilose-setulose. Inflorescences with 20-40 flowers in a dichasial cyme; pistillate flowers with tepals 3, glabrous; ovary coniform. Capsules ovoid. Fl. Feb-Jun. Fr. Apr-Jun. Limestone slopes in dense broad-leaved forests at streamsides at 200-300 m. Distributed in Yunnan (Gejiu, Hekou County).

长柱秋海棠 *Begonia longistyla*

## 歪叶秋海棠

**Begonia augustinei** Hemsl.

草本。叶片极浅裂，不对称，卵形至广卵形，边缘不规则锯齿状，基部深心形；叶柄密具长柔毛。雄花花被片4，浅粉色，外轮2个长圆形至卵形，内轮2个椭圆形；雌花花被片5；子房疏具短柔毛，2室。蒴果弯头，具不等3翅。花期

崇左秋海棠 *Begonia chongzuoensis*

6-9月，果期7月开始。生海拔1000-1800米的灌丛下或潮湿处岩石上。产云南南部。

Herbs. Leaves very shallowly lobed, asymmetric, ovate to broadly ovate, margin irregularly serrulate, deeply cordate at base; petioles densely villous. Staminate flowers with tepals 4, pinkish, outer 2 oblong to ovate, inner 2 elliptic; pistillate flowers with tepals 5; ovary sparsely pilose, 2-loculed. Capsules nodding, unequally 3-winged. Fl. Jun-Sep. Fr. since Jul. Shrubbery or damp rocks at 1000-1800 m. Distributed in S Yunnan.

## 崇左秋海棠

**Begonia chongzuoensis** Yan Liu, S. M. Ku et C. I Peng

多年生草本。叶不对称，宽卵形，长(6-)7-11(-13)厘米，先端渐尖或短渐尖，基部深心形，上面被细刚毛，下面沿脉被长柔毛。二歧聚伞花序具4-8朵花；总花梗长5-12厘米，无毛；雄花外轮2枚花被片无毛；雄蕊25-35；雌花子房无毛，具3翅。蒴果三角状椭圆体形，长7.5-10毫米。花期5-9月，果期6-10月。生石灰岩山坡上。产广西西南部。

Perennial herbs. Leaves asymmetric, broadly ovate, (6-)7-11(-13) cm long, apex acuminate or shortly so, base deeply cordate, adaxially setulose, abaxially pilose along veins. Dichasial cyme 4-8-flowered; peduncle 5-12 cm long, glabrous; staminate flowers, outer 2 tepals glabrous; stamens 25-35; carpellate flowers, ovary glabrous, 3-winged. Capsules trigonous-ellipsoid, 7.5-10 mm long. Fl. May-Sep. Fr. Jun-Oct. Limestone slopes. Distributed in SW Guangxi.

## 一口血秋海棠

**Begonia picturata** Y. Liu, S. M. Ku et C. I Peng

多年生草本。叶片卵圆形至广卵圆形，厚纸质，下面具长柔毛，上面具长毛状刚毛或绒毛状刚毛，基部歪斜，深心形。雄花花被片4；雌花被片3；子房具红色长柔毛状刚毛或硬毛状刚毛，具3翅，1室具侧膜胎座。蒴果弯头。花期2-5月，果期3-6月。生海拔700-800米的阴石灰岩坡洞穴。产广西(靖西)。

Perennial herbs. Leaves ovate to broadly ovate, thickly papery, abaxially villous, adaxially villous-setose or tomentose-setose, base obliquely, deeply cordate. Staminate flowers with 4 tepals; pistillate flowers with 3 tepals; ovary red villous-setose or hispid-setose, 3-winged, 1-loculed with parietal placentation. Capsules nodding. Fl. Feb-May. Fr. Mar-Jun. Caves of shaded rocky limestone slopes at 700-800 m. Distributed in Guangxi (Jingxi).

一口血秋海棠 *Begonia picturata*

彭氏秋海棠 *Begonia pengii*

## 彭氏秋海棠

**Begonia pengii** S. M. Ku et Yan Liu

多年生草本。叶不对称，狭卵形，长(11-)17-27厘米，上面密被长柔毛，下面被绒毛状长柔毛。二歧聚伞花序具4-12朵花；总花梗长17-37厘米，密被长柔毛或硬毛状绒毛；雄花外轮2枚花被片外面密被长柔毛；花药黄色，药囊边缘红色。蒴果三角状椭圆体形，长18-25毫米，具不等3翅。花期3-6月，果期6-12月。生石灰岩山上。特产广西西部。

Perennial herbs. Leaves asymmetric, narrowly ovate, (11-)17-27 cm long, adaxially densely pilose, abaxially villose-piloses. Dichasial cyme 4-12-flowered; peduncle 17-37 cm long, densely pilose or hispid-villous; staminate flowers, outer 2 tepals abaxially densely pilose; anthers yellow with red margins along anther sacs. Capsules trigonous-ellipsoid, 18-25 mm long, unequally 3-winged. Fl. Mar-Jun. Fr. Jun-Dec. Limestone hill. Endemic to W Guangxi.

长柄秋海棠 *Begonia smithiana*

红斑秋海棠 *Begonia rubropunctata*

## 红斑秋海棠

**Begonia rubropunctata** S. H. Huang et Y. M. Shui

草本。叶基生；叶圆形，不对称，疏具糙毛，基部稍斜，心形，边缘5或6浅裂；叶柄具红色条状斑点。雄花花被片4，外面2枚宽倒卵状，里面2枚倒卵状矩圆形；子房2室。蒴果下垂，椭圆状。花期11月。生海拔600-1100米的林中、石灰岩石上。产云南(西双版纳)。

Herbs. Leaves basal; leaves orbicular, asymmetric, sparsely setulose, base slightly oblique, cordate, margin 5- or 6-lobed; petioles with red linear dots. Staminate flowers with 4 tepals, pink, outer 2 broadly obovate, inner 2 obovate-oblong; ovary 2-loculed. Capsules nodding, ellipsoid. Fl. Nov. Forests, on limestone rocks at 600-1100 m. Distributed in Yunnan (Sipsongpanna).

## 长柄秋海棠

**Begonia smithiana** T. T. Yu

草本。茎极短。叶多基生，卵形至阔卵形，不对称；叶柄有时淡红色，达25厘米长。雄花花被片4，淡粉红色；雌花花被片5，淡粉红色；子房2室。蒴果弯头，有不等3翅。花期8月，果期9月。生海拔700-1300米的林中、阴湿环境的岩石上或沟谷中。产四川、贵州、湖北和湖南。

Herbs. Stems very short. Leaves mostly basal, ovate to broadly ovate, asymmetric; petioles sometime s reddish, to 25 cm long. Flowers paired on stem apices; staminate flowers with 4 tepals, pinkish; pistillate flowers with 5 tepals, pinkish; ovary 2-loculed. Capsules nodding, unequally 3-winged. Fl. Aug. Fr. Sep. Forests, on rocks in shaded moist environments, or in valleys at 700-1300 m. Distributed in Sichuan, Guizhou, Hubei and Hunan.

桂南秋海棠 *Begonia austroguangxiensis*

## 桂南秋海棠

**Begonia austroguangxiensis** Y. M. Shui et W. H. Chen

草本。具根茎。叶全基生；叶片阔卵形或近圆形，5-9 × 6-8厘米，先端圆形或钝。花序腋生，5-12朵花组成二歧聚伞状；雌花花被片3。蒴果先端具宿存花柱。花期5-10月。生海拔200-600米的石灰林中、阴处岩石上。产广西(龙州县)。

Herbs. Rhizomatous. Leaves all basal; blade broadly ovate or suborbicular, 5-9 × 6-8 cm, apex rounded or obtuse. Inflorescences axillary, flowers 5-12 in a dichasial cyme; pistillate flowers with tepals 3. Capsules apex with persistent styles. Fl. May-Oct. Limestone forests, on rocks in shaded environments at 200-600 m. Distributed in Guangxi (Longzhou County).

## 大王秋海棠 (虾蟆秋海棠)

**Begonia rex** Putz.

多年生草本。叶基生，下面淡紫色，卵形至狭卵形，不对称，上面疏具毛，基部倾斜，心形。花2朵生于茎顶；雄花花被片4；雌花花被片5；子房无毛，2室。蒴果弯头，有3个不等的翅。花期5月，果期8月。生海拔400-1100米的林中、岩石上或沟谷中。产云南、贵州和广西。印度东北部和越南亦有。

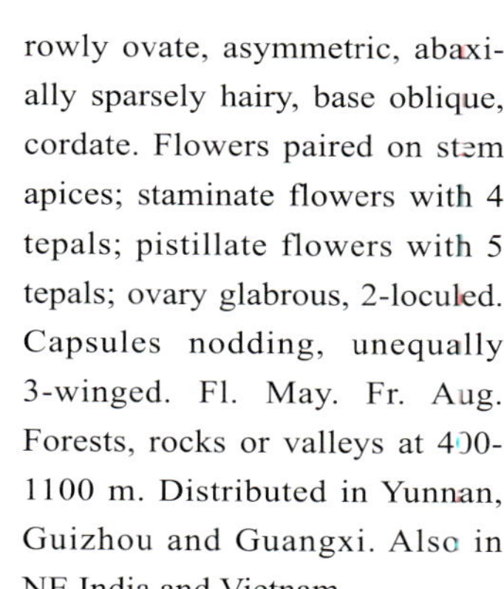

Perennial herbs. Leaves basal, abaxially purplish, ovate to narrowly ovate, asymmetric, abaxially sparsely hairy, base oblique, cordate. Flowers paired on stem apices; staminate flowers with 4 tepals; pistillate flowers with 5 tepals; ovary glabrous, 2-loculed. Capsules nodding, unequally 3-winged. Fl. May. Fr. Aug. Forests, rocks or valleys at 400-1100 m. Distributed in Yunnan, Guizhou and Guangxi. Also in NE India and Vietnam.

大王秋海棠 (虾蟆秋海棠) *Begonia rex*

## 弯果秋海棠

**Begonia curvicarpa** S. M. Ku, C. I Peng et Yan Liu

草本。具根茎。根茎伸长，疏具毛。叶全基生；叶片斜阔卵形至圆形，纸质，下面具长柔毛，上面具柔毛。花序腋生，3-5朵花组成1-3次二歧分枝聚伞花序；雌花花被片3。蒴果下垂，明显弯曲。花期8-11月，果期10-12月。生石灰岩上或潮湿洞穴洞口半阴至阴处。产广西(融安县、永福县)。

Herbs. Rhizomatous. Rhizome elongate, sparsely hairy. Leaves all basal; blade obliquely broadly ovate to orbicular, papery, abaxially villous, adaxially pilose. Inflorescences axillary, flowers 3-5 in 1-3 times branched dichasial cyme; pistillate flowers with tepals 3. Capsules nodding, manifestly crooked. Fl. Aug-Nov. Fr. Oct-Dec. limestone rocks or entrances to semishaded to shaded, slightly moist caves. Distributed in Guangxi (Rong'an County, Yongfu County).

弯果秋海棠 *Begonia curvicarpa*

半侧膜秋海棠 *Begonia semiparietalis*

## 半侧膜秋海棠

**Begonia semiparietalis** Yan Liu, S. M. Ku et C. I Peng

草本。具根茎。叶基生；叶片沿主脉具白色斑点，阔卵形或近圆形，3.5-15 × 3-13厘米，具皱，上面疏具刚毛状柔毛。6-25朵花组成一个二歧聚伞状花序；雌花花被片3；子房淡红色，椭圆体形，1室中部以上具侧生胎座，以下为中轴胎座。花期5月。生石灰岩山。产广西(扶绥县)。

Herbs. Rhizomatous. Leaves basal; blade with white spots along major veins, broadly ovate or suborbicular, 3.5-15 × 3-13 cm, rugose, adaxially sparsely setulose-pilose. Inflorescences a dichasial cyme with 6-25 flowers; pistillate flowers with tepals 3; ovary reddish, ellipsoid, 1-loculed with parietal placentation above middle, lower half with axile placentation. Fl. May. Limestone hills. Distributed in Guangxi (Fusui County).

## 掌裂叶秋海棠

**Begonia pedatifida** Lévl.

草本。叶均基生，轮廓扁圆形至宽卵形，5-6深裂，中间3裂片再中裂。花葶高7-15厘米；花白色至淡粉色；雄花花被片4；雌花花被片5，不等大；子房2室。蒴果弯头，倒卵球形，具不等大3翅。花期1-7月，果期10月。生海拔300-1700米的林下潮湿处、常绿阔叶林山坡沟谷或阴湿林下石壁上。产四川、贵州、湖北和湖南。

Herbs. Leaves all basal, oblate-orbicular or broadly ovate in outline, 5-6-parted, middle 3 lobes again divided into 1/2. Scapes 7-15 cm tall; flowers white to pinkish; staminate flowers with tepals 4; pistillate flowers with tepals 5, unequal; ovary 2-loculed. Capsules nodding, obovoid, unequally 3-winged. Fl. Jan-Jul. Fr. Oct. Moist places under forests, evergreen broad-leaved forests by streams, or shady stony cliffs at 300-1700 m. Distributed in Sichuan, Guizhou, Hubei and Hunan.

## 掌叶秋海棠

**Begonia hemsleyana** Hook. f.

草本。叶茎生或基生兼茎生，掌状复叶，小叶7(或8)，下面疏具粗毛。花粉色；雄花花被

掌裂叶秋海棠 *Begonia pedatifida*

片4；雌花花被片5；子房无毛，2室。蒴果弯头，倒卵球形或椭圆体形，常具不等大3翅。花期12月，果期翌年6月。生海拔1000-1300米的潮湿处、林中、岩石上或水边。产云南东南部和广西。越南亦有。

Herbs. Leaves cauline, or basal and cauline, palmately compound, leaflets 7(or 8), abaxially sparsely hirsute. Flowers pink; staminate flowers with tepals 4; pistillate flowers with tepals 5; ovary glabrous, 2-loculed. Capsules nodding, obovoid or ellipsoid, unequally 3-winged. Fl. Dec. Fr. Jun next year. Wet places, woods, rocks or by waters at 1000-1300 m. Distributed in SE Yunnan and Guangxi. Also in Vietnam.

## 方氏秋海棠

**Begonia fangii** Y. M. Shui et C. I Peng

草本。具根茎。根茎伸长。叶为掌状复叶；小叶3-6，下面红褐色，上面深绿色，披针形，近革质。花序腋生；雄花花被片4，淡粉色；雌花花被片3。蒴果下垂。花期12月至翌年4月，果期2-6月。生海拔400-700米的石灰岩林中。产广西(龙州县)。

Herbs. Rhizomatous. Rhizomes elongate. Leaves palmately compound; leaflets 3-6, abaxially reddish brown, adaxially dark green, lanceolate, subleathery. Inflorescences axillary; staminate flowers with tepals 4, pinkish; pistillate flowers with tepals 3. Capsules nodding. Fl. Dec to next Apr. Fr. Feb-Jun. Limestone rocks in forests at 400-700 m. Distributed in Guangxi (Longzhou County).

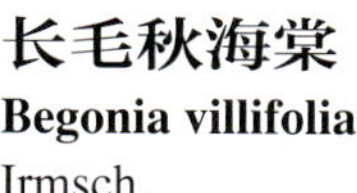

## 长毛秋海棠

**Begonia villifolia** Irmsch.

草本。叶茎生，阔卵形至卵形，不对称，密具长柔毛，基部倾斜，心形，边缘浅裂至叶长的1/3。雄花花被片4，白色；雌花花被片5；子房密具长柔毛。蒴果弯头，椭圆形，具不等大3翅。花期5-7月，果期7月开始。生海拔1600-1700米的林中潮湿地、河边灌丛或岩石上。产云南东南部。缅甸和越南亦有。

Herbs. Leaves cauline, broadly ovate to ovate, asymmetric, densely long villous, base oblique, cordate, margin shallowly lobed, divided to 1/3 of leaf length. Staminate flowers with tepals 4, white; pistillate flowers with tepals 5; ovary densely villous. Capsules nodding, ellipsoid, unequally 3-winged. Fl. May-Jul. Fr. since Jul. Damp places in forests, shrubbery by rivers, or rocks at 1600-1700 m. Distributed in SE Yunnan. Also in Myanmar and Vietnam.

长毛秋海棠 *Begonia villifolia*

掌叶秋海棠 *Begonia hemsleyana*

方氏秋海棠 *Begonia fangii*

# 钩枝藤科 Ancistrocladaceae

### 钩枝藤

**Ancistrocladus tectorius** (Lour.) Merr.

攀援灌木。侧枝具弯曲至螺旋状钩。叶常聚生枝顶，革质。花序疏松至紧密，圆锥状；花小，直径7-8毫米；萼片5；花瓣5，基部贴生，常内卷；3心皮，1室。坚果红色，倒圆锥形。花期4-6月，果期6月。生海拔500-700米的山坡、山谷密林或山地森林。产广西西南部和海南。印度、缅甸、老挝、泰国、越南和柬埔寨亦有。

Climbing shrubs. Lateral branches with recurved to spiraling hooks. Leaves clustered at top of branchlets, leathery. Inflorescences lax to congested, paniculate; flowers small, 7-8 mm diam; sepals 5; petals 5, connate basally, usually involute; carpels 3, 1-locular. Nuts red, obconical. Fl. Apr- Jun. Fr. Jun. Mountain slopes, dense forests in valleys or forests on mountains at 500-700 m. Distributed in SW Guangxi and Hainan. Also in India, Myanmar, Laos, Thailand, Vietnam and Cambodia.

# 仙人掌科 Cactaceae

### 仙人掌

**Opuntia dillenii** (Ker Gawl.) Haw.

肉质灌木。多数小窠具刺1-12(-20)枚，开展，黄色，稍具褐色带，钻形。叶钻形。外轮花被片绿色，具黄色边缘；内轮花被片开展，亮黄色。浆果紫色，陀螺状或倒卵状。花期6-12月。生近海平面的灌丛，岩地或沙地。在广西、广东和海南归化。原产加勒比；广泛引进或归化于热带地区。

Succulent shrubs. Spines 1-12 (-20) per areole on most areoles, spreading, yellow, ± brown banded, subulate. Leaves subulate. Outer petals greenish with yellow margin; inner tepals spreading, bright yellow. Berries purple, turbinate to obovoid. Fl. Jun-Dec. Thickets, rocks or sandy soils near sea level. Naturalized in Guangxi, Guangdong and Hainan. Native to Caribbean; widely introduced and naturalized in tropical regions.

### 梨果仙人掌

**Opuntia ficus-indica** (L.) Mill.

仙人掌 *Opuntia dillenii*

肉质灌木或小乔木。每小窠具1-6枚刺，有时无刺，开展或反折，刚毛状或针状。倒刺刚毛黄色，早落。叶圆锥状，早落。外轮花被片黄色，中央淡红色或绿色；内轮花被片开展，黄色至橘黄色。浆果黄色、橙色或淡紫色，倒卵球形。花期5-6月。生海拔600-2900米的干热山谷或岩石上。产云南、四川、西藏、贵州和广西。原产墨西哥；广泛引入；在全球热带及亚热带地区归化。

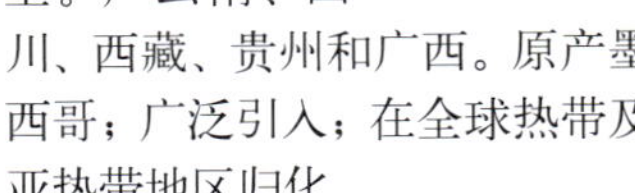

Succulent shrubs or small trees. Spines 1-6 per areole, sometimes absent, spreading or deflexed, bristlelike or acicular. Glochids yellow, early deciduous. Leaves conic, early deciduous. Outer petals yellow with reddish or green center; inner petals spreading, yellow to orange. Berries yellow, orange or purplish, obovoid. Fl. May-Jun. Hot dry valleys or rocks at 600-2900 m. Distributed in Yunnan, Sichuan, Xizang, Guizhou and Guangxi. Native to Mexico; widely introduced as a hedge or for its edible fruits; also naturalized in tropical and subtropical regions of the world.

### 单刺仙人掌

**Opuntia monacantha** (Willd.) Haw.

肉质灌木或乔木状。茎圆柱状。刺在分枝每个小窠1或2(或3)枚，在主干上多至12枚，直立或开展，浅灰色，先端深褐

钩枝藤 *Ancistrocladus tectorius*

梨果仙人掌 *Opuntia ficus-indica*

单刺仙人掌 *Opuntia monacantha*

色，针状。叶锥形，早落。外轮花被片具红色中肋和黄色边缘；内轮花被片开展，黄色至橘黄色。果红紫色，倒卵球形。花期4-8月。生海拔2000米以下的山坡或海边。栽培于云南、广西、广东、台湾和福建。热带地区广泛栽培；原产南美洲。

Succulent shrubs or treelike. Trunk terete. Spines 1 or 2(or 3) per areole on branches, but on main trunk to 12 per areole, erect or spreading, grayish, dark brown tipped, acicular. Leaves conical, deciduous. Outer petals with red midrib and yellow margin; inner petals spreading, yellow to orange. Fruits reddish purple, obovoid. Fl. Apr-Aug. Slopes or seashores below 2000 m. Cultivated in Yunnan, Guangxi, Guangdong, Taiwan and Fujian. Widely cultivated in tropical areas; native to South America.

### 量天尺
**Hylocereus undatus** (Haw.) Britt. et Rose

攀援肉质灌木。茎3棱状或翅状。每小窠具1-3(-6)刺，刺以不同方向开展，灰褐色，圆锥形至近钻形。外轮花被片具浅绿色中肋，多具白色边缘；内轮花被片直立或开展，白色。浆果红色。花期7-12月。生海拔300米以下的岩地或沙地。栽培或归化广西、广东、海南、台湾和福建。原产墨西哥和中美洲，广泛引进或归化于热带亚洲、东澳大利亚和南美洲。

Climbing succulent shrubs. Stems 3-angled or winged. Spines 1-3(-6) per areole, spreading in various directions, gray-brown, conic to subulate. Outer petals with greenish midrib and mostly white margin; inner petals erect to spreading, white. Berries red. Fl. Jul-Dec. Rocks or sandy soils below 300 m. Cultivated or naturalized in Guangxi, Guangdong, Hainan, Taiwan and Fujian. Native to Mexico and Central America, widely introduced or naturalized in tropical Asia, E Australia and South America.

量天尺 *Hylocereus undatus*

# 瑞香科 Thymelaeaceae

### 土沉香
**Aquilaria sinensis** (Lour.) Spreng.

乔木，高5-15米。叶圆形或椭圆形至长圆形。花夜间芳香；花萼黄绿色；花瓣状附属物10，鳞片状，生花冠筒喉部。外果皮薄，干后光滑。种子1或2粒，被白色绢毛，具1.5厘米长的附属体。花期春夏，果期夏秋。生低海拔森林、向阳山坡或路边。产广西、广东、海南和福建。

Trees, 5-15 m tall. Leaves orbicular or elliptic to oblong. Flowers fragrant at night; calyx yellowish green; petaloid appendages 10, scalelike, inserted at throat of tube. Pericarps thin, smooth when dried. Seeds 1 or 2, white sericeous, with 1.5 cm long appendages. Fl. spring and summer. Fr. summer and autumn. Lowland forests, sunny places on slopes or along roadsides. Distributed in Guangxi, Guangdong, Hainan and Fujian.

土沉香 *Aquilaria sinensis*

## 丽江荛花

**Wikstroemia lichiangensis** W. W. Sm.

灌木。叶互生，倒披针形或长圆形，纸质，两面散生灰白色绒毛，侧脉3或4对。头状花序顶生，具5-15花；花萼黄绿色，外部有时淡紫色；萼筒外部密具灰白色柔毛；裂片4。核果狭椭圆状。花期夏秋季。生海拔2600-3500米的林中。产云南西北部和四川西南部。

Shrubs. Leaves alternate, oblanceolate or oblong, papery, both surfaces scattered grayish white pubescent, lateral veins 3 or 4 pairs. Capitulums terminal, 5-15-flowered; calyx yellow-green, exterior sometimes purplish; tubes exterior densely grayish white pubescent; lobes 4. Drupes narrowly ellipsoid. Fl. summer-autumn. Forests at 2600-3500 m. Distributed in NW Yunnan and SW Sichuan.

了哥王 *Wikstroemia indica*

## 了哥王

**Wikstroemia indica** (L.) C. A. Mey.

灌木。叶对生，纸质或近革质，倒卵形、椭圆状长圆形或披针形，纸质至薄革质，两面无毛。花萼黄绿色，外面渐无毛；裂片4；雄蕊8，2轮。核果红色至深紫色，椭圆体形。花果期夏秋间。生海拔1500米以下的开阔林下或石山坡。产中国西南、华南和东南。南亚、东南亚和太平洋岛屿亦有。

Shrubs. Leaves opposite, papery or subleathery, obovate, elliptic-oblong or lanceotate papery to thinly leathery, both surfaces glabrous. Calyx yellow-green, exterior glabrescent; lobes 4; stamens 8, 2-whorled. Drupes red to dark purple, ellipsoid. Fl. and fr. summer to autumn. Open forests or rocky slopes below 1500 m. Distributed in SW, S and SE China. Also in S and SE Asia, and Pacific Islands.

## 北江荛花

**Wikstroemia monnula** Hance

灌木。叶对生至互生，卵状椭圆形至椭圆形或椭圆状披针形，纸质或硬纸质，侧脉4或5对，下面至少沿脉具柔毛。总状花序顶生，具8-12花；花萼紫红色；萼筒细弱，外部具白色柔毛；裂片4；雄蕊8，2列。蒴果干燥，卵球形。花期4-8月，果期8月开始。生海拔650-1100米的山坡、灌丛或路旁。产贵州、广西、广东、浙江和湖南。

Shrubs. Leaves opposite to alternate, ovate-elliptic to elliptic or elliptic-lanceolate, papery or stiffly so, lateral veins 4 or 5 per side, abaxially puberulous at least on veins. Racemes terminal, with 8-12 flowers; calyx purplish red; tubes slender, exterior white pubescent; lobes 4; stamens 8, 2-ranked. Drupes dry, ovoid. Fl. Apr-Aug. Fr. since Aug. Slopes, thickets or roadsides at 650-1100 m. Distributed in Guizhou, Guangxi, Guangdong, Zhejiang and Hunan.

北江荛花 *Wikstroemia monnula*

丽江荛花 *Wikstroemia lichiangensis*

## 河朔荛花

**Wikstroemia chamaedaphne** (Bunge) Meisn.

灌木。多分枝。叶披针形至长圆状披针形，薄革质，两面无毛。穗状花序或圆锥花序顶生或腋生；花萼绿黄色或乳黄色，外部具灰色丝状柔毛，裂片4。核果干燥，卵球形。花期6-8月，果期9月。生海拔500-2400的林中、灌丛、山坡或路边。产中国西南、东南、华北和华西。

Shrubs. Much branched. Leaves lanceolate to oblong-lanceolate, thinly leathery, both surfaces glabrous. Spikes or panicles terminal and axillary; calyx greenish yellow or creamy-yellow, exterior gray sericeous-pubescent; lobes 4 . Drupes dry, ovoid. Fl. Jun-Aug. Fr. Sep. Forests, bushes, slopes or roadsides at 500-2400 m. Distributed in SW, SE, N and W China.

## 一把香

**Wikstroemia dolichantha** Diels

灌木。叶互生，长圆形或倒披针状长圆形，纸质，疏具柔

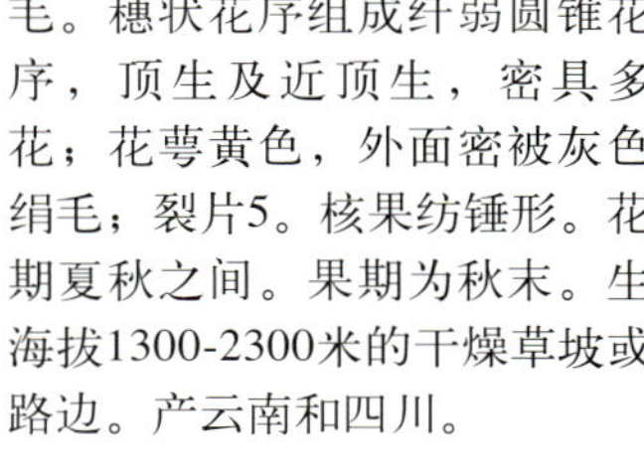

毛。穗状花序组成纤弱圆锥花序，顶生及近顶生，密具多花；花萼黄色，外面密被灰色绢毛；裂片5。核果纺锤形。花期夏秋之间。果期为秋末。生海拔1300-2300米的干燥草坡或路边。产云南和四川。

Shrubs. Leaves alternate, oblong to oblanceolate-oblong, papery, sparsely puberulous. Spikes forming a gracile panicle, terminal and subterminal, densely many flowered; calyx yellow, densely gray-sericeous outside; lobes 5. Drupes fusiform. Fl. between summer and autumn. Fr. end of autumn. Dry grassy slopes or roadsides at 1300-2300 m. Distributed in Yunnan and Sichuan.

河朔荛花 *Wikstroemia chamaedaphne*

一把香 *Wikstroemia dolichantha*

荛花 *Wikstroemia canescens*

丝毛瑞香 *Daphne holosericea*

## 荛花

**Wikstroemia canescens** Wall. ex Meisn.

灌木。叶披针形，下面具长柔毛。头状花序顶生或在上部腋生，具4-10花；花萼黄色，内部具8棱，外部具灰色长柔毛；子房棍棒状，具柄，具柔毛。核果干燥。花期秋季。生海拔1000-3500米的山坡或岩石周围。产西藏。印度、尼泊尔、孟加拉国、巴基斯坦、阿富汗和日本亦有。

Shrubs. Leaves lanceolate, abaxially villous. Capitulums terminal or axillary at upper parts; 4-10-flowered; calyx yellow, 8-ribbed inside, exterior gray villous; ovary clavate, stipitate, pubescent. Drupes dry. Fl. autumn. Slopes or among rocks at 1000-3500 m. Distributed in Xizang. Also in India, Nepal, Bangladesh, Pakistan, Afghanistan and Japan.

## 芫花

**Daphne genkwa** Siebold et Zucc.

落叶灌木。小枝黄绿色或紫褐色，密具淡黄色绢毛。叶卵形、卵状披针形或椭圆状长圆形。具3-15花，簇生叶腋或侧生；花萼蓝紫色，浅紫色或紫罗兰色；萼筒圆柱状，外面具绢毛；裂片4。果肉质，白色至淡红色，椭圆体形。花期3-5月，果期6-7月。生海拔300-1000米的林中或灌丛坡地。产中国西南、东南、华中、华北、华西和华东。朝鲜半岛亦有。

Deciduous shrubs. Branches yellowish green or purplish brown, densely yellowish sericeous. Leaves ovate, ovate-lanceolate, or elliptic-oblong. Flowers 3-15-clustered in axillary or lateral; calyx bluish purple, lilac, or lavender; tubes cylindric, exterior sericeous; lobes 4. Drupes fleshy, white to reddish, ellipsoid. Fl. Mar-May. Fr. Jun-Jul. Forests or shrubby slopes at 300-1000 m. Distributed in SW, SE, C, N, W and E China. Also in Korean Peninsula.

## 丝毛瑞香

**Daphne holosericea** (Diels) Hamaya

常绿灌木。多分枝。叶革质或近革质，条形或条状长圆形，两面

芫花 *Daphne genkwa*

东北瑞香 *Daphne pseudomezereum*

被绢毛。花序顶生，具多花；花萼外面白色，内部黄色；萼筒筒状，外被绢毛；雄蕊10，2轮。核果干，圆锥形。花期7-8月，果期9-10月。生海拔3000-3600米的干旱河谷或灌丛草坡。产云南西部、四川和西藏东南部。

Evergreen shrubs. Many-branched. Leaves leathery or subleathery, linear or linear-oblong, sericeous on both surfaces. Inflorescences terminal, many flowered; calyx exterior white, interior yellow; calyx tubes tubular, abaxially sericeous; stamens 10, 2-whorled. Drupes dry, conic. Fl. Jul-Aug. Fr. Sep-Oct. Dry valleys or thickets on grassy slopes at 3000-3600 m. Distributed in W Yunnan, Sichuan and SE Xizang.

## 东北瑞香
**Daphne pseudomezereum** A. Gray

落叶灌木。老茎具大叶痕。叶披针形或长圆披针形，膜质，两面光滑，先端钝，侧脉8-12对。花序顶生，具2-10花；花萼黄绿色；萼筒外面无毛；裂片4；雄蕊8。核果红色，卵圆形。花期2-4月，果期7-8月。生海拔800-1600米的林中。产吉林和辽宁。朝鲜半岛和日本亦有。

Deciduous shrubs. Older stems with large leaf scars. Leaves lanceolate or oblong-lanceolate, membranous, both surfaces glabrous, apex obtuse; veins 8-12 pairs. Inflorescences terminal, 2-10-flowered; calyx yellowish green; tubes exterior glabrous; lobes 4; stamens 8. Drupes red, ovoid. Fl. Feb-Apr. Fr. Jul-Aug. Forests at 800-1600 m. Distributed in Jilin and Liaoning. Also in Korean Peninsula and Japan.

凹叶瑞香 *Daphne retusa*

## 橙黄瑞香
**Daphne aurantiaca** Diels

常绿灌木。小枝褐色。叶椭圆形、倒卵形或长倒卵形。花序顶生，常生缩短的侧枝上，具2-5花；花芳香；花萼深黄色至橘黄色；萼筒外部无毛；裂片4；雄蕊8。核果黄褐色，椭圆体形。花期5-6月，果期8月。生海拔2600-3500米的高山灌丛或灌丛山坡。产云南西北部和四川西南部。

Evergreen shrubs. Branchlets brown. Leaves elliptic, obovate or long obovate. Inflorescences terminal, often on reduced lateral shoots, 2-5-flowered; flowers fragrant; calyx deep yellow to orange; tubes exterior glabrous; lobes 4; stamens 8. Drupes brownish yellow, ellipsoid. Fl. May-Jun. Fr. Aug. Alpine thickets or shrubby slopes at 2600-3500 m. Distributed in NW Yunnan and SW Sichuan.

橙黄瑞香 *Daphne aurantiaca*

## 凹叶瑞香
**Daphne retusa** Hemsl.

常绿灌木。叶革质或纸质，两面无毛，中脉在上面下陷。花序顶生，常具数花；花芳香；花萼内面紫红色，具浅紫红色或白色裂片；萼筒圆柱形，外部具柔毛；裂片4。核果红色，近球形或卵球形。花期4-5月，果期6-7月。生海拔3000-3900米的高山草地、草坡或灌丛中。产中国西南、华北和西北。印度、克什米尔地区、尼泊尔和不丹亦有。

Evergreen shrubs. Leaves leathery or papery, both surfaces glabrous, midrib adaxially impressed. Inflorescences terminal, usually several flowered; flowers fragrant; calyx purplish red abaxially with paler purple-pink or white lobes; tubes cylindric, exterior pubescent; lobes 4. Drupes red, subglobose or ovoid. Fl. Apr-May. Fr. Jun-Jul. Alpine glasslands, herbaceous slopes or thickets at 3000-3900 m. Distributed in SW, N and NW China. Also in India, Kashmir, Nepal and Bhutan.

滇瑞香 *Daphne feddei*

唐古特瑞香 *Daphne tangutica*

## 唐古特瑞香
**Daphne tangutica** Maxim.

常绿灌木。叶具光泽暗绿色，披针形至倒披针形。花序顶生，头状，具3-12花；花萼内面粉色至白色带紫红色；萼筒圆柱状，外部无毛；裂片4。核果红色，近球形或卵球形。花期4-6月，果期5-7月。生海拔1000-3800米的灌丛或开阔林中。产中国西南和华西。

Evergreen shrubs. Leaves lustrous dark green, lanceolate to oblanceolate. Inflorescences terminal, capitate, 3-12-flowered; calyx pink to white flushed purplish red abaxially; tubes cylindric, exterior glabrous; lobes 4. Drupes red, subglobose or ovoid, Fl. Apr-Jun. Fr. May-Jul. Thickets or open forests at 1000-3800 m. Distributed in SW and W China.

## 白瑞香
**Daphne papyracea** Wall. ex Steud.

常绿灌木。叶互生；叶卵形、披针形、椭圆形或倒披针形，薄革质，膜质或纸质，两面无毛。花序顶生、束生，具3-10(-12)花；总花梗长约2毫米；花不芳香；花药长圆形，长1-2毫米。核果红色，卵球状梨形。花期11月至翌年1月，果期4-5月。生海拔700-3100米的林中、灌丛或草坡。产中国西南、华南和华中。印度和尼泊尔亦有。

Shrubs evergreen. Leaves alternate; leaves ovate, lanceolate, elliptic, or oblanceolate, thinly leathery, membranous, or papery, both surfaces glabrous. Inflorescences terminal, fasciculate, 3-10(-12)-flowered; peduncles ca. 2 mm long; flowers not fragrant; anthers oblong, 1-2 mm long. Drupes red, ovoid-pyriform. Fl. Nov to next Jan. Fr. Apr-May. Forests, shrubby and herbaceous slopes at 700-3100 m. Distributed in SW, S and C China. Also in India and Nepal.

## 滇瑞香
**Daphne feddei** Lévl.

常绿灌木。叶上面深亮绿色，倒披针形或长圆状披针形至倒卵状披针形或倒卵形。花序顶生，常具8-12花；花萼白色，萼筒圆柱状，外面密具柔毛。核果橘红色，球形或卵球形。花期2-4月，果期5-6月。生海拔1800-2600米的林中或灌丛山坡。产云南、贵州和四川。

Evergreen shrubs. Leaves dark lustrous green adaxially, oblanceolate or oblong-lanceolate to obovate-lanceolate or obovate. Inflorescences terminal, usually 8-12-flowered; calyx white, tubes cylindric, exterior densely pubescent. Drupes orange-red, globose or ovoid. Fl. Feb-Apr. Fr. May-Jun. Forests or shrubby slopes at 1800-2600 m. Distributed in Yunnan, Guizhou and Sichuan.

白瑞香 *Daphne papyracea*

## 毛花瑞香(毛管花)
**Eriosolena composita** (L. f.) Tiegh.

灌木。叶椭圆形至椭圆状披针形。头状花序腋生，具(4-)8-12花；花芳香；花萼浅黄色或白色；萼筒外部密具绢毛；裂

毛花瑞香(毛管花) *Eriosolena composita*

片4；子房椭圆体形，先端具白色粗毛。核果黑紫色或红色，卵球形。花期春季。生海拔1300-1800米的林中或山坡灌丛中。产云南南部。南亚和东南亚亦有。

Shrubs. Leaves elliptic to elliptic-lanceolate. Inflorescences axillary, capitate, (4-)8-12-flowered; flowers fragrant; calyx pale yellow or white; tubes exterior densely sericeous; lobes 4; ovary ellipsoid, apex white hirsute. Drupes blackish violet or red, ovoid. Fl. spring . Forests or thickets on slopes at 1300-1800 m. Distributed in S Yunnan. Also in S and SE Asia.

## 鼠皮树

**Rhamnoneuron balansae** (Drake) Gilg

灌木或小乔木。叶互生，卵形或长圆形至披针形。花序顶生及腋生，头状，排成圆锥状，具4花；花萼红色，外面被绢毛，萼片4，卵形；无花瓣；雄蕊8，2轮；子房被白色长硬毛；柱头近球形。花期3月。生海拔900-1200米的林中。产云南(屏边县)。越南亦有。

Shrubs or small trees. Leaves alternate, ovate or oblong to lanceolate. Inflorescences terminal and axillary, capitate, in panicles, 4-flowered; calyx red, abaxially sericeous, calyx lobes 4, ovate; petals absent; stamens 8, 2-whorled; ovary white-hirsute; stigmas subglobose. Fl. Mar. Forests at 900-120C m. Distributed in Yunnan (Pingbian County). Also in Vietnam.

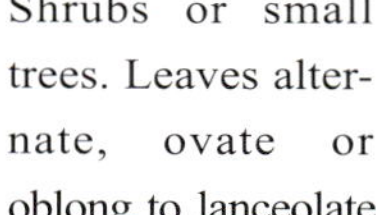

## 结香

**Edgeworthia chrysantha** Lindl.

落叶灌木。枝常具绒毛。叶常绿或二年生。花先叶开放；花黄色，芳香；花萼内部黄色；萼筒外部密具白色绢毛；裂片4；子房卵球形，先端具绢毛；花柱无毛；柱头球形。核果椭圆体形。花期晚冬和早春，果期春季至夏季。生林中或山坡灌丛，亦栽培。产中国西南、东南和华中；许多省有栽培。日本亦有。

Deciduous shrubs. Branching usually trichotomous. Leaves evergreen or biennial. Blossom before leaves appearing; flowers yellow, fragrant; calyx yellow inside; tubes exterior densely white sericeous; lobes 4; ovary ovoid, apex sericeous; styles glabrous; stigma globose. Drupes ellipsoid. Fl. late winter and early spring. Fr. spring-summer. Forests or shrubby slopes, also cultivated. Distributed in SW, SE and C China; cultivated in many provinces. Also in Japan.

结香 *Edgeworthia chrysantha*

鼠皮树 *Rhamnoneuron balansae*

## 草瑞香

**Diarthron linifolium** Turcz.

一年生草本。多分枝。小枝纤细，淡绿色，无毛。叶互生，线形至线状披针形。总状花序顶生；花小，绿色，先端常紫色。果实卵形或圆锥状，黑色。花期5-7月，果期6-8月。生海拔500-1400米的沙质荒地。产华北、西北和东北。俄罗斯和蒙国亦有。

Annual herbs. Multi-branched. Branchlets slender, greenish, glabrous. Leaves, alternate, linear to linear-lanceolate. Racemes terminal; flowers small, green, apex tinted with purple. Fruits ovate or conical, black. Fl. May-Jul. Fr. Jun-Aug. Sandy wastelands at 500-1400 m. Distributed in N, NW and NE China. Also in Russia and Mongolia.

草瑞香 *Diarthron linifolium*

## 狼毒

**Stellera chamaejasme** L.

多年生草本。叶下面灰色或浅绿色，披针形或长圆状披针形。花序顶生，头状，球形，具多花，具绿色的叶状苞片形成的总苞；花芳香；花萼白色、黄色或红紫色，花萼裂片5；雄蕊10。核果圆锥形。花期4-5月，果期7-9月。生海拔2600-4200米的高山草甸、沙地、废弃地或疏林中。产中国西南、华北、华西、西北和东北。不丹、尼泊尔、俄罗斯和蒙古亦有。

Perennial herbs. Leaves pale or grayish green abaxially, lanceolate or oblong-lanceolate. Inflorescences terminal, capitate, globose, many flowered, with involucre of green leaflike bracts; flowers fragrant; calyx white, yellow, or reddish purple, calyx lobes 5; stamens 10. Drupes conic. Fl. Apr-May. Fr. Jul-Sep. Alpine meadows, sandy places, wastelands or open forests at 2600-4200 m. Distributed in SW, N, W, NW and NE China. Also in Bhutan, Nepal, Russia and Mongolia.

狼毒 *Stellera chamaejasme*

# 胡颓子科
# Elaeagnaceae

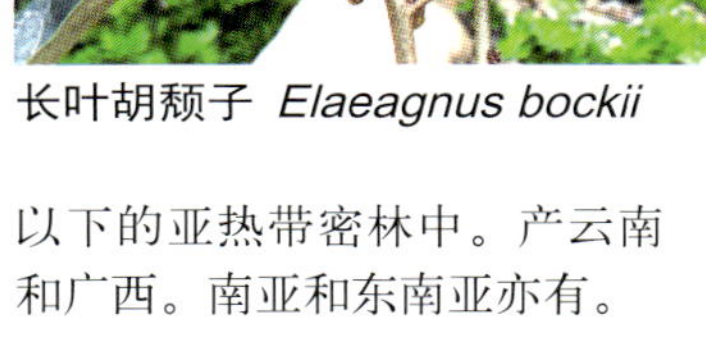

长叶胡颓子 *Elaeagnus bockii*

## 密花胡颓子
**Elaeagnus conferta** Roxb.

常绿攀援灌木。腋生枝条密被白盾状毛。叶椭圆形至椭圆状长圆形。花银白色；萼筒坛状，近无梗，成伞形短总状花序生于叶腋。核果狭椭圆体形或长圆形。花期10-11月，果期翌年2-3月。生海拔2100米以下的亚热带密林中。产云南和广西。南亚和东南亚亦有。

Evergreen scandent shrubs. Axillary branches densely with scales. Leaves elliptic to elliptic-oblong. Flowers silver-white; calyx tubes urceolate, subsessile, in umbellate short racemes. Drupes narrowly ellipsoid or oblong. Fl. Oct-Nov. next Fr. Feb-Mar. Subtropical dense forests below 2100 m. Distributed in Yunnan and Guangxi. Also in S and SE Asia.

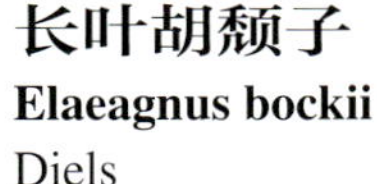

## 长叶胡颓子
**Elaeagnus bockii** Diels

直立灌木，具刺。幼枝密具锈色或褐色盾状毛。叶背密被白色盾状毛和稀疏锈色或褐色盾状毛，狭长圆椭圆形至狭披针形。花常5-7朵腋生；花梗长3-8毫米；花白色，密被盾状毛。核果红色，阔圆柱形。花期10-11月，果期翌年3-4月。生海拔350-2900米的向阳山坡或路旁灌丛中。产中国西南、华南、华中和西北。

Erect shrubs, spiny. Young branches with dense rust-colored or brown scales. Leaves adaxially with dense white and sparsely brown scales, narrowly oblong-elliptic to narrowly lanceolate. Flowers 5-7 in axils; pedicels 3-8 mm long; flowers white, outside densely scaly. Drupes red, broadly cylindric. Fl. Oct-Nov. Fr. next Mar-Apr. Sunny slopes or among roadside shrubs at 350-2900 m. Distributed in SW, S, C and NW China.

密花胡颓子 *Elaeagnus conferta*

## 披针叶胡颓子
**Elaeagnus lanceolata** Warb. ex Diels

常绿灌木。幼枝密被淡黄色和锈色盾状毛。叶常披针形，革质，背面密被银色盾状毛。花常3-5朵生于叶腋；花淡黄色，外面密被银色和锈色盾状毛；萼筒管状，于子房之上急收缩。核果红色，椭球体，密被锈色盾状毛。花期8-10月，果期翌年4-5月。生海拔300-2900米的沟边、山谷林下、山坡灌木丛。产中国西南、华中和华西。

Evergreen shrubs. Young branches densely yellowish white and rust-colored scales. Leaves usually lanceolate, coriaceous, abaxially densely silvery scaly. Flowers often 3-5 in axils; flowers yellowish white, outside densely silvery and rust-colored scaly; calyx tube tubular, abruptly constricted above ovary. Drupes red, ellipsoid, densely rust-colored scaly. Fl. Aug-Oct. Fr. next Apr-May. Streams, valley forests, bushy slopes at 300-2900 m. Distributed in SW, C and W China.

披针叶胡颓子 *Elaeagnus lanceolata*

绿叶胡颓子 *Elaeagnus viridis*

## 绿叶胡颓子

**Elaeagnus viridis** Serv.

常绿灌木，具刺。幼枝密被黄白色或锈色盾状毛。叶椭圆形至矩圆形，纸质或薄革质，叶背密被银白色或锈色盾状毛。花常多枚组成短的伞形总状花序，白色，外被银白色和稀疏的红棕色盾状毛，萼筒宽圆筒形。果实椭球形。花期10-11月，果期翌年3-5月。生海拔500-3150米的溪流沟边和山坡林下。产湖北西部和山西南部。

Evergreen shrubs, spiny. Young branches densely with yellowish white and rust-colored scales. Leaves elliptic to oblong, papery or thinly leathery, abaxially with dense silvery scales. Flowers often several clustered in a shortened umbellate-raceme, white, outside with silvery and sparse reddish brown scales, calyx broadly tubular. Drupes ellipsoid. Fl. Oct-Nov. Fr. next Mar-May. Stream and thickets on slope at 500-3150 m. Distributed in W Hubei and S Shanxi.

## 蔓胡颓子

**Elaeagnus glabra** Thunb.

常绿蔓生或攀援灌木。分枝处常微微膨大呈三角状，幼枝密被锈色盾状毛。叶椭圆形至倒卵形，纸质。花常2-8朵簇生成伞形总状花序；花淡黄色，密被白色和锈色盾状毛。核果橘红色，阔椭球形至长圆球形。花期9-11月，果期翌年4-5月。生海拔约2200米的溪边、山谷林中、灌丛。产中国西南、华南、东南、华中和华东。朝鲜半岛南部和日本南部亦有。

Evergreen shrubs, twinning or climbing. Branche nodes often slightly triangular, young branches densely covered with rust-colored scales. Leaves elliptic to obovate, papery. Flowers 2-8 usually clustered to umbellate racemes; flowers yellowish white, densely with white and rust-colored scales. Drupes red-orange, broadly ellipsoid to ellipsoid. Fl. Sep-Nov. Fr. next Apr-May. Streams, thickets, bushy slopes at ca. 2200 m. Distributed in SW, S, SE, C and E China. Also in S Korean Peninsula and S Japan.

## 胡颓子

**Elaeagnus pungens** Thunb.

常绿乔木，高3-4米，多刺。叶革质，长圆形至狭长圆形，侧脉7-9对，背面密被银白色和少数褐色盾状毛。花少，簇生于叶腋；花萼管漏斗形。核果长圆形，具褐色盾状毛。花期9-12月，果期翌年4-6月。生海拔1000米以下的向阳山坡、开阔山坡、灌丛或路边。产中国西南、东南、华中和华东。日本亦有。

Evergreen shrubs, 3-4 m tall, spines frequent. Leaves coriaceous, oblong to narrowly so, lateral nerves 7-9 pairs, abaxially covered with dense silvery scales and sparse brown scales. Flowers few, clustered in axils; calyx tube funnel form. Drupes oblong, brown scaly. Fl. Sep-Dec. Fr. next Apr-Jun. Sunny slopes, open slopes, thickets or roadsides

蔓胡颓子 *Elaeagnus glabra*

胡颓子 *Elaeagnus pungens*

below 1000 m. Distributed in SW, SE, C and E China. Also in Japan.

## 宜昌胡颓子

**Elaeagnus henryi** Warb. ex Diels

常绿灌木，具刺。幼枝密被锈色叠生盾状毛。叶宽披针形至卵圆形，革质，叶背密被黄白色和少许锈色盾状毛。花黄白色，外面密被银白色和散生锈色盾状毛，萼筒倒圆锥状管形。果实长圆球形，成熟时红色。花期10-11月，果期翌年4-5月。生海拔200-2700米的水沟边灌丛、河谷林中、山坡密林中。产中国西南、华南、华中和华东。

Evergreen shrubs, spiny. Young branches densely with overlapping rubiginous scales. Leaves broadly lanceolate to ovate, leathery, abaxially with yellowish white and sparse rust-colored scales. Flower yellowish white, outside densely with silvery and sparse rust-colored scales, calyx tube obconic-tubular. Drupes oblong, red when ripe. Fl. Oct-Nov. Fr. next Apr-May. Streamsides, valley forests, and thickets on slope at 200-2700 m. Distributed in SW, S, C and E China.

宜昌胡颓子 *Elaeagnus henryi*

## 巴东胡颓子

**Elaeagnus difficilis** Ser.

常绿灌木，有刺。幼枝密被锈色盾状毛。叶卵状椭圆形，稀狭披针形，纸质或薄革质，叶背散生或密被锈色盾状毛。花常5朵组成短的总状花序，黄色，萼筒常短钟形。果实椭球形，成熟时粉红或橘红色。花期12月至翌年3月，果期5-6月。生海拔300-1900米的溪边和林下灌丛。产中国西南、华南、华中和华东。

Evergreen shrubs, spiny. Young branches densely with rust-colored scales. Leaves ovate-elliptic, rarely tolanceolate, papery or thinly leathery, abaxially with dense or sparse rust-colored scales. Flower 5 clustered in a shortened raceme, yellow, calyx tube short campanulate. Drupes ellipsoid, salmon or orange-red when ripe. Fl. Dec to next Mar. Fr. May-Jun. Streamsides and valley forests at 300-1900 m. Distributed in SW, S, C and E China.

巴东胡颓子 *Elaeagnus difficilis*

沙枣 *Elaeagnus angustifolia*

## 沙枣

**Elaeagnus angustifolia** L.

落叶乔木。树皮红棕色，具刺。叶薄纸质。花1-3朵生于老叶腋；萼筒钟形，芳香，外面密被银白色盾状毛，内部黄色。核果黄褐色，卵球形、球形或近球形，果肉粉质。花期5-6月，果期8-10月。生山地、平原、沙滩或荒地。产华北、华西和西北。西南亚、中亚、俄罗斯、蒙古和欧洲东部亦有。归化于北美洲。

Deciduous trees. Bark reddish brown, spines sharp. Leaves thin papery. Flowers 1-3 in axils of older leaves; flowers fragrant, calyx tube campanulate, outside densely with silvery white scales, inside yellow. Drupes yellowish brown, globose-ovoid, globose, or subglobose, and flesh mealy. Fl. May-Jun. Fr. Aug-Oct. Mountains, plains, sandy coasts or wastelands. Distributed in N, W and NW China. Also in SW and C Asia, Russia, Mongolia and E Europe. Naturalized in North America.

## 翅果油树

**Elaeagnus mollis** Diels

落叶灌木或小乔木。树皮灰褐色。叶背密具灰色星状绒毛。花通常1-5朵生于叶腋，芳香；花黄绿色，密具灰白色星状绒毛。核果卵球形或近球形，明显具8个翅状棱脊。花期4-5月，果期8-9月。生海拔700-1300米的山谷或阴处。产山西南部和陕西(户县)。

Deciduous shrubs or small trees. Bark grayish brown. Leaves abaxially densely grayish stellate-tomentose. Flowers often 1-5 in leaf axils, fragrant; flowers yellowish green, densely with grayish white stellate-tomentose. Drupes globose-ovoid or subglobose, conspicuoly 8-ribbed, winged. Fl. Apr-May. Fr. Aug-Sep. Valleys or shaded areas at 700-1300 m. Distributed in S Shanxi and Shaanxi (Hu County).

## 尖果沙枣

**Elaeagnus oxycarpa** Schltdl.

落叶乔木。树皮红褐色，具锐刺。叶长圆状线形或线状披针形。花通常1-3朵生于叶腋，黄色，密被银白色盾状毛。核果卵球形或近球形，成熟时黄褐色。花期5-6月，果期9-10月。生海拔400-1500米的沙漠、田中或路边。产甘肃和新疆。俄罗斯亦有。

Deciduous trees. Bark reddish brown, spines sharp. Leaves oblong-linear or linear-lanceolate. Flowers 1-3 in leaf axils, yellow, outside densely with silvery white scales. Drupes globose-ovoid or subglobose, yellowish brown when mature. Fl. May-Jun. Fr. Sep-Oct. Deserts, fields or roadsides at 400-1500 m. Distributed in Gansu and Xinjiang. Also in Russia.

翅果油树 *Elaeagnus mollis*

## 星毛羊奶子

**Elaeagnus stellipila** Rehd.

半常绿灌木，高达4米，具刺。叶片宽卵形或卵状椭圆形，叶背密被叠生白色或黄白色星状毛。花常1-3朵生于叶腋，淡黄色；萼筒倒圆锥状管形，不明显4棱。核果成熟时红色，狭椭球形或矩圆形。花期3-4月，果

尖果沙枣 *Elaeagnus oxycarpa*

星毛羊奶子 *Elaeagnus stellipila*

期翌年7-8月。生海拔140-2400米的向阳山坡、溪边和农田。产中国西南、湖北和江西。

Semi-evergreen shrubs, to 2 m tall, spiny. Leaves broadly ovate or ovate-elliptic, abaxially with densely overlapping white or yellowish white stellate hairs. Flowers often 1-3 together in leaf axils, light yellow. Calyx tubes obconic-tubular, faintly 4-ribbed. Drupes red when ripe, narrowly ellipsoid or oblong. Fr. Mar-Apr. Fr. next Jul-Aug. Sunny slopes, by streams and farmlands, at 140-2400 m. Distributed in SW China, Hubei and Jiangxi.

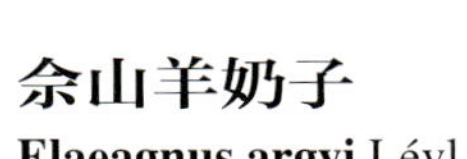

### 佘山羊奶子
**Elaeagnus argyi** Lévl.

半常绿灌木，高2-3米，具刺。叶薄纸质或膜质，随季节二型，叶背具白色和褐色点及重叠的盾状毛。花常5-7朵簇生于新生长的枝条基部；花黄色。核果成熟后红色，倒卵状长圆形，被银色盾状毛。花期1-3月，果期4-5月。生海拔100-300米的林下或路边。产浙江、湖北、湖南、安徽和江苏。

Semi-evergreen shrubs, 2-3 m tall, usually spiny. Leaves thinly papery or membranous, dimorphic by season, abaxially white with brown dots, scales overlapping. Flowers often 5-7 in a fascicle at base of new growing branches; flowers yellow. Drupes red when ripe, obovoid-oblong, covered by silver scales. Fl. Jan-Mar. Fr. Apr-May. Forests or roadsides at 100-300 m, Distributed in Zhejiang, Hubei, Hunan, Anhui and Jiangsu.

佘山羊奶子 *Elaeagnus argyi*

### 牛奶子
**Elaeagnus umbellata** Thunb.

落叶灌木。叶椭圆形、卵状椭圆形或倒卵状披针形，叶背密被银白色盾状毛。花1-3(-7)簇生于叶腋；花白色，萼筒漏斗状。核果红色，近球形。花期4-5月，果期7-8月。生海拔(100-)500-3000米的疏林、灌丛或荒地。产中国西南、华北、华西和华东。印度、尼泊尔、不丹、阿富汗、朝鲜半岛和日本亦有。归化于北美洲。

Deciduous shrubs. Leaves elliptic, ovate-elliptic or obovate-lanceolate, abaxially densely silvery white scaly. Flowers 1-3(-7)-fasciculate in axils; flowers white; calyx tube funnel-shaped. Drupes red, nearly globose. Fl. Apr-May. Fr. Jul-Aug. Sparse forests, thickets or wastelands at (100-)500-3000 m. Distributed in SW, N. W and E China. Also in India, Nepal, Bhutan, Afghanistan, Korean Peninsula and Japan. Naturalized in North America.

牛奶子 *Elaeagnus umbellata*

银果牛奶子 *Elaeagnus magna*

## 银果牛奶子
**Elaeagnus magna** (Servett.) Rehd.

落叶灌木，具刺。幼枝、花和果实密被银色盾状毛。叶纸质或膜质。花1-3朵生于新生枝基部，黄色，外被银白色盾状毛；萼筒管状，四棱状；核果粉红色，圆柱形或狭椭圆体形至纺锤形，长1.2-1.6厘米。花期4-5月，果期6月。生海拔100-2300米的山地沙壤、路边、林缘或河岸。产中国西南、华南和华中。

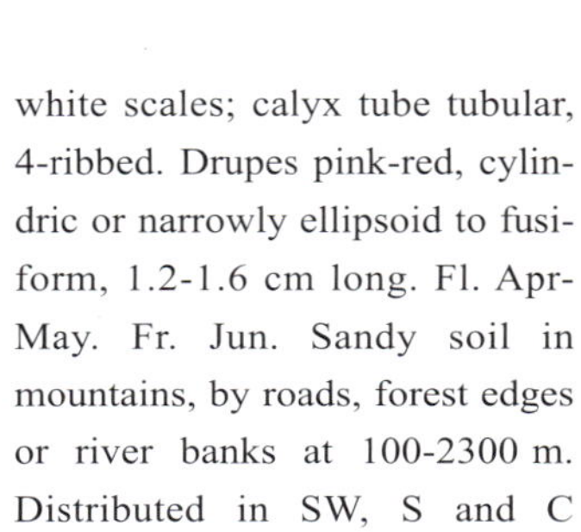

Deciduous shrubs, spines present. Young branches, flowers and fruits densely silvery scaly. Leaves papery or membranous. Flowers 1-3 at base of new grown shoots, yellow, outside densely with silvery white scales; calyx tube tubular, 4-ribbed. Drupes pink-red, cylindric or narrowly ellipsoid to fusiform, 1.2-1.6 cm long. Fl. Apr-May. Fr. Jun. Sandy soil in mountains, by roads, forest edges or river banks at 100-2300 m. Distributed in SW, S and C China.

## 木半夏
**Elaeagnus multiflora** Thunb.

落叶灌木。幼枝被锈色盾状毛。叶背密被叠生白色和散生棕色盾状毛。花1(-2)朵腋生，白色，密被白色和疏被棕色盾状毛；萼筒于基部收缩。核果长圆形、卵球形或椭圆体形，果梗纤细。花期4-5月，果期6-7月。生海拔1800米以下的灌丛或低地和山坡的开阔林地。产中国西南、东南、华北和华东。朝鲜半岛和日本亦有。

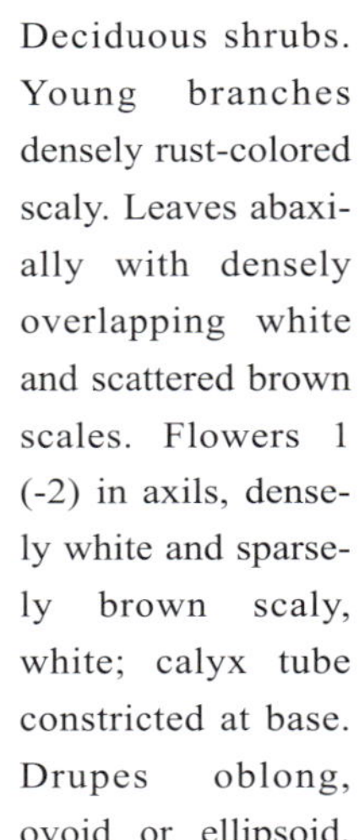

Deciduous shrubs. Young branches densely rust-colored scaly. Leaves abaxially with densely overlapping white and scattered brown scales. Flowers 1 (-2) in axils, densely white and sparsely brown scaly, white; calyx tube constricted at base. Drupes oblong, ovoid or ellipsoid, fruits pedicels slender. Fl. Apr-May. Fr. Jun-Jul. Thickets or open woodland in lowlands and mountains below 1800 m. Distributed in SW, SE, N and E China. Also in Korean Peninsula and Japan.

## 西藏沙棘
**Hippophae tibetana** Schltdl.

落叶灌木。叶多3枚轮生，叶背散生近全缘的红褐色盾状毛，中脉红褐色，线状长圆形，密被盾状毛。果实橘黄色，顶端具黑色星状条纹，球形至椭圆体形。花期5-6月，果期9月。生海拔3600-4700米的干旱碎石或多石地，特别是河床和河漫滩。产西藏、甘肃和青海。印度北部、尼泊尔和不丹亦有。

Deciduous shrubs. Leaves mostly in whorls of 3, leaves abaxially

木半夏 *Elaeagnus multiflora*

西藏沙棘 *Hippophae tibetana*

柳叶沙棘 *Hippophae salicifolia*

云南沙棘 *Hippophae rhamnoides* subsp. *yunnanensis*

with scattered subentire, reddish brown scales and reddish brown midrib, linear-oblong, densely scaly. Fruits orange, tip with black star-stripe, globose to ellipsoid. Fl. May-Jun. Fr. Sep. Dry gravelly or stony places, especially on riverbeds or flood plains at 3600-4700 m. Distributed in Xizang, Gansu and Qinghai. Also in N India, Nepal and Bhutan.

## 柳叶沙棘

**Hippophae salicifolia** D. Don

落叶灌木或乔木，2-3(-10)米高。枝顶端具短刺。叶纸质，线状长圆形或披针形，叶背常具白色星状毛，中脉红褐色。果实成熟时橘黄色，球形或近球形。花期6月，果期10月。生海拔2800-3500米的潮湿砾石地、河边或溪边。产西藏南部。印度北部、尼泊尔和不丹亦有。

Deciduous shrubs or trees, 2-3 (-10) m tall. Branch top with short spine. Leaves papery, linear-oblong or lanceolate, abaxially whitish stellate hairs and reddish brown midrib. Fruits orange-yellow when ripe, globose or almost globose. Fl. Jun. Fr. Oct. Moist gravels, by rivers or streams at 2800-3500 m. Distributed in S Xizang. Also in N India, Nepal and Bhutan.

## 中国沙棘

**Hippophae rhamnoides** L. subsp. **sinensis** Rousi

落叶灌木或乔木，高1-15(-18)米。树皮淡褐色，枝近黑色，粗糙，侧生刺较少，粗壮。叶多对生或近对生，叶背常银白色，狭披针形。果实球状，黄色至深红色。花期4-5月，果期9-10月。生海拔800-3600米的河岸、干河床、林缘或灌丛中。产中国西南、华北和华西。

Deciduous shrubs or trees, 1-15 (-18) m tall. Bark brownish; branches almost black, rough; lateral spines relatively few, stout. Leaves mostly opposite or subopposite; leaves abaxially usually silvery, narrowly lanceolate. Fruits globose, yellow to dark red. Fl. Apr-May. Fr. Sep-Oct. River banks, dry river beds, forest edges or thickets at 800-3600 m. Distributed in SW, N and W China.

中国沙棘 *Hippophae rhamnoides* subsp. *sinensis*

## 云南沙棘

**Hippophae rhamnoides** L. subsp. **yunnanensis** Rousi

落叶灌木或乔木。新枝常密被红褐色星状毛。叶互生，狭披针形或长圆状披针形，基部常为圆形或楔形，背面灰褐色，具多而大的褐色盾状毛。果实黄色，圆球形，直径5-7毫米。种子强烈压扁。花期4月，果期8-9月。生海拔2200-3700米的河谷沙地、石砾地或山坡密林。产云南、四川西部和西藏东南部。

Deciduous shrubs or trees. New shoots with dense reddish brown stellate hairs. Leaves alternate, narrowly lanceolate or oblong-lanceolate, often round or cuneate at base, abaxially gray-brown, covered with numerous large brown scales. Fruits yellow, globose, 5-7 mm diam. Seeds strongly flattened. Fl. Apr. Fr. Aug-Sep. Sandy or gravelly places in valleys or in montane dense forests at 2200-3700 m. Distributed in Yunnan, W Sichuan and SE Xizang.

# 千屈菜科
# Lythraceae

耳基水苋 *Ammannia auriculata*

多花水苋菜 *Ammannia multiflora*

### 水苋菜
**Ammannia baccifera** L.

一年生草本。叶在茎下部对生，狭椭圆形或倒披针形至条形，基部渐狭至截形。花3至数朵形成密集腋生聚伞状；萼片4；无外花萼；无花瓣；雄蕊4。蒴果1/4-1/2伸出。花期8-10月，果期9-12月。生海拔800-1800米的草地、湿地、田间或路边。产中国西南、华南、东南、华中、华北和华东。南亚、东南亚、西南亚、澳大利亚、热带非洲和加勒比亦有。

Annual herbs. Leaves opposite on basal stem portion, narrowly elliptic or oblanceolate to linear, basally attenuate to truncate. Flowers 3 to many in dense axillary cymes; sepals 4; epicalyx absent; petals absent; stamens 4. Capsules 1/4-1/2 exserted. Fl. Aug-Oct. Fr. Sep-Dec. Grasslands, wet places, fields or roadsides at 800-1800 m. Distributed in SW, S, SE, C, N and E China. Also in S, SE and SW Asia, Australia, tropical Africa and Caribbean.

### 耳基水苋
**Ammannia auriculata** Willd.

草本。叶对生，狭披针形或长圆状披针形，膜质，无柄，叶基部心状耳形。聚伞花序腋生，具(1-)3-15花；花瓣4，玫瑰紫色，近圆形。蒴果扁球形，紫红色，成熟时不规则周裂。花果期8-12月。生海拔1200-2000米的湿地或水稻田中。产中国西南、东南、华北、华西和华东。泛热带地区亦有。

Herbs. Leaves opposite, narrowly lanceolate or oblong-lanceolate, membranous, sessile, cordate-auriculate at base. Cymes axillary, flowers (1-)3-15; petals 4, rose-purple, suborbicular. Capsules flat-globose, purple-red, irregularly circumscissile when mature. Fl. and fr. Aug-Dec. Wet places or paddy fields at 1200-2000 m. Distributed in SW, SE, N, W and E China. Also in pantropical.

### 多花水苋菜
**Ammannia multiflora** Roxb.

一年生草本。叶对生，狭椭圆形、宽条形或披针状长圆形。花3-7(-20)朵形成密集腋生聚伞花序；萼片4；花瓣4，粉色至白色；雄蕊4。蒴果红褐色或带酒红色，约1/2伸出。花期7-8月，果期9月。生潮湿地或水田中。产华南。热带和亚热带亚洲、大洋洲和非洲亦有。

Annual herbs. Leaves opposite,

水苋菜 *Ammannia baccifera*

节节菜 *Rotala indica*

narrowly elliptic, broadly linear, or lanceolate-oblong. Flowers 3-7(-20) in dense axillary cymes; sepals 4; petals 4, pink to whitish; stamens 4. Capsules red-brown or red-wine colored, ca. 1/2 exserted. Fl. Jul-Aug. Fr. Sep. Wet places or fields. Distributed in S China. Also in tropical and subtropical Asia, Oceania and Africa.

## 节节菜

**Rotala indica** (Willd.) Koehne

一年生草本。叶交互对生，倒卵状椭圆形或倒卵状长圆形。腋生穗状花序或，无柄生主茎苞片内；花小，淡红色；花瓣4，倒卵形，粉色。蒴果椭圆体形，2瓣开裂。花期9-10月，果期10月至翌年4月。生海拔1400-1800米的稻田或潮湿地方。产中国西南、华南、东南、华中和华东。南亚、东南亚和东北亚亦有。非洲(刚果)、欧洲(意大利、葡萄牙)和北美洲(美国)引入稻田栽培。

Annual herbs. Leaves decussate, obovate-elliptic or obovate-oblong. Flowers in axillary spikes or sessile in bracts on main stem; flowers small, pale red; petals 4, obovate, pink. Capsules ellipsoidal, 2-valved. Fl. Sep-Oct. Fr. Oct to next Apr. Paddy fields or moist places at 1400-1800 m. Distributed in SW, S, SE, C and E China. Also in S, SE and NE Asia. Introduced in rice fields in Africa (Congo), Europe (Italy, Portugal), and North America (USA).

## 圆叶节节菜

**Rotala rotundifolia** (Buch.-Ham. ex Roxb.) Koehne

多年生或偶为一年生草本。叶交互对生，倒卵状椭圆形至圆形或椭圆形。花1-8顶生，穗状；萼片4；花瓣4，亮玫红色，超出萼片；雄蕊4。蒴果球形，4瓣裂。花果期11月至翌年6月。生海拔2700米以下的山坡、水边田地或潮湿处。产中国西南、华南、东南、华中和华东。南亚和日本亦有。

Perennial herbs or possibly annual. Leaves decussate, obovate-elliptic to orbicular or elliptic. Flowers in 1-8 terminal, spikes; sepals 4; petals 4, bright rose, surpassing sepals; stamens 4. Capsules globose, 4-valved. Fl. and fr. Nov to next Jun. Mountains, water fields or damp places below 2700 m. Distributed in SW, S, SE, C and E China. Also in S Asia and Japan.

圆叶节节菜 *Rotala rotundifolia*

千屈菜 *Lythrum salicaria*

## 千屈菜

**Lythrum salicaria** L.

多年生草本或亚灌木，粗糙或具灰色柔毛(或绒毛)，有时渐无毛。叶对生或三叶轮生，披针形或宽披针形，基部圆形、截形或半抱茎。穗状花序顶生；花1至多朵在叶腋形成聚伞状；花瓣6，紫红色或淡紫色。蒴果扁球形。花期7-9月，果期10月。生河岸、溪边或潮湿草地。广布中国各地。亚洲、澳大利亚东南部、非洲(阿尔及利亚)、欧洲和美洲亦有。

Perennial herbs or subshrubs, scabrous or gray pubescent (or tomentose), sometimes glabrescent. Leaves opposite or 3-whorled, lanceolate or broadly lanceolate, base rounded, truncate, or semi-clasping. Inflorescences terminal, spicate; flowers in 1- to multi-flowered whorled axillary cymes; petals 6, purple-red or pale-purple. Capsules depressed globose. Fl. Jul-Sep. Fr Oct. River banks, by streams, or moist grasslands. Widespread in China. Also in Asia, SE Australia, Africa (Algeria), Europe and America.

## 帚枝千屈菜

**Lythrum virgatum** L.

多年生草本。叶对生，有时上部互生，狭披针形至条状披针形，基部狭楔形。穗状花序顶生；具1-3(-7)花组成腋生聚伞花序，单生或少数轮生；花瓣紫红色；雄蕊12；花柱长短不等。蒴果圆柱形或长球形。花期4-8月，果期7-9月。生湿地。产河北和新疆。欧洲东部至西伯利亚东南部亦有。

Perennial herbs. Leaves opposite, sometimes alternate above, narrowly lanceolate to linear-lanceolate, base narrowly cuneate. Inflorescences terminal, spicate; flowers in 1-3(-7)-flowered axillary cymes, solitary or in sparse whorls; petals purple-red; stamens 12; styles unequal. Capsules terete or long-globose. Fl. Apr-Aug. Fr. Jul-Sep. Wet places. Distributed in Hebei and Xinjiang. Also in E Europe to SE Siberia.

## 小瓣萼距花

**Cuphea micropetala** Kunth

直立灌木，高达1米。小枝无毛。叶近对生，薄革质，披针形，长5-12厘米。花单生，腋生或腋外生，组成总状花序；花萼筒阔管状，被绒毛，下部深红色，有距，上部黄色；雄蕊红色。花期秋冬季。广东和云南栽培。原产墨西哥。

Erect shrubs, to 1 m tall. Branchlets glabrous. Leaves sub-opposite, thinly coriaceous, lanceolate, 5-12 cm long. Flowers solitary, axillary or extra-axillary, formed in racemes; calyx tube broadly tabulate, tomentose, lower part crimson, calcarate, upper part yellow; stamens red. Fl. autumn to winter. Cultivated in Guangdong and Yunnan. Native to Mexico.

帚枝千屈菜 *Lythrum virgatum*

小瓣萼距花 *Cuphea micropetala*

火红萼距花 *Cuphea platycentra*

## 火红萼距花

**Cuphea platycentra** Lem.

亚灌木，全株无毛。叶对生，披针形至卵状披针形，长2.5-6厘米。花单生叶腋或近腋生，具细长的花梗；萼筒细长，无毛，有距，火焰红色，末端有紫黑色的环。花期8月。北京温室有栽培。原产墨西哥。

Semishrubs, wholly glabrous. Leaves opposite, lanceolate to ovate-lanceolate, 2.5-6 cm long. Flowers solitary, axillary or extra-axillary, long pedicellate; calyx tube slender, glabrous, calcarate, flammate, apex with dark purple ring. Fl. Aug. Cultivated in greenhouses of Beijing. Native to Mexico.

## 细叶萼距花

**Cuphea hyssopifolia** Kunth

常绿小灌木。分枝多而细密。叶对生，线状披针形，长1-2厘米。花单生叶腋，小而多，紫色、淡紫色或白色；花萼花冠状，高脚蝶状，具5齿，齿间具退化的花瓣。花期几全年。中国南方城市广泛栽培。原产中美洲，世界各地广泛栽培。

Evergreen small shrubs. Branchlets dense, very slender. Leaves opposite, linear-lanceolate, 1-2 cm long. Flowers solitary, axillary, small, numerous, purple, purplish or white; calyx corolla-like, salverform, 5-teethed, with regressive petals between teeth. Fl. almost all year. Widely cultivated in southern cities of China. Native to Central America, widely cultivated worldwide.

细叶萼距花 *Cuphea hyssopifolia*

哥伦比亚萼距花 *Cuphea carthagenensis*

## 哥伦比亚萼距花

**Cuphea carthagenensis** (Jacq.) J. F. Mac Br.

小灌木，高30-60厘米。小枝密被长腺毛。叶对生，纸质，椭圆形或长圆形，长2-3厘米，侧脉弧形上升。花腋生或生于短枝顶端，淡紫色。花期9-12月。西藏和海南有归化。原产中南美洲。

Small shrubs, 30-60 cm tall. Branchlets densely covered with long glandular hairs. Leaves opposite, chartaceous, elliptic or oblong, 2-3 cm long, laternal veins arched ascending. Flowers axillary or terminal, purplish. Fl. Sep-Dec. Naturalized in Xizang and Hainan. Native to Central and South America.

## 虾子花

**Woodfordia fruticosa** (L.) Kurz

灌木。茎与枝下垂。叶披针形或卵状披针形，革质。1-15花密集于枝腋；苞叶亮红色、橙红色或深红色，狭佛焰苞状；花瓣6；雄蕊12。蒴果室背开裂，开裂成2-4果瓣。花期1-5月(主要3-4月)，果期4-5月。生海拔300-2000米的干热河谷、山坡或阳坡灌丛。产云南、广西和广东。南亚、东南亚和热带非洲亦有。

Shrubs. Stems and branches pendulous. Leaves lanceolate or ovate-lanceolate, leathery. Inflorescences condensed axillary shoots of 1-15 flowers; floral tubes light red, orange-red, or deep red, narrowly cyathiform; petals 6; stamens 12. Capsules loculicidal, dehiscing into 2-4 valves. Fl. Jan-May (mainly Mar-Apr). Fr. Apr-May. Dry and hot valleys, slopes, or sunny thickets at 300-2000 m. Distributed in Yunnan, Guangxi and Guangdong. Also in S and SE Asia, and tropical Africa.

虾子花 *Woodfordia fruticosa*

## 水芫花

**Pemphis acidula** J. R. et G. Forst.

多分枝小灌木，被灰色短柔毛。叶对生，厚，肉质，椭圆形或线状披针形。花腋生，二型，花萼具12棱，6浅裂；花瓣6，白色或粉红色；雄蕊12，6长6短。蒴果革质，倒卵形。花期6月。生热带海边。产台湾南部。东半球热带海岸广布。

Small shrubs, multi-branched, gray pubescent. Leaves opposite, thick, carnose, elliptic or linear-lanceolate. Flowers axillary, dichotypic, calyx 12-redged, 6-lobed; petals 6, white or pink; stamens 12, 6 longer than the rest. Capsules coriaceous, obovate. Fl. Jun. Tropical seacoast. Distributed in S Taiwan. Widely distributed in tropical seacoast of

水芫花 *Pemphis acidula*

黄薇 *Heimia myrtifolia*

Eastern Hemisphere.

## 黄薇

**lia** Cham. et Schlechtend.

小灌木。小枝略四棱。叶对生，椭圆形至线形，近无柄。花单生叶腋；花萼半球形，果时包围蒴果；花瓣5-7，黄色。蒴果球形。花果期4-7月。广东、广西和云南栽培。原产巴西。

Small shrubs. Branchlets somewhat tetragonous. Leaves opposite, elliptic to linear, subsessile. Flowers axillary, solitary; calyx hemispherical, surrounding Capsules at fruit; petals 5-7, yellow. Capsules globose. Fl. and fr. Apr-Jul. Cultivated in Guangdong, Guangxi and Yunnan. Native to Brazil.

紫薇 *Lagerstroemia indica*

## 紫薇

**Lagerstroemia indica** L.

灌木或小乔木。叶椭圆形、长圆形、倒卵形或近圆形。圆锥花序近金字塔形，密具花；花冠筒6数，壁光滑或具不明显的6棱，无毛；萼片上面无毛；花瓣紫色、紫红色、粉色或白色，边缘具皱波纹；子房无毛。蒴果椭圆体形。花期6-9月，果期9-11月。生海拔1100-2800米的开阔林中、路边或栽培。产中国西南、华南、东南、华中、华北、华东和东北。南亚和东南亚亦有；常栽培于热带、亚热带和暖温带地区。

Shrubs or small trees. Leaves elliptic, oblong, obovate, or suborbicular. Panicles subpyramidal, densely flowered; floral tubes 6-merous, smooth walled or obscurely 6-ribbed, glabrous; sepals adaxially glabrous; petals purple, fuchsia, pink, or white, margin crisped; ovary glabrous. Capsules ellipsoidal. Fl. Jun-Sep. Fr. Sep-Nov. Semishaded places, rich fields, open forests, roadsides, or cultivated at 1100-2800 m. Distributed in SW, S, SE, C, N, E and NE China. Also in S and SE Asia; commonly cultivated in tropical, subtropical and warm-temperate areas.

## 大花紫薇

**Lagerstroemia speciosa** Pers.

乔木。叶对生，椭圆形或长圆状椭圆形。圆锥花序顶生，长10-15厘米；花蕾顶端明显尖，花冠管6数且明显具棱，密具黄褐色柔毛；萼片上面无毛；花瓣干后紫色至蓝紫色；雄蕊75-130，成2-3轮。蒴果近球形，6裂。花期5-6月，果期10-12月。生海拔800-1500米的林中或路边。广西、广东和福建有栽培。产印度、斯里兰卡、越南、马来西亚和菲律宾。

Trees. Leaves opposite, elliptic or oblong-elliptic. Panicles terminal, 10-15 cm long; buds are typically apiculate, floral tubes 6-merous, deeply and conspicuously ribbed, densely yellow-brown puberulous; sepals adaxially glabrous; petals purple to bluish purple when dry; stamens 75-130, 2-3-whorled. Capsules subglobose, 6-lobed. Fl. May-Jun. Fr. Oct-Dec. Forests or roadsides at 800-1500 m. Cultivated in Guangxi, Guangdong and Fujian. Distributed in India, Sri Lanka, Vietnam, Malaysia and the Philippines.

大花紫薇 *Lagerstroemia speciosa*

云南紫薇 *Lagerstroemia intermedia*

福建紫薇 *Lagerstroemia limii*

## 云南紫薇
**Lagerstroemia intermedia** Koehne

乔木。叶对生，椭圆形或长圆状椭圆形。圆锥花序顶生，长10-15厘米；花冠管6数，壁光滑或具12个浅的阔棱，密具黄褐色柔毛；萼片上面无毛；花瓣干后紫色至蓝紫色；雄蕊75-130，成2-3轮；子房无毛。蒴果近球形，6裂。花期5-6月，果期10-12月。生海拔800-1500米的林中或路边。产云南西南部。缅甸亦有。

Trees. Leaves opposite, elliptic or oblong-elliptic. Panicles terminal, 10-15 cm long; floral tubes 6-merous, smooth walled or with 12 shallow, broad ribs, densely yellow-brown puberulous; sepals adaxially glabrous; petals purple to bluish purple when dry; stamens 75-130, 2-3-whorled; ovary glabrous. Capsules subglobose, 6-lobed. Fl. May-Jun. Fr. Oct-Dec. Forests or roadsides at 800-1500 m. Distributed in SW Yunnan. Also in Myanmar.

## 福建紫薇
**Lagerstroemia limii** Merr.

灌木或小乔木。叶互生至近对生，椭圆形或长圆状椭圆形，革质或近革质。圆锥花序顶生；花冠筒(5或)6数，具12-14深棱至浅脊；萼片上面无毛；花瓣淡红色至粉色；雄蕊约35；子房无毛。蒴果4或5裂。花期5-6月，果期7-8月。生低山区杂木林。产福建、浙江和湖北。

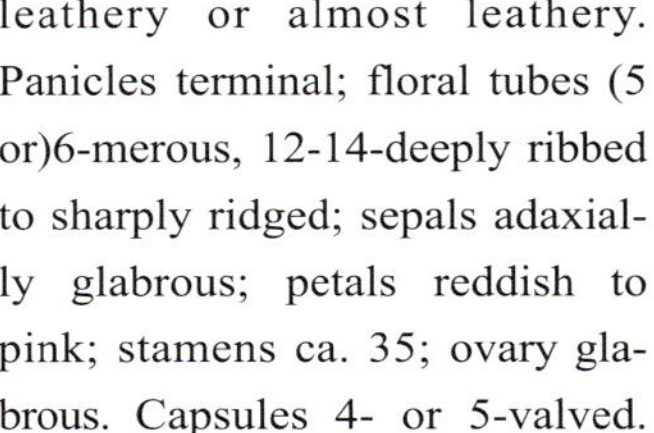

Shrubs or small trees. Leaves alternate to opposite, elliptic or oblong-elliptic, leathery or almost leathery. Panicles terminal; floral tubes (5 or)6-merous, 12-14-deeply ribbed to sharply ridged; sepals adaxially glabrous; petals reddish to pink; stamens ca. 35; ovary glabrous. Capsules 4- or 5-valved. Fl. May-Jun. Fr. Jul-Aug. Mixed forests of low mountains. Distributed in Fujian, Zhejiang and Hubei.

## 南紫薇(九芎)
**Lagerstroemia subcostata** Koehne

落叶乔木或灌木。叶长圆形，卵状披针形或椭圆形。圆锥花序锥状，密具花；花冠管6数，

南紫薇(九芎) *Lagerstroemia subcostata*

广东紫薇 *Lagerstroemia fordii*

具(10或)12(或14)条黑色脉或(10或)12(或14)棱；萼片上面无毛；花瓣白色、粉色或紫色；雄蕊15-30；子房无毛。蒴果球形至矩圆状。花期6-8月，果期7-10月。生低海拔至中海拔的林缘或溪边。产中国西南、华南、东南、华中、华东和华西。菲律宾和日本亦有。

Deciduous trees or shrubs. Leaves oblong, ovate-lanceolate or elliptic. Panicles pyramidal, densely flowered; floral tubes 6-merous, with (10 or)12(or 14) dark veins or obscurely (10 or)12(or 14)-ribbed; sepals adaxially glabrous; petals white, pink, or purple; stamens 15-30; ovary glabrous. Capsules globose to oblong. Fl. Jun-Aug. Fr. Jul-Oct. Forest edges or streamsides at low to medium elevations. Distributed in SW, S, SE, C, E and W China. Also in the Philippines and Japan.

## 广东紫薇

**Lagerstroemia fordii** Oliv. et Koehne

灌木或乔木。叶阔披针形或椭圆状披针形至椭圆状卵形，互生，纸质。圆锥花序顶生；花筒6数，明显具10-12棱，具灰白色绒毛；萼片上面前端稍具柔毛；雄蕊25-30；子房无毛。蒴果卵球状，褐色。果期8月。生低山疏林中。产香港和福建。

Shrubs or trees. Leaves broadly lanceolate or elliptic-lanceolate to elliptic-ovate, alternate, papery. Panicles terminal; floral tubes 6-merous, distinctly 10-12-ribbed, gray-white pubescent; sepals adaxially slightly pubescent on distal half; stamens 25-30; ovary glabrous. Capsules ovoid, brown. Fr. Aug. Sparse forests on low mountains. Distributed in Hong Kong and Fujian.

绒毛紫薇 *Lagerstroemia tomentosa*

## 绒毛紫薇

**Lagerstroemia tomentosa** C. Presl

乔木。叶长圆状披针形或卵形。圆锥花序相对开放；花冠筒6数，具12棱，密具金色绒毛；萼片上面完全无毛；花瓣白色、浅粉色或紫色；雄蕊24-70；子房密具金色绒毛。蒴果球形至卵球状长圆形。花期5月，果期8-11月。生海拔600-1200米的疏林或溪边。产云南。缅甸、老挝、泰国和越南亦有。

Trees. Leaves oblong-lanceolate or ovate. Panicles relatively open; floral tubes 6-merous, 12-ribbed, densely golden tomentose; sepals adaxially completely glabrous; petals white, pale pink, or purple; stamens 24-70; ovary densely golden tomentose. Capsules globose to ovoid-oblong. Fl. May. Fr. Aug-Nov. Open forests or by streams at 600-1200 m. Distributed in Yunnan. Also in Myanmar, Laos, Thailand and Vietnam.

## 散沫花

**Lawsonia inermis** L.

大灌木。小枝略呈四棱形。叶交互对生，薄革质，椭圆形或椭圆状披针形。圆锥花序顶生，塔状；花4基数，极香，白色、玫瑰红色或朱红色。蒴果扁球形。花期6-10月，果期12月。中国南部各地栽培。原产东非和东南亚。

Large shrubs. Branchlets somewhat tetragonous. Leaves decussate, thinly coriaceous, elliptic or elliptic-lanceolate. Panicles terminal, pyramidal; flowers 4-merous, very fragrant, white, rose-colored or vermilion. Capsules compressed globose. Fl. Jun-Oct. Fr. Dec. Cultivated in southern parts of China. Native to E Africa and SE Asia.

散沫花 *Lawsonia inermis*

# 海桑科
# Sonneratiaceae

海桑 *Sonneratia caseolaris*

## 无瓣海桑
**Sonneratia apetala** Buch.-Ham.

乔木。植株具呼吸根。小枝下垂。叶对生；叶片狭椭圆形至披针形，长5-13厘米，宽1.5-4厘米，先端钝，侧脉5-8对，不明显。聚伞花序腋生或顶生，具花3-7朵；花萼筒高4-5毫米；无花瓣；雄蕊多数，花丝白色，长约1厘米；柱头盾状，直径可达7毫米。浆果近球形，直径1.5-3.5厘米。种子“U”形或镰形，长8-9.5毫米。花期5-12月，果期8月至翌年4月。红树林植物。生海边滩涂。广东和海南有引进栽培。原产印度、缅甸、孟加拉国和斯里兰卡。

Trees. Plants with pneumatophores. Branches pendulous. Leaves opposite; leaves narrowly elliptic to lanceolate, 5-13 cm long, 1.5-4 cm wide, apex obtuse, lateral veins 5-8 pairs, inconspicuous. Cyme axillary or terminal, 3-7 flowered; calyx tube 4-5 mm tall; petal absent; stamens numerous, filament white, ca. 1 cm long; stigma peltate, 7 mm diam. Berry spheroidal, 1.5-3.5 cm diam. Seeds U-shaped or falcate, 8-9.5 mm long. Fl. May-Dec. Fr. Aug to next Apr. Mangrove plant. Intertidal mudflat by the sea. Introduced and cultivted into Guangdong and Hainan. Native to India, Myanmar, Bangladesh and Sri Lanka.

无瓣海桑 *Sonneratia apetala*

## 海桑
**Sonneratia caseolaris** (L.) Engl. et Prantl

乔木。植株具呼吸根。小枝通常下垂。叶对生；叶片椭圆形或长圆形，长4-11厘米，宽2-7厘米，先端钝并具小尖头，侧脉7-10对，不明显。花单生或5-7朵排成聚伞花序，顶生；花萼筒高1.2-1.5厘米；花瓣条形，深红色；雄蕊多数，花丝粉红色至红色，有时上部白色，长2.5-3厘米；柱头头状。浆果球形，直径4-5厘米。种子不规则角状，长约7毫米。花期冬春季，果期春夏季。红树林植物。生海边滩涂。产海南。东南亚和澳大利亚亦有。

Trees. Plants with pneumatophores. Branches usually pendulous. Leaves opposite; leaves elliptic or oblong, 4-11 cm long, 2-7 cm wide, apex obtuse with minute, lateral veins 7-10 pairs, inconspicuous. Flower solitary or 5-7 in a cyme, terminal; calyx tube 1.2-1.5 cm tall; petal linear, dark red; stamens numerous, filament pink to red, sometimes white distally, 2.5-3 cm long; stigma capitate. Berry globular, 4-5 cm diam. Seeds irregularly angular, ca. 7 mm long. Fl. winter-spring. Fr. spring-summer. Mangrove plant. Intertidal mudflat by the sea. Distributed in Hainan. Also in SE Asia and Australia.

## 卵叶海桑
**Sonneratia ovata** Backer

乔木。叶宽卵形至近圆形，长4-10厘米，先端圆，基部宽圆形或近心形；叶柄长5-6毫米；萼筒具细疣和6棱，棱沿叶柄基部下延，萼片内面明显带红色；花瓣无，稀残留，白色，线形；花丝白色。果直径3-4.5厘米，约与萼筒等宽。花期3-10月，果期4-10月。生淡海水和淤泥中的红树林沼泽地的向陆方向的边缘。产海南。东南亚和新几内亚岛亦有。

Trees. Leaves broadly ovate to suborbicular, 4-10 cm long, apex rounded, base broadly rounded or subcordate; Petioles 5-6 mm long;

卵叶海桑 *Sonneratia ovata*

线形；花丝白色。果实直径2-4.5厘米，约与萼筒等宽。花期10-11月，果期翌年2月。生静海的浅水地带、海滨或退潮的海湾。产海南。南亚、东南亚，热带东非、澳大利亚北部、西太平洋岛屿和塞舌尔群岛亦有。

Small trees. Leaves elliptic to ovate or obovate, 5-11 cm long; Petioles 5-15 mm long. Calyx tube campanulate or obconiform, often 6-ribbed, sepals adaxially tinged red, recurved in fruit; petals white, linear; filaments white. Fruit 2-4.5 cm diam, ca. equal to width of calyx tube. Fl. Oct-Nov. Fr. Feb next year. Shallow parts of calm seas, seashores or tidal creeks. Distributed in Hainan. Also in S and SE Asia, tropical E Africa, N Australia, W Pacific Islands and Seychelles.

calyx tube finely verruculose, 6-ribbed, ribs decurrent along stipitate base; sepals adaxially strongly tinged red; petals generally absent, rarely vestigial, white, linear; filaments white. Fruit 3-4.5 cm diam, ca. equal to width of floral tube. Fl. Mar-Oct. Fr. Apr-Oct. Landward edge of mangrove swamps in brackish water and muddy soil. Distributed in Hainan. Also in SE Asia and New Guinea.

## 海南海桑

**Sonneratia × hainanensis** W. C. Ko et al.

小乔木。叶宽椭圆形或近圆形，稀宽卵形，长6.5-8厘米，宽6-8厘米，先端圆形或钝，基部短渐狭；叶柄长2-7毫米。萼筒长1.2-1.5厘米，具6棱，萼片内面红色，直立至开展，部分包被成熟果实；花瓣白色，披针形，长2.5-3厘米；花丝白色。果实直径5-6厘米，约与萼筒等宽。生红树林中。产海南(文昌县)。

Small trees. Leaves broadly elliptic or suborbicular, rarely broadly ovate, 6.5-8 cm long, 6-8 cm wide, apex rounded or obtuse, base shortly attenuate; Petioles 2-7 mm long. Calyx tube 1.2-1.5 cm long, 6-ribbed, sepals adaxially red, erect to spreading, partially enclosing mature fruit; petals white, lanceolate, 2.5-3 cm long; filaments white. Fruit 5-6 cm diam, ca. equal to width of calyx tube. Mangrove communities. Distributed in Hainan (Wenchang County).

海南海桑 *Sonneratia × hainanensis*

## 杯萼海桑

**Sonneratia alba** J. Sm.

小乔木。叶椭圆形至卵形或倒卵形，长5-11厘米；叶柄长5-15毫米。萼筒钟形或倒圆锥形，常具6棱，萼片内面红色，果期外卷；花瓣白色，

杯萼海桑 *Sonneratia alba*

拟海桑 *Sonneratia × gulngai*

## 拟海桑

**Sonneratia × gulngai** N. C. Duke et B. R. Jackes

乔木。叶倒卵形，长5-9(-11)厘米，先端急尖至钝，基部宽渐狭；叶柄长3-5毫米。萼筒无棱，萼片内面明显红色，花期直立，果期直立至部分开展；花瓣红色，条形，长4-5厘米；花丝红色。果实直径3-5厘米，约与萼筒等宽。花期12月和翌年3月，果期翌年3月和8月。生红树林中。产海南。印度尼西亚、马来西亚和澳大利亚东部亦有。

Trees. Leaves obovate, 5-9(-11) cm long, apex acute to obtuse, base broadly attenuate; Petioles 3-5 mm long. Calyx tube not ribbed, sepals adaxially usually strongly tinged red, erect at anthesis, erect to partially spreading in fruit; petals red, linear, 4-5 cm long; filaments red. Fruit 3-5 cm diam, ca. equal to width of calyx tube. Fl. Dec and next Mar. Fr. next Mar and Aug. Mangrove communities. Distributed in Hainan. Also in Indonesia, Malaysia and E Australia.

## 八宝树

**Duabanga grandiflora** (Roxb. ex DC.) Walp.

乔木。植株具小的板状根。枝条下垂，螺旋状或轮生于树干上，幼时具4棱。叶排成二列，大，长12-20厘米，宽5-7厘米，侧脉20-24对。伞房花序顶生，具花3-20朵；花萼筒高约1厘米；花瓣6，倒卵形，白色；雄蕊多数，排成2轮；花柱长3-4厘米；柱头微裂。蒴果近球形，6-9瓣裂。花期3-4月，果期5-8月。生海拔900-1500米的开阔地或溪岸。产云南南部。南亚和东南亚亦有。

Trees. Plants with small buttresses. Branches pendulous, spirally or whorled on trunk, tetragonous when young. Leaves distichous, large, 12-20 cm long, 5-7 cm wide, lateral veins 20-24 pairs. Corymb terminal, 3-20 flowered; calyx tube ca. 1 cm tall; petals 6, obovate, white; stamens numerous, 2-ranked; styles 3-4 cm long, stigmas slightly lobed. Capsules subglobose, 6-9-valved. Fl. Mar-Apr. Fr. May-Aug. Open places or river banks at 900-1500 m. Distributed in S Yunnan. Also in S and SE Asia.

八宝树 *Duabanga grandiflora*

# 隐翼科 Crypteroniaceae

## 隐翼木

**Crypteronia paniculata** Blume

乔木。小叶椭圆形至矩圆形或卵状矩圆形。圆锥花序腋生或顶生，稍下垂，无明显总梗；轴上苞片常早落；花多数，多达150朵，密生；雄蕊5，多退化且长期反折于雌花中；子房2(或3)室。蒴果两侧压扁，2瓣裂。花期7-8月，果期9-11月。生海拔300-1300米的潮湿热带雨林。产云南南部。南亚和东南亚亦有。

Trees. Leaflets elliptic to oblong, or ovate-oblong. Panicles axillary or terminal, ± pendulous, without definite peduncles; bracts of axes usually caducous; flowers many, up to 150, dense; stamens 5, reduced and mostly permanently inflexed in pistillate flowers; ovary 2(or 3)-loculed. Capsules laterally compressed, valves 2. Fl. Jul-Aug. Fr. Sep-Nov. Humid rain forests at 300-1300 m. Distributed in S Yunnan. Also in S and SE Asia.

# 安石榴科 Punicaceae

## 石榴

**Punica granatum** L.

灌木或小乔木。叶对生，披针形、椭圆状倒披针形或长圆形，纸质。花大；花瓣5-9，红色、黄色或白色。果实球形，革质浆果，红色至黄绿色或红褐色，不规则开裂。种子倒圆锥体形，具多汁囊状层。花期3-7月。常见果树。中国各地均有栽培；归化于华西。可能汉朝从中亚或欧洲引进中国。

Shrubs or small trees. Leaves opposite, lanceolate, elliptic-oblanceolate, or oblong, papery. Flowers large; petals 5-9, red, yellow or white. Fruits globose, leathery berries, red to yellow-green or red-brown, irregularly dehiscent. Seeds obpyramidal within juicy sarcotestal layer. Fl. Mar-Jul. Common fruits trees. Widely cultivated in China; naturalized in W China. Probably introduced to China from C Asia or Europe during the Han dynasty.

隐翼木 *Crypteronia paniculata*

石榴 *Punica granatum*

# 玉蕊科 Lecythidaceae

## 滨玉蕊(棋盘角)

**Barringtonia asiatica** (L.) Kurz

常绿乔木。叶大，丛生枝顶，有短柄，近革质，狭倒卵形，全缘。总状花序直立；花瓣4，椭圆形；雄蕊多数6轮，花丝长8-12厘米，粉红色或白色。果实卵形或近圆锥形，具4棱。生滨海地区。花果期全年。产台湾南部。亚洲、东非和澳大利亚的热带地区广布。

Evergreen trees. Leaves very large, clustered near top of branchlets, shortly petiolate, subcoriaceous, narrowly obovate, margin entire. Racemes erect; petals 4, elliptic; stamens numerous, in 6 whorls, filaments 8-12 cm long, pinkish or white. Fruits ovoid or subconical, 4-ridged. Coastal regions. Fl. and fr. all year. Distributed in S Taiwan. Widely distributed in tropical areas of Asia, E Africa and Australia.

滨玉蕊(棋盘角) *Barringtonia asiatica*

## 玉蕊

**Barringtonia racemosa** (L.) Spreng.

灌木或乔木。分枝下垂。叶倒卵状长圆形，先端锐尖或渐尖。总状花序常顶生或腋生于落叶，下垂，多花；花梗长达2.5厘米；花瓣4，绿色或带红色或黄色。果实卵球状圆柱形，具4棱。花果期几全年。生溪岸、潮汐河流或河口湾。产海南和台湾。日本和澳大利亚亦有。

Shrubs or trees. Branches pendulous. Leaves obovate-oblong, apex acute or acuminate. Racemes usually terminal or in axils of fallen leaves, pendulous, many flowered; pedicels 2.5 cm long; petals 4, green or tinged red or yellow. Fruits ovoid-cylindric, 4-angled. Fl. and fr. almost all year. Seashores, along tidal rivers or estuaries. Distributed in Hainan and Taiwan. Also in Japan and Australia.

玉蕊 *Barringtonia racemosa*

## 梭果玉蕊

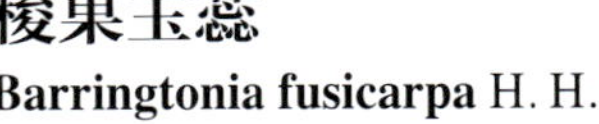

**Barringtonia fusicarpa** H. H. Hu

常绿大乔木。叶倒卵状长圆形至长圆形，纸质，通常长15-30(-36)厘米；侧脉14-15对。穗状花序顶生或腋生于老枝，长达100厘米，下垂；花萼花期2-4浅裂；花瓣4，白色或带粉红色。果棕色，梭形，外果皮具粉。花果期全年。生海拔100-800米的密林中或潮湿地。产云南南部和东南部。

Evergreen large trees. Leaves obovate-oblong to oblong, papery, usually 15-30(-36) cm long, lateral veins 14-15 pairs. Spikes terminal or axillary on old branches, to 100 cm long, pendulous; calyx becoming 2-4-lobed at anthesis; petals 4, white or pink-white. Fruits brown, fusiform, exocarps pulverulent. Fl. and fr. all year. Dense forests or moist places at 100-800 m. Distributed in S and SE Yunnan.

梭果玉蕊 *Barringtonia fusicarpa*

# 红树科
# Rhizophoraceae

## 红树
**Rhizophora apiculata** Blume

乔木或灌木。叶具柄，革质，交互对生，椭圆形，全缘。聚伞花序腋生，有花2朵；花萼4深裂，革质；花瓣4，膜质，早落；雄蕊约12。果实倒梨形，略粗糙；胚轴圆柱状棒形，长20-40厘米。花果期几全年。生海湾红树林内。产广东和海南。亚洲、澳大利亚和太平洋岛屿热带海岸亦有。

Trees or shrubs. Leaves petiolate, coriaceous, decussate, elliptic, margin entire. Cymes axillary, with 2 flowers; calyx 4-divided, coriaceous; petals 4, membranous, caducous; stamens ca. 12. Fruits obpyriform, slightly coarse; hypocotyl cylindric-clavate, 20-40 cm long. Fl. and fr. almost all year. Mangrove forests. Distributed in Guangdong and Hainan. Also in tropical seacoast of Asia, Australia and Pacific Islands.

红树 *Rhizophora apiculata*

## 角果木
**Ceriops tagal** (Perr.) C. B. Rob.

灌木或乔木。叶交互对生，集生枝端，革质，倒卵形，全缘。聚伞花序腋生，具总花梗；花小，2-4朵，花萼5-6深裂，革质；花瓣白色。果实圆锥状卵形；胚轴长棒状，长15-30厘米。花期秋冬季，果期冬季。生海湾红树林内。产广东、海南和台湾。非洲、亚洲和澳大利亚热带海岸亦有。

角果木 *Ceriops tagal*

Shrubs or trees. Leaves decussate, clustered near the top, coriaceous, obovate, margin entire. Cymes axillary, pedunculate; flowers 2-4, small; calyx 5-6 divided, coriaceous; petals white. Fruits conical-ovoid; hypocotyl long cylindrical, 15-30 cm. long. Fl. autumn-winter. Fr. winter. Mangrove forests. Distributed in Guangdong, Hainan and Taiwan. Also in tropical seacoast of Africa, Asia and Australia.

秋茄树 *Kandelia obovata*

## 秋茄树

**Kandelia obovata** Sheue, H. Y. Liu et J. Yong

灌木或小乔木，高1-3(-8)米。叶对生；叶片椭圆形、长圆形或倒卵状长圆形，长4-12厘米，宽2-5厘米，边缘全缘，先端钝、圆或微凹，质厚。花序腋生，2或3次二歧分枝，具花4-9朵；花萼乳白色，无毛；花萼裂片5或6，线形，先端渐尖，花后反折；花瓣5(或6)，白色；雄蕊多数，长6-13毫米；子房下位，1室。果实卵球形，长1.5-2.5厘米，直径约1厘米，不开裂。种子1，胎生。胚轴棒状，长15-23厘米。花果期几全年。红树林植物。生海边滩涂。产广西、广东、海南、台湾和福建。日本南部亦有。

Shrubs or small trees, 1-3(-8) m tall. Leaves opposite, leaves elliptic, oblong, or obovate-oblong, 4-12 cm long, 2-5 cm wide, margin entire, apex obtuse, rounded, or sometimes slightly emarginated, thick. Inflorescences axillary, 2 or 3 times dichotomously branched, 4-9 flowered; calyx cream colored, glabrous; lobes 5 or 6, linear, apex acuminate, reflexed after anthesis; petals 5 (or 6), white; stamens numerous, 6-13 mm long; ovary inferior, 1-loculed. Fruits ovoid, 1.5-2.5 cm long, ca. 1 cm daim, indehiscent. Seed 1, viviparous. Hypocotyl clavate, 15-23 cm long. Fl. and fr. almost all year. Mangrove plant. Intertidal mudflat by the sea. Distributed in Guangxi, Guangdong, Hainan, Taiwan and Fujian. Also in S Japan.

## 木榄

**Bruguiera gymnorrhiza** (L.) Savigny

乔木。植株具支柱根及膝状呼吸根。叶对生；叶片椭圆状长圆形，长8-21厘米，宽4-9厘米，边缘全缘，先端急尖，革质。花单生，腋生；花萼筒状，粉红色至紫红色；花萼裂片10-14，线形，表面无毛；花瓣12-14，外部边缘具白色丝状毛，先端具刺毛；雄蕊24-28，长8-11毫米；子房下位；花柱线形，柱头3-4裂。果实与萼筒贴生。种子1，胎生。胚轴圆柱状，长15-25厘米。花期5-6月。红树林植物。生海边滩涂。产广西、广东、海南、台湾和福建。南亚、东南亚、东非和太平洋岛屿亦有。

Trees. Plants with prop roots and kneelike pneumatophores. Leaves opposite; leaves elliptic-oblong, 8-21 cm long, 4-9 cm wide, margin entire, apex acute, leathery. Flowers solitary, axillary; calyx tubular pinkish to purplish red; calyx lobes 10-14, linear, glabrous; petals 12-14, outer margin with white silky hairs, apex with bristles; stamens 24-28, 8-11 mm long; ovary inferior; style filiform; stigma lobes 3-4. Fruit adnate to calyx tube. Seed 1, viviparous. Hypocotyl terete, 15-25 cm long. Fl. May-Jun. Mangrove plant. Intertidal mudflat by the sea. Distributed in Guangxi, Guangdong, Hainan, Taiwan and Fujian. Also in S and SE Asia, E Africa and Pacific Islands.

## 竹节树

**Carallia brachiata** (Lour.) Merr.

乔木，高达10米。叶对生；叶片椭圆形、倒卵形或倒披针形，长5-15厘米，宽2-10厘米，边缘全缘，稀具细齿，先端急尖或具短尖头，革质或近革质。花序为二歧聚伞花序，腋生；花具短梗或无梗；花萼钟状；花萼裂片6或7，三角形；花瓣6或7，白色，近圆形，直径约1.5毫米，边缘不规则撕裂状；雄蕊12-14；子房球状；柱头圆盘状；果粉色至红色，球形，直径约5毫米。花期冬季至春季，果期春季至夏季。生海拔900米以下的常绿阔叶林、灌丛或沼泽地。产云南、广西、广东、海南和福建。南亚、东南亚、马达加斯加、澳大利亚北部和太平洋岛屿亦有。

Trees, up to 10 m tall. Leaves opposite; leaves elliptic, obovate or oblanceolate, 5-15 cm long, 2-10 cm wide, margin entire or rare serrate, apex acute to shortly acuminate, leathery or subleathery. Inflorescences dichasial

木榄 *Bruguiera gymnorrhiza*

竹节树 *Carallia brachiata*

cymes, axillary; flowers shortly pedicellate or sessile; calyx campanulate, calyx lobes 6 or 7, deltoid; petals 6 or 7, white, suborbiculate, 1.5 mm diam, margin unevenly lacerate; stamens 12-14; ovary bulbous; stigma discoid. Fruits pink to red, globose, ca. 5 mm diam. Fl. winter-spring. Fr. spring-summer. Evergreen forests, thickets or swamps below 900 m. Distributed in Yunnan, Guangxi, Guangdong, Hainan and Fujian. Also in S and SE Asia, Madagascar, N Australia and Pacific Islands.

## 锯叶竹节树
**Carallia diplopetala** Hand.-Mazz.

乔木或灌木。枝及小枝上的皮孔显著。叶对生；叶片长圆形，长9-11厘米，宽2-3厘米，边缘具明显的锯齿，先端急尖或具小尖头，纸质。花序为二歧聚伞花序，腋生；花梗长约5毫米；花萼钟状；花萼裂片7，三角状卵形；花瓣14，排成二轮，玫瑰红色，卵圆形，边缘不规则撕裂状；雄蕊7或14；花柱短于花萼；柱头盘状。果实红色、紫红色或黑色，球形至椭圆体形。花期秋冬季，果期春季。生海拔300-1000米的林中或灌丛。产云南、广西和广东。越南亦有。

Trees or shrubs. Lenticels conspicuous on branches and branchlets. Leaves opposite; leaves oblong, 9-11 cm long, 2-3 cm wide, margin finely denticulate, apex acute to shortly acuminate, papery. Inflorescences dichasial cymes, axillary; pedicel ca. 5 mm long; calyx campanulate; calyx lobes 7, triangular-ovate; petals 14, in 2 whorls, rose red, ovate, margin unevenly lacerate; stamens 7 or 14; style shorter than calyx; stigma platelike. Fruits red, purplish red, or black, globose to ellipsoid. Fl. autumn-winter. Fr. spring. Forests or shrublands at 300-1000 m. Distributed in Yunnan, Guangxi and Guangdong. Also in Vietnam.

锯叶竹节树 *Carallia diplopetala*

## 山红树
**Pellacalyx yunnanensis** H. H. Hu

乔木，高达15米。小枝空心。叶片倒卵状披针形或椭圆形，下部具柔毛，边缘具细锯齿，干后稍反折。果实单生，近球形，宿存花萼裂片6或7，生果实顶端，披针形。果期冬季。生海拔800-1200米的林中。产云南南部。

Trees, up to 15 m tall. Branchlets hollowed. Leaves obovate-lanceolate to elliptic, abaxially pubescent, margin serrulate and slightly reflexed when dried. Fruits solitary, subglobose; persistent calyx lobes 6 or 7, at fruits apex, lanceolate. Fr. winter. Forests at 800-1200 m. Distributed in S Yunnan.

山红树 *Pellacalyx yunnanensis*

# 紫树科 Nyssaceae

蓝果树 *Nyssa sinensis*

## 喜树
**Camptotheca acuminata** Decne.

落叶乔木。幼枝紫色。叶长圆状卵形或长圆状椭圆形，纸质，全缘，侧脉11-15对。头状花序近球形；花瓣5，浅绿色；雄蕊10，外轮5枚稍长于花瓣。翅果长圆形。种子1粒。花期5-7月，果期9月。生海拔1000米以下的林缘或溪边。产中国西南、华南、东南、华中和华东。

Deciduous trees. Young branchlets purplish. Leaves oblong-ovate or oblong-elliptic, papery, entire, lateral veins 11-15 pairs. Capitulum subglobose; petals 5, light green; stamens 10, outer 5 longer than petals. Samaras oblong. Seeds 1. Fl. May-Jul. Fr. Sep. Forest edges or along streams below 1000 m. Distributed in SW, S, SE, C and E China.

喜树 *Camptotheca acuminata*

## 蓝果树
**Nyssa sinensis** Oliv.

乔木，高达20米。小枝、叶柄和花梗幼时具柔毛，老时近无毛。叶纸质至近革质，具6-10对侧脉。花序伞形或总状；花具柄。核果椭圆体形或长圆状倒卵球形，幼时紫绿，成熟时蓝色后变褐色。花期4月，果期9月。生海拔300-1700米的湿润杂木林。产中国西南、华南、东南、华中和华东。越南亦有。

Trees, up to 20 m tall. Branchlets, petioles and pedicels pilose when young, subglabrous when old. Leaves papery to subleathery, lateral veins 6-10 pairs. Inflorescences umbels or racemes; flowers pedicellate. Drupes ellipsoid or oblong-obovoid, purple-green when young, blue and then becoming brown when mature. Fl. Apr. Fr. Sep. Wet mixed forests at 300-1700 m. Distributed in SW, S, SE, C and E China. Also in Vietnam.

## 华南蓝果树
**Nyssa javanica** (Blume) Wanger.

落叶乔木。当年生小枝密具绒毛，渐无毛。叶常簇生于近枝末端；叶柄长1.5-3.5厘米；叶倒披针形或长圆状倒卵形，革质。花为腋生头状，近球形；花瓣4或5，淡黄色或淡绿色。核果紫色，椭圆体形。花期4-5月，果期10月。生海拔100-2500米的常绿林中。产云南、广西、广东和海南。南亚和东南亚亦有。

Deciduous trees. Current year twigs densely tomentose, glabrescent. Leaves often crowded near ends of branches; petioles 1.5-3.5 cm long; leaves oblance-

华南蓝果树 *Nyssa javanica*

云南蓝果树 *Nyssa yunnanensis*

olate or oblong-obovate, leathery. Flowers in ± globose axillary heads; petals 4 or 5, yellowish or greenish. Drupes purple, ellipsoid. Fl. Apr-May. Fr. Oct. Evergreen forests at 100-2500 m. Distributed in Yunnan, Guangxi, Guangdong and Hainan. Also in S and SE Asia.

## 云南蓝果树

**Nyssa yunnanensis** W. Q. Yin ex H. N. Qin et Phengklai

乔木。树皮深棕色。小枝、花梗和叶下始终具微绒毛。叶厚革质，椭圆形，长15-22厘米，背面密被微绒毛。花单性，异株。核果4或5个聚成头状，紫褐色，微被绒毛。花期3月，果期9月。生海拔500-1100米的山谷密林。产云南南部。

Trees. Bark dark brown. Branchlets, pedicels and leaves abaxially persistently pilose. Leaves thickly leathery, elliptic, 15-22 cm long, abaxially densely tomentulose. Flowers unisexual, dioecious. Drupes 4 or 5 in head, purple-brown, tomentulose. Fl. Mar. Fr. Sep. Dense forests in valleys at 500-1100 m. Distributed in S Yunnan.

## 珙桐

**Davidia involucrata** Baill.

乔木。叶纸质，背面密被柔毛，边缘有三角形锐重锯齿。花序头状，具长的总梗，下面有总苞，总苞片2-3，膜质，椭圆形或长圆状卵形，幼时淡绿色，成熟时黄褐色。核果狭卵球形。花期4月，果期10月。生海拔1500-2200米的杂木林。产云南、四川、贵州、湖北和湖南。

Trees. Leaves papery, abaxially densely pubescent, sharply and doubly triangular-serrate at margin. Inflorescences capitate, with long peduncle and an involucre below it, the involucral bracts 2-3, membranous, elliptic or oblong-ovate, light green to white when young, yellow-brown when mature. Drupes narrowly ovoid. Fl. Apr. Fr. Oct. Mixed forests at 1500-2200 m. Distributed in Yunnan, Sichuan, Guizhou, Hubei and Hunan.

珙桐 *Davidia involucrata*

光叶珙桐 *Davidia involucrata* var. *vilmoriniana*

### 光叶珙桐
**Davidia involucrata** Baill. var. **vilmoriniana** (Dode) Wanger.

本变种与原变种的区别在于本变种的下面幼时无毛或疏具柔毛，有时下面被白霜。花期4-5月，果期8-10月。生海拔1500-2000米的林中。产四川、贵州和湖北西部。

This variety differs from the typical variety in its abaxially glabrous or scarcely pubescent when young, sometimes abaxially glaucous. Fl. Apr-May. Fr. Aug-Oct. Forests at 1500-2000 m. Distributed in Sichuan, Guizhou and W Hubei.

# 八角枫科 Alangiaceae

### 土坛树
**Alangium salviifolium** (L. f.) Wanger.

落叶灌木或乔木，高可达20米。小枝有时具刺。叶长圆形至长圆状披针形，纸质或近革质，近光滑，基部楔形，全缘。花序近无梗，常4-8朵花簇生，花少，有时单花，密被锈色绒毛；萼齿5-10；花瓣4-6(-10)，外被绒毛；雄蕊10-30；子房一室，下垂。核果成熟时红色，近球形。花期2月，果期3月。生海拔1200米以下的林中。产广西南部、广东和海南。南亚、东南亚和非洲东南部亦有。

Deciduous shrubs or trees, to 20 m tall. Branchlets sometimes spinescent. Leaves oblong to oblong-lanceolate, papery to subcoriaceous, glabrescent, base cuneate, margin entire. Inflorescences sessile, often clusters of 4-8 flowers, usually fewer, sometimes only a solitary flower, densely rusty tomentose; calyx lobes 5-10 toothed; petals 4-6 (-10), outside tomentose; stamens 10-30; ovule 1 per locule, pendulous. Drupes red when ripe, subglobose. Fl. Feb. Fr. Mar. Forests below 1200 m. Distributed in S Guangxi, Guangdong and Hainan. Also in S and SE Asia, and SE Africa.

### 高山八角枫
**Alangium alpinum** (C. B. Clarke) W. W. Sm. et Cave

落叶乔木，高可达12米。叶近圆形、卵形或长圆状卵形，纸质，叶背被稀疏毛。花序腋生，3花；花萼6-8裂；花瓣6-8，线形；花序腋生，3花；花萼6-8裂；花瓣6-8，线形；花丝无毛。核果椭圆形。种子1。花期1-8月，果期10-11月。生海拔1800-3000米的林中。产西藏南部和云南西部。印度北部、尼泊尔东部、不丹和缅甸北部亦有。

Deciduous trees, to 12 m tall. Leaves suborbicular, ovate, or oblong-ovate, papery, abaxially sparsely hairy. Inflorescences axillary, 3-flowered; calyx lobes 6-8; petals 6-8, linear; filaments glabrous. Drupes ellipsoid. Seed 1. Fl. Jan-Aug. Fr. Oct-Nov. Forests at 1800-3000 m. Distributed in S Xizang, W Yunnan. Also in N India, E Nepal, Bhutan and N Myanmar.

土坛树 *Alangium salviifolium*

高山八角枫 *Alangium alpinum*

八角枫 *Alangium chinense*

## 三裂瓜木

**Alangium platanifolium** (Siebold et Zucc.) Harms var. **trilobum** (Miq.) Ohwi

灌木。叶腹面灰色，背面绿色，心状圆形，薄膜质，腹面明显具短柔毛，背面常稍具短柔毛，3-5 (-7)浅裂。花序疏散少花；花瓣白色，花期反卷，条形；雄蕊12。核果蓝色，椭圆体形，无毛。花期3-6月，果期7-9月。生海拔2000米以下的山地林中或疏松和肥沃的土壤。产中国西南、东南、华中、华北、华西、华东和东北。朝鲜半岛和日本亦有。

Shrubs. Leaves abaxially pale, adaxially green, cordate-orbicular, thinly membranous, abaxially usually prominently shortly pubescent, adaxially usually slightly shortly pilose, shallowly 3-5(-7) -lobed. Inflorescences loosely few-flowered; petals white, revolute at anthesis, linear; stamens 12. Drupes blue, ellipsoid, glabrous. Fl. Mar-Jun. Fr. Jul-Sep. Woods in mountains or loose and fertile soils below 2000 m. Distributed in SW, SE, C, N, W, E and NE China. Also in Korean Peninsula and Japan.

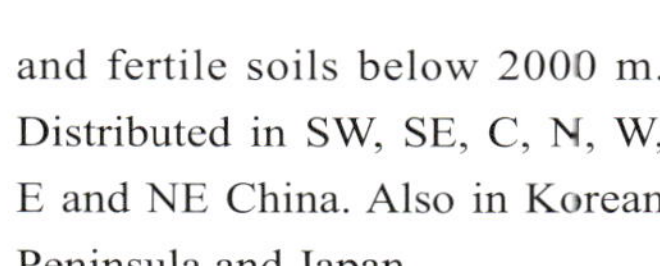

## 八角枫

**Alangium chinense** (Lour.) Harms

灌木或小乔木。叶近圆形、卵形或椭圆形，基部歪斜，全缘或3-7 (-9)裂。腋生聚伞花序具3-15花；花瓣镊合状，6(-8)片，披针形，长1-1.5厘米；雄蕊附属物无毛。核果卵球形。种子1。花期5-7月和9-10月，果期7-11月。生海拔2500米以下的山地或疏林。产中国大部分地区。东南亚和东非亦有。

Shrubs or small trees. Leaves suborbicular, ovate or elliptic, base oblique, entire or 3-7(-9)-lobed. Inflorescences axillary cymes, 3-15-flowered; petals valvate, 6(-8), lanceolate, 1-1.5 cm long; connectives of stamens glabrous. Drupes ovoid. Seed 1. Fl. May-Jul and Sep-Oct. Fr. Jul-Nov. Mountains or sparse forests below 2500 m. Distributed in most parts of China. Also in SE Asia and E Africa.

三裂瓜木 *Alangium platanifolium* var. *trilobum*

毛八角枫 *Alangium kurzii*

## 毛八角枫
**Alangium kurzii** Craib

直立乔木或灌木，高可达28米。叶近圆形或阔卵形，纸质，腹面浅绿并具柔毛；叶柄长2.5-4厘米。花5-7朵聚生，具强烈芳香；花瓣黄色、深黄、橙色或砖红色，极少白色；子房2室。成熟核果深紫色至近黑色，椭圆体形。花期5-6月，果期9月。生海拔600-1600米的林中或灌丛。产中国西南、华南、东南、华中、华北和华东。南亚、东南亚和东北亚亦有。

Erect trees or shrubs, to 28 m tall. Leaves suborbicular or broadly ovate, papery, abaxially light green and pubescent; petioles 2.5-4 cm long. Inflorescences 5-7-flowered, flowers strongly fragrant; petals yellow, dark yellow, orange, or brick red, rarely white; ovary 2-loculed. Mature drupes dark violet to nearly black, ellipsoid. Fl. May-Jun. Fr. Sep. Forests or thickets at 600-1600 m. Distributed in SW, S, SE, C, N and E China. Also in S, SE and NE Asia.

## 小花八角枫
**Alangium faberi** Oliv.

落叶灌木，1-4米高。树皮光滑，灰褐色或深褐色。叶薄纸质至膜质，3裂或全缘，长圆形或披针形，幼时具柔毛，渐无毛。聚伞花序具花5-20朵；花瓣条形，里外被毛。核果成熟时浅紫色，近卵球形或卵球状椭圆体形。花期6月，果期9月。生海拔1600米以下的林中或灌丛。产中国西南、华南和华中。

Deciduous shrubs, 1-4 m tall. Bark smooth, greyish brown or dark brown. Leaves thinly papery to membranous, 3-lobed or entire, oblong or lanceolate, pubescent when young, glabrescent. Cymes 5-20-flowered; petals linear, outside and inside with hairs. Drupes light purple when mature, subovoid or ovoid-ellipsoid. Fl. Jun. Fr. Sep. Forests or thickets below 1600 m. Distributed in SW, S and C China.

小花八角枫 *Alangium faberi*

云南八角枫 *Alangium yunnanense*

## 云南八角枫

**Alangium yunnanense** C. Y. Wu ex W. P. Fang et al.

灌木或小乔木。小枝浅紫色。小叶近圆形，纸质，两面均具柔毛，具有掌状5脉。花序腋生，具有7-15花；花瓣6-10，长约7毫米；花柱具柔毛。核果椭圆体形。花期4-5月，果期8-9月。生海拔约1400米的林中。产云南中部。

Shrubs or small trees. Branchlets light purple. Leaves suborbicular, papery, both surfaces pubescent, palmately 5-veined. Inflorescences axillary, 7-15-flowered; petals 6-10, ca. 7 mm long; styles pubescent. Drupes ellipsoid. Fl. Apr-May. Fr. Aug-Sep. Forests at ca. 1400 m. Distributed in C Yunnan.

## 髯毛八角枫

**Alangium barbatum** Baill. ex Kuntze

灌木或小乔木。叶阔椭圆形或卵状长圆形，两面被黄色硬毛和微绒毛，基部心形或圆形，全缘稀1裂或具1锯齿。花序二叉分枝，3-15花或更多；花白色或黄色，花瓣4-7；花药内部具软毛。核果卵球形或椭圆体形。种子1。花期6-8月，果期10月至翌年3月。生海拔1000米以下的林中。产云南、广东和广西。印度、缅甸、老挝、泰国和越南亦有。

Shrubs or small trees. Leaves broadly elliptic or ovate-oblong, yellow-wirehaired and yellow-tomentulose on both surfaces, base cordate or rounded, margin entire, rarely 1-lobed or 1-toothed. Inflorescences branched binate, 3-15 or more flowered; flowers white or yellow, petals 4-7, anthers inside with soft hairs. Drupes ovate or ellipsoid. Seeds 1. Fl. Jun-Aug. Fr. Oct to next Mar. Forests below 1000 m. Distributed in Yunnan, Guangdong and Guangxi. Also in India, Myanmar, Laos, Thailand and Vietnam.

髯毛八角枫 *Alangium barbatum*

# 使君子科 Combretaceae

### 榆绿木
**Anogeissus acuminata** (Roxb. ex DC.) Guillem. et Perr.

乔木，高达20米。小枝稍下垂，细弱，幼时与叶柄和叶被金色长柔毛。叶披针形至狭披针形。花集成头状花序；苞片易早落，条形；花无柄；萼筒下面具黄色柔毛。坚果具翅及宿存萼管。花期2-3月。生海拔700米以下的石灰山林下。产云南。南亚亦有。

Trees, to 20 m tall. Branchlets slightly pendent, slender, together with petioles and leaves golden villous when young. Leaves lanceolate to narrowly lanceolate; flowers in a capitulum; bracts easily deciduous, linear; flowers sessile; calyx tubes abaxially yellow pubescent. Nuts with wings and persistent calyx tube. Fl. Feb-Mar. Forests on limestone hills below 700 m. Distributed in Yunnan. Also in S Asia.

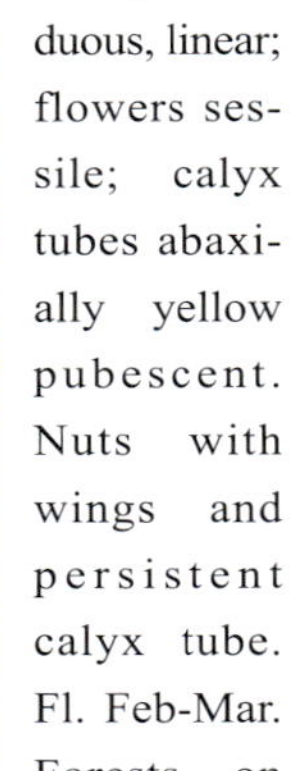

榆绿木 *Anogeissus acuminata*

### 千果榄仁
**Terminalia myriocarpa** Van. Huerck et Müe. Arg.

常绿乔木，高达35米。叶对生，长圆状椭圆形或长圆状披针形。圆锥花序大型，长18-26厘米，由密集多数的穗状花序组成。坚果有3翅，其中2翅等大，1翅特小，翅膜质，被疏毛。花期8-9月，果期10月至翌年1月。生海拔600-2100米的热带雨林。产云南、西藏和广西。南亚和东南亚亦有。热带雨林沟谷优势树种。

Evergreen trees, to 35 m tall. Leaves opposite, oblong-elliptic or oblong-lanceolate. Panicles very large, 18-26 cm long, composed of many clustered spikes. Nuts with 3 wings, 2 large and equal, 1 very small, wings membranous, sparsely hairy. Fl. Aug-Sep. Fr. Oct to next Jan. Tropical rain forests at 600-2100 m. Distributed in Yunnan, Xizang and Guangxi. Also in S and SE Asia. A dominant tree species in tropical rain forests in valleys.

千果榄仁 *Terminalia myriocarpa*

### 榄仁树
**Terminalia catappa** L.

乔木，高达20米。叶倒卵形至倒

榄仁树 *Terminalia catappa*

披针形，大，常密集于枝顶。花序腋生，单生，细弱穗状；花芳香；萼筒长7-8毫米。果实椭圆体形，压扁，明显2棱至具2狭翅。花期3-6月，果期7-9月。生海边沙滩或村庄附近。产云南、广东、海南和台湾。南亚、东南亚、马达加斯加、澳大利亚北部和太平洋岛屿亦有。

Trees, to 20 m tall. Leaves obovate to oblanceolate, big, usually clustered at branch tops. Inflorescences axillary, simple, long, slender spikes; flowers fragrant; calyx tubes 7-8 mm long. Fruits ellipsoid, compressed, strongly 2-ridged to narrowly 2-winged. Fl. Mar-Jun. Fr. Jul-Sep. Beaches or near villages. Distributed in Yunnan, Guangdong, Hainan and Taiwan. Also in S and SE Asia, Madagascar, N Australia and Pacific Islands.

## 毗黎勒

**Terminalia bellirica** (Gaertn.) Roxb.

落叶乔木，高达35米。叶倒卵形。花序腋生，单穗状，常聚生于小枝顶端形成圆锥状；萼筒上部浅杯状。果实具短柄，近球形至阔椭圆体形或倒卵球形，具浅至明显5棱。花期3-4月，果期5-7月。生海拔500-1400米的林中或山坡。产云南南部。南亚、东南亚和澳大利亚北部亦有。东非有引种。

Deciduous trees, to 35 m tall. Leaves obovate. Inflorescences axillary, simple spikes, often grouped at branchlet apex and forming a panicle; calyx tubes distally shallowly cupular. Fruits shortly stipitate, subglobose to broadly ellipsoid or ovoid, weakly to strongly 5-ridged. Fl. Mar-Apr. Fr. May-Jul. Forests or slopes at 500-1400 m. Distributed in S Yunnan. Also in S and SE Asia, and N Australia. Introduced in E Africa.

毗黎勒 *Terminalia bellirica*

诃子 *Terminalia chebula*

红榄李 *Lumnitzera littorea*

## 诃子

**Terminalia chebula** Retz.

乔木，高达30米。幼枝被绒毛。叶互生或对生；叶椭圆形。花序腋生或顶生，单穗状，具多花；花稍芳香，两性。果实不具柄，成熟后黑褐色，卵球形或阔卵球形、椭圆体形或圆柱状卵球形，具钝5棱。花期5-6月，果期7-10月。生海拔500-1800米的疏林。产云南西部；广西、广东、台湾和福建有栽培。南亚亦有。

Trees, to 30 m tall. Branchlets tomentose. Leaves alternate or opposite; leaves elliptic. Inflorescences axillary or terminal, simple spikes, numerous flowered; flowers slightly fragrant, bisexual. Fruits not stipitate, blackish brown when ripe, ovoid or broadly so, ellipsoid, or cylindric-ovoid, obtusely 5-ridged. Fl. May-Jun. Fr. Jul-Oct. Sparse forests at 500-1800 m, or cultivated. Distributed in W Yunnan; cultivated in Guangxi, Guangdong, Taiwan and Fujian. Also in S Asia.

## 红榄李

**Lumnitzera littorea** (Jack) Voigt

乔木。叶互生，常聚生枝顶。叶片肉质而厚，侧脉4-5对。总状花序顶生；花多数；小苞片2，三角形；花瓣5，红色；雄蕊5-10(常7)，约为花瓣长的2倍。果实木质，纺锤形，黑褐色。花期11-12月，5月，果期翌年6-8月。生海岸边。产海南南部。热带亚洲和太平洋岛屿亦有。

榄李 *Lumnitzera racemosa*

Trees. Leaves alternate, often clustered at top of branches. Leaves fleshy and thick, lateral veins 4-5-paired. Inflorescences terminal, racemes; flowers numerous; bracteoles 2, triangular; petals 5, red; stamens 5-10 (usually 7), ca. 2 × as long as petals. Fruits woody, fusiform, black-brown. Fl. Nov-Dec, May. Fr. next Jun-Aug. Seashores. Distributed in S Hainan. Also in tropical Asia and Pacific Islands.

## 榄李

**Lumnitzera racemosa** Willd.

常绿灌木或小乔木。叶匙形至倒披针形或倒卵形。花序腋生；花芳香，小苞片裂片三角形，花瓣白色，椭圆形。果实成熟后黑褐色，椭圆体形或卵球形，一侧稍压扁，具2或3棱。花期11月至翌年8月，果期8月至翌年4月。生红树林海岸边或咸水沼泽地。产广西、广东、海南和台湾。热带亚洲、热带东非和大洋洲亦有。

Evergreen shrubs or small trees. Leaves spatulate to oblanceolate or obovate. Inflorescences axillary; flowers fragrant, bracteoles lobes deltoid, petals white, elliptic. Fruits blackish brown when ripe, ellipsoid or ovoid, slightly compressed on one side, 2- or 3-ridged. Fl. Nov to next Aug. Fr. Aug to next Apr. Mangrove forests along sea shores or saltwater swamps. Distributed in Guangxi, Guangdong, Hainan and Taiwan. Also in tropical Asia, tropical E Africa and Oceania.

## 使君子

**Quisqualis indica** L.

藤本。叶多长圆状椭圆形或椭圆形；叶柄长，无关节。穗状花序疏生，组成伞房式花序；花芳香；花瓣开放时白色，后下面为浅黄色，上面为浅红色。果实幼时红色，成熟后黑绿色或褐色，纺锤形或狭倒卵球形，具锐5棱。花期3-11月，果期6-11月。生海拔1500米以下的疏林中、灌丛、山坡、路边或河岸边；或栽培。产中国西南、华南和东南。南亚、东南亚、东非海岸和太平洋岛屿亦有。广泛栽培于热带地区。

Lianas. Leaves mostly oblong-elliptic or elliptic; petioles long, exarticulated. Spikes sparse, in a

使君子 *Quisqualis indica*

corymb; flowers fragrant; petals opening white, later turning yellowish abaxially and reddish adaxially. Fruits red when young, greenish black or brown when ripe, fusiform or narrowly ovoid, sharply 5-ridged. Fl. Mar-Nov. Fr. Jun-Nov. Sparse forests, thickets, slopes, roadsides or riversides below 1500 m; or cultivated. Distributed in SW, S and SE China. Also in S and SE Asia, coastal E Africa and Pacific Islands. Commonly cultivated in tropical areas.

## 石风车子
**Combretum wallichii** DC.

藤本。小枝与叶柄具柔毛及密褐色鳞片，渐无毛。叶对生或互生，坚纸质。花具强烈气味；萼筒长3.5-5毫米；裂片4；花瓣4，黄色至绿色；花盘密被黄白色长硬毛。果实具4翅，具白色或金色鳞片。花期3-8月，果期7-11月。生海拔(500-) 800-2200(-3200)米的山坡、路边、沟边的杂木林中或灌丛中。产云南、四川、贵州和广西。印度、尼泊尔、不丹、孟加拉国、缅甸北部和越南北部亦有。

Lianas. Branchlets together with petioles puberulous and densely brown scaly, glabrescent. Leaves opposite or alternate, hardy papery. Flowers strongly scented; calyx tubes 3.5-5 mm long; lobes 4; petals 4, yellow to green; disk densely white-hirsute. Fruits 4-winged, white or golden scaly. Fl. Mar-Aug. Fr. Jul-Nov. Slopes, roadsides, by streams, mixed forests or thickets at (500-)800-2200(-3200) m. Distributed in Yunnan, Sichuan, Guizhou and Guangxi. Also in India, Nepal, Bhutan, Bangladesh, N Myanmar and N Vietnam.

## 西南风车子
**Combretum griffithii** Van Huerck et Müe. Arg.

木质藤本。叶对生或互生，稀三叶轮生，叶片纸质，两面无毛但背面被锈色鳞片。花序顶生及腋生，单生，狭圆柱状穗状；花瓣4，白至黄或黄绿色。果实褐色，球形，具4翅。花期4-5月，果期9-12月。生海拔(600-)1100-1600米的林下、山谷或山坡。产云南南部。南亚和东南亚亦有。

Lianas woody. Leaves opposite or alternate, rarely 3 in a whorl, leaves papery, both surfaces glabrous, but with ferruginous scales abaxially. Inflorescences terminal and axillary, simple, narrowly cylindric spikes; petals 4, white to yellow or yellowish green. Fruits brown, globose, 4-winged. Fl. Apr-May. Fr. Sep-Dec. Forests, valleys or slopes at (600-)1100-1600 m. Distributed in S Yunnan. Also in S and SE Asia.

石风车子 *Combretum wallichii*

西南风车子 *Combretum griffithii*

云南风车子 *Combretum griffithii* var. *yunnanense*

### 云南风车子
**Combretum griffithii** Van Huerck et Müe. Arg. var. **yunnanense** (Exell) Turland et C. Chen

本变种与原变种的区别在于：叶两面有柔毛，随生长渐无毛，仅在脉上具柔毛，具橙黄色鳞片。花期4-6月，果期7-12月。生海拔500-1600(-2000)米的山谷、河边或林下。产云南。缅甸和泰国亦有。

This variety differs from *Combretum griffithii* in its leaves pilose on both surfaces, glabrescent with age but remaining pilose on veins. Fl. Apr-Jun. Fr. Jul-Dec. Valleys, by rivers or forests at 500-1600(-2000) m. Distributed in Yunnan. Also in Myanmar and Thailand.

阔叶风车子 *Combretum latifolium*

### 阔叶风车子
**Combretum latifolium** Blume

大型藤本。小枝与叶柄常无毛，具鳞片。叶对生，革质，阔椭圆形或卵状椭圆形，两面无毛，背面无鳞片。花极芳香；萼筒裂片4，反折；花瓣4，绿白至黄绿色。坚果淡黄色至淡褐色，光亮，稍倒卵球形，具4翅，无鳞片。花期1-4月，果期6-10月。生海拔500-1000米的林下。产云南南部。南亚和东南亚亦有。

Lianas large. Branchlets together with petioles usually glabrous, scaly. Leaves opposite, leathery, broadly elliptic or ovate-elliptic, both surfaces glabrous, without scales abaxially. Flowers very fragrant; calyx tubes lobes 4, reflexed; petals 4, greenish white to yellowish green. Nuts yellowish to brownish, glossy, ± obovoid, 4-winged, without scales. Fl. Jan-Apr. Fr. Jun-Oct. Forests at 500-1000 m. Distributed in S Yunnan. Also in S and SE Asia.

# 桃金娘科 Myrtaceae

### 柠檬桉
**Eucalyptus citriodora** Hook.

大乔木，高可达28米。树皮光滑，灰白色。幼态叶披针形；成熟叶狭披针形，宽约1厘米，具柠檬气味。圆锥花序腋生；花白色，雄蕊多数，长6-7毫米。蒴果壶形。花期4-9月。广西、广东、福建和海南常见栽培。原产澳大利亚。

Large trees, to 28 m tall. Bark smooth, offwhite-gray. Juvenile

柠檬桉 *Eucalyptus citriodora*

赤桉 *Eucalyptus camaldulensis*

leaves lanceolate; climax leaves narrowly lanceolate, ca. 1 cm wide, scented like lemon. Inflorescences paniculate, axillary; flowers white, stamens numerous, 6-7 mm long. Capsules urceolate. Fl. Apr-Sep. Commonly cultivated in Guangxi, Guangdong, Fujian and Hainan. Native to Australia.

## 赤桉

**Eucalyptus camaldulensis** Dehnh.

大乔木。树皮平滑，暗灰色，片状脱落。幼态叶对生，阔披针形；成熟叶薄革质，狭披针形至披针形，宽1-2厘米。伞形花序腋生，有花5-8朵；花蕾卵形；帽状体尖锐。蒴果近球形，果瓣4。花期12月至翌年8月。中国南部广泛栽培。原产澳大利亚。

Large trees. Bark smooth, dark gray, exfoliating in flakes. Juvenile leaves opposite, broadly lanceolate; climax leaves thinly coriaceous, narrowly lanceolate to lanceolate, 1-2 cm wide. Inflorescences axillary, umbelliform, with 5-8 flowers; buds ovate; calyptras acute. Capsules subglobose, valves 4. Fl. Dec to next Aug. Widely cultivated in southern China. Native to Australia.

## 蓝桉

**Eucalyptus globulus** Labill.

大乔木。树皮灰蓝色，片状剥落。幼态叶对生，卵形，无柄，有白粉；成熟叶革质，披针形，镰状，宽1-2厘米。花大，宽4毫米，单生或2-3朵聚生于叶腋内；帽状体稍扁平。蒴果半球形，有4棱，宽2-2.5厘米。花期12月至翌年5月，果期冬季。云南、四川和广西栽培。原产澳大利亚。

Large trees. Bark bluish-gray, exfoliating in flakes. Juvenile leaves opposite, ovate, sessile, glaucous; climax leaves coriaceous, lanceolate, falcate, 1-2 cm wide. Flowers large, 4 mm wide, axillary, solitary or clustered in 2-3; calyptras somewhat flat. Capsules hemispheric, 4-ridged, 2-2.5 cm wide. Fl. Dec to next May. Fr. winter. Cultivated in Yunnan, Sichuan and Guangxi. Native to Australia.

## 直杆蓝桉

**Eucalyptus globulus** subsp. **maidenii** (F. Muell.) Kirkpatr.

本种与原变种的区别区别在于：伞形花序有花7朵。蒴果直径0.6-1厘米。花期12月至翌年5月，果期冬季。云南、四川、广西和江西栽培。原产澳大利亚。

This subspecies differs from *Eucalyptus globulus* in its umbels 7-flowered. Capusules 0.6-1 cm diam. Fl. Dec to next May. Fr. winter. Cultivated in Yunnan, Sichuan, Guangxi and Jiangxi. Native to Australia.

蓝桉 *Eucalyptus globulus*

直杆蓝桉 *Eucalyptus globulus* subsp. *maidenii*

红千层 *Callistemon rigidus*

垂枝红千层 *Callistemon viminalis*

## 红千层

**Callistemon rigidus** R. Br.

小乔木。小枝直立。叶互生，坚革质，线形，全缘，油腺点明显。穗状花序生于枝顶，直立；花瓣绿色，卵形；雄蕊长2.5厘米，鲜红色，花药暗紫色。蒴果半球形，先端平截。花期6-8月。云南、广西和广东栽培。原产澳大利亚。

Small trees. Branchlets erect. Leaves alternate, coriaceous, linear, margin entire, oil glands distinct. Racemes terminal, erect; petals green, ovate; stamens 2.5 cm long, scarlet, anthers dark purple. Capsules hemispheric, apex truncate. Fl. Jun-Aug. Cultivated in Yunnan, Guangxi and Guangdong. Native to Australia.

## 垂枝红千层

**Callistemon viminalis** (Sol. ex Gaertn.) G. Don ex Loudon

小乔木。小枝细长，下垂。叶互生，坚革质，细线形，灰绿色，全缘，油腺点明显。穗状花序生于枝顶，下垂；雄蕊长1.5厘米，绯红色。蒴果半球形，先端平截。花期4-9月。云南、广西和广东栽培。原产澳大利亚。

Small trees. Branchlets slender, pendulous. Leaves alternate, thick coriaceous, narrowly linear, grayish-green, margin entire, oil glands distinct. Racemes terminal, pendulous; stamens 1.5 cm long, scarlet. Capsules hemispheric, apex truncate. Fl. Apr-Sep. Cultivated in Yunnan, Guangxi and Guangdong. Native to Australia.

## 白千层

**Melaleuca cajuputi** subsp. **cumingiana** (Turcz.) Barlow

乔木。树皮灰白色，厚而松软，呈薄层状剥落。叶互生，革质，披针形或狭长圆形，具油腺点。花白色，密集

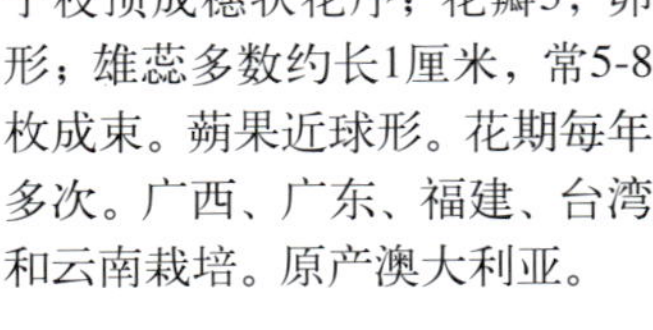

于枝顶成穗状花序；花瓣5，卵形；雄蕊多数约长1厘米，常5-8枚成束。蒴果近球形。花期每年多次。广西、广东、福建、台湾和云南栽培。原产澳大利亚。

Trees. Brak offwhite-gray, very thick, soft, peeling in many thin flakes. Leaves alternate, coriaceous, lanceolate or narrowly oblong, with numerous oil glands. Flowers white, terminal, clustered, formed spikes; petals 5, ovate; stamens numerous, ca. 1 cm long, in 5-8 bundles. Capsules subglobose. Fl. several times in a year. Cultivated in Guangxi, Guangdong, Fujian, Taiwan and Yunnan. Native to Australia.

## 岗松

**Baeckea frutescens** L.

灌木或小乔木。叶条形，下面有突起的油腺点，上面具槽。花单个腋生，小型；苞片早落；花瓣白色，离生，圆形，基部具爪；子房下位，2或3室。蒴果小。种子扁平，具棱。花期夏季。生灌丛、山

白千层 *Melaleuca cajuputi* subsp. *cumingiana*

岗松 *Baeckea frutescens*

肖蒲桃 *Syzygium acuminatissimum*

坡、低山丘或草地。产广西、广东、海南、福建、浙江和江西。南亚、东南亚和大洋洲亦有。

Shrubs or small trees. Leaves linear, abaxially with raised oil glands, adaxially grooved. Flowers axillary, solitary, small; bracts caducous; petals white, distinct, rounded, base clawed; ovary inferior, 2- or 3-loculed. Capsules small. Seeds flattened, angular. Fl. summer. Thickets, slopes, low hills or grasslands. Distributed in Guangxi, Guangdong, Hainan, Fujian, Zhejiang and Jiangxi. Also in S and SE Asia, and Oceania.

## 红果仔

**Eugenia uniflora** L.

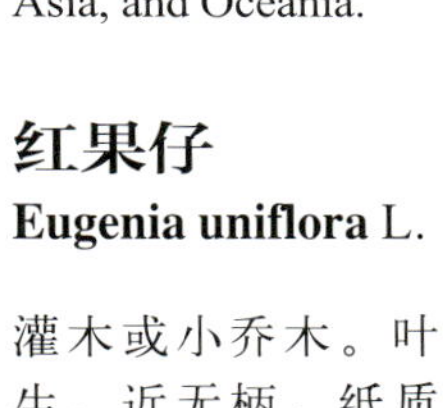

灌木或小乔木。叶对生，近无柄，纸质，卵形至卵状披针形，全缘。花白色，稍芳香，单生或数朵聚生于叶腋；萼片4；花瓣4。浆果球形，直径1-2厘米，有8棱，熟时深红色。花期春季。福建、云南、广西、广东和海南栽培。原产巴西。

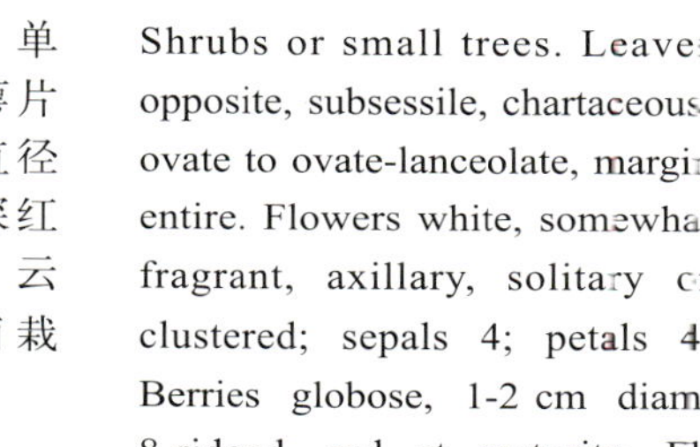

Shrubs or small trees. Leaves opposite, subsessile, chartaceous, ovate to ovate-lanceolate, margin entire. Flowers white, somewhat fragrant, axillary, solitary or clustered; sepals 4; petals 4. Berries globose, 1-2 cm diam, 8-ridged, red at maturity. Fl. spring. Cultivated in Fujian, Yunnan, Guangxi, Guangdong and Hainan. Native to Brazil.

## 肖蒲桃

**Syzygium acuminatissimum** (Blume) DC.

乔木。叶革质，卵状披针形至狭披针形，上面多油点，侧脉多而密。聚伞花序排成圆锥状，萼管倒圆锥形；花瓣小，白色，离生，长约1毫米。浆果球形，成熟时黑紫色。花期7-10月。生海拔100-600米的常绿阔叶林中。产广西、广东、海南和台湾。南亚、东南亚和太平洋岛屿亦有。

Trees. Leaves leathery, ovate-lanceolate to narrowly lanceolate, adaxially with glandular dots, lateral veins dense and many. Panicled cymes; calyx-tube obconical; petals small, white, distinct, ca. 1 mm long. Berries globose, black-purple when ripe. Fl. Jul-Oct. Evergreen broad-leaved forests at 100-600 m. Distributed in Guangxi, Guangdong, Hainan and Taiwan. Also in S and SE Asia, and Pacific Islands.

红果仔 *Eugenia uniflora*

黑嘴蒲桃 *Syzygium bullockii*

蒲桃 *Syzygium jambos*

## 黑嘴蒲桃

**Syzygium bullockii** (Hance) Merr. et Perry

灌木至乔木。叶片革质，椭圆形至卵状长圆形，基部圆形或微心形；叶柄极短，近于无柄。圆锥花序顶生；萼管倒圆锥形；花瓣连成帽状体；花柱与雄蕊同长。果实红色至黑色椭圆形。花期3-8月。生海拔100-400米的林下。产广西、广东和海南。老挝和越南亦有。

Shrubs to trees. Leaves leathery, elliptic to ovate-oblong, base rounded or slightly cordate; petioles very short to none. Inflorescences terminal, paniculate cymes; hypanthium obconic; petals coherent; style as long as stamens. Fruits red to black, ellipsoid. Fl. Mar-Aug. Forests at 100-400 m. Distributed in Guangxi, Guangdong and Hainan. Also in Laos and Vietnam.

## 轮叶蒲桃

**Syzygium grijsii** (Hance) Merr. et Perry

灌木。叶片革质，细小，常3叶轮生，狭窄长圆形或狭披针形。聚伞花序顶生，少花；花梗长3-4毫米，花白色；花瓣4，分离，近圆形，长约2毫米；雄蕊长约5毫米；花柱与雄蕊同长。果实球形。花期5-6月。生海拔200-1500米的山坡灌丛、林下、山谷或溪边。产广西、广东、海南、福建、江西、浙江和安徽。

Shrubs. Leaves leathery, thin and small, usually 3 leaves verticillate, narrowly oblong or lanceolate. Inflorescences terminal, few flowered; peduncles 3-4 mm, flowers white; petals 4, distinct, suborbicular, ca. 2 mm long; stamens ca. 5 mm long; style as long as stamens. Fruits globose. Fl. May-Jul. Forests, valley or streamsides at 200-1500 m. Distributed in Guangxi, Guangdong, Hainan. Fujian, Jiangxi, Zhejiang and Anhui.

## 蒲桃

**Syzygium jambos** (L.) Alston

乔木。茎非常短，多分枝。叶披针形、卵状披针形、长圆形或条形，基部狭至阔楔形。聚伞花序顶生；花白色或粉红色；花瓣离生。果实球状或椭圆状，具油腺，成熟时淡黄色或红色。花期3-4月，果期5-6月。生海拔100-1500米的林中、山坡、溪边或山谷。栽培于中国西南和华南。菲律宾亦有。

Trees. Stems very short, broadly

轮叶蒲桃 *Syzygium grijsii*

南洋蒲桃（莲雾） *Syzygium samarangense*

branched. Leaves lanceolate, ovate-lanceolate, oblong, or linear, base narrow to broadly cuneate. Inflorescences terminal, cymes; flowers white or pink; petals distinct. Fruits globose or ellipsoid, with oil glands, pale yellow or red when mature. Fl. Mar-Apr. Fr. May-Jun. Forests, slopes, riversides or valleys at 100-1500 m. Cultivated in SW and S China. Also in the Philippines.

## 南洋蒲桃（莲雾）

**Syzygium samarangense** (Blume) Merr. et L. M. Perry

乔木。叶椭圆形至长圆形，薄革质，下面有小腺点，侧脉14-19对。聚伞花序顶生或腋生；花白色，花瓣4。果实红色，梨形至圆锥形，肉质，直径4-6厘米，光亮，顶端压扁。花期3-4月，果期5-6月。云南、四川、广西、广东、台湾和福建栽培。原产泰国、马来西亚和印度尼西亚。

Trees. Leaves elliptic to oblong, thinly leathery, abaxially with small glands, lateral veins 14-19 per side. Inflorescences terminal or axillary, cymes; flowers white; petals 4. Fruits red, pyriform to conic, fleshy, 4-6 cm diam, glossy, apex impressed. Fl. Mar-Apr. Fr. May-Jun. Cultivated in Yunnan, Sichuan, Guangxi, Guangdong, Taiwan and Fujian. Native to Thailand, Malaysia and Indonesia.

## 硬叶蒲桃

**Syzygium sterrophyllum** Merr. et L. M. Perry

灌木或小乔木。叶片革质，狭披针形，基部狭楔形；叶柄长3-6毫米。聚伞花序腋生或生枝顶叶腋，花白色；萼管倒圆锥形，长3毫米；花瓣连合；雄蕊长3-4毫米；花柱约与雄蕊等长。果实椭圆体形，顶部冠宿存长1-1.5毫米的萼檐。花期6-10月，果期11月至翌年1月。生海拔600-1200米的山谷或河边。产云南、广西和海南。越南亦有。

Shrubs or small trees. Leaves narrowly leathery, lanceolate, base narrowly cuneate; petioles 3-6 mm long. Inflorescences axillary or in leaf axils apically on branches, cymes; flowers white; hypanthium obconic, 3 mm long; petals coherent; stamens 3-4 mm long; style equal length to stamens. Fruits ellipsoid, apically covered with 1-1.5 mm long persistent calyx limb. Fl. Jun-Oct. Fr. Nov to next Jan. Mountain valleys or riversides at 600-1200 m. Distributed in Yunnan, Guangxi and Hainan. Also in Vietnam.

## 短药蒲桃

**Syzygium globiflorum** (Craib) P. Chantar. et J. Parnell

灌木或小乔木。叶椭圆形至狭椭圆形，薄革质。圆锥花序或聚伞花序顶生，具3-11花；花序梗长1-1.5厘米；花瓣离生，阔卵形，长7-8毫米；雄蕊大小可变，花药长约0.6毫米。果实近球形。花期4-8月。生海拔200-1000(-2400)米的山谷密林中。产云南、广西和海南。泰国亦有。

Shrubs or small trees. Leaves elliptic to narrowly elliptic, thinly leathery. Inflorescences terminal, panicles or cymes, 3-11-flowered; peduncles 1-1.5 cm long; petals distinct, broadly ovate, 7-8 mm long; stamens size variable, anthers ca. 0.6 mm long. Fruits subglobose. Fl. Apr-Aug. Dense forests in valleys at 200-1000(-2400) m. Distributed in Yunnan, Guangxi and Hainan. Also in Thailand.

硬叶蒲桃 *Syzygium sterrophyllum*

短药蒲桃 *Syzygium globiflorum*

马六甲蒲桃 *Syzygium malaccense*

## 马六甲蒲桃

**Syzygium malaccense** (L.) Merr. et L. M. Perry

乔木。叶狭椭圆形至椭圆形，革质。花序侧生无叶老枝上，聚伞状，4-9朵花成簇；花序梗极短；花红色；花瓣圆形，约1 × 1厘米，离生；雄蕊完全离生。果实卵球形至坛形。花期1-2月，果期4-5月。栽培或逸生云南和台湾。可能原产于马来西亚；广泛栽培于东南亚。

Trees. Leaves narrowly elliptic to elliptic, leathery. Inflorescences lateral on older leafless branches, cymes, in 4-9-flowered clusters; peduncles very short; flowers red; petals rounded, ca. 1 × 1 cm, distinct; stamens completely distinct. Fruits ovoid to pot-shaped. Fl. Jan-Feb. Fr. Apr-May. Cultivated or escaped in Yunnan and Taiwan. Propably native to Malaysia; commonly cultivated in SE Asia.

香胶蒲桃 *Syzygium balsameum*

## 香胶蒲桃

**Syzygium balsameum** (Wight) Wall. ex Walp.

灌木或乔木。小枝稍扁。叶椭圆形至狭长圆形。圆锥花序侧生无叶的枝上；萼裂片不明显；花瓣合生；雄蕊长2-3毫米。果实球状，红色，直径5-6毫米。花期11-12月，果期翌年1-2月。生海拔500-1300米的林中或溪边。产云南南部和西藏。印度、缅甸、泰国和越南亦有。

Shrubs or trees. Branchlets slightly compressed. Leaves elliptic to narrowly oblong. Inflorescences panicles, lateral on leafless branches; calyx lobes inconspicuous; petals coherent; stamens 2-3 mm long. Fruits globose, red, 5-6 mm diam. Fl. Nov-Dec. Fr. Jan-Feb next year. Forests or riversides at 500-1300 m. Distributed in S Yunnan and Xizang. Also in India, Myanmar, Thailand and Vietnam.

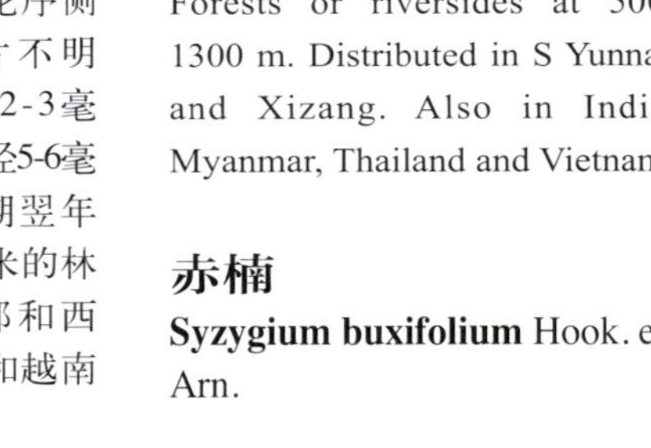

## 赤楠

**Syzygium buxifolium** Hook. et Arn.

灌木或小乔木。叶革质，下面有腺点，侧脉多而密。聚伞花序顶生，有花数朵；萼管倒圆锥形；萼裂片浅波状；花瓣白色，4，离生，长约2毫米。果红色，后为紫黑色，球形。花期6-8月，果期10-12月。生海拔200-1200米的低山疏林或灌丛中。产中国西南、华南、东南、华中和华东。越南和琉球群岛亦有。

Shrubs or small trees. Leaves leathery, abaxially glandular, lateral veins numerous and dense. Inflorescences terminal, cymes, several-flowered; hypanthium obconical; calyx lobes shallow wavy; petals white, 4, free, ca. 2 mm long. Fruits red turning purplish black, globose. Fl. Jun-Aug. Fr. Oct-Dec. Sparse forests in low mountains or thickets at 200-1200 m. Distributed in SW, S, SE, C and E China. Also in Vietnam and Ryukyu Islands.

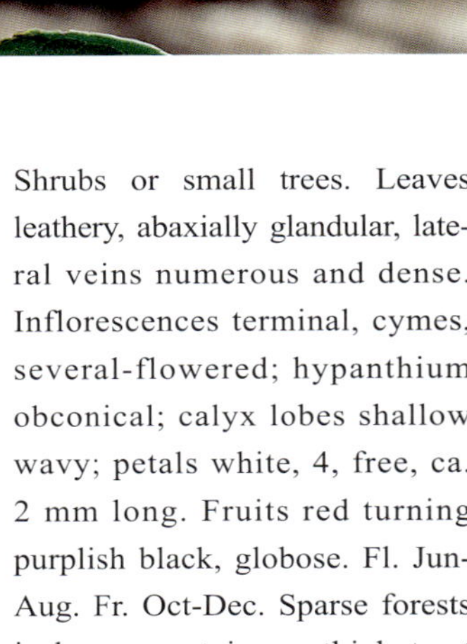

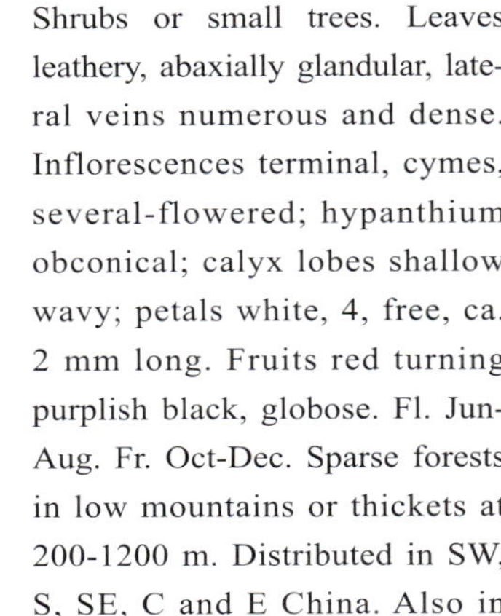

## 阔叶蒲桃

**Syzygium megacarpum** (Craib) Rathakr. et N. C. Nair

乔木。叶片狭长椭圆形至椭圆形，基部圆形至浅心形。花序顶生，聚伞状，具2-6花；花序梗极短；花白色，大；萼管长，为倒圆锥形，长1.5-2厘米；花瓣离生，圆形，长约2厘米。果实卵球形。花期4-10月，果期7-10月。生海拔300-1200米的河边密林下。产云南、广西和海南。孟加拉国、缅甸、泰国和越南亦有。

赤楠 *Syzygium buxifolium*

阔叶蒲桃 *Syzygium megacarpum*

贵州蒲桃 *Syzygium handelii*

乌墨 *Syzygium cumini*

Trees. Leaves narrowly long elliptic to elliptic, base round to shallowly cordate. Inflorescences terminal, cymes, 2-6-flowered; peduncles very short; flowers white, large; hypanthium long obconic, 1.5-2 cm long; petals distinct, rounded, ca. 2 cm long. Fruits ovoid. Fl. Apr-Oct. Fr. Jul-Oct. Dense forests by rivers at 300-1200 m. Distributed in Yunnan, Guangxi and Hainan. Also in Bangladesh, Myanmar, Thailand and Vietnam.

## 贵州蒲桃

**Syzygium handelii** Merr. et L. M. Perry

灌木。叶柄长3-4毫米；叶披针形至狭长圆形。圆锥花序顶生，花序轴有棱；花萼管倒圆锥状，长约3毫米，光滑；萼裂片不明显；花瓣常4，分离，阔倒卵形。果实球形。花期5-6月。生海拔500-1000米的林中、灌丛、溪边或山谷。产贵州、广西、广东、湖北和湖南。

Shrubs. Leaves lanceolate to narrowly oblong; petioles 3-4 mm long. Inflorescences terminal, panicles, axis ridged; hypanthium obconic, ca. 3 mm long, smooth; calyx lobes inconspicuous; petals usually 4, distinct, broadly obovate. Fruits globose. Fl. May-Jun. Forests, thickets, streamsides or valleys at 500-1000 m. Distributed in Guizhou, Guangxi, Guangdong, Hubei and Hunan.

## 华南蒲桃

**Syzygium austrosinense** (Merr. et L. M. Perry) Hung T. Chang et R. H. Miao

灌木或乔木。叶椭圆形，革质。聚伞花序顶生或近顶生；花萼管倒圆锥形，长2.5-3厘米；萼裂片4，短三角形；花瓣离生，倒卵形。果实球形，黑色。花期6-8月。生海拔200-800(-2300)米的常绿阔叶林中。产中国西南、华南、东南和华中。

Shrubs or trees. Branchlets 4-angled. Leaves elliptic, leathery. Inflorescences terminal, cymes or subterminal; hypanthium obconic, 2.5-3 mm long; calyx lobes 4, shortly triangular; petals distinct, obovate. Fruits globose, black. Fl. Jun-Aug. Broad leaved evergreen forests at 200-800 (-2300) m. Distributed in SW, S, SE and C China.

## 乌墨

**Syzygium cumini** (L.) Skeels

乔木。叶对生，革质，广椭圆形至窄椭圆形，两面被细小腺点。花序腋生，圆锥聚伞状，长达11厘米；花萼管倒圆锥状或长梨形；萼裂片不明显；花瓣4，白色或淡紫色，合生，卵形且稍圆。果实红色至黑色，椭圆体形至坛形。花期4-5月，果期6-9月。生海拔1200米以下的低海拔次生林或荒地上。产云南、广西、广东、海南和福建。南亚、东南亚和澳大利亚亦有。

Trees. Leaves opposite, broadly elliptic to narrowly elliptic, leathery, both surfaces minutely glandular-punctate. Inflorescences axillary, paniculate cymes, 11 cm long; hypanthium obconic or long pyriform; calyx lobes inconspicuous; petals 4, white or light purple, coherent, ovate and slightly rounded. Fruits red to black, ellipsoid to pot-shaped. Fl. Apr-May. Fr. Jun-Sep. Secondary forests on lowlands or wastelands below 1200 m. Distributed in Yunnan, Guangxi, Guangdong, Hainan and Fujian. Also in S and SE Asia, and Australia.

华南蒲桃 *Syzygium austrosinense*

水竹蒲桃 *Syzygium fluviatile*

## 水竹蒲桃

**Syzygium fluviatile** (Hemsl.) Merr. et L. M. Perry

灌木。小枝圆柱状。叶条状披针形，革质。聚伞花序腋生；花瓣白色或紫色；花萼管倒圆锥状，长约3.5毫米；萼裂片4，极短；花瓣白色或紫色，分离，圆形。果实球状，成熟时黑色。花期4-7月，果期9-12月。生海拔100-1000米的林中溪边。产贵州、广西和海南。

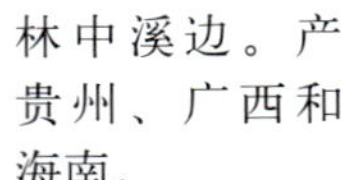

Shrubs. Branchlets terete. Leaves linear-lanceolate, leathery. Inflorescences axillary, cymes; petals white or purple; hypanthium obconic, ca. 3.5 mm long; calyx lobes 4, very short; petals white or purple, distinct, rounded. Fruits globose, black when mature. Fl. Apr-Jul. Fr. Sep-Dec. Streamsides in forests at 100-1000 m. Distributed in Guizhou, Guangxi and Hainan.

## 水翁蒲桃

**Cleistocalyx nervosum** DC.

乔木。叶薄革质，长圆形至椭圆形。圆锥花序生无叶的老枝上；花萼管半球形，长约3毫米，不具柄；花瓣退化；雄蕊长5-8毫米。果实成熟后紫色至黑色，阔卵球形。花期5-6月。生海拔200-600米的林中或水边。产云南、西藏、广西、广东、香港和海南。南亚、东南亚和大洋洲亦有。

桃金娘 *Rhodomyrtus tomentosa*

水翁蒲桃 *Cleistocalyx nervosum*

Trees. Leaves oblong to ellipsoid, thinly leathery. Panicles on leafless old branches; hypanthium hemispheric, ca. 3 mm long, not stipitate; petals obsolete; stamens 5-8 mm long. Fruits violet to black when mature, broadly ovoid. Fl. May-Jun. By waters or in forests at 200-600 m. Distributed in Yunnan, Xizang, Guangxi, Guangdong, Hong Kong and Hainan. Also in S and SE Asia, and Oceania.

## 桃金娘

**Rhodomyrtus tomentosa** (Ait.) Hassk.

灌木。叶椭圆形至倒卵形，对生，革质，三出脉。花单生，紫色，直径2-4厘米；花萼管倒卵球形，长约6毫米，具灰色绒毛；萼裂片5，近圆形，宿存；花瓣5，紫色，倒卵形；子房3(或4)室。浆果成熟时黑紫色，坛形。花期4-5月。生海拔200-1000米的山坡、低丘陵或草地。产中国西南、华南和东南。南亚、东南亚和日本南部亦有。

Shrubs. Leaves elliptic to obovate, opposite, leathery, ternate. Flowers solitary, purple, 2-4 cm diam; hypanthium obovoid, ca. 6 mm long, gray tomentose; calyx lobes 5, subrounded, persistent; petals 5, violet, obovate; ovary 3(or 4)-loculed. Berries purplish black when mature, urceolate. Fl. Apr-May. Mountain slopes, low hills or grasslands at 200-1000 m. Distributed in SW, S and SE China. Also in S and SE Asia, and S Japan.

## 番石榴

**Psidium guajava** L.

乔木。嫩枝有棱。叶片长圆形至椭圆形，背面有毛。花单生或2或3朵成聚伞状；花萼管钟形；萼帽近圆形，不规则开展；花瓣白色；子房贴生花萼管。浆果球形、卵球形或梨形。花期夏季。广泛栽培或逸生于中国西南和华南。原产热带美洲。

Trees. Branchlets ribbed. Leaves oblong to elliptic, abaxial surfaces hairy. Flowers solitary or 2 or 3 in cymes; hypanthium campanulate; calyx cap nearly rounded, irregularly opening; petals white; ovary adnate to hypanthium. Berries globose, ovoid, or pyriform. Fl. summer. Commonly cultivated or escaped in SW and S China. Native to tropical America.

番石榴 *Psidium guajava*

## 子楝树

**Decaspermum gracilentum**

(Hance) Merri. et L. M. Perry

灌木。叶片薄革质，椭圆形或卵状长圆形，基部阔楔形或略圆；叶柄长3-5毫米。花为腋生单花或4-6朵排成聚伞花序；花梗长7-10毫米，有短柔毛；花白，直径1厘米，萼管被毛，萼片4；花瓣4，卵形；雄蕊比花瓣略长，花丝红色；花柱比雄蕊短。浆果球形。生海拔200-800米的石灰岩山地林下。产贵州、广西、广东、台湾和湖南。越南亦有。

Shrubs. Leaves thinly leathery, elliptic or oblong, base broadly cuneate to round; petioles 3-5 mm long. Inflorescences axillary solitary or 4-6-flowered cluster; peduncles 7-10 mm long, short tomentose; flowers white, 1 cm diam; hypanthium tomentose, calyx lobes 4; petals 4, ovate; stamens slightly longer than petals, filaments red; style shorter than stamens. Berries globose. Limestone montane forests at 200-800 m. Distributed in Guizhou, Guangxi, Guangdong, Taiwan and Hunan. Also in Vietnam.

子楝树 *Decaspermum gracilentum*

## 五瓣子楝树

**Decaspermum parviflorum**

(Lam.) A. J. Scott

灌木或乔木。叶披针形或长圆披针形，两面密被黑色腺点，侧脉12-15对。花序腋生或顶生，腋生者聚伞圆锥状，顶生者圆锥状；花两性或雄性；花瓣5，白色或粉红色，圆形；花丝粉色或白色；子房4-6室。每果实具3-12粒种子。花期春夏间。生海拔2000米以下的林中。产云南、西藏、贵州、广西、广东和海南。南亚、东南亚和太平洋岛屿亦有。

Shrubs or trees. Leaves lanceolate or oblong-lanceolate, both surfaces densely black glandular-punctate, lateral veins 12-15 pairs. Inflorescences axillary or terminal, when axillary then thyrses, when terminal then paniculately arranged; flowers bisexual or staminate; petals 5, white or pink, orbicular; filaments pink or white; ovary 4-6-loculed. Seeds 3-12 per fruit. Fl. between spring and summer. Forests below 2000 m. Distributed in Yunnan, Xizang, Guizhou, Guangxi, Guangdong and Hainan. Also in S and SE Asia, and Pacific Islands.

五瓣子楝树 *Decaspermum parviflorum*

# 野牡丹科
# Melastomaceae

## 菲油果(南美稔)
**Acca sellowiana** (O. Berg) Burret

常绿小乔木。叶对生，革质，椭圆形或倒卵状椭圆形，顶端圆形，下面密被灰白色短绒毛。花单生叶腋，有长梗，直径2.5-5厘米；花瓣白色，有时内面带紫色；雄蕊与花柱红色。浆果卵圆形或长圆形，顶部有宿存萼片。中国南部栽培。原产南美洲。

Evergreen small trees. Leaves opposite, coriaceous, elliptic or obovate-elliptic, apex round, abaxially densely grayish-white tomentose. Flowers solitary, axillary, pedicellate, 2.5-5 cm diam; petals white, sometimes adaxially tinted with purple; stamens and style red. Berries ovate or oblong, apex with persistent calyx lobes. Cultivated in S China. Native to South America.

菲油果(南美稔) *Acca sellowiana*

## 金锦香
**Osbeckia chinensis** L.

草本或灌木。叶片坚纸质。头状花序顶生，具2-8(-10)花；花萼裂片4(或5)，三角状披针形，边缘具缘毛，裂片间具刚毛；花瓣4，淡紫红色或粉红色；宿存萼坛状。蒴果紫红色，卵球形。花期7-9月，果期9-11月。生海拔2800米以下的山坡草地、疏林下或路边。产中国长江流域以南各省区。日本、南亚、东南亚和澳大利亚亦有。

Herbs or shrubs. Leaves hard-papery. Inflorescences terminal, capitulum, 2-8(-10)-flowered; calyx lobes 4(or 5), triangular-lanceolate, margin ciliate, setose between lobes; petals 4, pale purplish red or pink; persistent calyx urceolate. Capsules purplish red, ovoid. Fl. Jul-Sep. Fr. Sep-Nov. Grasslands on mountain slopes, sparse forests or roadsides below 2800 m. Distributed throughout the provinces on south of the Yangtze River. Also in Japan, S and SE Asia, and Australia.

金锦香 *Osbeckia chinensis*

## 朝天罐
**Osbeckia opipara** C. Y. Wu et C. Chen

灌木。叶坚纸质，卵形至卵状披针形，长5.5-11.5厘米，全缘，具缘毛，两面被糙伏毛、微柔毛及透明腺点。花萼长坛状，中部略上缢缩，长约2.3厘米，外面被刺毛状有柄星状毛和微柔毛，裂片长三角形或卵状三角形。蒴果长卵形，为宿存萼所包。花果期7-9月。生海拔250-800米的山坡、路旁、疏林或灌丛中。产长江流域以南各省。越

朝天罐 *Osbeckia opipara*

星毛金锦香 *Osbeckia stellata*

南和泰国亦有。

Shrubs. Leaves rigidly papery, ovate to ovate-lanceolate, 5.5-11.5 cm long, margin entire, ciliate, strigose, pubescent and transparent glandular on both surfaces. Calyx long-urceolate, contracted slightly above middle, ca. 2.3 cm long, with echinulate and stalked stellate hairs and pubescence outside, lobes long-triangular or ovate-triangular. Capsules long-ovate, included in persistent calyx. Fl. and fr. Jul-Sep. Slopes, roadsides, sparse forests or shrubs at 250-800 m. Distributed in the south of Yangtze River. Also in Vietnam and Thailand.

## 星毛金锦香

**Osbeckia stellata** Buch.-Ham. ex Ker Gawl.

草本或小灌木。叶对生，有时3叶轮生，坚纸质，两面具糙毛或亦有绒毛。花4数；花瓣倒卵形，粉红色至紫色；萼片线状披针形或钻形。蒴果长坛状，中部收缩。花期7-11月，果期10-12月。生海拔200-2300米的草坡、路边或灌丛。产中国西南、华南、东南和华中。南亚和东南亚亦有。

蚂蚁花 *Osbeckia nepalensis*

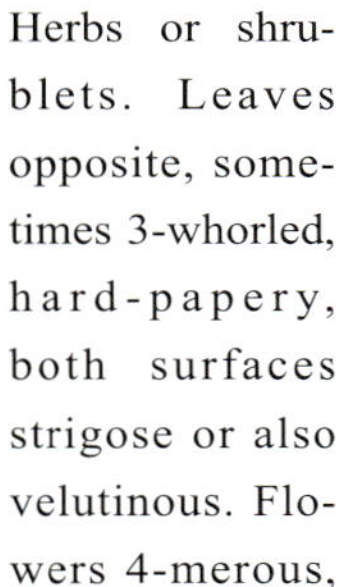
Herbs or shrublets. Leaves opposite, sometimes 3-whorled, hard-papery, both surfaces strigose or also velutinous. Flowers 4-merous, 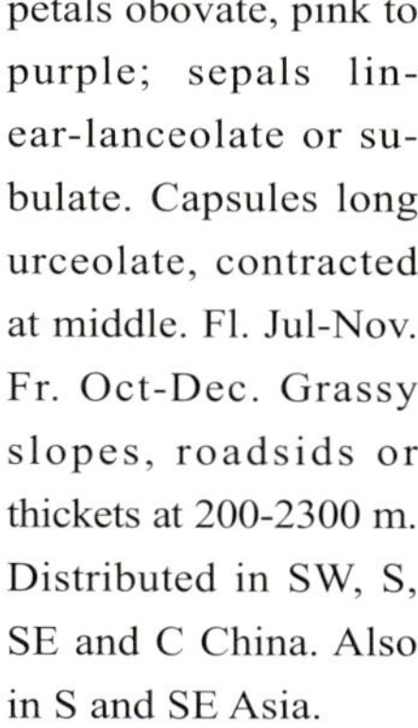
petals obovate, pink to purple; sepals linear-lanceolate or subulate. Capsules long urceolate, contracted at middle. Fl. Jul-Nov. Fr. Oct-Dec. Grassy slopes, roadsids or thickets at 200-2300 m. Distributed in SW, S, SE and C China. Also in S and SE Asia.

## 蚂蚁花

**Osbeckia nepalensis** Hook. f.

灌木。叶长圆状披针形至卵状披针形，硬纸质，两面密具糙毛。花5数；萼裂片5，长卵形，与花萼管等长，两面无毛；花瓣宽倒卵形，粉红色或红色；花丝较花药长，花药具短喙。蒴果卵球形。花期8-10月，果期9-12月。生海拔500-1900米的草坡、灌丛、林缘、路边或田间。产云南、西藏和广西。南亚和东南亚亦有。

Shrubs. Leaves oblong-lanceolate to ovate-lanceolate, stiffly papery, both surfaces densely strigose. Flowers 5-merous; calyx lobes 5, long ovate, as long as hypanthium, both surfaces glabrous; petals broadly obovate, pink or red; filaments longer than anthers, anthers shortly beaked. Capsules ovoid. Fl. Aug-Oct. Fr. Sep-Dec. Grassy slopes, thickets, forest edges, roadsides or fields at 500-1900 m. Distributed in Yunnan, Xizang and Guangxi. Also in S and SE Asia.

野牡丹 *Melastoma malabathricum*

## 野牡丹

**Melastoma malabathricum** L.

直立灌木。叶片披针形、卵状披针形或椭圆形，基出脉5。萼片宽披针形。花序近头状伞房状，顶生，具3-7花，基部具2个叶状苞片；花瓣红紫色。果实坛状球形，肉质。花期2-8月，果期7-12月。生海拔100-2800米的开阔地、山坡、林中、草地或灌丛。产中国西南、华南和东南。南亚、东南亚、日本和太平洋岛屿亦有。

Shrubs erect. Leaves lanceolate, ovate-lanceolate or elliptic, 5-basal nerved. Sepals broadly lanceolate. Inflorescences subcapitate corymbose, terminal, 3-7-flowered, with 2 leaflike bracts at base; petals reddish purple. Fruits urceolate-globular, succulent. Fl. Feb-Aug. Fr. Jul-Dec. Open fields, slopes, forests, grasslands or thickets at 100-2800 m. Distributed in SW, S and SE China. Also in S and SE Asia, Japan and Pacific Islands.

毛菍 *Melastoma sanguineum*

## 毛菍

**Melastoma sanguineum** Sims

灌木。叶坚纸质，基出脉5，两面具糙毛。伞房花序顶生，常仅有1花；苞片戟形，膜质；花瓣粉红色或紫红色；子房半下位，密被刚毛。蒴果坛状陀螺形，肉质。花期几全年，果期8-10月。生海拔400米以下的山坡、草丛、灌丛或山脚下林缘。产广西、广东、海南和福建。印度、马来西亚和印度尼西亚亦有。

Shrubs. Leaves hard-papery, pentanervious, both surfaces strigose. Inflorescences terminal, corymb, often solitary; bracts halberd-shaped, membranous; petals pink or purple-red; ovary half-inferior, densely bristly. Capsules urceolate-turbinate, succulent. Fl. almost all year. Fr. Aug-Oct. Slopes, grasslands, thickets, or forest edges on low hills below 400 m. Distributed in Guangxi, Guangdong, Hainan and Fujian. Also in India, Malaysia and Indonesia.

## 地菍

**Melastoma dodecandrum** Lour.

小灌木。茎常匍匐；小枝多数，匍匐，幼时具硬毛，后无毛。叶卵形至椭圆形；叶柄长0.2-0.6(-1.5)厘米。花序顶生，聚伞状，具(1-)3花；花盘具硬毛；花瓣紫罗兰色至紫色。果实坛状球形，肉质，具硬毛。花期5-7月，果期7-9月。生海拔1300米以下的开阔林地、灌木、草地或铁道边。产中国西南、华南和东南。越南亦有。

长穗花 *Styrophyton caudatum*

Shrublets. Stems often repent; branchlets numerous, procumbent, strigose when young, later glabrous. Leaves ovate to elliptic; Petioles 0.2-0.6(-1.5) cm long. Inflorescences terminal, cymose, (1-)3-flowered; hypanthium strigose; petals lavender to purple. Fruits urceolate-globular, succulent, strigose. Fl. May-Jul. Fr. Jul-Sep. Open fields, thickets, grasslands or trailsides below 1300 m. Distributed in SW, S and SE China. Also in Vietnam.

## 长穗花

**Styrophyton caudatum** (Diels) S. Y. Hu

直立灌木。叶卵形至阔卵形，幼时密具硬毛，后粗糙。花序顶生，穗状，密具长柔毛；花单生或3-5朵成簇，较小；花瓣粉色或白色。蒴果卵球形，明显具纵向8条肋纹。花期5-6月，果期10月至翌年1月。生海拔400-1500米的密林、溪岩、灌丛、山谷或湿处。产云南和广西。

Erect shrubs. Leaves ovate to broadly ovate, densely strigose when young but later scabrous. Inflorescences terminal, spicate, densely villous; flowers solitary or in clusters of 3-5, small; petals pink or white. Capsules ovoid, conspicuously longitudinally 8-ribbed. Fl. May-Jun. Fr. Oct to next Jan. Dense forests, stream banks, thickets, valleys or damp places at 400-1500 m. Distributed in Yunnan and Guangxi.

地菍 *Melastoma dodecandrum*

尖子木 *Oxyspora paniculata*

## 尖子木
**Oxyspora paniculata** (D. Don) DC.

灌木。茎具秕糠状星状毛及疏具柔刚毛。叶卵形、狭椭圆状卵形或近圆形，下面常沿脉具秕糠状星状毛。圆锥花序大，宽约10厘米或更宽；花瓣粉色、红色或深红色，卵形。蒴果倒卵球形。花期7-9月，果期翌年1-3月。生海拔500-2000米的林下、山谷或湿地。产云南、西藏、贵州和广西。南亚和东南亚亦有。

Shrubs. Stems furfuraceous stellate and sparsely puberulous-setose. Leaves ovate, narrowly ellipticovate, or suborbicular, abaxially usually furfuraceous stellate on veins. Panicles large, about 10 cm broad or more; petals pink, red, or dark red, ovate. Capsules obovoid. Fl. Jul-Sep. Fr. next Jan-Mar. Forests, valleys or moist places at 500-2000 m. Distributed in Yunnan, Xizang, Guizhou and Guangxi. Also in S and SE Asia.

## 滇尖子木
**Oxyspora yunnanensis** H. L. Li

灌木。茎幼时明显具糙毛或无。叶披针状长圆形至长圆状倒卵形，5基出脉。花序顶生，伞房圆锥状，长10-20厘米；花萼窄漏斗状；花瓣粉红色至红色，卵圆形。蒴果漏斗状。花期8月，果期10-11月。生海拔1300-2800米的密林下或沟边岩石缝中。产云南和贵州。

Shrubs. Stems patently setose or not when young. Leaves lanceolate-oblong or oblong-obovate, 5 basal nerved. Inflorescences terminal, a cymose panicle, 10-20 cm long; calyx narrowly infundiformis; petals pink to red, ovate. Capsules infundibular. Fl. Aug. Fr. Oct-Nov. Dense forests or rock crevices by streams at 1300-2800 m. Distributed in Yunnan and Guizhou.

滇尖子木 *Oxyspora yunnanensis*

## 偏瓣花
**Plagiopetalum esquirolii** (Lévl.) Rehd.

小灌木。茎幼时具狭翅。叶披针形或卵状披针形，膜质或纸质。疏松的伞房花序或复伞花序顶生或腋生；花瓣红色至紫色，稀粉红色，倒卵形，倾斜。蒴果卵球形，具4棱。花期8-10月，果期12月至翌年2月。生海拔500-3500米的林中、草地、山谷、湿地或岩石缝中。产云南、四川、贵州和广西，缅甸和越南亦有。

Shrublets. Stems narrowly winged when young. Leaves lanceolate or ovate-lanceolate, membranous to papery. Remotely cymoses or compound umbels terminal or axillary; petals red to purple, rarely pink, obovate, oblique. Capsules ovoid, 4-ribbed. Fl. Aug-Oct. Fr. Dec to next Feb. Forests, grasslands, valleys, moist places or rock crevices at 500-3500 m. Distributed in Yunnan, Sichuan, Guizhou and Guangxi. Also in Myanmar and Vietnam.

偏瓣花 *Plagiopetalum esquirolii*

棱果花 *Barthea barthei*

## 棱果花

**Barthea barthei** (Hance ex Benth.) Krass.

灌木。叶坚纸质或近革质，基出脉5。聚伞花序顶生，具3花，仅1朵可育；花瓣白色至粉红色或紫色；子房梨形或四棱形。蒴果长圆形，具四面，秕糠状，肋上翅1-2毫米宽。花期1-5月或10-12月，果期10-12月或5月。生海拔400-2500(-2800)米的山坡、山谷、林中或水旁。产广西、广东、台湾、福建和湖南。

Shrubs. Leaves hard-papery or subleathery, 5-nerved. Inflorescences terminal, cymes, 3-flowered but usually only 1 fertile; petals white to pink or purple; ovary pyriform or tetragonous. Capsules oblong, 4-sided, furfuraceous, wings 1-2 mm wide on ribs. Fl. Jan-May or Oct-Dec. Fr. Oct-Dec or May. Slopes, valleys, forests or by rivers at 400-2500(-2800) m. Distributed in Guangxi, Guangdong, Taiwan, Fujian and Hunan.

## 少花柏拉木

**Blastus pauciflorus** (Benth.) Guillaumin

灌木。茎圆柱状，幼时具柔毛及黄色腺点。叶片卵状披针形至卵形，纸质。聚伞圆锥状伞房花序顶生；花萼管漏斗形，具(3或)4棱，密具带柄腺体；花瓣(3-)4，粉红至淡紫红色。蒴果椭圆体形。花期6-8月，果期8-11月。生海拔100-1600米的林下、山谷、溪边或路旁。产中国西南、华南和东南。

Shrubs. Stems terete, puberulous and yellow glandular when young. Leaves ovate-lanceolate to ovate, papery. Inflorescences terminal, panicled cymose; hypanthium funnelform, (3 or)4-sided, densely stipitate glandular; petals (3-)4, pink to reddish lavender. Capsules ellipsoid. Fl. Jun-Aug. Fr. Aug-Nov. Under forests, valleys, riversides or roadsides at 100-1600 m. Distributed in SW, S and SE China.

少花柏拉木 *Blastus pauciflorus*

过路惊 *Bredia quadrangularis*

## 过路惊

**Bredia quadrangularis** Cogn.

灌木或小灌木。叶硬纸质，三出脉，卵状椭圆形，基部圆形至楔形。伞房状花序顶生于小枝顶端，具3-9花或有时具更多花；花瓣玫红色至紫色。蒴果四棱状杯形，先端截形。花期6-10月，果期8-10月。生海拔300-1500米的山谷林下、山坡、潮湿处或路边。产华南、东南和华东。

Shrubs or shrublets. Leaves stiffly papery, ternate, ovate-elliptic, base rounded to cuneate. Inflorescences terminal at tips of branchlets, cymose, 3-9-flowered or sometimes with more flowers; petals rose red to purple. Capsules tetragonous-cuplike, apex truncate. Fl. Jun-Oct. Fr. Aug-Oct. Forests in valleys, slopes, wet places or roadsides at 300-1500 m. Distributed in S, SE and E China.

小叶野海棠 *Bredia microphylla*

## 小叶野海棠
**Bredia microphylla** H. L. Li

小灌木或草本，匍匐。茎圆柱状。叶卵形至卵状圆形，下面仅沿脉疏具糙毛。聚伞花序顶生，具1-3花；花萼管钟形，4棱；萼裂片线形；花瓣淡紫红色，稍倾斜。蒴果杯状。花期10月。生林中、山坡或阴湿处。产广西、广东和江西。

Shrublets or herbs, creeping. Stems terete. Leaves ovate to ovate-orbicular, abaxially only rarely strigose on veins. Inflorescences terminal, cymes, 1-3-flowered; hypanthium campanulate, 4-sided; calyx lobes linear; petals lavender red, slightly oblique. Capsules cupulate. Fl. Oct. Forests, slopes or moist shaded places. Distributed in Guangxi, Guangdong and Jiangxi.

## 叶底红
**Bredia fordii** (Hance) Diels

灌木或小灌木。茎浅褐色或红色，4棱。叶片硬纸质，下面紫红色，仅疏具糙毛和柔毛。伞房聚伞花序；花萼杯状，裂片4；花瓣4，粉红色至紫红色；子房半下位。蒴果为宿存萼所包被；宿存萼被平展红色长刚毛、腺毛和微柔毛。花期6-8月，果期8-10月。生海拔100-1400米的坡地、湿润的草丛中或水旁。产中国西南、华南和东南。

Shrubs or shrublets. Stems pale brown or red, 4-sided. Leaves stiffly papery, abaxially purplish red and only sparsely strigose and puberulous. Inflorescences corymbose cymes; sepals cup-like, 4; petals 4, pink to purple-red; ovary half-inferior. Capsules not exserted; persistent sepals spreading red bristly, glandular and pilose. Fl. Jun-Aug. Fr. Aug-Oct. Hillsides, wet grasslands or by rivers at 100-1400 m. Distributed in SW, S and SE China.

叶底红 *Bredia fordii*

## 大叶熊巴掌
**Phyllagathis longiradiosa** C. Chen

草本或小灌木。茎圆柱形。叶阔卵形至近椭圆形，纸质或近膜质；密被细糠秕及疏短糙伏毛；叶柄具槽，槽两侧具长柔毛。伞形花序顶生；萼筒被刺毛；花瓣玫瑰色，具缘毛；花药基部具2小瘤，药隔基部膨大成短距。蒴果被基部膨大的刺毛。花期5-7月或11月，果期12月至翌年1月。生海拔300-2200米的常绿阔叶林下。产云南、贵州和广西。

Herbs or shrubs. Stems terete. Leaves broadly ovate to subelliptic, papery to submembranous, densely furfuraceous and sparsely strigose; petioles sulcate, lateral sides villous. Inflorescences terminal, umbellate; hypanthium setose; petals rose-colored, margin ciliate; anther bases 2 tuberculate; connective base inflated forming a short spur. Capsules setose with inflated bases. Fl. May-Jul or Nov. Fr. Dec to next Jan. Broad-leaved evergreen forests at 300-2200 m. Distributed in Yunnan, Guizhou and Guangxi.

大叶熊巴掌 *Phyllagathis longiradiosa*

丽萼熊巴掌 *Phyllagathis longiradiosa* var. *pulchella*

异药花 *Fordiophyton faberi*

## 丽萼熊巴掌

**Phyllagathis longiradiosa** C. Chen var. **pulchella** C. Chen

本变种与原变种大叶熊巴掌的主要区别是：叶柄的槽两侧不具长柔毛，叶缘具疏缘毛；萼筒及萼片两面被微柔毛。花期5月或11月。特产广西(大新县)。

This variety differs from var. *longiradiosa* in its Petioles not villous on lateral sides; leaves margin sparsely ciliate; hypanthium and sepals puberulous on both surfaces. Fl. May or Nov. Endemic to Guangxi (Daxin County).

## 锦香草

**Phyllagathis cavaleriei** (Lévl. et Vaniot) Guillaumin

草本。叶纸质至近硬纸质。伞形花序顶生；花萼管漏斗形，4棱，被秕糠，有时具长糙毛；花瓣粉色或紫色；药隔下延成短距；子房杯形，顶端具一膜质冠，冠边缘具8锯齿。蒴果杯状。花期5-8月，果期7-10月。生海拔300-3100米的林中、山谷、山坡、溪边或阴湿处。产中国西南和东南。

Herbs. Leaves papery to substiffly papery. Inflorescences terminal, umbels; hypanthium funnel-shaped, 4-sided, furfuraceous, sometimes long setose; petals pink or purple; connectives decurrent, forming a short spur; ovary cup-shaped, apex with a membranous crown, crown margin 8-dentate. Capsules cupulate. Fl. May-Aug. Fr. Jul-Oct. Forests, valleys, slopes, streamsides or moist shaded places at 300-3100 m. Distributed in SW and SE China.

## 异药花

**Fordiophyton faberi** Stapf

草本或亚灌木。叶膜质，大小差别很大，下面无毛，密被白色小腺点。花序顶生，伞房着圆锥花序或伞形，多达50花；花瓣粉色至紫红色；药隔微膨大呈小距；子房顶端具膜质冠，冠檐具缘毛。蒴果钟形，先端4裂。花期6-10月，果期8-11月。生海拔500-1800米的山谷、疏密林中、阴湿处或水边或山坡草地、土质肥厚和湿润处。产中国西南、华南和东南。

Herbs or shrublets. Leaves membranous, unequal in size, abaxially glabrous, with white glands. Inflorescences terminal, cymose paniculate or umbellate, to 50-flowered; petals pink to reddish purple; connectives inflated in spurs; ovary apex with membranous hat, limps ciliate. Capsules campanulate, apex 4-lobed. Fl. Jun-Oct. Fr. Aug-Nov. Valleys, sparse or dense forests, wet places, by stream, or grassy slopes, on wet and humid soil at 500-1800 m. Distributed in SW, S and SE China.

## 肉穗草

**Sarcopyramis bodinieri** Lévl. et Vaniot

草本。叶卵形至披针形，纸质，下面常紫红色且常无毛，边缘近波状且具牙齿。聚伞花序顶生；花萼管4棱，沿棱具狭翅；花瓣紫色或粉红色；花药黄色。蒴果杯状。花期5-7月，果期10-12月或翌年1月。生海拔300-2800米的密林下、山谷、阴湿的地方或石隙。产中国西南和东南。

锦香草 *Phyllagathis cavaleriei*

肉穗草 *Sarcopyramis bodinieri*

Herbs. Leaves ovate to elliptic, papery, abaxially usually purplish red and usually glabrous, margin subrepand and armed with teeth. Inflorescences terminal, cymes; hypanthium 4-sided, narrowly winged on angles; petals purple or pink; anthers yellow. Capsules cup-shaped. Fl. May-Jul. Fr. Oct-Dec or Jan next year. Dense forests, valleys, shaded and wet places or rocky crevices at 300-2800 m. Distributed in SW and SE China.

## 楮头红

**Sarcopyramis napalensis** Wall.

直立草本。叶阔卵形或卵形，膜质，边缘具细锯齿，被疏糙伏毛。聚伞花序顶生，具1-3花，基部具2苞片；花萼管4棱，沿棱具狭翅；花瓣粉红色；药隔下延，形成一短矩或小脊。蒴果杯状。花期8-10月，果期9月至翌年1月。生海拔1000-3200米的林下潮湿地或溪边。产中国西南、东南和华中。南亚和东南亚亦有。

楮头红 *Sarcopyramis napalensis*

Erect herbs. Leaves broadly ovate or cvate, membranous, margin serrulate, strigose on both surfaces. Inflorescences terminal, cymes, 1-3-flowered, with 2 bracts at base; hypanthium 4-sided, narrowly winged on angles; petals pink; connective decurrent, forming a short spur or minute bulge. Capsules cup-shaped. Fl. Aug-Oct. Fr. Sep to next Jan. Damp places in forests or by streams at 1000-3200 m. Distributed in SW, SE and C China. Also in S and SE Asia.

## 直立蜂斗草

**Sonerila erecta** Jack

草本。叶狭椭圆形至卵形，腹面被秕糠状微柔毛及短刺毛。花序为顶生蝎尾状聚伞，具1-5(-11)花；花萼管管状，近3棱，具6肋；花瓣粉红色至紫色。蒴果管状。花期7-10月，果期10-12月。生海拔500-1800米的林下、石灰质山腰、密竹林或多草地。产云南、广西、广东、湖南和江西。南亚和东南亚亦有。

Herbs. Leaves narrowly elliptic to ovate, adaxially scurfy-puberulous and short-setose. Inflorescences terminal, scorpioid cymes, 1-5(-11)-flowered; hypanthium tubular, almost 3-sided, 6-ribbed; petals pink to purple. Capsules tubular. Fl. Jul-Oct. Fr. Oct-Dec. Forests, limestone hillsides, bamboo thickets or grassy areas at 500-1800 m. Distributed in Yunnan, Guangxi, Guangdong, Hunan and Jiangxi. Also in S and SE Asia.

直立蜂斗草 *Sonerila erecta*

海棠叶蜂斗草 *Sonerila plagiocardia*

## 海棠叶蜂斗草
**Sonerila plagiocardia** Diels

草本。叶卵形，基部斜心形，掌状(6-)8(-9-11)条脉。聚伞花序顶生，蝎尾状；花萼管管状钟形，中部稍收缩，3棱，6肋；花瓣粉红色至红色；萼外被微柔毛及疏腺毛。蒴果倒圆锥形。花期8-9月，果期10-11月。生海拔600-2500米的密林、山腰、山谷或阴湿处。产云南、广西、广东和江西。老挝、泰国、越南、柬埔寨和马来西亚亦有。

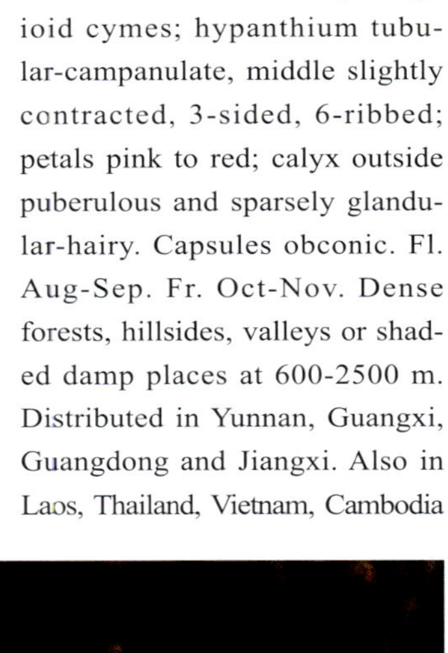

Herbs. Leaves ovate, base oblique-cordate, palmately (6-)8(-9-11)-nerved. Inflorescences terminal, scorpioid cymes; hypanthium tubular-campanulate, middle slightly contracted, 3-sided, 6-ribbed; petals pink to red; calyx outside puberulous and sparsely glandular-hairy. Capsules obconic. Fl. Aug-Sep. Fr. Oct-Nov. Dense forests, hillsides, valleys or shaded damp places at 600-2500 m. Distributed in Yunnan, Guangxi, Guangdong and Jiangxi. Also in Laos, Thailand, Vietnam, Cambodia and Malaysia.

## 溪边桑勒草
**Sonerila maculata** Roxb.

草本或小灌木。叶倒卵形至椭圆形，下面沿脉疏具糙毛。聚伞花序顶生，蝎尾状，具秕糠，渐无毛；花萼管漏斗形，3棱，6肋；萼裂片阔三角形；花瓣紫色。蒴果钟形。花期6-9月，果期8-12月。生海拔100-1300米的灌丛、山谷、溪边或阴湿处。产云南、西藏、广西、广东和福建。南亚和东南亚亦有。

Herbs or shrublets. Leaves obovate to elliptic, abaxially scattered setose on veins. Inflorescences terminal, scorpioid cymes, furfuraceous, glabrescent; hypanthium funnel-shaped, 3-sided, 6-ribbed; calyx lobes broadly triangular; petals purple. Capsules campanulate. Fl. Jun-Sep. Fr. Aug-Dec. Thickets, valleys, streamsides or shaded damp places at 100-1300 m. Distributed in Yunnan, Xizang, Guangxi, Guangdong and Fujian. Also in S and SE Asia.

溪边桑勒草 *Sonerila maculata*

## 藤牡丹
**Diplectria barbata** (Wall. ex C. B. Clarke) Franken et Roos

攀援灌木或藤本。幼枝钝四棱形。叶片长圆形或宽披针形，全缘，基出脉5。聚伞圆锥花序顶生，多花；花萼管状钟形；花瓣白色；雄蕊8，异型，极不相等，4长4短。浆果近球形，顶端平截。花期6月或11月，果

藤牡丹 *Diplectria barbata*

矮酸脚杆 *Medinilla nana*

附生美丁花 *Medinilla arboricola*

期6-7月。生海拔约400米的密林中。产海南。印度、越南和马来西亚亦有。

Climbing shrubs or vines. Branchlets obtusely tetragonous. Leaves oblong or broadly lanceolate, margin entire, basal veins 5. Cymose panicles terminal, multi-flowered; calyx tube tubiform-campanulate; petals white; stamens 8, heterotypic, very unequal, 4 long, and the rest 4 short. Berries subglobose, apex truncate. Fl. Jun or Nov. Fr. Jun-Jul. Dense forests at ca. 400 m. Distributed in Hainan. Also in India, Vietnam and Malaysia.

## 矮酸脚杆
**Medinilla nana** S. Y. Hu

灌木。叶片倒卵形至椭圆形，厚纸质或近革质，肉质至近革质，两面无毛。聚伞花序顶生，具1-3花；花萼管长漏斗形，具隆起；花瓣4，粉红色。果坛状，具小瘤。花期6月，果期11月。生海拔1100-2000米的林中树上。产云南和西藏东南部。越南亦有。

Shrubs. Leaves obovate to elliptic, stout papyraceous or subleathery, succulent to subleathery, both surfaces glabrous. Inflorescences terminal, cymes, 1-3-flowered; hypanthium long funnel-shaped, bulging; petals 4, pink. Berries urceolate, small tuberculate. Fl. Jun. Fr. Nov. On trees in forests at 1100-2000 m. Distributed in Yunnan and SE Xizang. Also in Vietnam.

## 北酸脚杆
**Medinilla septentrionalis** (W. W. Sm.) H. L. Li

灌木。叶片披针形、卵状披针形或阔卵形，下面稍秕糠状，基部钝或近圆形。聚伞花序腋生，(1-)3(-5)花；花瓣粉红色至紫红色；花萼被极疏的腺毛或几无。浆果坛形。花期6-9月，果期翌年2-5月。生海拔200-1800米的林中或潮湿地。产云南、广西和广东。缅甸、泰国和越南亦有。

Shrubs. Leaves lanceolate, ovate-lanceolate, or broadly ovate, abaxially ± furfuraceous, base obtuse or subrotund. Inflorescences axillary, cymes, (1-)3(-5)-flowered; petals pink to purplish red; calyx with very sparse glandular hairs or almost absent. Berries urceolate. Fl. Jun-Sep. Fr. Feb-May next year. Forests or damp places at 200-1800 m. Distributed in Yunnan, Guangxi and Guangdong. Also in Myanmar, Thailand and Vietnam.

北酸脚杆 *Medinilla septentrionalis*

## 附生美丁花
**Medinilla arboricola** F. C. How

攀援灌木。茎四棱形。叶3-5枚轮生，卵状椭圆形，全缘，离基3出脉，无毛。聚伞花序腋生；花3-5朵，4-5数，白色。浆果近球状壶形。花期6-7月，果期8-9月。生林中、阴处、水旁岩石上或附生于树上。产海南(保亭县)。

Climbing shrubs. Stems tetragonous. Leaves verticillate, 3-5 in a whorl, ovate-elliptic, margin entire, triplicostate, glabrous. Cymes axillary; flowers 3-5, 4-5-merous, white. Berries subglobose-urceolate. Fl. Jun-Jul. Fr. Aug-Sep. Forests, shaded places, rocks by streams or adnascent on trees. Distributed in Hainan (Baoting County).

沙巴酸脚杆 *Medinilla petelotii*

## 沙巴酸脚杆
**Medinilla petelotii** Merr.

灌木。小枝具狭翅。叶片倒卵形或椭圆形，下面秕糠状且具瘤，基部楔形且稍下延。花序生于无叶老茎上或匍匐茎上，聚伞状，具3-7花；花瓣粉红色，近圆形。浆果坛状。花期8-10月，果期翌年5-6月。生海拔800-1400米的林中或河边，生树上。产云南。越南亦有。

Shrubs. Branchlets with narrow wings. Leaves obovate or elliptic, abaxially furfuraceous and tuberculate, base cuneate and slightly decurrent. Inflorescences inserted on leafless old stems or stolons, cymose, 3-7-flowered; petals pink, suborbicular. Berries urceolate. Fl. Aug-Oct. Fr. next May-Jun. Forests or by rivers at 800-1400 m, on trees. Distributed in Yunnan. Also in Vietnam.

## 红花酸脚杆
**Medinilla rubicunda** (Jack) Blume

灌木。树皮鸡冠状，具皱。叶椭圆形至披针形，两面无毛，背面密布小窝点，离基三出脉。聚伞花序(1-)3-5花，生于叶腋或老枝叶痕；花瓣粉红色。浆果卵球形至坛形。花期8-9月，果期10月至翌年3月。生海拔800-1800米的林中、山坡或河边。产云南、西藏、海南和广西。南亚亦有。

Shrubs. Bark corky, wrinkled. Leaves elliptic to lanceolate, both surfaces glabrous, adaxially densely foveolate, triplinerved. Inflorescences axillary in axils of leaves or at leaf scars on older branches, cymose, (1-)3-5-flowered; petals pink. Berries ovoid to urceolate. Fl. Aug-Sep. Fr. Oct to next Mar. Forests, mountain slopes or river banks at 800-1800 m. Distributed in Yunnan, Xizang, Hainan and Guangxi . Also in S Asia.

## 酸脚杆
**Medinilla lanceata** (Nayar) C. Chen

灌木或乔木。树皮鸡冠状，纵向脱落。叶披针形或卵状披针形，3或5条脉。花序生于无叶的茎上或块根上，聚伞圆锥状；花瓣阔卵形。浆果坛状。花期8月，果期4月或10月。生海拔400-1000米的山谷、山坡或林中阴湿处。产云南和海南。

Shrubs or trees. Bark corky, longitudinally shedding. Leaves lanceolate or ovate-lanceolate, 3- or 5-veined. Inflorescences inserted on leafless stems or rhizomes, cymose paniculate; petals broadly ovate. Berries urceolate. Fl. Aug. Fr. Apr or Oct. Valleys, slopes or damp and shady places in forests at 400-1000 m. Distributed in Yunnan and Hainan.

## 谷木
**Memecylon ligustrifolium** Champ. ex Benth.

灌木或乔木。叶革质，长5.5-8厘米，宽2.5-3.5厘米；顶端渐尖，钝头。聚伞花序生叶腋；苞片卵形；花梗两边具髯毛；萼筒半球形；花瓣白色或淡黄

红花酸脚杆 *Medinilla rubicunda*

酸脚杆 *Medinilla lanceata*

绿色，或紫色，半圆形；雄蕊蓝色。浆果状核果球形，密布瘤状突起。花期5-8月，果期12月至翌年2月。生海拔100-1600米的密林、山坡、河谷及湿地。产云南、广西、广东、海南和福建。

Shrubs or trees. Leaves leathery, 5.5-8 cm long 2.5-3.5 cm long broad, apex acuminate with an obtuse tip. Inflorescences in axils of leaves, cymose; bracts ovate; pedicel both sides pubescent; hypanthium semiglobose; petals white or tinged yellowish green or purple, semiorbicular; stamens blue. Fruit a baccate drupe, globular, minutely tuberculate. Fl. May-Aug. Fr. Dec to next Feb. Dense forests, mountain slopes or valleys, damp places at 100-1600 m. Distributed in Yunnan, Guangxi, Guangdong, Hainan and Fujian.

谷木 *Memecylon ligustrifolium*

## 细叶谷木

**Memecylon scutellatum**

(Lour.) Hook. et Arn.

灌木，稀为小乔木。叶革质，椭圆形至卵状披针形；基部广楔形，顶端钝、圆形或微凹。聚伞花序腋生，花梗基部具刺毛；花梗无毛；萼筒浅杯形；花瓣紫色或蓝色，广卵形。浆果状核果球形，密布小疣状突起。花期(3-)6-8月，果期(11月至)翌年1-3月。生海拔300米的密林、灌木丛、草地及水边。产广西、广东和海南。缅甸、老挝、泰国、越南、柬埔寨和马来西亚亦有。

Shrubs or rarely small trees. Leaves leathery, elliptic to ovate-lanceolate, base broadly cuneate, apex obtuse, rounded, or retuse. Inflorescences axillary, cymose; peduncle base often setose; pedicel glabrous; hypanthium shallowly cup-shaped; petals purple or blue, broadly ovate. Fruit a baccate drupe, densely tuberculate. Fl. (Mar-)Jun-Aug. Fr. (Nov-) to next Jan-Mar. Dense forests, thickets, grassy areas, streamsides at 300 m. Distributed in Guangxi, Guangdong and Hainan. Also in Myanmar, Laos, Thailand, Vietnam, Cambodia and Malaysia.

细叶谷木 *Memecylon scutellatum*

# 菱科 Trapaceae

## 欧菱

**Trapa natans** L.

一年生水生草本。叶三角状菱形至扁球状菱形。花小，单生于叶腋；花瓣白色。果实三角状至菱状，具(0-)2-4角，具一明显突出的脊至一薄棱；果冠四棱形至圆形或房顶形。花期5-10月，果期7-11月。生海拔2700米以下的湖泊、沼泽或池塘。栽培于中国大部分地区。广泛栽培或逸生于世界各地。

Herbs, annual, aquatic. Leaves deltoid-rhombic to oblate-rhombic. Flowers small, solitary in leaf axils; petals white. Fruits triangular to rhombic, (0-)2-4-horned, crest a prominent bulge to a thin rib; crown tetragonal to rounded, or dome-shaped. Fl. May-Oct. Fr. Jul-Nov. Lakes, swamps, or ponds below 2700 m. Cultivated in most parts of China. Widely cultivated or escaped worldwide.

欧菱 *Trapa natans*

# 柳叶菜科 Onagraceae

## 毛草龙

**Ludwigia octovalvis** (Jacq.) P. H. Raven

多年生粗壮直立草本。茎多分枝，至少于上部茎密具开展柔毛，或具柔毛或近无毛。叶披针形或条状披针形。花瓣黄色；萼片4；雄蕊8。蒴果浅褐色，具8个色深的棱，圆柱形。花果期全年。生海拔2200米以下的田边、湖边、沟谷或开阔的湿地。产中国西南、华南和东南。亚洲、大洋洲、非洲、欧洲、北美洲、南美洲和太平洋岛屿亦有。

Herbs robust, erect, perennial. Stems well-branched, densely spreading pubescent at least on upper stem, or puberulous or subglabrous. Leaves lanceolate or linear-lanceolate. Petals yellow; sepals 4; stamens 8. Capsules pale brown with 8 darker ribs, cylindric. Fl. and fr. all year. Fields or lakes, valleys, or open wet places below 2200 m. Distributed in SW, S and SE China. Also in Asia, Oceania, Africa, Europe, North and South America and Pacific Islands.

毛草龙 *Ludwigia octovalvis*

## 细花丁香蓼

**Ludwigia perennis** L.

一年生直立草本。叶狭椭圆形至披针形。萼片4，三角形；花瓣黄色，椭圆形，长1-3毫米；雄蕊与萼片等数。蒴果常下垂，浅褐色，倒披针体形，长3-16(-19)毫米。种子每室多列。花果期7-11月。生海拔1200米以下的岸边或灌丛。产中国西南、华南和东南。南亚、东南亚、澳大利亚、热带

细花丁香蓼 *Ludwigia perennis*

非洲和太平洋岛屿(新喀里多尼亚)亦有。

Herbs erect, annual. Leaves narrowly elliptic to lanceolate. Sepals 4, deltate; petals yellow, elliptic, 1-3 mm long; stamens as many as sepals. Capsules often nodding, pale brown, oblanceoloid, 3-16(-19) mm long. Seeds multiserial in each locule. Fl. and fr. Jul-Nov. Banks or shrubbery below 1200 m. Distributed in SW, S and SE China. Also in S and SE Asia, Australia, tropical Africa and Pacific Islands (New Caledonia).

## 假柳叶菜

**Ludwigia epilobioides** Maxim.

一年生草本。茎直立，通常粗壮，分枝多。叶狭椭圆形至狭披针形。萼片4或5，稀6；花瓣黄色，倒卵形；雄蕊与萼片等数。蒴果淡褐色，近条形或圆柱状。每室种子1或2列。花期5-8月，果期6-10月。生海拔1600米以下的潮湿地。产中国大部分地区。越南、俄罗斯、朝鲜半岛和日本亦有。

Annual herbs. Stems erect, often stout, well-branched. Leaves narrowly elliptic to narrowly lanceolate. Sepals 4 or 5, rarely 6; petals yellow, obovate; stamens as many as sepals. Capsules light brown, sublinear or terete. Seeds in 1 or 2 rows per locule. Fl. May-Aug. Fr. Jun-Oct. Moist places below 1600 m. Distributed in most parts of China. Also in Vietnam, Russia, Korean Peninsula and Japan.

假柳叶菜 *Ludwigia epilobioides*

## 丁香蓼

**Ludwigia prostrata** Roxb.

一年生直立草本。茎常带红色。叶狭椭圆形；叶柄稍具翅。萼片4，盾形；花瓣黄色，狭匙形；雄蕊与萼片等数。蒴果浅褐色，近圆柱状，四棱形。种子每室1列。花果期6-11月。生海拔800米以下的水边、草地、田间或沼泽地。产云南、广西和海南。印度、尼泊尔、不丹、斯里兰卡、印度尼西亚和菲律宾亦有。

Annual herbs, erect. Stems often red tinged. Leaves narrowly elliptic; petioles slightly winged. Sepals 4, deltate; petals yellow, narrowly spatulate; stamens as many as sepals. Capsules pale brown, subcylindric, tetragonal. Seeds uniserial in each locule. Fl. and fr. Jun-Nov. Waters, grasslands, fields or marshes below 800 m. Distributed in Yunnan, Guangxi and Hainan. Also in India, Nepal, Bhutan, Sri Lanka, Indonesia and the Philippines.

丁香蓼 *Ludwigia prostrata*

水龙 *Ludwigia adscendens*

## 水龙
**Ludwigia adscendens** (L.) H. Hara

多年生草本。叶长圆形至匙状长圆形，无毛。萼片5，三角状渐尖；花瓣乳白色，基部黄色，倒卵形；雄蕊10。蒴果浅褐色，具深褐色肋纹，圆柱形。种子每室1列。花期4-11月，果期5-11月。生海拔1600米以下的水田或水塘。产中国西南、华南和东南。南亚、东南亚、澳大利亚和非洲亦有。

Perennial herbs. Leaves oblong to spatulate-oblong, glabrous. Sepals 5, deltoid-acuminate; petals creamy-white with yellow base, obovate; stamens 10. Capsules light brown with dark brown ribs, cylindric. Seeds in one row per locule. Fl. Apr-Nov. Fr. May-Nov. Water fields or ponds below 1600 m. Distributed in SW, S and SE China. Also in S and SE Asia, Australia and Africa.

## 倒挂金钟
**Fuchsia hybrida** Hort. ex Sieb. et Voss.

灌木或小乔木。叶对生，卵形或狭卵形。花单生，下垂，花梗纤细；花管红色，筒状，上部较大；萼片4，红色，开放时反折；花瓣色多变；花丝红色，伸出花管外；柱头棍棒状。果实紫红色，倒卵状长圆形。花期4-11月。中国广泛栽培。引自南美洲。

Shrubs or little trees. Leaves opposite, ovate or narrowly ovate. Flowers single, pendulous; pedicle slender; tubes red, cylindrical, larger at the upper part; calyx 4, red, reflex after anthesis; petals variable; filaments red, longer than tubes; stigma clavate. Fruits purplish-red, obovoid-oblong. Fl. Apr-Nov. Commonly cultivated in most parts of China. Introduced from South America.

## 露珠草(牛泷草)
**Circaea cordata** Royle

粗壮草本，植株密被柔毛。叶狭至阔卵形。总状花序不分枝或近基部分枝；花瓣反折，白色；蜜腺完全生于花筒内，不明显。果实2室，斜倒卵球形至凸镜状，下面扁平。花期6-8月，果期7-9月。生海拔3500米以下的林下湿地。产中国大部分地区。印度、尼泊尔、克什米尔地区、巴基斯坦、俄罗斯(远东地区)、朝鲜半岛和日本亦有。

Robust herbs, plants densely pubescent. Leaves narrowly to broadly ovate. Racemes simple or branched near base; petals reflexed, white; nectary wholly

倒挂金钟 *Fuchsia hybrida*

露珠草(牛泷草) *Circaea cordata*

高山露珠草 *Circaea alpina*

水珠草
*Circaea canadensis* var. *quadrisulcata*

within floral tube and inconspicuous. Fruits locules 2, obliquely obovoid to lenticular, abaxially flattened. Fl. Jun-Aug. Fr. Jul-Sep. Moist places under forests below 3500 m. Distributed in most parts of China. Also in India, Nepal, Kashmir, Pakistan, Russia (Far East), Korean Peninsula and Japan.

## 高山露珠草
**Circaea alpina** L.

多年生草本。根状茎顶部具结节。茎无毛或具短钩毛。叶形极度多变。总状花序顶生；花瓣白色；蜜腺完全生于花筒内，不明显。果实棍棒状至倒卵球形，1室，具1粒种子。花期6-8月，果期7-9月。生海拔5000米以下的林中、灌丛、高山多草地或潮湿地。产中国大部分地区。低纬度高海拔寒带广布。

Perennial herbs. Rhizomes with tuberous thickening at apex. Stems glabrous or pubescent with short falcate hairs. Leaves highly variably shaped. Racemes terminal; petals white; nectary wholly within floral tube and inconspicuous. Fruits clavate to obovoid, locule 1, seed 1. Fl. Jun-Aug. Fr. Jul-Sep. Forests, thickets, grassy alpine areas or moist or wet places below 5000 m. Distributed in most parts of China. Circumboreal in forests, but restricted to high elevations at lower latitudes.

## 水珠草
**Circaea canadensis** subsp. **quadrisulcata** (Maxim.) Boufford

多年生草本。茎无毛或稀具疏镰状毛。叶狭至阔卵形至长圆状卵形。总状花序不分枝或近基部分枝；花萼反折，大多紫色；花瓣多为粉色，顶端缺刻长为花瓣的1/3或稍长于1/2。果实梨形或近球形。花期6-8月，果期7-9月。生海拔1500米以下的凉温落叶林及混生极地落叶林。产河北、内蒙古、山东、黑龙江、吉林和辽宁。日本北部、朝鲜半岛、俄罗斯和欧洲东部亦有。

Herbs, perennial. Stem glabrous or rarely with sparse falcate hairs. Leaves narrowly to broadly ovate to oblong-ovate. Racemes simple or branched at base; sepals reflexed, most commonly purple; petals commonly pink, apical notch 1/3 to slightly more than 1/2 length of petal. Fruit pyriform or subglobose. Fl. Jun-Aug. Fr. Jul-Sep. Cool-temperate deciduous forests and mixed deciduous-boreal forests below 1500 m. Distributed in Hebei, Neimenggu, Shandong, Heilongjiang, Jilin and Liaoning. Also in N Japan, Korean Peninsula, Russia and E Europe.

## 小花山桃草
**Gaura parviflora** Dougl. ex Lehm.

一年生或短命两年生草本，密被长腺毛。叶互生，狭椭圆形至菱状卵形。花序顶生，穗状；花小，花瓣粉色至玫瑰色，长1.5-3毫米，傍晚开放。蒴果坚果状，纺锤形。花期7-8月，果期8-9月。生海拔1000米以下的荒地上。华北和华东地区归化。原产美国，世界各地归化。

Annual or short-lived biennial herbs, wholly covered with long glandular hairs. Leaves alternate, narrowly elliptic to rhombic-ovate. Inflorescences terminal, spicate; flowers small, petals pink to rose, 1.5-3 mm long, opening at dusk. Capsules nutlike, fusiform. Fl. Jul-Aug. Fr. Aug-Sep. Wastelands below 1000 m. Naturalized in N and E China. Native to USA, naturalized worldwide.

小花山桃草 *Gaura parviflora*

月见草 *Oenothera biennis*

## 月见草
**Oenothera biennis** L.

直立两年生草本。茎生叶狭倒披针形至椭圆形。穗状花序；苞片叶状；花瓣黄色，稀淡黄色；子房绿色，圆柱状，4棱。蒴果绿色，狭披针体形至披针体形，长2-4厘米，无柄。花期7-10月，果期7-11月。生海拔1500米以下的空旷、受干扰地区。中国西南、华南、华中、华北、华东和东北有栽培，后归化。原产美洲；早期引种欧洲，后传播到世界温带和亚热带地区。

Herbs erect, biennial. Cauline leaves narrowly oblanceolate to elliptic. Spikes; bracts foliaceous; petals yellow, rarely pale yellow; ovary green, terete, 4-ribbed. Capsules green, narrowly lanceoloid to lanceoloid, 2-4 cm long, sessile. Fl. Jul-Oct. Fr. Jul-Nov. Open, disturbed areas below 1500 m. Cultivated in SW, S, C, N, E and NE China, later naturalized. Native to America; introduced to Europe, later naturalized in temperate and subtropical regions of the world.

## 黄花月见草
**Oenothera glazioviana** Mich.

直立两年生至多年生短命草本。茎生叶狭椭圆形至披针形或倒披针形。花序为密生不分枝穗状；花瓣黄色，退至橘红色，长4-5厘米；柱头高过花药。蒴果绿色，狭披针体形。花期7-9月，果期8-10月。生海拔800米以下的荒地或田地。栽培或逸生于中国西南、华中、华北、华东和东北。引自美洲。全世界广泛栽培。

Herbs erect, biennial to short-lived perennial. Cauline leaves narrowly elliptic to lanceolate or oblanceolate. Inflorescences a dense unbranched spike; petals yellow, fading to reddish orange, 4-5 cm long; stigmas higher than anther. Capsules green, narrowly lanceoloid. Fl. Jul-Sep. Fr. Aug-Oct. Wastelands or fields below 800 m. Cultivated or escaped in SW, C, N, E and NE China. Introduced from America. Commonly planted worldwide.

## 海边月见草
**Oenothera drummondii** Hook.

草本。茎常平铺地面。叶互生，狭倒卵形至倒披针形，边缘疏生浅齿或全缘。花序顶生，穗状；花大，疏生，黄

黄花月见草 *Oenothera glazioviana*

海边月见草 *Oenothera drummondii*

四翅月见草 *Oenothera tetraptera*

色，花冠筒长25-50毫米；萼片长13-33毫米；花瓣长20-45毫米。蒴果圆柱状。花期5-8月，果期8-11月。生海边沙地。福建、广东和海南栽培或归化。原产美国大西洋海岸与墨西哥湾海岸。

Herbs. Stems usually procumbent. Leaves alternate, narrowly obovate to oblanceolate, margin sparsely and shallowly serrate or entire. Inflorescences terminal, spicate; flowers large, sparse, yellow, floral tubes 25-50 mm long; sepals 13-33 mm long; petals 20-45 mm long. Capsules cylindrical. Fl. May-Aug. Fr. Aug-Nov. Coastal sandy places. Cultivated or naturalized in Fujian, Guangdong and Hainan. Native to Atlantic seacoast of USA and seacoast of Mexico Gulf.

## 四翅月见草

**Oenothera tetraptera** Cav.

匍匐至斜升草本。茎具糙毛，常亦具少量长柔毛。茎生叶倒披针形至倒卵形或椭圆状披针形，边缘稀具锯齿至波状羽裂。花序总状；花瓣白色。蒴果倒卵球形，稀棍棒状，具4翅，顶端有喙，密被开展长柔毛。花期5-8月，果期6-10月。生海拔300-2200米的山坡、路边、田埂或草地。产云南、四川、贵州和台湾。原产北美洲南部；归化于西南亚、澳大利亚、欧洲、中美洲和南美洲北部。

Herbs decumbent to ascending. Stems strigillose, often also moderately villous. Cauline leaves oblanceolate to obovate or elliptic-lanceolate, margin weakly serrate to sinuate-pinnatifid. Inflorescences racemose; petals white. Capsules obovoid, rarely clavate, with 4 wings, apex beaked, densely spreading villose. Fl. May-Aug. Fr. Jun-Oct. Slopes, roadsides, field ridges or grasslands at 300-2200 m. Distributed in Yunnan, Sichuan, Guizhou and Taiwan. Native to S North America; naturalized in SW Asia, Australia, Europe, Central America and N South America.

## 粉花月见草

**Oenothera rosea** L'Hér. ex Ait.

多年生草本，斜升至匍匐。基生叶倒披针形，茎生叶椭圆形至倒披针形或长圆状卵形。花单生于茎顶；花瓣粉红色至紫红色。蒴果棍棒状或狭倒卵球形，果爿具棱或稍具翅。花期5-11月，果期6-12月。生海拔1000-2000米的荒地、山坡草地或沟边潮湿处。产云南、四川、贵州、浙江和江西。原产北美洲南部和南美洲北部，栽培或归化于西南亚、澳大利亚、欧洲和南美洲。

Herbs ascending to decumbent, perennial. Basal leaves oblanceolate, cauline leaves elliptic to oblanceolate or oblong-ovate. Flowers solitary at apical stems; petals pink to purple-red. Capsules clavate or narrowly obovoid, valves angled or weakly winged. Fl. May-Nov. Fr. Jun-Dec. Wastelands, grassy slopes, or wet places by streams at 1000-2000 m. Distributed in Yunnan, Sichuan, Guizhou, Zhejiang and Jiangxi. Native to S North America and N South America, cultivated or naturalized in SW Asia, Australia, Europe and South America.

粉花月见草 *Oenothera rosea*

柳兰 *Chamerion angustifolium*

毛脉柳兰 *Chamerion angustifolium* subsp. *circumvagum*

## 柳兰
**Chamerion angustifolium** (L.) Holub

多年生直立草本。茎与叶背面中脉无毛；茎生叶绿色，条形至披针形，近无柄。花蕾期下垂，花期近直立；花瓣浅粉色至紫色；花柱下部具长柔毛。蒴果密具平伏灰白毛。花期7-9月，果期8-10月。生海拔500-4700米的山坡、灌丛、高山草甸或岩石上。产中国西南、华北、华西、西北和东北。北温带地区亦有。

Perennial herbs, erect. Costae on lower surfaces of leaves and stems glabrous, cauline leaves green, linear to lanceolate, nearly sessile. Flowers nodding in bud, suberect at anthesis; petals pale pink to purple; styles lower part villous. Capsules densely appressed-canescent. Fl. Jul-Sep. Fr. Aug-Oct. Slopes, shrubbery, alpine meadows or rocky places at 500-4700 m. Distributed in SW, N, W, NW and NE China. Also in the north temperate zone.

## 毛脉柳兰
**Chamerion angustifolium** (L.) Holub subsp. **circumvagum** (Mosquin) Hoch

本亚种与柳兰的区别在于茎至少上部具糙毛；叶下常沿中脉具柔毛，(6-)9-23 × (0.7-)1.5-3.4厘米，基部楔形，边缘具小齿，叶柄长2-7毫米；花瓣14-25 × 7-15厘米。花期7-9月，果期8-10月。生海拔3600-4400米的潮湿及受干扰的山地。产中国西南、华北、华西、华东和东北。北温带地区亦有。

This variety differs from the in its stems strigillose at least above; leaves abaxially usually pubescent on midvein, (6-)9-23 × (0.7-)1.5-3.4 cm long, base cuneate, margin denticulate, petioles 2-7 mm long; petals 14-25 × 7-15 mm. Fl. Jul-Sep. Fr. Aug-Oct. Moist, often disturbed places in mountains at 3600-4400 m. Distributed in SW, N, W, E and NE China. Also in the north temperate zone.

## 柳叶菜
**Epilobium hirsutum** L.

多年生草本。叶无柄且抱茎；茎生叶披针状椭圆形至狭倒卵形或椭圆形。花序与花直立；花瓣玫瑰色、粉红或紫红色，宽倒心形；柱头4深裂。蒴果被卷曲柔毛和短腺毛。花期6-8月，果期7-9月。生海拔(100-)500-2800(-3500)米的河谷、沟边、湖边、路边或山坡湿地。产中国大部分地区。欧亚大陆与非洲温带亦有，归化于北美洲。

Perennial herbs. Leaves sessile and clasping stem; cauline leaves lanceolate-elliptic to narrowly obovate or elliptic. Inflorescences and flowers erect; petals rose, pink or purple-red, broadly obcordate; stigmas 4-parted. Capsules floccose-villose, and shortly glandular-hairy. Fl. Jun-Aug. Fr. Jul-Sep. Valleys, by streams, lake edges, roadsides or moist places

柳叶菜 *Epilobium hirsutum*

小花柳叶菜 *Epilobium parviflorum*

on slopes at (100-) 500-2800 (-3500) m. Distributed in most parts of China. Also in Eurasia and temperate Africa, naturalized in North America.

## 小花柳叶菜
**Epilobium parviflorum**
Schreb.

多年生粗壮草本。茎生叶披针状椭圆形至狭披针形或长圆状披针形。花序和花直立；花瓣亮粉色或深紫色，长4-8.5毫米；柱头4深裂。蒴果被柔毛或稀无毛。花期6-9月，果期7-10月。生海拔(300-)500-1800(-2500)米的水边、开阔荒坡或草甸。产中国西南、华中、华北和华西。西南亚、东北亚和非洲亦有。归化于新西兰和北美洲。

Perennial herbs, robust. Cauline leaves lanceolate-elliptic to narrowly lanceolate or oblong-lanceolate. Inflorescences and flowers erect; petals bright pink to dark purple, 4-8.5 mm long; stigma deeply 4-lobed. Capsules pubescent or rarely glabrescent. Fl. Jun-Sep. Fr. Jul-Oct. Watersides, open waste slopes or meadows at (300-)500-1800(-2500) m. Distributed in SW, C, N and W China. Also in SW and NE Asia, and Africa. Naturalized in New Zealand and North America.

## 南湖柳叶菜
**Epilobium nankotaizanense**
Yamamoto

多年生草本。茎高3-18厘米，常于上部分枝，全株具硬毛。叶簇生于上部茎，革质，极肉质；叶柄长1-3毫米。花序斜升；花于蕾期稍下垂；花瓣玫瑰紫色，长1.6-3.3厘米。蒴果；果梗长0.4-0.7厘米。种子密具网纹。花期7-8月，果期8-9月。生海拔2600-3800米的高山湿润开阔山坡。产台湾。

Herbs perennial. Stems 3-18 cm tall, usually branched above, strigillose throughout. Leaves crowded on upper stems, leathery, rather fleshy; petioles 1-3 mm long. Inflorescences ascending; flowers slightly nodding in bud; petals rose-purple, 1.6-3.3 cm long. Capsules; pedicels 0.4-0.7 cm long. Seeds finely reticulate. Fl. Jul-Aug. Fr. Aug-Sep. Local on moist open scree slopes in high mountains at 2600-3800 m. Distributed in Taiwan.

## 长籽柳叶菜
**Epilobium pyrricholophum**
Franch. et Sav.

多年生直立草本，基部具线形根茎。叶边缘每侧常具7-15个锐锯齿。花序与花直立；花瓣粉色至紫色，柱头棍棒状至近棍棒状，全缘。蒴果具糙毛，具腺；果梗长0.7-1.5厘米。种子褐色，长1.5-1.8毫米。生海拔(100-)300-1800米的溪边湿地和低地、湿润丘陵干扰区。产中国西南、华南、华东和华中。俄罗斯(远东地区)和日本亦有。

Herbs perennial, erect, with basal filiform stolons. Leaves margin usually sharply serrulate with 7-15 teeth per side. Inflorescences and flowers erect; petals pink to purple, stigma clavate to subcapitate, entire. Capsules strigillose, glandular; pedicels 0.7-1.5 cm long. Seeds brown, 1.5-1.8 mm long. Wet places along streams and low areas, disturbed moist hillsides in mountains at (100-) 300-1800 m. Distributed in SW, S, E and C China. Also in Russia (Far East) and Japan .

南湖柳叶菜 *Epilobium nankotaizanense*

长籽柳叶菜 *Epilobium pyrricholophum*

毛脉柳叶菜 *Epilobium amurense*

## 毛脉柳叶菜

**Epilobium amurense** Hausskn.

多年生直立草本。茎密具糙毛。叶卵形或长圆卵形，沿脉和边缘被卷曲柔毛和腺毛。花序与花直立或稍下垂；花瓣白色、粉红色或玫瑰紫色；柱头头状或阔头状，全缘。蒴果疏具短硬毛。花期(5-)7-8月，果期(6-)9-10月。生海拔600-4200米的山地溪边、沼泽地、草坡或林缘湿润处。产中国大部分地区。喜马拉雅、克什米尔地区、俄罗斯、朝鲜半岛和日本亦有。

Perennial herbs, erect. Stems densely strigillose. Leaves ovate or oblong-ovate, floccose-villose and glandular-hairy at margin and along nerves. Inflorescences and flowers erect to slightly nodding; petals white, pink or rose-purple; stigmas capitate or broadly capitate, entire. Capsules sparsely strigillose. Fl. (May-)Jul-Aug. Fr. (Jun-) Sep-Oct. Streams on mountains, marshlands, grassy slopes, or moist places of forest edges at 600-4200 m. Distributed in most parts of China. Also in Himalaya, Kashmir, Russia, Korean Peninsula and Japan.

## 光滑柳叶菜

**Epilobium amurense** subsp. **cephalostigma** (Hausskn.) C. J. Chen

此亚种与原变种的区别在于其茎具细弱、稀直立的硬毛；花序具硬毛且无腺；花冠筒具均匀硬毛，无簇毛；叶长圆状披针形至狭卵形。花期6-9月，果期8-10月。生海拔600-2100米的溪边湿地、低海拔沟渠或山坡南向。产中国西南、华南、华中、华东、华北和东北。日本、朝鲜半岛和俄罗斯(远东地区)亦有。

This subspecies differs to subsp. *amurense* in its stems with weak, barely raised strigillose lines; inflorescences strigillose but eglandular; floral tube evenly strigillose, without tufts; leaves oblong-lanceolate to narrowly ovate. Fl. Jun-Sep. Fr. Aug-Oct. Wet areas along streams, roadside ditches at low elevations or in mountains in south at 600-2100 m. Distributed in SW, S, C, E, N and NE China. Also in Japan, Korean Peninsula, Russia (Far East).

光滑柳叶菜 *Epilobium amurense* subsp. *cephalostigma*

## 鳞片柳叶菜

**Epilobium sikkimense** Hausskn.

多年生草本。茎不分枝或有时分枝，主体无毛，高(5-)10-25(-60)厘米。茎生叶卵形至椭圆形或长圆状披针形。花序及花下垂至近直立；花瓣粉色至玫紫色；柱头头状，全缘。蒴果疏被硬毛及腺体。花期7-8月，果期8-9月。生海拔(2400-)3200-4700米的高山草甸或溪边潮湿岩石坡。产云南、四川、西藏、陕西、甘肃和青海。印度、

鳞片柳叶菜 *Epilobium sikkimense*

尼泊尔、不丹和缅甸亦有。

Perennial herbs. Stems simple or sometimes branched, mainly glabrous, (5-)10-25(-60) cm long tall. Cauline leaves ovate to elliptic or oblong-lanceolate. Inflorescences and flowers nodding to suberect; petals pink to rose-purple; stigmas capitate, entire. Capsules sparsely strigillose and glandular. Fl. Jul-Aug. Fr. Aug-Sep. Alpine meadows or moist rocky slopes along streams at (2400-)3200-4700 m. Distributed in Yunnan, Sichuan, Xizang, Shaanxi, Gansu and Qinghai. Also in India, Nepal, Bhutan and Myanmar.

### 沼生柳叶菜
**Epilobium palustre** L.

多年生直立草本。根茎具新出顶生匍匐枝。叶近条形至狭披针形，近全缘至具不明显牙齿。花序直立或花蕾期稍下垂，密具糙毛，有时具腺毛；花直立；萼片长2.5-4.5毫米。花瓣白色至粉色。果梗长1-5厘米。种毛污白色。花期6-8月，果期8-9月。生海拔200-4500(-5000)米的溪边湿地、河边或沼泽。产中国西南、西北和东北。广布于中亚、北亚、西南亚、欧洲和北美洲(包括格陵兰岛)亦有。

Herbs perennial, erect. Stolons with fleshy terminal turions. Leaves sublinear to narrowly lanceolate, subentire to obscurely denticulate; Inflorescences erect or slightly nodding in bud, densely strigillose, sometimes with glandular hairs; flowers erect; sepals 2.5-4.5 mm long; petals white to pink. Fruiting pedicels 1-5 cm long. Seed comas dingy white. Fl. Jun-Aug. Fr. Aug-Sep. Wet places along streams, rivers or marshes at 200-4500(-5000) m. Distributed in SW, NW and NE China. Widespread in C, N and SW Asia, Europe and North America (including Greenland).

沼生柳叶菜 *Epilobium palustre*

# 小二仙草科 Haloragaceae

### 穗状狐尾藻
**Myriophyllum spicatum** L.

多年生沉水草本，雌雄同株。叶4或5枚轮生，篦齿状，裂片13-16对，线形。花两性或单性；穗状花序顶生，花常4朵轮生，雄花生于花序上部，雌花生下部；花瓣4(或5)，浅粉色；雄蕊8。果实4室，球形。花果期4-9月。生海拔4200米以下的池塘、河沟或沼泽。中国各地均有。亚洲、欧洲和北美洲亦有。

Perennial submerged herbs, monoecious. Leaves 4- or 5-whorled, pectinate, segments in 13-16 pairs, filiform. Flowers bisexual or unisexual; spikes terminal, 4-whorled flowers, the upper flowers male, the lower flowers female; petals 4(or 5), pale pink; stamens 8. Fruits 4-loculed, globose. Fl. and fr. Apr-Sep. Ponds, streams or in bogs below 4200 m. Distributed throughout China. Also in Asia, Europe and North America.

穗状狐尾藻
*Myriophyllum spicatum*

粉绿狐尾藻 *Myriophyllum aquaticum*

## 粉绿狐尾藻
**Myriophyllum aquaticum** (Vellozo) Verdc.

多年生水生植物。挺水叶5-7枚轮生，羽状分列，粉绿色；沉水叶丝状。雌花腋生，小，粉红白色。不结果，因除原产地外无雄性植株。花期4-9月。生池塘或溪流浅水处。北京、福建、广东和云南等地栽培或归化。原产南美洲，世界各地栽培或归化。

Perennial aquatic plants. Emergent leaves verticillate, 5-7 in a whorl, pinnate, glaucous; submerged leaves filamentous. Females axillary, small, pinkish-white. Seeds are not produced because there are no male plants found outside of South America. Fl. Apr-Sep. Ponds or shallow streams. Cultivated or naturalized in Beijing, Fujian, Guangdong and Yunnan. Native to South America, widely cultivated or naturalized worldwide.

## 黄花小二仙草
**Gonocarpus chinensis** (Lour.) Orchard

多年生小草本。茎四棱形。叶对生，近无柄，条状披针形至矩圆形，边缘具小锯齿，两面粗糙。总状花序纤细，组成顶生圆锥花序；花两性，极小；花瓣4，黄色。坚果极小，近球形。花期春夏秋季，果期夏秋季。生潮湿的荒山草丛中。产华中、华东、华南和西南。东南亚、澳大利亚和太平洋岛屿亦有。

Perennial small herbs. Stems tetragonous. Leaves opposite, subsessile, linear-lanceolate to oblong, margin minutely serrate, both surfaces coarse. Racemes slender, formed in terminal panicles; flowers small, bisexual; petals 4, yellow. Fruits small, nutlike, subglobose. Fl. spring-autumn. Fr. summer-autumn. Watery underbrushs in mountainous areas. Distributed in C, E, S and SW China. Also in SE Asia, Australia and Pacific Islands.

## 小二仙草
**Gonocarpus micranthus** Thunb.

多年生陆生草本。叶交互对生，卵状心形或椭圆形，通常无毛。圆锥花序顶生；花直立，后下垂；花两性，极小，直径约1毫米；花瓣4，淡红色；雄蕊8。坚果近球形，小，具8棱，无毛；种子1。花期4-8月，果期5-10月。生海拔100-1800米的湿润处或沼泽地，也生于开阔地或多草处。产中国西南、东南、华南、华中和华北。印度、泰国、越南、朝鲜半岛、日本和大洋洲亦有。

Perennial terrestrial herbs. Leaves decussate, ovate-cordate or elliptic, almost glabrous. Panicles terminal; flowers erect then nodding at anthesis; flowers bisexual, small, ca. 1 mm diam, petals reddish; stamens 8. Fruits small, nutlike, subglobose, 8-ribbed, glabrous. Seed 1 per fruit. Fl. Apr-Aug. Fr. May-Oct. Wet or boggy places, either in open or grassy situations at 100-1800 m. Distributed in SW, SE, S, C and N China. Also in India, Thailand, Vietnam, Korean Peninsula, Japan and Oceania.

黄花小二仙草 *Gonocarpus chinensis*

小二仙草 *Gonocarpus micranthus*

# 杉叶藻科 Hippuridaceae

## 杉叶藻

**Hippuris vulgaris** L.

多年生水生草本。茎长10-150厘米，或在流动水中更长。叶二型，具沉水叶及浮水叶；(4-)8-12轮生，常开展，披针形至条形，全缘至稍具锯齿，先端稍加厚，近锐尖；沉水叶长于浮水叶。花淡紫色。瘦果卵球状椭圆体形，光滑。花期4-9月，果期5-10月。生海拔5000米以下的水域。产中国各省温带地区。世界其他温带地区亦有。

Perennial herbs, aquatic. Stems 10-150 cm long or longer in running water. Leaves dimorphic, immersed and floating; leaves (4-)8-12-whorled, often spreading, lanceolate to linear, margin entire to weakly denticulate, apex somewhat thickened, subacute; submerged leaves longer than emergent leaves. Flowers purplish. Achenes ovoid-ellipsoid, smooth. Fl. Apr-Sep. Fr. May-Oct. Watery lands below 5000 m. Distributed in temperate areas of various provinces in China. Also in other temperate areas of the world.

杉叶藻 *Hippuris vulgaris*

# 假繁缕科 Theligonaceae

## 假繁缕

**Theligonum macranthum** Franch.

一年生草本。茎被锈色柔毛。叶卵形、卵状披针形或近椭圆形，长2-5厘米，先端渐尖，两面被疏柔毛或变无毛；托叶膜质，卵形或卵状三角形。雄花花被2深裂，裂片披针形，反折，具5-7脉；雄蕊20余枚，花丝纤细，下垂；雌花花被2浅裂。核果卵球形。花期4-7月，果期6-8月。生海拔1800-2400米的林下。产四川、湖北和浙江。

Annual herbs. Stem ferruginous pubescent. Leaves ovate, ovate-lanceolate or subelliptic, 2-5 cm long, apex acuminate, both surfaces sparsely pubescent or glabrate; stipules membranous, ovate or ovate-triangular. Perianth of staminate flowers 2-parted, lobes lanceolate, reflexed, 5-7-veined; stamens more than 20, filaments slender, nodding; perianth of pistillate flowers 2-lobed. Drupes ovoid. Fl. Apr-Jul. Fr. Jun-Aug. Forests at 1800-2400 m. Distributed in Sichuan, Hubei and Zhejiang.

假繁缕 *Theligonum macranthum*

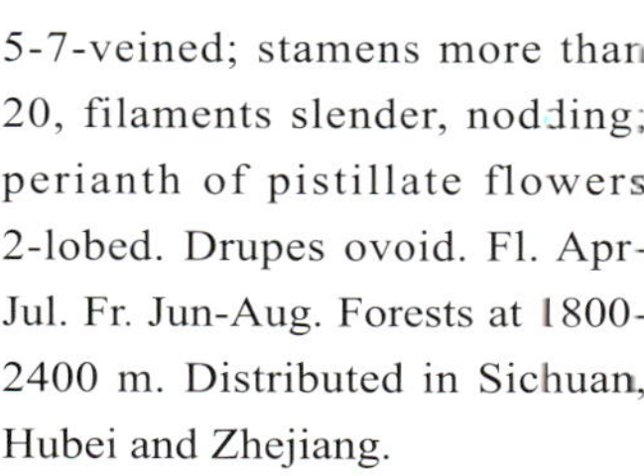

日本假繁缕
*Theligonum japonicum*

### 日本假繁缕

**Theligonum japonicum** Okubo et Makino

多年生草本。茎上部和叶均被短毛。叶卵形或近椭圆形，长0.7-3厘米，先端急尖；托叶膜质，卵形或卵状三角形。雄花花被3深裂，裂片倒披针形，反折，具3-5脉；雄蕊约20，花丝纤细、下垂；雌花花被3-4齿裂。核果斜卵球形。花期3-6月，果期6-8月。生海拔900-1200米的山谷阴湿处或溪边。产浙江、安徽和陕西。日本亦有。

Perennial herbs. Stem pubescent above as well as leaves. Leaves ovate or subelliptic, 0.7-3 cm long, apex acute; stipules membranous, ovate or ovate-triangular. Perianth of staminate flowers 3-parted, lobes oblanceolate, reflexed, 3-5-veined; stamens ca. 20, filaments slender, nodding; perianth of pistillate flowers 3-4-toothed. Drupes obliquely ovoid. Fl. Mar-Jun. Fr. Jun-Aug. Wet places in valleys or by streams at 900-1200 m. Distributed in Zhejiang, Anhui and Shaanxi. Also in Japan.

### 台湾假繁缕

**Theligonum formosanum** (Ohwi) Ohwi et Liu

多年生草本。茎稍被短柔毛。叶宽卵圆形，长1.2-2厘米，先端急尖，上面被毛，下面脉上被稀疏毛；托叶三角状卵形，边缘有睫毛。雄花花被3深裂，裂片倒披针形，反卷，有3-5脉；雄蕊3-8，花丝纤细；雌花花被2裂，裂片钝。核果斜卵球形，有毛，包藏于膜质的萼内。花期3-7月，果期6-8月。生海拔约2700米的林中或山路。产台湾。

Perennial herbs. Stem slightly pubescent. Leaves broadly ovate, 1.2-2 cm long, apex acute, pilose adaxially, puberulent on veins abaxially; stipules deltoid-ovate, margin ciliate. Perianth of staminate flowers 3-parted, lobes oblanceolate, reflexed, 3-5-veined; stamens 3-8, filaments slender; perianth of pistillate flowers 2-lobed, lobes obtuse. Drupes obliquely obovoid, hairy, included in membranous calyx. Fl. Mar-Jul. Fr. Jun-Aug. Forests and along mountain trails at ca. 2700 m. Distributed in Taiwan.

台湾假繁缕 *Theligonum formosanum*

# 锁阳科 Cynomoriaceae

### 锁阳

**Cynomorium songaricum** Rupr.

多年生肉质寄生草本。茎近地生，着生螺旋状排列的鳞片，基部稍加厚。叶鳞片状，卵状三角形。肉穗花序棍棒状，芳香，生于茎顶。果实多数，近白色，近球形；柱头宿存，黄色。花期5-7月，果期6-7月。生海拔500-700米的河边、湖边、池边等沙漠区。常寄生于白刺属、红砂属、猪毛菜属和柽柳属等灌木根部。产中国西北。中亚和西南亚亦有。

Perennial succulent parasitic herbs. Stems subterranean, with spiral scales, ± thickened at base. Leaves scalelike, ovate-deltoid. Spadix clavate, fragrant on stem top. Fruits numerous, whitish, subglobose; styles persistent, yellow. Fl. May-Jun. Fr. Jun-Jul. Lakes, bogs, streams, and rivers in deserts at 500-700 m. Usually parasitic on roots of *Nitraria*, *Reaumuria*, *Salsola*, and *Tamarix* shrubs. Distributed in NW China. Also in C and SW Asia.

锁阳 *Cynomorium songaricum*

# 五加科 Araliaceae

## 多蕊木
**Tupidanthus calyptratus** Hook. f. et Thoms.

小乔木，初直立，后渐为大藤本。叶具7-10小叶。花大，绿色，直径1.5-2.5厘米，雄蕊30-70，子房多室。核果扁球形，黄绿色。种子多粒，黄白色。花期2-3月，果期4-8月。生海拔900-1700米的常绿阔叶林或季雨林。产云南南部和西藏。南亚和东南亚亦有。

Small trees, at first erect, later becoming lofty climbers. Leaves 7-10-foliolate. Flowers large, green, 1.5-2.5 cm diam, stamens 30-70, ovary multilocular. Drupes depressed globose, yellow-green. Seeds many, yellow-white. Fl. Feb-Mar. Fr. Apr-Aug. Evergreen broad-leaved forests or monson forests at 900-1700 m. Distributed in S Yunnan and Xizang. Also in S and SE Asia.

## 刺通草
**Trevesia palmata** (Roxb. ex Lindl.) Visiani

小乔木。枝幼时密被黄棕色锈色绒毛，疏生短刺。单叶，直径60-90厘米，革质，5-9掌状深裂，边缘有大锯齿。伞形花序组成圆锥花序，幼时密具带粉星状柔毛，渐无毛；子房7-12室。果实近球形至扁球形。花期10月，果期翌年5-7月。生海拔600-2000米的密林或杂木林。产云南、贵州和广西。南亚和东南亚亦有。

Small trees. Densely brown-yellow-stellate-tomentose. Leaves simple, 60-90 cm diam, coriaceous, palmately 5-9-parted, grosse-serrate at margin. Inflorescences a panicle of umbels, densely farinose stellate pubescent when young, glabrescent; ovary 7-12-carpellate. Fruits subglobose to compressed-globose. Fl. Oct. Fr. next May-Jul. Dense or mixed forests at 600-2000 m. Distributed in Yunnan, Guizhou and Guangxi. Also in S and SE Asia.

## 通脱木
**Tetrapanax papyrifer** (Hook.) K. Koch

灌木或小乔木。单叶，掌状5-11裂，常再2-3浅裂，腹面密被白色绒毛。花序顶生，小伞形花序多花；花黄白色；花瓣4(或5)。果实成熟时深紫色，球形。花期10-12月，果期翌年1-2月。生海拔100-2800米的向阳肥沃土壤上或杂木灌丛中，有时庭园栽培。产中国西南、华南、东南、华中和华东。

Shrubs or small trees. Leaves simple, palmately 5-11-parted, each section often 2-3-lobed again, abaxially densely white-tomentose. Inflorescences terminal, umbels many flowered; flowers yellowish white; petals 4(or 5). Fruits dark purple at maturity, globose. Fl. Oct-Dec. Fr. next Jan-Feb. Sunny places with rich soils or mixed thickets 100-2800 m, sometimes cultivated in gardens. Distributed in SW, S, SE, C and E China.

多蕊木 *Tupidanthus calyptratus*

刺通草 *Trevesia palmata*

通脱木 *Tetrapanax papyrifer*

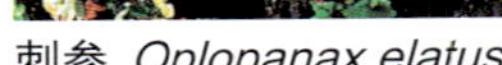
刺参 *Oplopanax elatus*

粗毛罗伞 *Brassaiopsis hispida*

## 刺参

**Oplopanax elatus** (Nakai) Nakai

多刺灌木。小枝粗壮，密具橘黄色刺。叶片圆至扁圆，两面具柔毛或于脉上具刚毛，5-7裂。伞形花序组成总状，顶生，具6-12花。果实倒卵球形，成熟后黄红色。花期6-7月，果期9月。生海拔1400-1600米的杂木林。产吉林东部。俄罗斯和朝鲜半岛亦有。

Spiny shrubs. Branches stout, with dense orange-yellow prickles. Leaves suborbicular to oblate, both surfaces pubescent or setose on veins, 5-7-lobed. Inflorescences terminal, a raceme of umbels, 6-12-flowered. Fruits obovoid, yellowish red at maturity. Fl. Jun-Jul. Fr. Sep. Mixed forests at 1400-1600 m. Distributed in E Jilin. Also in Russia and Korean Peninsula.

## 浅裂罗伞

**Brassaiopsis hainla** (Buch.-Ham.) Seem.

乔木。枝条具圆锥形刺。单叶，裂片5-7，下面具星状柔毛，渐无毛。花序顶生，明显直立至斜升，密被绒毛，花期后无毛，具散生皮刺；子房2室。果实近球形。花期12月至翌年3月，果期6-8月。生海拔1300-2100米的山谷林中。产云南。印度东北部、尼泊尔、不丹、缅甸和泰国亦有。

Trees. Branches with conic prickles. Leaves simple, 5-7-lobed, abaxially stellate pubescent, glabrescent. Inflorescences terminal, apparently erect to ascending, densely tomentose, glabrescent after anthesis, with scattered prickles; ovary 2-carpellate. Fruits subglobose. Fl. Dec to next Mar. Fr. Jun-Aug. Forests in valleys at 1300-2100 m. Distributed in Yunnan. Also in NE India, Nepal, Bhutan, Myanmar and Thailand.

浅裂罗伞 *Brassaiopsis hainla*

## 粗毛罗伞

**Brassaiopsis hispida** Seem.

灌木。枝条密具平伏刺和锈色星状毛。叶单生，具9-11裂片，革质，叶深裂至叶片长度的3/4-4/5，边缘具刺状锯齿。花序顶生，密生皮刺和棕色星状毛被。果实球形，稍压扁。花期6-12月，果期翌年1-2月。生海拔1400-2300米的山谷密林。产云南和西藏。印度(大吉岭、锡金)、不丹、缅甸和越南亦有。

Shrubs. Branches with dense, compressed prickles and ferruginous stellate. Leaves simple, 9-11-lobed, coriaceous, lobes divided 3/4-4/5 way to base, margin spinose-serrulate. Inflorescences terminal, with dense prickles and brown stellate indumentum. Fruits globose, slightly compressed. Fl. Jun-Dec. Fr. next Jan-Feb. Dense forests in valleys at 1400-2300 m. Distributed in Yunnan and Xizang. Also in India(Darjeeling, Sikkim), Bhutan, Myanmar and Vietnam.

## 穗序鹅掌柴

**Schefflera delavayi** (Franch.) Harms

乔木，雌雄同株。小叶(4-)5，椭圆形至卵状长圆形或卵状披针形，下面密被灰白色或黄褐色星状绒毛。圆锥花序组成穗状，顶生，具灰白色绒毛；花无柄；花柱合生成筒状。果实球形，干后具5棱。花期10-11月，果期翌年1月。生海拔600-3000米的常绿阔叶林，散生于沟谷或溪边。产中国西南、东南和华中。越南亦有。

Trees, hermaphroditic. Leaflets (4-)5, elliptic to ovate-oblong or

穗序鹅掌柴 *Schefflera delavayi*

白花鹅掌柴 *Schefflera leucantha*

ovate-lanceolate, abaxially densely gray-white or yellow-brown stellate tomentose. Inflorescences terminal, a panicle of spikes, gray-white tomentose. Flowers sessile; styles united into a column. Fruits globose, 5-ribbed when dry. Fl. Oct-Nov. Fr. next Jan. Evergreen broad-leaved forests, as scattered trees in valleys or stream banks at 600-3000 m. Distributed in SW, SE and C China. Also in Vietnam.

## 白花鹅掌柴 **Schefflera leucantha** R. Vig.

藤状灌木，有时附生。小叶5-7，倒卵形或椭圆形，两面无毛。伞形花序紧缩，顶生组成圆锥状，疏具柔毛；子房5室。果实球形至卵形，稀倒卵形，橙色，有红色腺点。花期1-2月，果期3-8月。生海拔1200-1700米的常绿阔叶林林缘。产云南西北部和广西西南部。泰国和越南北部亦有。

Vine-like shrubs, sometimes epiphytic. Leaflets 5-7, obovate or elliptic, both surfaces glabrous. Inflorescences a compact terminal panicle of umbels, sparsely pubescent; ovary 5-carpellate. Fruits globose to ovoid, rarely obovoid, orange, with red glands. Fl. Jan-Feb. Fr. Mar-Aug. Evergreen broad-leaved forest edges at 1200-1700 m. Distributed in NW Yunnan and SW Guangxi. Also in Thailand and N Vietnam.

## 鹅掌藤 **Schefflera arboricola** (Hayata) Merr.

藤状灌木。小枝有不规则纵皱纹；小叶片(5-)7-9(-10)，倒卵状长圆形至长圆形或椭圆形，两面无毛。顶生伞形花序组成圆锥状，疏具星状绒毛；子房5或6室。果实近球形。花期7-10月，果期8-12月。生海拔900米以下的山地密林下或溪边较湿润处，有时附生。产海南和台湾。

Vine-like shrubs. Branchlets with irregular longitudinal ruga; leaflets (5-)7-9(-10), obovate-oblong to oblong or elliptic, both surfaces glabrous. Inflorescences a terminal panicle of umbels, sparsely stellate tomentose; ovary 5- or 6-carpellate. Fruits subglobose. Fl. Jul-Oct. Fr. Aug-Dec. Dense forests on mountains or damp sites by rivers below 900 m, sometimes epiphytic. Distributed in Hainan and Taiwan.

鹅掌藤 *Schefflera arboricola*

密脉鹅掌柴 *Schefflera elliptica*

球序鹅掌柴 *Schefflera pauciflora*

## 密脉鹅掌柴

**Schefflera elliptica** (Blume) Harms

灌木或小乔木。小叶5-7，椭圆形至长圆形或倒卵形，两面无毛，侧脉5或6(-20)对。伞形花序组成的圆锥花序顶生，幼时疏生或密生星状毛，后无毛；子房5室。果实卵球形至椭球形或近球形。花期3-7月，果期2-7月和10月。生海拔900-2100米的山谷常绿阔叶林或附生于树上。产云南、西藏、贵州、广西和湖南。印度、泰国和越南亦有。

Shrubs or small trees. Leaflets 5-7, elliptic to oblong or obovate, both surfaces glabrous, secondary veins 5 or 6(-20) pairs. A terminal panicle of umbels, sparsely to densely stellate when young, later glabrescent; ovary 5-carpellate. Fruits ovoid to ellipsoid or subglobose. Fl. Mar-Jul. Fr. Feb-Jul and Oct. Evergreen broad-leaved forests in valleys or epiphytic on trees at 900-2100 m. Distributed in Yunnan, Xizang, Guizhou, Guangxi and Hunan. Also in India, Thailand and Vietnam.

## 球序鹅掌柴

**Schefflera pauciflora** R. Vig.

乔木或灌木，雌雄同株。小叶(3-)5-7，卵形至椭圆形或倒卵形，两面无毛。花序为顶生头状花序聚成圆锥形，疏被星状毛；每个头状花序具5-8花；子房5室。果实卵形至近球形或倒卵形，干后具5条肋纹。花期5月、7月、9月，果期6月、7月、9-12月。生海拔200-1700米的山谷常绿阔叶林中或山坡上。产云南东南部、贵州、广西和广东。印度、老挝和越南亦有。

Trees or shrubs, hermaphroditic. Leaflets (3-)5-7, ovate to elliptic or obovate, both surfaces glabrous. Inflorescences a terminal panicles of heads, sparsely stellate tomentose; flowers 5-8 per head; ovary 5 carpellate. Fruits ovoid to subglobose or obovoid, 5-ribbed when dry; disk conic-pentagonal. Fl. May, Jun, Sep. Fr. Jun, Jul, Sep-Dec. Broad-leaved evergreen forests in valleys or mountain slopes at 200-1700 m. Distributed in SE Yunnan, Guizhou, Guangxi and Guangdong. Also in India, Laos and Vietnam.

## 鹅掌柴

**Schefflera heptaphylla** (L.) Frodin

乔木，雄全同株。小叶6-9(-11)，椭圆形至长圆状椭圆形或倒卵状椭圆形；幼时密具星状柔毛，后几渐无毛。伞形花序顶生，组成圆锥状，密具星状绒毛，渐无毛；子房5-9(-10)室。果实球形，黑色。花期9-12月，果期12月至翌年2月。生海拔100-2100米的常绿阔叶林。产中国西南、华南和东南。印度、泰国、越南和日本亦有。

Trees, andromonoecious. Leaflets 6-9(-11), elliptic to oblong-elliptic or obovate-elliptic, densely stellate pubescent when young, almost glabrescent. Inflorescences a terminal panicle of umbels, densely stellate tomentose, glabrescent; ovary 5-9(-10)-carpellate. Fruits orbicular, black. Fl. Sep-Dec. Fr. Dec to next Feb. Evergreen broad-leaved forests at 100-2100 m. Distributed in SW, S and SE China. Also in India,

鹅掌柴 *Schefflera heptaphylla*

星毛鹅掌柴
*Schefflera minutistellata*

Thailand, Vietnam and Japan.

## 星毛鹅掌柴

**Schefflera minutistellata** Merr. ex H. L. Li

灌木或小乔木，髓白色。小叶7-15，椭圆形至卵状披针形或长圆状披针形，叶背稍具星状毛，后渐无毛至无毛。顶生伞形花序组成圆锥状，密具黄褐色或锈色星状柔毛；子房5室。果实球形。花期8-10月，果期10-12月。生海拔1000-1800米的山地林中。产中国西南、华南和东南。

Shrubs or small trees, marrow white. Leaflets 7-15, elliptic to ovate-lanceolate or oblong-lanceolate, abaxially minutely stellate pubescent, later glabrescent to glabrous. Inflorescences a terminal panicle of umbels, densely yellow-brown or ferruginous stellate pubescent; ovary 5-carpellate. Fruits globose. Fl. Aug-Oct. Fr. Oct-Dec. Montane forests at 1000-1800 m. Distributed in SW, S and SE China.

## 白背鹅掌柴

**Schefflera hypoleuca** (Kurz) Harms

小乔木。小叶片革质，常7，下面疏被星状绒毛，上面灰色无毛；小叶柄极不等长。圆锥花序顶生，有花多数，具星状柔毛或无毛；子房5室。果实球形至卵形。花期1-2月，果期4月。生海拔约1300米的密林。产云南南部和西藏(墨脱县)。印度、缅甸和越南亦有。

Small trees. Leaflets coriaceous, usually 7, abaxially sparsely stellate tomentose, adaxially glaucous and glabrous, petiolules very unequal. Panicles terminal, with many flowers, stellate pubescent or glabrous; ovary 5-carpellate. Fruits subglobose to ovoid. Fl. Jan-Feb. Fr. Apr. Dense forests at ca. 1300 m. Distributed in S Yunnan and Xizang (Mêdog County). Also in India, Myanmar and Vietnam.

白背鹅掌柴 *Schefflera hypoleuca*

## 中华鹅掌柴

**Schefflera chinensis** (Dunn) H. L. Li

乔木。小叶5-7，长圆状椭圆形或椭圆形，下面疏具星状柔毛或近无毛。花无梗或几无梗，密集成头状花序，再组成圆锥花序，密具星状绒毛，渐无毛；花柱基部合生；苞片宿存；子房5室。果实球形至倒卵球形。花期10-11月，果期翌年2-3月。生海拔1500-2700米的常绿阔叶林或沟谷湿润处。产云南和江西。

Trees. Leaflets 5-7, oblong-elliptic or elliptic, abaxially sparsely stellate pubescent or subglabrous. Flowers sessile or subsessile, aggregate into a capitulum, then the capitula arranged in a panicle, densely woolly stellate pubescent, glabrescent; styles connate at base; bracts persistent; ovary 5-carpellate. Fruits globose to obovoid. Fl. Oct-Nov. Fr. next Feb to Mar. Evergreen broad-leaved forests or wet places in valleys at 1500-2700 m. Distributed in Yunnan and Jiangxi.

中华鹅掌柴 *Schefflera chinensis*

谅山鹅掌柴 *Schefflera lociana*

## 谅山鹅掌柴

**Schefflera lociana** Grushv et Skvortsova

乔木或灌木。掌状复叶；小叶8-10，革质；托叶和叶柄基部合生成鞘状。圆锥花序；花梗无关节；花萼灰白色，全缘，被毛；花瓣5-11，镊合状排列；雄蕊和花瓣同数；子房常6室；花柱全部合生成柱状，柱头明显。果实球形、近球形或卵球形。种子扁平，被棕色毛。花期8-9月。生石山的密林中。产广西(龙州县)。

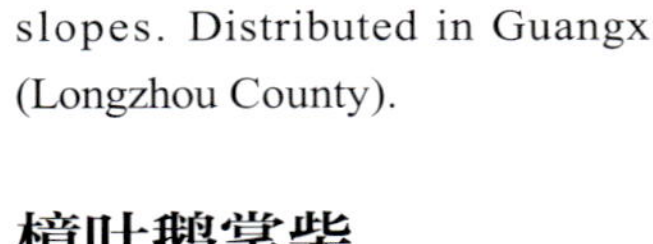

Trees or shubs. Leaves palmately compound; leaflets 8-10, leathery; stipules and petioles basally connate into a sheath. Inflorescences panicle; pedicels not articulate; calyx gray white pubescent, entire; petals 5-11, valvate; stamens 5-11; ovary 6-carpellate; styles united into a column; stigma obvious. Fruit globose to ovoid. Seeds compressed, brown pubescent. Fl. Aug-Sep. Dense forests on rocky mountain slopes. Distributed in Guangxi (Longzhou County).

## 樟叶鹅掌柴

**Schefflera pes-avis** R. Vig

乔木。掌状复叶，叶柄3-10厘米；小叶(3-)5(-7)，椭圆形，革质，两面无毛，全缘，先端锐尖。伞形花序组成圆锥花序，无毛；花萼无毛；子房5室；花柱合生成圆锥柱；柱头头状。果实近球形到椭圆形，直径3-5毫米，干时具5棱。花期8-9月，果期10月至翌年1月。生海拔600-800米的山坡和丘陵的顶部。产广西(靖西、龙州县、那坡县)。

Trees. Leaves palmately compound; petioles 3-10 cm long; leaflets (3-)5(-7), elliptic, leathery, glabrous, margin entire, acute acuminate. Inflorescences a terminal panicle of umbels, glabrous; calyx glabrous; ovary 5-carpellate; styles connate into a conic column; stigmas capitate. Fruit subglobose to ellipsoid, 3-5 mm diam, 5-ribbed when dry. Fl. Aug-Sep. Fr. Oct to next Jan. Rocky mountain slopes and hill tops at 600-800 m. Distributed in Guangxi (Jingxi, Longzhou County, Napo County).

## 树参

**Dendropanax dentiger** (Harms) Merr.

乔木或灌木。叶有时二型，厚纸质或革质，密生粗大半透明的红棕色腺点。花序顶生；伞形花序顶生，单花或2-3(-5)个生，具10-25(-50)花；子房5室。果实椭圆形至近球形，宿存花柱离生部分于顶端分开至反折。花期6月、8-9月，果期7月、10-12月。生海拔1800米以下的常绿阔叶林或灌丛。产中国西南、东南、华中和华东。老挝、泰国、越南和柬埔寨亦有。

Trees or shrubs. Leaflets sometimes dimorphic, thickly papery or coriaceous, densely clothed with broad reddish brown glands. Inflorescences terminal; umbels solitary or 2-3(-5), 10-25(-50)-flowered; ovary 5-carpellate. Fruits ellipsoid to subglobose, persistent styles free arms divergent to recurved apically. Fl. Jun, Aug-Sep. Fr. Jul, Oct-Dec. Evergreen broad-leaved forests or thickets below 1800 m. Distri-

樟叶鹅掌柴 *Schefflera pes-avis*

树参 *Dendropanax dentiger*

海南树参 *Dendropanax hainanensis*

buted in SW, SE, C and E China. Also in Laos, Thailand, Vietnam and Cambodia.

## 海南树参

**Dendropanax hainanensis** (Merr. et Chun) Merr. et Chun

乔木，雄花两性花同株。叶不具腺点，全缘。花序顶生，(2-)3-5个伞形花序组成圆锥状，大多具两性花，两侧伞形花序常为雄性；子房5室；花柱完全合生成筒状。果实球形，干后具肋纹。花期6-7月，果期10月。生海拔700-1500米的山谷林中或山坡上。产云南、贵州、广西、广东、海南和湖南。越南北部亦有。

Trees, andromonoecious. Leaves not glandular punctate, margin entire. Inflorescences terminal, a panicles of (2-)3-5 umbels, mostly with bisexual flowers, usually some lateral umbels with male flowers; ovary 5-carpellate; styles completely united into a column. Fruits globose, ribbed when dry. Fl. Jun-Jul. Fr. Oct. Forests in valleys or mountain slopes at 700-1500 m. Distributed in Yunnan, Guizhou, Guangxi, Guangdong, Hainan and Hunan. Also in N Vietnam.

变叶树参 *Dendropanax proteus*

## 变叶树参

**Dendropanax proteus** (Champ. ex Benth.) Benth.

灌木。小叶二型；叶片纸质至革质，常无腺点，三出脉。单个或2-3(-5)个伞形花序簇生于顶部，具(15-)20-40花；子房4或5室；花柱完全合生成筒状。果实卵球形至球形。花期7-9月，果期9-12月。生林中、溪边或山坡。产中国西南、华南和东南。

Shrubs. Leaflets dimorphic; leaves papery to coriaceous, usually without glandular punctate, 3-veined. Umbels terminal, solitary or 2-3(-5)-clustered, (15-)20-40-flowered; ovary 4- or 5-carpellate; styles completely united into a column. Fruits ovoid to globose. Fl. Jul-Sep. Fr. Sep-Dec. Forests, along streams or mountain slopes. Distributed in SW, S and SE China.

## 常春藤

**Hedera nepalensis** K. Koch var. **sinensis** (Tobl.) Rehd.

常绿藤本，有附生气根，攀援。单叶，二型，全缘或3浅裂，脉纹在两面明显。花序为顶生伞形或为一小总状花序，具锈色鳞片；子房5室。果实成熟时红色或黄色，球形。花期9-11月，果期翌年3-5月。生海拔3500米以下的林缘、多石山坡或林下路边。产中国西南、东南、华中、华西和华东。老挝和越南亦有。

Evergreen vine, with epiphytic air-roots, scandent. Leaves dimorphic, simple, entire or 3-lobed, venation distinct on both surfaces. Inflorescences a terminal umbel or a small raceme, with ferruginous scales; ovary 5-carpellate. Fruits red or yellow at maturity, globose. Fl. Sep-Nov. Fr. next Mar-May. Forest edges, rocky slopes or roadsides under forests below 3500 m. Distributed in SW, SE, C, W and E China. Also in Laos and Vietnam.

常春藤 *Hedera nepalensis* var. *sinensis*

梁王茶 *Metapanax delavayi*

刺楸 *Kalopanax septemlobus*

## 刺楸

**Kalopanax septemlobus** (Thunb.) Koidz.

落叶乔木。小枝劲直，具许多刺。小叶纸质，在长枝上互生，短枝上簇生，5-7裂。花序18-25 × 20-30厘米；花冠白色或黄绿色；花柱2，下面联合，顶端分枝反折。果实球形，成熟时深蓝色。花期7-8月，果期9-10月。生2500米以下的山地林中、灌丛或林缘。产中国西南、东南、华中、华北、华西和华东。俄罗斯、朝鲜半岛和日本亦有。

Deciduous trees. Branches stout, with numerous prickles. Leaves papery, alternate on long branches, clustered at short branches, 5-7-lobed. Inflorescences 18-25 × 20-30 cm; corolla white or yellowish green; styles 2, united below, apical branches recurved. Fruits dark blue at maturity, globose. Fl. Jul-Aug. Fr. Sep-Oct. Montane forests, thickets or forest edges below 2500 m. Distributed in SW, SE, C, N, W and E China. Also in Russia, Korean Peninsula and Japan.

## 异叶梁王茶

**Metapanax davidii** (Franch.) J. Wen et Frodin

小乔木。单叶，全缘或3裂，长圆状卵形至长圆状披针形，革质，基生三出脉，边缘疏具锯齿。伞房花序组成圆锥状，顶生。果实侧扁，球形。花期6-8月，果期9-10月。常见生海拔800-3000米的灌丛、河岸、林缘或路边。产云南、四川、贵州、湖北、湖南和陕西。越南北部亦有。

Small trees. Leaves simple, entire or 3-lobed, oblong-ovate to oblong-lanceolate, coriaceous, 3-veined from base, margin sparsely serrate. Inflorescences terminal, a panicle of umbels. Fruits laterally compressed, circular. Fl. Jun-Aug. Fr. Sep-Oct. Common in scrubs, stream banks, forest edges or roadsides at 800-3000 m. Distributed in Yunnan, Sichuan, Guizhou, Hubei, Hunan and Shaanxi. Also in N Vietnam.

## 梁王茶

**Metapanax delavayi** (Franch.) J. Wen et Frodin

灌木。掌状复叶，小叶 2-5，近无柄或柄长达1厘米，披针形或狭披针形，边缘具锯齿。伞形花序顶生，组成圆锥状。果实侧面扁，球形或稍扁球形。花期9-10月，果期12月至翌年1月。生海拔1500-3000米的沟谷灌丛或杂木林。产云南、四川和贵州。越南北部亦有。

Shrubs. Leaves palmately compound, leaflets 2-5, subsessile or petiolules to 1 cm long, lanceolate to narrowly lanceolate, margin serrulate. Inflorescences terminal, a panicle of umbels. Fruits laterally compressed, circular to slightly oblate. Fl. Sep-Oct. Fr. Dec to next Jan. Scrubs or mixed forests in ravines at 1500-3000 m. Distributed in Yunnan, Sichuan

异叶梁王茶 *Metapanax davidii*

红毛五加
*Eleutherococcus giraldii*

and Guizhou. Also in N Vietnam.

## 红毛五加
**Eleutherococcus giraldii**
(Harms) Nakai

灌木。小枝密被开展或反折的鬃毛状刺。小叶(3-)5，倒卵状长圆形，边缘具不规则重锯齿。伞形花序单生于顶部；花冠白色；子房具5室；花柱1/5-1/2联合。果实成熟时黑色，球形。花期6-7月，果期9-10月。生海拔1300-3500米的山坡灌丛中。产中国西南、华中和华西。

Shrubs. Branches with dense bristlelike spreading or reflexed prickles. Leaflets (3-)5, obovate-oblong, margin irregularly doubly serrate. Inflorescences terminal, a solitary umbel; corolla white; ovary 5-carpellate; styles united 1/5-1/2 their length. Fruits black at maturity, globose. Fl. Jun-Jul. Fr. Sep-Oct. Scrubs on mountain slopes at 1300-3500 m. Distributed in SW, C and W China.

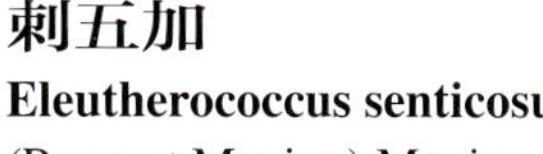

## 刺五加
**Eleutherococcus senticosus**
(Rupr. et Maxim.) Maxim.

一年生或二年生草本。枝上密生刺。小叶(3-)5，椭圆状倒卵形或长圆形，纸质。伞形花序单个顶生，或2-6个组成稀疏的圆锥花序；花冠紫黄色；子房5室；柱头联合成筒状。果实球形或卵球形。花期6-7月，果期8-10月。生海拔2000米以下的森林、沟谷或灌丛。产中国西南、华北和东北。俄罗斯、朝鲜半岛和日本亦有。

Annual or biennial herbs. Branches densely clothed with prickles. Leaflets (3-)5, elliptic-obovate or oblong, papery. Umbels solitary, terminal, or sparse panicles with flowers 2-6; corolla purple-yellow; ovary 5-carpellate; styles united into a column. Fruits orbicular or ovoid-crbicular. Fl. Jun-Jul. Fr. Aug-Oct. Forests, valleys or among bushes below 2000 m. Distributed in SW, N and NE China. Also in Russia, Korean Peninsula and Japan.

刺五加
*Eleutherococcus senticosus*

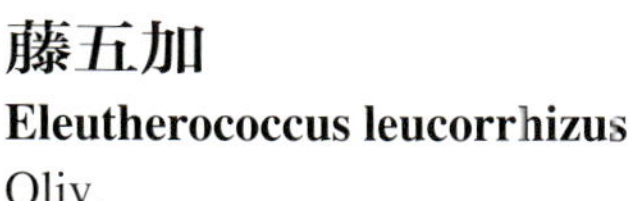

## 藤五加
**Eleutherococcus leucorrhizus**
Oliv.

灌木，有时攀援状。枝条无毛，于节上具少量细弱、圆柱的刺。小叶侧脉在两面隆起而明显。伞形花序单个顶生或数个组成圆锥状；子房5室，无毛。果实卵球形。花期6-8月，果期8-11月。生海拔100-3200米的林中、林缘或灌丛。产中国西南、东南、华中、华西和华东。不丹亦有。

Shrubs, sometimes climbers. Branches glabrous, with few, slender, terete prickles generally at nodes. Leaflets lateral nerves obviously convex at both sides. Inflorescences terminal, a corymbose panicle of umbels or a solitary umbel; ovary 5-carpellate, glabrous. Fruits ovoid-orbicular. Fl. Jun-Aug. Fr. Aug-Nov. Forests, forest edges or scrubs at 100-3200 m. Distributed in SW, SE, C, W and E China. Also in Bhutan.

藤五加 *Eleutherococcus leucorrhizus*

糙叶五加 *Eleutherococcus henryi*

## 糙叶五加
**Eleutherococcus henryi** Oliv.

灌木。树枝幼时密被糙毛，具散生劲直刺。小叶(3-)5，腹面稍具糙毛，上部边缘具细齿。花序顶生，数个伞形花序组成圆锥状；子房2-5室；花柱贴生成筒状；花柱宿存。果实成熟时黑色，椭圆状球形。花期7-9月，果期9-10月。生海拔800-3200米的灌丛、林缘、路边或山坡。产中国东南、华北、华西和华东。

Shrubs. Branches densely and roughly pubescent when young, with scattered, stout prickles. Leaflets(3-)5, adaxially ± scabrous-pubescent, margin serrate apically. Inflorescences terminal, a panicle of umbels, with several umbels; ovary 2-5-carpellate; styles united into a column; styles persistent. Fruits black at maturity, ellipsoid-globose. Fl. Jul-Sep. Fr. Sep-Oct. Scrub fields, forest edges, roadsides, or mountain slopes at 800-3200 m. Distributed in SE, N, W, and E China.

## 白簕 (三叶五加)
**Eleutherococcus trifoliatus** (L.) S. Y. Hu

藤状或攀援灌木。叶为3小叶，稀4-5小叶；萼片具5齿，无毛。顶生伞形花序组成总状或3-10个伞形花序组成复伞状；子房2室；柱头合生至中部；花柱2裂。果实球形。花期8-11月，果期9-12月。生海拔3200米以下的林中、林缘、灌丛、路边或山坡。产华南、东南、华中和华东。印度、泰国、越南、菲律宾和日本亦有。

Shrubs, scandent or climbers. Leaves ternate, rarely 4-5-foliolate; calyx with 5 teeth, glabrous. Inflorescences a terminal raceme of umbels or a compound umbel, with 3-10 umbels; ovary 2-carpellate; styles united to middle; styles bifid. Fruits globose. Fl. Aug-Nov. Fr. Sep-Dec. Forests, forest edges, bushes, roadsides or mountain slopes below 3200 m. Distributed in S, SE, C and E China. Also in India, Thailand, Vietnam, the Philippines and Japan.

白簕(三叶五加) *Eleutherococcus trifoliatus*

## 细柱五加
**Eleutherococcus nodiflorus** (Dunn) S. Y. Hu

灌木。小叶柄短；叶为3-4(-5)小叶；叶柄长3-8厘米，无毛，有疏刺。伞形花序生于短枝叶腋，单生或2至3个聚生；花冠微黄绿色；子房2-3室。果实成熟时黑色，近球形。花期4-7月，果期6-10月。生海拔3000米以下的灌丛、山坡、路边、林缘或河边。产中国黄河以南大部分地区。

Shrubs. Petiolules very short; leaves 3-4 (-5) -foliolate; petioles 3-8 cm long, glabrous, with small scattered prickles. Inflorescences born in axils of leaves on short shoots, a solitary umbel or sometimes 2-3 umbels together; corolla yellowish green; ovary 2-3-carpellate. Fruits black at maturity, subglobose. Fl. Apr-Jul. Fr. Jun-Oct. Scrubs, mountain slopes, roadsides, forest edges or along rivers below 3000 m. Distributed in most parts of S Yellow River in China.

## 无梗五加
**Eleutherococcus sessiliflorus** (Rupr. et Maxim.) S. Y. Hu

灌木或小乔木。刺粗壮，直或弯曲。小叶3-5，纸质，下面无毛或稍粗糙。3-6个头状伞形花序聚成总状，顶生；花冠浅暗紫色；子房具2室；花柱基部联合成筒状。果实倒卵球形。花期8-9月，果期9-11月。生海拔200-1000米的林中或灌丛。产

细柱五加 *Eleutherococcus nodiflorus*

河北、山西、黑龙江、吉林和辽宁。朝鲜半岛亦有。

Shrubs or small trees. Spines robust, erect or curved. Leaflets 3-5, papery, adaxially glabrous or slightly scabrous. Inflorescences terminal, a raceme of umbels, with 3-6 capitate umbels; corolla dull purplish; ovary 2-carpellate; styles united basally into a column. Fruits obovoid. Fl. Aug-Sep. Fr. Sep-Nov. Forests or bushes at 200-1000 m. Distributed in Hebei, Shanxi, Heilongjiang, Jilin and Liaoning. Also in Korean Peninsula.

## 人参木

**Chengiopanax fargesii** (Franch.) C. B. Shang et J. Y. Huang

落叶乔木，枝、叶和花序幼时密被短的锈色星状毛，渐无毛。掌状复叶；小叶5-7(-9)，两面无毛，边缘稍具细锯齿。伞房花序组成圆锥状，顶生；小伞形花序具8-20花；花柱贴生成筒状。果实具宿存花柱。花期9月，果期11-12月。生海拔1000-2000米的山坡杂木林。产重庆(万县)和湖南(新宁县)。

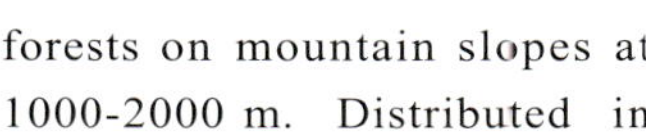

Deciduous trees, branches, leaves and inflorescences densely shortly ferruginous stellate tomentose when young, glabrescent. Leaves palmately compound; leaflets 5-7(-9), both surfaces glabrous, margin minutely serrulate. Inflorescences terminal, a corymbose panicle; umbels 8-20-flowered; styles united into a column. Fruits with persistent styles. Fl. Sep. Fr. Nov-Dec. Mixed forests on mountain slopes at 1000-2000 m. Distributed in Chongqing (Wan County) and Hunan (Xinning County).

无梗五加 *Eleutherococcus sessiliflorus*

人参木 *Chengiopanax fargesii*

罗伞 *Brassaiopsis glomerulata*

## 罗伞

**Brassaiopsis glomerulata** (Blume) Regel

乔木。枝条被红锈色绒毛。掌状复叶具小叶5-9；小叶柄长2-9厘米。花序顶生，下垂，无刺；花白色，芳香，密被锈色绒毛。果实球形或扁球形。花期6-8月，果期翌年1-2月。生海拔400-2400米的山坡林中或山谷林中。产云南、四川、贵州、广西和广东。南亚和东南亚亦有。

Trees. Branchlets rust-red-tomentose. Leaves palmately compound, with 5-9 leaflets; petiolules 2-9 cm long. Inflorescences terminal, pendent, unarmed; flowers white, fragrant, densely rusted tomentose. Fruits globose or compressed globose. Fl. Jun-Aug. Fr. next Jan-Feb. Forests on mountain slopes or in valleys at 400-2400 m. Distributed in Yunnan, Sichuan, Guizhou, Guangxi and Guangdong. Also in S and SE Asia.

## 短梗大参

**Macropanax rosthornii** (Harms) C. Y. Wu ex G. Hoo

常绿灌木或小乔木，约8米高。小叶纸质，3-7。花序圆锥状，光滑；花序梗长1-3厘米；花梗长3-8毫米；花萼边缘不明显；花柱宿存，顶部分裂。果实卵球形至球形。花期7-9月，果期10-12月。生海拔1500米的林中、灌丛、林缘或路边。产中国西南、东南、华中和华西。

Evergreen shrubs or small trees, ca. 8 m tall. Leaflets papery, 3-7. Inflorescences paniculate, glabrous throughout; peduncles 1-3 cm long; pedicels 3-8 mm long; calyx rim inconspicuous; styles persistent, divided apically. Fruits ovoid to globose. Fl. Jul-Sep. Fr. Oct-Dec. Forests, among bushes, forest edges or roadsides at 1500 m. Distributed in SW, SE, C and W China.

## 波缘大参

**Macropanax undulatus** (Wall. ex G. Don) Seem.

乔木。掌状复叶，小叶3-5，近革质，无毛，边缘有疏离的小锯齿。花序圆锥状，无毛；花小，长约3毫米；花序梗长0.5-2厘米；花梗长3-5毫米。果实卵球形至椭球形。花期9月。生海拔400-2200米的杂木林。产云南、贵州和广西。南亚和东南亚亦有。

Trees. Leaves palmately compound, 3-5-foliolate, subleathery, glabrous, margin remotely serrulate. Inflorescences paniculate, glabrous throughout; flowers

短梗大参 *Macropanax rosthornii*

波缘大参 *Macropanax undulatus*

显脉大参 *Macropanax chienii*

small, ca. 3 mm long; peduncles 0.5-2 cm long; pedicels 3-5 mm long. Fruits ovoid to ellipsoid. Fl. Sep. Mixed forests at 400-2200 m. Distributed in Yunnan, Guizhou and Guangxi. Also in S and SE Asia.

## 显脉大参
**Macropanax chienii** G. Hoo

乔木。掌状复叶，小叶3-4，长圆状椭圆形至长圆形，边缘有不明显疏离锯齿，脉网在两面明显。圆锥花序较窄，分枝斜上升，密具棕色绒毛；花萼具5齿，无毛。花期11月，果期不详。生海拔800-900米的山坡灌丛。产云南南部。

Trees. Leaves palmately compound, leaflets 3-4, oblong-elliptic or oblong, inconspicuously and remotely serrate at margin, with conspicuous reticulation on both surfaces. Panicles narrow, with ascendent branchlets, densely brown pubescent; calyx 5-toothed, glabrous. Fl. Nov. Fr. unknown. Thickets on mountain slopes at 800-900 m. Distributed in S Yunnan.

## 大参
**Macropanax dispermus** (Blume) Kuntze

乔木。掌状复叶，两面无毛，小叶5-7，边缘有粗齿状锯齿，脉网在背面不明显。圆锥花序较大，分枝平展，密具锈色短星状柔毛；花萼具5齿。果实椭圆形至长圆形。花期8-9月，果期翌年1-2月。生海拔300-2300米的山谷混交林或密林或灌丛中。产云南。南亚和东南亚亦有。

Trees. Leaves palmately compound, both surfaces glabrous, leaflets 5-7, grossly serrate at margin, abaxially with inconspicuous reticulation. Panicles ample, with horizontal branchlets, densely shortly ferruginous stellate pubescent; calyx 5-toothed. Fruits ellipsoid to oblong. Fl. Aug-Sep. Fr. next Jan-Feb. Mixed or dense forests in valleys or thickets at 300-2300 m. Distributed in Yunnan. Also in S and SE Asia.

马蹄参 *Diplopanax stachyanthus*

大参 *Macropanax dispermus*

## 马蹄参
**Diplopanax stachyanthus** Hand.-Mazz.

常绿乔木。单叶，革质，全缘，椭圆形、倒卵状椭圆形或倒卵状披针形。圆锥状花序顶生。果实长圆状卵形至长圆状椭圆形。胚弯曲，横切面呈马蹄形。花期6-8月，果期8-10月。生海拔1300-1900米的常绿阔叶林。产云南(屏边县、喜洲)、贵州、广西、广东(阳春、阳江、余元)和湖南(莽山)。越南北部亦有。

Evergreen trees. Leaves simple, coriaceous, entire, elliptic, obovate-elliptic or obovate-lanceolate. Inflorescences terminal, paniculate. Fruits oblong-ovoid to oblong-ellipsoid. Embryo curved, hippocrepiform in transverse section. Fl. Jun-Aug. Fr. Aug-Oct. Evergreen broad-leaved forests at 1300-1900 m. Distributed in Yunnan (Pingbian County, Xizhou), Guizhou, Guangxi, Guangdong (Yangchun, Yangjiang, Yuyuan) and Hunan (Mang Mountain). Also in N Vietnam.

华幌伞枫 *Heteropanax chinensis*

## 华幌伞枫
**Heteropanax chinensis** (Dunn.) H. L. Li

灌木。三至五回羽状复叶，小叶小，椭圆状披针形，长2.5-6厘米，先端长渐尖，两面无毛。花序密具锈色绒毛；主轴长70厘米；次级主轴长20厘米；花柱反折。果实极扁。花期10-11月，果期翌年1-2月。生海拔800米以下的山坡密林或沟谷。产云南(思茅)和广西(南宁、上思县)。越南北部亦有。

Shrubs. Leaves 3-5-pinnate; leaflets small, elliptic-lanceolate, 2.5-6 cm long, apexacuminate, both surfaces glabrous. Inflorescences densely ferruginous tomentose; primary axis to 70 cm long; secondary axes to 20 cm long, styles recurved. Fruits very flattened. Fl. Oct-Nov. Fr. next Jan to Feb. Dense forests on slopes or valleys below 800 m. Distributed in Yunnan (Simao) and Guangxi (Nanning, Shangsi County). Also in N Vietnam.

## 羽叶参
**Pentapanax fragrans** (D. Don) T. D. Ha

小乔木或攀援灌木。一回羽状复叶；小叶边缘具缘毛或锯齿，有时全缘。伞形花序组成顶生的伞房花序，具1个两性花组成的顶生伞形花序及2-6个侧生轮生雄性伞形花序；子房5室；花柱合生成筒状。果实卵球形。花期6-8月，果期8-10月。生海拔2000-3600米的湿润森林、林缘、沟谷或山坡。产云南、四川和西藏。南亚和东南亚亦有。

Small trees or climbing shrubs. Leaves 1-pinnately compound; leaflets margin ciliate or serrate, sometimes entire. Terminal corymb of umbels, with a terminal umbel of bisexual flowers and 2-6 lateral, verticillate umbels of male flowers; ovary 5-carpellate; styles united into a column. Fruits ovoid. Fl. Jun-Aug. Fr. Aug-Oct. Moist forests, forest edges, ravines or mountain slopes at 2000-3600 m. Distributed in Yunnan, Sichuan and Xizang. Also in S and SE Asia.

羽叶参 *Pentapanax fragrans*

## 长刺楤木
**Aralia spinifolia** Merr.

灌木。叶柄、叶轴、次级叶轴和小叶均具散生的圆锥状刺；复叶二至三回，每回5-9小叶，卵形至狭卵形。顶生伞形花序组成圆锥状，多刚毛。果实卵球形，黑褐色。花期8-10月，果期10-12月。生海拔200-800米

长刺楤木 *Aralia spinifolia*

野楤头 *Aralia armata*

的山坡、开阔林地、峡谷、路边或林缘阳光充足处。产中国西南、华南和东南。

Shrubs. Petioles, rachises, secondary rachises, and leaflets with scattered conic prickles; leaves 2-3-pinnately compound, leaflets 5-9 per pinna, ovate to narrowly ovate. Inflorescences a terminal panicle of umbels, setose. Fruits ovoid-orbicular, black brown. Fl. Aug-Oct. Fr. Oct-Dec. Mountain slopes, open woods, ravines, roadsides or sunny sites of forest edges at 200-800 m. Distributed in SW, S and SE China.

## 野楤头

**Aralia armata** (Wall. ex G. Don) Seem.

灌木。枝条具圆锥状、反曲的刺；二或三回羽状复叶；每个羽片具小叶5-9，两面多毛。顶生伞形花序组成圆锥状，具反曲刺。果实球形，具5棱，柱头宿存，辐射状至反折。花期8-10月，果期9-12月。生海拔1600米以下的林中或林缘。产云南南部和西部、贵州、广西、广东和江西(武功山)。印度、缅甸、泰国、越南和马来西亚北部亦有。

Shrubs. Branches with conic, often recurved prickles; 2- or 3-pinnately compound; leaflets 5-9 per pinna, both surfaces pilose. Inflorescences a terminal panicle of umbels, with recurved prickles; styles persistent, radiating to recurved. Fruits globose, with 5 ribs, Fl. Aug-Oct. Fr. Sep-Dec. Forests or forest edges below 1600 m. Distributed in S and W Yunnan, Guizhou, Guangxi, Guangdong and Jiangxi (Wugong Mountain). Also in India, Myanmar, Thailand, Vietnam and N Malaysia.

## 黄毛楤木

**Aralia chinensis** L.

灌木或小乔木。枝多刺。二回羽状复叶，具柔毛及刺；每个羽片具小叶5-13，下面灰绿色，具绒毛。顶生伞形花序组成圆锥状；花柱宿存，花柱反曲。果实球形至卵球形。花期9-12月，果期11月至翌年1月。生溪边林中或山坡灌丛。产贵州、广西、广东、海南、福建和江西。

Shrubs or small trees. Branches prickly. Leaves 2-pinnately compound, pubescent and prickly; leaflets 5-13 per pinna, abaxially pale grayish green, tomentose. Terminal panicle of umbels; styles persistent, recurved. Fruits globose to ovoid. Fl. Sep-Dec. Fr. Nov to next Jan. Forests by stream banks or thickets on slopes. Distributed in Guizhou, Guangxi, Guangdong, Hainan, Fujian and Jiangxi.

黄毛楤木 *Aralia chinensis*

楤木 *Aralia elata*

## 楤木

**Aralia elata** (Miq.) Seem.

灌木或小乔木，雄全同株。小枝具稀疏刺。二(三)回羽状复叶；羽片具小叶5-11(-13)，下面无毛或具浅黄色或灰色柔毛。伞形花序组成圆锥状，顶生；子房5室；花柱5，分离或联合至中部；花柱宿存。果实球形。花期7-9月，果期9-12月。生海拔2700米以下的林中、林缘、灌丛或路边。产中国除华南和西北外的广大地区。俄罗斯东部、朝鲜半岛和日本亦有。

Shrubs or small trees, andromonoecious. Branches armed with sparse prickles. Leaves 2(-3)-pinnately compound; leaflets 5-11(-13) per pinna, abaxially glabrous or light yellow or gray pubescent. Inflorescences a terminal panicle of umbels; ovary 5-carpellate; styles 5, free or united to middle; styles persistent. Fruits globose. Fl. Jul-Sep. Fr. Sep-Dec. Forests, forest edges, scrubs fields or roadsides below 2700 m. Widely distributed in China except S and NW parts. Also in E Russia, Korean Peninsula and Japan.

## 食用土当归

**Aralia cordata** Thunb.

多年生草本。具有地下长圆柱状根茎。二或三回羽状复叶；每个羽片具小叶3-5，下面沿脉疏具柔毛。花序为稀疏的顶生或少分枝的腋生伞形花序，排列成圆锥状；柱头宿存。果实球形。花期7-8月，果期9-10月。生海拔1300-1600米的山坡草丛或林下。产中国西南、东南和华东。

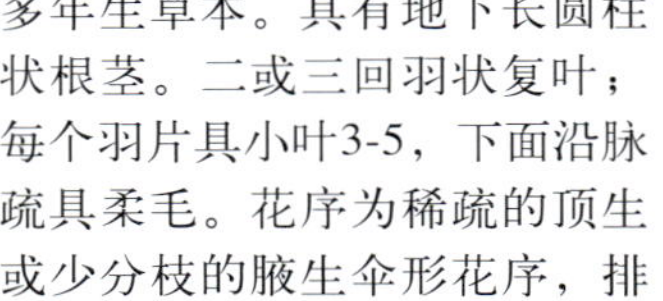

Perennial herbs. With underground long cylindric rhizomes. Leaves 2- or 3-pinnately compound; leaflets 3-5 per pinna, abaxially sparsely pubescent on veins. Inflorescences alax terminal or axillary few branched panicle of umbels; styles persistent. Fruits globose. Fl. Jul-Aug. Fr. Sep-Oct. Grasslands on slopes or under forests at 1300-1600 m. Distributed in SW, SE and E China.

## 东北土当归

**Aralia continentalis** Kitag.

多年生草本。具粗壮的根状茎。二或三回羽状复叶；每个羽片具3-7小叶，两面具灰色柔毛；紧缩的顶生或腋生伞形花序组成圆锥状，被灰色短柔毛；伞形花序多花；花柱宿

食用土当归 *Aralia cordata*

东北土当归 *Aralia continentalis*

存，离生花柱反曲。果球形。花期7-8月，果期8-9月。生海拔800-3200米的林中或山坡草地。产中国西南、华北、华东和东北。俄罗斯和朝鲜半岛亦有。

Perennial herbs. With stout rhizome. Leaves 2-or 3-pinnately compound; leaflets 3-7 per pinna, both surfaces gray pubescent; a compact terminal or axillary panicle of umbels, gray pubescent; umbels many flowers; styles persistent, free styles recurved; Fruits globose. Fl. Jul-Aug. Fr. Aug-Sep. Forests or grasslands on slopes at 800-3200 m. Distributed in SW, N, E and NE China. Also in Russia and Korean Peninsula.

## 西藏土当归

**Aralia tibetana** G. Hoo

多年生草本。二回羽状复叶；羽片具3-5小叶，两面疏具柔毛，异形，叶缘具锯齿。顶生伞形花序组成圆锥状；每个二级分枝上具1-3个伞形花序，多花；花柱5，基部联合，顶部分离。果实卵球形。花期8月，果期9月。生海拔3200-3500米的林中或灌丛。产西藏。

Perennial herbs. Leaves 2-pinnately compound; leaflets 3-5 per pinna, both surfaces sparsely pubescent, heteromorphic, margin serrulate. Inflorescences a terminal panicle of umbels; umbels 1-3 per secondary axis, many flowered; styles 5, basally united, apically free. Fruits ovoid. Fl. Aug. Fr. Sep. Forests or scrub fields at 3200-3500 m. Distributed in Xizang.

西藏土当归 *Aralia tibetana*

## 龙眼独活

**Aralia fargesii** Franch.

多年生草本。叶为一至三回羽状复叶；每羽片具3-5小叶，膜质，两面粗糙，背面脉上具柔毛。花序为少数分枝的伞房状聚伞花序；聚伞花序总状排列，具10-20花；柱头宿存，辐射状。果实近球形。花期7-8月，果期10-11月。生海拔1800-2700米的林中或溪岸。产云南(鹤庆县、昆明和嵩明县)、四川和陕西(太白山)。

Perennial herbs. Leaves 1-3-pinnately compound; leaflets 3-5 per pinna, membranous, both surfaces scabrous, pubescent on veins abaxially. Inflorescences a few branched corymb of umbels; umbels racemosely arranged, 10-20-flowered; styles persistent, radiating. Fruits subglobose. Fl. Jul-Aug. Fr. Oct-Nov. Forests or stream banks at 1800-2700 m. Distributed in Yunnan (Heqing County, Kunming, Songming County), Sichuan and Shaanxi (Taibai Mountain).

龙眼独活 *Aralia fargesii*

柔毛龙眼独活 *Aralia henryi*

## 柔毛龙眼独活
**Aralia henryi** Harms

多年生草本。叶为二至三回至二回羽状复叶；每分枝具3小叶，两面沿脉具柔毛。伞形花序集合为伞房状，顶生，具柔毛；每个伞形花序具3-10花；子房(3-)5室；花柱(3-)5，分离。果实近球形。花期7-8月，果期9-11月。生海拔1500-2300米的林中。产四川、重庆、湖北、陕西和安徽。

Perennial herbs. Leaves 2-ternately to 2-pinnately compound; leaflets 3 per rachis, both surfaces villous on veins. Inflorescences a terminal corymb of umbels, villous; umbels 3-10-flowered; ovary (3-)5-carpellate; styles (3-)5, free. Fruits subglobose. Fl. Jul-Aug. Fr. Sep-Nov. Forests at 1500-2300 m. Distributed in Sichuan, Chongqing, Hubei, Shaanxi and Anhui.

## 浓紫龙眼独活
**Aralia atropurpurea** Franch.

多年生草本。一或二回羽状复叶；羽片具小叶3-7，膜质，两面疏生糙毛。伞形花序组成伞房状，顶生；子房5室；花柱5，分离；花柱宿存，反折。果实球形。花期6-7月，果期8-9月。生海拔2700-3300米的散生林中、草地、山坡或路边。产云南(德钦县)、四川和西藏(波密县)。

Perennial herbs. Leaves 1-or 2-pinnately compound; leaflets 3-7 per pinna, membranous, both surfaces sparsely setose-scabrous. Inflorescences a terminal corymb of umbels; ovary 5-carpellate; styles 5, free; styles persistent, recurved. Fruits globose. Fl. Jun-Jul. Fr. Aug-Sep. Scattered trees, grasslands on slopes or roadsides at 2700-3300 m. Distributed in Yunnan (Dêqên County), Sichuan and Xizang (Bomê County).

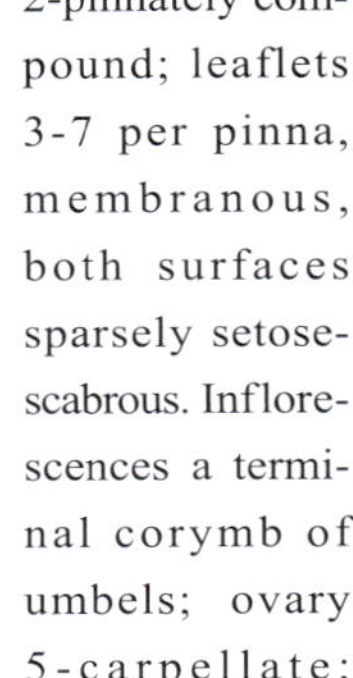

浓紫龙眼独活 *Aralia atropurpurea*

人参 *Panax ginseng*

## 人参
**Panax ginseng** C. A. Mey.

多年生草本。主根肥大，纺锤形或圆柱形。叶片3-6，轮生于茎顶，掌状复叶；小叶3-5，膜状，离轴无毛，近轴稀少硬毛。花序为单生、顶生的伞形花序，具30-50花。果实扁球形，鲜红色。生杂木林或落叶阔叶林。产中国东北。河北和山西有栽培。俄罗斯东部和朝鲜半岛亦有。

Perennial herbs. Taproots hypertrophic, fusiform or cylindric. Leaves 3-6, verticillate at apex of stem, palmately compound; leaflets 3-5, membranous, abaxially glabrous, adaxially sparsely setose. Inflorescences a solitary, terminal umbel 30-50-flowered. Fruits oblate, scarlet. Mixed forests or deciduous broad-leaved forests. Distributed in NE China. Cultivated in Hebei and Shanxi. Also in E Russia and Korean Peninsula.

三七 *Panax notoginseng*

## 三七

**Panax notoginseng** (Burk.) F. H. Chen ex C. Chow et W. G. Huang

多年生草本，主根肉质。掌状复叶3-6；叶柄基部无托叶或托叶状附属物。顶生伞形花序单生，具80-100至更多花；花序梗长7-25厘米；子房2室；花柱2，至少联合至中部。果实红色，扁球状肾形。花期7-8月，果期8-10月。生海拔1200-1800米的林中。产云南东南部。广西、福建、浙江和江西有栽培。越南北部亦有。

Perennial herbs, main roots fleshy. Leaves 3-6, palmately compound; petioles base without stipule or stipulelike appendage. Inflorescences a solitary, terminal umbel 80-100(or more)-flowered; peduncles 7-25 cm long; ovary 2-carpellate; styles 2, united at least to middle. Fruits red, compressed globose-nephroid. Fl. Jul-Aug. Fr. Aug-Oct. Forests at 1200-1800 m. Distributed in SE Yunnan. Cultivated in Guangxi, Fujian, Zhejiang and Jiangxi. Also in N Vietnam.

竹节参 *Panax japonicus*

## 竹节参

**Panax japonicus** (T. Nees) C. A. Mey.

多年生草本。根状茎横走，鞭状，似竹根。叶3-5，于茎顶轮生，掌状复叶；小叶5，不为二回羽状分裂，倒卵状椭圆形至长圆形。花序为单生的伞形花序，顶生，具50-80(或更多)花；子房2-5室。果实红色，近球形。花期5-6月，果期7-9月。生海拔1200-3600米的林中或山谷林中。产中国西南、东南、华中、华西和华东。越南、朝鲜半岛和日本亦有。

Perennial herbs. Rootstocks horizontal, flagellate, like rhizome of bamboo. Leaves 3-5, verticillate at apex of stem, palmately compound; leaflets 5, not 2-pinnatifid, obovate-elliptic to oblong. Inflorescences a solitary, terminal umbels 50-80 (or more)-flowered; ovary 2-5-carpellate. Fruits red, subglobose. Fl. May-Jun. Fr. Jul-Sep. Forests or forests in valleys at 1200-3600 m. Distributed in SW, SE, C, W and E China. Also in Vietnam, Korean Peninsula and Japan.

## 狭叶竹节参

**Panax japonicus** (T. Nees) C. A. Mey. var. **angustifolius** (Burk.) C. Y. Cheng et C. Y. Chu

本变种与原变种的区别在于：叶狭披针形，长约为宽的5倍，顶部具长尾状渐尖。生海拔1600-3600米的林中。产云南、四川和贵州。尼泊尔、印度、不丹和泰国亦有。

This variety differs from var. *japonicus* in its leaflets narrowly lanceolate, ca. 5 × as long as wide, apex long caudate-acuminate. Forests at 1600-3600 m. Distributed in Yunnan, Sichuan and Guizhou. Also in Nepal, India, Bhutan and Thailand.

狭叶竹节参 *Panax japonicus* var. *anyustifolius*

珠子参 *Panax japonicus* var. *major*

# 伞形花科 Umbelliferae

### 珠子参
**Panax japonicus** (T. Nees) C. A. Mey. var. **major** (Burk.) C. Y. Wu et K. M. Feng

本变种与原变种的区别在于：根状茎串珠状。小叶倒卵状椭圆形至椭圆形，先端渐尖。生海拔1700-3600米的林下或灌丛草坡。产中国西南、华北和华西。尼泊尔、缅甸北部和越南北部亦有。

This variety differs from var. *japonicus* in its: rhizomes moniliform. Leaflets obovate-elliptic to elliptic, apex acuminate. Under forests or among bushes on grassy slopes at 1700-3600 m. Distributed in SW, N and W China. Also in Nepal, N Myanmar and N Vietnam.

### 姜状三七
**Panax zingiberensis** C. Y. Wu et K. M. Feng

多年生草本。主根肉质姜块状，根茎具短而增厚的节间；掌状复叶，小叶3-5，边缘具锯齿或微重锯齿。单个伞形花序顶生；花梗长24-26厘米；子房2室；花柱2，合生至中部。果实红色，球状肾形。花期7-8月，果期8-10月。生常绿阔叶林。产云南东南部。越南北部亦有。

Perennial herbs. Main roots fleshy, like zingber-tuber, rhizomes with short and thicken nodes. Leaves palmately compound, leaflets 3-5, serrate or inconspicuously doubly serrate at margin. Inflorescences a solitary, terminal umbel; peduncles 24-26 cm long; ovary 2-carpellate; styles 2, united to middle. Fruits red, globose-nephroid. Fl. Jul-Aug. Fr. Aug-Oct. Broad-leaved evergreen forests. Distributed in SE Yunnan. Also in N Vietnam.

### 红马蹄草
**Hydrocotyle nepalensis** Hook.

草本。叶圆形或肾形。花序数个生于枝条顶端，密被柔毛；伞形花序束生，每个伞形花序均密头状，具20-60花；花瓣白色或具紫红色斑点。果实黄褐色或紫黑色。花果期5-11月。生海拔300-3600米的山坡、阴处或草地。产中国南方大部分地区。南亚亦有。

Herbs. Leaves orbicular or reniform. Umbels several to numerous, inserted on the top of branchlets, densely villose; umbels fascicled; each umbel densely capitate, 20-60-flowered; petals white or with purplish red stains. Fruits yellow-brown or purple-black. Fl. and fr. May-Nov. Slopes, shady places or grasslands at 300-3600 m. Distributed

姜状三七 *Panax zingiberensis*

红马蹄草 *Hydrocotyle nepalensis*

天胡荽 *Hydrocotyle sibthorpioides*

in most parts of S China. Also in S Asia.

## 天胡荽

**Hydrocotyle sibthorpioides** Lam.

草本。茎软，细丝状，匍匐。叶肾状圆形，膜质。花序梗短于叶柄，单生，无毛或上部有毛；伞形花序单生于叶腋，具5-18花；花瓣绿白色，卵形至卵状披针形，有黄色腺点。果实阔球形。花果期4-9月。生海拔100-3000米的湿润草地、溪边或杂木林。产中国南方大部分地区。南亚、东南亚、东北亚和非洲热带亦有。

Herbs. Stems weak, slender, filiform, creeping. Leaves reniform-rounded, membranous. Peduncles shorter than petioles, solitary, glabrous or hairy on the upper part; umbels solitary in axils, 5-18-flowered; petals green-white, ovate to ovate-lanceolate, yellow-punctate. Fruits broadly globose. Fl. and fr. Apr-Sep. Wet grasslands, by streams or mixed forests at 100-3000 m. Distributed in most parts of S China. Also in S, SE and NE Asia, and tropical Africa.

中华天胡荽 *Hydrocotyle hookeri* subsp. *chinensis*

## 中华天胡荽

**Hydrocotyle hookeri** (C. B. Clarke) Craib subsp. **chinensis** (Dunn ex R. H. Shan et S. L. Liou) M. F. Watson et M. L. Sheh

多年生草本植物。茎匍匐至长1.5米。叶柄和叶片密被或疏被白色或紫色柔毛；叶片轮廓圆形，边缘浅5-7裂，裂片先端钝。伞形花序具30-55花，具紫褐色柔毛；花瓣白色。果实具褐色斑点，近球形，基部浅心形或截形。花果期7-8月。生海拔1000-2900米的山坡、溪岸或阴湿路边。产云南、四川和湖南。

Perennial herbs. Stems creeping to 1.5 m long. Petioles and leaves densely or sparsely white or purple pubescent; leaves round in outline, margin shallowly 5-7-lobed, lobe apex obtuse. Umbels 30-55-flowered, with purple-brown pubescent; petals white Fruits brown-spotted, subglobose, base shallowly cordate or truncate. Fl. and fr. Jul-Aug. Mountain slopes, stream banks, shady wet roadsides at 1000-2900 m. Distributed in Yunnan, Sichuan and Hunan.

## 肾叶天胡荽

**Hydrocotyle wilfordii** Maxim.

草本。叶圆形或肾圆形，基部心形，两面无毛或下面沿脉疏具粗毛。伞形花序多花，总花梗纤细，单生，长于叶柄或近等长；花白色至浅黄色。果实浅棕略带紫斑。花果期5-9月。生海拔300-1400米的阴湿山谷、田野或溪边。产中国西南、华南和东南。越南、朝鲜半岛和日本亦有。

Herbs. Leaves orbicular or reniform-rounded, cordate at base, glabrous on both surfaces or abaxially sparsely hirsute on veins. Umbels many-flowered, peduncles slender, solitary; peduncles longer than or equaling petioles; petals white to pale yellow. Fruits light brown with purplish stains. Fl. and fr. May-Sep. Moist shady valleys, fields or by streams at 300-1400 m. Distributed in SW, S and SE China. Also in Vietnam, Korean Peninsula and Japan.

肾叶天胡荽 *Hydrocotyle wilfordii*

南美天胡荽 *Hydrocotyle verticillata*

野鹅脚板 *Sanicula orthacantha*

## 南美天胡荽

**Hydrocotyle verticillata** Thunb.

多年生草本。植株低矮，全株无毛。茎匍匐，节上生根。叶片圆形至宽椭圆形，具8-13辐射状叶脉；叶柄盾状着生，纤细。花在花序轴上间断性轮生，呈穗状轮伞花序；花瓣淡黄色。果实宽椭圆形，两侧压扁，果体扁平，基部宽楔形，棱丝状，突出。中国有引种和逸生。原产北美洲(美国)、南美洲、中美洲、澳大利亚和非洲南部。

Perennial herbs. Plants dwarf, glabrous. Stems prostrate, rooting at the nodes. Blade round to broadly elliptic, with 8-13 radiate veins; petioles shield insertion, slender. Inflorescences axis with whorled flowers formed spike verticillaster; petals yellow. Fruit broadly oblong, flattening on both sides, base wide wedge, ribs prominent, silky. Introduction in China and escaping into the wild. Native to North America (USA), South and Central America, Australia and S Africa.

## 积雪草

**Centella asiatica** (L.) Urban

草本。茎匍匐。叶单生圆形或肾形，掌状脉5-7，两面无毛或叶背脉上疏被柔毛，基部宽心形，具粗齿。伞形花序头状，具3-4花；花瓣白色或带玫瑰红色，覆瓦状；苞片宿存。果实背棱及侧棱明显。花果期4-10月。生海拔200-1900米的湿草地或溪边。产中国长江以南地区。南亚、东南亚、中亚、东北亚、南美洲和非洲亦有。

Herbs. Stems repent. Leaves simple, orbicular or reniform, palmate veins 5-7, both surfaces glabrous or abaxially sparsely pubescent on the veins, base broadly cordate, coarsely toothed. Umbels 3-4-flowered, capitate; petals white or rose-tinged, imbricate; bracts present. Dorsal and lateral fruit ribs prominent. Fl. and fr. Apr-Oct. Moist grassy places or along streams at 200-1900 m. Widely distributed in South of Yangtze River. Also in S, SE, C and NE Asia, South America and Africa.

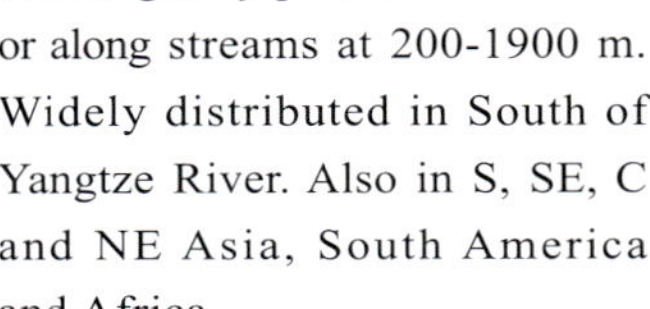

## 野鹅脚板

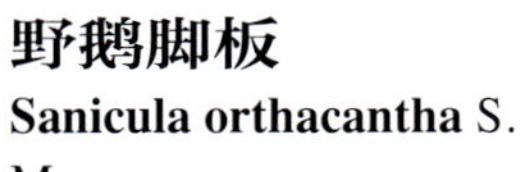

**Sanicula orthacantha** S. Moore

多年生草本。根短，木质。叶掌状3裂，侧裂片斜倒卵形，深裂至基部。花序2-3分枝；小伞形花序3-8；小苞片5；小伞形花序具6或7花；花瓣白色、淡蓝色或紫红色，卵形。果实卵球形，有皮刺。花果期4-9月。生海拔200-3200米的林中、山坡或溪边。产中国西南、华南、华中和华东。印度、老挝、越南和柬埔寨亦有。

Perennial herbs. Roots short, woody. Leaves palmately 3-parted, lateral segments of leaves oblique-obovate, deeply divided to base. Inflorescences 2-3-branched; umbels 3-8; bracteoles 5; umbellules 6 or 7-flowered; petals white, pale blue or purplish red, obovate. Fruits ovoid, with spiny. Fl. and fr. Apr-Sep. Forests, slopes or streamsides at 200-3200 m. Distributed in SW, S, C and E China. Also in India, Laos, Vietnam and Cambodia.

## 变豆菜

**Sanicula chinensis** Bunge

多年生草本。基生叶极少，3-5

积雪草 *Centella asiatica*

变豆菜 *Sanicula chinensis*

裂；茎生叶3裂。花序常二至三回叉状分枝。小伞形花序具6-10花；每个小伞形花序具雄花3-7朵，雌花3或4朵；萼齿窄线形；花瓣白色。果实卵球形；油管5，合生面油管2，稍大。花果期4-10月。生海拔200-2300米的杂木林、河岸、路边或山坡阴处。中国广泛分布。俄罗斯(东西伯利亚)、朝鲜半岛北部和日本亦有。

Perennial herbs. Basal leaves few, 3-5-parted; cauline leaves 3-parted. Inflorescences usually bi- or trichotomously branched; umbellules 6-10-flowered; staminate flowers 3-7, fertile flowers 3 or 4 per umbellule; calyx teeth linear petals white. Fruits ovoid; vittae 5, commissural vittae 2, larger. Fl. and fr. Apr-Oct. Mixed forests, stream banks, roadsides or shady slopes at 200-2300 m. Widely distributed in China. Also in Russia (E Siberia), N Korean Peninsula and Japan.

## 刺芹

**Eryngium foetidum** L.

草本。基生叶长披针形、披针形或倒披针形，不分裂。复头状花序二歧状三叉分枝；侧枝继续发育形成单歧聚伞状，头状花序多数，具短梗；总苞片4-7；花两性。果实卵球形。花果期4-12月。生海拔100-1500米的山地林下、路边湿润处或溪边。产云南、贵州、广西和广东。原产美洲中部，现为热带和亚热带区域广布杂草。

Herbs. Basal leaves long-elliptic, lanceolate or oblanceolate, undivided. Inflorescences divaricately trifurcate; lateral branches often continuing to form a monochasium, heads numerous, short-pedunculate; involucral bracts 4-7; flowers bisexual. Fruits ovoid. Fl. and fr. Apr-Dec. Forests on mountains, moist places by roads or by streams at 100-1500 m. Distributed in Yunnan, Guizhou, Guangxi and Guangdong. Native to C America, now a widespread weed in tropical and subtropical regions.

刺芹 *Eryngium foetidum*

## 扁叶刺芹

**Eryngium planum** L.

多年生草本。茎灰白色至紫色。基生叶数个，叶片狭椭圆状卵形，具7-9掌状脉，两面突出，基部心形，边缘具粗齿、短尖的牙齿至短微刺。花序1-4回三叉分枝，分枝顶端生头状花序；花瓣浅蓝色。果实长椭圆形或近球形，背面扁压，被白色、狭窄的长鳞片。花果期7-8月。生海拔500-1500米的杂草丛。产

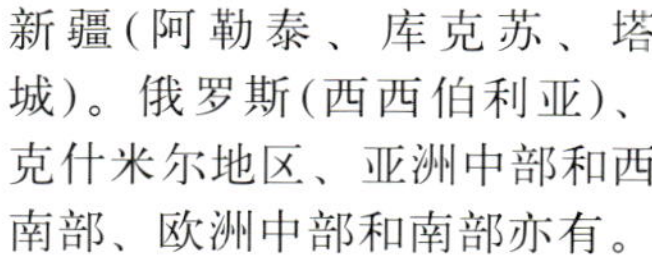

新疆(阿勒泰、库克苏、塔城)。俄罗斯(西西伯利亚)、克什米尔地区、亚洲中部和西南部、欧洲中部和南部亦有。

Perennial herbs. Stem gray-white to purple. Basal leaves several; blade narrowly elliptic-ovate, palmately 7-9-nerved, prominent on both surfaces, base cordate, margin coarsely toothed, teeth mucronate to short spinulose. Inflorescences 1-4-trifurcate, heads terminal on branches; petals pale blue. Fruit long-ellipsoid or subglobose, flattened dorsally, clothed with white, narrow long scales. Fl. and fr. Jul-Aug. Ruderal of disturbed habitats at 500-1500 m. Distributed in Xinjiang (Altay, Kukesu, Tacheng). Also in Russia (W Siberia), Kashmir, C and SW Asia, C and S Europe.

扁叶刺芹 *Eryngium planum*

迷果芹 *Sphallerocarpus gracilis*

## 迷果芹

**Sphallerocarpus gracilis**
(Besser ex Trevir.) Koso-Pol.

多年生草本。根块状或圆锥状，基部叶早落。茎生叶具柄。叶鞘边缘膜质，具白色柔毛，叶片末端3裂或具齿。伞幅6-13；小伞形花序具15-25花。果实线状长圆形；每侧具2-4油管。花果期7-10月。生海拔500-2800米的山坡、可耕地或弃荒地。产中国西南、华北、华西、西北和东北。俄罗斯(西伯利亚)、蒙古和日本亦有。

Perennial herbs. Roots tuberous or conic, basal leaves caduceus. Cauline leaves petiolate. Sheath margin with white pubescent; ultimate segments 3-lobed or toothed. Rays 6-13; umbellules 15-25-flowered. Fruits linear-oblong; vittae 2-4 in each furrow. Fl. and fr. Jul-Oct. Mountain slopes, arablelands or waste places at 500-2800 m. Distributed in SW, N, W, NW and NE China. Also in Russia (Siberia), Mongolia and Japan.

## 峨参

**Anthriscus sylvestris** (L.)
Hoffm.

多年生草本。二回羽状复叶。伞辐4-15，不等长；小苞片5-8，卵形至披针形，渐尖；花梗顶端于果期常具白色刚毛。果实表面光滑或疏生小瘤点；果喙短于果体。花果期4-5月。生海拔高达4500米的林中、沟边或山坡草地。产中国除华南以外的大部分地区。东亚和东欧亦有；北美洲引进。

Perennial herbs. Leaves bipinnate. Rays 4-15, unequal; bracteoles 5-8, ovate to lanceolate, acuminate; pedicels apex usually surrounded by white bristles in fruit. Fruits surface smooth or sparsely verruculose; fruit beaks shorter than the body. Fl. and fr. Apr-May. Forests, valley sides or grassy places on mountain slopes up to 4500 m. Widely distributed in China, except S part. Also in E Asia and E Europe; introduced in North America.

峨参 *Anthriscus sylvestris*

## 窃衣

**Torilis scabra**
(Thunb.) DC.

一年生或多年生草本。叶片轮廓卵形，羽片披针形至狭卵形。总苞片常无；苞片常无，稀1；伞幅2-4(-5)；小苞片2-6；小伞形花序具2-6花。果实常深绿色，偶尔浅紫色，长圆形，4-7 × 2-3毫米。花果期4-11月。生海拔200-2400米的山坡、林下、路旁、河边和空

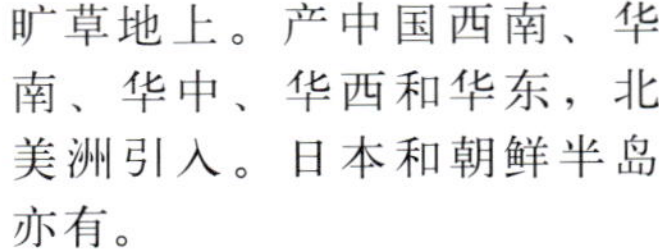

旷草地上。产中国西南、华南、华中、华西和华东，北美洲引入。日本和朝鲜半岛亦有。

Annual or perennial herbs. Leaves ovate in outline; pinnae lanceolate to narrowly ovate. Involucres usually absent, bracts usually absent, rarely 1; rays 2-4(-5); bracteoles 2-6; umbellules 2-6-flowered. Fruits usually dark

窃衣 *Torilis scabra*

green, occasionally tinged dark purple, oblong, 4-7 × 2-3 mm. Fl. and fr. Apr-Nov. On mountain slopes, forests, by roads, rivers, on fields at 200-2400 m. Distributed in SW, S, C, W and E China, introduced in North America. Also in Japan and Korean Peninsula.

## 小窃衣

**Torilis japonica** (Houtt.) DC.

一年生或多年生草本。叶片轮廓三角卵形。苞片3-6；伞幅4-12；小苞片5-8；小伞形花序具4-12花；复伞形花序顶生或腋生；花瓣白色、紫红或紫蓝色。果实卵球形，成熟时紫黑色，1.5-5 × 1-2.5厘米。花果期4-10月。生海拔100-3800米的林中、山谷、草地或受干扰地区。产中国除内蒙古、新疆和黑龙江以外的大部分地区。亚洲、欧洲和非洲亦有。

Annual or perennial herbs. Leaves triangular-ovate to ovate-lanceolate in outline. Bracts 3-6; rays 4-12; bracteoles 5-8; umbellules 4-12-flowered; compound umbels terminal or axillary; petals white, purplish red or purplish blue. Fruits ovoid, blackish purple when mature, 1.5-5 × 1-2.5 mm. Fl. and fr. Apr-Oct. Forests, valleys, grasslands or disturbed areas at 100-3800 m. Distributed in most parts of China, except Neimenggu, Xinjiang and Heilongjiang. Also in Asia, Europe and Africa.

小窃衣 *Torilis japonica*

芫荽 *Coriandrum sativum*

## 芫荽

**Coriandrum sativum** L.

一年生或二年生草本。茎圆柱形，通常光滑。茎中部和上部叶二至三回羽状全裂。伞幅2-8；小苞片2-5，条形，全缘；小伞形花序具3-9花。果实圆球形，外果皮质硬。花果期4-11月。中国大部分地区有栽培。原产欧洲。

Annual or biennial herbs. Stems cylindric, usually smooth. Mid and upper cauline leaves ternate 2-3-pinnatisect. Rays 2-8; bracteoles 2-5, linear, entire; umbellules 3-9-flowered. Fruits globose, pericarps hard. Fl. and fr. Apr-Nov. Cultivated in most parts of China. Native to Europe.

## 矮棱子芹

**Pleurospermum nanum** Franch.

多年生小草本，5-15厘米高，无茎。全株平滑无毛。叶片卵状长圆形，二至三回羽状分裂。伞幅5-15，稍不等长；小苞片6-10；花瓣白色或紫红色，倒卵形；花柱基压扁，花柱短。果实宽卵球形。花期7-9月，果期10-11月。生海拔(2600-)3500-4600米的高山或亚高山草甸。产云南西北部和西藏东南部。

Perennial small herbs, 5-15 cm tall, acaulescent. Whole plant smooth and glabrous. Leaves ovate-oblong, 2-3-pinnate. Rays 5-15, slightly unequal; bracteoles 6-10; petals white or purplish red, obovate; stylopodium short-conic, styles short. Fruits broad-ovoid. Fl. Jul-Sep. Fr. Oct-Nov. Alpine or subalpine meadows at (2600-) 3500-4600 m. Distributed in NW Yunnan and SE Xizang.

## 棱子芹

**Pleurospermum uralense** Hoffm.

多年生草本。根芳香。茎中空具肋纹。叶为二至三回三出式羽状复叶，上部叶不规则羽裂。伞形花序大，长10-20厘米；伞幅20-40(-60)；花瓣白色。果实宽卵球形；棱厚鸡冠状突起；每棱槽有油管1，合生面2。花期6-7月，果期7-8月。生林中河岸边或山地峡谷。产河北、山西、内蒙古、陕西、吉林和辽宁。俄罗斯东南部、蒙古和日本亦有。

Perennial herbs. Roots aromatic. Stems hollow, ribbed. Leaves 2-3-ternate-pinnate, ultimate segments irregular-pinnatifid. Umbels large, 10-20 cm long; rays 20-40(-60); petals white. Fruits broad-ovoid; ribs all thickly cristate-winged; vittae 1 in each furrow, 2 on commissure. Fl. Jun-Jul. Fr. Jul-Aug. Stream banks in forests or mountain ravines. Distributed in Hebei, Shanxi, Neimenggu, Shaanxi, Jilin and Liaoning. Also in SE Russia, Mongolia and Japan.

## 美丽棱子芹

**Pleurospermum amabile** Craib et W. W. Sm.

多年生草本。叶为三至四回羽状

矮棱子芹 *Pleurospermum nanum*

棱子芹 *Pleurospermum uralense*

复叶，基生叶和茎中部叶轮廓三角状卵形，长4-8厘米。叶鞘边缘啮蚀状，有紫色脉纹，上部叶和苞片包被花序。伞幅20-30；小苞片约12，膜质。果实卵球状长圆形；果棱具狭波状翅。花期7-9月，果期9-10月。生海拔(3000-)4000-5100米的山坡草地或灌丛。产云南西北部和西藏东南部。印度北部和不丹亦有。

Perennial herbs. Leaves 3-4-pinnate, basal and middle leaves triangular-ovate in outline, 4-8 cm long; leaf sheaths erose at margin, with purple-nerved, upper leaves and bracts enveloping the inflorescences. Rays 20-30; bracteoles ca. 12, membranous. Fruits ovoid-oblong; ribs very narrowly sinuolate-winged. Fl. Jul-Sep. Fr. Sep-Oct. Grassy slopes cr bushes at (3000-)4100-5100 m. Distributed in NW Yunnan and SE Xizang. Also in N India and Bhutan.

美丽棱子芹 *Pleurospermum amabile*

归叶棱子芹 *Pleurospermum angelicoides*

## 归叶棱子芹

**Pleurospermum angelicoides**
(Wall. ex DC.) C. B. Clarke

多年生草本。茎棱稀疏，无毛。基生叶具长叶柄，叶片长圆形，三至四回三出式羽状分裂。茎生叶向上渐简化，具非常膨大的鞘，膜质。花瓣卵形，白色或微带紫红色。果实长圆形，暗褐色，具突出的背棱，侧棱狭翅。种子表面凹。花果期6-9月。生海拔3000-4000米的森林、高山草甸的溪岸。产云南西北部、四川西南部和西藏东南部。印度北部、不丹、克什米尔地区、缅甸和尼泊尔亦有。

Perennial herbs. Stem thinly ribbed, glabrous. Basal leaves with long petioles, blades oblong, 3-4-ternate-pinnate. Stem leaves reduced upwards, sheaths strongly inflated and auriculate, membranous. Petals ovate, white or tinged purplish red. Fruit oblong, dark brown. dorsal ribs prominent, lateral ribs narrowly winged. Seed face concave. Fl. and fr. Jun-Sep. Forests, stream banks in alpine meadows at 3000-4000 m. Distributed in NW Yunnan. SW Sichuan and SE Xizang. Also in N India, Bhutan, Kashmir, Myanmar and Nepal.

宝兴棱子芹 *Pleurospermum benthamii*

翼叶棱子芹 *Pleurospermum decurrens*

## 宝兴棱子芹

**Pleurospermum benthamii** (Wall. ex DC.) C. B. Clarke

多年生草本。茎空，稍具棱，常微染紫色。叶片宽三角状卵形，二至三回三出式羽状，无毛。苞片5-9，倒披针形，边缘白色干膜质，先端羽状；小苞片6-9，倒披针形，边缘白色干膜质，先端3浅裂。花瓣倒卵形，白色，锐尖。果实卵球状椭圆形，棱全部微波状翅；花果期6-9月。生海拔2200-4300米的稀疏灌丛、高山牧场、河边。产云南西北部、四川西部和西藏东南部。印度北部、不丹、缅甸北部和尼泊尔东部亦有。

Perennial herbs. Stem hollow, thinly ribbed, often tinged purple. Blades broadly triangular-ovate, 2-3-ternate-pinnate, glabrous. Bracts 5-9, oblanceolate, margin white-scarious, apex pinnate. bracteoles 6-9, oblanceolate, margin white-scarious, apex 3-lobed; petals obovate, white, acute at tip. Fruits ovoid-ellipsoid, ribs all sinuolate-winged; Fl. and fr. Jun-Sep. Open scrubs, alpine pastures, riversides at 2200-4300 m. Distributed in NW Yunnan, W Sichuan and SE Xizang. Also in N India, Bhutan, N Myanmar and E Nepal.

## 翼叶棱子芹

**Pleurospermum decurrens** Franch.

多年生草本。茎纤细，具棱。叶片宽卵形，二回三出式或羽状分裂，两面沿着主脉具微糙硬毛，或无毛。苞片6-10，苍绿色，长圆状披针形，边缘白色膜质，先端锐尖或3浅裂；小苞片6-8，线状披针形，全缘或偶尔3浅裂；具多花；花瓣卵状披针形，白色。果实卵球形，棱具微波状狭翅。花果期7-9月。生海拔3000-4000米的松林和混交林、高山草甸的阴湿处。产云南西北部。

Perennial herbs. Stem slender, ribbed. Blades broad-ovate, 2-ternate or pinnate, hirtellous along the main veins on both surfaces, otherwise glabrous. Bracts 6-10, pale green, oblong-lanceolate, margin white membranous, apex acute or 3-lobed; bracteoles 6-8, linear-lanceolate, entire or occasionally 3-lobed; flower numerous; petals ovate-lanceolate, white. Fruitsovoid, ribs narrowly sinuolate-winged. Fl. and fr. Jul-Sep. Shady areas in *Pinus* and mixed forests, alpine grasslands at 3000-4000 m. Distributed in NW Yunnan.

## 丽江棱子芹

**Pleurospermum foetens** Franch.

多年生草本，具较明显臭味。茎短缩，具棱，粗糙。叶片长圆形，三至四回三出式羽裂。苞片6-8，倒卵形，先端羽状；小苞片约10，倒卵形，先端羽状，苍绿色，边缘宽，白色，膜质；具多花；花瓣倒卵形，白色的或带粉红色，锐尖，基部有爪。果实卵球形，紫黑色，暗红棕色，棱全部具宽深波状翅。花果期6-9月。生海拔3600-4500米的高山草甸、岩石坡或松散的山麓碎石。产云南西北部、四川、西藏东南部和甘肃。

Perennial herbs, plants with characteristic strong unpleasant odor. Stems reduced, ribbed, scabrous. Blades oblong, 3-4-ternate-pinnate. Bracts 6-8, obovate, apex pinnate; bracteoles ca. 10, obovate, apex pinnatifid, pale green, margin broad, white, membranous; flower numerous; petals obovate, white or pinkish, acute at tip, clawed at base. Fruits-ovoid, purple-black, dark red-brown, ribs all broadly sinuate winged. Fl. and fr. Jun-Sep. Open alpine meadows, rocky slopes or loose screes at 3600-4500 m. Distributed in NW Yunnan, Sichuan, SE Xizang and Gansu.

丽江棱子芹 *Pleurospermum foetens*

太白棱子芹 *Pleurospermum giraldii*

## 太白棱子芹
**Pleurospermum giraldii** Diels

多年生草本。茎微带紫色，具条棱。叶片三角状心形，三至四回三出式羽状分裂。苞片5-7，卵状椭圆形或倒卵形，白色或微带紫色，膜质；小苞片5-7，倒卵形，边缘白色膜质，先端羽状全裂；花瓣倒心形，白色。果实长圆形，棱具翅。花果期7-10月。生海拔3000-3600米的草质山坡。产四川、湖北、陕西和贵州。

Perennial herbs. Stems tinged purple, ribbed. Blades triangular-ovate, 3-4-ternate-pinnate. bracts 5-7, ovate-elliptic or obovate, white or tinged purple, membranous; bracteoles 5-7, obovate, margin white membranous, apex pinnatisect; petals obcordate, white. Fruits oblong, ribs winged; Fl. and fr. Jul-Oct. Grassy mountain slopes at 3000-3600 m. Distributed in Sichuan, Hubei, Shaanxi and Gansu.

## 垫状棱子芹
**Pleurospermum hedinii** Diels

多年生莲座状草本。茎非常短，肉质。叶片长圆形，二回羽裂。伞形花序密集头状，顶生；苞片多数，叶状；小苞片8-12，倒卵形或倒披针形，绿色，边缘宽，白色，顶3浅裂；花瓣圆形，白色至略带紫色。果实宽卵球形，棱宽，微波状翅。花果期7-9月。生海拔4200-5000米的高山草甸。产云南西北部、西藏东部、青海南部和西部。

Perennial herbs, rosette. Stem very short, fleshy. Blades oblong, 2-pinnate. Umbel densely capitate, terminal; bracts numerous, leaf-like; bracteoles 8-12, obovate or oblanceolate, pale green, margin broad, white, apex 3-lobed; petals rounded, white to purplish red. Fruit broad-ovoid, ribs broadly sinuolate-winged. Fl. and fr. Jul-Sep. Alpine grasslands at 4200-5000 m. Distributed in NW Yunnan, E Xizang, S and W Qinghai.

## 云南棱子芹
**Pleurospermum yunnanense** Franch.

多年生草本。茎中空，上部具条棱和分枝。叶片宽三角状卵形，二至三回三出式或羽状分裂，无毛。苞片6-8，长圆形至宽披针形，顶端全裂；小苞片6-10，长圆状倒卵形，除中脉均膜质，楔形，顶端3-5浅裂至羽状半裂；花瓣倒卵形，绿白色。果实宽卵球形，棱具狭翅。花果期6-10月。生海拔3600-4100米的林地边缘、矮小的杜鹃花灌丛、山谷边、岩石坡。产云南西北部和四川西部。缅甸东北部亦有。

Perennial herbs. Stem hollow, distally ribbed and branched. Blades broadly triangular-ovate, 2-3-ternate or pinnate, glabrous. Bracts 6-8, oblong to broadly lanceolate, divided at apex; bracteoles 6-10, oblong-obovate, membranous except midribs, cuneate, 3-5-lobed to pinnatifid at apex; petals obovate, greenish-white. Fruit broad-ovoid, ribs narrowly winged. Fl. and fr. Jun-Oct. Woodland margins, dwarf *Rhododendron* scrubs, valley sides, rocky slopes at 3600-4100 m. Distributed in NW Yunnan and W Sichuan, Also in NE Myanmar.

垫状棱子芹 *Pleurospermum hedinii*

云南棱子芹 *Pleurospermum yunnanense*

西藏棱子芹 *Pleurospermum hookeri* var. *thomsonii*

## 西藏棱子芹

**Pleurospermum hookeri** C. B. Clarke var. **thomsonii** C. B. Clarke

多年生草本。茎具条棱。叶片三角状心形，三至四回三出式羽状分裂。苞片5-7，倒卵状披针形或线状披针形，边缘膜质，白色或微染棕色，先端长尾状或偶羽状半裂；小苞片披针状线形，边缘狭窄，微带褐色，先端常羽状半裂；花瓣圆形，白色。果实卵球形，棱具狭翅。花果期8-10月。生海拔2700-4500米的草坡。产云南西北部、四川、西藏东南部、甘肃和青海。

Perennial herbs. Stem ribbed. Blades triangular-ovate, 3-4-ternate-pinnate. Bracts 5-7, obovate-lanceolate or linear-lanceolate, margin membranous, white or tinged brown, apex long-caudate or occasionally pinnatifid; bracteoles linear-lanceolate, margin narrow, tinged brown, apex usually pinnatifid; petals rounded, white. Fruit ovoid, ribs narrowly winged. Fl. and fr. Aug-Oct. Grassy slopes at 2700-4500 m. Distributed in NW Yunnan, Sichuan, SE Xizang, Gansu and Qinghai.

## 瘤果棱子芹

**Pleurospermum wrightianum** de Boiss.

多年生草本。叶柄有狭翅，基部扩展但不呈鞘状；叶片轮廓狭长圆状卵形，二至三回羽状分裂。总苞片7-9，基部边缘狭膜质；伞辐10-20，中间的较周围的短；小总苞片与总苞片同形；小伞形花序有花10-15；花瓣白色或紫红色，倒卵形。果实椭圆状卵形，常具小瘤状突起，果棱有明显的鸡冠状翅，每棱槽有油管1，合生面2。花期7-8月，果期9-10月。生海拔3600-4600米的山坡草地上。产中国西南。

Perennial herbs. Petioles winged, extending at base but not clasping; blades narrowly oblong-ovate, 2-3-ternate-pinnate. Bracts 7-9, linear-lanceolate, apex pinnatifid, magin narrow membranous; rays 10-20; bracteoles similar to bract; umbellules 10-15-flowered. petals obovate, white or purplish reddish. Fruits narrowly elliptic-ovoid, usually tuberculate, ribs all broadly cristate-winged; vittae 1 in each furrow, 2 on commissure. Fl. Jul-Aug. Fr. Sep-Oct. Alpine grasslands at 3600-4600 m. Distributed in SW China.

瘤果棱子芹 *Pleurospermum wrightianum*

## 裂苞舟瓣芹

**Sinolimprichtia alpina** H. Wolff var. **dissecta** R. H. Shan et S. L. Liou

多年生草本。叶片卵状长圆形或长圆形，末回裂片线形，顶端圆形，全缘或具齿。总苞片1-4厘米，为伞辐的一半；伞辐15-35(-50)；小苞片的轮廓宽倒卵形，超过小伞形花序，二至三回羽状分裂。花萼宿存，花药深紫色。花果期5-10月。生海拔3500-4800米的高山草坡、山麓碎石或岩石缝中。产云南西北部、四川西南部和西藏东南部。

Perennial herbs. Blade ovate-oblong or oblong, ultimate segments linear, apex rounded, entire or toothed. Bracts 1-4 cm long, about half as long as rays; rays 15-35(-50); bracteoles broadly obovate in outline, exceeding umbellule, 2-3-pinnate; calyx teeth persistent in fruit; anthers dark purple. Fl. and fr. May-Oct. Alpine grassy slopes, screes, rock crevices at 3500-4800 m. Distributed in NW Yunnan, SW Sichuan and SE Xizang.

裂苞舟瓣芹 *Sinolimprichtia alpina* var. *dissecta*

## 密瘤瘤果芹

**Trachydium subnudum** C. B. Clarke ex H. Wolff

多年生草本。茎常退化，植株几乎呈莲座状。叶片长圆状披针形，二回三出式羽状分裂，末回

密瘤瘤果芹 *Trachydium subnudum*

裂片卵形或披针形，边缘锐裂，两面具糙硬毛。伞形花序铺散；萼齿不明显；花瓣白色，基部楔形。果实宽卵形，果棱丝状，密被具瘤，特别是在果棱上。种子表面深凹。花果期7-9月。生海拔3000-4500(-5000)米的高山草甸。产四川西南部、西藏南部。印度东北部亦有。

Perennial herbs. Stem often reduced, plants almost rosette. Blade oblong-lanceolate, ternate-2-pinnate, ultimate segments ovate or lanceolate, margins incised, both surfaces moderately hispid. Umbels lax; calyx teeth obsolete; petals white, base cuneate. Fruits broadly ovoid, ribs filiform, densely tuberculate especially on ribs. Seeds face deeply concave. Fl. and fr. Jul-Sep. Alpine meadows at 3000-4500 (-5000) m. Distributed in SW Sichuan, S Xizang. Also in NE India.

## 大叶柴胡

**Bupleurum longiradiatum** Turcz.

多年生高大草本。根茎长圆柱形，坚硬。叶宽卵状椭圆形或披针形，8-17 × 2.5-5(-8)厘米。伞幅3-9，不等长；小苞片5-6；小伞形花序具5-16花；花瓣黄色或紫色。果实长圆状椭圆体形。花果期8-10月。生海拔200-900米的林下、阴湿溪边或山坡。产内蒙古、甘肃、辽宁、吉林和黑龙江。俄罗斯、

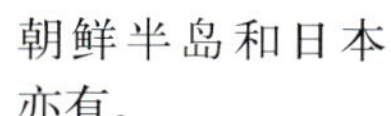
朝鲜半岛和日本亦有。

Perennial tall herbs. Rhizomes cylindric, rigid. Leaves broadly ovate-elliptic or lanceolate, 8-17 × 2.5-5(-8) cm. Rays 3-9, unequal; bracteoles 5-6; umbellules 5-16-flowered; petals yellow or purple. Fruits oblong-ellipsoid. Fl. and fr. Aug-Oct. Forests, shady river banks or mountain slopes at 200-900 m. Distributed in Neimenggu, Gansu, Liaoning, Jilin and Heilongjiang. Also in Russia, Korean Peninsula and Japan.

大叶柴胡 *Bupleurum longiradiatum*

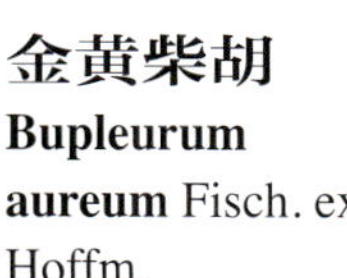

## 金黄柴胡

**Bupleurum aureum** Fisch. ex Hoffm.

多年生草本。茎分枝少。茎生叶阔卵形。顶部复伞形花序宽6-10厘米，侧生复伞形花序宽3-5厘米；伞幅6-10；小伞形花序具15-20花；花瓣黄色，倒卵形，中脉深黄色。果实长圆形，深棕色；棱突出；每棱槽内油管3，合生面4。花果期7-9月。生海拔1300-1900米的开阔林中、林缘、灌丛、山坡或河岸。产新疆(天山)。哈萨克斯坦、吉尔吉斯斯坦、俄罗斯和蒙古亦有。

Perennial herbs. Stems little-branched. Cauline leaves broadly ovate. Terminal umbels 6-10 cm across, lateral umbels 3-5 cm across; rays 6-10; umbellules 15-20-flowered; petals yellow, obovate, midvein dark yellow. Fruits oblong, dark brown; ribs prominent; vittae 3 in each furrow, 4 on commissure. Fl. and fr. Jul-Sep. Open forests, forest edges, among shrubs, mountain slopes or river banks at 1300-1900 m. Distributed in Xinjiang (Tianshan Mountain). Also in Kazakhstan, Kyrgyzstan, Russia and Mongolia.

金黄柴胡 *Bupleurum aureum*

兴安柴胡 *Bupleurum sibiricum*

### 兴安柴胡
**Bupleurum sibiricum** Vest ex Spreng. (C. B. Clarke) Ridley

多年生草本。茎基部常紫红色。基部叶片狭披针形，叶脉7-9，先端短渐尖，基部渐狭成叶柄；上部叶片披针形，抱茎，先端渐尖。小苞片7-12，椭圆状披针形，浅黄色，常具5脉；花瓣黄色。果实宽椭圆形，暗褐色，稍灰绿色；果棱突出，狭翅状。花果期7-9月。生海拔300-800米的山坡。产内蒙古、黑龙江和辽宁。俄罗斯东南部和蒙古亦有。

Perennial herbs. Stems base often purplish red. Blade narrowly lanceolate, 7-9 nerved, apex short-acuminate, apiculate, base tapering into petioles; upper leaves sessile; blades lanceolate, embracing, apex acuminate. Bracteoles 7-12, elliptic-lanceolate, pale yellow, 5-nerved; petals yellow. Fruit broad-ellipsoid, dark brown, slightly glaucous; ribs prominent, narrowly winged. Fl. and fr. Jul-Sep. Mountain slopes at 300-800 m. Distributed in Neimenggu, Heilongjiang and Liaoning. Also in SE Russia and Mongolia.

雾灵柴胡 *Bupleurum sibiricum* var. *jeholense*

### 雾灵柴胡
**Bupleurum sibiricum** Vest ex Spreng. var. **jeholense** (Nakai) Y. C. Chu ex R. H. Shan et Y. Li

本变种与原变种的区别在于：叶片卵状披针形。小苞片5。花果期7-9月。生海拔1500-2000米的山坡。产河北(雾灵山)。

This variety differs from var. *sibiricum* in its: leaves ovate-lanceolate. Bracteoles 5. Fl. and fr. Jul-Sep. Mountain slopes at 1500-2000 m. Distributed in Hebei (Wuling Mountain).

### 大苞柴胡
**Bupleurum euphorbioides** Nakai

一年生或二年生草本。茎常带紫色。伞形花序数个；伞幅4-11，极不等长；小苞片5(-7)，阔椭圆形或倒卵形，超出花和果；小伞形花序具16-24花；花瓣黄色，背面浅紫色。果实广卵球形，紫棕色。花期7-8月果期8-9月，生海拔1200-2500米的林缘或高山草地。产吉林南部。朝鲜半岛亦有。

Annual or biennial herbs. Stems often tinged purple. Rays 4-11, very unequal; bracteoles 5(-7), broadly elliptic or obovate, exceeding flowers and fruits; umbellules 16-24-flowered; petals yellow, abaxially light purplish. Fruits wide ovoid, purplish brown. Fl. Jul-Aug. Fr. Aug-Sep. Forest edges or alpine grasslands at 1200-2500 m. Distributed in S Jilin. Also in Korean Peninsula.

大苞柴胡 *Bupleurum euphorbioides*

### 黑柴胡
**Bupleurum smithii** H. Wolff

多年生草本植物。根状茎深褐色。茎多数，簇状，粗壮，基部多叶。叶柄微紫红色，叶片窄长圆形至倒披针形。伞幅4-9，不等长；小苞片6-9，超出花约1.5倍；花瓣黄色。果实卵球形，褐色。花果期7-9月。生海拔1400-3700米的山坡、草地或河边。产华北、华西和西北。

黑柴胡 *Bupleurum smithii*

Perennial herbs. Rhizomes dark brown. Stems several, tufted, stout, basal leaves many; petioles often purplish red. Leaves narrow-oblong or oblanceolate. Rays 4-9, unequal; bracteoles 6-9, exceeding (to × 1.5) flowers; petals yellow. Fruits ovoid, brown. Fl. and fr. Jul-Sep. Slopes, grasslands or riversides at 1400-3700 m. Distributed in N, W and NW China.

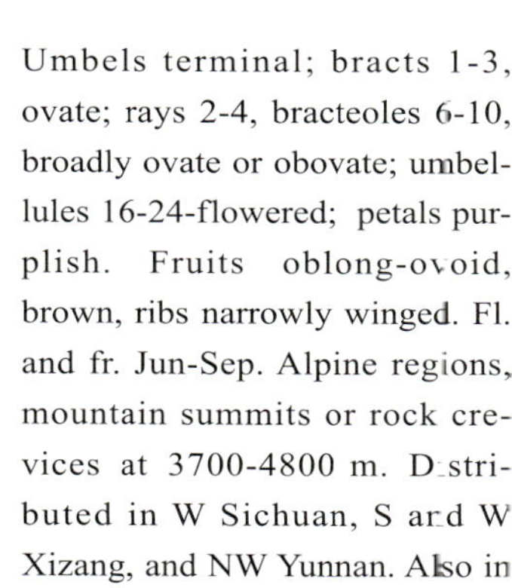

## 匍枝柴胡

**Bupleurum dalhousieanum** (C. B. Clarke) Koso-Pol.

多年生草本。茎紫红色，常外倾。叶片披针形或卵形，基部圆形，抱茎，先端渐尖。伞形花序顶生；苞片1-3，卵形；伞辐2-4；小苞片6-10，宽卵形或倒卵形；小伞形花序有花16-24；花瓣略带紫色。果实长圆状卵球形，褐色。果棱狭翅状。花果期6-9月。生海拔3700-4800米的高山地区、山顶或岩石裂缝中。产四川西部、西藏南部和西部、云南西北部。中亚和南亚亦有。

Perennial herbs. Stems purplish red, usually decumbent. Leaves lanceolate or ovate, base rounded, clasping, apex acuminate. Umbels terminal; bracts 1-3, ovate; rays 2-4, bracteoles 6-10, broadly ovate or obovate; umbellules 16-24-flowered; petals purplish. Fruits oblong-ovoid, brown, ribs narrowly winged. Fl. and fr. Jun-Sep. Alpine regions, mountain summits or rock crevices at 3700-4800 m. Distributed in W Sichuan, S and W Xizang, and NW Yunnan. Also in C and S Asia.

匍枝柴胡 *Bupleurum dalhousieanum*

## 空心柴胡

**Bupleurum longicaule** DC. var. **franchetii** H. Boiss.

多年生草本。根木质化，灰褐色。茎多丛生，下部叶披针形，无柄，基部抱茎。苞片1-2，早落，小伞形花序具8-15花；花瓣黄色；小苞片常10-12。果矩圆形，棱具狭翅。生海拔1000-4000米的林中或山坡草地。产中国西南、华中和华西。

Perennial herbs. Roots woody, gray-brown. Stems usually several, lower leaves lanceolate, sessile, clasping. Bracts 1-2, deciduous, umbellules 8-15-flowered; petals yellow; bracteoles usually 10-12. Fruits oblong, ribs narrowly winged. Forests or grasslands or mountain slopes at 1000-4000 m. Distributed in SW, C and W China.

空心柴胡 *Bupleurum longicaule* var. *franchetii*

紫花鸭跖柴胡 *Bupleurum commelynoideum*

抱茎柴胡 *Bupleurum longicaule* var. *amplexicaule*

## 紫花鸭跖柴胡

**Bupleurum commelynoideum** de Boiss. (C. B. Clarke) Ridley

多年生草本。茎数个。基生叶无梗，叶片线状披针形，背面微紫色，叶脉5，基部圆形，抱茎。苞片1或2，或无，卵状披针形；小苞片略带紫蓝色，卵形或披针形，大大超过花；小伞形花序具花16-30朵；花瓣正面紫色或微淡黄色，背面紫色。果实长圆形，红棕色，果棱淡褐色，突出或狭翅状。花果期8-10月。生海拔3000-4300米的高山草甸。产云南西北部、四川西部和西藏。

Perennial herbs. Stems several. Basal leaves sessile; blade linear-lanceolate, abaxially tinged purple, 5-nerved, base rounded, clasping. Bracts 1 or 2, or absent, ovate-lanceolate, bracteoles purplish blue, ovate or lanceolate, greatly exceeding the flowers; umbellules 16-30-flowered; petals adaxially purple or yellowish-tinged, abaxially purple. Fruits oblong, reddish-brown, ribs pale brown, prominent or narrowly winged. Fl. and fr. Aug-Oct. Alpine meadows at 3000-4300 m. Distributed in NW Yunnan, W Sichuan and Xizang.

## 抱茎柴胡

**Bupleurum longicaule** DC. var. **amplexicaule** C. Y. Wu ex R. H. Shan et Yin Li

多年生草本。茎直立，数个，不分枝或很少上面分枝。下部叶线形，无梗，抱茎；中间叶长披针形，无梗，基部圆形或心形。上部叶片狭卵形，基部深心形。总苞片2-3或无，披针形或卵形；伞辐7-9；小苞片10-12；小伞形花序有花20；花瓣黄色。果实卵球形或椭圆状卵球形，灰棕色，果棱突出，锐尖。花果期7-9月。生海拔2500-2700米的山坡上的森林中。产云南西北部。

Perennial herbs. Stems several, erect, unbranched or few-branched above. Lower leaves linear, sessile, clasping; middle leaves long-lanceolate, sessile, base rounded or cordate; upper leaves narrow-ovate, base deep cordate. Bracts 2-3 or 0, lanceolate or ovate; rays 7-9; bracteoles 10-12; umbellules 20-flowered; petals yellow. Fruits ovoid or ellipsoid-ovoid, gray-brown, ribs prominent, acute. Fl. and fr. Jul-Sep. Forests on mountain slopes at 2500-2700 m. Distributed in NW Yunnan.

## 阿尔泰柴胡

**Bupleurum krylovianum** Schischk. ex G. V. Krylov

多年生草本。茎多数，上部分枝。复伞形花序宽3-7厘米；顶生复伞形花序具伞幅10-20，侧生复伞形花序具伞幅6-8；小伞形花序具18-22花；花瓣黄色，先端2裂。果实圆柱状长圆形，深棕色；棱突出。花果期7-9月。生海拔1200-2000米的灌丛下或干旱石山坡。产新疆。哈萨克斯坦、吉尔吉斯斯坦和俄罗斯亦有。

Perennial herbs. Stems numerous, branched above. Umbels 3-7 cm across; rays of terminal umbels 10-20, lateral umbels 6-8-rayed; umbellules 18-22-flowered; petals yellow, tips 2-lobed. Fruits terete-oblong, dark brown; ribs prominent. Fl. and fr. Jul-Sep. Under shrubs or dry stony mountain slopes at 1200-2000 m. Distributed in Xinjiang. Also in Kazakhstan, Kyrgyzstan and Russia.

## 竹叶柴胡

**Bupleurum marginatum** Wall. ex DC.

多年生草本。单叶，长披针形，全缘。复伞形花序，多数，顶生花序常短于侧生花序，伞辐3-4(-7)，不等长；小苞片5；小伞形花序具(6-)8-10(-12)花；花瓣浅黄色。果实长圆形。花期6-9月，果期9-11月。生海拔700-4000米的山坡草地或林中。产中国西南和西北。南亚亦有。

阿尔泰柴胡 *Bupleurum krylovianum*

竹叶柴胡 *Bupleurum marginatum*

Perennial herbs. Leaves simple, long lanceolate, entire. Compound umbels many, terminal ones often shorter than lateral ones; rays 3-4(-7), unequal in length; bracteoles 5; umbellules (6-)8-10 (-12)-flowered; petals pale yellow. Fruits oblong. Fl. Jun-Sep. Fr. Sep-Nov. Grassy slopes or forests at 700-4000 m. Distributed in SW and NW China. Also in S Asia.

## 红柴胡

**Bupleurum scorzonerifolium** Will.

多年生草本。主根粗壮，深红棕色，分枝。茎二歧状分枝。基生叶线形，厚纸质，稍硬，叶脉3-5，背面突起，边缘白色软骨质，基部轻微变窄，抱茎。伞形花序多数，小伞形花序有花(6-)9-11(-15)；花瓣黄色。果实椭圆体形，暗褐色，果棱浅，突出。花果期7-9月。生海拔100-2300米的灌木丛中林缘、向阳的山坡、干燥草地。

产中国东北、华北、西北、华中、华东等地。俄罗斯、蒙古、朝鲜半岛和日本亦有。

Perennial herbs. Taproot stout, dark reddish-brown, branched. Stems greatly dichotomously branched. Basal leaves linear, thick-papery, rigid, nerves 3-5, prominent abaxially, margin white cartilaginous, base slightly narrowed and clasping. Umbels numerous, umbellules (6-)9-11(-15)-flowered; petals yellow. Fruits ellipsoid, dark brown, ribs pale, prominent. Fl. and fr. Jul-Sep. Shrub forest edges, sunny mountain slopes, dry grasslands at 100-2300 m. Distributed in NE, N, NW, C and E China. Also in Russia, Mongolia, Korean Peninsula and Japan.

## 细叶旱芹

**Cyclospermum leptophyllum** (Pers.) Sprague ex Britton et P. Wils.

多年生草本，高25-45厘米。基部叶柄长2-5(-11)厘米；叶末端裂片线形至丝形；茎生叶多回三出式羽状分裂。复伞形花序顶生或腋生；伞幅2-3(-5)；小伞形花序有花5-23朵。果实球形。花期5-6月，果期6-7月。生溪边或荒地。产广东、台湾、福建和江苏。原产南美洲，现热带和温带地区归化。

Plants perennial, 25-45 cm tall. Basal petioles 2-5(-11) cm long; leaves ultimate segments linear to filiform; cauline leaves ternate-pinnately decompound. Compound umbels terminal or lateral; rays 2-3(-5); umbellules 5-23-flowered. Fruits globose. Fl. May-Jun. Fr. Jun-Jul. Streamsides or wastelands. Distributed in Guangdong, Taiwan, Fujian and Jiangsu. Native to South America, widely naturalized in tropical and temperate regions.

红柴胡 *Bupleurum scorzonerifolium*

细叶旱芹 *Cyclospermum leptophyllum*

毒芹 *Cicuta virosa*

## 毒芹
**Cicuta virosa** L.

多年生草本。根状茎粗2-4厘米，切割后流出黄色汁液。茎单生。基生叶羽片3裂或羽裂；上部叶一至二回羽裂。复伞形花序长5-15厘米；伞幅6-25；小伞形花序具15-35花；每棱槽内油管1，合生面2。花果期7-9月。生海拔300-3300米的林缘、沼泽区域、沼泽地或溪边，常见于浅水。产中国西南、华北、华西、西北和东北。印度、俄罗斯、蒙古、朝鲜半岛、日本和欧洲亦有。

Perennial herbs. Rootstocks 2-4 cm long thick, exudes yellow sap when cut. Stems solitary. Basal leaves pinnae 3-lobed or pinnatifid; upper leaves 1-2-pinnate. Umbels 5-15 cm across; rays 6-25; umbellules 15-35-flowered; vittae 1 in each furrow, 2 on commissure. Fl. and fr. Jul-Sep. Forest edges, marshy areas, bogs or streamsides, often emergent in shallow water at 300-3300 m. Distributed in SW, N, W, NW and NE China. Also in India, Russia, Mongolia, Korean Peninsula, Japan and Europe.

## 宽叶毒芹
**Cicuta virosa** L. var. **latisecta** Celak.

本变种与原变种的区别在于：叶片三角形或卵形三角形，羽片3浅裂或羽状半裂；叶末回裂片长椭圆形或椭圆状卵形。种子表面平。花果期7-9月。生海拔300-500米的沼泽。产山西和吉林。俄罗斯东南部和日本亦有。

This variety differs from var. *virosa* in its: Blade triangular or ovate-triangular, pinnae 3-lobed or pinnatifid; ultimate segments of leaves long-elliptic or elliptic-ovate. Seed face plane. Fl. and fr. Jul-Sep. Marshy places at 300-500 m. Distributed in Shanxi and Jilin. Also in SE Russia and Japan.

宽叶毒芹 *Cicuta virosa* var. *latisecta*

鸭儿芹 *Cryptotaenia japonica*

## 鸭儿芹
**Cryptotaenia japonica** Hassk.

多年生草本。主根短，侧根多数。基部和下部叶柄具长圆形叶鞘；叶三角形至宽卵形；中部小叶菱状倒卵形或心形。复伞形花序呈圆锥状；伞幅2-3，极不等长；小苞片1-3。花果期2-10月。生海拔200-2400米的山地、山谷或林下。产中国西南、华南、华中、

华北、华西和华东。朝鲜半岛和日本亦有。

Perennial herbs. Taproots short, branch roots many. Basal and lower petioles with oblong sheaths; leaves triangular to broad-ovate; middle leaflets rhombic-obovate or cordate. Compound umbels coniform; rays 2-3, very unequal; bracteoles 1-3. Fl. and fr. Feb-Oct. Mountains, valleys or under forests at 200-2400 m. Distributed in SW, S, C, N, W and E China. Also in Korean Peninsula and Japan.

## 田葛缕子

**Carum buriaticum** Turcz.

多年生草本。基部及下部叶片三至四回羽状分裂，末回裂片线形。伞幅9-15，稍不等长；小苞片5-8，披针形；小伞形花序具10-30花；花瓣白色。果实长圆状椭圆体形。花期5-7月，果期8-10月。生海拔1500-3600米的林中、草甸、田中或路边。产华西、华北、西北和东北。俄罗斯和蒙古亦有。

Perennial herbs. Basal and lower leaves 3-4-pinnate; ultimate segments linear. Rays 9-15, slightly unequal; bracteoles 5-8, lanceolate; umbellules 10-30-flowered; petals white. Fruits oblong-ellipsoid. Fl. May-Jul. Fr. Aug-Oct. Forests, meadows, fields or roadsides at 1500-3600 m. Distributed in W, N, NW and NE China. Also in Russia and Mongolia.

田葛缕子 *Carum buriaticum*

## 河北葛缕子

**Carum bretschneideri** H. Wolff

多年生草本。茎单生或2-3分枝。基生叶轮廓卵状披针形，二至三回羽状；末回裂片披针形。苞片1-6，线形，在边缘上具缘毛；伞辐8-12，小苞片5-8，与苞片同形，等于或长于小伞形花序；小伞形花序具15-25花；花瓣白色，基部具短爪。果实长圆状椭圆形；棱槽油管1，合生面油管2。花果期6-9月。生海拔1500-2000米的阴湿处。产河北和山西。

河北葛缕子 *Carum bretschneideri*

Perennial herbs. Stem solitary or 2-3 branched. Basal leaves ovate-lanceolate in outline, 2-3-pinnate; ultimate segments lanceolate. Bracts 1-6, linear, ciliate on the margins; rays 8-12; bracteoles 5-8, similar to bracts, as long as or longer than the umbellules; umbellules 15-25-flowered; petals white, base shortly clawed. Fruit oblong-ellipsoid; vittae solitary in each furrow, 2 on commissure. Fl. and fr. Jun-Sep. Shady moist places at 1500-2000 m. Distributed in Hebei and Shanxi.

长柄小芹 *Sinocarum dolichopodum*

## 长柄小芹

**Sinocarum dolichopodum** (Diels) H. Wolff ex R. H. Shan et F. T. Pu

草本。茎单生，淡紫色，无分枝。叶片轮廓三角形，三至四回羽状分裂。复伞形花序，无总梗，伞辐4-6；小苞片2-6；小伞形花序具10-15花；花瓣白色或紫色。幼果长卵球形，果每棱槽内油管3条。花果期7-9月。生海拔3000-4000米高山草垫或山地岩石上。产云南西北部和四川西部。

Herbs. Stems solitary, with purplish, usually unbranched. Leaves triangular in outline, 3-4- pinnatipartite. Compound umbel, expedunculate; rays 4-6; bracteoles 2-6; umbellules 10-15- flowered; petals white or purplish. Young fruits oblong-ovoid; each rib of fruits with 3 oil tubes. Fl. and fr. Jul-Sep. Alpine meadows or rocks of mountains at 3000-4000 m. Distributed in NW Yunnan and W Sichuan.

## 五匹青

**Pternopetalum vulgare** (Dunn) Hand.-Mazz.

多年生草本。茎单生或2-3分枝。叶片轮廓三角状卵形，三出式分裂；小伞形花序具2-5朵花；花瓣白色或紫色，果实长椭圆形，果棱具小齿。花果期4-9月。生海拔1400-3500米的森林或阴处溪边。产云南、四川、贵州、湖南、湖北、陕西和甘肃南部。印度东北部、尼泊尔和缅甸北部亦有。

Perennial herbs. Stems solitary or 2-3 branches. Blade triangular-ovate in outline, ternate. Umbellules 2-5-flowered; petals white or purple-white. Fruits oblong-ovoid, ribs denticulate. Fl. and fr. Apr-Sep. Forests or shady streamsides at 1400-3500 m. Distributed in Yunnan, Sichuan, Guizhou, Hunan, Hubei, Shaanxi and S Gansu. Also in NE India, Nepal and N Myanmar.

## 川鄂囊瓣芹

**Pternopetalum rosthornii** (Diels) Hand.-Mazz.

多年生草本。基生叶为一至二回三出复叶。复伞形花序花期宽1.5-3厘米；伞幅(7-)15-30(-40)；小伞形花序具2-3花；花瓣白色。果实卵球形；棱粗糙，每棱槽具油管1-3，合生面2-4。花期4-6月，果期7-8月。生海拔1300-2100米的林中、沟边或潮湿岩缝。产四川和湖北。

Perennial herbs. Basal leaves, 1-2-ternate. Umbels 1.5-3 cm across in flower; rays (7-)15-30 (-40); umbellules 2-3-flowered; petals white. Fruits ovoid; ribs scabrid, vittae 1-3 in each furrow, 2-4 on commissure. Fl. Apr-Jun. Fr. Jul-Aug. Forests, valley sides or moist rock crevices at 1300-2100 m. Distributed in Sichuan and Hubei.

## 异叶囊瓣芹

**Pternopetalum heterophyllum** Hand.-Mazz.

多年生草本，高15-30厘米。叶为三出复叶，小叶3(-5)；叶片三角形，裂片扇形或菱形；茎生叶1-3，一至二回三出复叶；末回裂片条形。伞幅10-20；小伞形花序具2(-3)花；花瓣白色。果实卵球形；棱线形，每棱槽具油管2，合生面4。花期3-5月，果期6-8月。生海拔1200-3400米的林下、灌丛、草地或溪边。产四川西部、湖北西部、湖南、陕西南部、甘肃和青海东部。

Perennial herbs, 15-30 cm tall. Leaves ternate; leaflets 3(-5), blade triangular in outline; ultimate segments flabelliform or rhomboidal; cauline leaves 1-3, 1-2-ternate; ultimate segments strip-type. Rays 10-20; umbellules 2(-3) -flowered; petals white. Fruits ovoid; ribs filiform; vittae 2 in each furrow, 4 on commissure. Fl. Mar-May. Fr. Jun-Aug. Under forests, shrubs, grasslands or streamsides at 1200-3400 m. Distributed in W Sichuan, W

五匹青 *Pternopetalum vulgare*

川鄂囊瓣芹 *Pternopetalum rosthornii*

异叶囊瓣芹 *Pternopetalum heterophyllum*

膜蕨囊瓣芹 *Pternopetalum trichomanifolium*

Hubei, Hunan, S Shaanxi, Gansu and E Qinghai.

## 膜蕨囊瓣芹

**Pternopetalum trichomanifolium** (Franch.) Hand.-Mazz.

多年生草本。叶几乎全部基生，叶片外形三角状卵形，末回裂片条形。伞幅(6-)15-30(-40)；小苞片2-4；小伞形花序具2-4花；花瓣白色，顶端微凹。果实狭长卵球形。花期3-5月，果期6-8月。生海拔600-2400米的林下、沟边或阴湿岩石上。产中国西南、华南和华中。

Perennial herbs. All leaves almost basilar, leaves triangular-ovate in outline; ultimate segments linear. rays (6-)15-30(-40); bracteoles 2-4; umbellules 2-4-flowered; petals white, apex slightly concave. Fruits narrow long-ovoid. Fl. Mar-May. Fr. Jun-Aug. Under the forests, by the valleys or damp and shady rocks at 600-2400 m. Distributed in SW, S and C China.

## 异叶茴芹

**Pimpinella diversifolia** DC.

多年生草本。全株被短柔毛，无匍匐茎。叶片三出式分裂，小叶卵状心形；茎生叶向上退化，一回羽状分裂或3浅裂，披针形，经常撕裂。小伞形花序具6-20花；花瓣白色，倒卵形，先端缺刻，具弯曲的小裂片。果实卵球形，基部心形，表面具小乳突状的短柔毛。种子表面平。花果期5-11月。生海拔200-3300米的森林、林缘、山地灌丛和草地或溪边。产中国西北、华南、华中、华北和西南。东亚、东南亚和中亚等地区亦有。

Perennial herbs. Plants pubescent throughout, without stolons. Blade ternate, leaflets ovate-cordate. Cauline leaves reduced upwards, 1-pinnate or 3-lobed, lanceolate, often lacerate. Umbellules 6-20-flowered; petals white, obovate, apex notched with small incurved lobule. Fruit ovoid, base cordate, surface shortly papillose-pubescent. Seed face plane. Fl. and fr. May-Nov. Forests, forest edges, montane scrub and grasslands or streamsides at 200-3300 m. Distributed in NW, S, C, N and SW China. Also in S, SE and C Asia.

## 短果茴芹

**Pimpinella brachycarpa** (Kom.) Nakai

多年生草本。顶端小叶宽卵形。茎生叶基生，无柄；通常无总苞片；伞幅7-15；小苞片2-5，短于花梗；小伞形花序具15-20花，杂性；花瓣白色。果实卵球形，约长2 × 1.8毫米。花果期6-9月。生海拔500-900米的河边或林缘。产贵州、河北、山西、辽宁和吉林。俄罗斯和朝鲜半岛亦有。

Perennial herbs. Terminal leaflets broad-ovate. Cauline leaves basal, sessile; usually without common perianth; rays 7-15; bracteoles 2-5, shorter than pedicels; umbellules 15-20-flowered, polygamous; petals white. Fruits ovoid, ca. 2 × 1.8 mm. Fl. and fr. Jun-Sep. Riversides or forest edges at 500-900 m. Distributed in Guizhou, Hebei, Shanxi, Liaoning and Jilin. Also in Russia and Korean Peninsula.

异叶茴芹 *Pimpinella diversifolia*

短果茴芹 *Pimpinella brachycarpa*

山茴香 *Carlesia sinensis*

天山泽芹 *Berula erecta*

## 山茴香

**Carlesia sinensis** Dunn

多年生矮小草本。叶片通常三回羽状全裂，末回裂片线形，上部叶3深裂。伞幅7-12(-20)；小伞形花序多花；萼齿明显，卵状三角形。果实长倒卵球形至长圆状卵球形，基部圆形。花果期7-9月。生海拔300-1000米的干旱山坡或岩石缝中。产山东和辽宁。朝鲜半岛亦有。

Perennial herbs short. Leaves usually pinnately 3-parted, ultimate segments linear, upper leaves deep 3-parted. Rays 7-12 (-20); umbellules many-flowered; calyx teeth conspicuous, ovate-triangular. Fruits long obovoid to oblong-ovoid, base rounded. Fl. and fr. Jul-Sep. Dry slopes or rock crevices at 300-1000 m. Distributed in Shandong and Liaoning. Also in Korean Peninsula.

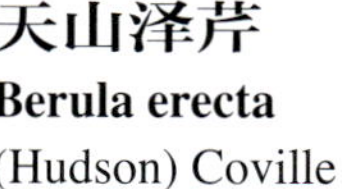

## 天山泽芹

**Berula erecta** (Hudson) Coville

多年生草本。茎中空。沉水叶三至四回羽状分裂；陆生叶一回羽状分裂；卵状披针形或长圆形，边缘有锯齿或不规则锐裂。苞片3-6，长圆形或披针形；小伞形花序具花10-20；小苞片5-8；花瓣白色，广卵形。果广卵形，木栓质增厚。花果期5-8月。生海拔约1500米的溪边、平原或低山河边。产新疆。南亚、西南亚、中亚、欧洲和北非亦有；美洲和澳大利亚有引入。

Perennial herbs. Stem hollow. Submerged leaves 3-4-pinnate; aerial leaves 1-pinnate, ovate-lanceolate or oblong, margins serrate or irregularly incised. Bracts 3-6, oblong or lanceolate; umbellules 10-20-flowered, bracteoles 5-8; petals white, widely ovate. Fruits broadly ovate, corky thickened. Fl. and fr. May-Aug. Streamsides, other riparian habitats on plains or hills at ca. 1500 m. Distributed in Xinjiang. Also in S, SW and C Asia, Europe and N Africa; introduced in America and Australia.

泽芹 *Sium suave*

## 泽芹

**Sium suave** Walter

多年生草本。根纺锤形。羽片3-9对。复伞形花序宽4-8厘

密花岩风 *Libanotis condensata*

米。伞幅(8-)10-20；小伞形花序具10-20花；萼齿三角状披针形或稍三角形。果实卵球形；棱突出，鸡冠状加厚，具狭翅；每棱槽具油管1-3，合生面2-6。花期7-8月，果期9-10月。生湿草地、沼泽地或溪边。产华南、华北、华西、华东和东北。朝鲜半岛、日本、俄罗斯和北美洲亦有。

Perennial herbs. Roots fusiform. Pinnae 3-9 pairs. Umbels 4-8 cm across; rays (8-)10-20; umbellules 10-20-flowered; calyx teeth triangular-lanceolate or minute triangular. Fruits ovoid; ribs prominent, corky thickened, narrowly winged; vittae 1-3 in each furrow, 2-6 on commissure. Fl. Jul-Aug. Fr. Sep-Oct. Damp grasslands, marshlands or streamsides. Distributed in S, N, W, E and NE China. Also in Korean Peninsula, Japan, Russia and North America.

## 密花岩风

**Libanotis condensata** (L.) Crantz

多年生草本。叶长圆形，二至三回羽状全裂；末回裂片线形。复伞形花序顶生，通常不分枝；小伞形花序具15-20花；花瓣白色。果实椭圆体形，每棱内2-4油管。花果期7-9月。生海拔1400-2400米的林缘、草地或溪边。产河北、山西、内蒙古和新疆。哈萨克斯坦、俄罗斯和蒙古亦有。

Perennial herbs. Leaves oblong, 2-3-pinnatisect; ultimate segments linear. Compound umbels terminal, usually unbranched; umbellules 15-20-flowered; petals white. Fruits ellipsoid, vittae 2-4 in each furrow. Fl. and fr. Jul-Sep. Forest edges, grasslands or streamsides at 1400-2400 m. Distributed in Hebei, Shanxi, Neimenggu and Xinjiang. Also in Kazakhstan, Russia and Mongolia.

## 地岩风

**Libanotis depressa** R. H. Shan et M. L. Sheh

多年生草本。矮小，无茎，莲座状。叶片长圆形，二回羽状分裂全裂；羽片2-4对；末回裂片线状披针形，基部和边缘具毛，先端具细尖，无毛或稍被微柔毛。小伞形花序有花10-20；花瓣白色，中脉淡黄色，无毛。果实长圆形或近圆形，稍微背压扁，密被鳞片状糙硬毛；果棱丝状，突出。花果期7-9月。生海拔3400-4100米的山坡草丛中，河岸边。产青海(玉树)、四川(德格县)和西藏(贡觉县)。

Perennial herbs. Plants dwarf, acaulescent. Rosette leaf; oblong, 2-pinnatisect; pinnae 2-4 pairs; ultimate segments linear-lanceolate, base and margins pilose, apex apiculate, glabrous or minutely puberulent. Umbellules 10-20-flowered; petals white, midrib yellowish, glabrous. Fruit oblong or suborbicular, slightly dorsally compressed, densely scaly-hispid, ribs filiform, prominent. Fl. and fr. Jul-Sep. Grassy places, river banks at 3400-4100 m. Distributed in Qinghai (Yushu), Sichuan (Dege County) and Xizang (Gongjue County).

地岩风 *Libanotis depressa*

水芹 *Oenanthe javanica*

## 水芹

**Oenanthe javanica** (Blume) DC.

多年生草本。叶片轮廓三角形，一至二回羽状分裂，叶裂片长1.5-3厘米。复伞形花序顶生；伞幅6-16（-30），小苞片2-8，条形；小伞形花序约具20花。果实近球形或卵球形，果棱稍木质加厚。生海拔600-4000米的浅水低洼地区、池沼或水沟边。产中国大多数省区。南亚、东南亚、东北亚和新几内亚岛亦有。

Perennial herbs. Leaves triangular in outline, 1-2- pinnatipartite, lobes 1.5-3 cm long. Compound umbels terminal; rays 6-16 (-30); bracteoles 2-8, linear; umbellules ca. 20-flowered. Fruits subglobose or ovoid, ribs slightly corky-thickened. Swamps, ponds or streams at 600-4000 m. Distributed in most parts of China. Also in S, SE and NE Asia, and New Guinea.

## 卵叶水芹

**Oenanthe javanica** (Blume) DC. subsp. **rosthornii** (Diels) F. T. Pu

本亚种与水芹的区别在于伞幅不等长，长2-6厘米；小苞片披针形。果实卵球形。花期8-9月，果期10-11月。生海拔1400-4000米的林缘草地、沼泽地、湿草甸或河边。产中国西南、华南、东南和华中。泰国亦有。

This subspecies differs from subsp. *javanica* in its rays unequal, 2-6 cm long; bracteoles lanceolate. Fruits ovoid. Fl. Aug-Sep. Fr. Oct-Nov. Grassland at forest edges, marshes, water meadows or river banks at 1400-4000 m. Distributed in SW, S, SE and C China. Also in Thailand.

## 线叶水芹

**Oenanthe linearis** Wall. ex DC.

多年生草本。根茎短。茎直立或匍匐。叶三角状卵形，多为一回羽状分裂，末回裂片线形。伞幅3-12；小苞片3-8；小伞形花序具8-20花。果实长圆形或近圆球形。花期5-7月，果期7-8月。生海拔800-3000米的山坡、山谷林下或溪旁。产中国西南、东南和华中。印度、尼泊尔、缅甸、越南和印度尼西亚亦有。

Perennial herbs. Rhizomes short. Stems erect or creeping. Leaves triangular-ovate, mostly 1-pinnate; ultimate segments linear.

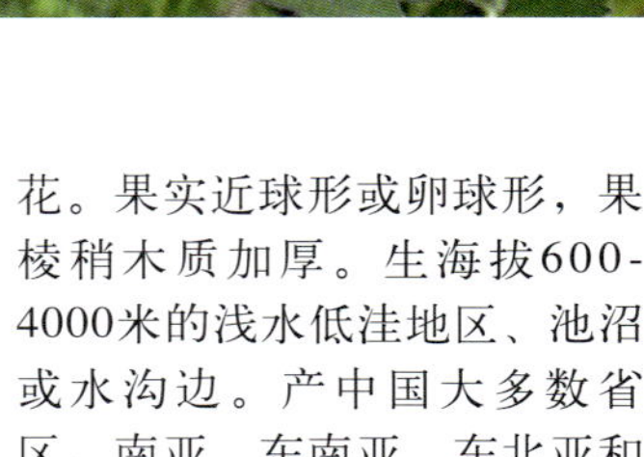

卵叶水芹 *Oenanthe javanica* subsp. *rosthornii*

线叶水芹 *Oenanthe linearis*

多裂叶水芹 *Oenanthe thomsonii*

Rays 3-12; bracteoles 3-8; umbellules 8-20-flowered. Fruits oblong or almost globose. Fl. May-Jul. Fr. Jul-Aug. Slopes, valley forests or by rivers at 800-3000 m. Distributed in SW, SE and C China. Also in India, Nepal, Myanmar, Vietnam and Indonesia.

## 多裂叶水芹

**Oenanthe thomsonii** C. B. Clarke

多年生匍匐草本。根簇生或纤维状。叶同形，三至四(五)回羽状分裂；初级羽片5-7对；末回羽片短，条形。复伞形花序宽3-8厘米，常与叶对生；伞幅4-12；小伞形花序具15-20花。果实近球形；背棱及中棱凸出，丝状。花期7-8月，果期9-10月。生海拔1800-3500米的沼泽草甸、湿草地或溪边。产中国西南、华中和华东。印度、尼泊尔、不丹和缅甸亦有。

Perennial creeping herbs. Roots fascicled or fibrous. Leaves homomorphic, 3-4(-5)-pinnate; primary pinnae 5-7 pairs; ultimate segments short, linear. Compond umbels 3-8 cm across, frequently leaf-opposed; rays 4-12; umbellules 15-20-flowered. Fruits subglobose; dorsal and intermediate ribs protruding, filiform. Fl. Jul-Aug. Fr. Sep-Oct. Marshy meadows, moist grasslands or streamsides at 1800-3500 m. Distributed in SW, C and E China. Also in India, Nepal, Bhutan and Myanmar.

## 白花苞裂芹

**Schulzia albiflora** (Kar. et Kir.) Popov

多年生草本。叶长圆状披针形，三回羽状分裂，末回条状披针形或条形。复伞形花序多数，宽3-7厘米；伞幅10-20 (-30)。果实长圆状卵球形；每棱槽具油管3，合生面8。花果期7-8月。生海拔2700-4600米的高寒草甸或草坡。产新疆。哈萨克斯坦、吉尔吉斯斯坦、塔吉克斯坦和俄罗斯(西伯利亚)亦有。

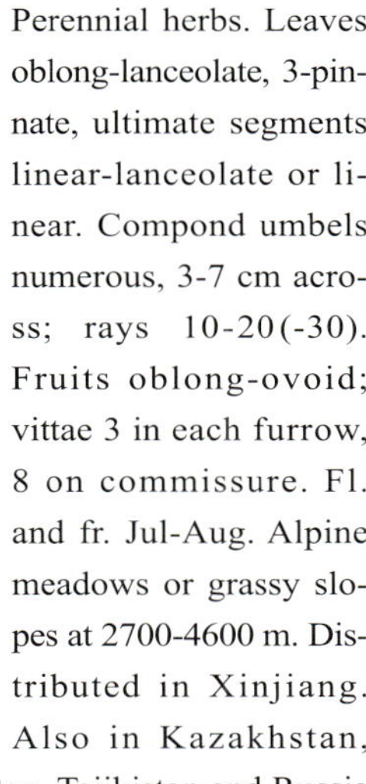
Perennial herbs. Leaves oblong-lanceolate, 3-pinnate, ultimate segments linear-lanceolate or linear. Compond umbels numerous, 3-7 cm across; rays 10-20(-30). Fruits oblong-ovoid; vittae 3 in each furrow, 8 on commissure. Fl. and fr. Jul-Aug. Alpine meadows or grassy slopes at 2700-4600 m. Distributed in Xinjiang. Also in Kazakhstan, Kyrgyzstan, Tajikistan and Russia (Siberia).

白花苞裂芹 *Schulzia albiflora*

茴香 *Foeniculum vulgare*

## 茴香
**Foeniculum vulgare** (L.) Mill.

多年生草本，全株具强烈香气。叶片轮廓宽三角形，末回裂片线形，宽不及1毫米。伞幅6-29(-40)，不等长；小伞形花序具14-39小花；花梗纤细。果实长圆形，圆柱状；肋等长。花期5-6月，果期7-9月。海拔200-2600米的中国大部分地区有栽培。原产地中海地区；全世界广泛栽培。

Perennial herbs, strongly aromatic throughout. Leaves broadly triangular in outline, ultimate segments filiform, less than 1 mm wide. Rays 6-29 (-40), unequal; umbellules 14-39-flowered; pedicels slender. Fruits oblong, terete; ribs equal. Fl. May-Jun. Fr. Jul-Sep. Cultivated in most parts of China at 200-2600 m. Native to Mediterranean regions; cultivated worldwide.

## 蛇床
**Cnidium monnieri** (L.) Cusson

一年生草本，高10-80厘米。茎单生。苞片宿存；伞幅8-20(-30)，不等长；小苞片5-9，条形，与花梗近等长；小伞形花序具15-20花。萼无齿。果实卵球形，1.5-3 × 1-2毫米。花期4-7月，果期7-10月。在河边草地或田边。产中国大部分地区。南亚、东南亚、东北亚亦有，在北美洲为外来种。

Annual herbs, 10-80 cm tall. Stems solitary. Bracts persistent; rays 8-20 (-30), unequal; bracteoles 5-9, linear, nearly equal pedicels; umbellules 15-20-flowered; calyx teeth obsolete. Fruits ovoid, 1.5-3 × 1-2 mm. Fl. Apr-Jul. Fr. Jul-Oct. Riparian grasslands or field edges. Distributed in most parts of China. Also in S, SE and NE Asia, and adventive in North America.

蛇床 *Cnidium monnieri*

## 岩茴香
**Ligusticum tachiroei** (Franch. et Sav.) M. Hiroe et Constance

多年生草本，高10-30厘米。根圆柱形。叶为三回羽状复叶，基部羽片5-7对。复伞形花序宽2-4厘米。伞幅5-10；花瓣白色或粉色，基部具短爪。果实长圆形；棱突出；每棱槽具油管1，合生面2。种子表面平滑至稍凹。花期7-8月，果期8-9月。生海拔1200-2500米的砾石山坡、潮湿河岸或岩缝。产河南、河北、山西、吉林和辽宁。蒙古、朝鲜半岛和日本亦有。

Perennial herbs, 10-30 cm tall. Roots cylindrical. Leaves ovate, 3-pinnate, primary pinnae 5-7 pairs. Umbels 2-4 cm across; rays 5-10; petals white or pinkish, base shortly clawed. Fruits oblong; ribs prominent; vittae 1 in each furrow, 2 on commissure. Seeds face plane to slightly con-

岩茴香 *Ligusticum tachiroei*

cave. Fl. Jul-Aug. Fr. Aug-Sep. Pebbly slopes, damp river banks or rock crevices at 1200-2500 m. Distributed in Henan, Hebei, Shanxi, Jilin and Liaoning. Also in Mongolia, Korean Peninsula and Japan.

## 抽葶藁本

**Ligusticum scapiforme** H. Wolff

多年生草本。茎2-3条，不分枝，近花葶状，基部具枯萎的叶鞘纤维。叶片长圆状披针形，二至三回羽状分裂；茎生叶无或偶有1，退化。伞形花序基部被柔毛；花瓣白色或略带紫色，倒卵形，基部具短爪。果实长圆状卵球形，背棱和中棱丝状，侧棱具翅。花果期6-10月。生海拔2700-4800米的针叶林、山地的灌丛、林缘草地、高山灌丛和草甸或河岸。产云南西北部、四川西部和西藏南部。

Perennial herbs. Stems 2-3, unbranched, subscapose, base clothed in fibrous remnant sheaths. Leaves oblong-

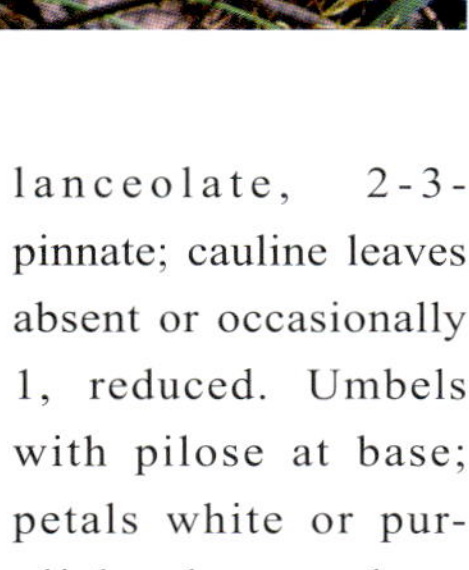

lanceolate, 2-3-pinnate; cauline leaves absent or occasionally 1, reduced. Umbels with pilose at base; petals white or purplish, obovate, base shortly clawed. Fruit oblong-ovoid, dorsal and intermediate ribs filiform, lateral ribs winged. Fl. and fr. Jun-Oct. Coniferous forests, montane thickets, grassland at forest edges, alpine scrub and meadows or river banks at 2700-4800 m. Distributed in NW Yunnan, W Sichuan and S Xizang.

## 尖叶藁本

**Ligusticum acuminatum** Franch.

多年生草本。根茎粗。茎略带淡紫色，中空。叶片三角状心形，三回三出式羽状分裂。花瓣白色，倒卵形，基部楔形。果实长圆状卵球形，背棱和中棱狭翅状，侧棱翅更宽。种子合生面平或稍凹。花果期7-10月。生海拔1500-4000米的森林、林缘、高山灌丛或草甸。产云南、四川、湖南、湖北、河南、陕西和甘肃。

Perennial herbs. Rootstock thick. Stems purplish tinged, hollow. Blades triangular-ovate, ternate-3-pinnate. Petals white, obovate, base cuneate. Fruits oblong-

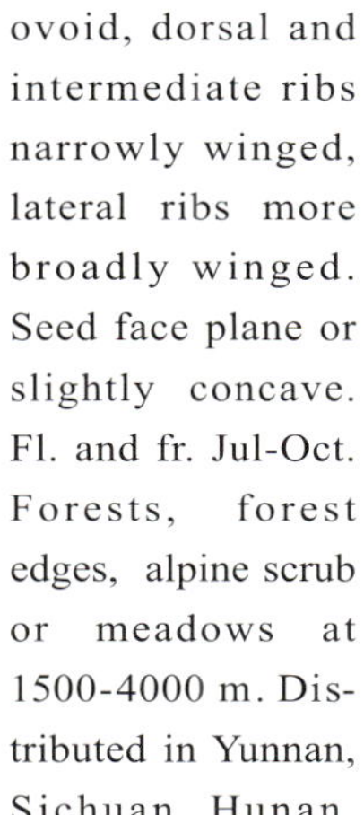

ovoid, dorsal and intermediate ribs narrowly winged, lateral ribs more broadly winged. Seed face plane or slightly concave. Fl. and fr. Jul-Oct. Forests, forest edges, alpine scrub or meadows at 1500-4000 m. Distributed in Yunnan, Sichuan, Hunan, Hubei, Henan, Shaanxi and Gansu.

抽葶藁本 *Ligusticum scapiforme*

尖叶藁本 *Ligusticum acuminatum*

川滇藁本 *Ligusticum sikiangense*

高山芹 *Coelopleurum saxatile*

## 川滇藁本

**Ligusticum sikiangense** M. Hiroe

多年生草本。茎单生或2-3分枝。叶片长圆形或披针形，二至三回羽状分裂。伞形花序顶生和侧生；苞片2-3，线形，全缘；小苞片5-7，线状披针形；小伞形花序多花；花瓣白色，倒卵形，基部具短爪。果实长圆状卵球形，果棱均具狭翅。种子合生面微凹。花果期7-9月。生海拔3400-4500米的针叶林、高山灌丛和草甸或高山岩斜坡。产云南西北部和四川西部。

Perennial herbs. Stems single or 2-3 branched. Blades oblong or lanceolate, 2-3-pinnate. Umbels terminal and lateral; bracts 2-3, linear, entire; bracteoles 5-7, linear-lanceolate; umbellules many-flowered; petals white, obovate, base shortly clawed. Fruit oblong-ovoid, ribs all narrowly winged. Seed face slightly concave. Fl. and fr. Jul -Sep. Coniferous forests, alpine scrub and meadows or alpine talus slopes at 3400-4500 m. Distributed in NW Yunnan and W Sichuan.

## 细裂藁本

**Ligusticum tenuisectum** de Boiss.

多年生草本。茎直立，分枝。叶片三角状卵形，三出式三至四回羽状分裂；上部叶片退化，一至二回羽状分裂。伞形花序顶生和侧生，小伞形花序具多花；花瓣白色，倒卵形，基部楔形。果实长圆状卵球形，背棱和中棱突出，侧棱具翅，种子表面稍凹。花果期8-9月。生海拔2000-4500米的灌丛、草坡、高山草甸。产云南西北部、四川东北部和湖北西部。

Perennial herbs. Stem erect, branched. Blades triangular-ovate, ternate-3-4-pinnate; upper leaves reduced, 1-2-pinnate. Umbels terminal and lateral, umbellules many-flowered; petals white, obovate, base cuneate. Fruits oblong-ovoid, dorsal and intermediate ribs prominent, lateral ribs winged. Seeds face slightly concave. Fl. and fr. Aug-Sep. Scrubs, grassy slopes, alpine meadows at 2000-4500 m. Distributed in NW Yunnan, NE Sichuan and W Hubei.

## 高山芹

**Coelopleurum saxatile** (Turcz. ex Ledeb.) Drude

二年生草本。根圆锥形。茎单生，紫绿色。叶为二至三回三出羽状复叶。伞状花序直径5-9厘米。伞幅20-27，密具柔毛；小苞片7-8，条形。果实椭圆体形，果棱为较厚的三角形翅状。花期7-8月，果期8-9月。生海拔1900米以上的高山岩石间或林下。产吉林(长白山)。俄罗斯和朝鲜半岛亦有。

Biennial herbs. Roots coniform. Stems solitary, purplish green. Leaves 2-3-ternate-pinnate. Umbels 5-9 cm diam; rays 20-27, densely pubescent; bracteoles 7-8, linear. Fruits ellipsoid, ribs triangular winged. Fl. Jul-Aug. Fr. Aug-Sep. Alpine rocky crevices or under forests above 1900 m. Distributed in Jilin (Changbai Mountain). Also in Russia and Korean Peninsula.

细裂藁本 *Ligusticum tenuisectum*

## 黑水当归

**Angelica amurensis** Schischk.

多年生草本。茎上部具微柔毛。叶为二至三回三出式羽状复叶，有一回羽片2-3对。伞幅20-45；小伞形花序具30-45花；花白色。果实椭圆体形至近球形；背棱隆起，侧棱宽翅状；棱槽中有油管1，全生面油管(2-3-)4。花期7-8月，果期8-9月。生海拔500-1000米的林缘、草坡或溪边。产内蒙古、黑龙江、吉林和辽宁。俄罗斯(西伯利亚)、朝鲜半岛和日本亦有。

Perennial herbs. Stems puberulous above. Leaves 2-3-ternate-pinnate, pinnae 2-3 pairs. Rays 20-45; umbellules 30-45-flowered; petals white. Fruits ellipsoid to subglobose; dorsal ribs prominent, lateral ribs broad-winged; vittae 1 in each furrow, (2-3-)4 on commissure. Fl. Jul-Aug. Fr. Aug-Sep. Forest edges, grassy slopes or streamsides at 500-1000 m. Distributed in Neimenggu, Heilongjiang, Jilin and Liaoning. Also in Russia (Siberia), Korean Peninsula and Japan.

黑水当归 *Angelica amurensis*

## 东当归

**Angelica acutiloba** (Siebold et Zucc.) Kitag.

多年生草本。根黄褐色，强烈芳香。茎坚实，略带紫色，具细棱。叶片三角状卵形，一至二回三出式羽状。小伞形花序具30花；花瓣白色，倒卵形至长圆形。果实狭长圆形，背棱丝状，侧棱具狭翅。花果期7-9月。栽培于海拔约400米。产吉林。原产日本和朝鲜半岛。

Perennial herbs. Root yellow-brown, strongly aromatic. Stem solid, purplish, thinly ribbed. Blade triangular-ovate, 1-2-ternate-pinnate. Umbellules 30- flowered; petals white, obovate to oblong. Fruit narrow-oblong, dorsal ribs filiform, lateral ribs narrow-winged. Fl. and fr. Jul-Sep. Cultivated at ca. 400 m. Distributed in Jilin. Native to Japan and Korean Peninsula.

东当归 *Angelica acutiloba*

紫花前胡 *Angelica decursiva*

## 紫花前胡

**Angelica decursiva** (Miq.) Franch. et Sav.

多年生草本。根圆锥形。茎单生，紫色。伞幅10-22，具柔毛；小苞片3-8，条形至披针形，绿色或紫色；花瓣暗紫色，倒卵形或椭圆状披针形，先端内折但不具凹口；花药深紫色。果实椭圆体形，扁平。花期8-9月，果期9-11月。生海拔200-800米的山地林下或溪边。产中国西南、华南、华中、华北、华东和东北。越南、俄罗斯、朝鲜半岛和日本亦有。

Perennial herbs. Roots coniform. Stems solitary, purple. Rays 10-22, pubescent; bracteoles 3-8, linear to lanceolate, green or purple; petals dark purple, obovate or ellipsoid-lanceolate, apex incurved but not notched; anthers dark purple. Fruits ellipsoid, depressed. Fl. Aug-Sep. Fr. Sep-Nov. Montane forests or streamsides at 200-800 m. Distributed in SW, S, C, N, E and NE China. Also in Vietnam, Russia, Korean Peninsula and Japan.

## 白芷

**Angelica dahurica** (Fisch. ex Hoffm.) Benth. et Hook. f. ex Franch. et Sav.

多年生草本。小叶无柄，长圆状椭圆形长至圆状披针形。伞幅18-40(-70)，具短毛；小苞片多数，条状披针形，干膜质；花瓣白色，倒卵形且具凹口；子房无毛或少毛。果实近球形。花期7-8月，果期8-9月。生海拔500-1000米的林缘、河谷草丛或溪边。产河北、陕西、黑龙江、吉林和辽宁。俄罗斯、朝鲜半岛和日本亦有。

Plants perennial. Leaflets sessile, oblong-elliptic to oblong-lanceolate. Rays 18-40(-70), short-hairy; bracteoles many, linear-lanceolate, scarious; petals white, obovate and notched; ovary glabrous or pubescent. Fruits subglobose. Fl. Jul-Aug. Fr. Aug-Sep. Forest edges, valley grasslands or streamsides at 500-1000 m. Distributed in Hebei, Shaanxi, Heilongjiang, Jilin and Liaoning. Also in Russia, Korean Peninsula and Japan.

白芷 *Angelica dahurica*

重齿当归 *Angelica biserrata*

## 重齿当归

**Angelica biserrata** (R. H. Shan et C. Q. Yuan) C. Q. Yuan et R. H. Shan

多年生草本。茎上部具硬毛。叶鞘膨大，无毛或腹面稍被柔毛；叶二回三出式羽状复叶。伞幅10-25，密被硬毛；小伞形花序具17-28 (-36)花；花白色。果实椭圆体形；背棱突出，侧棱宽翅状；棱槽中有油管2-3。花期8-9月，果期9-10月。生海拔1000-1700米的稀疏灌丛或湿山坡。产四川、浙江、湖北、安徽和江西。

Perennial herbs. Stems hispid above. Sheaths inflated, glabrous or slightly pubescent abaxially. Leaves 2-ternate-pinnate. Rays 10-25, densely hispidulous; umbellules 17-28 (-36)-flowered; petals white. Fruits ellipsoid; dorsal ribs prominent, lateral ribs broad-winged; vittae 2-3 in each furrow. Fl. Aug-Sep. Fr. Sep-Oct. Sparse thickets or damp slopes at 1000-1700 m. Distributed in Sichuan, Zhejiang, Hubei, Anhui and Jiangxi.

## 拐芹

**Angelica polymorpha** Maxim.

多年生草本。小叶卵形至菱状长卵形。花梗、伞幅和花梗均密被硬毛；伞幅10-20；小苞片7-10，狭条形，浅紫色，具缘毛；花瓣白色，匙形。果实长圆状椭圆体形。花期8-9月，果期9-10月。生海拔1000-1500米的林中、草地或溪边。产中国东南、华中、华北、华东和东北。朝鲜半岛和日本亦有。

Perennial herbs. Leaflets ovate or rhombic-oblong. Peduncles, rays and pedicels densely hispidulous; rays 10-20; bracteoles 7-10, narrow-linear, purplish, ciliate; petals white, spatulate. Fruits oblong-ellipsoid. Fl. Aug-Sep. Fr. Sep-Oct. Forests, grasslands or streamsides at 1000-1500 m. Distributed in SE, C, N, E and NE China. Also in Korean Peninsula and Japan.

拐芹 *Angelica polymorpha*

山芹 *Ostericum sieboldii*

## 山芹
**Ostericum sieboldii** (Miq.) Nakai

多年生草本，高0.5-1.5米。叶为二至三回三出式羽状复叶。复伞形花序宽4-8厘米；伞幅7-13；小伞形花序约具20花；花瓣白色。果实椭圆体形；背棱突出，侧棱宽翅状；每棱槽具油管1-3，合生面具4-6(-8)。花期8-9月，果期9-10月。生海拔600-1200米的林中、峡谷、草坡或草地。产黑龙江、吉林、辽宁、内蒙古、陕西和山东。俄罗斯、朝鲜半岛和日本亦有。

Perennial herbs, 0.5-1.5 m tall. Leaves 2-3-ternate-pinnate. Umbels 4-8 cm across; rays 7-13; umbellules ca. 20-flowered; petals white. Fruits ellipsoid; dorsal ribs prominent, lateral ribs broad-winged; vittae 1-3 in each furrow, 4-6(-8) on commissure. Fl. Aug-Sep. Fr. Sep-Oct. Forests, ravines, grassy slopes or grasslands at 600-1200 m. Distributed in Heilongjiang, Jilin, Liaoning, Neimenggu, Shaanxi and Shandong. Also in Russia, Korean Peninsula and Japan.

珊瑚菜 *Glehnia littoralis*

## 珊瑚菜
**Glehnia littoralis** F. Schmidt ex Miq.

多年生草本。全株被白色柔毛。叶宽卵形，一至二回复叶，沿脉处粗糙，叶边缘具白色骨质缺刻状锯齿；小叶厚纸质。伞幅8-16，不等长。果实球形或椭圆体形，表面密具粗毛及绒毛。花果期6-8月。生海拔50-100米的海边沙滩，并栽培于沙地中。产中国沿海地区。俄罗斯、朝鲜半岛和日本亦有。

Perennial herbs. Whole plant white pubescent. Leaves broad-ovate, 1-2-ternate, scabrous along nerves, incised-serrate with white cartilaginous margins; leaflets thickly papery. Rays 8-16, unequal. Fruits globose or ellipsoid, densely hirsute and velutinous on the surface. Fl. and fr. Jun-Aug. Distributed on sandy beaches of costal areas in China, also cultivated in sandy soils at 50-100 m. Also in Russia, Korean Peninsula and Japan.

## 石防风
**Peucedanum terebinthaceum** (Fisch. ex Trevir.) Ledeb.

多年生草本。基部生二回羽状全裂；羽片3-5对。复伞形花序多分枝，宽3-10(-15)厘米；伞幅8-20(至更多)，带4棱角；花白色。果实椭圆体形；背棱突出，侧棱翅状，宽约为果的1/3；每棱槽具油管1，合生面2。花期7-9月，果期9-10月。生海拔200-1200米的杂木林、灌丛或草坡。产河北、内蒙古(大兴安岭)、黑龙江、吉林和辽宁。俄罗斯、朝鲜半岛和日本亦有。

Perennial herbs. Basal leaves 2-pinnate pinnatisect; pinnae 3-5-paired. Umbels much branched; umbels 3-10 (-15) cm across; rays 8-20 (or more), 4-angled; petals white. Fruits ellipsoid; dorsal ribs prominent, lateral ribs winged, ca. 1/3 width of body; vittae 1 in each furrow, 2 on commissure. Fl. Jul-Sep. Fr. Sep-Oct. Mixed forests, scrubs or grassy slopes at 200-1200 m. Distributed in Hebei, Neimenggu (Daxing'an Mountain), Heilongjiang, Jilin and Liaoning. Also in Russia, Korean Peninsula and Japan.

## 华中前胡
**Peucedanum medicum** Dunn

多年生草本。叶为二至三回三出式复叶，羽片3对。复伞形花序宽7-15厘米；伞幅15-30；小

石防风 *Peucedanum terebinthaceum*

伞形花序具10-30花；花瓣白色。果实卵球状椭圆体形，背棱稍突出，侧棱翅状，翅约占宽的1/3；每棱槽具油管3，合生面8-10。花期7-9月，果期10-11月。生海拔700-2000米的多石湿润山坡或草原地带。产中国西南、华南、华中和华东。

Perennial herbs. Leaves 2-3-ternate; pinnae 3 pairs. Umbels 7-15 cm across; rays 15-30; umbellules 10-30-flowered; petals white. Fruits ovoid-ellipsoid, dorsal ribs slightly prominent, lateral ribs winged, wings ca. 1/3 width of body; vittae 3 in each furrow, 8-10 on commissure. Fl. Jul-Sep. Fr. Oct-Nov. Wet rocky slopes or grassy places at 700-2000 m. Distributed in SW, S, C and E China.

## 中亚阿魏

**Ferula jaeschkeana** Vatke

多年生草本。一次结果，无强烈香味。茎单生，粗壮，红棕色，圆锥状分枝，下部分枝互生，上部分枝轮生。叶片轮廓宽三角形，二回三出式多裂，正面无毛，背面短柔毛，早枯萎。小伞形花序有花15-20；花瓣长椭圆形，先端渐尖，弯曲；花黄色。果实椭圆体形。花期6月，果期7月。生海拔约3600米的草坡或灌木中。产西藏西部(阿里、札达县)。印度东北部、巴基斯坦西部、不丹、阿富汗和中亚亦有。

Perennial herbs. Monocarpic, not strongly scented. Stem solitary, thick, robust, reddish brown, paniculate-branched, lower branches alternate, upper branches verticillate. Blade broadly triangular in outline, 2-ternately dissected, glabrous adaxially, pubescent abaxially, soon wilting. Umbellules 15-20-flowered; petals yellow, long-elliptic, apex acuminate, incurved. Fruits ellipsoid. Fl. Jun. Fr. Jul. Grassy slopes or among shrubs at ca. 3600 m. Distributed in W Xizang (Ngari, Zanda County) . Also in NE India, W Pakistan, Bhutan, Afghanistan and C Asia.

华中前胡 *Peucedanum medicum*

中亚阿魏 *Ferula jaeschkeana*

准噶尔阿魏 *Ferula songarica*

## 准噶尔阿魏

**Ferula songarica** Pall. ex Spreng.

多年生草本。茎带紫红色，圆锥状分枝，下部分枝互生，上部分枝轮生。叶片宽三角形，三至四回三出式羽状全裂，末回裂片线形。无总苞片；侧生花序常(1-)2-4，超出中央花序；小伞形花序有花15-20；花瓣黄色，椭圆形。果实椭圆体形，背腹极扁压，背棱丝状，侧棱呈狭翅状。花果期6-7月。产新疆(阿勒泰、塔城)。生山地草坡和灌丛中。哈萨克斯坦和俄罗斯(西西伯利亚)亦有。

Perennial herbs. Stems purplish red with age, paniculate-branched, lower branches alternate, upper branches verticillate. Blades broadly triangular, 3-4-ternate-pinnatisect; ultimate segments linear. Bracts absent; terminal umbel short-pedunculate, lateral umbels (1-)2-4 or absent, exceeding terminal; umbellules 15-20-flowered; petals yellow, elliptic. Fruits ellipsoid, dorsal strongly compression, dorsal ribs filliform, lateral ribs narrowly winged. Fl. and fr. Jun-Jul. Distributed in Xinjiang (Altay, Tacheng). Grassy places on mountain slopes or scrubs. Also in Kazakhstan and Russia (W Siberia).

硬阿魏 *Ferula bungeana*

## 硬阿魏

**Ferula bungeana** Kitag.

多年生草本。叶宽卵形或三角形，顶端具三角状齿，近革质，坚硬，密具柔毛。伞幅4-15，不等长，开展；小伞形花序具5-12花；花瓣黄色。果实椭圆体形或宽椭圆体形。花果期5-7月。生海拔200-2500米的砾石质山坡或沙地。产华北、华西和东北。

Perennial herbs. Leaves broadly ovate or triangular, apex 3-triangular-toothed, sub-leathery, rigid, densely pubescent. Rays 4-15, unequal, spreading; umbellules 5-12-flowered; petals yellow. Fruits ellipsoid or broadly ellipsoid. Fl. and fr. May-Jul. Gravelly slopes or sandy places at 200-2500 m. Distributed in N, W and NE China.

## 铜山阿魏

**Ferula licentiana** Hand.-Mazz. var. **tunshanica** (S. W. Su) R. H. Shan et Q. X. Liu

多年生草本，高60-120厘米。伞辐3-7，且伞幅较短，仅1.5-3厘米。小伞形花序具5-12花；花瓣黄色。果实长7-10毫米；每棱槽内油管1-3，合生面4-6。生海拔100-200米的山坡。产安徽、山东和江苏。

Perennial herbs, 60-120 cm tall. Rays 3-7, short, 1.5-3 cm long; umbellules 5-12-flowered; petals yellow. Fruits 7-10 mm long; vittae 1-3 in each furrow,

铜山阿魏 *Ferula licentiana* var. *tunshanica*

兴安独活 *Heracleum dissectum*

4-6 on commissure. Mountain slopes at 100-200 m. Distributed in Anhui, Shandong and Jiangsu.

## 兴安独活

**Heracleum dissectum** Ledeb.

多年生草本。基生叶和下部叶羽状分裂，小叶羽状半裂，上部叶3浅裂。伞幅20-30；花瓣白色，复伞形花序外缘辐射状。果实倒卵球形；每棱槽具油管1，超出分生果长的2/3。花期7-8月，果期8-9月。生海拔2200米以下的山地林中、林缘或湿草地。产新疆、黑龙江和吉林。中亚和东北亚亦有。

Perennial herbs. Basal and lower leaves pinnate, pinnatifid; upper leaves 3-lobed. Rays 20-30; petals white, outer flowers in umbels radiant. Fruits obovoid; vittae solitary in each furrow, extending to 2/3 length of mericarps. Fl. Jul-Aug. Fr. Aug-Sep. Montane forests, forest edges or moist grasslands below 2200 m. Distributed in Xinjiang, Heilongjiang and Jilin. Also in C and NE Asia.

## 短毛独活

**Heracleum moellendorffii** Hance

草本。根圆锥形，粗大，多分枝。茎上部叶具有显著宽展的叶鞘。伞幅12-30；小苞片5-10，披针形；每个小伞形花序花多于20朵；花瓣白色，伞形花序外轮花辐射状，增大。果实倒卵球形，疏被柔毛或近无毛。花期7-8月，果期8-9月。生海拔3200米以下的阴坡山沟旁、林缘或草甸。产中国除华南以外大部分地区。日本和朝鲜半岛亦有。

Herbs. Roots coniform, large, many-branched. Leaves on upper part of stem with obvious conspicucous extended sheath. Rays 12-30; bracteoles 5-10, lanceolate; flowers more than 20 per umbellule; petals white, on outer flowers of umbels radiant, enlarged. Fruits obovoid, sparsely hispidulous or almost glabrous. Fl. Jul-Aug. Fr. Aug-Sep. Shady slopes by valleys, forest edges or marshland below 3200 m. Distributed in most parts of China, except S China. Also in Japan and Korean Peninsula.

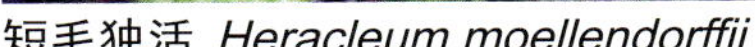

短毛独活 *Heracleum moellendorffii*

## 白亮独活

**Heracleum candicans** Wall. ex DC.

多年生草本。全体被灰白色柔毛或绒毛。伞幅15-25(-35)，不等长；小苞片5-8，条形；小伞形花序具20-25花；花瓣白色，外轮伞形花序花辐射状。果实倒卵球形。花期5-7月，果期8-9月。生海拔1800-4500米的山坡林中、路旁、高山草甸或溪边。产云南、四川和西藏。印度、尼泊尔、不丹和巴基斯坦亦有。

白亮独活 *Heracleum candicans*

Perennial herbs. With gray-white villose or tomentum. Rays 15-25 (-35), unequal; bracteoles 5-8, linear; umbellules 20-25-flowered; petals white, outer flowers of umbels radiant. Fruits obovoid. Fl. May-Jul. Fr. Aug-Sep. Hillsides forests, roadsides, alpine meadows or streamsides at 1800-4500 m. Distributed in Yunnan, Sichuan and Xizang. Also in India, Nepal, Bhutan and Pakistan.

## 裂叶独活

**Heracleum millefolium** Diels

多年生草本。叶多基生，狭长圆形或披针形，三至四回羽状分裂。复伞形花序顶生或侧生；花瓣白色、淡黄色或淡紫色，伞形花序外轮花辐射状，增大花瓣极明显，深2裂。果实宽卵球形。花期6-8月，果期9-10月。生海拔2800-5000米的林中、林缘、高山草甸或田边。产云南、四川、西藏、甘肃和青海。不丹亦有。

Perennial herbs. Leaves mostly basal, narrowly oblong or lanceolate, 3-4-pinnate. Compound umbels terminal or lateral; petals white, yellowish or purplish, outer flowers in umbel radiant, enlarged petals very conspicuous, deeply 2-lobed. Fruits broad ovoid. Fl. Jun-Aug. Fr. Sep-Oct. Forests, forest edges, alpine meadows or fieldsides at 2800-5000 m. Distributed in Yunnan, Sichuan, Xizang, Gansu and Qinghai. Also in Bhutan.

裂叶独活 *Heracleum millefolium*

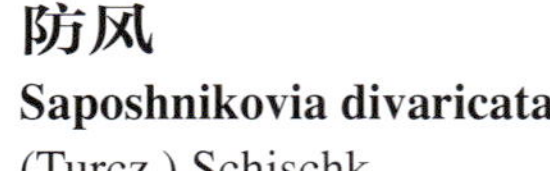

## 防风

**Saposhnikovia divaricata** (Turcz.) Schischk.

多年生草本，全体无毛。叶长圆状卵形至阔卵形，二回羽状分裂；叶柄基部有宽叶鞘。伞形花序多数，宽约6厘米；伞幅5-7；小苞片4-6；小伞形花序具4-5花。果实幼时具瘤，成熟后光滑。花期8-9月，果期9-10月。生海拔400-800米的丘陵、多砾石山坡或草地。产华北、华东、西北和东北。俄罗斯、蒙古和朝鲜半岛亦有。

Perennial herbs, whole plant glabrous. Leaves oblong-ovate to

防风 *Saposhnikovia divaricata*

broad-ovate, 2-pinnate; petioles with wide sheathes at base. Umbels numerous, ca. 6 cm across; rays 5-7; bracteoles 4-6; umbellules 4-5-flowered. Fruits tuberculate when young, becoming smooth when mature. Fl. Aug-Sep. Fr. Sep-Oct. Hills, stony mountain slopes or grasslands at 400-800 m. Distributed in N, E, NW and NE China. Also in Russia, Mongolia and Korean Peninsula.

## 野胡萝卜

**Daucus carota** L.

二年生草本。根细，多分枝，木质，不分枝，常褐色。叶长条形，二至三回羽状全裂。伞幅不等长；小苞片5-7，条形；花瓣白色，有时黄色或粉红色。果实卵球形，棱上有白色刺毛。花期5-7月。生海拔2000-3000米的山坡、路旁或田间。产中国西南、东南、华中和华东。亚洲西南部、北非和欧洲亦有。

Biennial herbs. Roots slender, many-branched, woody, not fleshy, usually brown. Leaves oblong, 2-3-pinnatisect. Rays unequal; bracteoles 5-7, linear; petals white, sometimes yellow or pinkish. Fruits ovoid, ribs with white bristles. Fl. May-Jul. Mountain slopes, by roads or fields at 2000-3000 m. Distributed in SW, SE, C and E China. Also in SW Asia, N Africa and Europe.

野胡萝卜 *Daucus carota*

### 胡萝卜
**Daucus carota** L. var. **sativa** Hoffm.

本变种与野胡萝卜的区别在于根加粗，长圆柱状或长棍棒状，肉质，浅红色、橙红色或黄色。花期5-7月。中国大部分地区有栽培，世界广泛栽培。

This variety differs from var. *carota* in its taproots thickened, elongate terete or clavate, fleshy, reddish, reddish-yellow, or yellow. Fl. May-Jul. Cultivated in most parts of China, cultivated worldwide.

# 山茱萸科 Cornaceae

### 八蕊单室茱萸
**Mastixia euonymoides** Prain

常绿乔木。叶革质，长圆状椭圆形至长圆形，顶端长尾尖。圆锥花序顶生，疏具花；花盘圆形，稍4裂，具4个雄蕊的痕迹，花萼和柱头宿存。果实干后棕褐色，长圆形。花蕾期3月，果期10-12月。生海拔1400-1600米的亚热带森林中或沟谷湿润疏林中。产云南(西双版纳南糯山)。

Evergreen trees. Leaves coriaceous, oblong-elliptic to oblong, apex long-caudate. Panicles terminal, loosely flowered; disks circular, slightly 4-lobed, with scars of 4 stamens; calyx and styles persistent. Fruits dark brown after drying, oblong. Fl. (buds) Mar. Fr. Oct-Dec. Subtropical forests or sparse wet forests in valleys at 1400-1600 m. Distributed in Yunnan (Nannuo Mountain of Sipsongpanna).

### 桃叶珊瑚
**Aucuba chinensis** Benth.

乔木或灌木。叶椭圆形至宽椭圆形，宽3-8厘米，革质；侧脉6-8(-10)对。雄花序与雌花序圆锥状；花序分枝密被柔毛；花萼4齿裂；花瓣4；雄蕊4。果亮红色或深红色，圆柱状或卵球形。花期1-2月，果期2月，常与一二年生果序同存于枝上。生海拔300-1000米的林中。产中国西南、华南和东南。缅甸和越南北部亦有。

Trees or shrubs. Leaves elliptic to broadly elliptic, 3-8 cm wide, leathery; lateral veins 6-8(-10) pairs. male and female inflorescences paniculate; branches densely pubescent; calyx 4-dentate; petals 4; stamens 4. Fruits bright red or dark red, cylindrical or ovoid. Fl. Jan-Feb. Fr. Feb, often with fruits from preceding two years. Forests at 300-1000 m. Distributed in SW, S and SE China. Also in Myanmar and N Vietnam.

### 密花桃叶珊瑚
**Aucuba confertiflora** W. P. Fang et T. P. Soong

常绿小乔木。叶片厚革质，宽3-5.5厘米，下面密被柔毛，侧脉及网脉在背面微突出；叶柄长2-2.5厘米。雄花序为近圆柱状的圆锥花序，花密集；花瓣4，深紫红色。花期2-3月。生海拔1000-1600米的林中。产云南东南部。

胡萝卜 *Daucus carota* var. *sativa*

八蕊单室茱萸 *Mastixia euonymoides*

桃叶珊瑚 *Aucuba chinensis*

Small trees, evergreen. Leaves thickly leathery, 3-5.5 cm wide, abaxially densely pubescent, lateral veins and veins slightly raised adaxially; petioles 2-2.5 cm long. Staminate inflorescences almost cylindric paniculate, densely flowered; petals 4, dark purplish red. Fl. Feb-Mar. Forests at 1000-1600 m. Distributed in SE Yunnan.

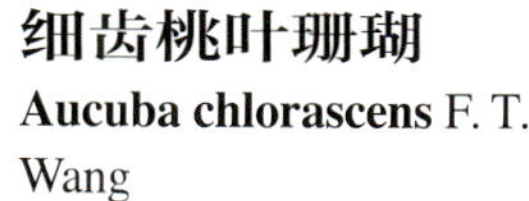

## 细齿桃叶珊瑚

**Aucuba chlorascens** F. T. Wang

灌木或乔木。叶纸质至薄革质，被柔毛，边缘反卷，中段以上具8-14对不明显细齿。雄花序为圆锥花序，长5-6厘米，雄花绿色；雌花序长1-2厘米，花瓣尖端具尖尾，长1-1.5毫米。果柄直，与果实的连接处不膨大。花期2-3月，果期10-12月。生海拔1400-2800米的林中。产云南。

Shrubs or trees. Leaves papery to thinly leathery, pubescent, margin revolute, with 8-14 pairs of inconspicuous teeth on upper half. Staminate inflorescences paniculate, 5-6 cm, staminate flowers green; carpellate inflorescences 1-2 cm long, petals with a tail-like appendage, 1-1.5 mm. Fruits stalks straight, not expanded at junction with fruit. Fl. Feb-Mar. Fr. Oct-Dec. Forests at 1400-2800 m. Distributed in Yunnan.

密花桃叶珊瑚 *Aucuba confertiflora*

## 纤尾桃叶珊瑚

**Aucuba filicauda** Chun et F. C. How

灌木。叶阔椭圆形至倒卵状椭圆形，纸质至革质，先端具急尖尾，长1-1.5厘米。雄花组成总状圆锥花序，长7-15厘米，被紧贴的粗伏毛；雌花序较短，长2-5厘米。果实椭圆体形、卵球形或卵球状长圆形。花期4-5月，果期7月以后。生海拔900-1900米的林中。产云南、贵州、广西和江西。

Shrubs. Leaves broadly elliptic or obovate-elliptic, papery to leathery, apex long cuspidate for 1-1.5 cm long. Staminate inflorescences racemose panicles, 7-15 cm long, with appressed thick trichomes; carpellate inflorescences stout, 2-5 cm. Fruits ellipsoid, ovoid, or ovoid-oblong. Fl. Apr-May. Fr. after Jul. Forests at 900-1900 m. Distributed in Yunnan, Guizhou, Guangxi and Jiangxi.

细齿桃叶珊瑚 *Aucuba chlorascens*

纤尾桃叶珊瑚 *Aucuba filicauda*

喜马拉雅珊瑚 *Aucuba himalaica*

## 喜马拉雅珊瑚

**Aucuba himalaica** Hook. f. et Thoms.

常绿小乔木或灌木。叶脉在上面明显下凹，下面凸起，叶下表面具柔毛，脉上具粗毛。雄花序长8-10(-13)厘米；雌花序长3-5厘米。核果卵球状长圆形，长1-1.2厘米。花期3-5月，果期10月至翌年5月。生海拔500-1200米的林中。产中国西南、东南、华中和华北。印度、不丹和缅甸亦有。

Small trees or shrubs, evergreen. Leaf veins raised abaxially, conspicuously impressed adaxially, abaxially pubescent with soft trichomes, with thick trichomes on veins. Staminate inflorescences 8-10(-13) cm long; carpellate inflorescences 3-5 cm. Drupes ovoid-oblong, 1-1.2 cm long. Fl. Mar-May. Fr. Oct to next May. Forests at 500-1200 m. Distributed in SW, SE, C and N China. Also in India, Bhutan and Myanmar.

## 长叶珊瑚

**Aucuba himalaica** Hook. f. et Thoms. var. **dolichophylla** W. P. Fang et T. P. Soong

本变种与原变种的区别在于本变种的叶狭披针形或披针形，无毛或下面中脉上具短的软绒毛。花期3-5月，果期10月至翌年5月。生海拔约1000米的林中。产四川、贵州、广西、广东、浙江、湖北西部和湖南。

This variety differs from the typical variety in its leaves narrowly lanceolate or lanceolate, glabrous or only midvein of abaxial surface with short, soft trichomes. Fl. Mar-May. Fr. Oct to next May. Forests at ca. 1000 m. Distributed in Sichuan, Guizhou, Guangxi, Guangdong, Zhejiang, W Hubei and Hunan.

## 倒披针叶珊瑚

**Aucuba himalaica** Hook. f. et Thoms. var. **oblanceolata** W. P. Fang et T. P. Soong

本变种与原变种的区别在于本变种的叶片倒披针形，叶上表面密被短柔毛。生海拔约700米的林中。产四川和湖南北部。

This variety differs from the typical variety in its leaves oblanceolate, abaxially densely pubescent with short, soft trichomes. Forests at ca. 700 m. Distributed in Sichuan and N Hunan.

## 密毛桃叶珊瑚

**Aucuba himalaica** Hook. f. et Thoms. var. **pilossima** W. P. Fang et T. P. Soong

本变种与原变种的区别在于本变种的叶下密被柔毛，混生短毛、软毛和硬毛。雄花序长12厘米；果期花序长2-3厘米。果实近椭圆体形，长约1.5厘米。花期3-5月，果期10月至翌年5月。生海拔1000-1300米的林中。产四川、湖北、湖南和陕西。

This variety differs from the typical variety in its leaves abaxially surface densely pubescent with short, soft, and stiff trichomes.

长叶珊瑚 *Aucuba himalaica* var. *dolichophylla*

倒披针叶珊瑚 *Aucuba himalaica* var. *oblanceolata*

密毛桃叶珊瑚
*Aucuba himalaica* var. *pilossima*

Shrubs evergreen. Leavess thickly papery, narrowly lanceolate, margin serrate, 11-15 cm long, 1.5-3 cm wide, with irregular white or light yellow spots adaxially. Inflorescences panicles, terminal, loosely flowered; flowers dark purple. Fruits ovoid, shiny red at maturity. Under forests at 1300-2100 m. Distributed in Sichuan, Zhejiang and Hunan.

Staminate inflorescences 12 cm long; fruiting inflorescences 2-3 cm. Fruits subellipsoid, ca. 1.5 cm long. Fl. Mar-May. Fr. Oct to next May. Forests at 1000-1300 m. Distributed in Sichuan, Hubei, Hunan and Shaanxi.

## 窄斑叶珊瑚

**Aucuba albopunctifolia** F. T. Wang var. **angustula** W. P. Fang et T. P. Soong

常绿灌木。叶厚纸质，狭披针形，边缘具锯齿，长11-15厘米，宽1.5-3厘米，背面常具白色或淡黄色不规则斑点。顶生圆锥花序，花较疏散；花深紫色。果实卵球形，熟后亮红色。生海拔1300-2100米的林下。产四川、浙江和湖南。

## 倒心叶珊瑚

**Aucuba obcordata** (Rehd.) Fu ex W. K. Hu et T. P. Soong

灌木或乔木，高1-4米。叶常倒心形或倒卵形，厚纸质，不具斑纹，长(4-)8-14厘米，宽(2-)4.5-8厘米，厚纸质，边缘具缺刻状粗锯齿。雄花序为总状圆锥花序；花较稀疏，花瓣紫红色；花药2室。核果较密集，卵球形。花期3-4月，果期9-11月或更晚。生海拔约1300米的林中。产中国西南、华南、华中和华西。

Shrubs or trees, 1-4 m tall. Leaves often obcordate or obovate, not variegated, (4-)8-14 cm long, (2-)4.5-8 cm wide, thickly papery, margin coarsely dentate. Staminate inflorescences racemose-paniculate, sparsely flowered, petals purplish red; anthers 2-locular. Drupes clustered, ovoid. Fl. Mar-Apr. Fr. Sep-Nov or later. Forests at ca. 1300 m. Distributed in SW, S, C and W China.

窄斑叶珊瑚 *Aucuba albopunctifolia* var. *angustula*

倒心叶珊瑚 *Aucuba obcordata*

Fruits subglobose to oblong. Seeds 3 or 4. Fl. Apr-May. Fr. Aug-Oct. Thickets on slopes or forest edges at (1000-)1700-3000 m. Distributed in SW, S and C China. Also in S and SE Asia.

青荚叶 *Helwingia japonica*

## 青荚叶

**Helwingia japonica** (Thunb.) Dietr.

落叶灌木。叶卵形或卵圆形，下面浅绿色，纸质，边缘具刺状细齿；托叶条形或线形，分裂。伞形花序着生在叶中脉1/3处；雄花序具3-12(-18)花；雌花序具1-3花。果实近球形或椭圆体形。种子3-5粒。花期4-5月，果期8-10月。生海拔100-3400米的林中、灌丛、沟谷、溪边或路边。产中国西南、华南、东南、华北和华东。不丹、缅甸、朝鲜半岛和日本亦有。

Shrubs deciduous. Leaves ovate or ovate-rounded, abaxially light green, papery, margin spiculate serrate; stipules linear or filiform, divided. Umbels between middle and lower 1/3 of midvein; staminate umbels 3-12(-18)-flowered; carpellate umbels 1-3-flowered. Fruits subglobose or ellipsoid. Seeds 3-5. Fl. Apr-May. Fr. Aug-Oct. Forests, thickets, valleys, streamsides or roadsides at 100-3400 m. Distributed in SW, S, SE, N and E China. Also in Bhutan, Myanmar, Korean Peninsula and Japan.

## 西域青荚叶

**Helwingia himalaica** Hook. f. et Thoms. ex C. B. Clarke

常绿灌木。叶厚纸质，长椭圆形或长椭圆披针形，边缘具腺状细齿，叶脉(5-)7-9条，长5-11(-18)厘米，宽2.5-4(-5)厘米；托叶常2-3裂，稀不裂。伞形花序位于叶面中下部1/3处。果实近球形至长圆形。种子3或4粒。花期4-5月，果期8-10月。生海拔(1000-)1700-3000米的山坡灌丛或林缘。产中国西南、华南和华中。南亚和东南亚亦有。

Shrubs evergreen. Leaves thickly papery, oblong or oblong-lanceolate, margin glandular serrulate, veins (5-)7-9, 5-11(-18) cm long 2.5-4(-5) cm wide; stipules often 2-3-divided, rarely not. Umbels between middle and lower 1/3 of midvein on adaxial surface.

## 中华青荚叶

**Helwingia chinensis** Batal.

常绿灌木。叶条状披针形或狭披针形，革质，边缘具稀疏腺状锯齿。伞形花序生于叶面中脉近中部或幼枝上端；雄花4-5朵成伞形花序，花3-5数；雌花1-3朵成伞形花序。果实长圆形。种子3-5。花期4-5月，果期8-10月。生海拔1000-2600米的山坡杂木林。产中国西南、华中和华西。缅甸和泰国亦有。

Shrubs evergreen. Leaves linear-lanceolate or narrowly lanceolate, leathery, margins sparsely glandular serrate; umbels near mid point of midvein on adaxial surface of leaves or on upper part of young branches; flowers of staminate umbels 4-5, 3-5-merous; flowers of carpellate umbels 1-3. Fruits oblong. Seeds 3-5. Fl. Apr-May. Fr. Aug-Oct. Mixed woods on slopes at 1000-2600 m. Distributed in SW, C and W China. Also in Myanmar and Thailand.

## 峨眉青荚叶

**Helwingia omeiensis** (W. P. Fang) H. Hara et S. Kuros.

常绿乔木或灌木。叶长圆形或倒卵状长圆形，革质，边缘上部2/3具针状腺齿。雄花多朵簇生，常5-20(-30)朵成密伞花序或伞形花序；雌花1-4(-6)朵组成伞形花序；雄花浅紫色；雌花绿色。果

西域青荚叶 *Helwingia himalaica*

中华青荚叶
*Helwingia chinensis*

实近球形或狭椭圆体形。种子3或4(-5)。花期4-5月，果期7-8月。生海拔600-1700米的林中或山坡湿润处。产中国西南、华中和华西。

Trees or shrubs evergreen. Leaves oblong or obovate-oblong, leathery, margin glandular spiculate-serrate on apical 2/3. Staminate umbels many, 5-20(-30)-flowered; carpellate umbels 1-4(-6)-flowered; staminate flowers purplish white; carpellate flowers green. Fruits subglobose or narrowly ellipsoid. Seeds 3 or 4(-5). Fl. Apr-May. Fr. Jul-Aug. Moist habitats in woods or slopes at 600-1700 m. Distributed in SW, C and W China.

## 鞘柄木
**Toricellia tiliifolia** DC.

落叶乔木。叶纸质，长圆状卵形至宽卵形，边缘的粗锯齿有须头，有时有波状棱角，掌状脉7-9条；叶柄向下扩展成鞘。花序长12-22厘米，稍被柔毛；花瓣白色。果实卵球形，无毛。花期11月至翌年3月，果期3-4月。生海拔1600-2600米的阔叶林或林缘。产云南西部和西藏东南部。印度、尼泊尔和不丹亦有。

Trees deciduous. Leaves papery, oblong-ovate to broadly ovate, margin grossly apiculate-serrate and sometimes undulate-angulate, veins palmate 7-9; petioles gradually vaginate at lower part. Inflorescences 12-22 cm long, slightly pubescent; petals white. Fruits ovoid, glabrous. Fl. Nov to next Mar. Fr. Mar-Apr. Broad-leaved forests or forest edges at 1600-2600 m long. Distributed in W Yunnan and SE Xizang. Also in India, Nepal and Bhutan.

峨眉青荚叶 *Helwingia omeiensis*

鞘柄木 *Toricellia tiliifolia*

## 灯台树
**Cornus controversa** Hemsl.

乔木。叶互生，下面色浅或浅绿色，疏被平伏柔毛，具乳突，具6-7(-9)脉，在下面稍紫色。聚伞状伞房花序，花白色。核果幼时紫红色，成熟时蓝黑色，球形。花期5-6月，果期7-9月。生海拔200-2600米的常绿阔叶或针阔混交林。产中国大部分地区。南亚和东北亚亦有。

灯台树 *Cornus controversa*

Trees. Leaves alternate, abaxially light or grayish green, sparsely pubescent with appressed trichomes, papillate, veins 6-7(-9), abaxially slightly purplish. Cymose-corymb with white flowers. Drupes purple-red when

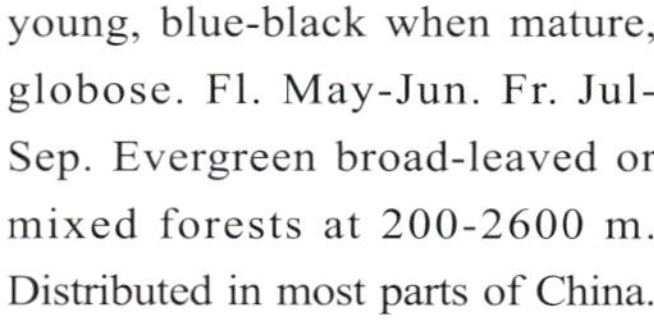

young, blue-black when mature, globose. Fl. May-Jun. Fr. Jul-Sep. Evergreen broad-leaved or mixed forests at 200-2600 m. Distributed in most parts of China. Also in S and NE Asia.

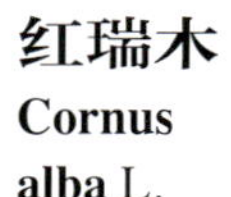

## 红瑞木
**Cornus alba** L.

落叶灌木，开展。树皮紫红色。叶背粉绿色，侧脉(4-)5(-6)条。伞房状聚伞花序较密；花白色或黄白色。核果长圆形，成熟时乳白色或蓝白色，果核侧扁。花期6-7月，果期8-10月。生海拔600-1700(-2700)米的针阔混交林或溪边杂灌丛。产华中、华东、华北和东北。俄罗斯、蒙古、朝鲜半岛和欧洲亦有。

Shrubs deciduous, spreading. Bark purplish red. Leaves abaxially glaucous green, veins (4-)5(-6). Corymbose cymes dense; flowers white or yellowish white. Drupes oblong, creamy white or bluish white at maturity, stones of fruit laterally compressed. Fl. Jun-Jul. Fr. Aug-Oct. Mixed broad-leaved and coniferous forests or mixed thickets by streams at 600-1700 (-2700) m. Distributed in C, E, N and NE China. Also in Russia, Mongolia, Korean Peninsula and Europe.

红瑞木 *Cornus alba*

## 长圆叶梾木
**Cornus oblonga** Wall.

常绿乔木。茎褐色或紫褐色。叶椭圆形至长圆形，下面被乳突。伞房状聚伞花序，花白色。花序、叶背面及叶柄具灰色贴伏的短柔毛。果实成熟时黑色，椭圆体形。花期9月至翌年1月，果期翌年4-6月。生海拔800-3700米的常绿阔叶林、常绿落叶阔叶混交林或溪边灌丛。产云南、四川、西藏、贵州和湖北。南亚、东南亚和西南亚亦有。

Trees evergreen. Stems brown or purple-brown. Leaves elliptic to oblong, abaxially papillate. Corymbose-cyme, flowers white. Inflorescences, abaxial surface of leaves, and petioles pubescent with short grayish appressed trichomes. Fruits black at maturity, ellipsoid. Fl. Sep to next Jan. Fr. next Apr-Jun. Broad-leaved evergreen, mixed broad-leaved evergreen-deciduous forests or thickets along streams at 800-3700 m. Distributed in Yunnan, Sichuan, Xizang, Guizhou and Hubei. Also in S, SE and SW Asia.

## 沙梾
**Cornus bretschneideri** L.

灌木或小乔木。树皮紫红色。

长圆叶梾木 *Cornus oblonga*

沙梾 *Cornus bretschneideri*

叶背灰白色，密被不明显的乳头状突起及白色贴生的短柔毛，侧脉5或6(-7)条，侧脉细，细脉不显明。伞房状聚伞花序顶生；花白色。核果近球形，蓝黑色至黑色。花期6-7月，果期8-9月。生海拔600-2300米的林中、灌丛中或山坡。产中国西南、华中、华北、西北和东北。

Shrubs or small trees. Bark purplish red. Leaves abaxially grayish white, densely papillose and pubescent with appressed white trichomes, lateral veins 5 or 6(-7), thin, smaller veins inconspicuous. Corymbose cymes terminal; flowers white. Drupes subglobose, bluish black to black. Fl. Jun-Jul. Fr. Aug-Sep. Forests, thickets or slopes at 600-2300 m. Distributed in SW, C, N, NW and NE China.

## 红椋子

**Cornus hemsleyi** C. K. Schneid. et Wanger.

灌木或小乔木。小枝初为红色或绿色，后转红。叶对生，纸质，下面密被白色贴生短柔毛及乳头状突起，叶脉6-8条，下面无黑色条纹。花白色或带黄色。果实紫红色或黑色，球形。花期6-7月，果期8-9月。生海拔1000-4000米的杂木林、灌丛或溪边。产中国西南、华北、华西和西北。

Shrubs or small trees. Young branches red or green, later red. Leaves opposite, papery, abaxially densely papillate, pubescent with appressed short white trichomes, veins 6-8, abaxial surface without blackish streaks. Flowers white or yellowish. Fruits purplish red or black, globose. Fl. Jun-Jul. Fr. Aug-Sep. Mixed forests, thickets or streamsides at 1000-4000 m. Distributed in SW, N, W and NW China.

红椋子 *Cornus hemsleyi*

光皮梾木 *Cornus wilsoniana*

## 光皮梾木
**Cornus wilsoniana** Wanger.

落叶灌木或乔木。树皮灰色或灰绿色，块状剥落。叶对生；叶下面灰绿色，纸质，密被白色乳头状突起及平贴短柔毛，脉3或4条。顶生圆锥或聚伞花序近塔形；花白色。核果球形，紫黑色或黑色。花期5月，果期9-11月。生海拔100-1100米的林中。产中国西南、华南、东南、华北、华西和华东。

Shrubs or trees deciduous. Bark gray or greenish gray, rectangularly splitting. Leaves opposite, abaxially grayish green, papery, abaxially densely pubescent with white short appressed trichomes and papillae, veins 3 or 4. Paniculate to corymbose cymes almost turreted, terminal; flowers white. Drupes globose, purplish black or black. Fl. May. Fr. Sep-Nov. Forests at 100-1100 m. Distributed in SW, S, SE, N, W and E China.

## 梾木
**Cornus macrophylla** Wall.

乔木。幼枝粗壮，有4棱。叶对生，下面被乳突和灰白色或棕色平伏短柔毛，早落。花序圆锥状或有时为伞房状聚伞花序；花白色，芳香。果实紫黑色或蓝黑色，近球形；核扁球形，具6或8条肋纹。花期6-7(-8)月，果期8-9(-10)月。生海拔3600米以下的密林、杂木林、林缘、山坡、溪边。产中国西南、华南、东南、华中、华西和华东。南亚和西南亚亦有。

Trees. Young branches stout, 4-angled. Leaves opposite, abaxially papillate, with appressed grayish white or brown short deciduous trichomes. Inflorescences paniculate or sometimes corymbose cymes; flowers fragrant, white. Fruits purplish black or bluish black, subglobose; stones compressed globose, ribs 6 or 8. Fl. Jun-Jul. (-Aug). Fr. Aug-Sep(-Oct). Dense forests, mixed woods, margins of woods, slopes or streamsides below 3600 m. Distributed in SW, S, SE, C, W and E China. Also in S and SW Asia.

梾木 *Cornus macrophylla*

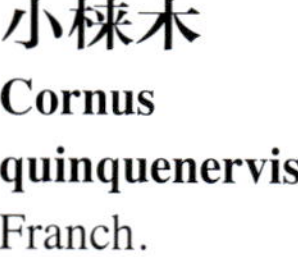

## 小梾木
**Cornus quinquenervis** Franch.

灌木。幼枝绿色或紫红色，4棱。叶对生，下面被较少白色的平贴短柔毛，早落，脉(2或)3(或4)。伞房状聚伞花序宽3.5-8厘米；花白色或黄白色。果实成熟时黑色，球形；核近球形，具不明显6条肋纹。花期6-7月，果期10-11月。生海拔2500米以下的山地

小梾木 *Cornus quinquenervis*

毛梾 *Cornus walteri*

林中、溪边灌丛、灌丛或山腰。产中国西南、华南、华中、华西和华东。

Shrubs. Young branches green or purplish red, 4-angled. Leaves opposite, abaxially with sparse white appressed deciduous short trichomes, veins (2 or)3(or 4). Corymbose cymes 3.5-8 cm wide; flowers white or yellowish white. Fruits black at maturity, globose; stones subglobose, inconspicuously 6-ribbed. Fl. Jun-Jul. Fr. Oct-Nov. Montane forests, thickets by streams, scrubs or hillsides below 2500 m. Distributed in SW, S, C, W and E China.

## 毛梾

**Cornus walteri** Wanger.

乔木。树皮厚，深灰色。叶对生，背面浅绿色，纸质，背面被灰白色短伏毛，叶脉4(-5)条。伞房状聚伞花序花密，紧缩，分枝直；萼齿与花盘等长。核果球形，成熟时黑色。花期5-6月，果期8-10月。生海拔300-2500米的杂木林。产中国大部分地区。

Trees. Bark thick, dark gray. Leaves opposite, light green abaxially, papery, abaxially with grayish white short appressed trichomes, veins 4(-5). Corymbose cymes dense, compact, branches straight; calyx lobes equal to disk. Drupes globose, black after mature. Fl. May-Jun. Fr. Aug-Oct. Mixed forests at 300-2500 m. Distributed in most parts of China.

## 川鄂山茱萸

**Cornus chinensis** Wanger.

乔木。单轴分枝。叶卵状披针形至狭椭圆形，下面微被灰白色贴生短柔毛，脉腋有明显的灰色丛毛。伞形花序侧生。果实成熟后红色或黑色，长圆形；核狭椭圆体形，具几条肋纹。花期4月，果期9月。生海拔700-2500(-3500)米的密林、杂木林林缘或山坡。产中国西南、华南、东南、华中、华北和华西。缅甸亦有。

Trees. Branches monopodial. Leaves ovate-lanceolate to narrowly elliptic, abaxially sparsely pubescent with grayish white appressed trichomes and a cluster of conspicuous gray long trichomes in axils of veins. Umbellate inflorescences lateral. Fruits red or black (when ripe), oblong; stones narrowly ellipsoid, with few ribs. Fl. Apr. Fr. Sep. Dense forests, mixed forest edges or slopes at 700-2500 (-3500) m. Distributed in SW, S, SE, C, N and W China. Also in Myanmar.

山茱萸 *Cornus officinalis*

## 山茱萸

**Cornus officinalis** Siebold et Zucc.

常绿乔木或灌木。合轴分枝，树皮灰褐色。叶下面浅绿色，脉腋密生淡褐色丛毛，叶脉6或7条。伞形花序顶生；花梗粗壮，长约2毫米；花瓣反折。核果红色至紫红色，狭椭圆体形。花期3-4月，果期9-10月。生海拔400-1500(-2100)米的林中或林缘。产华南、东南、华中、华西、华东和华北。朝鲜半岛和日本亦有。

Trees or shrubs evergreen. Branches sympodial. Bark grayish brown. Leaves abaxially light green, with clusters of light brown trichomes in axils of lateral veins, veins 6 or 7. Umbellate inflorescences terminal; pedicels robust, ca. 2 mm long; petals reflexed. Drupes red to purple, narrowly ellipsoid. Fl. Mar-Apr. Fr. Sep-Oct. Forests or forest edges at 400-1500(-2100) m. Distributed in S, SE, C, W, E and N China. Also in Korean Peninsula and Japan.

川鄂山茱萸 *Cornus chinensis*

香港四照花 *Cornus hongkongensis*

## 香港四照花

**Cornus hongkongensis** Hemsl.

常绿小乔木或灌木。树皮深灰色或深棕色。叶革质或厚革质，两面初被柔毛，渐无毛，下面脉腋无Y型绒毛，叶脉2-3(-4)条。头状花序具50-70花；苞片白色；花瓣淡黄色。花期5-6月，果期11-12月。生海拔600-1800米的常绿阔叶林。产贵州、广西、广东和湖南。老挝和越南亦有。

Small trees or shrubs, evergreen. Bark dark gray or dark brown. Leaves leathery or thickly leathery, pubescent on both surfaces when young, gradually glabrous, without Y-shaped trichomes in axils of veins, veins 2-3(-4); capitulums 50-70-flowered; bracts white; petals light yellow. Fl. May-Jun. Fr. Nov-Dec. Evergreen broad-leaved forests at 600-1800 m. Distributed in Guizhou, Guangxi, Guangdong and Hunan. Also in Laos and Vietnam.

## 秀丽四照花

**Cornus hongkongensis** Hemsl. subsp. **elegans** (W. P. Fang et Y. T. Hsieh) Q. Y. Xiang

本变种与香港四照花的区别在于本变种的侧叶脉，尤其是次级侧脉不明显。花期5-6月，果期11月。生海拔200-1200米的林中或溪边。产福建、浙江和江西。

This variety differs from the typical variety in its lateral leaf veins, particularly secondary lateral veins, inconspicuous. Fl. May-Jun. Fr. Nov. Forests or streamsides at 200-1200 m. Distributed in Fujian, Zhejiang and Jiangxi.

## 尖叶四照花

**Cornus elliptica** (Pojark.) Q. Y. Xiang ex Boufford

常绿乔木或灌木。幼枝、叶背面和苞片均密被白色平伏毛。叶两面灰白色，脉3或4，先端渐尖状尾尖。头状花序，总苞片白色，狭卵形至倒卵形。果序球形，成熟时红色。花期6-7月，果期10-11月。生海拔300-2200米的灌丛、溪边或杂木林。产中国西南、华南、东南和华中。

Trees or shrubs evergreen. Young branches, leaves abaxial surface, and bracts all pubescent with white appressed trichomes. Leaves grayish green on both surfaces, veins 3 or 4, apex acuminate-caudate. Inflorescences capitate, involucral bracts white, narrowly ovate to obovate. Infructescence globose, red at maturity. Fl. Jun-Jul. Fr. Oct-Nov. Shrubs, streamsides or mixed forests at 300-2200 m. Distributed in SW, S, SE and C China.

## 头状四照花 (鸡嗉子)

**Cornus capitata** Wall.

常绿小乔木。叶对生，侧脉3或4对，两面均灰绿色，常被柔毛混有短的浅灰或白色的绒毛，手感粗糙。头状花序，扁球

秀丽四照花 *Cornus hongkongensis* subsp. *elegans*

尖叶四照花 *Cornus elliptica*

形；总苞4，白色；花瓣4，白色至淡黄色。果序成熟时紫红色。花期5-7月，果期9-11月。生海拔1000-3200米的常绿林或杂木林。产云南、四川、西藏和贵州。印度、尼泊尔、不丹和巴基斯坦亦有。

Small trees evergreen. Leaves opposite, lateral veins 3- or 4-paired, grayish green on both surfaces, typically pubescent with short, light gray or white trichomes, scabrous. Inflorescences capitate, compressed globose; involucral bracts 4, white; petals 4, white to yellowish; infructescences purple red at maturity. Fl. May-Jul. Fr. Sep-Nov. Evergreen and mixed forests at 1000-3200 m. Distributed in Yunnan, Sichuan, Xizang and Guizhou. Also in India, Nepal, Bhutan and Pakistan.

头状四照花（鸡嗉子） *Cornus capitata*

## 四照花

**Cornus kousa** F. Buerger ex Hance subsp. **chinensis** (Osborn) Q. Y. Xiang

落叶乔木或灌木。叶纸质或厚纸质，背面粉绿色，密被乳头状突起和平伏柔毛，叶脉4或5对。头状聚伞花序球形，基部常具一明显的环状加厚；花萼内侧有一圈褐色短柔毛。果实成熟时紫红色，球形。花期5-7月，果期9-10月。生海拔400-2200米的林中、沟谷或溪边。产中国除西北外大部分地区。

Trees or shrubs deciduous. Leaves papery or thickly papery, pink green abaxially, often densely papillate and pubescent with appressed trichomes, veins 4 or 5 pairs. Capitate cymes globose, often with a conspicuously thickened ring at base; calyx interior with brown pubescence. Fruits purple when mature, globose. Fl. May-Jul. Fr. Sep-Oct. Forests valleys or streamsides at 400-2200 m. Distributed in most parts of China, except NW China.

四照花 *Cornus kousa* subsp. *chinensis*

多脉四照花 *Cornus multinervosa*

## 草茱萸

**Cornus canadensis** L.

多年生草本。叶常6枚轮生枝顶，纸质，倒卵形至椭圆形，全缘，侧脉3对，弧形。伞形状聚伞花序顶生；总苞片4，白色，花瓣状，宽卵形；花小，绿白色。核果球形，红色。花期7-8月，果期8-9月。生海拔约1200米的针叶林下。产吉林(长白山)。朝鲜半岛、俄罗斯和北美洲亦有。

Perennial herbs. Leaves usually terminal, verticillate, 6 in a whorl, chartaceous, obovate to elliptic, margin entire, lateral veins 3 pairs, arcuate. Cymes terminal, umbrella-like; bracts 4, white, petal-like, broadly ovate; flowers small, greenish-white. Drupes globose, red. Fl. Jul-Aug. Fr. Aug-Sep. Under coniferous forests at ca. 1200 m. Distributed in Jilin (Changbai Mountain). Also in Korean Peninsula, Russia and North America.

## 多脉四照花

**Cornus multinervosa** (Pojark.) Q. Y. Xiang

落叶乔木。树皮黑褐色。叶纸质，下面浅绿色，被较密的白色贴生短柔毛，叶脉(5-)6(-7)条，几乎延伸至顶部。头状花序球形，具27-45花；花药深蓝色。核果球形，成熟时红色，基部无环状加厚；果核具散生红点。花期5-6月，果期10-11月。生海拔900-2700米的林中。产云南和四川。

Trees deciduous. Bark blackish brown. Leaves papery, abaxially light green, densely pubescent with white appressed trichomes, veins (5-)6(-7), nearly extending to apex. Capitates cymes globose, 27-45-flowered; anthers dark blue. Drupes globose, red at maturity, usually without a thickened ring at base; stones of fruit with scattered red spots. Fl. May-Jun. Fr. Oct-Nov. Forests at 900-2700 m. Distributed in Yunnan and Sichuan.

草茱萸 *Cornus canadensis*

# 中文名索引

## Y



## Z

# 拉丁学名索引

**J**



**K**



**L**



**M**

## T



## U



## V



## W



## X



## Y



## Z